Photocatalytic Systems by Design

Materials, Mechanisms and Applications

Photocatalytic Systems by Design

Materials, Mechanisms and Applications

Edited by

Mohan Sakar

Centre for Nano and Material Sciences, Jain University, Bengaluru, India

R. Geetha Balakrishna

Centre for Nano and Material Sciences, Jain University, Bengaluru, India

Trong-On Do

Department of Chemical Engineering, Laval University, Quebec, QC, Canada

ELSEVIER

Elsevier
Radarweg 29, PO Box 211, 1000 AE Amsterdam, Netherlands
The Boulevard, Langford Lane, Kidlington, Oxford OX5 1GB, United Kingdom
50 Hampshire Street, 5th Floor, Cambridge, MA 02139, United States

British Library Cataloguing-in-Publication Data
A catalogue record for this book is available from the British Library

Library of Congress Cataloging-in-Publication Data
A catalog record for this book is available from the Library of Congress

ISBN: 978-0-12-820532-7

For Information on all Elsevier publications
visit our website at https://www.elsevier.com/books-and-journals

Publisher: Susan Dennis
Acquisitions Editor: Kostas KI Marinakis
Editorial Project Manager: Emerald Li
Production Project Manager: Debasish Ghosh
Cover Designer: Greg Harris

Typeset by MPS Limited, Chennai, India

Working together
to grow libraries in
developing countries

www.elsevier.com • www.bookaid.org

Contents

Chapter 3: Anion-modified photocatalysts **55**

Shwetharani R., Bindu K., Laveena P. D'Souza, R. Mithun Prakash and R. Geetha Balakrishna

Chapter 8: Cocatalyst-integrated photocatalysts for solar-driven hydrogen and oxygen production

Chinh Chien Nguyen, Thi Hong Chuong Nguyen and Minh Tuan Nguyen Dinh

Chapter 9: Materials and features of ferroelectric photocatalysts: the case of multiferroic BiFeO₃

U. Bharagav, N. Ramesh Reddy, K. Pratap, K.K. Cheralathan,
M.V. Shankar, P.K. Ojha and M. Mamatha Kumari

Chapter 10: Metal organic framework-based photocatalysts for hydrogen production *275*

Lakshmana Reddy Nagappagari, Kiyoung Lee and
Shankar Muthukonda Venkatakrishnan

Chapter 11: Transition metal chalcogenide—based photocatalysts for small-molecule activation *297*

Afsar Ali and Arnab Dutta

Chapter 12: MXene-based photocatalysts *333*

Yolice P. Moreno Ruiz, William Leonardo da Silva and João H. Zimnoch dos Santos

Chapter 16: Oxyhalides-based photocatalysts: the case of bismuth oxyhalides 441

Y.N. Teja, K. Gayathri, C. Ningaraju, Adhigan Murali and Mohan Sakar

Chapter 17: Design of photocatalysts for the decontamination of emerging pharmaceutical pollutants in water....................................475

K. Rokesh, Mohan Sakar and Trong-On Do

Chapter 18: Magnetic photocatalytic systems 503

Jagadeesh Babu Sriramoju, Chitrabanu C. Paramesh, Guddappa Halligudra, Dinesh Rangappa and Prasanna D. Shivaramu

List of contributors

Afsar Ali   Chemsitry Discipline, IIT Gandhinagar, Gandhinagar, India

R. Geetha Balakrishna   Centre for Nano and Material Sciences, Jain University, Bangalore, India

U. Bharagav   Nanocatalysis and Solar Fuels Research Laboratory, Department of Materials Science & Nanotechnology, Yogi Vemana University, Kadapa, India

Vassilios Binas   Institute of Electronic Structure and Laser, Foundation for Research and Technology-Hellas (FORTH-IESL), Irákleion, Greece; Department of Physics, University of Crete, Herakleio, Greece

K.K. Cheralathan   Department of Chemistry, School of Advanced Sciences, Vellore Institute of Technology, Vellore, India

G.L. Colpani   Post-Graduation Program in Technology and Innovation Management, Community University of Chapecó Region, Servidão Anjo da Guarda, Chapecó, Brazil; Department of Chemical Engineering, Community University of Chapecó Region, Servidão Anjo da Guarda, Chapecó, Brazil

Laveena P. D'Souza   Centre for Nano and Material Sciences, Jain University, Bangalore, India

William Leonardo da Silva   Universidade Franciscana, Santa Maria, Brazil

Maria Vittoria Diamanti   Department of Chemistry, Materials and Chemical Engineering "G. Natta," Polytechnic of Milan, Milan, Italy

Minh Tuan Nguyen Dinh   The University of Da Nang, University of Science and Technology, Faculty of Chemical Engineering, Da Nang, Viet Nam

Trong-On Do   Department of Chemical Engineering, Laval University, Quebec, QC, Canada

Arnab Dutta   Chemistry Department, IIT Bombay, Mumbai, India

Vijayakumar Elayappan   Department of Materials science and Engineering, Korea University, Seoul, Republic of South Korea

M.A. Fiori   Post-Graduation Program in Technology and Innovation Management, Community University of Chapecó Region, Servidão Anjo da Guarda, Chapecó, Brazil; Department of Chemical Engineering, Community University of Chapecó Region, Servidão Anjo da Guarda, Chapecó, Brazil

K. Gayathri   Centre for Nano and Material Sciences, Jain University, Bangalore, India

Guddappa Halligudra   Department of Applied Sciences, Visvesvaraya Technological University, Center for Postgraduate Studies, Bengaluru Region, Muddenahalli, Chikkaballapur, India

Jaeyeong Heo   Department of Materials Engineering, Optoelectronics Convergence Research Center, Chonnam National University, Gwangju, Republic of Korea

Bindu K.   Department of Inorganic and Physical Chemistry, Indian Institute of Science, Bangalore, India

Mohammad Mansoob Khan Chemical Sciences, Faculty of Science, Universiti Brunei Darussalam, Gadong, Brunei Darussalam

George Kiriakidis Institute of Electronic Structure and Laser, Foundation for Research and Technology-Hellas (FORTH-IESL), Irákleion, Greece

Dimitrios Kotzias European Commission-Joint Research Centre, Institute for Health and Consumer Protection, Ispra, Italy

M. Mamatha Kumari Nanocatalysis and Solar Fuels Research Laboratory, Department of Materials Science & Nanotechnology, Yogi Vemana University, Kadapa, India

Hai-Gun Lee Department of Materials science and Engineering, Korea University, Seoul, Republic of South Korea

Kiyoung Lee School of Nano & Materials Science and Engineering, Kyungpook National University, Sangju, South Korea

Francisco José Maldonado-Hódar Carbon Materials Research Group, Department of Inorganic Chemistry, Faculty of Sciences, University of Granada, Granada, Spain

Saikumar Manchala Department of Chemistry, National Institute of Technology, Warangal, India; Centre for Advanced Materials, National Institute of Technology, Warangal, India

J.M.M. Mello Post-Graduation Program in Technology and Innovation Management, Community University of Chapecó Region, Servidão Anjo da Guarda, Chapecó, Brazil; Department of Chemical Engineering, Community University of Chapecó Region, Servidão Anjo da Guarda, Chapecó, Brazil

Maryam Mokhtarifar Department of Chemistry, Materials and Chemical Engineering "G. Natta," Polytechnic of Milan, Milan, Italy; Department of Chemical Engineering, Laval University, Quebec, QC, Canada

Sergio Morales-Torres Carbon Materials Research Group, Department of Inorganic Chemistry, Faculty of Sciences, University of Granada, Granada, Spain

Yolice P. Moreno Ruiz Centro de Tecnologias Estratégicas do Nordeste, Recife, Brazil

Adhigan Murali School for Advanced Research in Polymers (SARP)-ARSTPS, Central Institute of Plastics Engineering & Technology (CIPET), Chennai, India

Shankar Muthukonda Venkatakrishnan Nano Catalysis and Solar Fuels Research Laboratory, Department of Materials Science & Nanotechnology, Yogi Vemana University, Kadapa, India

Lakshmana Reddy Nagappagari School of Nano & Materials Science and Engineering, Kyungpook National University, Sangju, South Korea

KrishnaRao Neerugatti Eswar Department of Materials Engineering, Optoelectronics Convergence Research Center, Chonnam National University, Gwangju, Republic of Korea

Chinh Chien Nguyen Institute of Research and Development, Duy Tan University, Da Nang, Viet Nam; Faculty of Environmental and Chemical Engineering, Duy Tan University, Da Nang, Viet Nam

Thi Hong Chuong Nguyen Institute of Research and Development, Duy Tan University, Da Nang, Viet Nam; Faculty of Environmental and Chemical Engineering, Duy Tan University, Da Nang, Viet Nam

C. Ningaraju Centre for Nano and Material Sciences, Jain University, Bangalore, India

P.K. Ojha Naval Materials Research Laboratory (NMRL), Ambernath, India

Chitrabanu C. Paramesh Department of Applied Sciences, Visvesvaraya Technological University, Center for Postgraduate Studies, Bengaluru Region, Muddenahalli, Chikkaballapur, India

Luisa M. Pastrana-Martínez Carbon Materials Research Group, Department of Inorganic Chemistry, Faculty of Sciences, University of Granada, Granada, Spain

MariaPia Pedeferri Department of Chemistry, Materials and Chemical Engineering "G. Natta," Polytechnic of Milan, Milan, Italy

Álvaro Pérez-Molina Carbon Materials Research Group, Department of Inorganic Chemistry, Faculty of Sciences, University of Granada, Granada, Spain

R. Mithun Prakash Centre for Nano and Material Sciences, Jain University, Bangalore, India

K. Pratap Centre for Advanced Studies in Electronics Science and Technology (CASEST), School of Physics, University of Hyderabad, Hyderabad, India

Shwetharani R. Centre for Nano and Material Sciences, Jain University, Bangalore, India

Dinesh Rangappa Department of Applied Sciences, Visvesvaraya Technological University, Center for Postgraduate Studies, Bengaluru Region, Muddenahalli, Chikkaballapur, India

Nur Aqilah Mohd Razali Advanced Membrane Technology Research Centre (AMTEC), Universiti Teknologi Malaysia, Skudai, Malaysia; School of Chemical and Energy Engineering, Faculty of Engineering, Universiti Teknologi Malaysia, Skudai, Malaysia

N. Ramesh Reddy Nanocatalysis and Solar Fuels Research Laboratory, Department of Materials Science & Nanotechnology, Yogi Vemana University, Kadapa, India

K. Rokesh Department of Chemical Engineering, Laval University, Quebec, QC, Canada

Nurafiqah Rosman Advanced Membrane Technology Research Centre (AMTEC), Universiti Teknologi Malaysia, Skudai, Malaysia; School of Chemical and Energy Engineering, Faculty of Engineering, Universiti Teknologi Malaysia, Skudai, Malaysia

Mohan Sakar Department of Chemical Engineering, Laval University, Quebec, QC, Canada; Centre for Nano and Material Sciences, Jain University, Bangalore, India

Wan Norharyati Wan Salleh Advanced Membrane Technology Research Centre (AMTEC), Universiti Teknologi Malaysia, Skudai, Malaysia; School of Chemical and Energy Engineering, Faculty of Engineering, Universiti Teknologi Malaysia, Skudai, Malaysia

M.V. Shankar Nanocatalysis and Solar Fuels Research Laboratory, Department of Materials Science & Nanotechnology, Yogi Vemana University, Kadapa, India

Vishnu Shanker Department of Chemistry, National Institute of Technology, Warangal, India; Centre for Advanced Materials, National Institute of Technology, Warangal, India

Prasanna D. Shivaramu Department of Applied Sciences, Visvesvaraya Technological University, Center for Postgraduate Studies, Bengaluru Region, Muddenahalli, Chikkaballapur, India

L.L. Silva Post-Graduation Program in Technology and Innovation Management, Community University of Chapecó Region, Servidão Anjo da Guarda, Chapecó, Brazil; Department of Chemical Engineering, Community University of Chapecó Region, Servidão Anjo da Guarda, Chapecó, Brazil

Jagadeesh Babu Sriramoju Department of Applied Sciences, Visvesvaraya Technological University, Center for Postgraduate Studies, Bengaluru Region, Muddenahalli, Chikkaballapur, India

Y.N. Teja Centre for Nano and Material Sciences, Jain University, Bangalore, India

Danae Venieri School of Environmental Engineering, Technical University of Crete, Chania, Greece

M. Zanetti Department of Chemical Engineering, Community University of Chapecó Region, Servidão Anjo da Guarda, Chapecó, Brazil

R.C.F. Zeferino   Department of Chemical Engineering, Community University of Chapecó Region, Servidão Anjo da Guarda, Chapecó, Brazil

João H. Zimnoch dos Santos   Instituto de Química–Universidade Federal do Rio Grande do Sul, Porto Alegre, Brazil

CHAPTER 1

Principles and mechanisms of photocatalysis

Mohammad Mansoob Khan

Chemical Sciences, Faculty of Science, Universiti Brunei Darussalam, Gadong, Brunei Darussalam

1.1 Introduction and historical developments

From history, we have learned that the human race has been motivated by nature to discover intelligent answers for complicated problems in their everyday life. Similarly, natural photosynthesis drives chemical pathways using solar radiations. Hence, photocatalysis has arisen as a nature-inspired method for harvesting and converting sun energy to facilitate challenging synthetic conversions for various applications. In 1972 Fujishima and Honda (1972) first reported "photocatalysis" by water splitting under UV irradiation since then research in this field has increased many folds. Photocatalysis has been attracting extra attention as it finds applications in a variety of products across a broad range of research areas, mainly environmental and energy-related fields. Recently, metal oxides, such as TiO_2, ZnO, SnO_2, and CeO_2, have been the main choice for most studies in basic research and practical applications because of their high activity, low cost, high stability, nontoxicity, and chemical inertness, which make them suitable for applications in water and air purification, sterilization, hydrogen evolution, etc. (Albini & Fagnoni, 2008; Chen, Nanayakkara, & Grassian, 2012; Khan, Adil, & Mayouf, 2015; Ravelli, Dondi, Fagnoni, & Albini, 2009).

The "photocatalysis" shows the development of some dynamic concepts of photochemistry. The turning point that allowed photochemistry to become a science on its own was distinguishing the difference with thermal chemistry. In fact, till the beginning of the 20th century, many scientists felt that irradiation or illumination was one of the several ways existing for catalyzing a reaction, that is, making it faster, such as treating with some chemicals and heating. Giacomo Ciamician was the first scientist who put a systematic effort to understand the chemical effect of light on ascertaining whether "light and light alone," and not, for instance, heat, caused the reactions to accelerate (Albini & Fagnoni, 2008). He properly assigned these reactions as "photochemical reaction" while the term "photocatalytic"-tagged reactions were accelerated by light, but maintaining the same course as the thermal reactions. Later, researchers, recognized that photochemical reactions involve electronically excited states, "electronic isomers" of ground states that have a reactivity (and thermodynamics) of their own. Bodenstein in 1914 observed it, however, becomes a common concept after several years (Albini & Fagnoni, 2008; Ravelli et al., 2009).

The word "photocatalysis" is of Greek origin and includes two terms (prefix "photo" derived from phos means light and "catalysis" derived from katalyo means break apart or decompose). However, there is no agreement in the scientific community as to have a proper definition of photocatalysis. It is generally used to describe a process in which light is used to trigger a substance, the photocatalyst, which modifies the rate of a chemical reaction without being involved itself in the chemical reactions. Thus, the main difference between a conventional thermal catalyst and a photocatalyst is that the latter is activated by photons of appropriate energy whereas the former is activated by heat. Photocatalytic reactions may take place homogeneously or heterogeneously. However, heterogeneous photocatalysis is more deeply studied in recent years because of its potential usage in a variety of environmental and energy-related applications as well as in organic syntheses (Choi et al., 2016). In heterogeneous photocatalysis, the reaction pathway involves the formation of an interface among photocatalysts and fluid containing the reactants and products. The processes involving irradiation adsorbate—semiconductors interfaces are generally categorized as a branch of photochemistry. Therefore, the term "heterogeneous photocatalysis" is mainly used in the cases where a light-absorbing semiconductors photocatalyst is utilized, which is in contact with either a liquid phase or a gas phase.

A material is considered to be a *catalyst* when it accelerates a chemical reaction without being consumed in a reaction, denatured or unaltered and lowers the free activation enthalpy of the reaction. *Photocatalysis* is defined as the acceleration of a reaction in the presence of light and a suitable catalyst. *Photocatalysts* are defined as a material that accelerates a chemical reaction in the presence of appropriate light and a suitable catalyst without being utilized that lowers the free activation enthalpy of a chemical reaction (Ravelli et al., 2009). *Photocatalysis* is a fast-developing field with a great possibility of a widespread range of industrial applications, such as water disinfection, air disinfection, mineralization of organic pollutants, production of renewable fuels, and organic syntheses (Chen et al., 2012; Choi et al., 2016; Khan et al., 2015; Ravelli et al., 2009). In this chapter, the discussion will be on semiconductor-mediated photocatalysis, basic principles, classification, mechanism, limitations, and operating parameters of photocatalytic processes. This chapter also covers theory and fundamentals essential for understanding heterogeneous photocatalysis. Emphasis has been given to the electronic and optical properties of the photocatalysts. This is followed by the discussion of the principles of photocatalysis and thermodynamics as well as kinetic aspects, which determine the photocatalytic efficiency. This chapter is designed to understand fundamentals, theories, and concepts necessary for understanding the semiconductors, such as metal oxides (TiO_2, ZnO, SnO_2, CeO_2, etc.) and chalcogenides (ZnS, CdS, MoS, $CdSe$, $CdTe$, etc.) as photocatalysts (Ansari et al., 2013; Ansari, Khan, Ansari, Lee, & Cho, 2014b; Chen et al., 2012; Choi et al., 2016; Kalathil, Khan, Ansari, Lee, & Cho, 2013; Khan et al., 2015; Khan, Ansari, Pradhan, Ansari, et al., 2014; Khan, Ansari, Pradhan, Han, et al., 2014; Ravelli et al., 2009). This chapter also emphasizes the optical and electronic properties of the semiconductors, which will be described with the use of the band model, etc. Photocatalysis has been discussed in detail by giving several examples.

1.2 Semiconductors and photocatalysis

Several semiconductors are useful as photocatalysts because of the promising amalgamated light absorption properties, electronic structures, excited-state lifetimes, and charge transport characteristics. A semiconductor is nonconductive in its undoped ground state because an energy gap, that is, a wide band gap, exists between the top of the filled valence band (VB) and the bottom of the vacant conduction band (CB). Accordingly, electron transport between these bands must occur only with considerable energy change. In semiconductor photocatalysis, the excitation of an electron from the VB to CB is achieved by absorption of a photon of energy equal to or higher than the band gap energy of the semiconductor. This leads to the formation of an electron–hole pair ($e-/h^+$), which is a prerequisite for all the semiconductor-assisted photocatalytic reactions. The photo-generated species formed tend to recombine and dissipate energy as heat because the kinetic barrier for the electron–hole recombination step is low. However, holes in VB and electrons in CB can be separated well in the presence of an electric field, such as the one generated spontaneously in the space charge layer of a metal-semiconductor or a fluid–semiconductor interface. Hence, the lifetime of photo generated carriers' ($e-/h^+$) increases and the possibility offered to these species to exchange charge with substrates adsorbed on the photocatalyst surface and persuade primary and secondary chemical reactions.

Transfer of an electron to or from a substrate adsorbed onto the light-activated semiconductor, that is, interfacial electron transfer, is probably the most critical step in photocatalytic reactions, and its efficiency controls to a large extent the ability of the semiconductor to serve as a photocatalyst for a given redox reaction. The efficiency of electron transfer reactions is, in turn, a function of the position of the semiconductor's VB and CB edges relative to the redox potentials of the adsorbed substrates. For an anticipated electron transfer step to occur, the potential of the electron donor species should be located above (more negative than) the VB of the semiconductor, whereas the potential of the electron acceptor species should be located below (more positive than) the CB of the semiconductor. Interfacial electron transfer processes are then initiating subsequent redox reactions to form free radicals for primary and secondary reactions. The free radicals formed, such as hydroxyl radicals ($^{\bullet}$OH) and superoxide radicals ($^{\bullet}O_2$), will be used as strong oxidizing agents for decomposing or degrading the organic pollutants, etc.

1.3 Fundamentals of photocatalysis

In short, photocatalysis is defined as the "acceleration of a reaction in the presence of a suitable catalyst and suitable light." A catalyst does not change or being used up during a chemical reaction and accelerates the rate of reaction by lowering the activation energy. It includes photosensitization, which is a process by which a photochemical reaction takes place in one molecular unit as a result of the initial absorption of suitable light energy by another molecular unit, called the photosensitized. Photocatalysis assist in forming strong reducing and oxidizing agents, which help in breaking down the organic pollutants to CO_2 and H_2O in the presence of light, photocatalyst, and water (Ansari, Khan, Ansari, & Cho, 2016; Ansari, Khan, Ansari, Lee, & Cho, 2014a; Chen et al., 2012; Fujishima, Zhang, & Tryk, 2008; Khan, Ansari, Pradhan, Ansari, et al., 2014; Khan, Ansari, Pradhan, Han, et al., 2014).

1.3.1 Mechanism

When photocatalyst (such as CdS, ZnS, TiO_2, ZnO, SnO_2, CeO_2, etc.) absorbs suitable light, it produces electron hole pairs (e-/h$^+$) in the CB and VB respectively. The electrons in the VB of semiconductors become excited when irradiated by light. The excess energy of this excited electron promoted the electron to the CB of semiconductors, therefore, creating the positive-hole (h$^+$) and negative-electron (e$^-$) pairs. This stage is referred to as the semiconductor's "photo-excitation" state. The energy difference between the VB and CB is known as the "band gap" energy (E_g). Fig. 1.1 shows the band gap energy of insulators, semiconductors, and conductors (Hernandez-Ramırez & Medina-Ramırez, 2015; Saravanan et al., 2016).

The photoactivation of semiconductors photocatalyst is based on its electronic excitation by photons (light) with energy (hv) greater than the band gap energy (E_g). The electrons migrate after the excitation generating vacancies in the VB (holes, h+)

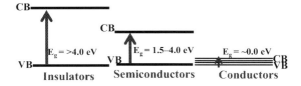

Figure 1.1
Band gap energies of the insulators, semiconductors, and conductors.

and forming regions with high electron density (e$^-$) in the CB (Hernandez-Ramırez & Medina-Ramırez, 2015; Hoffmann, Martin, Choi, & Bahnemann, 1995; Khataee, Zarei, & Ordikhani-Seyedlar, 2011; Kumar & Devi, 2011; Ni, Leung, Leung, & Sumathy, 2007; Nogueira & Jardim, 1998; Ziolli & Jardim, 1998). These holes are pH dependent and have high positive electrochemical potentials, in the range between +2.0 and +3.5 V, measured against a saturated calomel electrode (Khataee et al., 2011). This potential is sufficiently positive to generate hydroxyl radicals ($^\bullet$OH) from water molecules adsorbed on the surface of the semiconductor photocatalysts [Eqs. (1.i), (1.ii), (1.iii)]. The photocatalytic efficiency depends on the competition between the formation of e$^-$/h$^+$ pairs and the recombination of these pairs Eq. (1.iv) on the semiconductor photocatalysts surfaces (Hoffmann et al., 1995; Ni et al., 2007; Ziolli & Jardim, 1998).

$$\text{Photocatalysts} + \text{hv} \rightarrow \text{photocatalysts}\left(e_{CB}^- + h_{VB}^+\right) \tag{1.i}$$

$$h^+ + H_2O_{ads.} \rightarrow {}^\bullet OH + H^+ \tag{1.ii}$$

$$h^+ + OH_{ads.}^- \rightarrow {}^\bullet OH \tag{1.iii}$$

$$\text{Photocatalysts}\left(e_{CB}^- + h_{VB}^+\right) \rightarrow \text{photocatalysts} + \Delta \tag{1.iv}$$

Though the oxidation reactions caused by the generated holes occur at the VB, the electrons transferred to the CB are responsible for reduction reactions, such as the formation of gaseous hydrogen and the generation of other important oxidizing species, such as superoxide anion radicals ($^\bullet O_2$). In case of semiconductors, such as chalcogenides and metal oxides, the E_g is between 1.8−2.7 and 3.00−3.40 eV, respectively (Hernandez-Ramırez & Medina-Ramırez, 2015; Hoffmann et al., 1995; Kumar & Devi, 2011). The whole procedure is shown schematically in Fig. 1.2.

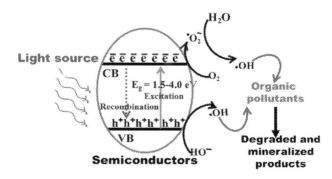

Figure 1.2
Tentative photocatalysis mechanism occurs during photocatalytic reaction at the surface of the semiconductors.

The positive hole formed in semiconductor photocatalysts dissociates the H_2O molecules to form H_2 gas and hydroxyl radicals ($^{\bullet}OH$). The negative electron reacts with adsorbed oxygen molecules to form superoxide anions ($O_2^{\bullet-}$) (Ansari et al., 2016; Chen et al., 2012; Fujishima et al., 2008; Hernandez-Ramırez & Medina-Ramırez, 2015; Khan, Ansari, Pradhan, Ansari, et al., 2014; Khan, Ansari, Pradhan, Han, et al., 2014). This cycle continues till suitable light of appropriate intensity and wavelength is available. The complete mechanism of photocatalytic reaction that happens at its surface of semiconductors, in the presence of suitable light, is shown in Fig. 1.2.

The ultimate leading and advanced oxidation reactions are based on the formation of hydroxyl radicals ($^{\bullet}OH$), which are extremely powerful oxidizing agents, second only to fluorine in power (2.23 in relative oxidizing power). Through the use of the strong oxidation strength of $^{\bullet}OH$ radicals, photocatalytic oxidation can efficiently decompose, degrade, disinfect, deodorize, and purify the air, water, and different types of surfaces. Table 1.1 shows the common chemical oxidants, placed in the order of their oxidizing strength.

1.3.1.1 Photocatalysis mechanism

The photocatalysis method based on semiconductors basically involves the following stages:

- The light energy of a certain wavelength is made to fall onto a semiconductor. If the energy of incident light is equivalent to the band gap energy of the semiconductor, the electron would be excited from the VB to CB of the semiconductor.
- Holes would be left in the valance band of the semiconductor. The holes in the VB can oxidize donor molecules.

The electrons and holes could undergo subsequent reduction and oxidation reaction, respectively, with any dye molecules, which might be adsorbed on the surface of the

Table 1.1: Common chemical oxidants, placed in the order of their oxidizing strength.

Compounds/radicals	Oxidation potentials (V)	Relative oxidizing power ($Cl_2 = 1.0$)
Hydroxyl radicals ($^{\bullet}OH$)	2.8	2.1
Sulfate radicals ($SO_4^{\bullet-}$)	2.6	1.9
O_3	2.1	1.5
H_2O_2	1.8	1.3
MnO_4^{-}	1.7	1.2
ClO_2	1.5	1.1
Cl_2	1.4	1.0
O_2	1.2	0.90
Br_2	1.1	0.80
I_2	0.76	0.54

semiconductor to give the degradation products. Fig. 1.2 shows the reaction mechanism of photocatalysis.

CB electrons and VB holes are generated when aqueous semiconductor suspension is irradiated with light energy greater than its band gap energy (E_g). Photo-generated electrons can react with water or OH^- and oxidize them to produce $^\cdot OH$ radicals. The photo-generated electrons can react with electron acceptors, such as O_2, adsorbed on the catalysts' surface or dissolved in water, reducing it to the superoxide radical anion $O_2^{\cdot-}$. The powerful hydroxyl radicals ($^\cdot OH$) are accountable for the decomposition of organic pollutants. The excited electron and hole pairs can recombine and release the input energy as thermal energy (heat) with no chemical effect.

1.3.1.2 Oxidation mechanism

The surface of a semiconductor photocatalyst contains water, which is referred to as "adsorbed water." When this water is oxidized by positive holes, hydroxyl radicals ($^\cdot OH$), which have strong oxidative decomposing power, are formed. Then, the hydroxyl radicals react with organic matter. If oxygen is present when this process takes place, the intermediate radicals in the organic compounds and oxygen molecules can undergo radical chain reactions and consume oxygen in some cases. In such case, the organic matter eventually decomposes and ultimately ends up with the end product to CO_2 and H_2O. Under some conditions, organic pollutants can directly react with the positive holes, resulting in oxidative decomposition products (Fig. 1.3) (Khan et al. 2015).

1.3.1.3 Reduction mechanism

The reduction of oxygen contained in the air occurs as a pairing reaction. As oxygen is an easily reducible substance, if oxygen is present, the reduction of oxygen takes place instead

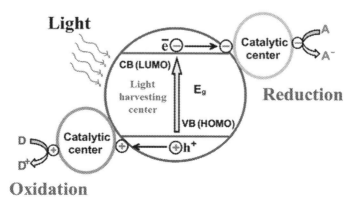

Figure 1.3
Oxidation and reduction centers at the surface of the semiconductor.

of hydrogen generation. The reduction of oxygen results in the generation of superoxide anions ($O_2^{\bullet-}$). Superoxide anions react with the intermediate in the oxidative reaction, forming peroxide or changing to H_2O_2 and then to H_2O. As reduction tends to occur more easily in organic solvent than in water, when the concentration of organic matter becomes high, the possibility of positive holes being used in the oxidative reactions with organic matter increases, thus reducing the rate of charge carrier recombination.

1.3.2 Major advantages of photocatalysis

Following are the advantages of photocatalysis:

1. Photocatalysis offers a better substitute for the energy-intensive conventional treatment methods with the capacity for using renewable and pollution-free solar energy.
2. Photocatalysis leads to the formation of harmless products, unlike conventional treatment measures that transfer pollutants from one phase to another.
3. This process can be used to destroy a variety of hazardous compounds in different wastewater streams.
4. The reaction conditions for photocatalysis are mild, the reaction time is modest, and a lesser chemical input is required.
5. Minimum secondary waste generation.
6. Photocatalysis can also be applied to the gaseous phase and aqueous treatments as well as solid-phase treatments to some extent.

1.3.3 Limitations of photocatalysis

Following are the rate-limiting steps in photocatalytic reactions, which involve:

1. interfacial charge transfer,
2. improving the charge separation, and
3. inhibition of charge carrier recombination.

These steps are important for enhancing the efficiency of the photocatalytic process.

1.3.4 Operating parameters in photocatalytic processes

The rate of photo mineralization of an organic compound by photocatalysis method depends mainly on the following parameters: reaction temperature, pH, light intensity, and effect of photocatalysts (Pelaez et al., 2012).

1. *Effect of reaction temperature*
    Usually, photocatalysis is not temperature dependent. However, an increase in temperature can affect the amount of adsorption and enhances the recombination of

charge carriers, resulting in a decrease in the photocatalytic activity. To determine the reaction rate, the temperature depending on the kinetic parameter k_{obs} in terms of the Arrhenius equation can be expressed as follows:

$$k_{obs} = A\,e^{-\frac{E_a}{RT}} \tag{1.1}$$

where k_{obs} is the kinetic parameter (1/min), A is the frequency factor (1/min), E_a is the activation energy (kcal/mol), T is the temperature (K), and R is the gas constant $(1.987 \times 10^{-3}$ kcal/mol/K). According to this equation, the reaction rate should increase linearly with $e^{(-1/T)}$.

2. *pH effect*

   In photocatalytic degradation reactions, the pH value is an important parameter because of the photocatalytic activity that takes place on the surface of photocatalysts. The pH of an aqueous solution affects the surface charge on the photocatalyst, the oxidation potential of the valance band, and other physiochemical properties of the system. In accordance with Nernst's law, varying the pH of the solution shifts the energies of the valence and CB edges by 0.059 V per pH unit (at ambient temperature). This results in the VB electrons becoming more potent and the CB holes becoming less potent at higher pH.

3. *Effect of light intensity*

   The photocatalytic reaction rate largely depends on the light-harvesting ability of the photocatalyst. In general, the decomposition rate increases with increasing the light intensity during the photocatalytic reaction. Recent reports confirmed that during photocatalytic degradation reactions, increase in the degradation rate has been observed with increase in light intensity. However, excessive light intensity results in more electron−hole recombination. Another factor that limits the photonic efficiency is the thermal recombination between electrons and holes (Amani-Ghadim et al., 2019; Wang & Domen, 2020). When considering the electron−hole recombination, the appropriate light intensity I_a can be predicted by the following equation:

$$I_a = mI \tag{1.2}$$

where m is an excess coefficient and I is the light intensity. This shows that the light intensity had a great importance on the electron−hole pair's recombination. Occasionally with the increase in light intensity, the increment in reaction rate would reduce, indicating that the utilization ratio of light energy might drop.

4. *Effect of dosage of photocatalyst*

   The amount of photocatalyst loaded may also affect the process of photocatalytic degradation. Initially, the increase in the amount of photocatalyst increases the number of active sites on the semiconductor surface that in turn increases the number of $^{\bullet}OH$ and $O_2^{\bullet-}$ radicals. As a result, the rate of photocatalytic degradation is increased.

In the review of TiO_2-assisted photocatalytic degradation of azo dyes in an aqueous solution, Konstantinou and Albanis reported that the initial rates are directly proportional to catalyst concentration in any reactor system. Furthermore, they observed that there is a limit of catalyst concentration, which is used for the degradation of particular organic pollutants from wastewater. Moreover, the effect of loading photocatalyst content beyond this catalyst concentration may result in the agglomeration of catalyst; hence, for photon absorption, part of the catalyst surface will be unavailable.

5. *Concentration of pollutants in wastewater*

Another main factor to determine the degradation rate is the pollutant type and its concentration. Many researchers have reported the photocatalytic activity under similar operating conditions and using similar catalysts, but the variation in the initial concentration of water contaminants results in different irradiation times necessary to achieve complete mineralization.

1.4 Semiconductors that are mainly used as photocatalysts

1.4.1 Metal oxides as photocatalysts

Photocatalysis is a promising, green, environmentally friendly method for the transformation of solar energy to chemical energy or chemical conversion using suitable semiconductor nanostructures, such as the degradation of organic and inorganic pollutants, removal of inorganic pollutants, and H_2 production. In the last three to four decades, considerable research efforts have been made to understand efficient, economical, and green sources for the treatments of organic and inorganic pollutants for environmental remediation processes, energy production, and optoelectronic devices. Suitable semiconductor nanomaterials have played a chief role in this effort because they have an excellent combination of photochemical activity, major light-harvesting, mechanical and thermal stability. They have long been pursued photocatalytic applications, such as H_2 production through water splitting or decomposition of water pollutants. The catalytic activities of semiconductor nanomaterials are influenced significantly by the reactive sites present on the surface, due to different types of defects in the crystal structures.

Among the semiconductors, transition metal oxide nanomaterials have attracted considerable attention owing to their potential applications, such as photocatalysis, H_2 production and storage, environmental remediation, and energy. Therefore these nanomaterials are expected to be the key nanomaterials for further developments of nanoscience and nanotechnology (Ansari et al., 2016; Chen et al., 2012; Fujishima et al., 2008; Hernandez-Ramırez & Medina-Ramırez, 2015). A large diversity of semiconducting materials, mainly metal oxides and chalcogenides, have been explored with respect to their photocatalytic behavior, but only a few of them are considered to be effective photocatalysts. Among the metal oxide nanostructures available, TiO_2 has attracted specific attention due to its extraordinary properties, such as low cost, high stability, high chemical inertness, biocompatibility, and nontoxicity. TiO_2 has been examined widely as an efficient photocatalyst for the purification of water, and the degradation of dyes, pesticides, etc., since the discovery of its photocatalytic properties by Honda—Fujishima (Choi et al., 2016; Fujishima & Honda, 1972; Khan et al., 2015; Khataee et al., 2011; Ni et al., 2007; Ziolli & Jardim, 1998). Due to the attractive properties, metal oxides such as ZnO, TiO_2, SnO_2, CeO_2, $Fe2O_2$ and V_2O_2 have attracted significant interest in several fields of research, such as materials science, physics and chemistry. In general, wide-band gap metal oxides, such as TiO_2, prove to be better photocatalysts than low-band gap materials, such as cadmium sulfide (CdS), mainly due to the higher free energy of photo-generated charge carriers of the former and the inherently low chemical and photochemical stability of the latter (Chen et al., 2012; Ravelli et al., 2009). However, low-band gap metal oxides are better altered to the solar spectrum, thus posing the significant benefit of potential utilization of a continuous and readily available power supply, that is, the sunlight. A considerable amount of work has been reported in recent years for the development of more efficient photocatalysts considered by improving light harvesting and increased quantum efficiency. Promising results in this way have been obtained with the use of several approaches targeting the optical and/or modification of electronic properties of different metal oxides, which includes metal deposition, dye sensitization, doping with transition metals or nonmetallic elements, use of composite semiconductor photocatalysts, etc. (Ansari et al., 2016; Chen et al., 2012; Khan, Ansari, Pradhan, Ansari, et al., 2014; Khan, Ansari, Pradhan, Han, et al., 2014).

The possible applications of heterogeneous photocatalysis mainly depend on the progress of the light-harvesting ability and reduced recombination efficiency of electron—hole pairs scaled-up reactor designs. The main task in the design of a photocatalytic reactor is the mass transfer optimization and efficient light-harvesting ability of the catalyst, especially in liquid-phase reactions. Mass transfer restrictions can be dealt with using monolithic reactors, spinning disk reactors, and microreactors, which have been proven to be much more efficient than conventional reactors. Photon transfer can be optimized using light-emitting diodes (LEDs) and optical fibers, however, key developments in this area are still lacking. Successively the artificial formation of photons needed for photocatalytic reactions is the most significant basis of operating costs in practical applications, and a considerable amount of research effort has been made for the development of solar photoreactors. Among the various types of solar reactor configurations evaluated so far, compound parabolic collectors are the most promising and have been successfully scaled up for applications related to water cleaning, water disinfection, and wastewater treatment

(Fujishima, Rao, & Tryk, 2000; Ibhadon & Fitzpatrick, 2013; Ni et al., 2007; Pelaez et al., 2012; Spasiano, Marotta, Malato, Fernandez-Ibanez, & Di Somma, 2015). Metal oxide-based photocatalysis is currently one of the most active interdisciplinary research areas and has been examined from the standpoint of catalysis, photochemistry, electrochemistry, organic, inorganic, physical, polymer, and environmental chemistry. Owing to a large number of research in these core areas, the fundamental processes of photocatalysis are now much well understood. The applicability of photocatalysis has been proven in a laboratory scale for a large number of different processes, such as deodorizing and air cleaning, water treatment, disinfection applications, and production of fuels from water and atmospheric gases, selective organic synthesis, and metal recovery (Chen et al., 2012; Fujishima et al., 2008; Khan et al., 2015). However, industrial applications remain incomplete and restricted. The present insufficiency of extensive industrial applications is mainly due to the low photocatalytic efficiency of the semiconductor photocatalysts and the lack of efficient as well as large-scale photochemical reactor setups.

1.4.2 Chalcogenides as photocatalysts

The chalcogens are the chemical elements in group VIA of the periodic table. It consists of oxygen (O), sulfur (S), selenium (Se), and tellurium (Te). Chalcogenide is composed of elements from groups II−VI. Then, it can be modified into several types of chalcogenides, such as I−III−VI$_2$, I$_2$−V−VI$_4$, I$_2$−II−IV−VI$_4$, and I$_2$−IV−VI$_3$. This modification is carried out to obtain material properties as expected. With this wide variety of chalcogenides, it has varied physical and chemical properties. The compounds of the heavier chalcogens (the sulfides, selenides, and tellurides) are collectively known as chalcogenides. A unique property shown by chalcogenide semiconductor material is photochemical reaction. When photons from sunlight is absorbed by the chalcogenide semiconductor material, it generates a pair of electron and hole simultaneously. The electrons have a negative charge, and the holes have a positive charge. These electrons and holes can be used for a redox reaction. Therefore, chalcogenide compounds can be used as photovoltaics or as photocatalyst materials.

A chalcogenide is a chemical compound consisting of at least one chalcogen anion, such as sulfur (S), selenium (Se), and tellurium (Te), and at least one more electropositive element. Materials based on group VI elements are generally termed chalcogenide compounds and semiconducting chalcogenide binary compounds include oxides, sulfides, selenides, and tellurides of metals like Cd, Zn, Pb, Ag, Cu, etc. Nevertheless, metal chalcogenides would undergo photocorrosion when irradiated in the absence of sacrificial electron donors. Thus it processes a suitable band engineering to prohibit rapid recombination of h + /e − pairs and backward reactions. To improve the photocatalytic activity, modifying the surface of the particles is a necessary stage steps to prevent the electron—hole recombination. There are various ways to improve the photocatalytic activities of the metal chalcogenide materials such as, metal and non-metal doping, coupling with various metal and metal oxides, surface sensitization by polymers, dye sensitization, composite formation using semiconductors, etc.

1.4.2.1 Significance of chalcogenide nanomaterials

Metal chalcogenide nanostructures are attracting significant attention in a variety of energy devices, including sensors, fuel cells, Li-ion batteries, supercapacitors, solar cells, LEDs, thermoelectric devices, and memory devices. Many metal chalcogenides have become popular choices for solar absorber materials and device architectures. When the size of the metal chalcogenides is reduced to the nanometer scale, new physical and chemical properties emerge owing to the well-known quantum size effect.

In addition, nanostructured metal chalcogenides can provide a much higher specific surface area as compared with their bulk counterparts, which is beneficial to energy devices because the reaction/interaction between the devices and the interacting media can be significantly enhanced. The wide applications of metal chalcogenides are profited from the progress in the synthesis of new nanostructured materials of different sizes and new morphologies. When used as catalysts or electrode materials for electrochemical storage systems, nanostructured materials are found to be more effective than their bulk counterparts due to the following several advantages:

- higher specific surface areas and more active sites,
- longer cycle life,
- higher electric conductivity, and
- shorter path lengths for the transport of electrons.

Also, their unique physical and chemical properties are due to the "nanoscale effects."

1.5 Applications of the photocatalysis

Since last few decades, metal oxide nanomaterials, such as Fe_2O_3, TiO_2, SnO_2, ZnO, WO_3, and CeO_2, have been largely studied for their photocatalytic properties (Chen et al., 2012; Fujishima & Honda, 1972; Hernandez-Ramirez & Medina-Ramirez, 2015; Hoffmann et al., 1995; Khataee et al., 2011; Kumar & Devi, 2011; Ni et al., 2007; Nogueira & Jardim, 1998; Saravanan et al., 2016; Ziolli & Jardim, 1998). Metal oxide nanomaterials used as a photocatalyst showed the excellent degradation of organic and toxic pollutants due to their high reactivity at low concentrations, low toxicity, and high stability. Metal oxide photocatalysis has attracted significant attention because of its promising applications in various fields, such as environmental remediation by the photodecomposition of hazardous dyes in polluted water, industrial effluents, and solar energy conversion. Therefore metal oxides, such as TiO_2, ZnO, SnO_2, and CeO_2, have been the prime choice for basic research and practical applications owing to their high activity, good stability, easy availability, low cost, nontoxicity, and chemical inertness.

In recent years, emergent alarms about energy and environmental problems have encouraged extensive research on solar energy utilization. Dyes are used widely in a range of fields, but their discharge into the water can cause environmental pollution. In addition, most dyes are toxic, carcinogenic, and harmful, resulting in adverse impacts on human and animal health. Dyes find considerable applications in several industries, including textile, plastic, rubber, paper, concrete, and medicine with the textile industry as the main user. Unfortunately, approximately 10% of dyes used in industry are discharged directly into the environment as a harmful pollutant, which is environmentally unsafe and esthetically unacceptable. Heterogeneous photocatalysis involves the application of metal oxide catalyst (such as TiO_2, ZnO, SnO_2, WO_3, and CeO_2) irradiated with light of an appropriate wavelength to generate highly reactive transitory oxidative species (i.e., $^{\bullet}OH$ and $^{\bullet}O_2{}^-$) for the mineralization of organic contaminants, impurities, and pollutants (Choi et al., 2016; Khan et al., 2015). Therefore a range of approaches has been explored for the photocatalytic degradation of organic dyes using semiconductor photocatalysts (Ansari et al., 2016; Chen et al., 2012; Fujishima et al., 2008; Hernandez-Ramırez & Medina-Ramırez, 2015; Hoffmann et al., 1995; Khan, Ansari, Pradhan, Ansari, et al., 2014; Khan, Ansari, Pradhan, Han, et al., 2014; Kumar & Devi, 2011; Nogueira & Jardim, 1998; Saravanan et al., 2016). Generally, semiconductor catalysts show relatively low quantum degradation efficiency because of the high recombination rate of light-induced $e-/h+$ pairs at or near the surface of the photocatalysts, which is considered as one of the major limitations to hindering the photocatalytic efficiency (Fujishima et al., 2008; Khan, Ansari, Pradhan, Ansari, et al., 2014; Khan, Ansari, Pradhan, Han, et al., 2014).

The chemistry that occurs at the surfaces of metal oxides has attracted considerable attention for a range of industrial applications (Fig. 1.4), including catalysis, photocatalysis, water purification, deodorization, air purification, self-cleaning, self-sterilizing, antifogging surfaces optical display technology, chemical synthesis, solar energy devices, antibacterial activities, batteries, energy production, and storage (Ansari et al., 2014b; Banerjee, Dionysiou, & Pillai, 2015; Chen et al., 2012; Fujishima & Zhang, 2006; Fujishima et al., 2000, 2008; Hernandez-Ramırez & Medina-Ramırez, 2015; Hoffmann et al., 1995; Ibhadon & Fitzpatrick, 2013; Khataee et al., 2011; Kumar & Devi, 2011; Nakata & Fujishima, 2012; Ni et al., 2007; Nogueira & Jardim, 1998; Pelaez et al., 2012; Saravanan et al., 2016; Spasiano et al., 2015; Wang et al., 1997; Ziolli & Jardim, 1998).

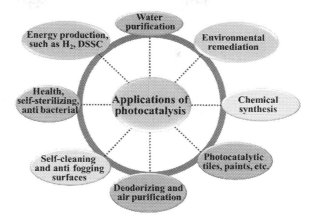

Figure 1.4
Photocatalysis-based possible applications using semiconductors as photocatalysts.

1. *Water purification*

   Water is well thought-out as one of the necessities of life because the world would not have any life without water. Water cleansing/treatment has become a worldwide problem, mostly, in industrialized countries, where wastewater normally contains organic pollutants, such as organic dyes from the textile industries, organic waste materials from the paper industry, leather and tanning industries, food industry, agricultural research, and pharmaceutical industries. The release of these colored and toxic compounds in the environment has raised substantial concern because of their toxic effects on the environment, plants, and human beings. In addition, two classes of dyes, azo dyes, and thiazine dyes can cause serious health risks. Some of the azo dyes are highly carcinogenic. The conventional wastewater treatment plants cannot degrade the majority of these pollutants. Hence, there has been increasing interest in methods for decontamination over the past few decades (Chen et al., 2012; Hoffmann et al., 1995; Saravanan et al., 2016).

   The waste from the textile industry, paper industry, leather and tanning industries, food industry, and agricultural and pharmaceutical industries has long been considered a serious environmental issue. Most of the paper mills and textile industries produce wastewater that contains aromatic dyes, which are highly toxic and difficult to decompose due to their relatively stable chemical structures. Many dyes [methylene blue (MB), methyl orange] are used in industries for a range of purposes. Among them, MB is one of the most frequently used dyes for agriculture, textile, paper-making, cosmetic, and pharmaceutical purposes (Chen et al., 2012; Hoffmann et al., 1995; Khataee et al., 2011; Kumar & Devi, 2011; Nogueira & Jardim, 1998; Saravanan et al., 2016; Ziolli & Jardim, 1998).

   Photocatalysts combined with ultraviolet lights can oxidize organic pollutants into nontoxic by-products, such as water and CO_2, and can sterilize certain bacteria. This technology is very effective for removing hazardous organic compounds, such as dyes, volatile organic compounds (VOCs), and total organic carbons (TOCs), and killing a variety of bacteria and some viruses in the secondary wastewater treatments. Pilot projects demonstrated that photocatalytic detoxification systems could effectively kill fecal coliform bacteria in the secondary wastewater treatment (Chen et al., 2012; Fujishima et al., 2008; Hernandez-Ramırez & Medina-Ramırez, 2015; Hoffmann et al., 1995; Kumar & Devi, 2011; Nogueira & Jardim, 1998; Saravanan et al., 2016).

The excessive use of organic chemicals in both industrial manufacturing and normal household uses has led to their leaching into the environment, causing a shocking environmental contamination. Organic chemicals are also present as pollutants in groundwater and surface water, such as wells, ponds, and lakes. To achieve drinking water quality, pollutants need to be removed to protect the water resources. Numerous processes, such as biological and chemical oxidation reactions, adsorption onto supported substrates, ultrasonic irradiation, and electrochemical, have been used widely to destroy or remove these toxins. Among these techniques, the photocatalytic detoxification of organic pollutants has attracted considerable attention because of its numerous advantageous features, such as use of very small amounts of catalysts, regeneration of catalysts, utilization of natural sunlight for environmental remediation, and clean energy production.

2. *Deodorizing and air purification*

During deodorizing process, the hydroxyl radicals accelerate the breakdown of any VOCs by breaking the molecular bonds. This will help combine the organic gases to form a single molecule that is not much harmful to the environment, plants, and human beings and thus enhance the air-cleaning efficiency. Some of the examples of odor molecules are tobacco odor, formaldehyde, nitrogen dioxide, gasoline, urine and fecal odor, and many other hydrocarbon molecules in the atmosphere (Fujishima et al., 2008; Hoffmann et al., 1995; Pelaez et al., 2012).

The air purifier developed with TiO_2 can avoid soil and smoke, pollen, bacteria, virus, and harmful gases and halt the free bacteria in the air by filtering $\sim 99.9\%$ with the help of the highly oxidizing effect of photocatalysts, that is, metal oxides (TiO_2, etc.). The photocatalytic activity of metal oxides can be applied for the elimination or reduction of polluted compounds in the air, such as cigarette smoke, automobile smoke, NO_x, and volatile compounds arising from various industries and construction sites. Also, high photocatalytic reactivity can be applied to protect lamp-houses and walls in tunneling, as well as to prevent white tents from becoming sooty and dark. Atmospheric constituents, such as greenhouse gases, chlorofluorocarbons (CFCs), substituted CFCs, and nitrogenous and sulfurous compounds, undergo photochemical reactions either indirectly or directly in the presence of sunlight. In a polluted region, these pollutants can ultimately be efficiently removed using suitable metal oxides. The deoderization and air purification may help to keep the air pure and free from bad odor. This will be beneficial for plants and human beings to make them fit and healthy (Fujishima et al., 2008; Pelaez et al., 2012).

3. *Self-cleaning, self-sterilizing, and antifogging surfaces*

In most of industrial countries, most of the exterior walls of buildings become soiled from automotive exhaust fumes and smokes coming out from industries, which contain oily components. When the original building materials are coated with a photocatalyst, such as TiO_2, a protective layer of titanium offers self-cleaning the building by becoming antistatic, superoxidative, and superhydrophilic. The hydrocarbon from the automotive exhaust is oxidized, and the dirt on the walls washes away with rainfall, keeping the building exterior clean and shining at all times (Fujishima et al., 2008; Hernandez-Ramírez & Medina-Ramírez, 2015; Pelaez et al., 2012).

1.5.1 Superhydrophilic

When the surface of the photocatalytic layer is exposed to light, the contact angle of the photocatalyst surface with H_2O is decreased slowly. After sufficient exposure to light, the surface reaches superhydrophilic, that is, it does not repel water at all, so water cannot exist in the shape of a drop but spreads flatly on the surface of the substrate. Finally, the water acquired the form of a highly uniform thin film, which acts optically like a clear sheet of glass. The hydrophilic nature of metal oxides, coupled with gravity, will enable the dust particles to be cleaned away following the water stream, thus making the product self-cleaning.

TiO_2 can degrade organic contaminants at the surface level with the assistance of UV light. This characteristic has led to the new application of TiO_2 photocatalysis in the "self-cleaning" technique where TiO_2-coated surfaces can maintain themselves clean under UV light utilizing readily available sunlight or ultraviolet emission from fluorescent lamps, which in turn saves maintenance costs, and reduce the reliability of detergents for cleaning (Spasiano et al., 2015). Fujishima et al. (2000) confirmed this self-cleaning concept on a titania-coated ceramic tile in 1992. The first commercialized product using this method was the self-cleaning cover glass for highway tunnel lamps (Ibhadon & Fitzpatrick, 2013; Spasiano et al., 2015). An example is the sodium lamp in Japan, which emits UV light through the cover glass and is mainly used to decompose the contaminants from automobile exhaust since the UV light is not useful for lighting purposes. Therefore the cover glass maintains its transparency for long-term use. Also, Wang et al. (1997) added that the self-cleaning effect could be aided with flowing water (e.g., rainfall) on the TiO_2 surfaces. This enhancing phenomenon of the flowing water, such as rainfall, was ascribed to the superhydrophilic property of TiO_2 surface, that is, water was able to penetrate the molecular-level space between the stain and the superhydrophilic TiO_2 surface (Banerjee et al., 2015; Fujishima & Zhang, 2006; Nakata & Fujishima, 2012; Wang et al., 1997). The outdoor materials benefit from the combined self-cleaning effect of TiO_2 photocatalytic and superhydrophilic nature. Apart from lamps, other materials on the road, such as spray coatings for cars, rearview mirror and windshield, tunnel wall, reflectors, and traffic signs, have been made to use TiO_2 surface coating for the self-cleaning and antifogging advantage. Also, the application could be found in materials for residential and office buildings, such as exterior tiles, kitchen and bathroom components, interior furnishings, plastic surfaces, aluminum siding, tent material, building stone and curtains, glass window, and window blinds (Ibhadon & Fitzpatrick, 2013).

4. *Antibacterial agents*

   The advantage of using photocatalyst as antibacterial materials is that it does not only kill bacterial cells but also completely decompose the cells. The TiO_2 photocatalyst has been found to be more active than any other antibacterial agent because the photocatalytic process acts even when the cell surface is covered and bacteria are actively propagating. The end toxins produced after the death of the cell are also expected to be decomposed by photocatalytic action, that is, decomposing using $^{\bullet}OH$ and $^{\bullet}O_2$ radicals. TiO_2 does not deteriorate and demonstrates a long-term antibacterial effect. As per the researchers, decontaminations by TiO_2 are threefold stronger than chlorine and 1.5-fold stronger than ozone as well as much safer and efficient than others (Fujishima et al., 2008; Hernandez-Ramırez & Medina-Ramirez, 2015).

5. *Chemical synthesis*

   Photocatalytic organic synthesis using metal oxides is well known and in the progress. Choi et al. recently reported the photocatalytic synthesis of 2-hydroxyterephthalic acid (HTPA) using terephthalic acid (TPA). In this synthesis, HTPA was treated with novel photocatalysts ZnO and ZnS under visible light irradiation for 6 hours, which in situ generated $^{\bullet}OH$ radicals and reacted with HTPA leading to the formation of TPA (Choi et al., 2016; Fujishima et al., 2008; Hernandez-Ramırez & Medina-Ramırez, 2015; Hoffmann et al., 1995).

6. *Energy production*

   Recent advancement in the development of novel nanomaterial for energy production and storage with improved energy efficiencies, lower costs, and energy savings contributes significantly to the global energy sustainability as demand is continuously increasing. The nanostructured solar cell, such as TiO_2 and ZnO nanotube-based dye-sensitized solar cells, has potential to be cheaper and easier to install with recent advancement in their print-like manufacturing process and can be made into flexible rolls rather than discrete panels (Sharma, Siwach, Ghoshal, & Mohan, 2017; Siwach, Sharma, & Mohan, 2017). It is also inevitable that the abundant water resource and sunlight can be used to have hydrogen as the future source of clean energy. The advancement in research on improving the efficiency of hydrogen production and light harvesting using different metal oxide nanostructures, such as TiO_2, ZnO, Ce/TiO_2, and CeO_2-graphene, is expanding very fast (Albini & Fagnoni, 2008; Chen et al., 2012; de Lima et al., 2015; Fujishima & Honda, 1972; Khan et al., 2015; Khan, Khan, & Cho, 2017; Patrocinio, Paterno, & Murakami Iha, 2010; Ravelli et al., 2009). This would serve as a major source of clean energy for the world and help in energy sustainability for future generations.

1.6 Future prospects

This chapter presents an overview of the recent studies related to the heterogeneous photocatalysis using metal oxides as photocatalysts. These studies provide an important basis for its understanding; however, we believe that many questions remain unanswered. These questions can be answered through a combined effort of critical field studies, thorough laboratory studies, and modeling using novel tools. Some of the future directions are as follows:

1. Fabricating photocatalysts with crystal defects on the surface.
2. Ion doping in semiconductor photocatalysts.
3. Semiconductor-based photocatalytic reactions carried out under atmospherically relevant conditions of gas-phase concentrations, relative humidity, solar flux, etc.
4. Semiconductor-based photocatalytic studies that probe in situ both surface and gas-phase species formed under pertinent environmental conditions, as these are best suited for mechanistic studies of photocatalytic reactions.
5. Investigation of the effects of the physicochemical properties of semiconductors on heterogeneous photocatalysis that include shape, size, coatings, and also use as nanocomposites.
6. Hands-on field measurements that provide and validate the importance of these photocatalytic reactions for better understanding.
7. Proper modeling to get a better understanding of the relative impact of photocatalysis compared to other heterogeneous reactions.
8. Utilization of strong electrolytes releases high mobility ions and improves the conductivity of the electrolysis solution.
9. Addition of the sacrificial chemical reagents prevents the recombination of electrons and holes that produces hydrogen gas only.

Hence, it is believed that with further research and development on photocatalysis, various types of novel photocatalysts and other relevant products can be fabricated and used not only to overcome the existing challenges but also to improve the environmental effects and processing cost of photocatalysis.

1.7 Conclusions

This chapter discussed semiconductor-based heterogeneous photocatalysis and focused on basic concepts involved, such as theory and background necessary to understand the heterogeneous photocatalysis of metal oxides and chalcogenides. This chapter also includes a discussion on the principles of semiconductor-based photocatalysis, thermodynamics, and kinetic aspects, which regulate photocatalytic performance. Subsequently, different types of photocatalysts are discussed along with issues related to response to light, photochemical stability, and environmental issues. Later, several methods discussing semiconductor-mediated photocatalytic processes are described mainly focusing on applications related to the environment and energy. Finally, future prospects have been suggested. Through additional research and developments in this field, novel photocatalysts can be evolved and contribute to a cleaner environment and healthy life.

References

Albini, A., & Fagnoni, M. (2008). 1908: Giacomo Ciamician and the concept of green chemistry. *ChemSusChem, 1*, 63−66.

Amani-Ghadim, A. R., Khodam, F., & Dorraji, M. S. S. (2019). ZnS quantum dot intercalated layered double hydroxide semiconductors for solar water splitting and organic pollutant degradation. *Journal of Materials Chemistry A, 7*(18), 11408−11422.

Ansari, S. A., Khan, M. M., Ansari, M. O., & Cho, M. H. (2016). Nitrogen-doped titanium dioxide (N-doped TiO_2) for visible light photocatalysis. *New Journal of Chemistry, 40*, 3000−3009.

Ansari, S. A., Khan, M. M., Ansari, M. O., Lee, J., & Cho, M. H. (2014a). Band gap engineering of CeO_2 nanostructure by electrochemically active biofilm for visible light applications. *RSC Advances, 4*, 16782−16791.

Ansari, S. A., Khan, M. M., Ansari, M. O., Lee, J., & Cho, M. H. (2014b). Highly photoactive SnO_2 nanostructures engineered by electrochemically active biofilm. *New Journal of Chemistry, 38*, 2462−2469.

Ansari, S. A., Khan, M. M., Kalathil, S., Nisar, A., Lee, J., & Cho, M. H. (2013). Oxygen vacancy induced band gap narrowing of ZnO nanostructure by electrochemically active biofilm. *Nanoscale, 5*, 9238−9246.

Banerjee, S., Dionysiou, D. D., & Pillai, S. C. (2015). Self-cleaning applications of TiO_2 by photo-induced hydrophilicity and photocatalysis. *Applied Catalysis B: Environmental, 176*, 396−428.

Chen., Nanayakkara, C. E., & Grassian, V. H. (2012). Titanium dioxide photocatalysis in atmospheric chemistry. *Chemical Reviews, 112*, 5919−5948.

Choi, Y. I., Lee, S., Kim, S. K., Kim, Y. I., Cho, D. W., Khan, M. M., & Sohn, Y. (2016). Fabrication of ZnO, ZnS, Ag-ZnS, and Au-ZnS microspheres for photocatalytic activities, CO oxidation and 2-hydroxyterephthalic acid synthesis. *Journal of Alloys and Compounds, 675*, 46−56.

de Lima, J. F., Harunsani, M. H., Martin, D. J., Kong, D., Dunne, P. W., Gianolio, D., ... Walton, R. I. (2015). Control of chemical state of cerium in doped anatase TiO_2 by solvothermal synthesis and its application in photocatalytic water reduction. *Journal of Materials Chemistry A, 3*, 9890−9898.

Fujishima, A., & Honda, K. (1972). Electrochemical photolysis of water at a semiconductor electrode. *Nature, 238*, 37−38.

Fujishima, A., Rao, T. N., & Tryk, D. A. (2000). Titanium dioxide photocatalysis. *Journal of Photochemistry and Photobiology C: Photochemistry Reviews, 1*(1), 1−21.

Fujishima, A., & Zhang, X. (2006). Titanium dioxide photocatalysis: Present situation and future approaches. *Comptes Rendus Chimie, 9*(5), 750−760.

Fujishima, A., Zhang, X., & Tryk, D. A. (2008). TiO_2 photocatalysis and related surface phenomena. *Surface Science Reports, 63*, 515–582.

ISBN: 978-3-319-10999-2 (eBook); doi:10.1007/978-3-319-10999-2 Hernandez-Ramırez, A., & Medina-Ramırez, I. (Eds.), (2015). *Photocatalytic semiconductors: Synthesis, characterization, and environmental applications.* Switzerland: Springer International Publishing.

Hoffmann, M. R., Martin, T., Choi, W., & Bahnemann, D. W. (1995). Environmental applications of semiconductor photocatalysis. *Chemical Reviews, 95*(1), 69–96.

Ibhadon, A. O., & Fitzpatrick, P. (2013). Heterogeneous photocatalysis: Recent advances and applications. *Catalysts, 3*(1), 189–218.

Kalathil, S., Khan, M. M., Ansari, S. A., Lee, J., & Cho, M. H. (2013). Band gap narrowing of titanium dioxide (TiO_2) nanocrystals by electrochemically active biofilm and their visible light activity. *Nanoscale, 5*, 6323–6326.

Khan, M. E., Khan, M. M., & Cho, M. H. (2017). Ce^{3+}-ion, surface oxygen vacancy, and visible light-induced photocatalytic dye degradation and photocapacitive performance of CeO_2-graphene nanostructures. *Scientific Reports, 7*(7), 5928.

Khan, M. M., Adil, S. F., & Mayouf, A. A. (2015). Metal oxides as photocatalysts. *Journal of Saudi Chemical Society, 19*, 462–464.

Khan, M. M., Ansari, S. A., Pradhan, D., Ansari, M. O., Han, D. H., Lee, J., & Cho, M. H. (2014). Band gap engineered TiO_2 nanoparticles for visible light induced photoelectrochemical and photocatalytic studies. *Journal of Materials Chemistry A, 2*, 637–644.

Khan, M. M., Ansari, S. A., Pradhan, D., Han, D. H., Lee, J., & Cho, M. H. (2014). Defect-induced band gap narrowed CeO_2 nanostructures for visible light activities. *Industrial & Engineering Chemistry Research, 53*, 9754–9763.

Khataee, A. R., Zarei, M., & Ordikhani-Seyedlar, R. (2011). Heterogeneous photocatalysis of a dye solution using supported TiO_2 nanoparticles combined with homogeneous photoelectrochemical process: Molecular degradation products. *Journal of Molecular Catalysis A: Chemical, 338*(1–2), 84–91.

Kumar, S. G., & Devi, L. G. (2011). Review on modified TiO_2 photocatalysis under UV/visible light: Selected results and related mechanisms on interfacial charge carrier transfer dynamics. *The Journal of Physical Chemistry A, 115*, 13211–13241.

Nakata, K., & Fujishima, A. (2012). TiO_2 photocatalysis: Design and applications. *Journal of Photochemistry and Photobiology C: Photochemistry Reviews, 13*(3), 169–189.

Ni, M., Leung, M. K. H., Leung, D. Y. C., & Sumathy, K. (2007). A review and recent developments in photocatalytic water-splitting using TiO_2 for hydrogen production. *Renewable & Sustainable Energy Reviews, 11*(3), 401–425.

Nogueira, R. F. P., & Jardim, W. F. (1998). Fotocatálise heterogênea e sua aplicação ambiental. *Quimica Nova, 21*(1), 69–72.

Patrocinio, A. O. T., Paterno, L. G., & Murakami Iha, N. Y. (2010). Role of polyelectrolyte for layer-by-layer compact TiO_2 films in efficiency enhanced dye-sensitized solar cells. *The Journal of Physical Chemistry C, 114*, 17954–17959.

Pelaez, M., Nolan, N. T., Pillai, S. C., Seery, M. K., Falaras, P., Kontos, A. G., ... Entezari, M. H. (2012). A review on the visible light active titanium dioxide photocatalysts for environmental applications. *Applied Catalysis B: Environmental, 125*, 331–349.

Ravelli, D., Dondi, D., Fagnoni, M., & Albini, A. (2009). Photocatalysis. A multi-faceted concept for green chemistry. *Chemical Society Reviews, 38*, 1999–2011.

Saravanan, R., Khan, M. M., Gracia, F., Qin, J., Gupta, V. K., & Stephen, A. (2016). Ce^{3+}-ion-induced visible-light photocatalytic degradation and electrochemical activity of ZnO/CeO_2 nanocomposite. *Scientific Reports, 6*, 31641.

Sharma, S., Siwach, B., Ghoshal, S. K., & Mohan, D. (2017). Dye sensitized solar cells: From genesis to recent drifts. *Renewable and Sustainable Energy Reviews, 70*, 529–537.

Siwach, B., Sharma, S., & Mohan, D. (2017). Structural, optical and morphological properties of ZnO/MWCNTs nanocomposite photoanodes for Dye Sensitized Solar Cells (DSSCs) application. *Journal of Integrated Science and Technology*, *5*, 1–4.

Spasiano, D., Marotta, R., Malato, S., Fernandez-Ibanez, P., & Di Somma, I. (2015). Solar photocatalysis: Materials, reactors, some commercial, and pre-industrialized applications. A comprehensive approach. *Applied Catalysis B: Environmental*, *170*, 90–123.

Wang, Q., & Domen, K. (2020). Particulate photocatalysts for light-driven water splitting: Mechanisms, challenges, and design strategies. *Chemical Reviews*, *120*(2), 919–985. Available from https://doi.org/10.1021/acs.chemrev.9b00201.

Wang, R., Hashimoto, K., Fujishima, A., Chikuni, M., Kojima, E., Kitamura, A., . . . Watanabe, T. (1997). Light-induced amphiphilic surfaces. *Nature*, *388*(6641), 431.

Ziolli, R. L., & Jardim, W. F. (1998). Mechanism reactions of photodegradation of organic compounds catalyzed by TiO_2. *Quimica Nova*, *21*(3), 319–325.

CHAPTER 2

Cation-modified photocatalysts

G.L. Colpani[1,2], R.C.F. Zeferino[2], M. Zanetti[2], J.M.M. Mello[1,2], L.L. Silva[1,2] and M.A. Fiori[1,2]

[1]*Post-Graduation Program in Technology and Innovation Management, Community University of Chapecó Region, Servidão Anjo da Guarda, Chapecó, Brazil,* [2]*Department of Chemical Engineering, Community University of Chapecó Region, Servidão Anjo da Guarda, Chapecó, Brazil*

2.1 Fundamentals

The heterogeneous photocatalysis is a promising method to degrade a wide range of contaminants due to its ability to generate strongly oxidative species using a radiation source and a semiconductor, such as TiO_2 and ZnO (Zhua & Zhou, 2019). However, the photocatalysts possess some limitations, such as the fast recombination of the electron/vacancy pair, that occur due to impurities or other imperfections in the crystal surface (Hassan, Zhao, & Xie, 2016; Zhang, Wang, Liu, Sang, & Liu, 2017), and the short lifetime and nonselectivity of the formed radicals, what cause a reduction in the photocatalytic efficiency (Exposito et al., 2017; Fakhouri, Arefi-Khonsari, Jaiswal, & Pulpytel, 2015). To overcome these limitations, different strategies can be proposed to tune the optical, chemical, electrical, and structural properties, such as modification of semiconductors by adding cationic dopants like transition metals, noble metals and rare earth (RE) metals (Din, Najeeb, & Ahmad, 2018; Khaki, Shafeeyan, Raman, & Daud, 2017).

The cationic dopants have been widely evaluated, and these elements have been found to be able to reduce the recombination rate, shift the wavelength absorption and decrease the band gap energy, mainly because metal dopants introduce new energy levels, act as electron traps and absorb on the visible light due to the localized surface plasmon resonance (SPR) (Kumaravel, Mathew, Bartlett, & Pillai, 2019; Verbruggen, 2015). However, these features rely on several factors, including the nature and concentration of dopant, as well as the synthetic route applied in the doping process (Ghosh & Das, 2015; Shayegan, Lee, & Haghighat, 2018).

It has been demonstrated that the photocatalytic performance and stability of cation-doped semiconductors can be significantly affected by different synthesis methods, since they can change the nature, size, shape and surface area of the semiconductor materials. Among the various synthesis methods, the most commons are sol−gel, coprecipitation, hydrothermal, solvothermal, impregnation, chemical precipitation, photoreduction, and microwave (Khaki et al., 2017; Shayegan et al., 2018).

Depending on the ionic radius of the cationic dopant, it can usually be incorporated in substitutional or interstitial positions into photocatalyst lattice when the dopant has an ionic radius similar or smaller than those of semiconductors ions, respectively. On the other hand, when the photocatalysts are modified by metals with an ionic radius larger than their ionic radii, they are most likely to be found as highly dispersed metal oxide "clusters" (Coronado, Fresno, Hernández-Alonso, & Portela, 2013; Zou et al., 2019).

2.1.1 Electronic states

Cation-modified photocatalysts can extend the optical absorption-range to the visible light region when cations are incorporated in the bulk lattice. The absorption edge shifts to visible light because cationic dopants introduce additional states between the valence band (VB) and the conduction band (CB) of the semiconductors and these impurity levels reduce the band gap energy (Etacheri, Valentin, Schneider, Bahnemann, & Pillai, 2015; Verbruggen, 2015). In general, the transition metals incorporate 3d orbitals into the band gap and these energy levels are capable to transfer electrons from these orbitals to the semiconductor CB (Abdullah, Khan, Ong, & Yaakob, 2017), as presented in Fig. 2.1. For example, Fe^{3+} has been studied as a suitable cationic dopant for TiO_2 due to its similar radius to that of Ti^{4+}, where it is easy to substitute into TiO_2 lattice. The presence of this cationic dopant generally enhances the visible light absorption, which may be ascribed to the presence of 3d orbitals of Fe^{3+} that lead to the a variation of valence band and then reduces the bandgap. Moreover, the d−d transition in Fe^{3+} ions and the charge transfer transition between Fe^{3+}/Fe^{2+} ions can expand the photoresponse of TiO_2 into the visible light region (Komaraiaha et al., 2019). Furthermore, some cationic dopants can displace the wavelength absorption due to the substitution of ions into semiconductor lattice by cations of the dopants, which have absorption in greater wavelength and/or create oxygen vacancies, such as vanadium, chromium, and copper (Zhu & Zhou, 2019).

2.1.2 Electron traps

Semiconductors' electronic structure generally provides a fast recombination of the electronic charge carriers (electron/hole) generated by energy absorption supplied from light radiation (Wang, Huang, Yu, & Wong, 2015). Recombination mechanisms are related to the impurities or other surface imperfections of the crystal lattice and reduce the photocatalytic efficiencies (Singh, Mahalingam, & Singh, 2013). Cation doping improves the trapping of the photogenerated electrons at the surface while minimizing charge carrier recombination and leads to an increased rate in the formation of radicals. However, the electron trap

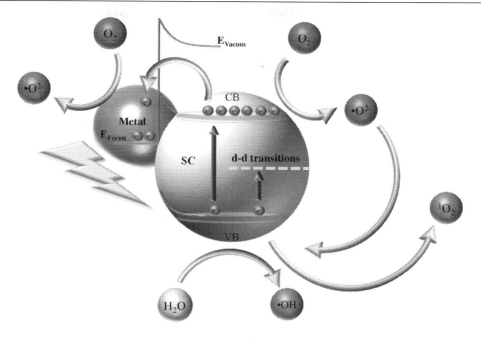

Figure 2.1
Schematic representation of the energy level and Schottky barrier formation by metal doping under UV-vis light. Source: *Modified from Prakash, J., Sun, S., Swart, H.C., Gupta, R.K. (2018). Noble metals-TiO$_2$ nanocomposites: From fundamental mechanisms to photocatalysis, surface enhanced Raman scattering and antibacterial applications. Applied Materials Today, 11 82−135. https://doi.org/10.1016/j. apmt.2018.02.002.*

mechanism relies on dopant nature and concentration. The transition metals incorporated in the substitutional or interstitial positions into photocatalyst lattice can induce the oxygen vacancies in the bulk and on the surface of the semiconductor. This enhances the adsorption of water and formation of surface hydroxyl groups, which promotes the photocatalytic activity. Furthermore, these transitions metals can act as electron scavengers to prevent the recombination process as well (Crişan et al., 2018; Khaki et al., 2017).

At the interfaces, cationic dopants like noble metals retard the electron−hole pair recombination due to the Schottky barrier formed between the semiconductor CB and the metal Fermi level, which assure an efficient electron trap (Luo, Zhang, Zheng, Geng, & Debliquy, 2017). Fermi level is an important parameter for establishing the electronic properties of a solid, which is related to the chemical potential of the electrons (Coronado et al., 2013). Since the Fermi level of metal dopants is lower than the n-type semiconductor, photogenerated electrons are entrapped by the metal nanoparticles in the interface. However, the Fermi level is higher for a p-type semiconductor and the migration of holes from the semiconductor to the metal is made easy by the migration of charge carriers (Adekoya, Tahir,

& Amin, 2019; Kumaravel et al., 2019). In both cases, there is an accumulation of electrons or holes at the interface, which are unable to flow back to their initial level due to the Schottky barrier (Ponraj, Vinitha, & Daniel, 2017), as depicted in Fig. 2.1.

The RE dopants can decorate the semiconductors' surface as highly dispersed metal oxide "clusters," which interact with the photocatalyst ions, such as Ti^{4+} and Zn^{2+} (Samadi, Zirak, Naseri, Khorashadizade, & Moshfegh, 2016; Shayegan et al., 2018). The semiconductor ions (SC) replace lanthanide ions (Ln^{3+}) in the RE oxides lattice on the interface, producing an $SC-O-Ln$ bond. For example, when a lanthanide is incorporated on the TiO_2 surface, the development of $Ti-O-Ln$ bonds could induce a charge surface imbalance in the photocatalyst lattice due to the reduction of Ti^{4+} ions to Ti^{3+} by electronegativity effects (Colpani, Zanetti, Cecchin, et al., 2018; Samadi et al., 2016). The presence of negative charges would disturb the surface energy of the structure and make it necessary to remove one O^{2-} ion for every two Ti^{3+} ions produced, causing the formation of oxygen and titanium vacancies capable of capturing electrons and reducing the recombination rate of the photogenerated charges, and leading to an increase in the photocatalytic activity (Colpani, Zanetti, Cecchin, et al., 2018).

However, when cation loading is beyond optimum value, the dopant can behave as a mediator for interfacial charge transfer or as a recombination center for electron-hole pairs, which will be adverse to the photocatalytic activity (Jiang et al., 2017). Therefore the concentration of dopant should be controlled and an optimum quantity of metal should be used to improve the efficiency of the photocatalytic activity.

2.2 Transition metals

Transition metals are the elements from groups 3 to 12 of the Periodic Table, which have been used as dopants to improve the photocatalytic performance of the semiconductors, since their valence-shell electrons are located in the d-orbital (Gao & Guo, 2017). The incompletely filled d-shells provide many energy levels for the effective optical transitions to take place in the semiconductor (Dramićanin, 2018). Many transition metals like Fe, Pd, Cd, V, Ni, Cr, Cu, Mn, and Co are employed as cationic dopants by introducing an energy level between CB and VB, which reduces the band gap energy and shifts the wavelength absorption of the photocatalyst into the visible region (Varma et al., 2020).

2.2.1 Iron

The incorporation of an appropriate amount of Fe^{3+} into semiconductors, such as TiO_2, ZnO, ZnS, WO_3, and CdS, can modify the optical properties of these photocatalysts due to its extrinsic band gap ($\sim 2.6\,eV$) (Chen, Qiu, Xu, Cao, & Zhu, 2015). Since the ionic radius of iron (0.64 Å) is similar or smaller than that of Ti^{4+} (0.68 Å), Zn^{2+} (0.74 Å), W^{6+} (0.74 Å), and Cd^{2+}

(0.95 Å), Fe^{3+} can easily enter into the semiconductor crystal lattice either substitutionally or interstitially and it can introduce an energy level that enhances the visible-light absorption (Khosroshahi & Mehrizad, 2019; Komaraiah, Radha, Sivakumar, Reddy, & Sayanna, 2019; Shamsipur, Rajabi, & Khani, 2013; Yazdanbakhsh, Eslami, Massoudinejad, & Avazpour, 2020). Therefore, among the transition metals, iron is most popular and widely employed due to its similar ionic radius to those of many semiconductors.

The substitution of iron cations in photocatalyst lattice introduces many crystal defects, such as oxygen vacancies, which may act as efficient electron traps. Besides, when Fe^{3+} is into the lattice it can turn into Fe^{2+} and Fe^{4+} ions by trapping photogenerated electrons and hole traps, which reduce the recombination rate (Kaur, Shahi, Shahi, Sandhu, & Sharma, 2020). When dispersed on the semiconductors' surface, the iron oxide clusters can trap the excited electrons in the CB and slow down the recombination rate of photogenerated e^-/h^+ pairs (Muramatsu, Jin, Fujishima, & Tada, 2012).

A summary of iron dopant for the photocatalytic performance is given in Table 2.1.

2.2.2 Copper

Copper ions, analogous to iron, can easily be incorporated into substitutional or interstitial positions in the semiconductor lattice because they have an ionic radius (0.73 Å) closer than that of SC (Khaki, Shafeeyan, Raman, & Daud, 2018). Accordingly, the ions replaced into the lattice of the photocatalyst can generate some defects by forming Cu−O−SM bonds, which reduce electron/hole pair recombination, such as above mentioned (Shayegan et al., 2018).

Table 2.1: Summary of iron-doped photocatalysts.

Photocatalyst	Doping method	Target pollutant	Light source	Degradation/ time	References
TiO_2	Sol−gel	Acid orange 7	UV light— 30 W Visible light—30 W	49.6%/360 min 53.2%/360 min	Shen, Chuang, Jiang, Liu, and Horng (2020)
TiO_2	Hydrothermal	Diazinon	UV light— 15 W	85%/100 min	Tabasideh, Maleki, Shahmoradi, Ghahremani, and McKay (2017)
ZnO	Sol−gel	Rhodamine B	Visible light	100%/80 min	Bousslama, Elhouichet, and Férid (2017)
WO_3	Coprecipitation	Methyl red	Visible light— 500 W	94%/120 min	Wang et al. (2018)
ZnS	Hydrothermal	P-nitrophenol	Visible light— 500 W	83.8%/180 min	Mehmood, Iqbal, Jan, and Mansoor (2017)
CdS	Chemical precipitation	Methylene blue	Visible light— 400 W	100%/60 min	Junaid et al. (2019)

This cationic dopant creates additional levels into band gap, which mainly exists in the form of CuO (Cu^{2+}) and Cu_2O (Cu^+). The photogenerated electrons are transferred from VB of the photocatalyst to Cu^{2+} ions, reducing the recombination rate. Then, Cu^{2+} is reduced into Cu^+, which acts as a multielectron oxygen reduction catalyst. Afterward, Cu^+ will be reduced into Cu^0 due to trap electrons to the VB or could be reoxidized to Cu^{2+} by O_2 that is present in the system (Abdullah et al., 2017; Etacheri et al., 2015; Kumaravel et al., 2019). These sequential reactions could effectively reduce the electron−hole recombination rate. The impact of Cu oxidation species, due to the high positive redox potentials of 0.52 V (Cu^{2+}/Cu) and 0.16 V (Cu^{2+}/Cu^+), makes copper a suitable modifier for various visible-light-responsive photocatalysts (Khaki et al., 2017).

Moreover, copper-doped TiO_2 shows disinfection property against both Gram-positive and Gram-negative bacteria under visible light in addition to the increased photocatalytic capacity (Yadava & Jaiswar, 2017). Among many transition metals, copper has been widely used as a cationic dopant due to these features (Khaki et al., 2018) and Table 2.2 shows some applications of Cu-doped photocatalysts.

2.2.3 Manganese

Manganese has several oxidation states with different ionic radii, such as Mn^{2+} (0.80 Å), Mn^{3+} (0.66 Å), and Mn^{4+} (0.60 Å), which produce oxides with different physical and chemical properties. Manganese oxides are made from the basic MnO_6 octahedrons that are connected in a variety of ways, which suggests that the incorporation of Mn into semiconductor crystal lattice could produce an octahedral coordination (Islam, Morton, Johnson, Mainali, & Angove, 2018; Pérez-Larios et al., 2016). This is essenital because the

Table 2.2: Comparison of photocatalytic efficiency under various systems.

Photocatalyst	Doping method	Target pollutant	Light source	Degradation/ time	References
TiO_2	Sol−gel	2-Chlorophenol	Visible light−5 W	100%/360 min	Lin, Sopajaree, Jitjanesuwan, and Lu (2018)
ZnO	Solvothermal	Methylene blue	UV light−500 W Visible light−500 W	100%/7 min > 90%/60 min	Xu, Bai, Li, Guo, and Bai (2018)
WO_3	Sol−gel	Bisphenol A	Visible light−150 W	80%/480 min	Goulart, Alves, and Mascaro (2019)
CdS	Hydrothermal	Methylene blue	Visible light	95.5%/240 min	Wang, Yang, et al. (2017)
ZnS	Coprecipitation	Methylene blue	IR light−150 WVisible light−20 W	32.9%/60 min 73.5%/60 min	Prasad and Balasubramanian (2017)

visible-light absorption in Mn-doped photocatalysts is usually due to the crystal d–d transition field of octahedral Mn^{2+} and Mn^{3+} (Feng et al., 2016; Pérez-Larios et al., 2016; Rao et al., 2015).

Therefore the role of Mn as a cationic dopant in semiconductors increases the visible-light absorption due to the mixing and interaction between the s and p electrons of the host photocatalyst and the 3d electrons of manganese cations that make the forbidden transition partially allowed (Kaur, Sharma, & Pandey, 2015; Putri et al., 2018). These dopant ions create new energy levels and they may occupy the new localized electronic states within the band gap near the CB, which can promote the narrowing of band gap (Jiang et al., 2016; Putri et al., 2018). Then, under the light irradiation source, the electrons on the semiconductor VB migrate to the impurity band and after either to the CB of the photocatalyst or directly to adsorbed oxygen, water molecules, or hydroxyl groups (Abhudhahir & Kandasamy, 2015).

The presence of cations Mn^{2+}, Mn^{3+}, and Mn^{4+} enhances the photocatalytic activity because they easily change the oxidation state, which can suppress the electron and hole recombination and improve the charge separation efficiency. All of these oxidation states can easily be oxidized until Mn^+ due to trap electrons or combine with hydroxyl groups. On the other hand, these cations can be reduced until Mn^{4+} by photogenerated holes or adsorbed oxygen. Eqs. (2.1)–(2.4) summarize some reactions (Binas, Venieri, Kotzias, & Kiriakidis, 2017; Pérez-Larios et al., 2016; Putri et al., 2018).

$$Mn^{2+} + e^- \rightarrow Mn^+ \tag{2.1}$$

$$Mn^+ + O_2 \rightarrow Mn^{2+} + \bullet O_2^- \tag{2.2}$$

$$Mn^{2+} + h^+ \rightarrow Mn^{3+} \tag{2.3}$$

$$Mn^{3+} + OH^- \rightarrow Mn^{2+} + \bullet OH \tag{2.4}$$

Furthermore, the incorporation of dopants, besides the achieved red shift in optical absorption edges, induces crystal defects, especially oxygen vacancies, which themselves act as electron traps that effectively suppress the recombination of the photogenerated carriers (Kaur et al., 2015; Samadi et al., 2016). However, the excessive amount of manganese ions can reduce the photocatalytic activity due to the dopant sites acting as an efficient recombination center (Jiang et al., 2016). Table 2.3 shows an overview of some Mn-doped photocatalysts. Compared with other transition metals, iron and manganese are more abundant and cheaper (Chen et al., 2020)

2.2.4 Nickel

Nickel naturally can exist in many oxidation states, including N^0, Ni^+, Ni^{2+}, Ni^{3+}, and Ni^{4+}, but the prevalent one under environmental conditions is Ni^{2+} (0.72 Å). These

Table 2.3: Examples of pollutants degraded by Mn-doped photocatalysts.

Photocatalyst	Doping method	Target pollutant	Light source	Degradation/ time	References
TiO_2	Sol—gel	Methylene blue Methyl orange	UV light— 8 W	MB—94%/ 60 min MO—85%/ 60 min	Gnanasekaran et al. (2016)
			Visible light— 300 W	MB—73%/ 240 min MO—65%/ 240 min	
ZnO	Combustion	Phenol Chromium VI	Visible light— 250 W	Ph—79%/ 210 min Ch—90%/ 210 min	Kumar, Amanchi, Sreedhar, Ghosal, and Subrahmanyam (2017)
ZnS	Solvothermal	Chromium VI	Visible light— 300 W	>98%/20 min	Wang, Wang, et al. (2017)
CdS	Coprecipitation	Methylene blue	Visible light— 300 W	>85%/90 min	Deka and Kalita (2018)
WO_3	Microwave irradiation	Methylene blue	Visible light— 125 W	79%/120 min	Abhudhahir and Kandasamy (2015)

oxidation states provide different redox pathways, such as Ni^{2+}/Ni ($E^0 = -0.23$ V vs SHE) (Cai, Geng, & Wang, 2016; Khake & Chatani, 2020; Wang, Liu, Zhong, & Lu, 2019), which can enhance the photocatalytic activity of Ni-doped semiconductors.

The incorporation of nickel ions into semiconductors lattice is possible, without significantly distorting the crystal structure, due to the similar ionic radius. Then, impurity energy levels are introduced and shifted the light absorption edge to the visible-light region when photocatalysts are doped with nickel ions, which are probably attributed to the interaction of the semiconductor 3d orbitals and 3d level of Ni^{2+} (Kaur et al., 2015; Khojasteh, Salavati-Niasari, & Mortazavi-Derazkola, 2016; Nsib, Saafi, Rayes, Moussa, & Houas, 2016).

Furthermore, Ni^{2+} also plays an important role in a semiconductor structure, such as ZnO, TiO_2, and ZnS, because it greatly suppresses the recombination of electron—hole pairs on the surface of the photocatalyst by trapping the photogenerated electrons at CB and transferring the excited electrons to the adsorbed oxygen. Similar to manganese ions, nickel oxidation states can easily be reduced by trapping electrons or be oxidized by trapping holes, as summarized in Eqs. (2.5)—(2.10) (Devi, Kottam, Kumar,

& Rajashekhar, 2010; Kaur et al., 2015; Shayegan et al., 2018; Tripathi et al., 2015).

$$Ni^{2+} + e^- \rightarrow Ni^+ \qquad (2.5)$$

$$Ni^{3+} + e^- \rightarrow Ni^{2+} \qquad (2.6)$$

$$Ni^+ + O_2 \rightarrow Ni^{2+} + \bullet O_2^- \qquad (2.7)$$

$$Ni^+ + h^+ \rightarrow Ni^{2+} \qquad (2.8)$$

$$Ni^{2+} + h^+ \rightarrow Ni^{3+} \qquad (2.9)$$

$$Ni^{3+} + OH^- \rightarrow Ni^{2+} + \bullet OH \qquad (2.10)$$

The photocatalytic activity enhancement can also be attributed to the formation of Schottky barriers at the interface, which acts as electron traps to suppress photogenerated charge carrier recombination (Blanco-Vega et al., 2017). Table 2.4 represents the comparative analysis of Ni-doped semiconductors in terms of photocatalysis.

2.3 Noble metals

Noble metals include silver (Ag), gold (Au), platinum (Pt), osmium (Os), palladium (Pd), ruthenium (Ru), rhodium (Rh), and iridium (Ir). These metals have been reported as powerful cationic dopants in photochemical processes due to their high efficiency in trapping photogenerated electron via the Schottky barrier since their Fermi level is typically lower than that of photocatalysts and because they are able to adsorb light energy over a

Table 2.4: Examples of pollutants degraded by Ni-doped photocatalysts.

Photocatalyst	Doping method	Target pollutant	Light source	Degradation/time	References
TiO_2	Microwave Sol—gel	Bisphenol A	Visible light—25 W	100%/180 min	Blanco-Vega et al. (2017)
ZnO	Hydrothermal	Tartrazine	UV light—8 W	>90%/30 min	Türkyılmaz, Güy, and Özacar (2017)
ZnS	Wet chemical precipitation	Methylene blue	Visible light	94%/150 min	Priyadharsini, Elango, Vairam, and Thamilselvan (2016)
WO_3	Coprecipitation	Methyl red	Visible light	96%/120 min	Mehmood, Iqbal, Ismail, and Mehmood (2018)
CdS	Solvothermal	Methylene blue	UV light—8 W Visible light	50%/80 min 70%/80 min	Ahmed, Ojha, and Kumar (2017)

wide range of the solar spectrum (Etacheri et al., 2015; Liu et al., 2017; Wang, Ye, Iocozzia, Lin, & Lin, 2016).

Noble metal dopants can red-shift the wavelength absorption of the photocatalysts due to the SPR, which is a phenomenon that occurs on the surface of noble metals and is related to the collective oscillation of surface electrons when they are exposed to electromagnetic radiation, especially at visible wavelengths (Chen et al., 2015; Dal'toé et al., 2018). The energy generated by the electron oscillation can be transferred to the semiconductor as electromagnetic energy or as excited electrons, improving its activity in visible light (Reddy et al., 2016). In noble metals, the SPR energy is between 1.0 and 4.0 eV in respect to the metal Fermi level and the electron formed in the SPR excitation process will be in this energy range, having enough energy to overcome the Schottky barrier, which allows the transfer of electrons from the metal to the semiconductor (Linic, Christopher, & Ingram, 2011). A schematic representation of the charge carrier separation and transfer via the Schottky barrier and SPR is shown in Fig. 2.2.

2.3.1 Silver

Silver (Ag) is a monovalent noble metal belonging to the transition metals group, with atomic number 47 and electronic distribution $[Kr]4d^{10}5s^1$. Among the noble metals, Ag is the cheapest and has the highest thermal and electrical conductivity (Li & Meng, 2020). Moreover, silver has a great potential in photocatalytic applications due to its low cost, compared to other candidates as gold and platinum, and high efficiency in the separation of charge carriers. This noble metal shows intense surface plasmon in the range of 320−450 nm. Nevertheless, the resonant wavelength and the SPR intensity depend on the nature, size and shape of metal NPs (Dal'Toé et al., 2018).

However, since the ionic radius of silver (1.26 Å) is higher than that of many semiconductors, the metallic silver species (Ag^0, Ag_2O, and AgO) cannot be easily incorporated in substitutional or interstitial positions into the crystal lattice. In this way, these species are generally decorated onto the surface of photocatalysts at the grain boundaries or like separated particles and clusters, although some discussions have been presented about changes in the crystal structure of semiconductors (Cui et al., 2017; Jaramillo-Páez, Navío, & Hidalgo, 2018; Samadi et al., 2016).

Doping silver on photocatalysts substantially inhibits the electron−hole recombination, enhancing the photocatalytic activity, due to the difference in the semiconductor CB potential and the Ag Fermi level potential (0.4 eV vs NHE), which arise in a space-charge layer, the Schottky barrier energy. Accordingly, if the CB energy of the photocatalyst is higher than the silver Fermi level, under the UV radiation, the photogenerated electrons on the n-type semiconductor CB will be transferred to the noble metal until to reach an

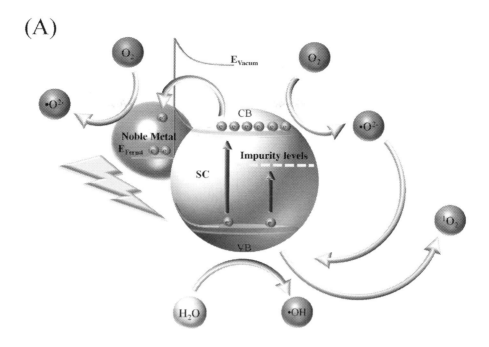

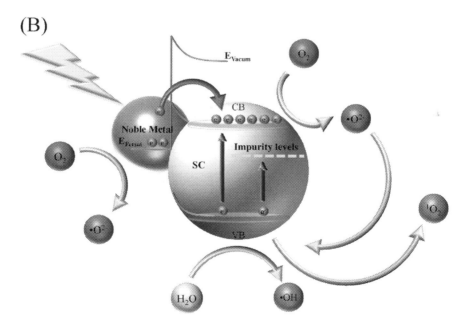

Figure 2.2

Schematic representation of electron transfer via Schottky barrier and SPR under (A) UV light and (B) visible light. Source: *Modified from Prakash, J., Sun, S., Swart, H.C., Gupta, R.K. (2018). Noble metals-TiO₂ nanocomposites: From fundamental mechanisms to photocatalysis, surface enhanced Raman scattering and antibacterial applications. Applied Materials Today, 11 82–135. https://doi.org/10.1016/j. apmt.2018.02.002.*

interfacial charge equilibrium. Thus, as the reversible process can be extremely difficult, Ag-trapped electrons react with the adsorbed oxygen to form superoxide radicals (Devi & Kavitha, 2016; Jin et al., 2020; Qi, Cheng, Yu, & Ho, 2017).

On the other hand, under visible light, the nanoparticles of this dopant strongly interact with incident electromagnetic radiation in a wide range of visible light, mentioned above. As a result, in n-type semiconductors, the electromagnetic excitations get coupled with collective oscillations of those free electrons in the silver, which overcome the Schottky barrier and are transferred to the CB of the semiconductor (Prakash et al., 2018).

The performance of silver-modified semiconductors in photocatalytic reduction of contaminants is tabulated in Table 2.5.

2.3.2 Gold

The catalytic activity of gold was unknown until the 1980s when studies demonstrated that the catalytic performance of this noble metal is dependent on the size and shape (Tang, Zhao, Long, Liu, & Li, 2016). Then, the properties of Au nanoparticles have received considerable attention, such as in photocatalysis. The electron energy levels of gold correspond to the completely filled $5d^{10}$ orbital with one electron on the $6s^1$ level, being

Table 2.5: Examples of pollutants degraded by Ag-doped photocatalysts.

Photocatalyst	Doping method	Target pollutant	Light source	Degradation/time	References
TiO_2	Sol—gel	Methylene blue	Visible light—500 W	96%/60 min	Ali et al. (2018)
ZnO	Precipitation	Methyl orange Rhodamine BPhenol	UV light—300 W	MO—100%/60 min RB—100%/60 min Ph—100%/60 min	Jaramillo-Páez et al. (2018)
			Visible light—300 W	RB—~65%/120 min Ph—~10%/120 min	
ZnS	Chemical	Strychnine	UV light—25 W	92.8%/60 min	Gupta, Fakhri, Azad, and Agarwal (2018)
WO_3	Hydrothermal	Sulfanilamide	Visible light—200 W	96.1%/300 min	Zhu, Liu, Yu, Zhou, and Yan (2016)
CdS	Sol—gel Ultrasound	Rhodamine BPhenol	Visible light—300 W	RB—100%/120 min Ph—80%/150 min	Khan et al. (2018)

demonstrated in the electronic distribution $[Kr]4f^{14}5d^{10}6s^1$, which favors electron losing and the formation of the oxidation state Au^+ (Li & Meng, 2020). Besides this oxidation states, gold can exist as Au^0 or Au^{3+} ions when incorporated into semiconductors and the control of these species is very important because high concentrations of Au^0 or Au^{3+} can act as hole trap or recombination centers, respectively, and reduce the photocatalytic activity (Devi & Kavitha, 2016; Kavitha & Kumar, 2019). Because of the difference between ionic radius of Au^+ (1.37 Å) and Au^{3+} (0.85 Å) (Terzioglu, Aydin, Koc, & Terzioglu, 2019), while carrying out the gold doping it is possible that these noble metal ions decorate the semiconductors' surface or exist into the crystal lattice in substitutional or interstitial positions, respectively.

In photochemical processes, gold nanoparticles can efficiently increase the charge carrier separation and electron trapping due to their higher work function (~ 5.31 eV) (Bora, Myint, Al-Harthi, & Dutta, 2015). In metal-doped semiconductors, under the UV light illumination, the photogenerated electrons on the CB of the semiconductor are transferred to the metal when the work function of the metal is lower (more positive) than the potential of the CB edge of the semiconductor (Li & Meng, 2020). Since the work function of Au is more positive than the CB of semiconductors, such as ZnO ($E_{CB} = -0.5$ eV vs SHE), TiO_2 ($E_{CB} = -0.24$ eV vs SHE), and WO_3 ($E_{CB} = 0.1$ eV vs SHE), the transfer of the photoexcited electrons from the semiconductor to the metal occurs until the Fermi level of the semiconductor reaches equilibrium with that of the metal and a potential barrier is formed, the Schottky barrier (Devi & Kavitha, 2016; Khan, Chuan, Yousuf, Chowdhury, & Cheng, 2015; Li, Yu, Quan, Chen, & Zhang, 2016; Sun et al., 2016). Therefore, the high work function of gold may enhance the Schottky barrier effect, facilitating electron transfer into Au compared to other metals and reducing the electron and hole recombination (Gupta, Melvin, Matthews, Dash, & Tyagi, 2016).

The SPR feature of silver also is observed in Au-doped semiconductors, although Ag nanoparticles are more intense than that of gold. However, gold has strong visible-light absorption over a wide range, with the absorption peak usually lies around 520 nm, demonstrating efficient harvesting of visible light and increasing the photocatalytic activity (Chava, Do, & Kang, 2018; Gupta et al., 2016).

Pertinent results on Au-doped semiconductor photocatalysis under visible light are summarized in Table 2.6.

2.3.3 Platinum

Among noble metals, platinum (Pt) is one of the most employed elements in photocatalysis (Coronado et al., 2013). The electron configuration of Pt is $[Xe]4f^{14}5d^96s^1$ and the main oxidation states are Pt^{2+} and Pt^{4+} with ionic radius equal to 0.94 and 0.75 Å, respectively (Dorothy, Subramaniam, & Panigrahi, 2019; Hu, Song, Jiang, & Wei, 2015). Consequently,

Table 2.6: Comparison of Au-doped photocatalysts efficiency under various systems.

Photocatalyst	Doping method	Target pollutant	Light source	Degradation/ time	References
TiO_2	Deposition Precipitation	Methylene blue	Visible light—150 W	>60%/40 min	Pal and Kryschi (2016)
ZnO	Photodeposition	Methylene blue	Visible light—500 W	>90%/90 min	Bora et al. (2015)
ZnS	Chemical precipitation	Methyl orange	Visible light—150 W	76%/60 min	Misra, Gupta, Paul, and Singla (2015)
WO_3	Sputtering	Methylene blue	Visible light—80 W	>50%/360 min	Choi, Kim, and Hahn (2010)
CdS	Photodeposition	Cyanide	Visible light—150 W	100%/40 min	Aazam (2014)

when semiconductors are doped with platinum, these oxidation states are predominant and the ion Pt^{4+} could be introduced as an impurity in substitutional or interstitial positions into the crystal lattice of photocatalysts due to the similar ionic radius than that of Ti^{4+}, Zn^{2+}, W^{6+}, and Cd^{2+}. On the other hand, the insertion of Pt^{2+} into the semiconductor structure is difficult and this ion is usually on the catalyst surface in the form of PtO. The presence of these cationic species into the lattice or on the surface could enhance charge carrier separation by trapping electrons or holes, as shown in Eqs. (2.11) and (2.12). Furthermore, the charge transfer between Pt^{2+} and Pt^{4+} can contribute to increasing the light absorption (Abdennouri et al., 2015; Hu et al., 2015; Zhao, Liu, Gong, Wu, & Wu, 2019).

$$Pt^{4+} + e^- \rightarrow Pt^{3+} \tag{2.11}$$

$$Pt^{2+} + h^+ \rightarrow Pt^{3+} \tag{2.12}$$

Platinum, with the highest work function (~ 5.6 eV) in comparison to Au and Ag (consequently the lower Fermi level), can create the largest Schottky barrier at the interface of semiconductor/metal junction, which results in an intense depletion layer. Because of this energetic difference, it is expected that Pt can promote one of the most efficient and faster traps of photoexcited electrons and charge separation (Devi & Kavitha, 2016; Liu et al., 2018; Zare, Mortezaali, & Shafiekhani, 2016).

Another commonly ascribed effect of platinum-doped semiconductors is the ability to shift the photocatalyst light absorption into visible light due to SPR, such as observed in gold and silver. The strongest visible-light plasmon absorption peak generally appears near 540 nm when platinum nanoparticles are employed in the doping of semiconductors (Antolín, Contreras, Medina, & Tichit, 2017; Qin, Wang, & Tan, 2018; Zhang et al., 2016).

Table 2.7 summarizes some of the platinum-doped semiconductors developed for photocatalysis degradation.

Table 2.7: A summary of the key findings of Pt-doped photocatalysts.

Photocatalyst	Doping method	Target pollutant	Light source	Degradation/ time	References
TiO_2	Photoreduction	Methyl orange	Visible light— 250 W	84.3%/ 120 min	Hu et al. (2015)
ZnO	Photodeposition	Malachite green	Visible light— 300 W	100%/50 min	Mohamed, McKinney, Kadi, Mkhalid, and Sigmund (2016)
WO_3	Hydrothermal Reduction	Toluene	Visible light— 300 W	98%/90 min	Li et al. (2020)
CdS	Photodeposition	Tetracycline	Visible light— 300 W	66%/100 min	Fang et al. (2017)

2.3.4 Palladium

Palladium (Pd) is another noble metal that contributes to the photocatalytic activity of semiconductors due to the surface plasmonic resonance effect and because it can suppress the recombination of electron/hole pair. The electron configuration of Pd is equal to $[Kr]4d^{10}$ and palladium primarily exists as Pd^0 and Pd^{2+} oxidation states (Li & Meng, 2020). Accordingly, when semiconductors are doped with palladium, these oxidation states can coexist as Pd^0 clusters loaded on the surface of the photocatalysts and as Pd^{2+}, which usually enters into semiconductor lattice because the ionic radius of Pd^{2+} (0.86 Å) is similar to that of several semiconductor ions (Chang et al., 2019; Güy, Çakar, & Özacar, 2016).

Therefore dopping with palladium ions has been considered as a promising candidate to improve the photocatalytic activity due to the creation of impurity levels of Pd^{2+} inside the conduction and VBs of semiconductors, which can capture the photogenerated electrons and holes and change to Pd^+ or Pd^{3+}, respectively. On the other hand, since the stability of Pd^+ is lower than the Pd^{2+} species, it can be easily oxidized to Pd^{2+} after trapping electrons on the photocatalyst surface and, then, this charge transfer reduces the recombination of photogenerated electrons (Channei, Nakaruk, Jannoey, & Phanichphant, 2019; Güy et al., 2016).

Another distinguishing feature of Pd that could collaborate to increase the photocatalytic activity is related to work function. The large work function of palladium (~ 5.6 eV) produces a Schottky barrier with a higher potential difference between the Fermi level of this noble metal and the CB of semiconductors, such as TiO_2, which provides an effective transfer and trap of photogenerated electrons from the CB of semiconductor to the palladium. Besides, Pd has a much lower electron affinity and hence electron trapping capability than Pt and Au, which may enable more facile electron transfer from Pd to donor species (Chang et al., 2019; Li, Yu, Liu, Sun, & Huang, 2016; Al-Azri et al., 2015).

Palladium also acts as an antenna to capture resonance photons due to the SPR effect, intensifying the light absorption and photocatalytic activity. However, the particle size is an important parameter in the plasmonic characteristics. Palladium has a plasmonic absorption peak in UV light wavelengths when the particles dispersed on the semiconductor surface are smaller than 10 nm, while larger particle size and Pd clusters contribute to the enhanced absorption and resonance in visible light (Devi and Kavitha, 2016; Leong et al., 2018; Hosseini-Sarvari & Bazyar, 2018; Zhou, Jiang, Xue, & Li, 2019).

Table 2.8 shows an overview of some Pd-doped photocatalysts.

2.4 Rare earths

RE elements comprise 17 elements of the periodic table, including lanthanides (La, Ce, Pr, Nd, Pm, Sm, Eu, Gd, Tb, Dy, Ho, Er, Tm, Yb, and Lu), with the atomic number from 57 to 71, scandium (Sc) and yttrium (Y). The electron configuration of these elements is described as $[Xe]4f^n6s^2$ or $[Xe]4f^n5d^16s^2$ ($0 \le n \le 14$), which usually have an oxidation state equal to RE^{3+}, having partly filled 4f and empty 5d orbitals (Colpani et al., 2018; Dramićanin, 2018; Shayegan et al., 2018). These 4f electrons are shielded from the environment by the outer orbitals 5s and 5p, resulting in a low interaction of 4f electrons with the chemical environment. This effective shielding is responsible for the optical properties of REs since there are almost no perturbations of the f—f transitions within 4f orbitals. Due to the different arrangements of these orbitals, it is possible to generate different levels of energy, allowing the absorption in a broad spectrum of light radiation, from near-infrared to

Table 2.8: Photocatalytic degradation of contaminants by Pd-doped semiconductors.

Photocatalyst	Doping method	Target pollutant	Light source	Degradation/ time	References
TiO_2	Chemical reduction	Perfluorooctanoic acid	UV light—125 W	94%/420 min	Li et al. (2016)
ZnO	Microwave Photoreduction $NaBH_4$ reduction	Congo red	UV light—100 W	MW = 75%/ 60 min PR = 95%/ 60 min NR = 98%/ 60 min	Güy et al. (2016)
$ZnWO_4$	Hydrothermal Chemical deposition	Nitric oxide	Visible light—300 W	55%/ < 10 min	Chang et al. (2019)
CeO_2	Wet impregnation	Methyl orange	Visible light—54 W	92%/120 min	Channei et al. (2019)

ultraviolet (Li & Yan, 2019; Mazierski et al., 2018; Sousa Filho, Lima, & Serra, 2015). Therefore, electronic structures of lanthanides are unique among the elements and the RE-doped semiconductors are of particular interest for photocatalytic applications.

Moreover, REs have a strong ability to form complexes by coordination bonding between their f orbitals and lone electron pairs of various Lewis bases (e.g., acids, amines, aldehydes, alcohols, and thiols), which could concentrate the organic molecules at the semiconductor enhancing the adsorption at the surface, where the reactive oxygen species (ROS) are formed (Dal'Toé et al., 2018; Al-Mamun et al., 2019). A large amount of these species return to a thermodynamically stable state without reacting with organic molecules due to their short lifetimes and because the molecules are dispersed in the solution. Thus, when the molecules are adsorbed, the transfer of electrons between the ROS (•OH, $^1O_2$, and •O_2^-) and the contaminant becomes more probable, increasing the efficiency of the process (Colpani, Zanetti, Cecchin, et al., 2018). The incorporation of REs in semiconductors reduces the particle size and can inhibit the transition from anatase to rutile phase, in the case of RE-doped TiO_2 (Al-Mamun, Kader, Islam, & Khan, 2019; Shayegan et al., 2018). Both features may be associated with the fact that the lanthanides are dispersed interstitially in the grain boundaries, producing SC$-$O$-$Ln bonds, which will stabilize the surface energy (Colpani et al., 2018; Dal'Toé et al., 2018).

The charge surface imbalance produced in the photocatalyst lattice due to the presence of RE elements was discussed in Section 2.1.2, which is an important feature to generate oxygen vacancies that capture electrons and reduce the recombination rate.

2.4.1 Lanthanum

Lanthanum (La) is the first element in the lanthanide group with an electron configuration equal to [Xe]$5d^16s^2$, where the 4f level is empty, which implies the absence of f$-$f transitions and consequently poor visible-light absorption (Dramićanin, 2018; Kaczmarek, Liu, Voort, & Deun, 2013). Due to the large difference between the ionic radii of La^{3+} (1.15 Å) and some semiconductor ions, lanthanum is commonly dispersed on the surface of the photocatalysts, in the form of La_2O_3, rather than into the crystal lattice (Meksi et al., 2016; Mrabet, Kamoun, Boukhachem, Amlouk, & Manoubi, 2015). Then, the substitutional or interstitial inclusion of this ion into the semiconductor structure is intrinsically associated with the doping method. The incorporation into the crystal structure will lead to the lattice deformation, thus causing an impurity state in the band gap. The existence of an impurity state improves the absorption of visible light by narrowing the band gap.

However, ions such as Ti^{4+}, Zn^{2+}, W^{6+}, and Cd^{2+}, can substitute La^{3+} in the lattice of La_2O_3 clusters, producing SC$-$O$-$La bonds and a charge imbalance, which reduces the electron/hole recombination and increases the photocatalytic activity (Li & Feng, 2016;

Mrabet et al., 2015; Murtaza, Osama, Saleem, Hassan, & Watoo, 2018; Zhu et al., 2017). Furthermore, due to lower energy of the empty 4f orbitals of La^{3+} dopant ions, lanthanum shows the faster transfer of photogenerated electrons among REs (Ayawanna & Sato, 2019). The charge imbalance can result in the higher adsorption of OH^- ions and water molecules. Both bound hydroxyl groups and adsorbed water molecules can interact with vacancies in the electronic structures of the nanoparticles and also reduce the recombination rates of the photogenerated charges (Colpani et al., 2018; Samadi et al., 2016).

Lanthanum can also improve the photocatalytic activity of semiconductors in the visible-light irradiation due to the charge transfer transition between the f-electrons of the RE ions and the CB of photocatalysts, which introduce electronic states into the band gap (Coronado et al., 2013; Dal'Toé et al., 2018). Because of these shielded f−f transitions, the lanthanides can act as luminescence centers and increase the visible-light absorption of RE-doped semiconductors (Lee, Lai, Ngai, & Juan, 2016; Mrabet et al., 2015).

The efficiency application of La-doped semiconductors in photodegradation to remove contaminants is presented in Table 2.9.

2.4.2 Cerium

Cerium (Ce) exhibits an electron configuration equal to $[Xe]4f^15d^16s^2$ with incompletely 4f level, which could result in different optical and catalytic properties (Fifere et al., 2018). The oxidation states of cerium are Ce^{3+} ($4f^15d^0$) and Ce^{4+} ($4f^05d^0$), with an ionic radii equal to 1.03 and 0.92 Å, respectively (Chouchene et al., 2016; Touati, Hammedi, Najjar, Ksibi, & Sayadi, 2016). Ce-doped photocatalysts can enhance the photocatalytic activity probably due to their ability to reduce the electron/hole recombination rate and because of their capacity to shift the absorption edge value to visible light or near-infrared wavelengths (Vieira et al., 2018), similar to the lanthanum dopping.

Table 2.9: Photocatalytic degradation of contaminants by La-doped semiconductors.

Photocatalyst	Doping method	Target pollutant	Light source	Degradation/ time	References
TiO_2	Sol−gel	Acetone	Visible light—300 W	47%/30 min	Ho, Kang, Chang, You, and Wang (2019)
ZnO	Precipitation	Paracetamol	Visible light—60 W	99%/180 min	Thi and Lee (2017)
ZnS	Coprecipitation	Turquoise blue H5G	Visible light—Sunlight	60%/180 min	Suganthi and Pushpanathan (2018)
WO_3	Precipitation	Rhodamine B	Visible light—1000 W	>80%/600 min	Zhu et al. (2017)
CdS	Chemical precipitation	Rhodamine B	Visible light—300 W	95%/45 min	Abdelhameed and El Radaf (2018)

Because of the ionic radii mentioned above, it is difficult to introduce cerium ions into the lattice of semiconductors and thus Ce^{3+} probably occurs like CeO_2 clusters dispersed on the surface (Meshram, Adhyapak, Pardeshi, Mulla, & Amalnerkar, 2017; Touati et al., 2016). Cerium on the photocatalyst surface can change between CeO_2 and Ce_2O_3 (Ce^{4+}) because Ce^{3+} helps in the transport of photogenerated electrons to the oxygen to form superoxide and because Ce^{4+} can act as a trapping site for the photoexcited electron, according to Eqs. (2.13) and (2.14), respectively (Fifere et al., 2018; Sumathi & Kavipriya, 2017; Touati et al., 2016). These cerium oxides can also form SC—O—Ce bonds on the semiconductor surface, which promote charge imbalance and oxygen vacancies (Vieira et al., 2018), as described above.

$$Ce^{4+} + e^- \rightarrow Ce^{3+} \tag{2.13}$$

$$Ce^{3+} + O_2 \rightarrow \cdot O_2^- + Ce^{4+} \tag{2.14}$$

However, depending on the doping method, cerium can be incorporated into the crystal structure, which leads to structural disorder in the formation of lattice and impurity states (Samadi et al., 2016). In this case, the cerium oxide acts as an "antenna" for the visible-light photons and allows some electron transfer from the discrete 4f levels to the bulk semiconductor (Calza et al., 2017). Furthermore, the red shift of absorption spectra for Ce-doped photocatalysts is assigned to a new energy level in the band gap and thus the charge transition between the host VB or CB and the cerium-shielded 4f level, which allows various well-defined narrow optical transitions (Yu et al., 2010).

In addition, Ce doping reduces the particle size and increases the surface area, which favors photodegradation reactions (Raj, Raj, & Das, 2015). Table 2.10 presents

Table 2.10: Photocatalytic degradation of contaminants by Ce-doped semiconductors.

Photocatalyst	Doping method	Target pollutant	Light source	Degradation/ time	References
TiO$_2$	Hydrothermal	Diclofenac	Visible light	100%/80 min	Thiruppathi et al. (2018)
ZnO	Hydrothermal	Acelsulfame K	Visible light—1500 W	>60%/90 min	Calza et al. (2017)
ZnS	Coprecipitation	Turquoise blue H5G	Visible light—sunlight	78%/180 min	Suganthi and Pushpanathan (2019)
WO$_3$	Coprecipitation	Methylene blue	Visible light—sunlight	89%/180 min	Saleem, Iqbal, Nawaz, Islam, and Hussain (2020)
CdS	Hydrothermal	Rhodamine B	Visible light—500 W	76%/120 min	Ranjith, Krishnakumar, Boobas, Venkatesan, and Jayaprakash (2018)

the effects of the Ce-doped photocatalyst on the degradation of some molecules under visible light.

2.4.3 Neodymium

Neodymium (Nd) is considered one of the most reactive lanthanides, having an electronic configuration equal to $[Xe]4f^46s^2$, with quarter-filled 4f shells. Furthermore, neodymium shows a very interesting emission in the near-infrared among the RE metals (Atwood, 2012).

Because of the large mismatch in ionic radii between Nd^{3+} (0.98 Å) and some semiconductor ions, it is difficult for doped neodymium ions to replace the host crystal lattice and probably the neodymium ions could occupy the interstitial spaces or can be presented as clusters of neodymium oxides (Nd_2O_3) on the semiconductor surface, forming Nd—O—SC bonds (Parnicka et al., 2017; Shayegan et al., 2018; Yurtsever & Çifçioğlu, 2018). Thereby, the presence of neodymium can result in the formation of oxygen vacancies and surface defects, which effectively restrain the recombination of photogenerated electron/hole pairs and enhance the photocatalytic efficiency (Dhandapani et al., 2016; Parnicka et al., 2017). Moreover, Nd^{3+} could create some electronic states into the band gap of semiconductors by neodymium 4f orbitals. Then, based on the energy redox potentials of Nd^{3+}/Nd^{2+} ($E^0 = -0.40$ V vs NHE) and because Nd^{3+} ions can act as a Lewis acid due to the presence of partially filled f-orbital, the electrons photoexcited from the VB are easily transferred to the Nd^{3+}/Nd^{2+} dopping energy level (Mazierski et al., 2018; Poongodi, Kumar, & Jayavel, 2015; Sin & Lam, 2018).

Despite the fact that neodymium usually is dispersed on the grain boundaries or interstitial lattice, depending on the doping method some Nd^{3+} can enter into semiconductor lattice, which can shift the light absorption range at the wavelengths of visible light. The red shift is attributed to the formation of new energy levels within the semiconductor band gap, which are related to the absorption bands of Nd^{3+} ion resulting from 4f–4f transitions and play a crucial role in efficient photosensitization by visible light (Mazierski et al., 2018; Oppong, Anku, Shukla, & Govender, 2017; Yurtsever & Çifçioğlu, 2018).

However, high concentrations of neodymium in semiconductor dopping suppress the photocatalytic activity due to a higher number of recombination centers on the semiconductor surface that capture the photogenerated electrons and holes, thereby preventing their diffusion to the photocatalyst surface, which significantly reduces the photocatalytic activity (Mazierski et al., 2018; Vieira et al., 2018).

Table 2.11 summarizes some Nd-doped photocatalysts synthesized to enhance the photocatalytic degradation performance.

Table 2.11: Photocatalytic degradation of contaminants by Nd-doped semiconductors.

Photocatalyst	Doping method	Target pollutant	Light source	Degradation/ time	References
TiO_2	Coprecipitation Hydrothermal	Methylene blue	Visible light— 350 W	99%/120 min	Liang et al. (2020)
ZnO	Solvothermal	Resorcinol	Visible light— sunlight	98%/60 min	Sin and Lam (2018)
SnO_2	Sol—gel	Phenol	UV light— 8 W	95%/120 min	Al-Hamdi, Sillanpää, and Dutta (2014)
ZnSe	Hydrothermal	Acid orange 7	Visible light—6 W	84%/120 min	Khataee, Hosseini, Hanifehpour, Safarpour, and Joo (2014)

References

Aazam, E. S. (2014). Environmental remediation of cyanide solutions by photocatalytic oxidation using Au/CdS nanoparticles. *Journal of Industrial and Engineering Chemistry*, 20, 2870—2875. Available from https://doi.org/10.1016/j.jiec.2013.11.020.

Abdelhameed, R. M., & El Radaf, I. M. (2018). Self-cleaning lanthanum doped cadmium sulfide thin films and linear/nonlinear optical properties. *Materials Research Express*, 5, 066402. Available from https://doi.org/10.1088/2053-1591/aac638.

Abdennouri, M., Elhalil, A., Farnane, M., Tounsadi, H., Mahjoubi, F. Z., Elmoubarki, R., . . . Barka, N. (2015). Photocatalytic degradation of 2,4-D and 2,4-DP herbicides on Pt/TiO_2 nanoparticles. *Journal of Saudi Chemical Society*, 19(5), 485—493. Available from https://doi.org/10.1016/j.jscs.2015.06.007.

Abdullah, H., Khan, Md. M. R., Ong, H. R., & Yaakob, Z. (2017). Modified TiO_2 photocatalyst for CO_2 photocatalytic reduction: An overview. *Journal of CO_2 Utilization*, 22, 15—32. Available from https://doi.org/10.1016/j.jcou.2017.08.004.

Abhudhahir, M. H. S., & Kandasamy, J. (2015). Photocatalytic effect of manganese doped WO3 and the effect of dopants on degradation of methylene blue. *Journal of Materials Science: Materials in Electronics*, 26, 8307—8314. Available from https://doi.org/10.1007/s10854-015-3496-z.

Adekoya, D., Tahir, M., & Amin, N. A. S. (2019). Recent trends in photocatalytic materials for reduction of carbon dioxide to methanol. *Renewable and Sustainable Energy Reviews*, 116, 109389. Available from https://doi.org/10.1016/j.rser.2019.109389.

Ahmed, B., Ojha, A. K., & Kumar, S. (2017). One-pot synthesis of Ni doped CdS nanosheets for near infrared emission and excellent photocatalytic materials for degradation of MB dye under UV and sunlight irradiation. *Spectrochimica Acta Part A: Molecular and Biomolecular Spectroscopy*, 179, 144—154. Available from https://doi.org/10.1016/j.saa.2017.02.007.

Al-Azri, Z. H. N., Chen, W., Chan, A., Jovic, V., Ina, T., Idriss, H., & Waterhouse, G. I. N. (2015). The roles of metal co-catalysts and reaction media in photocatalytic hydrogen production: Performance evaluation of M/TiO_2 photocatalysts (M = Pd, Pt, Au) in different alcohol—water mixtures. *Journal of Catalysis*, 329, 355—367. Available from https://doi.org/10.1016/j.jcat.2015.06.005.

Al-Hamdi, am, Sillanpää, M., & Dutta, J. (2014). Photocatalytic degradation of phenol in aqueous solution by rare earth-doped SnO2 nanoparticles. *Journal of Materials Science*, 49, 5151—5159. Available from https://doi.org/10.1007/s10853-014-8223-2.

Al-Mamun, M. R., Kader, S., Islam, M. S., & Khan, M. Z. H. (2019). Photocatalytic activity improvement and application of UV-TiO$_2$ photocatalysis in textile wastewater treatment: A review. *Journal of Environmental Chemical Engineering, 7*, 103248. Available from https://doi.org/10.1016/j.jece.2019.103248.

Ali, T., Ahmed, A., Alam, U., Uddin, I., Tripathi, P., & Muneer, M. (2018). Enhanced photocatalytic and antibacterial activities of Ag-doped TiO$_2$ nanoparticles under visible light. *Materials Chemistry and Physics, 212*, 325−335. Available from https://doi.org/10.1016/j.matchemphys.2018.03.052.

Antolín, A. M., Contreras, S., Medina, F., & Tichit, D. (2017). Silver/platinum supported on TiO$_2$ p25 nanocatalysts for nonphotocatalytic and photocatalytic denitration of water. *Topics in Catalysis, 60*, 1156−1170. Available from https://doi.org/10.1007/s11244-017-0793-1.

Atwood, D. A. (Ed.), (2012). *The rare earth elements* (1st ed.). London: John Wiley & Sons Ltd, ISBN: 978-1-118-63263-5.

Ayawanna, J., & Sato, K. (2019). Photoelectrodeposition effect of lanthanum oxide-modified ceria particles on the removal of lead (II) ions from water. *Catalysis Today, 321−322*, 128−134. Available from https://doi.org/10.1016/j.cattod.2017.11.010.

Binas, V., Venieri, D., Kotzias, D., & Kiriakidis, G. (2017). Modified TiO$_2$ based photocatalysts for improved air and health quality. *Journal of Materiomics, 3*, 3−16. Available from https://doi.org/10.1016/j.jmat.2016.11.002.

Blanco-Vega, M. P., Guzmán-Mar, J. L., Villanueva-Rodríguez, M., Maya-Treviño, L., Garza-Tovar, L. L., Hernández-Ramírez, A., & Hinojosa-Reyes, L. (2017). Photocatalytic elimination of bisphenol A under visible light using Ni-doped TiO$_2$ synthesized by microwave assisted sol-gel method. *Materials Science in Semiconductor Processing, 71*, 275−282. Available from https://doi.org/10.1016/j.mssp.2017.08.013.

Bora, T., Myint, M. T. Z., Al-Harthi, S. H., & Dutta, J. (2015). Role of surface defects on visible light enabled plasmonic photocatalysis in Au−ZnO nanocatalysts. *RSC Advances, 5*, 96670−96680. Available from https://doi.org/10.1039/c5ra16569e.

Bousslama, W., Elhouichet, H., & Férid, M. (2017). Enhanced photocatalytic activity of Fe doped ZnO nanocrystals under sunlight irradiation. *Optik, 134*, 88−98. Available from https://doi.org/10.1016/j.ijleo.2017.01.025.

Cai, W., Geng, D., & Wang, Y. (2016). Assessment of cathode materials for Ni(II) reduction in microbial electrolysis cells. *RSC Advances, 6*, 31732. Available from https://doi.org/10.1039/c6ra02082h.

Calza, P., Gionco, C., Giletta, M., Kalaboka, M., Sakkas, V. A., Albanis, T., & Paganini, M. C. (2017). Assessment of the abatement of acelsulfame K using cerium doped ZnO as photocatalyst. *Journal of Hazardous Materials, 323*, 471−477. Available from https://doi.org/10.1016/j.jhazmat.2016.03.093.

Chang, L., Zhu, G., Hassan, Q., Cao, B., Li, S., Jia, Y., ... Wang, Q. (2019). Synergetic effects of Pd^0 metal nanoparticles and Pd^{2+} ions on enhanced photocatalytic activity of ZnWO$_4$ nanorods for nitric oxide removal. *Langmuir, 35*, 11265−11274. Available from https://doi.org/10.1021/acs.langmuir.9b01323.

Channei, D., Nakaruk, A., Jannoey, P., & Phanichphant, S. (2019). Preparation and characterization of Pd modified CeO$_2$ nanoparticles for photocatalytic degradation of dye. *Solid State Sciences, 87*, 9−14. Available from https://doi.org/10.1016/j.solidstatesciences.2018.10.016.

Chava, R. K., Do, J. Y., & Kang, M. (2018). Smart hybridization of Au coupled CdS nanorods with few layered MoS$_2$ nanosheets for high performance photocatalytic hydrogen evolution reaction. *ACS Sustainable Chemistry & Engineering, 6*, 6445−6457. Available from https://doi.org/10.1021/acssuschemeng.8b00249.

Chen, J., Qiu, F., Xu, W., Cao, S., & Zhu, H. (2015). Recent progress in enhancing photocatalytic efficiency of TiO$_2$-based materials. *Applied Catalysis A: General, 495*, 131−140. Available from https://doi.org/10.1016/j.apcata.2015.02.013.

Chen, Y., Liu, Y., Li, Y., Wu, Y., Chen, Y., Liu, Y., ... Li, L. (2020). Synthesis, application and mechanisms of Ferro-Manganese binary oxide in water remediation: A review. *Chemical Engineering Journal, 388*, 124313. Available from https://doi.org/10.1016/j.cej.2020.124313.

Choi, H. W., Kim, E. J., & Hahn, S. H. (2010). Photocatalytic activity of Au-buffered WO$_3$ thin films prepared by RF magnetron sputtering. *Chemical Engineering Journal, 161*, 285−288. Available from https://doi.org/10.1016/j.cej.2010.01.050.

Chouchene, B., Chaabane, T. B., Balan, L., Girot, E., Mozet, K., Medjahdi, G., & Schneider, R. (2016). High performance Ce-doped ZnO nanorods for sunlight-driven photocatalysis. *Beilstein Journal of Nanotechnolgy*, *7*, 1338–1349. Available from https://doi.org/10.3762/bjnano.7.125.

Colpani, G. L., Zanetti, M., Zeferino, R. C. F., Silva, L. L., Mello, J. M. M., Riella, H. G., ... Soares, C. (2018). Lanthanides effects on TiO_2 photocatalysts. In S. B. Khan, & K. Akhtar (Eds.), *Photocatalysts: Applications and attributes* (pp. 81–98). London: Intechopen. Available from http://doi.org/10.5772/intechopen.80906.

Colpani, G. L., Zanetti, J. T., Cecchin, F., Dal'Toé, A., Fiori, M. A., Moreira, R. F. P. M., & Soares, C. (2018). Carboxymethyl-β-cyclodextrin functionalization of TiO_2 doped with lanthanum: Characterization and enhancement of photocatalytic activity. *Catalysis Science & Technology*, *8*, 2636–2647. Available from https://doi.org/10.1039/C7CY02115A.

Coronado, J. M., Fresno, F., Hernández-Alonso, M. D., & Portela, R. (2013). *Design of advanced photocatalytic materials for energy and environmental applications. Green energy and technology* (1st ed.). London: Springer-Verlag. Available from https://doi.org/10.1007/978-1-4471-5061-9.

Crişan, M., Mardare, D., Ianculescu, A., Drăgan, N., Niţoi, I., Crişan, D., ... Vasile, B. (2018). Iron doped TiO_2 films and their photoactivity in nitrobenzene removal from water. *Applied Surface Science*, *455*, 201–215. Available from https://doi.org/10.1016/j.apsusc.2018.05.124.

Cui, Y., Ma, Q., Deng, X., Meng, Q., Cheng, X., Xie, M., ... Liu, H. (2017). Fabrication of Ag-Ag_2O/reduced TiO_2 nanophotocatalyst and its enhanced visible light driven photocatalytic performance for degradation of diclofenac solution. *Applied Catalysis B: Environmental*, *206*, 136–145.

Dal'Toé, A. T. O., Colpani, G. L., Padoin, N., Fiori, M. A., & Soares, C. (2018). Lanthanum doped titania decorated with silver plasmonic nanoparticles with enhanced photocatalytic activity under UV-visible light. *Applied Surface Science*, *441*, 1057–1071. Available from https://doi.org/10.1016/j.apsusc.2018.01.291.

Deka, K., & Kalita, M. P. C. (2018). Structural phase controlled transition metal (Fe, Co, Ni, Mn) doping in CdS nanocrystals and their optical, magnetic and photocatalytic properties. *Journal of Alloys and Compounds*, *757*, 209–220. Available from https://doi.org/10.1016/j.jallcom.2018.04.323.

Devi, L. G., & Kavitha, R. (2016). A review on plasmonic metal–TiO_2 composite for generation, trapping, storing and dynamic vectorial transfer of photogenerated electrons across the Schottky junction in a photocatalytic system. *Applied Surface Science*, *360*, 601–622. Available from https://doi.org/10.1016/j.apsusc.2015.11.016.

Devi, L. G., Kottam, N., Kumar, S. G., & Rajashekhar, K. E. (2010). Preparation, characterization and enhanced photocatalytic activity of Ni^{2+} doped titania under solar light. *Central European Journal of Chemistry*, *8*(1), 142–148. Available from https://doi.org/10.2478/s11532-009-0115-y.

Dhandapani, C., Narayanasamy, R., Karthick, S. N., Hemalatha, K. V., Selvam, S., Hemalatha, P., ... Kim, H. (2016). Drastic photocatalytic degradation of methylene blue dye by neodymium doped zirconium oxide as photocatalyst under visible light irradiation. *Optik.*, *127*, 10288–10296. Available from https://doi.org/10.1016/j.ijleo.2016.08.048.

Din, M. I., Najeeb, J., & Ahmad, G. (2018). Recent advancements in the architecting schemes of zinc oxide based photocatalytic assemblies. *Separation & Purification Reviews*, *47*, 267–287. Available from https://doi.org/10.1080/15422119.2017.1383918.

Dorothy, A. A., Subramaniam, N. G., & Panigrahi, P. (2019). Tuning electronic and optical properties of TiO_2 with Pt/Ag doping to a prospective photocatalyst: A first principles DFT study. *Materials Research Express*, *6*, 045913. Available from https://doi.org/10.1088/2053-1591/aafc56.

Dramićanin, M. (2018). Lanthanide and transition metal ion doped materials for luminescence temperature sensing. In M. Dramićanin (Ed.), *Luminescence thermometry: Methods, materials, and applications* (pp. 113–157). Oxford: Elsevier. Available from https://doi.org/10.1016/B978-0-08-102029-6.00006-3.

Etacheri, V., Valentin, C., Schneider, J., Bahnemann, D., & Pillai, S. C. (2015). Visible-light activation of TiO_2 photocatalysts: Advances in theory and experiments. *Journal of Photochemistry and Photobiology C: Photochemistry Reviews*, *25*, 1–29. Available from https://doi.org/10.1016/j.jphotochemrev.2015.08.003.

Exposito, A. J., Patterson, D. A., Mansor, W. S. W., Monteagudo, J. M., Emanuelsson, E., Sanmartín, I., & Duran, A. (2017). Antipyrine removal by TiO_2 photocatalysis based on spinning disc reactor technology.

Journal of Environmental Management, 187, 504−512. Available from https://doi.org/10.1016/j.jenvman.2016.11.012.

Fakhouri, H., Arefi-Khonsari, F., Jaiswal, A. K., & Pulpytel, J. (2015). Enhanced visible light photoactivity and charge separation in TiO$_2$/TiN bilayer thin films. *Applied Catalysis A: General, 492*, 83−92. Available from https://doi.org/10.1016/j.apcata.2014.12.030.

Fang, J., Wang, W., Zhu, C., Fang, L., Jin, J., Ni, Y., . . . Xu, Z. (2017). CdS/Pt photocatalytic activity boosted by high-energetic photons based on efficient triplet−triplet annihilation upconversion. *Applied Catalysis B: Environmental, 217*, 100−107. Available from https://doi.org/10.1016/j.apcatb.2017.05.069.

Feng, L., Jiang, H., Zou, M., Xiong, F., Ganeshraja, A. S., Pervaiz, E., . . . Yang, M. (2016). Enhanced photocatalytic degradation of dye under visible light on mesoporous microspheres by defects in manganese- and nitrogen-co-doped TiO$_2$. *Journal of Nanoparticles Research, 18*, 278. Available from https://doi.org/10.1007/s11051-016-3591-y.

Fifere, N., Airinei, A., Timpu, D., Rotaru, A., Sacarescu, L., & Ursu, L. (2018). New insights into structural and magnetic properties of Ce doped ZnO nanoparticles. *Journal of Alloys and Compounds, 757*, 60−69. Available from https://doi.org/10.1016/j.jallcom.2018.05.031.

Gao, X., & Guo, Z. (2017). Biomimetic superhydrophobic surfaces with transition metals and their oxides: A review. *Journal of Bionic Engineering, 14*, 401−439. Available from https://doi.org/10.1016/S1672-6529(16)60408-0.

Ghosh, S., & Das, A. P. (2015). Modified titanium oxide (TiO$_2$) nanocomposites and its array of applications: A review. *Toxicological & Environmental Chemistry, 97*, 491−514. Available from https://doi.org/10.1080/02772248.2015.1052204.

Gnanasekaran, L., Hemamalini, R., Saravanan, R., Ravichandran, K., Gracia, F., & Gupta, V. K. (2016). Intermediate state created by dopant ions (Mn, Co and Zr) into TiO$_2$ nanoparticles for degradation of dyes under visible light. *Journal of Molecular Liquids, 223*, 652−659. Available from https://doi.org/10.1016/j.molliq.2016.08.105.

Goulart, L. A., Alves, S. A., & Mascaro, L. H. (2019). Photoelectrochemical degradation of bisphenol A using Cu doped WO$_3$ electrodes. *Journal of Electroanalytical Chemistry, 839*, 123−133. Available from https://doi.org/10.1016/j.jelechem.2019.03.027.

Gupta, B., Melvin, A. A., Matthews, T., Dash, S., & Tyagi, A. K. (2016). TiO2 modification by gold (Au) for photocatalytic hydrogen (H$_2$) production. *Renewable and Sustainable Energy Reviews, 58*, 1366−1375. Available from https://doi.org/10.1016/j.rser.2015.12.236.

Gupta, V. K., Fakhri, A., Azad, M., & Agarwal, S. (2018). Synthesis and characterization of Ag doped ZnS quantum dots for enhanced photocatalysis of Strychnine as a poison: Charge transfer behavior study by electrochemical impedance and time-resolved photoluminescence spectroscopy. *Journal of Colloid and Interface Science, 510*, 95−102. Available from https://doi.org/10.1016/j.jcis.2017.09.043.

Güy, N., Çakar, S., & Özacar, M. (2016). Comparison of palladium/zinc oxide photocatalysts prepared by different palladium doping methods for congo red degradation. *Journal of Colloid and Interface Science, 466*, 128−137. Available from https://doi.org/10.1016/j.jcis.2015.12.009.

Hassan, M., Zhao, Y., & Xie, B. (2016). Employing TiO$_2$ photocatalysis to deal with landfill leachate: Current status and development. *Chemical Engineering Journal, 285*, 264−275. Available from https://doi.org/10.1016/j.cej.2015.09.093.

Ho, C., Kang, F., Chang, G., You, S., & Wang, Y. (2019). Application of recycled lanthanum doped TiO$_2$ immobilized on commercial air filter for visible-light photocatalytic degradation of acetone and NO. *Applied Surface Science, 465*, 31−40. Available from https://doi.org/10.1016/j.apsusc.2018.09.136.

Hosseini-Sarvari, M., & Bazyar, Z. (2018). Visible light driven photocatalytic cross coupling reactions on nano Pd/ZnO photocatalyst at room-temperature. *ChemistrySelect, 3*, 1898−1907. Available from https://doi.org/10.1002/slct.201702219.

Hu, Y., Song, X., Jiang, S., & Wei, C. (2015). Enhanced photocatalytic activity of Pt-doped TiO$_2$ for NO$_x$ oxidation both under UV and visible light irradiation: A synergistic effect of lattice Pt^{4+} and surface PtO. *Chemical Engineering Journal, 274*, 102−112. Available from https://doi.org/10.1016/j.cej.2015.03.135.

Islam, M. A., Morton, D. W., Johnson, B. B., Mainali, B., & Angove, M. J. (2018). Manganese oxides and their application to metal ion and contaminant removal from wastewater. *Journal of Water Process Engineering*, *26*, 264−280. Available from https://doi.org/10.1016/j.jwpe.2018.10.018.

Jaramillo-Páez, C., Navío, J. A., & Hidalgo, M. C. (2018). Silver-modified ZnO highly UV-photoactive. *Journal of Photochemistry and Photobiology A: Chemistry*, *356*, 112−122. Available from https://doi.org/10.1016/j.jphotochem.2017.12.044.

Jiang, L., Yuan, X., Pan, Y., Liang, J., Zeng, G., Wu, Z., & Wang, H. (2017). Doping of graphitic carbon nitride for photocatalysis: A review. *Applied Catalysis B: Environmental*, *217*, 388−406. Available from https://doi.org/10.1016/j.apcatb.2017.06.003.

Jiang, T., Kong, J., Wang, Y., Meng, D., Wang, D., & Yu, M. (2016). Optical and Photocatalytic properties of Mn-doped CuO nanosheets prepared by hydrothermal method. *Crystal Research & Technology*, *51*(1), 58−64. Available from https://doi.org/10.1002/crat.201500152.

Jin, X., Chen, F., Jia, D., Cao, Y., Duan, H., & Long, M. (2020). Facile strategy for the fabrication of noble metal/ZnS composites with enhanced photocatalytic activities. *RSC Advances*, *10*, 4455−4463. Available from https://doi.org/10.1039/c9ra07163f.

Junaid, M., Imran, M., Ikram, M., Naz, M., Aqeel, M., Afzal, H., . . . Ali, S. (2019). The study of Fe-doped CdS nanoparticle-assisted photocatalytic degradation of organic dye in wastewater. *Applied Nanoscience*, *9*, 1593−1602. Available from https://doi.org/10.1007/s13204-018-0933-3.

Kaczmarek, am, Liu, Y., Voort, P. V. D., & Deun, R. V. (2013). Tuning the architecture and properties of microstructured yttrium tungstate oxide hydroxide and lanthanum tungstate. *Dalton Transactions*, *42*, 5471−5479. Available from https://doi.org/10.1039/C3DT32699C.

Kaur, J., Sharma, M., & Pandey, O. P. (2015). Photoluminescence and photocatalytic studies of metal ions (Mn and Ni) doped ZnS nanoparticles. *Optical Materials*, *47*, 7−17. Available from https://doi.org/10.1016/j.optmat.2015.06.022.

Kaur, N., Shahi, S. K., Shahi, J. S., Sandhu, S., Sharma, R., & Singh, V. (2020). Comprehensive review and future perspectives of efficient N-doped, Fe-doped and (N, Fe)-codoped titania as visible light active photocatalysts. *Vacuum*, *178*, 109429. Available from https://doi.org/10.1016/j.vacuum.2020.109429.

Kavitha, R., & Kumar, S. G. (2019). A review on plasmonic Au-ZnO heterojunction photocatalysts: Preparation, modifications and related charge carrier dynamics. *Materials Science in Semiconductor Processing*, *93*, 59−91. Available from https://doi.org/10.1016/j.mssp.2018.12.026.

Khake, S. M., & Chatani, N. (2020). Nickel-catalyzed C-H functionalization using a non-directed strategy. *Chem*, *6*, 1056−1081. Available from https://doi.org/10.1016/j.chempr.2020.04.005.

Khaki, M. R. D., Shafeeyan, M. S., Raman, A. A. A., & Daud, W. M. A. W. (2017). Application of doped photocatalysts for organic pollutant degradation − A review. *Journal of Environmental Management*, *198*, 78−94. Available from https://doi.org/10.1016/j.jenvman.2017.04.099.

Khaki, M. R. D., Shafeeyan, M. S., Raman, A. A. A., & Daud, W. M. A. W. (2018). Evaluating the efficiency of nano-sized Cu doped TiO_2/ZnO photocatalyst under visible light irradiation. *Journal of Molecular Liquids*, *258*, 354−365. Available from https://doi.org/10.1016/j.molliq.2017.11.030.

Khan, M. R., Chuan, T. W., Yousuf, A., Chowdhury, M. N. K., & Cheng, C. K. (2015). Schottky barrier and surface plasmonic resonance phenomena towards the photocatalytic reaction: Study of their mechanisms to enhance photocatalytic activity. *Catalysis Science & Technology*, *5*, 2522−2531. Available from https://doi.org/10.1039/C4CY01545B.

Khan, U. A., Liu, J., Pan, J., Zuo, S., Ma, H., Yu, Y., . . . Li, B. (2018). One step fabrication of novel Ag-CdS@EP floating photocatalyst for superiorly efficient degradation of organic pollutants under visible light illumination. *Dalton Transactions*, *47*, 12253−12263. Available from https://doi.org/10.1039/C8DT02555J.

Khataee, A. R., Hosseini, M., Hanifehpour, Y., Safarpour, M., & Joo, S. W. (2014). Hydrothermal synthesis and characterization of Nd-doped ZnSe nanoparticles with enhanced visible light photocatalytic activity. *Research on Chemical Intermediates*, *40*, 495−508. Available from https://doi.org/10.1007/s11164-012-0977-z.

Khojasteh, H., Salavati-Niasari, M., & Mortazavi-Derazkola, S. (2016). Synthesis, characterization and photocatalytic properties of nickel-doped TiO_2 and nickel titanate nanoparticles. *Journal of Materials*

Science: Materials in Electronics, *27*, 3599–3607. Available from https://doi.org/10.1007/s10854-015-4197-3.

Khosroshahi, A. G., & Mehrizad, A. (2019). Optimization, kinetics and thermodynamics of photocatalytic degradation of Acid Red 1 by Sm-doped CdS under visible light. *Journal of Molecular Liquids*, *275*, 629–637. Available from https://doi.org/10.1016/j.molliq.2018.11.122.

Komaraiah, D., Radha, E., Sivakumar, J., Reddy, M. V. R., & Sayanna, R. (2019). Structural, optical properties and photocatalytic activity of Fe^{3+} doped TiO_2 thin films deposited by sol-gel spin coating. *Surfaces and Interfaces*, *17*, 100368. Available from https://doi.org/10.1016/j.surfin.2019.100368.

Komaraiaha, D., Radha, E., Kalarikkal, N., Sivakumar, J., Reddy, M. V. R., & Sayanna, R. (2019). Structural, optical and photoluminescence studies of sol-gel synthesized pure and iron doped TiO_2 photocatalysts. *Ceramics International*, *45*, 25060–25068. Available from https://doi.org/10.1016/j.ceramint.2019.03.170.

Kumar, K. V. A., Amanchi, S. R., Sreedhar, B., Ghosal, P., & Subrahmanyam, C. (2017). Phenol and Cr(VI) degradation with Mn ion doped ZnO under visible light photocatalysis. *RSC Advances*, *7*, 43030–43039. Available from https://doi.org/10.1039/C7RA08172C.

Kumaravel, V., Mathew, S., Bartlett, J., & Pillai, S. C. (2019). Photocatalytic hydrogen production using metal doped TiO_2: A review of recent advances. *Applied Catalysis B: Environmental*, *244*, 1021–1064. Available from https://doi.org/10.1016/j.apcatb.2018.11.080.

Lee, K. M., Lai, C. W., Ngai, K. S., & Juan, J. C. (2016). Recent developments of zinc oxide based photocatalyst in water treatment technology: A review. *Water Research*, *88*, 428–448. Available from https://doi.org/10.1016/j.watres.2015.09.045.

Leong, K. H., Aziz, A. A., Sim, L. C., Saravanan, P., Jang, M., & Bahnemann, D. (2018). Mechanistic insights into plasmonic photocatalysts in utilizing visible light. *Beilstein Journal of Nanotechnology*, *9*, 628–648. Available from https://doi.org/10.3762/bjnano0.9.59.

Li, H., & Feng, B. (2016). Visible-light-driven composite La_2O_3/TiO_2 nanotube arrays: Synthesis and improved photocatalytic activity. *Materials Science in Semiconductor Processing*, *43*, 55–59. Available from https://doi.org/10.1016/j.mssp.2015.11.021.

Li, H., Yu, H., Quan, X., Chen, S., & Zhang, Y. (2016). Uncovering the key role of the fermi level of the electron mediator in a Z-scheme photocatalyst by detecting the charge transfer process of WO_3-metal-gC_3N_4 (Metal = Cu, Ag, Au). *ACS Applied Materials & Interfaces*, *8*, 2111–2119. Available from https://doi.org/10.1021/acsami.5b10613.

Li, J., Zhang, M., Weng, B., Chen, X., Chen, J., & Jia, H. (2020). Oxygen vacancies mediated charge separation and collection in Pt/WO3 nanosheets for enhanced photocatalytic performance. *Applied Surface Science*, *507*, 145133. Available from https://doi.org/10.1016/j.apsusc.2019.145133.

Li, M., Yu, Z., Liu, Q., Sun, L., & Huang, W. (2016). Photocatalytic decomposition of perfluorooctanoic acid by noble metallic nanoparticles modified TiO_2. *Chemical Engineering Journal*, *286*, 232–238. Available from https://doi.org/10.1016/j.cej.2015.10.037.

Li, Q., & Yan, B. (2019). Multi-component assembly of luminescent rare earth hybrid materials. *Journal of Rare Earths*, *37*, 113–123. Available from https://doi.org/10.1016/j.jre.2018.10.001.

Li, Z., & Meng, X. (2020). Recent development on palladium enhanced photocatalytic activity: A review. *Journal of Alloys and Compounds*, *830*, 154669. Available from https://doi.org/10.1016/j.jallcom.2020.154669.

Liang, J., Wang, J., Yu, K., Song, K., Wang, X., Liu, W., . . . Liang, C. (2020). Enhanced photocatalytic performance of Nd^{3+}-doped TiO_2 nanosphere under visible light. *Chemical Physics*, *528*, 110538. Available from https://doi.org/10.1016/j.chemphys.2019.110538.

Lin, J. C., Sopajaree, K., Jitjanesuwan, T., & Lu, M. (2018). Application of visible light on copper-doped titanium dioxide catalyzing degradation of chlorophenols. *Separation and Purification Technology*, *191*, 233–243. Available from https://doi.org/10.1016/j.seppur.2017.09.027.

Linic, S., Christopher, P., & Ingram, D. B. (2011). Plasmonic-metal nanostructures for efficient conversion of solar to chemical energy. *Nature Materials*, *10*(12), 911–921. Available from https://doi.org/10.1038/nmat3151.

Liu, X., Iocozzia, J., Wang, Y., Cui, X., Chen, Y., Zhao, S., ... Lin, Z. (2017). Noble metal − metal oxide nanohybrids with tailored nanostructures for efficient solar energy conversion, photocatalysis and environmental remediation. *Energy & Environmental Science, 10,* 402−434. Available from https://doi.org/10.1039/C6EE02265K.

Liu, Y., Guo, J., Zhu, E., Liao, L., Lee, S., Ding, M., ... Duan, X. (2018). Approaching the Schottky−Mott limit in Van der Waals metal−semiconductor junctions. *Nature, 557,* 696−700. Available from https://doi.org/10.1038/s41586-018-0129-8.

Luo, Y., Zhang, C., Zheng, B., Geng, X., & Debliquy, M. (2017). Hydrogen sensors based on noble metal doped metal-oxide semiconductor: A review. *International Journal of Hydrogen Energy, 42,* 20386−20397. Available from https://doi.org/10.1016/j.ijhydene.2017.06.066.

Mazierski, P., Mikolajczyk, A., Bajorowicz, B., Malankowska, A., Zaleska-Medynska, A., & Nadolna, J. (2018). The role of lanthanides in TiO_2-based photocatalysis: A review. *Applied Catalysis B: Environmental, 233,* 301−317. Available from https://doi.org/10.1016/j.apcatb.2018.04.019.

Mehmood, F., Iqbal, J., Jan, T., & Mansoor, Q. (2017). Structural, Raman and photoluminescence properties of Fe doped WO_3 nanoplates with anti cancer and visible light driven photocatalytic activities. *Journal of Alloys and Compounds, 728,* 1329−1337. Available from https://doi.org/10.1016/j.jallcom.2017.08.234.

Mehmood, F., Iqbal, J., Ismail, M., & Mehmood, A. (2018). Ni doped WO_3 nanoplates: An excellent photocatalyst and novel nanomaterial for enhanced anticancer activities. *Journal of Alloys and Compounds, 746,* 729−738. Available from https://doi.org/10.1016/j.jallcom.2018.01.409.

Meksi, M., Turki, A., Kochkar, H., Bousselmi, L., Guillard, C., & Berhault, G. (2016). The role of lanthanum in the enhancement of photocatalytic properties of TiO2 nanomaterials obtained by calcination of hydrogenotitanate nanotubes. *Applied Catalysis B: Environmental, 181,* 651−660. Available from https://doi.org/10.1016/j.apcatb.2015.08.037.

Meshram, S. P., Adhyapak, P. V., Pardeshi, S. K., Mulla, I. S., & Amalnerkar, D. P. (2017). Sonochemically generated cerium doped ZnO nanorods for highly efficient photocatalytic dye degradation. *Powder Technology, 318,* 120−127. Available from https://doi.org/10.1016/j.powtec.2017.05.044.

Misra, M., Gupta, R. K., Paul, A. K., & Singla, M. (2015). Influence of gold core concentration on visible photocatalytic activity of gold−zinc sulfide core−shell nanoparticle. *Journal of Power Sources, 294,* 580−587. Available from https://doi.org/10.1016/j.jpowsour.2015.06.099.

Mohamed, R. M., McKinney, D., Kadi, M. W., Mkhalid, I. A., & Sigmund, W. (2016). Platinum/zinc oxide nanoparticles: Enhanced photocatalysts degrade malachite green dye under visible light conditions. *Ceramics International, 42,* 9375−9381. Available from https://doi.org/10.1016/j.ceramint.2016.02.147.

Mrabet, C., Kamoun, O., Boukhachem, A., Amlouk, M., & Manoubi, T. (2015). Some physical investigations on hexagonal-shaped nanorods of lanthanum-doped ZnO. *Journal of Alloys and Compounds, 648,* 826−837. Available from https://doi.org/10.1016/j.jallcom.2015.07.009.

Muramatsu, Y., Jin, Q., Fujishima, M., & Tada, H. (2012). Visible-light-activation of TiO_2 nanotube array by the molecular iron oxide surface modification. *Applied Catalysis B: Environmental, 119−120,* 74−80. Available from https://doi.org/10.1016/j.apcatb.2012.02.012.

Murtaza, G., Osama, S. M. A., Saleem, M., Hassan, M., & Watoo, N. R. K. (2018). Structural, optical, and photocatalytic properties of $Cd_{1-x}S:La_x$ nanoparticles for optoelectronic applications. *Applied Physics A., 124,* 778. Available from https://doi.org/10.1007/s00339-018-2199-8.

Nsib, M. F., Saafi, S., Rayes, A., Moussa, N., & Houas, A. (2016). Enhanced photocatalytic performance of Ni−ZnO/Polyaniline composite for the visible-light driven hydrogen generation. *Journal of the Energy Institute, 89,* 694−703. Available from https://doi.org/10.1016/j.joei.2015.05.001.

Oppong, S. O. B., Anku, W. W., Shukla, S. K., & Govender, P. P. (2017). Synthesis and characterization of neodymium doped-zinc oxide−graphene oxide nanocomposite as a highly efficient photocatalyst for enhanced degradation of indigo carmine in water under simulated solar light. *Research on Chemical Intermediates, 43,* 481−501. Available from https://doi.org/10.1007/s11164-016-2636-2.

Pal, N. K., & Kryschi, C. (2016). Improved photocatalytic activity of gold decorated differently doped TiO_2 nanoparticles: A comparative study. *Chemosphere, 144,* 1655−1664. Available from https://doi.org/10.1016/j.chemosphere.2015.10.060.

Parnicka, P., Mazierski, P., Grzyb, T., Wei, Z., Kowalska, E., Ohtani, B., . . . Nadolna, J. (2017). Preparation and photocatalytic activity of Nd-modified TiO_2 photocatalysts: Insight into the excitation mechanism under visible light. *Journal of Catalysis, 353*, 211−222. Available from https://doi.org/10.1016/j.jcat.2017.07.017.

Pérez-Larios, A., Hernández-Gordillo, A., Morales-Mendoza, G., Lartundo-Rojas, L., Mantilla, Á., & Gómez, R. (2016). Enhancing the H_2 evolution from water−methanol solution using Mn^{2+}−Mn^{+3}−Mn^{4+} redox species of Mn-doped TiO_2 sol−gel photocatalysts. *Catalysis Today, 266*, 9−16. Available from https://doi.org/10.1016/j.cattod.2015.12.029.

Ponraj, C., Vinitha, G., & Daniel, J. (2017). A review on the visible light active $BiFeO_3$ nanostructures as suitable photocatalyst in the degradation of different textile dyes. *Environmental Nanotechnology, Monitoring & Management, 7*, 110−120. Available from https://doi.org/10.1016/j.enmm.2017.02.001.

Poongodi, G., Kumar, R. M., & Jayavel, R. (2015). Structural, optical and visible light photocatalytic properties of nanocrystalline Nd doped ZnO thin films prepared by spin coating method. *Ceramics International, 41*, 4169−4175. Available from https://doi.org/10.1016/j.ceramint.2014.12.098.

Prakash, J., Sun, S., Swart, H. C., & Gupta, R. K. (2018). Noble metals-TiO_2 nanocomposites: From fundamental mechanisms to photocatalysis, surface enhanced Raman scattering and antibacterial applications. *Applied Materials Today, 11*, 82−135. Available from https://doi.org/10.1016/j.apmt.2018.02.002.

Prasad, N., & Balasubramanian, K. (2017). Optical, phonon and efficient visible and infrared photocatalytic activity of Cu doped ZnS micro crystals. *Spectrochimica Acta Part A: Molecular and Biomolecular Spectroscopy, 173*, 687−694. Available from https://doi.org/10.1016/j.saa.2016.10.014.

Priyadharsini, N., Elango, M., Vairam, S., & Thamilselvan, M. (2016). Effect of nickel substitution on structural, optical, magnetic properties and photocatalytic activity of ZnS nanoparticles. *Materials Science in Semiconductor Processing, 49*, 68−75. Available from https://doi.org/10.1016/j.mssp.2016.03.033.

Putri, N. A., Fauzia, V., Iwan, S., Roza, L., Umar, A. A., & Budi, S. (2018). Mn-doping-induced photocatalytic activity enhancement of ZnO nanorods prepared on glass substrates. *Applied Surface Science, 439*, 285−297. Available from https://doi.org/10.1016/j.apsusc.2017.12.246.

Qi, K., Cheng, B., Yu, J., & Ho, W. (2017). Review on the improvement of the photocatalytic and antibacterial activities of ZnO. *Journal of Alloys and Compounds, 727*, 792−820. Available from https://doi.org/10.1016/j.jallcom.2017.08.142.

Qin, L., Wang, G., & Tan, Y. (2018). Plasmonic Pt nanoparticles—TiO_2 hierarchical nano-architecture as a visible light photocatalyst for water splitting. *Scientific Reports, 8*, 16198. Available from https://doi.org/10.1038/s41598-018-33795-z.

Raj, D. J. V., Raj, C. J., & Das, S. J. (2015). Synthesis and optical properties of cerium doped zinc sulfide nano particles. *Superlattices and Microstructures, 85*, 274−281. Available from https://doi.org/10.1016/j.spmi.2015.04.029.

Ranjith, R., Krishnakumar, V., Boobas, S., Venkatesan, J., & Jayaprakash, J. (2018). An efficient photocatalytic and antibacterial performance of Ni/Ce−codoped CdS nanostructure under visible light irradiation. *ChemistrySelect, 3*, 9259−9267. Available from https://doi.org/10.1002/slct.201801485.

Rao, G. T., Stella, R. J., Babu, B., Ravindranadh, K., Reddy, C. V., Shim, J., & Ravikumar, R. V. S. S. N. (2015). Structural, optical and magnetic properties of Mn^{2+} doped ZnO-CdS composite nanopowder. *Materials Science and Engineering B, 201*, 72−78. Available from https://doi.org/10.1016/j.mseb.2015.06.012.

Reddy, P. A. K., Reddy, P. V. L., Kwon, E., Kim, K., Akter, T., & Kalagara, S. (2016). Recent advances in photocatalytic treatment of pollutants in aqueous media. *Environment International, 91*, 94−103. Available from https://doi.org/10.1016/j.envint.2016.02.012.

Saleem, M., Iqbal, J., Nawaz, A., Islam, B., & Hussain, I. (2020). Synthesis, characterization, and performance evaluation of pristine and cerium-doped WO_3 nanoparticles for photodegradation of methylene blue via solar irradiation. *International Journal of Applied Ceramic Technology, 17*, 1918−1929. Available from https://doi.org/10.1111/ijac.13496.

Samadi, M., Zirak, M., Naseri, A., Khorashadizade, E., & Moshfegh, A. Z. (2016). Recent progress on doped ZnO nanostructures for visible-light photocatalysis. *Thin Solid Films, 605*, 2−19. Available from https://doi.org/10.1016/j.tsf.2015.12.064.

Shamsipur, M., Rajabi, H. R., & Khani, O. (2013). Pure and Fe^{3+}-doped ZnS quantum dots as novel and efficient nanophotocatalysts: Synthesis, characterization and use for decolorization of Victoria blue R. *Materials Science in Semiconductor Processing, 16*, 1154−1161. Available from https://doi.org/10.1016/j.mssp.2013.02.010.

Shayegan, Z., Lee, C., & Haghighat, F. (2018). TiO_2 photocatalyst for removal of volatile organic compounds in gas phase − A review. *Chemical Engineering Journal, 334*, 2408−2439. Available from https://doi.org/10.1016/j.cej.2017.09.153.

Shen, J., Chuang, H., Jiang, Z., Liu, X., & Horng, J. (2020). Novel quantification of formation trend and reaction efficiency of hydroxyl radicals for investigating photocatalytic mechanism of Fe-doped TiO_2 during UV and visible light-induced degradation of acid orange 7. *Chemosphere, 251*, 126380. Available from https://doi.org/10.1016/j.chemosphere.2020.126380.

Sin, J., & Lam, S. (2018). One-dimensional ZnO nanorods doped with neodymium for enhanced resorcinol degradation under sunlight irradiation. *Chemical Engineering Communications, 205*(3), 311−324. Available from https://doi.org/10.1080/00986445.2017.1387855.

Singh, S., Mahalingam, H., & Singh, P. K. (2013). Polymer-supported titanium dioxide photocatalysts for environmental remediation: A review. *Applied Catalysis A: General, 462−463*, 178−195. Available from https://doi.org/10.1016/j.apcata.2013.04.039.

Sousa Filho, P. C., Lima, J. F., & Serra, O. A. (2015). From lighting to photoprotection: Fundamentals and applications of rare earth materials. *Journal of Brazilian Chemical Society, 26*(12), 2471−2495. Available from https://doi.org/10.5935/0103-5053.20150328.

Suganthi, N., & Pushpanathan, K. (2018). Photocatalytic degradation and ferromagnetism in mesoporous La doped ZnS nanoparticles. *Journal of Materials Science: Materials in Electronics, 29*, 13970−13983. Available from https://doi.org/10.1007/s10854-018-9530-1.

Suganthi, N., & Pushpanathan, K. (2019). Cerium doped ZnS nanorods for photocatalytic degradation of turquoise blue H5G dye. *Journal of Inorganic and Organometallic Polymers and Materials, 29*, 1141−1153. Available from https://doi.org/10.1007/s10904-019-01077-4.

Sumathi, S., & Kavipriya, A. (2017). Structural, optical and photocatalytic activity of cerium doped zinc aluminate. *Solid State Sciences, 65*, 52−60. Available from https://doi.org/10.1016/j.solidstatesciences.2017.01.003.

Sun, Y., Sun, Y., Zhang, T., Chen, G., Zhang, F., Liu, D., . . . Li, C. (2016). Complete Au@ZnO core-shell nanoparticles with enhanced plasmonic absorption enabling significantly improved photocatalysis. *Nanoscale, 8*, 10774−10782. Available from https://doi.org/10.1039/C6NR00933F.

Tabasideh, S., Maleki, A., Shahmoradi, B., Ghahremani, E., & McKay, G. (2017). Sonophotocatalytic degradation of diazinon in aqueous solution using irondoped TiO_2 nanoparticles. *Separation and Purification Technology, 189*, 186−192. Available from https://doi.org/10.1016/j.seppur.2017.07.065.

Tang, Y., Zhao, S., Long, B., Liu, J., & Li, J. (2016). On the nature of support effects of metal dioxides MO_2 (M = Ti, Zr, Hf, Ce, Th) in single-atom gold catalysts: Importance of quantum primogenic effect. *The Journal of Physical Chemistry C, 120*, 17514−17526. Available from https://doi.org/10.1021/acs.jpcc.6b05338.

Terzioglu, R., Aydin, G., Koc, N. S., & Terzioglu, C. (2019). Investigation of the structural, magnetic and electrical properties of the Au doped YBCO superconductors. *Journal of Materials Science: Materials in Electronics, 30*, 2265−2277. Available from https://doi.org/10.1007/s10854-018-0497-8.

Thi, V. H., & Lee, B. (2017). Effective photocatalytic degradation of paracetamol using La-doped ZnO photocatalyst under visible light irradiation. *Materials Research Bulletin, 96*, 171−182. Available from https://doi.org/10.1016/j.materresbull.2017.04.028.

Thiruppathi, M., Kumar, P. S., Devendran, P., Ramalingan, C., Swaminathan, M., & Nagarajan, E. R. (2018). Ce@TiO_2 nanocomposites: An efficient, stable and affordable photocatalyst for the photodegradation of diclofenac sodium. *Journal of Alloys and Compounds, 735*, 728−734. Available from https://doi.org/10.1016/j.jallcom.2017.11.139.

Touati, A., Hammedi, T., Najjar, W., Ksibi, Z., & Sayadi, S. (2016). Photocatalytic degradation of textile wastewater in presence of hydrogen peroxide: Effect of cerium doping titania. *Journal of Industrial and Engineering Chemistry, 35*, 36−44. Available from https://doi.org/10.1016/j.jiec.2015.12.008.

Tripathi, A. K., Mathpal, M. C., Kumar, P., Agrahari, V., Singh, M. K., Mishra, S. K., . . . Agarwal, A. (2015). Photoluminescence and photoconductivity of Ni doped titania nanoparticles. *Advanced Materials Letters, 6* (3), 201−208. Available from https://doi.org/10.5185/amlett.20150.5663.

Türkyılmaz, Ş. Ş., Güy, N., & Özacar, M. (2017). Photocatalytic efficiencies of Ni, Mn, Fe and Ag doped ZnO nanostructures synthesized by hydrothermal method: The synergistic/ antagonistic effect between ZnO and metals. *Journal of Photochemistry and Photobiology A: Chemistry, 341*, 39−50. Available from https://doi.org/10.1016/j.jphotochem.2017.03.027.

Varma, K. S., Tayade, R. J., Shah, K. J., Joshi, P. A., Shukla, A. D., & Gandhi, V. G. (2020). Photocatalytic degradation of pharmaceutical and pesticide compounds (PPCs) using doped TiO_2 nanomaterials: A review. *Water-Energy Nexus, 3*, 46−61. Available from https://doi.org/10.1016/j.wen.2020.03.008.

Verbruggen, S. W. (2015). TiO2 photocatalysis for the degradation of pollutants in gas phase: From morphological design to plasmonic enhancement. *Journal of Photochemistry and Photobiology C: Photochemistry Reviews, 24*, 64−82. Available from https://doi.org/10.1016/j.jphotochemrev.2015.07.001.

Vieira, G. B., José, H. J., Peterson, M., Baldissarelli, V. Z., Alvarez, P., & Moreira, R. F. P. M. (2018). CeO_2/TiO_2 nanostructures enhance adsorption and photocatalytic degradation of organic compounds in aqueous suspension. *Journal of Photochemistry and Photobiology A: Chemistry, 353*, 325−336. Available from https://doi.org/10.1016/j.jphotochem.2017.11.045.

Wang, J., Liu, W., Zhong, D., & Lu, T. (2019). Nickel complexes as molecular catalysts for water splitting and CO_2 reduction. *Coordination Chemistry Reviews, 378*, 237−261. Available from https://doi.org/10.1016/j.ccr.2017.12.009.

Wang, L., Wang, P., Huang, B., Ma, X., Wang, G., Dai, Y., . . . Qin, X. (2017). Synthesis of Mn-doped ZnS microspheres with enhanced visible light photocatalytic activity. *Applied Surface Science, 391*, 557−564. Available from https://doi.org/10.1016/j.apsusc.2016.06.159.

Wang, M., Ye, M., Iocozzia, J., Lin, C., & Lin, Z. (2016). Plasmon-mediated solar energy conversion via photocatalysis in noble metal/semiconductor composites. *Advanced Science, 3*, 1600024. Available from https://doi.org/10.1002/advs.201600024.

Wang, Q., Xu, P., Zhang, G., Zhang, W., Hu, L., & Wang, P. (2018). Characterization of visible-light photo-Fenton reaction using Fe-doped ZnS (Fex-ZnS) mesoporous microspheres. *Physical Chemistry Chemical Physics, 20*, 18601−18609. Available from https://doi.org/10.1039/c8cp02609b.

Wang, W., Huang, G., Yu, J. C., & Wong, P. K. (2015). Advances in photocatalytic disinfection of bacteria: Development of photocatalysts and mechanisms. *Journal of Environmental Sciences, 34*, 232−247. Available from https://doi.org/10.1016/j.jes.2015.05.003.

Wang, Y., Yang, X., Ye, T., Xu, C., Xia, F., & Meng, D. (2017). Comparative study of structural and photocatalytic properties of M-doped (M = Ce^{3+}, Zn^{2+}, Cu^{2+}) dendritic-like CdS. *Journal of Electronic Materials, 46*(3), 1598−1606. Available from https://doi.org/10.1007/s11664-016-5202-1.

Xu, D., Bai, Y., Li, Z., Guo, Y., & Bai, L. (2018). Enhanced photodegradation ability of solvothermally synthesized metallic copper coated ZnO microrods. *Colloids and Surfaces A, 548*, 19−26. Available from https://doi.org/10.1016/j.colsurfa.2018.03.057.

Yadava, S., & Jaiswar, G. (2017). Review on undoped/doped TiO2 nanomaterial; synthesis and photocatalytic and antimicrobial activity. *Journal of the Chinese Chemical Society, 64*, 103−116. Available from https://doi.org/10.1002/jccs.201600735.

Yazdanbakhsh, A., Eslami, A., Massoudinejad, M., & Avazpour, M. (2020). Enhanced degradation of sulfamethoxazole antibiotic from aqueous solution using $Mn-WO_3$/LED photocatalytic process: Kinetic, mechanism, degradation pathway and toxicity reduction. *Chemical Engineering Journal, 380*, 122497. Available from https://doi.org/10.1016/j.cej.2019.122497.

Yu, T., Tan, X., Zhao, L., Yin, Y., Chen, P., & Wei, J. (2010). Characterization, activity and kinetics of a visible light driven photocatalyst: Cerium and nitrogen co-doped TiO_2 nanoparticles. *Chemical Engineering Journal, 157*, 86−92. Available from https://doi.org/10.1016/j.cej.2009.10.051.

Yurtsever, H. A., & Çifçioğlu, M. (2018). The effect of powder preparation method on the artificial photosynthesis activities of neodymium doped titania powders. *International Journal of Hydrogen Energy, 43*(44), 20162−20175. Available from https://doi.org/10.1016/j.ijhydene.2018.08.185.

Zare, M., Mortezaali, A., & Shafiekhani, A. (2016). Photoelectrochemical determination of shallow and deep trap states of platinum-decorated TiO_2 nanotube arrays for photocatalytic applications. *Journal of Physical Chemystry C, 120*, 9017−9027. Available from https://doi.org/10.1021/acs.jpcc.5b11987.

Zhang, N., Han, C., Xu, Y., Foley, J. J., IV, Zhang, D., Codrington, J., . . . Sun, Y. (2016). Near-field dielectric scattering promotes optical absorption by platinum nanoparticles. *Nature Photonics, 10*, 473−482. Available from https://doi.org/10.1038/nphoton.2016.76.

Zhang, X., Wang, Y., Liu, B., Sang, Y., & Liu, H. (2017). Heterostructures construction on TiO_2 nanobelts: A powerful tool for building high-performance photocatalysts. *Applied Catalysis B: Environmental, 202*, 620−641. Available from https://doi.org/10.1016/j.apcatb.2016.09.068.

Zhao, Z., Liu, J., Gong, H., Wu, G., & Wu, H. (2019). $Pt/Ni_{0.17}Zn_{0.83}O$ hybrids with enhanced photocatalytic performance: Effect of reduction treatments. *Results in Physics, 14*, 102434. Available from https://doi.org/10.1016/j.rinp.2019.102434.

Zhou, W., Jiang, D., Xue, J., & Li, X. (2019). Selective growth of palladium nanocrystals on the (100) facets of truncated octahedral Cu_2O for UV plasmonic photocatalysis. *CrystEngComm, 21*, 30−33. Available from https://doi.org/10.1039/C8CE01697F.

Zhu, D., & Zhou, Q. (2019). Action and mechanism of semiconductor photocatalysis on degradation of organic pollutants in water treatment: A review. *Environmental Nanotechnology, Monitoring & Management, 12*, 100255. Available from https://doi.org/10.1016/j.enmm.2019.100255.

Zhu, W., Liu, J., Yu, S., Zhou, Y., & Yan, X. (2016). Ag loaded WO_3 nanoplates for efficient photocatalytic degradation of sulfanilamide and their bactericidal effect under visible light irradiation. *Journal of Hazardous Materials, 318*, 407−416. Available from https://doi.org/10.1016/j.jhazmat.2016.06.066.

Zhu, X., Zhang, P., Li, B., Hu, Q., Su, W., Dong, L., & Wang, F. (2017). Preparation, characterization and photocatalytic properties of La/WO_3 composites. *Journal of Materials Science: Materials in Electronics, 28*, 12158−12167. Available from https://doi.org/10.1007/s10854-017-7030-3.

Zhua, D., & Zhou, Q. (2019). Action and mechanism of semiconductor photocatalysis on degradation of organic pollutants in water treatment: A review. *Environmental Nanotechnology, Monitoring & Management, 12*, 100255. Available from https://doi.org/10.1016/j.enmm.2019.100255.

Zou, J. P., Chen, Y., Zhu, M., Wang, D., Luo, X. B., & Luo, S. L. (2019). *Semiconductor-based nanocomposites for photodegradation of organic pollutants. Nanomaterials for the Removal of Pollutants and Resource Reutilization*. Oxford: Elsevier. Available from https://doi.org/10.1016/b978-0-12-814837-2.00002-0.

CHAPTER 3

Anion-modified photocatalysts

Shwetharani R.[1], Bindu K.[2], Laveena P. D'Souza[1], R. Mithun Prakash[1] and R. Geetha Balakrishna[1]

[1]*Centre for Nano and Material Sciences, Jain University, Bangalore, India,* [2]*Department of Inorganic and Physical Chemistry, Indian Institute of Science, Bangalore, India*

3.1 Introduction

Solar energy is a renewable energy source that has been a great support to address various energy crises and environmental problems. Among a wide variety of renewable energy approaches, semiconductor photocatalysis is one of the clean, sustainable technologies thanks to its easy activation either under sunlight or artificial indoor illumination and it demonstrates a wide range of applications, such as hydrogen production (Shwetharani, Chandan, et al., 2019; Shwetharani, Sakar, Fernando, Binas, & Balakrishna, 2019; Xu, Ravi Anusuyadevi, Aymonier, Luque, & Marre, 2019), removal of organic/inorganic pollutants (Shwetharani, Poojashree, Balakrishna, & Jyothi, 2018), and bacterial disinfection (Liu, Wang, Xu, Marcel Veder, & Shao, 2019; Shwetharani & Geetha Balakrishna, 2014). Photocatalysis is mainly defined as the acceleration of the chemical reaction in the presence of a photoactive catalyst. The concept of photocatalysis came into light after the discovery of water splitting into hydrogen and oxygen using TiO_2 as the photoelectrocatalyst in 1972 by Fujishima and Honda (1972). The commonly used photocatalysts are the semiconductor-based metal oxides, which are due to their favorable light absorption ability, electronic structure, and charge transport features. The photocatalytic process can be typically explained in the following simple steps (Khan, Adil, & Al-Mayouf, 2015). First, the light illumination induces excitation of electrons from the valence band (VB) to the conduction band (CB) leaving the holes in VB. Second, the excited electrons and holes migrate to the surface; third, they react with the electron donors and electron acceptors, and thereby, it finally leads to the desired activities. Metal oxides, such as TiO_2, ZnO, SnO_2, and CeO_2, are extensively used as heterogeneous photocatalysts. The performance of a photocatalyst can be affected by its lattice structure, electronic structure, surface properties, defects, etc. (Hoffmann, Martin, Choi, & Bahnemann, 1995). The benefits of these photocatalysts include their high stability under various conditions, their biocompatibility, and their ability to generate charge carriers and reactive species using light energy. In addition, TiO_2 and CeO_2 are widely used as catalysts in energy conversion processes, such as solar-thermal conversion into chemical fuels. The photocatalytic activity of these materials can be further improved by modifying their electronic structure and ability of light absorption toward the visible light range, reducing their recombination processes, and enhancing the lifetime of their charge carriers.

Among all the developed photocatalysts, TiO_2 has been recognized as a renowned photocatalyst due to its properties, such as high refractive index, UV-light absorption, photostability, and chemical stability. Accordingly, TiO_2 has been extensively explored for various photocatalytic applications and eventually, it has been kept as a bench-mark photocatalyst in the field toward addressing environmental applications (Khlyustova et al., 2020). In TiO_2 photocatalyst, there are two reactions that occur simultaneously, which are the oxidation reactions by photogenerated holes on VB and the reduction reactions by photogenerated electrons on CB. However, there are few disadvantages that limit its large-scale applications, which include their large band gap energy (3.0 eV for rutile and 3.2 eV for anatase phase), where the photoexcitation largely takes place in the UV-light region, which is only 5% of the sunlight. Second, the high recombination rate of electron−hole pairs in TiO_2 results in low quantum yield as well as low photocatalytic activity (Khlyustova et al., 2020). To address such limitations, it is found that the incorporation of dopants, such as transitional metals, nonmetals, and rare earth metals, in TiO_2 can extend its light absorption to visible light (Cheng, Xu, Stadler, & Chen, 2015; Fu et al., 2014). The doping process essentially modifies the electronic structure of photocatalysts by introducing new impurity levels in the forbidden band gaps, which fundamentally causes the narrowing of band gap structures. The doping process also improves the electron trapping, inhibits electron−hole recombination, enhances the interfacial electron transfer rate, and thereby ultimately enhances the photocatalytic properties, as shown in Fig. 3.1 (Cheng et al., 2015; Fu et al., 2014; Sahoo, Martha, & Parida, 2016). For instance, a controlled doping induces the energy levels just below the CB or above the VB of TiO_2, which serves in prolonging the lifetime of electrons and holes, but at the increased dopant concentrations, they may act as charge recombination centers and decline the enhanced activity. Doping of TiO_2 with metals, such as Fe, Au, Cu, Cr, and Mn, and nonmetals, such as C, N, P, S, O, and Cl, has widely been done to obtain TiO_2 with modified optical absorption, band edge energies, and photoelectrical conductivity toward achieving the improved photocatalytic efficiencies (Cheng et al., 2015; Jyothi, D'Souza Laveena, Shwetharani, & Balakrishna, 2016; Khaki, Shafeeyan, Raman, & Daud, 2017; Muggli & Falconer, 2000; Sahoo et al., 2016).

In doping strategies, the introduction of dopants at anion sites leads to the manifestation of interesting properties in photocatalytic materials. In this direction, the incorporation of nonmetals, such as nitrogen (N), phosphorous (P) sulfur (S), oxygen (O), and halogens (F, Cl, Br) and their high ionization energies along with high electronegativity, are found to be competent enough toward enhancing the visible light absorption ability, band gap narrowing, and shifting of absorption edges and thereby increase their overall photocatalytic properties (Liu et al., 2019; Tan et al., 2011). Anions can be essentially doped into oxide lattice in three

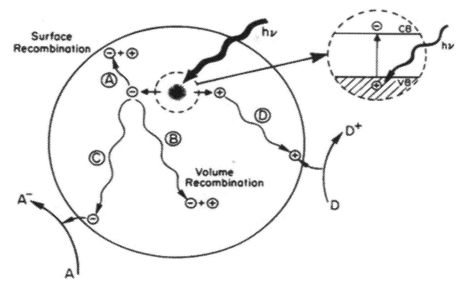

Figure 3.1
Various reactions and mechanisms of the photocatalytic process (Linsebigler, Lu, and Yates, 1995). Source: *Reprinted with permission from Linsebigler, A. L., Lu, G., & Yates Jr, J. T. (1995). Photocatalysis on TiO2 surfaces: Principles, mechanisms, and selected results.* Chemical Reviews, 95(3), 735−758.

ways, which include (1) replacement of lattice oxygen with nonoxygen anions, (2) substitution of two nonoxygen anions in place of one O^{2-} anion in the lattice, and (3) making them occupy the interstitial sites between different layers in the crystal structure (Liu et al., 2019; Muggli & Falconer, 2000). These nonmetal dopants fundamentally influence the VB of metal oxide through the formation of impurity levels above the O 2p electrons, and thereby, it improves photocatalytic activities in many ways. There are various reports available on anion-doped photocatalysts. Notably, in 2019 Liu et al. (2019) reviewed the recent advances in anion doping into metal oxides for catalytic applications, where it was mainly focused on the effect of anion doping in tuning the physical and chemical properties, and thereby catalytic performance in various applications, such as oxidative coupling of methane, oxidative dehydrogenation of ethane and ceramic membrane-based oxygen separation. In this context, this chapter discusses the synthesis, properties, mechanism, and various photocatalytic applications of the anions, such as N-, S-, F-, C-, Cl-, I-, and O-doped photocatalytic materials, and finally, it concludes with their prospects in future.

3.2 Synthesis methods of anion-doped photocatalysts

There are several ways to synthesize anion-doped photocatalysts, which include sol−gel, hydrothermal, coprecipitation, solid-state reactions, solution-based method, template-based method, and thermal treatment in the suitable gas environment. This section provides insights into these techniques and their mechanisms toward the synthesis of anion-doped photocatalysts.

3.2.1 Sol—gel method

The sol—gel method has been widely explored for the synthesis of oxide-based semiconductor photocatalysts (Marami, Farahmandjou, & Khoshnevisan, 2018). The main advantage of this process is the possibility of regulating the reaction parameters during the synthesis process. Recently, Ji, Liu, Xu, Gong, and Zhou (2020) have synthesized carbon-doped TiO_2 using polyacrylonitrile as a carbon precursor and also as a template for the carbon-doped TiO_2. The final C-doped TiO_2 microspheres were obtained after sintering the product at 450°C for 2 h under the nitrogen gas environment. Abraham and Devi (2020) have successfully fused W^{6+}, N^{3-}, C^{4+}, and S^{6+} ions into the TiO_2 lattice using sodium tungstate dihydrate (W) and L-cysteine (N, C, and S) is the dopant precursor that predissolved in concentrated H_2SO_4. The final product was examined for its efficiency of photocatalytic degradation of thymol. Furthermore, Tang, Chen, Zhang, Zhang, and Li (2020) have synthesized N-doped zinc oxide using zinc acetate ($Zn(CH_3COO)_2 \cdot 2H_2O$) and monoethanolamine as the starting precursors. They have achieved the N doping under N_2 treatment for 2 h at 600°C. Besides, Huang et al. (2021) have treated 1 g of sol—gel synthesized TiO_2 at 500°C under the flow of NH_3 for the successful incorporation of nitrogen into the TiO_2 lattice. Besides, Zhao, Liu, Zhang, Wang, and Wang (2019) investigated the photocatalytic activity of the N-doped TiO_2 by the sol—gel method taking urea as the precursor for N doping.

3.2.2 Hydrothermal/solvothermal method

The hydrothermal/solvothermal method is another method broadly explored for the synthesis of heterogeneous nanoparticles. The reaction is carried out in high temperature and pressure conditions accompanied by the aqueous/nonaqueous solvents termed as hydro and solvothermal methods, respectively. Recently, Qian et al. (2020) have synthesized honeycombed TiO_2 and $C-TiO_{2-x}$ using polystyrene as a template for the photocatalytic mineralization of Acid red 3R under visible light. Furthermore, Divyasri et al. (2021) synthesized the N-doped TiO_2 nanotubes through a hydrothermal method, where urea was used as a precursor for N doping and the product was annealed at 350°C for 3 h to obtain the final N-doped TiO_2 nanotubes. Also, Gundeboina, Venkataswamy, and Vithal (2020) have synthesized $K_2Al_2Ti_6O_{16}$ by a simple sol—gel method, where they doped carbon by taking glucose as a carbon source using the hydrothermal method. Similarly, Vitiello et al. (2021) hydrothermally synthesized the F-doped ZnO nano- and mesocrystals using zinc acetate dihydrate as Zn precursor and ammonium hydrogen fluoride as a precursor for F doping into ZnO. This synthesized material was investigated for the photocatalytic degradation of diclofenac molecules. Besides, Ai et al. (2020) have synthesized the defect-rich Cl-doped Bi_2S_3 homojunction nanorods for the photocatalytic reduction performance. More recently, Khazaee et al. (2020) have synthesized X-Bi_2MoO_6 (X: F, Cl, Br, I) nanoplates via template-confined growth technique and examined the synthesized materials for photocatalytic oxidation of RhB dye molecules. Zeng et al. (2020) have developed the F-doped Sn_3O_4 using the hydrothermal method and achieved excellent photocatalytic performance for the removal of Cr(VI) and organic pollutants. Also, Li, Lai, et al. (2020) have fabricated a wool ball-like F-doped $BiOCl_{0.4}Br_{0.3}I_{0.3}$ composite by a solvothermal method using ethylene glycol as a solvent, where they employed the material for the effective photocatalytic degradation of sulfamethazine molecules.

3.2.3 Coprecipitation method

The coprecipitation methods involve the manifestation of nucleation, particle growth, coarsening, and/or agglomeration processes during the process. Recently, Wafi et al. (2020) have synthesized N-doped TiO_2 using ammonium hydroxide as a nitrogen precursor using the coprecipitation method. Also, Kakhki, Mohammadpoor, Faridi, and Bahadori (2020) have developed the sulfur–nitrogen codoped on the surface of Fe_2O_3 nanostructures, where the thiourea was used as a precursor for both S and N and ammonium hydroxide was used as a precipitation agent.

3.2.4 Sonochemical method

The sonochemical methods utilize the potential of high-frequency ultrasound waves with wavelength ranges from 20 kHz to 1 GHz for the synthesis of materials (Shchukin, Radziuk, & Moehwald, 2010). Using this method, recently, Kadam, Salunkhe, Kim, and Lee (2020) explored the biogenic synthesis of mesoporous N-S-C tri-doped TiO_2 photocatalyst using the titanium (IV) butoxide and egg white (ovalbumin) extracted from the expired eggs as starting materials. The obtained gel was dried under air at 100°C for 24 h followed by the calcination at 500°C for 2 h. The resultant product was washed with ethanol and water and dried at 60°C and investigated for their photocatalytic degradation of RhB dye under visible light irradiation.

3.2.5 Microwave method

The microwave (MW) (frequency range from 0.3 to 300 GHz) radiation has been broadly explored for the synthesis of a wide variety of materials. It is known that MWs are not capable of breaking hydrogen or covalent bonds; thus they will not be able to induce chemical reactions since there is no bond breaking. However, the perturbation of MWs through the reaction components, such as solvents, reagents, and vessels, can convert the MW energy into heat.

Thus it is broadly accepted that the chemical reactions are happening because of the thermal effects of the MW radiation (Kappe, 2013). Recently, Reda, Khairy, and Mousa (2020) have synthesized nitrogen- and copper-doped TiO_2 nanoparticles by MW-assisted sol−gel process, where the ammonium nitrate and $Cu(NO_3)_2 0.3H_2O$ were employed as the nitrogen and copper dopants, respectively. Similarly, Chen, Chen, et al. (2020) have utilized the MW heating method to prepare the phosphorus-doped g-C_3N_4 with the assistance of Ag_3PO_4 nanoparticles, where the $NH_4H_2PO_4$ was used as a source for phosphorus. The obtained final powders were put into a corundum crucible and heated in an MW oven for 40 min to obtain the final P-doped g-C_3N_4 as depicted in Fig. 3.2. Besides, Payormhorm and Idem (2020) have synthesized C-doped TiO_2 by sol−gel-assisted MW method and demonstrated the photocatalytic conversion of glycerol to value-added chemicals under visible light irradiation.

3.2.6 Solution combustion method

Generally, solution combustion is a self-propagating high-temperature synthesis that evolves due to the exothermic reactions between the fuel (H, C) and the atoms of an oxidizer (e.g., O) (Merzhanov, 2004; Patil, 2008). Using this solution-combustion method, recently, Saroj, Singh, and Singh (2020) have synthesized iodine-doped TiO_2 nanoparticles using the presynthesized titaniumoxysulfate ($TiO(SO_4)$) and iodic acid as the precursors for TiO_2 and iodine, respectively. The synthesized material was investigated for the photocatalytic degradation of Direct Blue 199 dyes under visible light irradiation. Also, Kabir et al. (2020) have synthesized the N-doped ZnO through a controlled combustion method, where the urea was taken as a nitrogen precursor and ground along with the Zn paste to obtain the final N-doped ZnO photocatalyst. Mallikarjuna, Bari, Vattikuti, and Kim (2020) have synthesized carbon-doped SnO_2 nanostructures for visible-light-driven photocatalytic hydrogen production from water splitting, and Mani et al. (2015) have prepared two series of C-doped TiO_2 using titanyl nitrate ($TiO(NO_3)_2$) as TiO_2 precursor, and citric acid and ascorbic acid as carbon sources for the concurrent photocatalytic removal of phenol and Cr(VI) ions.

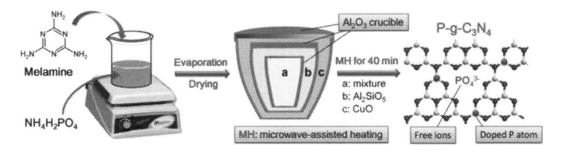

Figure 3.2
Schematic diagram of the microwave-assisted synthesis of P-doped g-C_3N (Chen, Chen, et al., 2020).

3.3 Concept and mechanism of anion doping

It is known that the effective properties of a photocatalyst fundamentally depend on the appropriate positioning of the band-edge potentials (Sakar, Prakash, Shinde, & Balakrishna, 2020). Accordingly, doping could be one of the effective strategies to perform band structure engineering in photocatalysts. In this direction, the anionic substitutions gain considerable importance due to the intrinsic modification in the photocatalysts toward tuning their band gap energy as well as band-edge potential. It was observed that doping of anions, such as N and F, into TiO_2 lattice will lead to the band gap narrowing due to the overlapping of N 2p states with the O 2p states in TiO_2 (Samsudin & Abd Hamid, 2017). As described, the impact of anion doping in photocatalysts can be at a substitutional or interstitial level, where both affect the electronic structure of the photocatalyst. Thus for a photocatalyst with ionic nature (e.g., TiO_2), the substitution of atoms is expected to affect only the VB, which is mainly made up of O 2p states. However, it is limited in the case of fluorine doping that simultaneously affects both VB and CB positions. Besides, the anion atoms of the first row are rather small, especially when they are neutral. So, they can diffuse through the lattice and make a bond at an interstitial position. The concept of substitutional and interstitial occupancy of some anions in TiO_2 is depicted in Fig. 3.3A and B, respectively.

For instance, in the case of TiO_2, it is observed that when the dopant atoms replace the oxygen atoms in the valance band, they create a new energy level just above the valance band maximum, which leads to the reduction in the overall band gap energy. However, anions with high electronegativities, such as fluorine (F), tend to form an energy level just

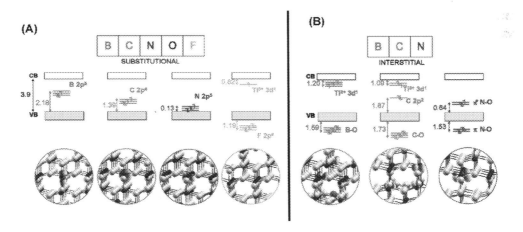

Figure 3.3
(A and B) Schematic representation of Kohn−Sham one-electron states and spin density plot for the substitutional and interstitial doping of anions (B, C, N, O, F) in anatase TiO_2 (Di Valentin & Pacchioni, 2013).

below the O 2p states in VB. This essentially influences the CB and leads to the formation of a Ti^{3+} state just below the CB. Similarly, the anions, such as boron (B), act as a net electron donor that forms B^{3+} and three Ti^{3+} ions, while the carbon (C) as an anion dopant donates only two electrons to the lattice with the formation of a C^{2+} ion. On the other hand, the dopant nitrogen (N) forms a direct bond with a lattice O and does not donate electrons to the host lattice (Di Valentin & Pacchioni, 2013). Among these dopants, boron donates all three valance electrons to lattice Ti ions; thus it is found to be the most oxidized system. Similarly, the interstitial-carbon forms three C–O bonds, so that the C ion can still hold a lone pair of an electron that makes them more electronegative than the B-doped system. Whereas the N is preferentially bound to only one lattice oxygen as compared to B and C. Typically, the N atom pushes the O atom from the Ti-O-Ti plane toward the interstitial cavity so that the NO species can be formed through lattice Ti. Therefore the interstitial nitrogen doping can enable visible light activity in photocatalysts as compared to B and C doping.

The band engineering can be further improved by doping with two or more anionic species in single-phase materials, which exhibit various functionalities that are normally unattainable with a single anion-doped material. Notably, the anions that are less electronegative than the oxygen can provide more negative energy to VB. On the other hand, in addition to doping, the construction of materials, such as oxynitrides and oxysulfides, can largely favor obtaining the photocatalytic systems with tunable band structures and band gap energies. This is because of the fact that these nitride phases possess d^0 or d^{10} electronic configurations, which are usually more negative than the reduction potential of water and favor for the effective photocatalytic process. Therefore besides mere anion doping, the formation of oxy-anion- or mixed-anion-based energy bands should be constructed and explored for the development of photocatalytic materials with improved properties and enhanced efficiencies.

3.4 Anion-doped photocatalysts and applications

3.4.1 Nitrogen-doped photocatalysts

Nitrogen is a relatively widely doped nonmetal in photocatalysts toward modifying their electronic states to obtain suitable properties for effective photocatalysis. The nitrogen doping at the anionic sublattices introduces new electronic states within the band gap near the VB edge of the semiconductor due to the matching energy difference between the N and O atomic valence orbitals (Zhang et al., 2020). Upon doping, nitrogen atoms are found to occupy either in interstitial sites (with N–O bonding as in TiO_2) or substitutional sites replacing O with N atoms in TiO_2. The interstitial doping shows increased visible light absorption, but it does not reduce the band gap as it forms a discrete energy state between

VB and CB. The active form of N-doped TiO_2 was found to be substitutional doping as it reduces the band gap energy via upward shifting of the edge of VB due to the hybridization of N 2p with O 2p orbitals (Asahi, Morikawa, Irie, & Ohwaki, 2014; Raizada et al., 2021).

It is demonstrated that the nitrogen-species can be introduced in the form of N, NO, and NO_2 into oxygen sites in anatase titanium oxide. All these modifications help in narrowing down the band gap of TiO_2, resulting in the visible light absorption in TiO_2. Also, it is found that the mode of nitrogen doping, that is, substitutional or interstitial doping, causes dipole–dipole interactions, electromagnetic interactions, or van der Waal's interactions in photocatalysts, which also lead to the surface modifications (Jyothi et al., 2016). Notably, the interstitial nitrogen doping changes the lattice structure of the photocatalysts. Dunnill and Parkin (2011) reported that nitrogen doping into TiO_2 leads to simultaneous occupancies in both substitutional and interstitial sites. Under such circumstances, N replaces the oxygen in the case of substitutional doping and it occupies the off-lattices of TiO_2 in the case of interstitial doping. As a result, the interstitial doping effectively reduces the band gap energy of TiO_2 than that of the substitutional doping. The interstitially nitrogen-doped TiO_2 showed a band gap of ~ 2.46 eV, while the substitutional TiO_2 showed a band gap of ~ 3.06 eV (Tang et al., 2020).

Lingling et al. synthesized the nitrogen-doped TiO_2/SnO_2 heterostructured microspheres using the conventional hydrothermal method. The band gap energy of the various synthesized materials, such as bare TiO_2, N-doped TiO_2, and N-doped TiO_2/SnO_2, was estimated to be 3.14, 3.06, and 2.97 eV, respectively. It was proposed that the interaction between 2p orbitals of nitrogen and 2p orbitals of oxygen are responsible for the band gap reduction in N-TiO_2 and N-TiO_2/SnO_2 materials (Zhou et al., 2016). Similarly, Xu, Steinmiller, and Skrabalak (2012) synthesized the N-TiO_2/SnO_2 composites and employed them to degrade rhodamine B (RhB) dye. The formation of heterojunction and proposed charge migration in the N-TiO_2/SnO_2 system toward the degradation of RhB dye is depicted in Fig. 3.4.

It is found that the nitrogen doping into the one-dimensional ZnO photocatalysts enhances the photocatalytic properties, where it is also found that the nitrogen doping converts ZnO toward p-type semiconductor by creating more holes in the VB of ZnO. Also, the N doping leads to the overlapping of 2p orbitals of N and O in ZnO and results in narrowing down the band gap structure. Furthermore, it is observed that when an atomic N replaces O, it acts as an acceptor, whereas, if molecular N occupies the O sites, then it acts as a donor. This is because of the fact that the molecular N has lower formation enthalpy than that of atomic N in oxygen sites, which as a result leads to the formation of n-type conduction in N-doped ZnO (Xu et al., 2012; Zhang et al., 2016). Bhawna et al. synthesized the N-doped SnO_2 nanoparticles using the solution-based one-step synthesis method, where the concentration of doped N was estimated to be 4.6%. The N-SnO_2 has a band gap of 3.45 eV and efficiently photo-catalyzed the water-splitting reaction under UV–visible light (Bhawna et al., 2020).

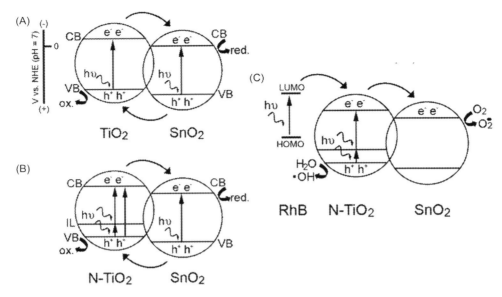

Figure 3.4

Schematic diagram of the formation of heterojunction in (A) TiO$_2$/SnO$_2$ system, (B) N-TiO$_2$/ SnO$_2$, and (C) anticipated charge migration for RhB degradation. *Source: Reprinted with permission from Xu, L., Steinmiller, E. M. P., & Skrabalak, S. E. (2012). Achieving synergy with a potential photocatalytic Z-scheme: Synthesis and evaluation of nitrogen-doped TiO$_2$/SnO$_2$ composites. The Journal of Physical Chemistry C, 116(1), 871–877. doi:10.1021/jp208981h.*

Zhou et al. prepared the N-doped g-C$_3$N$_4$ photocatalysts through thermo-induced polymerization and N doping was found to be responsible for the formation of improved and delocalized aromatic π-conjugated structure in g-C$_3$N$_4$ and accordingly, efficient charge transfer, and separation were also observed. As a result, the N-doped g-C$_3$N$_4$ showed enhanced hydrogen evolution attributed to the reduced recombination and enhanced visible light absorption (Zhou et al., 2016) (Fig. 3.5).

Qi et al. prepared CeO$_2$/N-doped carbon quantum dot/g-C$_3$N$_4$ Z-scheme photocatalyst for the degradation of tetracycline and hydrogen generation. The composite showed an improved photocatalytic activity attributed to the rapid charge separation and stability of CeO$_2$/N-doped CQDs/g-C$_3$N$_4$ as compared to the bare CeO$_2$ and g-C$_3$N$_4$. It was found that the N-doped-CQDs facilitated the enhanced photocatalytic efficiencies for the simultaneous degradation of H$_2$ production (Qi et al., 2020). As shown in Fig. 3.6, the NCQDs act as trapping centers for the holes and electrons generated in g-C$_3$N$_4$ and CeO$_2$, respectively. Through such trapping, the electrons in g-C$_3$N$_4$ are actively involved in the simultaneously degradation and hydrogen production; in addition, the holes in CeO$_2$ are involved in the degradation of tetracycline. In this process, NCQDs facilitated an enhanced trapping due to the discrete energy levels formed by N doping in the band structure CQDs.

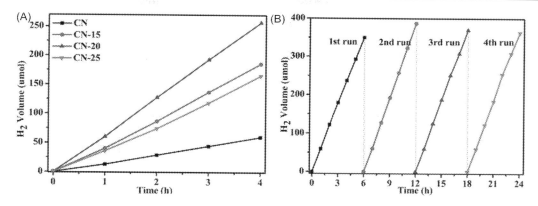

Figure 3.5

(A and B) Photocatalytic hydrogen evolution on N-doped g-C_3N_4 photocatalysts. Source: *Reprinted with permission from Zhou, Y., Zhang, L., Huang, W., Kong, Q., Fan, X., Wang, M., & Shi, J. (2016). N-doped graphitic carbon-incorporated g-C_3N_4 for remarkably enhanced photocatalytic H_2 evolution under visible light. Carbon, 99, 111−117, doi:https://doi.org/10.1016/j.carbon.2015.12.008.*

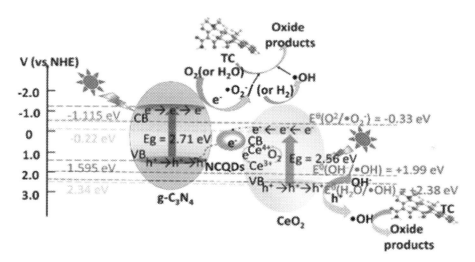

Figure 3.6

Band structure of Z-scheme photocatalyst CeO_2/N-doped carbon quantum dot/g-C_3N_4. Source: *Reprinted with permission from Qi, H., Shi, C., Jiang, X., Teng, M., Sun, Z., Huang, Z., . . . Guo, Z. (2020) Constructing CeO_2/nitrogen-doped carbon quantum dot/g-C_3N_4 heterojunction photocatalysts for highly efficient visible light photocatalysis. Nanoscale, 12(37), 19112−19120. doi:10.1039/D0NR02965C.*

3.4.2 Sulfur-doped photocatalysts

It is observed that the sulfur (S) doping distorts the crystal lattice in metal oxides due to its high ionic radius (1.8 Å) as compared to O (1.4 Å). For instance, in TiO_2, the sulfur doping takes place at O sites and also replaces Ti^{4+} in the bulk or at the surface (Kızılaslan, Kırkbınar, Cetinkaya, & Akbulut, 2020). The incorporation of S into TiO_2 extends the photocatalytic degradation of organic pollutants under visible irradiation. Studies also show that incorporation of sulfur into TiO_2 modifies the ratio of anatase and rutile phase and thereby improves the efficiency under visible irradiation (Khan, Swati, Younas, & Ullah, 2017). Die et al. studied the sulfur doping into the TiO_2 polymorph structure by first-principle calculations. From studies, it was concluded that the sulfur replaces Ti^{4+} ions in both anatase and rutile phase lattice and reduces the band gap energy. It also showed that sulfur can replace both Ti ions and O ions depending on whether the structure is O rich and Ti rich, respectively (Xu et al., 2010). Guo et al. prepared the S-doped $BiVO_4$ through the surfactant-free hydrothermal method. The S^{2-} ions were doped into the crystal lattice of $BiVO_4$ by replacing the O^{2-} ions, which enhanced the oxygen vacancies, increased the number of V^{4+} species, increased the lattice parameters of $BiVO_4$, and also led to the redshift of the diffuse reflectance spectra. The S-$BiVO_4$ exhibited enhanced photocatalytic activity for methylene blue degradation, ascribed to efficient charge separation, and reduced recombination (Guo et al., 2015). Gomathi Devi et al.(Gomathi Devi, 2014) doped the S^{6+} ions into the anatase TiO_2 lattice by sol−gel method to degrade phenol. The sulfur was doped at 0.15 at% by adding the stoichiometric amount of sulfur solution in benzene to anatase phase TiO_2 powders. The XRD of the sample showed the incorporation of sulfur ions into the lattices of TiO_2. Interestingly, it was described that the replacement of Ti^{4+} ions by S^{6+} ions caused the compressive strain in the system, which is confirmed from the observed shift in (101) peak in the XRD pattern. This also implied that the contraction of cell volume was occurred due to the sulfur doping in TiO_2. This is essentially ascribed to the ionic radius of the S^{2+} ion (0.17 nm), which is quite larger than that of O^{2-}, where the substitution of S^{6+} (0.029 nm) in the lattice of Ti^{4+} was more favored due to their comparable ionic radii. Agorku et al. (Agorku et al., 2014) performed photo-degradation of Indigo carmine using S/Gd^{3+} codoped TiO_2 synthesized by sol−gel method. The codoped TiO_2 showed enhanced activity for the degradation of Indigo carmine dye as compared to S-TiO_2 (Agorku, Mamba, Pandey, & Mishra, 2014). Pradhan et al. synthesized S and N codoped α-Fe_2O_3 nanostructures using the coprecipitation method. It was reported that the sulfur doping induced [110] plane-oriented growth in α-Fe_2O_3, which enhanced the photocatalytic activity of the system. The possible mechanism of S and N codoping in α-Fe_2O_3 is given in Eqs. (3.1)−(3.3) (Pradhan, Sahu, & Parida, 2013).

$$Fe(NO_3)_3 9H_2O + 6H_2O + NH_2CSNH_2 \rightarrow Fe(H_2O)_5 \left(S{=}C(NH_2)_2 \right)^{3+} + 2NO_3^- \qquad (3.1)$$

$$Fe(H_2O)_5 \left(S{=}C(NH_2)_2 \right)^{3+} \rightarrow S - FeOOH + NH_4^+ \qquad (3.2)$$

$$S{-}FeOOH + NH_4^+ \rightarrow SN{-}Fe_2O_3 \qquad (3.3)$$

Yu et al. studied the antibacterial effect of sulfur-doped titanium oxide under visible light. It was reported that the rapid formation of hydroxyl radicals due to sulfur doping was the key origin for the observed enhanced bactericidal photocatalytic activity of the S-TiO$_2$ under visible light irradiation (Yu et al., 2005). The photographic images of the visible light-induced photocatalytic disinfection of *Micrococcus lylae* and bacterial survival with increasing concentration of S-TiO$_2$ are shown in Fig. 3.7A and B (Yu et al., 2005).

In another study, Chen, Wang, Zhao, Gao, and Li (2020) doped S into WO$_3$ through the hydrothermal process and observed that the band gap of WO$_3$ was reduced with S doping along with increased oxygen vacancies in WO$_3$, which led to the improved visible light absorption and photocatalytic activity for methylene blue degradation. Xue et al. prepared S-doped Sb$_2$O$_3$ through hydrothermal synthesis using SbCl$_3$ and thioacetamide as precursors. The study reported that the doping amount of S can be modified by tuning the pH of the precursor solution, and the S^{2-} ions enter into the interstitial site of Sb$_2$O$_3$ rather than O sites, which in turn widened the optical absorption range of Sb$_2$O$_3$. The synthesized S-doped Sb$_2$O$_3$ showed enhanced photocatalytic activity for the decomposition of methyl orange and 4-phenylazophenol under visible light (Xue et al., 2018).

3.4.3 Fluoride-doped photocatalysts

It is found that halogen doping in photocatalysts can improve the surface properties as well as the optical properties, such as the increased absorption in the solar spectrum, surface

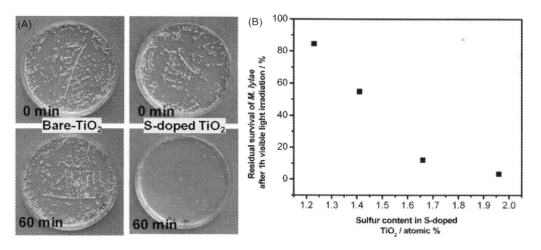

Figure 3.7
(A, B) Visible light-induced photocatalytic disinfection of *M. lylae* by S-doped TiO$_2$. *Source: Reprinted with permission from Yu, J. C., Ho, W., Yu, J., Yip, H., Wong, P. K., & Zhao, J. (2005). Efficient visible-light-induced photocatalytic disinfection on sulfur-doped nanocrystalline titania.* Environmental Science & Technology, 39(4), 1175–1179. doi:10.1021/es035374h.

reactivity, and efficient utilization of photoinduced electron/hole (e^-/h^+) pairs. In this direction, fluorine (F) is doped in TiO_2 and found to increase the percentage of the anatase phase, where the substitution of F atoms with O atoms in TiO_2 forms the Ti-F bond. Interestingly, it is found that the Ti-F bonds possess a stronger electron withdrawing ability and thus it facilitates the antirecombination of electron−hole pairs in the system (Yu, Yu, & Zhang, 2002). Furthermore, the pH of fluorine is found to show a significant role in determining the phase of TiO_2. The basic fluorine favors the formation of pure anatsae phase, while acidic F favors the formation of rutile/anatase or pure rutile phase TiO_2. Gasparotto et al. have reported F doping into p-type Co_3O_4 photocatalyst through the plasma-assisted process. F-doped Co_3O_4 films showed fivefold enhancement in H_2 production from water/ethanol mixture both under UV and visible light irradiation (Gasparotto et al., 2011). Gao et al. demonstrated the preparation of hydrogenated F-doped TiO_2, which effectively enhanced the UV-light absorption of TiO_2 and the further hydrogen treatment enhanced (H-F-TiO_2) the light absorption in visible light region. This system of H-F-TiO_2 showed higher photocatalytic activity for hydrogen generation and degradation of methylene blue attributed to the high charge carrier density and surface disorders created by F and H doping, respectively (Gao et al., 2019). Carroll et al. prepared F-doped ZnO photocatalysts through sol−gel method and investigated for antibacterial activity against *Escherichia coli* (Gram negative) and *Staphylococcus aureus* (Gram positive). The F-ZnO showed 99% antibacterial activity for *E. coli* and *S. aureus* ascribed to the rapid generation of high reactive oxygen species and the effective degradation of the bacterial cell membrane, leading to cell death due to the F doping-induced changes in ZnO nanostructures (Podporska-Carroll et al., 2017). Dozzi et al. doped S and F into TiO_2 using thiourea and NH_4F as dopant sources, respectively, and reported that F-doped TiO_2 exhibits enhanced photocatalytic activity for formic acid degradation in comparison to S-TiO_2 attributed to highly crystalline anatase structure and reduced recombination rates in the system. The high dopant concentrations limited the photocatalytic activity may be because of the increased charge recombination centers (Dozzi, Livraghi, Giamello, & Selli, 2011). Li et al. have designed the F-doped TiO_2/g-C_3N_4 heterostructure, and the photocatalytic activity was evaluated for the degradation of RhB under solar light. The heterostructure showed $\sim$97% degradation in 20 min attributed to the generation of higher superoxide radicals (O_2^-) and the reduced interfacial charge transfer resistance and increased photocurrent in the system (Li, Zhong, et al., 2020). Similarly, Kumar et al. prepared F-doped SnO_2 nanocrystals through low-temperature oxidation of Sn^{2+} containing fluoride complex $KSnF_3$ as single precursor with H_2O_2. This system showed an increased photocatalytic degradation toward RhB as compared to SnO_2, which was attributed to the very high concentration of oxygen vacancies in F-doped SnO_2 (Kumar, Govind, & Nagarajan, 2011). Yu et al. have prepared F-doped TiO_2 with anatase and brookite phase through the hydrolysis process. The F doping enhances the crystallinity of the material and also reduces phase transformation from anatase to rutile and also suppresses the brookite phase formation in addition to redshift in the band gap energy (Yu et al., 2005). Modak et al. have studied the role of F doping in enhancing the photoactivity of Rh-doped $SrTiO_3$ using density functional theory. Even though Rh doping alone can enhance the visible light activity, the photoconversion efficiency still was poor due to the introduction of localized unoccupied states above the VB, which improves the electron—hole recombination. Codoping of F was found to reduce these localized states, leading to band gap narrowing to 2.5 eV, and hence the codoping of F along with Rh increases the photocatalytic activity of $SrTiO_3$ (Modak & Ghosh, 2015). Torres et al. (Torres, 2018) have worked on C-, N-, S-, and F-doped TiO_2 (101) phase and analyzed through DFT calculations. It was predicted that the S doping can broaden the VB, N doping can generate the localized unoccupied states, and C doping on TiO_2 (101) can facilitate a cleaner band gap than the bulk and exhibit reduced recombination centers. The intensity of the visible light absorption of doped TiO_2 (101) followed the order of C > N > F > S dopant. Miao et al. have doped F into CeO_2 nanoparticles through combustion method and reported that F doping helps in the reduction of particle size and with a higher concentration of reactive facets exposed. The band gap was reduced from 3.16 to 2.88 eV, and the photocatalytic activity was enhanced with F doping by 9.5 times as compared to bare CeO_2 (Miao et al., 2016). Samsudin et al. have prepared N, F, and N—F codoped TiO_2 and reported the observation of reduced band gap energy in TiO_2 through the doping-induced formation of mid-band states and Ti^{3+} impurities states. The doped material showed a change in band gap structure as shown in

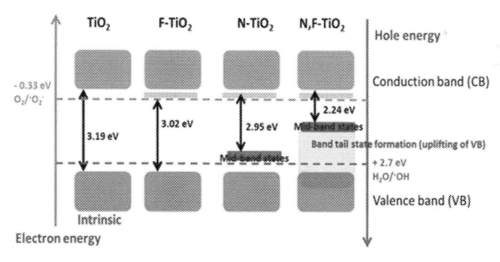

Figure 3.8

Band structure of TiO_2, F-TiO_2, N-TiO_2, and N-F-TiO_2. Source: *Reprinted with permission from Samsudin, E. M., & Abd Hamid, S. B. (2017). Effect of band gap engineering in anionic-doped TiO_2 photocatalyst.* Applied Surface Science, *391, 326—336, https://doi.org/10.1016/j.apsusc.2016.07.007.*

Fig. 3.8 attributed to the formation of oxygen vacancies, which contributed to better light absorption and improved optical properties in doped TiO_2 (Samsudin & Abd Hamid, 2017).

Considering halogen, iodine (I) is recognized as an effective dopant to extend the light absorption of a photocatalyst. Accordingly, a redshift in the band edge position of iodine-doped TiO_2 was reported, where it was attributed to the formation of Helmholtz potential on the surface of TiO_2 with the presence of iodine (Liu et al., 2009). Doping of iodine caused the reduction of TiO_2 band gap energy for 3.2−3.0 eV extending the photoresponse to the visible light region. In I-doped TiO_2, the I 5p states mix with Ti 3d while O 2p moves to high energy region. Iodine dopants were proven to increase the formation of surface hydroxyl groups and thus improved the photocatalytic activity of TiO_2 (Liu et al., 2009). In comparison to other nonmetal dopants (such as N, C, B, and S), iodine was predicted to be substituted in the lattice Ti due to closer ionic radii of I^{5+} and Ti^{4+}. Liu et al. doped the iodine on the surface of TiO_2 with I—O—I (unoccupied state below the CB) and I—O—Ti (occupied state above the VB) atomic configuration, which showed extended absorption edge up to 800 nm with enhanced RhB degradation (Liu et al., 2009). Zhang, Li, Ackerman, Gajdardziska-Josifovska, and Li, (2011) have prepared I-doped TiO_2 through the hydrothermal method and tested for photocatalytic reduction of CO_2 with H_2O resulting in CO as the major photoreduction product. I—TiO_2 exhibited high CO_2 reduction with a CO yield of 2.4 µmol/g/h. It is reported that the I doping induced the formation of mixed TiO_2 phases, such as anatase, rutile, and brookite. In addition, the introduction of I^{5+} leads to charge imbalances, leading to the formation of Ti^{3+} on the surface states, where it traps the excited electrons contributing to reduced recombination. Moreover, iodine atoms prefer to be doped at the near-surface of TiO_2 due to the I−O strong repulsion; hence, the surface-doped I^{5+} improves the overall electron transfer and reduces the recombination of charge carriers.

3.4.4 Carbon-doped photocatalysts

During the carbon doping in TiO_2, carbon atoms tend to occupy the interstitial positions and result in narrowing the band gap energy toward the absorption in the near-infrared region. Khan, Al-Shahry, and Ingler (2002) introduced the elemental C and carbonate species (C−O bond) as interstitial dopants in TiO_2, as well as the C substituted at O sites (Ti−C) in TiO_2 lattice. It was found that the substitution of carbon atoms in TiO_2 induced the new states (C 2p) close to the VB edge of TiO_2 (O 2p). Accordingly, the CB edge was shifted toward narrowing the band gap. Furthermore, the doping increased the high surface area of TiO_2 that increased the surface-adsorption efficiency and provided more reactive sites on the surface (Nowotny, Bak, & Alim, 2015). Similarly, Yang et al. (2015) prepared the carbon-doped TiO_2 using oleylamine-covered TiO_2 as a precursor. The formed C−TiO_2 exhibited visible light absorption, reduced recombination, and improved photocatalytic hydrogen evolution attributed to the presence of oxygen vacancies that facilitated rapid proton reduction to produce H_2 molecules as shown in Fig. 3.9.

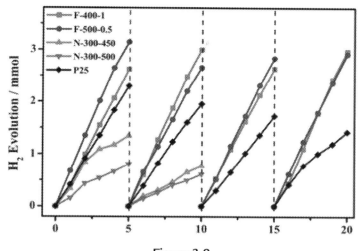

Figure 3.9
Simulated solar light-driven photocatalytic H₂ evolution from C—TiO₂ that calcined under different temperatures. Source: *Reprinted with permission from Yang, Y., Ni, D., Yao, Y., Zhong, Y., Ma, Y., & Yao, J. (2015). High photocatalytic activity of carbon doped TiO₂ prepared by fast combustion of organic capping ligands. RSC Advances, 5(113), 93635—93643. doi:10.1039/C5RA19058D.*

Wang, Chen, et al. (2021) have prepared C-doped boron nitride and used it for the reduction of uranium U(VI) ions. The doped boron nitride showed around 97.4% reduction efficiency toward U(VI), which attributed to the synergistic effect of photoinduced electrons and radicals along with the formation of the porous structure due to the carbon doping in boron nitride as shown in Fig. 3.10A and B (Wang, Chen, et al., 2021).

Shown et al. (2018) prepared C-SnS₂ and studied photocatalytic CO₂ reduction to hydrocarbons in the presence of solar light. It is reported that the doping induced the microstrain in SnS₂ lattice due to interstitial-carbon doping, which eventually enhanced the CO₂ reduction with a photochemical quantum yield of around 0.7%, where the doping induced changes in the band structure and subsequent CO₂ reduction mechanism is depicted in Fig. 3.11. Zhang et al. (2019) introduced carbon—hydrogen (CH) into WS₂, which led to reduction in band gap from 1.98 to 1.83 eV, and furthermore, their first principle calculation revealed that the CH group occupies S vacancies in WS₂. The pristine WS₂ exhibited n-type conductivity and carbon-hydrogen (CH) co-doping into WS₂ showed p-branch (p-type conductivity). Whereas, the increasing dopant concentration of carbon completely transformed the conduction band structure into p-type, which suggested that the tailoring of electronic and optical properties can be possible with CH doping in WS₂ (Zhang et al., 2019).

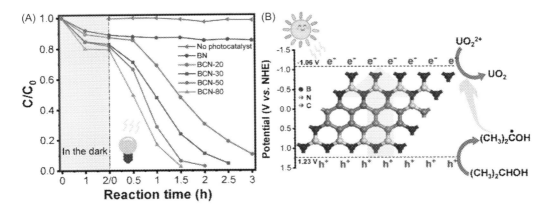

Figure 3.10

(A) Photocatalytic reduction of U(VI) over C-boron nitrides and (B) proposed mechanism for photocatalytic reduction of U(VI) over BCN under visible light irradiation. Source: *Reprinted with permission from Wang, Y., Chen, G., Weng, H., Wang, L., Chen, J., Cheng, S., . . . Lin, M.(2021). Carbon-doped boron nitride nanosheets with adjustable band structure for efficient photocatalytic U(VI) reduction under visible light.* Chemical Engineering Journal, 410, *128280. https://doi.org/10.1016/j.cej.2020.128280.*

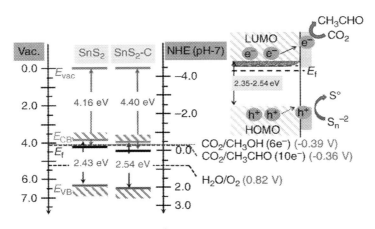

Figure 3.11

Band edge positions and photocatalytic reaction mechanism: Comparative band diagram of SnS_2-C and SnS_2, together with a proposed electron-hole separation of photo-excited electron-hole pairs in SnS_2-C. Source: *Reprinted with permission from Shown, I., Samireddi, S., Chang, Y.-C., Putikam, R., Chang, P.-H., Sabbah, A., . . . Chen, K.-H.(2018). Carbon-doped SnS_2 nanostructure as a high-efficiency solar fuel catalyst under visible light.* Nature Communications, 9(1), *169. doi:10.1038/s41467-017-02547-4.*

Tahir et al. doped the activated carbon onto tungsten trioxide (WO_3) nanotubes and studied the photocatalytic activity for the degradation of RhB (Tahir et al., 2020). WO_3 nanotubes showed band gap reduction from 2.76 to 2.26 eV with carbon doping and the 2% of activated carbon-doped WO_3 showed better degradation efficiency as compared to the undoped WO_3 (Tahir et al., 2020). Han et al. used Ti_3C_2 as a precursor to prepare C-doped TiO_2/g-C_3N_4 photocatalyst, which showed the enhanced hydrogen generation activity with 1409 μmol/h/g yield of hydrogen, ascribed to the formation of heterojunction structure between Ti_3C_2 MXene-derived C-doped TiO_2 and g-C_3N_4, which ultimately improved the charge carrier transport and reduced the recombination of electrons and holes (Han et al., 2020). Yu et al. reported a significantly enhanced photocatalytic hydrogen generation (211 μmol/g/h) using carbon-doped $KNbO_3$ photocatalyst prepared by hydrothermal and calcination processes. The improved activity was ascribed to the better separation of charge carriers due to carbon doping (Yu et al., 2018). Qi et al. prepared hydrophobic carbon-doped TiO_2/MCF-F (mesoporous cellular foam) using NH_4F as hydrophobic-modifying agent. Furthermore, the investigation reported that the carbon atoms were doped into the lattice by replacing titanium atoms in titania, leading to the narrowing of band gap to absorb visible light, resulting in improved photocatalytic activity. The TiO_2/MCF-F showed high absorption properties and high photo-activity for methyl orange degradation, where the hydrophobic property was induced by Si−F bonds (Qi, Xing, & Zhang, 2014). Zhou et al. prepared the carbon-doped carbon nitride (CCN)/$Bi_{12}O_{17}Cl_2$ composites through the in situ method. The CCN/$Bi_{12}O_{17}Cl_2$ showed higher activity for the degradation of tetracycline antibiotics as compared to the bare-$Bi_{12}O_{17}Cl_2$, CCN, and BiOCl (Zhou et al., 2018) (Fig. 3.12).

3.4.5 Cl- and O-doped photocatalysts

It is found that the Cl doping into the photocatalyst can cause alteration in the band gap structure. For example, Cl atoms intercalate into the g-C_3N_4 interlayer and cause band gap

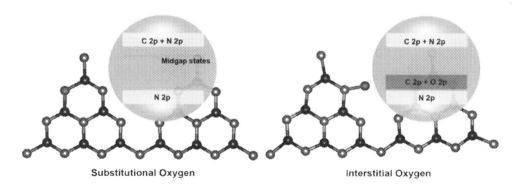

Figure 3.12

Structural site occupancy and band structure of oxygen-doped g-C_3N_4. Source: *Reprinted with permission from Putri, L. K., Ng, B.-J., Er, C.-C., Ong, W.-J., Chang, W. S., Mohamed, A. R., & Chai, S.-P. (2020). Insights on the impact of doping levels in oxygen-doped gC_3N_4 and its effects on photocatalytic activity.* Applied Surface Science, 504, 144427. https://doi.org/10.1016/j.apsusc.2019.144427.

tuning. The highly electronegative nature of the Cl atoms contributes to the easy trapping of positively charged holes leading to efficient electron–hole separation (Wang, Huang, et al., 2021). Ai et al. doped Cl into Bi_2S_3 homojunction nanorods, which created the defects in the material leading to the enhanced activity for Cr(VI) degradation. Cl doping induces surface sulfur defects in Bi_2S_3 mainly on the exposed facets and homojunction generated from the oriented attachment of nanoparticles. The sulfur defects acted as active sites for photocatalytic reaction and the homojunction structure improved the electron–hole separation, resulting in the increased photocatalytic degradation of Cr(VI) to Cr(III) (Ai et al., 2020). Wang et al. prepared Cl-doped g-C_3N_4 nanosheets and tested for the removal of RhB and tetracycline hydrochloride. The Cl-doped g-C_3N_4 exhibited higher photocatalytic activity attributed to the efficient charge carrier separation (Ai et al., 2020). Liang et al. doped Cl into SnO photocatalyst through mechanical alloying and studied for the photocatalytic degradation of methyl orange. The doped material showed around 99% degradation of methyl orange (Liang, Zhang, & Zhang, 2019). Guo et al. prepared Cl-doped porous g-C_3N_4 and studied for the degradation of tetracycline. The Cl-g-C_3N_4 exhibited higher activity attributed to a higher surface area, reduced recombination, and controlled electronic structure induced by Cl doping (Guo et al., 2019). Long et al. reported Mg- and Cl-doped g-C_3N_4 prepared by hydrothermal method, which reduced the band gap of g-C_3N_4, increased the surface area and visible light absorption, and improved the charge separation and transportation. The doped g-C_3N_4 showed enhanced hydrogen evolution of 8.4 μmol/h (Long, Diao, Rao, & Zhang, 2020). Phuruangrat, Dumrongrojthanath, Kuntalue, Thongtem, and Thongtem (2017) doped Cl into Bi_2MoO_6 through hydrothermal method. The doped Bi_2MoO_6 showed high crystallinity and excellent degradation of RhB. Putri et al. prepared the oxygen-doped g-C_3N_4 and studied the photocatalytic hydrogen evolution. The O-doped g-C_3N_4 showed increased porosity and mid-gap states in the electronic band structure as shown in Fig. 3.13, which enhanced the light-harvesting ability in O-doped g-C_3N_4 (Putri et al., 2020).

Jiang et al. prepared N- and O-doped g-C_3N_4 through the copyrolysis process. N doping induced defects and porous structure in g-C_3N_4, while the oxygen dopants changed the π band state and lone pair electrons in the system, leading to the change of electron transition in modified g-C_3N_4 as illustrated in Fig. 3.13. The modified g-C_3N_4 exhibited six times higher hydrogen generation as compared to pristine g-C_3N_4 attributed to the enhanced charge separation and improved specific surface area (Jiang et al., 2019).

Huang et al. doped the phosphorus and oxygen into g-C_3N_4 (CN) (POCN) through the thermal polymerization process. The P atoms were found to be substituted at the corner and bay carbon sites, while the O atoms were substituted at the nitrogen sites in the CN framework. Due to such structural modifications, the POCN showed improved activity toward the degradation of fluoroquinolones attributed to the narrow band gap, better charge carrier separation, and high specific surface area due to the doping (Huang et al., 2019). Li et al.

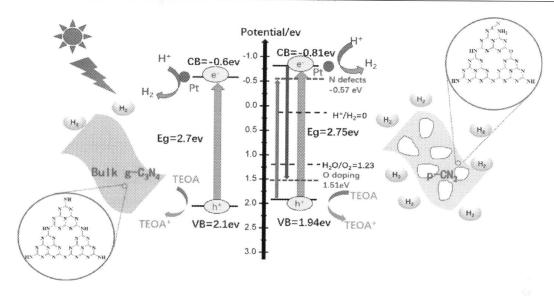

Figure 3.13
Band structure and photocatalytic H$_2$ evolution on N-O-doped [g-C$_3$N$_4$]. Source: *Reprinted with permission from Jiang, Y., Sun, Z., Tang, C., Zhou, Y., Zeng, L., & Huang, L.(2019). Enhancement of photocatalytic hydrogen evolution activity of porous oxygen doped g-C$_3$N$_4$ with nitrogen defects induced by changing electron transition.* Applied Catalysis B: Environmental, 240, 30—38. https://doi.org/10.1016/ j.apcatb.2018.08.059.

studied Ti^{3+} and oxygen doping toward achieving the improved properties of TiO$_2$/g-C$_3$N$_4$. The study suggested that the modified TiO$_2$/g-C$_3$N$_4$ composites showed improved RhB degradation, attributed to the synergistic effect of reduced electron transfer resistance, and improved charge carrier separation (Li et al., 2017). Zhao et al. prepared the oxygen-doped MoS$_2$ nanospheres/CdS quantum dots (QDs)/g-C$_3$N$_4$ nanosheet and reported the enhanced photocatalytic activity for hydrogen evolution and photodegradation of bisphenol A. The observed efficiency was ascribed to the longer lifetime of electron—hole charge carriers due to oxygen defects and more active sites, where the CdS QDs acted as a bridge between the intermediate charge transporters in the ternary composite (Zhao, Xing, et al., 2019).

3.5 Conclusion and outlook

This chapter is focused on the effect of anion doping into the photocatalysts, such as TiO$_2$, Co$_3$O$_3$, ZnO, g-C$_3$N$_4$, SnO$_2$, SrTiO$_3$, CeO$_2$, boronnitride, carbon-nitride/Bi$_{12}$O$_{17}$Cl$_2$, SnS$_2$, WS$_2$, WO$_3$, KNbO$_3$, and Bi$_2$S$_3$, using the anionic elements, such as N, S, F, C, Cl, I, and O. Anions occupy in a photocatalyst in two different ways: (1) substituting at the site of lattice oxygen and (2) introducing anion dopants into interstitial sites between different layers of the crystal. Accordingly, various methods have been proposed for the synthesis of anion-doped photocatalysts. The factors, such as concentration of the doped anions and doping sites, contribute to enhance many properties of photocatalysts, including the discrepancy in electronic structure, defect structure, surface area, and surface states. Similarly, the synthesis of anion-doped photocatalysts with controlled dopant concentration and doping sites is very important to optimize the activity of the photocatalyst, but still, there are more challenges to achieve such factors. Generally, anion doping decreases the surface area of the material, which is unfavorable for photocatalytic reactions. For instance, it was found from various studies that the N doping occupies either interstitial or substitution sites by replacing O and results in increased visible light absorption as well as the formation of oxygen vacancies, while the band gap reduction happens in the case of substitution-mediated doping. Carbon doping narrows the band gap by interstitial or substitution-mediated doping and results in higher visible light absorption. S doping enhances the visible light absorption and studies reported that S doping modifies the ratio of anatase and rutile to improve the efficiency of doped TiO_2 under visible irradiation. F doping narrows the band gap and also reduces the electron−hole recombination. Cl doping effectively channels the charge carrier separation and transport promoting photochemical reactions. Furthermore, anion doping modifies the concentration of oxygen vacancy and metal-oxygen bond strength that contribute to the high catalytic activity and oxygen ion transportation ability, which collectively improves the mobility and permeability of lattice oxygen. The property enhancments via anion doping make the anion-doped photocatalysts to be used in various photocatalytic applications, such as water splitting, toxic pollutant degradation, CO_2 reduction, and bacterial disinfection. However, further research is necessary towards producing more stable anion-doped photocatalysts and their mechanism should be explored to achieve further improved photocatalytic efficiencies.

Acknowledgment

One of the authors, Shwetharani R, acknowledges the TARE project (TAR/2019/000042) for funding support.

References

Abraham, C., & Devi, L. G. (2020). One-pot facile sol-gel synthesis of W, N, C and S doped TiO_2 and its application in the photocatalytic degradation of thymol under the solar light irradiation: Reaction kinetics and degradation mechanism. *Journal of Physics and Chemistry of Solids, 141*, 109350. Available from https://doi.org/10.1016/j.jpcs.2020.109350.

Agorku, E. S., Mamba, B. B., Pandey, A. C., & Mishra, A. K. (2014). Sulfur/gadolinium-codoped TiO_2 nanoparticles for enhanced visible-light photocatalytic performance. *Journal of Nanomaterials, 2014*, 289150. Available from https://doi.org/10.1155/2014/289150.

Ai, L., Jia, D., Guo, N., Xu, M., Zhang, S., Wang, L., & Jia, L. (2020). Cl-doped Bi_2S_3 homojunction nanorods with rich-defects for collaboratively boosting photocatalytic reduction performance. *Applied Surface Science, 529*, 147002. Available from https://doi.org/10.1016/j.apsusc.2020.147002.

Asahi, R., Morikawa, T., Irie, H., & Ohwaki, T. (2014). Nitrogen-doped titanium dioxide as visible-light-sensitive photocatalyst: Designs, developments, and prospects. *Chemical reviews*, *114*(19), 9824−9852. Available from https://doi.org/10.1021/cr5000738.

Bhawna., Gupta, A., Kumar, P., Tyagi, A., Kumar, R., Kumar, A., . . . Kumar, V. (2020). Facile synthesis of N-doped SnO_2 nanoparticles: A cocatalyst-free promising photocatalyst for hydrogen generation. *ChemistrySelect*, *5*(26), 7775−7782. Available from https://doi.org/10.1002/slct.202001301.

Chen, G., Wang, Q., Zhao, Z., Gao, L., & Li, X. (2020). Synthesis and photocatalytic activity study of S-doped WO_3 under visible light irradiation. *Environmental Science and Pollution Research*, *27*(13), 15103−15112. Available from https://doi.org/10.1007/s11356-020-07827-z.

Chen, P., Chen, L., Ge, S., Zhang, W., Wu, M., Xing, P., . . . He, Y. (2020). Microwave heating preparation of phosphorus doped g-C_3N_4 and its enhanced performance for photocatalytic H_2 evolution in the help of Ag_3PO_4 nanoparticles. *International Journal of Hydrogen Energy*, *45*(28), 14354−14367. Available from https://doi.org/10.1016/j.ijhydene.2020.03.169.

Cheng, G., Xu, F., Stadler, F. J., & Chen, R. (2015). A facile and general synthesis strategy to doped TiO_2 nanoaggregates with a mesoporous structure and comparable property. *RSC Advances*, *5*(79), 64293−64298. Available from https://doi.org/10.1039/C5RA11099H.

Di Valentin, C., & Pacchioni, G. (2013). Trends in non-metal doping of anatase TiO_2: B, C, N and F. *Catalysis Today*, *206*, 12−18. Available from https://doi.org/10.1016/j.cattod.2011.11.030.

Divyasri, Y. V., Reddy, N. L., Lee, K., Sakar, M., Rao, V. N., Venkatramu, V., . . . Reddy, N. C. G. (2021). Optimization of N doping in TiO_2 nanotubes for the enhanced solar light mediated photocatalytic H2 production and dye degradation. *Environmental Pollution*, *269*, 116170. Available from https://doi.org/10.1016/j.envpol.2020.116170.

Dozzi, M. V., Livraghi, S., Giamello, E., & Selli, E. (2011). Photocatalytic activity of S- and F-doped TiO_2 in formic acid mineralization. *Photochemical & Photobiological Sciences*, *10*(3), 343−349. Available from https://doi.org/10.1039/C0PP00182A.

Dunnill, C. W., & Parkin, I. P. (2011). Nitrogen-doped TiO_2 thin films: Photocatalytic applications for healthcare environments. *Dalton Transactions*, *40*(8), 1635−1640. Available from https://doi.org/10.1039/C0DT00494D.

Fu, R., Gao, S., Xu, H., Wang, Q., Wang, Z., Huang, B., & Dai, Y. (2014). Fabrication of $Ti3^+$ self-doped TiO_2(A) nanoparticle/TiO_2(R) nanorod heterojunctions with enhanced visible-light-driven photocatalytic properties. *RSC Advances*, *4*(70), 37061−37069. Available from https://doi.org/10.1039/C4RA06152G.

Fujishima, A., & Honda, K. (1972). Electrochemical photolysis of water at a semiconductor electrode. *Nature*, *238*(5358), 37−38. Available from https://doi.org/10.1038/238037a0.

Gao, Q., Si, F., Zhang, S., Fang, Y., Chen, X., & Yang, S. (2019). Hydrogenated F-doped TiO_2 for photocatalytic hydrogen evolution and pollutant degradation. *International Journal of Hydrogen Energy*, *44*(16), 8011−8019. Available from https://doi.org/10.1016/j.ijhydene.2019.01.233.

Gasparotto, A., Barreca, D., Bekermann, D., Devi, A., Fischer, R. A., Fornasiero, P., . . . Tondello, E. (2011). F-doped Co_3O_4 photocatalysts for sustainable H_2 generation from water/ethanol. *Journal of the American Chemical Society*, *133*(48), 19362−19365. Available from https://doi.org/10.1021/ja210078d.

Gomathi Devi., et al. (2014). Enhanced photocatalytic activity of sulfur doped TiO2 for the decomposition of phenol: A new insight into the bulk and surface modification. *Materials Chemistry and Physics*, *143*(3), 1300−1308.

Gundeboina, R., Venkataswamy, P., & Vithal, M. (2020). Hydrothermal synthesis of C-doped $K_2Al_2Ti_6O_{16}$ as a visible light−activated photocatalyst in the degradation of organic dyes. *Journal of the Australian Ceramic Society*, 1−8. Available from https://doi.org/10.1007/s41779-020-00472-2.

Guo, F., Li, M., Ren, H., Huang, X., Shu, K., Shi, W., & Lu, C. (2019). Facile bottom-up preparation of Cl-doped porous g-C_3N_4 nanosheets for enhanced photocatalytic degradation of tetracycline under visible light. *Separation and Purification Technology*, *228*, 115770. Available from https://doi.org/10.1016/j.seppur.2019.115770.

Guo, M., Wang, Y., He, Q., Wang, W., Wang, W., Fu, Z., & Wang, H. (2015). Enhanced photocatalytic activity of S-doped $BiVO_4$ photocatalysts. *RSC Advances*, 5(72), 58633−58639. Available from https://doi.org/10.1039/C5RA07603J.

Han, X., An, L., Hu, Y., Li, Y., Hou, C., Wang, H., & Zhang, Q. (2020). Ti3C2 MXene-derived carbon-doped TiO2 coupled with g-C_3N_4 as the visible-light photocatalysts for photocatalytic H_2 generation. *Applied Catalysis B: Environmental*, 265, 118539. Available from https://doi.org/10.1016/j.apcatb.2019.118539.

Hoffmann, M. R., Martin, S. T., Choi, W., & Bahnemann, D. W. (1995). Environmental applications of semiconductor photocatalysis. *Chemical Reviews*, 95(1), 69−96. Available from https://doi.org/10.1021/cr00033a004.

Huang, J., Dou, L., Li, J., Zhong, J., Li, M., & Wang, T. (2021). Excellent visible light responsive photocatalytic behavior of N-doped TiO_2 toward decontamination of organic pollutants. *Journal of Hazardous Materials*, 403, 123857. Available from https://doi.org/10.1016/j.jhazmat.2020.123857.

Huang, J., Li, D., Li, R., Zhang, Q., Chen, T., Liu, H., . . . Liu, G. (2019). An efficient metal-free phosphorus and oxygen co-doped g-C_3N_4 photocatalyst with enhanced visible light photocatalytic activity for the degradation of fluoroquinolone antibiotics. *Chemical Engineering Journal*, 374, 242−253. Available from https://doi.org/10.1016/j.cej.2019.05.175.

Ji, L., Liu, X., Xu, T., Gong, M., & Zhou, S. (2020). Technology, Preparation and photocatalytic properties of carbon/carbon-doped TiO_2 double-layer hollow microspheres. *Journal of Sol-Gel Science and Technology*, 93(2), 380−390. Available from https://doi.org/10.1007/s10971-019-05176-z.

Jiang, Y., Sun, Z., Tang, C., Zhou, Y., Zeng, L., & Huang, L. (2019). Enhancement of photocatalytic hydrogen evolution activity of porous oxygen doped g-C_3N_4 with nitrogen defects induced by changing electron transition. *Applied Catalysis B: Environmental*, 240, 30−38. Available from https://doi.org/10.1016/j.apcatb.2018.08.059.

Jyothi, M. S., D'Souza Laveena, P., Shwetharani, R., & Balakrishna, G. R. (2016). Novel hydrothermal method for effective doping of N and F into nano titania for both, energy and environmental applications. *Materials Research Bulletin*, 74, 478−484. Available from https://doi.org/10.1016/j.materresbull.2015.11.020.

Kabir, R., Saifullah, M., Khalid, A., Ahmed, A. Z., Masum, S. M., Molla, M., & Islam, A. (2020). Synthesis of N-doped ZnO nanocomposites for sunlight photocatalytic degradation of textile dye pollutants. *Journal of Composites Science*, 4(2), 49. Available from https://doi.org/10.3390/jcs4020049.

Kadam, A. N., Salunkhe, T. T., Kim, H., & Lee, S.-W. (2020). Biogenic synthesis of mesoporous N−S−C tri-doped TiO_2 photocatalyst via ultrasonic-assisted derivatization of biotemplate from expired egg white protein. *Applied Surface Science*, 518, 146194. Available from https://doi.org/10.1016/j.apsusc.2020.146194.

Kakhki, R. M., Mohammadpoor, M., Faridi, R., & Bahadori, M. (2020). The development of an artificial neural network−genetic algorithm model (ANN-GA) for the adsorption and photocatalysis of methylene blue on a novel sulfur−nitrogen co-doped Fe_2O_3 nanostructure surface. *RSC Advances*, 10(10), 5951−5960. Available from https://doi.org/10.1039/C9RA10349J.

Kappe, C. O. (2013). Unraveling the mysteries of microwave chemistry using silicon carbide reactor technology. *Accounts of Chemical Research*, 46(7), 1579−1587. Available from https://doi.org/10.1021/ar300318c.

Khaki, M. R. D., Shafeeyan, M. S., Raman, A. A. A., & Daud, W. M. A. W. (2017). Application of doped photocatalysts for organic pollutant degradation—A review. *Journal of Environmental Management*, 198, 78−94. Available from https://doi.org/10.1016/j.jenvman.2017.04.099.

Khan, H., Swati, I. K., Younas, M., & Ullah, A. (2017). Chelated nitrogen-sulphur-codoped TiO_2: Synthesis, characterization, mechanistic, and UV/visible photocatalytic studies. *International Journal of Photoenergy*, 2017, 7268641. Available from https://doi.org/10.1155/2017/7268641.

Khan, M. M., Adil, S. F., & Al-Mayouf, A. (2015). Metal oxides as photocatalysts. *Journal of Saudi Chemical Society*, 19(5), 462−464. Available from https://doi.org/10.1016/j.jscs.2015.04.003.

Khan, S. U. M., Al-Shahry, M., & Ingler, W. B. (2002). Efficient photochemical water splitting by a chemically modified n-TiO_2. *Science*, 297(5590), 2243−2245. Available from https://doi.org/10.1126/science.1075035.

Khazaee, Z., Khavar, A. H. C., Mahjoub, A. R., Motaee, A., Srivastava, V., & Sillanpää, M. (2020). Template-confined growth of X-Bi_2MoO_6 (X: F, Cl, Br, I) nanoplates with open surfaces for photocatalytic oxidation; experimental and DFT insights of the halogen doping. *Solar Energy*, *196*, 567−581. Available from https://doi.org/10.1016/j.solener.2019.12.061.

Khlyustova, A., Sirotkin, N., Kusova, T., Kraev, A., Titov, V., & Agafonov, A. (2020). Doped TiO_2: The effect of doping elements on photocatalytic activity. *Materials Advances*, *1*(5), 1193−1201. Available from https://doi.org/10.1039/D0MA00171F.

Kızılaslan, A., Kırkbınar, M., Cetinkaya, T., & Akbulut, H. (2020). Sulfur doped $Li_{1.3}Al_{0.3}Ti_{1.7}(PO_4)_3$ solid electrolytes with enhanced ionic conductivity and a reduced activation energy barrier. *Physical Chemistry Chemical Physics*, *22*(30), 17221−17228. Available from https://doi.org/10.1039/D0CP03442H.

Kumar, V., Govind, A., & Nagarajan, R. (2011). Optical and photocatalytic properties of heavily F-doped SnO_2 nanocrystals by a novel single-source precursor approach. *Inorganic Chemistry*, *50*(12), 5637−5645. Available from https://doi.org/10.1021/ic2003436.

Li, C., Zhong, W.-L., Gou, Q.-Z., Bai, X.-K., Zhang, G.-S., & Lei, C.-X. (2020). Facile design of F-doped TiO_2/ g-C_3N_4 heterojunction for enhanced visible-light photocatalytic activity. *Journal of Materials Science: Materials in Electronics*, *31*(4), 3681−3694. Available from https://doi.org/10.1007/s10854-020-02927-5.

Li, K., Huang, Z., Zeng, X., Huang, B., Gao, S., & Lu, J. (2017). Synergetic effect of Ti^{3+} and oxygen doping on enhancing photoelectrochemical and photocatalytic properties of TiO_2/g-C_3N_4 heterojunctions. *ACS Applied Materials & Interfaces*, *9*(13), 11577−11586. Available from https://doi.org/10.1021/acsami.6b16191.

Li, Y., Lai, J., Zheng, X., Lv, S., Yang, J., & Cui, S. (2020). Fabrication of wool ball-like F-doped $BiOCl_{0.4}Br_{0.3}I_{0.3}$ composite for effective sulfamethazine photocatalytic degradation. *Materials Research Bulletin*, *130*, 110937. Available from https://doi.org/10.1016/j.materresbull.2020.110937.

Liang, B., Zhang, L., & Zhang, W. (2019). Preparation of Cl doped SnO powder with excellent photocatalytic property by mechanical alloying. *Ceramics International*, *45*(7, Part A), 8908−8913. Available from https://doi.org/10.1016/j.ceramint.2019.01.220.

Linsebigler, A. L., Lu, G., & Yates, J. T., Jr (1995). Photocatalysis on TiO2 surfaces: Principles, mechanisms, and selected results. *Chemical Reviews*, *95*(3), 735−758.

Liu, G., Sun, C., Yan, X., Cheng, L., Chen, Z., Wang, X., . . . Cheng, H.-M. (2009). Iodine doped anatase TiO_2 photocatalyst with ultra-long visible light response: Correlation between geometric/electronic structures and mechanisms. *Journal of Materials Chemistry*, *19*(18), 2822−2829. Available from https://doi.org/10.1039/B820816F.

Liu, Y., Wang, W., Xu, X., Marcel Veder, J.-P., & Shao, Z. (2019). Recent advances in anion-doped metal oxides for catalytic applications. *Journal of Materials Chemistry A*, *7*(13), 7280−7300. Available from https://doi.org/10.1039/C8TA09913H.

Long, D., Diao, W., Rao, X., & Zhang, Y. (2020). Boosting the photocatalytic hydrogen evolution performance of Mg- and Cl-doped graphitic carbon nitride microtubes. *ACS Applied Energy Materials*, *3*(9), 9278−9284. Available from https://doi.org/10.1021/acsaem.0c01619.

Mallikarjuna, K., Bari, G. A. R., Vattikuti, S. P., & Kim, H. (2020). Synthesis of carbon-doped SnO_2 nanostructures for visible-light-driven photocatalytic hydrogen production from water splitting. *International Journal of Hydrogen Energy*, *45*(57), 32789−32796. Available from https://doi.org/10.1016/j.ijhydene.2020.02.176.

Mani, A. D. ;, Reddy, P. M. K., Srinivaas, M., Ghosal, P., Xanthopoulos, N., & Subrahmanyam, C. (2015). Facile synthesis of efficient visible active C-doped TiO_2 nanomaterials with high surface area for the simultaneous removal of phenol and Cr (VI). *Materials Research Bulletin*, *61*, 391−399. Available from https://doi.org/10.1016/j.materresbull.2014.10.051.

Marami, M. B., Farahmandjou, M., & Khoshnevisan, B. (2018). Sol−gel synthesis of Fe-doped TiO_2 nanocrystals. *Journal of Electronic Materials*, *47*(7), 3741−3748. Available from https://doi.org/10.1007/s11664-018-6234-5.

Merzhanov, A. G. (2004). The chemistry of self-propagating high-temperature synthesis. *Journal of Materials Chemistry*, *14*(12), 1779−1786. Available from https://doi.org/10.1039/B401358C.

Miao, H., Huang, G.-F., Liu, J.-H., Zhou, B.-X., Pan, A., Huang, W.-Q., & Huang, G.-F. (2016). Origin of enhanced photocatalytic activity of F-doped CeO_2 nanocubes. *Applied Surface Science*, *370*, 427−432. Available from https://doi.org/10.1016/j.apsusc.2016.02.122.

Modak, B., & Ghosh, S. K. (2015). Role of F in improving the photocatalytic activity of Rh-doped $SrTiO_3$. *The Journal of Physical Chemistry C, 119*(13), 7215−7224. Available from https://doi.org/10.1021/jp512948s.

Muggli, D. S., & Falconer, J. L. (2000). Role of lattice oxygen in photocatalytic oxidation on TiO_2. *Journal of Catalysis, 191*(2), 318−325. Available from https://doi.org/10.1006/jcat.2000.2821.

Nowotny, J., Bak, T., & Alim, M. A. (2015). Dual mechanism of indium incorporation into TiO_2 (Rutile). *The Journal of Physical Chemistry C, 119*(2), 1146−1154. Available from https://doi.org/10.1021/jp5112197.

Patil, K. (2008). *Chemistry of nanocrystalline oxide materials: Combustion synthesis, properties and applications*. World Scientific. Available from https://doi.org/10.1142/6754.

Payormhorm, J., & Idem, R. (2020). Synthesis of C-doped TiO_2 by sol-microwave method for photocatalytic conversion of glycerol to value-added chemicals under visible light. *Applied Catalysis A: General, 590*, 117362. Available from https://doi.org/10.1016/j.apcata.2019.117362.

Phuruangrat, A., Dumrongrojthanath, P., Kuntalue, B., Thongtem, S., & Thongtem, T. (2017). Synthesis and characterization of visible-light-driven Cl-doped Bi_2MoO_6 photocatalyst with enhanced photocatalytic activity. *Materials Letters, 196*, 256−259. Available from https://doi.org/10.1016/j.matlet.2017.03.073.

Podporska-Carroll, J., Myles, A., Quilty, B., McCormack, D. E., Fagan, R., Hinder, S. J., . . . Pillai, S. C. (2017). Antibacterial properties of F-doped ZnO visible light photocatalyst. *Journal of Hazardous Materials, 324*, 39−47. Available from https://doi.org/10.1016/j.jhazmat.2015.12.038.

Pradhan, G. K., Sahu, N., & Parida, K. M. (2013). Fabrication of S, N co-doped α-Fe_2O_3 nanostructures: Effect of doping, OH radical formation, surface area, [110] plane and particle size on the photocatalytic activity. *RSC Advances, 3*(21), 7912−7920. Available from https://doi.org/10.1039/C3RA23088K.

Putri, L. K., Ng, B.-J., Er, C.-C., Ong, W.-J., Chang, W. S., Mohamed, A. R., & Chai, S.-P. (2020). Insights on the impact of doping levels in oxygen-doped gC_3N_4 and its effects on photocatalytic activity. *Applied Surface Science, 504*, 144427. Available from https://doi.org/10.1016/j.apsusc.2019.144427.

Qi, D., Xing, M., & Zhang, J. (2014). Hydrophobic carbon-doped TiO_2/MCF-F composite as a high performance photocatalyst. *The Journal of Physical Chemistry C, 118*(14), 7329−7336. Available from https://doi.org/10.1021/jp4123979.

Qi, H., Shi, C., Jiang, X., Teng, M., Sun, Z., Huang, Z., . . . Guo, Z. (2020). Constructing CeO_2/nitrogen-doped carbon quantum dot/g-C_3N_4 heterojunction photocatalysts for highly efficient visible light photocatalysis. *Nanoscale, 12*(37), 19112−19120. Available from https://doi.org/10.1039/D0NR02965C.

Qian, H., Hou, Q., Duan, E., Niu, J., Nie, Y., Bai, C., . . . Ju, M. (2020). Honeycombed Au@ C-TiO_2-Xcatalysts for enhanced photocatalytic mineralization of Acid red 3R under visible light. *Journal of Hazardous Materials, 391*, 122246. Available from https://doi.org/10.1016/j.jhazmat.2020.122246.

Raizada, P., Soni, V., Kumar, A., Singh, P., Parwaz Khan, A. A., Asiri, A. M., . . . Nguyen, V.-H. (2021). Surface defect engineering of metal oxides photocatalyst for energy application and water treatment. *Journal of Materiomics, 7*(2), 388−418. Available from https://doi.org/10.1016/j.jmat.2020.10.009.

Reda, S., Khairy, M., & Mousa, M. A. (2020). Photocatalytic activity of nitrogen and copper doped TiO_2 nanoparticles prepared by microwave-assisted sol-gel process. *Arabian Journal of Chemistry, 13*(1), 86−95. Available from https://doi.org/10.1016/j.arabjc.2017.02.002.

Sahoo, P. C., Martha, S., & Parida, K. (2016). Solar fuels from CO_2 photoreduction over nano-structured catalysts. *Materials Science Forum, 855*, 1−19. Available from https://doi.org/10.4028/http://www.scientific.net/MSF.855.1.

Sakar, M., Prakash, R. M., Shinde, K., & Balakrishna, G. R. (2020). Revisiting the materials and mechanism of metal oxynitrides for photocatalysis. *International Journal of Hydrogen Energy, 45*(13), 7691−7705. Available from https://doi.org/10.1016/j.ijhydene.2019.04.222.

Samsudin, E. M., & Abd Hamid, S. B. (2017). Effect of band gap engineering in anionic-doped TiO_2 photocatalyst. *Applied Surface Science, 391*, 326−336. Available from https://doi.org/10.1016/j.apsusc.2016.07.007.

Saroj, S., Singh, L., & Singh, S. V. (2020). Solution-combustion synthesis of anion (iodine) doped TiO_2 nanoparticles for photocatalytic degradation of Direct Blue 199 dye and regeneration of used photocatalyst.

Journal of Photochemistry and Photobiology A: Chemistry, *396*, 112532. Available from https://doi.org/ 10.1016/j.jphotochem.2020.112532.

Shchukin, D. G., Radziuk, D., & Moehwald, H. (2010). Ultrasonic fabrication of metallic nanomaterials and nanoalloys. *Annual Review of Materials Research*, *40*, 345−362. Available from https://doi.org/10.1146/ annurev-matsci-070909-104540.

Shown, I., Samireddi, S., Chang, Y.-C., Putikam, R., Chang, P.-H., Sabbah, A., . . . Chen, K.-H. (2018). Carbon-doped SnS_2 nanostructure as a high-efficiency solar fuel catalyst under visible light. *Nature Communications*, *9*(1), 169. Available from https://doi.org/10.1038/s41467-017-02547-4.

Shwetharani, R., Poojashree, A., Balakrishna, G. R., & Jyothi, M. S. (2018). La activated high surface area titania float for the adsorption of Pb(ii) from aqueous media. *New Journal of Chemistry*, *42*(2), 1067−1077. Available from https://doi.org/10.1039/C7NJ03358C.

Shwetharani, R., Chandan, H. R., Sakar, M., Balakrishna, G. R., Reddy, K. R., & Raghu, A. V. (2019). Photocatalytic semiconductor thin films for hydrogen production and environmental applications. *International Journal of Hydrogen Energy*, *45*(36), 18289−18308. Available from https://doi.org/10.1016/j. ijhydene.2019.03.149.

Shwetharani, R., & Geetha Balakrishna, R. (2014). Comparative study of homogeneous and heterogeneous photo-oxidative treatment on bacterial cell via multianalytical techniques. *Journal of Photochemistry and Photobiology A: Chemistry*, *295*, 11−16. Available from https://doi.org/10.1016/j.jphotochem.2014.08.013.

Shwetharani, R., Sakar, M., Fernando, C. A. N., Binas, V., & Balakrishna, R. G. (2019). Recent advances and strategies to tailor the energy levels, active sites and electron mobility in titania and its doped/composite analogues for hydrogen evolution in sunlight. *Catalysis Science & Technology*, *9*(1), 12−46. Available from https://doi.org/10.1039/C8CY01395K.

Tahir, M. B., Ashraf, M., Rafique, M., Ijaz, M., Firman, S., & Mubeen, I. (2020). Activated carbon doped WO_3 for photocatalytic degradation of rhodamine-B. *Applied Nanoscience*, *10*(3), 869−877. Available from https://doi.org/10.1007/s13204-019-01141-y.

Tang, C., Chen, C., Zhang, H., Zhang, J., & Li, Z. (2020). Enhancement of degradation for nitrogen doped zinc oxide to degrade methylene blue. *Physica B: Condensed Matter*, *583*, 412029. Available from https://doi. org/10.1016/j.physb.2020.412029.

Torres., et al.3. Julio César González-Torres, Enrique Poulain, Víctor Domínguez-Soria, Raúl García-Cruz, and Oscar Olvera-Neria. (2018). C-, N-, S-, and F-Doped Anatase TiO_2 (101) with Oxygen Vacancies: Photocatalysts Active in the Visible Region. *International journal of photoenergy*7506151. doi:10.1155/ 2018/7506151.

Vitiello, G., Iervolino, G., Imparato, C., Rea, I., Borbone, F., De Stefano, L., . . . Vaiano, V. (2021). F-doped ZnO nano- and meso-crystals with enhanced photocatalytic activity in diclofenac degradation. *Science of the Total Environment*, *762*, 143066. Available from https://doi.org/10.1016/j.scitotenv.2020.143066.

Wafi, A., Szabó-Bárdos, E., Horváth, O., Pósfai, M., Makó, É., Juzsakova, T., & Fónagy, O. (2020). The photocatalytic and antibacterial performance of nitrogen-doped TiO_2: Surface-structure dependence and silver-deposition effect. *Nanomaterials*, *10*(11), 2261. Available from https://doi.org/10.3390/ nano10112261.

Wang, D., Huang, X., Huang, Y., Yu, X., Lei, Y., Dong, X., & Su, Z. (2021). Self-assembly synthesis of petal-like Cl-doped g-C_3N_4 nanosheets with tunable band structure for enhanced photocatalytic activity. *Colloids and Surfaces A: Physicochemical and Engineering Aspects*, *611*, 125780. Available from https://doi.org/ 10.1016/j.colsurfa.2020.125780.

Wang, Y., Chen, G., Weng, H., Wang, L., Chen, J., Cheng, S., . . . Lin, M. (2021). Carbon-doped boron nitride nanosheets with adjustable band structure for efficient photocatalytic U(VI) reduction under visible light. *Chemical Engineering Journal*, *410*, 128280. Available from https://doi.org/10.1016/j.cej.2020.128280.

Xu, C., Ravi Anusuyadevi, P., Aymonier, C., Luque, R., & Marre, S. (2019). Nanostructured materials for photocatalysis. *Chemical Society Reviews*, *48*(14), 3868−3902. Available from https://doi.org/10.1039/ C9CS00102F.

Xu, L., Steinmiller, E. M. P., & Skrabalak, S. E. (2012). Achieving synergy with a potential photocatalytic Z-scheme: Synthesis and evaluation of nitrogen-doped TiO_2/SnO_2 composites. *The Journal of Physical Chemistry C, 116*(1), 871−877. Available from https://doi.org/10.1021/jp208981h.

Xu, P., Xu, T., Lu, J., Gao, S., Hosmane, N. S., Huang, B., . . . Wang, Y. (2010). Visible-light-driven photocatalytic S- and C-codoped meso/nanoporous TiO_2. *Energy & Environmental Science, 3*(8), 1128−1134. Available from https://doi.org/10.1039/C001940M.

Xue, H., Lin, X., Chen, Q., Qian, Q., Lin, S., Zhang, X., . . . Xiao, L. (2018). S-doped Sb_2O_3 nanocrystal: An efficient visible-light catalyst for organic degradation. *Nanoscale Research Letters, 13*(1), 114. Available from https://doi.org/10.1186/s11671-018-2522-5.

Yang, Y., Ni, D., Yao, Y., Zhong, Y., Ma, Y., & Yao, J. (2015). High photocatalytic activity of carbon doped TiO_2 prepared by fast combustion of organic capping ligands. *RSC Advances, 5*(113), 93635−93643. Available from https://doi.org/10.1039/C5RA19058D.

Yu, J., Chen, Z., Zeng, L., Ma, Y., Feng, Z., Wu, Y., . . . He, Y. (2018). Synthesis of carbon-doped $KNbO_3$ photocatalyst with excellent performance for photocatalytic hydrogen production. *Solar Energy Materials and Solar Cells, 179*, 45−56. Available from https://doi.org/10.1016/j.solmat.2018.01.043.

Yu, J. C., Ho, W., Yu, J., Yip, H., Wong, P. K., & Zhao, J. (2005). Efficient visible-light-induced photocatalytic disinfection on sulfur-doped nanocrystalline titania. *Environmental Science & Technology, 39*(4), 1175−1179. Available from https://doi.org/10.1021/es035374h.

Yu, J. C., Yu, H., & Zhang, J. (2002). Effects of F-doping on the photocatalytic activity and microstructures of nanocrystalline TiO_2 powders. *Chemistry of Materials, 14*(9), 3808−3816. Available from https://doi.org/10.1021/cm020027c.

Zeng, D., Yu, C., Fan, Q., Zeng, J., Wei, L., Li, Z., . . . Ji, H. (2020). Theoretical and experimental research of novel fluorine doped hierarchical Sn_3O_4 microspheres with excellent photocatalytic performance for removal of Cr (VI) and organic pollutants. *Chemical Engineering Journal, 391*, 123607. Available from https://doi.org/10.1016/j.cej.2019.123607.

Zhang, F., Lu, Y., Schulman, D. S., Zhang, T., Fujisawa, K., Lin, Z., . . . Terrones, M. (2019). Carbon doping of WS_2 monolayers: Bandgap reduction and p-type doping transport. *Science Advances, 5*(5), eaav5003. Available from https://doi.org/10.1126/sciadv.aav5003.

Zhang, H., Kong, C., Li, W., Qin, G., Ruan, H., & Tan, M. (2016). The formation mechanism and stability of p-type N-doped Zn-rich ZnO films. *Journal of Materials Science: Materials in Electronics, 27*(5), 5251−5258. Available from https://doi.org/10.1007/s10854-016-4421-9.

Zhang, J., Deng, P., Deng, M., Shen, H., Feng, Z., & Li, H. (2020). Hybrid density functional theory study of native defects and nonmetal (C, N, S, and P) doping in a Bi_2WO_6 photocatalyst. *ACS Omega, 5*(45), 29081−29091. Available from https://doi.org/10.1021/acsomega.0c03685.

Zhang, Q., Li, Y., Ackerman, E. A., Gajdardziska-Josifovska, M., & Li, H. (2011). Visible light responsive iodine-doped TiO_2 for photocatalytic reduction of CO2 to fuels. *Applied Catalysis A: General, 400*(1), 195−202. Available from https://doi.org/10.1016/j.apcata.2011.04.032.

Zhao, T., Xing, Z., Xiu, Z., Li, Z., Yang, S., & Zhou, W. (2019). Oxygen-doped MoS_2 nanospheres/CdS quantum dots/g-C_3N_4 nanosheets super-architectures for prolonged charge lifetime and enhanced visible-light-driven photocatalytic performance. *ACS Applied Materials & Interfaces, 11*(7), 7104−7111. Available from https://doi.org/10.1021/acsami.8b21131.

Zhao, W., Liu, S., Zhang, S., Wang, R., & Wang, K. (2019). Preparation and visible-light photocatalytic activity of N-doped TiO_2 by plasma-assisted sol-gel method. *Catalysis Today, 337*, 37−43. Available from https://doi.org/10.1016/j.cattod.2019.04.024.

Zhou, C., Lai, C., Xu, P., Zeng, G., Huang, D., Li, Z., . . . Deng, R. (2018). Rational design of carbon-doped carbon nitride/$Bi_{12}O_{17}Cl_2$ composites: A promising candidate photocatalyst for boosting visible-light-driven photocatalytic degradation of tetracycline. *ACS Sustainable Chemistry & Engineering, 6*(5), 6941−6949. Available from https://doi.org/10.1021/acssuschemeng.8b00782.

Zhou, Y., Zhang, L., Huang, W., Kong, Q., Fan, X., Wang, M., & Shi, J. (2016). N-doped graphitic carbon-incorporated g-C$_3$N$_4$ for remarkably enhanced photocatalytic H$_2$ evolution under visible light. *Carbon, 99,* 111−117. Available from https://doi.org/10.1016/j.carbon.2015.12.008.

Further reading

Tan, Y. N., Wong, C. L., & Mohamed, A. R. (2011). An overview on the photocatalytic activity of nano-doped-TiO$_2$ in the degradation of organic pollutants. *ISRN Materials Science, 2011,* 261219. Available from https://doi.org/10.5402/2011/261219.

CHAPTER 4

Heterojunction-based photocatalyst

Nurafiqah Rosman[1,2], Wan Norharyati Wan Salleh[1,2] and Nur Aqilah Mohd Razali[1,2]

[1]*Advanced Membrane Technology Research Centre (AMTEC), Universiti Teknologi Malaysia, Skudai, Malaysia,* [2]*School of Chemical and Energy Engineering, Faculty of Engineering, Universiti Teknologi Malaysia, Skudai, Malaysia*

4.1 Introduction to heterojunction photocatalysts

In semiconductor photocatalysis, as the light is illuminated on a photocatalyst, the photon energy is absorbed to produce electrons and holes in the conduction band (CB) and the valence band (VB), respectively, which subsequently enables redox reactions. However, one of the drawbacks of the semiconductor photocatalysis is the recombination of electrons and holes in the bulk of semiconductors due to limited diffusion length (Sharma, Ruparelia, & Patel, 2011). A tight interface connection of the semiconductor to form heterojunction has become the key research due to the different properties in interfacial interactions at the nanoscale, which are not met in individual nanomaterials (Adams et al., 2003). A solid interfacial structure is a fundamental factor in establishing charge transfer behavior in different semiconductor heterostructures. By combining a semiconductor with a metal to form a Schottky junction with two or more semiconductors to produce semiconductor/semiconductor heterojunction, various applications were reviewed on the heterojunction photocatalysts for the degradation of pollutants, photocatalytic H_2 production, and photocatalytic water splitting. It should be noted that the recombination of excited state conduction-band electrons and VB holes is often facilitated by scavenger or crystalline defects that can trap the electron or the hole (Emeline, Kuznetsov, Ryabchuk, & Serpone, 2013). The electron—hole pairs should be efficiently separated and the charges should be quickly transported across the surface/interface to restrain the recombination. Hence, to recover the photocatalytic performance, an improved crystallinity with minimized defects would typically lessen the trapping states and recombination sites by forming a semiconductor heterojunction from coupling with a secondary substance (e.g., metal, other semiconductors), as discussed in the following section.

4.2 Categories of heterojunction photocatalysts

To enrich the visible-light photo effectiveness of a photocatalyst, the combination of narrow band gap semiconductors with suitable energy band edges in semiconductor nanoparticles

Photocatalytic Systems by Design.
DOI: https://doi.org/10.1016/B978-0-12-820532-7.00013-8

(NPs) for the development of binary and ternary composites could be an effective way (Lee, Lai, Ngai, & Juan, 2016). The appropriate band gap energy would harvest light as much as possible, which can directly increase the electron hole ($e^- - h^+$) conversion efficiency for the generation of hydroxyl radicals and active oxygen species (Kato et al., 2005). Instead of generating electron—hole conversion as much as possible, it increases the lifetime of the photogenerated hole (h^+_{VB}) and electron (e^-_{CB}) by suppressing the charge carrier recombination. The recombination of these $e^- - h^+$ pair lowers the quantum, consequently contributing to energy wastage. Forming a semiconductor with different interfaces or via the heterojunction technique will result in increased separation of photogenerated electron—hole pairs and contribute to improved stability (Narayanan & Viswanathan, 2019). The heterojunction can be described as the coupling of nanocrystals between two dissimilar crystalline semiconductors. The dissimilar crystalline semiconducting materials have different band gaps due to the difference in their electronic structure. Hence, the combination of the band gap and Fermi level will determine if there are holes or electrons that will be injected and which direction they are moving toward. Overall, it depends on how strong the local electric field that exists near the junction is, pointing from a semiconductor toward the other one due to the juxtaposition of high concentrations of negatively and positively charged ions (Khan, Pradhan, & Sohn, 2017).

These types of heterojunctions are designed to achieve good photocatalytic performance based on the mechanism that governs the heterogeneous photocatalysis. The mechanism of the heterojunction consists of four consecutive tandem steps: (1) light harvesting; (2) charge excitation/separation; (3) charge migration, transport, and recombination; and (4) charge utilization (surface electrocatalytic reduction and oxidation reactions) (see Fig. 4.1)

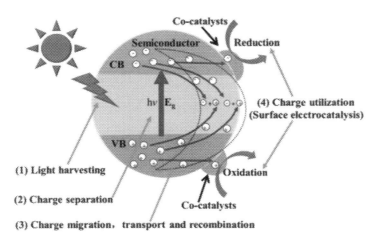

Figure 4.1
Four different stages in heterogeneous photocatalysis (Li et al., 2016c).

(Li, Yu, & Jaroniec, 2016c). Hence, to complete the steps, numerous factors must be assessed, including the micro- and nano-level structures, adsorption capacity, surface/interface morphology, cocatalysts, crystallinity, materials composition, and their band gap properties (Zheng, Li, Liu, Yin, & Li, 2011). A loss in the partial efficiency at each stage will contribute to the decrease in the overall photocatalytic efficiency. Besides that, these heterojunctions act as another step in the hierarchical structure photocatalyst, including morphology and porosity. The conduction and valence balance bands are hierarchically assembled nanomaterials that can increase the surface area and amount of active site, which can enhance light harvesting and improve molecular diffusion. Some distinct advantages of hierarchical structures in heterojunction that lead to the enhancement of the overall photocatalytic efficiency are discussed in the next subsection.

4.3 Semiconductor−semiconductor heterojunction

The semiconductor heterojunction can be designed in two different types: conventional heterojunction and p−n semiconductor heterojunction. In conventional heterojunction (non-p−n), the band alignment can be divided into three types of heterojunction structure, namely: (1) straddling gap (type I); (2) staggered gap (type II); and (3) broken gap (type III) (Low, Yu, Jaroniec, Wageh, & Al-Ghamdi, 2017), as shown in Fig. 4.2. Among these three types of heterojunctions, the architecture of heterojunction type III was found to be the most inefficient in separating photogenerated electron−hole due to its extremely staggering gap, where the band gaps' alignment of the two semiconductors did not overlap. Therefore electron−hole migration cannot occur, making it unsuitable for the enhancement of electron−hole pair separation. Meanwhile, for type-I heterojunction, the band gap alignment had overlapped with the CB and VB levels of the semiconductor A, which were respectively higher and lower than the bands of semiconductor B. This situation will result in the accumulation of electrons in CB, while the holes from semiconductor A had accumulated in the VB of semiconductor B when the light irradiated. Even though the electrons and holes can be separated, the accumulation of both electron−hole in the same semiconductor can potentially recombine the electron and hole, thus making it ineffective for a good separation mechanism. In fact, any band alignment that undergoes a redox reaction to generate more hydroxyl and oxygen radicals would be much higher and lower in CB and VB, respectively, in redox potential.

Type-II heterojunction is the most effective conventional heterojunction to improve photocatalytic activity due to its suitable structure for the electron−hole pair separation, which can effectively facilitate charge separation based on the mechanism proposed. Similar to type I, type-II heterojunction band alignment of both semiconductors is overlapped but the VB of semiconductor B is lower than that of semiconductor A. From the viewpoint of light energy harvesting, placing a narrow band gap semiconductor with a

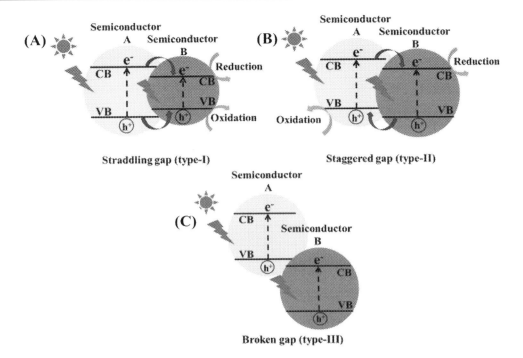

Figure 4.2
Schematic illustration of the three different types of separation of electron—hole pairs in the case of conventional light-responsive heterojunction photocatalysts: (A) type-I, (B) type-II, and (C) type-III heterojunctions (Low, Yu, Jaroniec, Wageh, & Al-Ghamdi, 2017).

strong visible light absorption (such as CdS, Bi_2S_3) can improve the solar energy utilization efficiency due to the synergic absorption of both semiconductors (Hou et al., 2019; Li et al., 2018b). However, the redox ability of type-II heterojunction photocatalyst is one of the limitations to counterpart the redox potential for generating oxygen and hydroxyl radicals. Moreover, the migration of electrons from A to electron-rich CB of semiconductor B or migration of holes from semiconductor B to the hole-rich VB of A is physically unfavorable due to electrostatic repulsion between electron and electron or hole and hole. In fact, the capability of type-II heterojunction includes hindering the ultrafast electron—hole recombination of the semiconductor.

This limitation was the purpose of the p—n heterojunction photocatalyst concept, in which it is capable of accelerating the electron—hole migration across the heterojunction to improve photocatalytic performance with additional electric field (Cao, Shen, Tong, Fu, & Yu, 2018). In general, once the p- and n-type semiconductors are in contact, the electrons on the n-type semiconductor near the p—n interface tend to diffuse into the p-type semiconductor, leaving a positively charged species (see Fig. 4.3). Meanwhile, the holes on

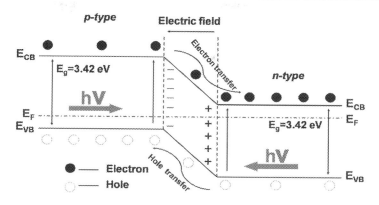

Figure 4.3
Schematic diagram showing the energy band structure and electron—hole pair separation in the p—n heterojunction (Wang et al., 2014).

the p-type semiconductor near the p—n interface tend to diffuse into the n-type semiconductor, leaving a negatively charged species. When the p—n heterojunction is irradiated by photons with energy higher or equal to the band gaps of the photocatalysts, there is an accelerated photogenerated electron—hole pair separation by the built-in electric field within the space charge region (Low, Yu, Jaroniec, Wageh, & Al-Ghamdi, 2017). This effective architecture forms a p—n junction with a space-charge region at the interface due to the diffusion of electrons and holes, thus generating a built-in electrical potential that allows the electron to travel in the opposite direction.

4.3.1 TiO$_2$-based heterojunction photocatalyst

Extensive efforts have been made on the synthetization of TiO$_2$ with other semiconductors over the past few decades to improve the photocatalytic efficiency. Typically, the anatase phase of TiO$_2$ is well known to have a higher photocatalytic activity than that in rutile and brookite phases. In fact, the mixed phase of anatase rutile, or called as the phase heterojunction, has displayed an enhanced photocatalytic performance, compared to the single phase. This is due to the synergistic effects between these two phases in the formation of the phase junctions.

Apart from that, TiO$_2$ is also known as an n-type semiconductor due to the loss of lattice oxygen (Nowotny, 2008). Through the combination of n-type TiO$_2$ with appropriate p-type semiconductor, a p—n heterojunction is formed. When the TiO$_2$ and some p-type semiconductor are in contact, a strong electric field can be created, leading to accelerated charge separation, thus impeding the recombination of the e/h$^+$ pairs. For example, Lv's group synthesized BiVO$_4$/TiO$_2$ heterostructure composites through a hydrothermal method (140°C for 6 h) exhibited high activity for the degradation of Rhodamine B (RhB). The

TiO$_2$ NPs were seeded with BiVO$_4$ from a solution of bismuth nitrate pentahydrate to grow the BiVO$_4$ particles on the surface of TiO$_2$. The large interface of BiVO$_4$/TiO$_2$ heterojunction resulted in the electron−hole pair separation and enhanced the visible light−light absorption ability. Similarly, according to Zhao et al., p-type BiOBr nanosheet can be grown over the n-type TiO$_2$ nanobelt to form p−n junction in the interface of BiOBr and TiO$_2$. The lattice fringes of BiOBr nanosheets and TiO$_2$ nanobelts (see Fig. 4.4C) had provided a large contact interface for creating p−n heterojunctions between BiOBr and TiO$_2$. It was found that the cross fringes that had occurred in the interface of BiOBr and TiO$_2$ (see Fig. 4.4D) signified a good combination of the two materials, making it beneficial for accelerating the charge transfer and separation process.

The accelerated charge transfer was basically initiated via a narrower band gap possessed by another heterojunction semiconductor. Upon visible light irradiation, the guest semiconductor can be excited due to its narrow band gap by absorbing incident photons to generate electrons and hole pairs so that the charge can be transferred between the guest and the TiO$_2$. The largely dispersed BiVO$_4$ narrow band gap (~ 2.4 eV) over TiO$_2$ has provided an extended spectral response to visible light and improves the quantum efficiency by the heterojunction between TiO$_2$ and BiVO$_4$, as reported by Wang et al. (2019). This

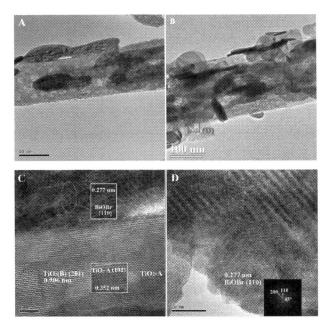

Figure 4.4
(A and B) Low-magnification TEM images of BiOBr@TiO$_2$−1. (C and D) High-resolution transmission electron microscopy (HRTEM) images of the designated square parts in (A) and (B) (Yang et al., 2016).

type-II heterojunction reached 79.3% of RhB photocatalytic degradation (BiVO$_4$ loading amounts: 1%; initial concentration: 10 mg/L; pH value: 5.3) after 300 min of sunlight irradiation, which was clearly much higher compared to the TiO$_2$ film. TiO$_2$ and BiVO$_4$ can restrain the recombination of electrons and holes, as well as expand the light absorption range. The higher intensification of light-harvesting efficiency was improved because of the fiddlehead-shaped morphology that resulted in a larger specific surface area (157.63 m^2/g). Moreover, the lower VB and CB location in TiO$_2$ in comparison with those in BiVO$_4$ had also promoted the separation of the photogenerated charge carriers.

It can be seen that a larger heterojunction interface area leads to faster charge migration at the interface, which can provide many reaction opportunities. Clearly, the well-controlled proportion of different compositions can enlarge the heterojunction interface area (Zheng, Li, Liu, Yin, & Li, 2011). Hao, Wang, Jiang, Tang, and Xu (2017) found that the photocatalytic performance of g-C$_3$N$_4$/TiO$_2$ composites was dependent on the content of melamine in the synthesis precursors, with an optimum melamine content of 3 g for 0.5 mL of TiCl$_4$. This heterojunction composition displayed the highest photocatalytic performance for the decomposition of RhB, which is superior to pure TiO$_2$ and g-C$_3$N$_4$ by a factor of 18.7 and 3.5, respectively. The high photocatalytic activity can be attributed to the high specific surface area (up to 115.6 m^2/g) of the g-C$_3$N$_4$/TiO$_2$ composites and a synergistic heterojunction structure between TiO$_2$ and g-C$_3$N$_4$.

However, the increased interface area that resulted from this approach remained restricted by microstructure component employment, such as the photocatalyst particle size, morphology, and surface area (Low, Yu, Jaroniec, Wageh, & Al-Ghamdi, 2017). Nevertheless, higher photocatalytic performance is generally associated with a high specific surface area, having a small particle size and band gap (E_g) energy. Specifically, it is very difficult to understand the exact control over different semiconductor components and contact structures within the interface (Li, 2018a; Shen, Jiang, Xiang, Xie, & Li, 2019). Rawool et al., (2018) reported that the exhibited grain boundaries to support the heterojunction mechanism are coexisted with a negative effect, which is attributed to the e − and h + recombination. The grain boundaries may be considered as strong links for the charge transport within these two semiconductors; however, it may also potentially lead to a higher rate of electron−hole recombination reaction when more grain boundaries exist.

Generally, the performance of TiO$_2$ in solar energy conversion can be influenced by its light-induced reactivity and chemical potential of electrons at the interface (Bai, Mora-Sero, Angelis, Bisquert, & Wang, 2014; Kohtani, Kawashima, & Miyabe, 2017). Bak's research in single crystal and polycrystalline TiO$_2$ has found that the grain boundaries are predictable to act as shortcuts in the charge transport in polycrystalline TiO$_2$ (Bak, Chu, Francis, Li, & Nowotny, 2014). Meanwhile, the defect disorder possessed by TiO$_2$ revealed a nonhomogeneous TiO$_2$ surface from the linear defect at the interaction between the grain boundaries and the external surfaces, in which cathodic behavior containing high electron concentration is expected (Bak et al., 2014). It has been documented that the reactivity and related semiconducting properties are determined by lattice imperfections (point defects) (Bak, Nowotny, Sucher, & Wachsman, 2011). Therefore the enhanced concentration of electrons at an interface should be related to the effect of segregation on the local defect disorder (Nowotny et al., 2015).

However, the synthesis of ZnO/TiO_2 core/shell system by Kwiatkowski et al. (2017) revealed that the defect behavior had increased through additional calcination processes. The EDX-TEM line scan analysis along the redlines is indicated in Fig. 4.5A, which demonstrated the increased amount of zinc in the outer part of TiO_2 layer after additional calcination of ZnO/TiO_2 was done. The outward diffusion of Zn ions into the TiO_2 layer allowed the formation of voids at ZnO/TiO_2 interface. From the inset TEM images in

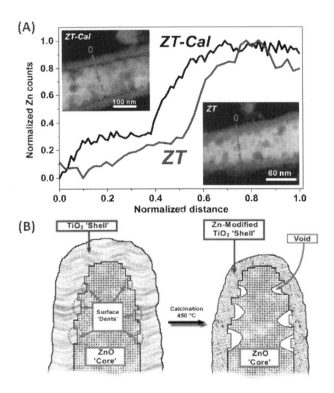

Figure 4.5

(A) Zinc concentration profiles obtained from linear EDX-TEM analysis across the ZnO/TiO_2 nanorod before and after calcination. Insets: HAADF-STEM images with arrows indicating the line scan direction. (B) A schematic illustration depicting presumable thermal diffusion process of Zn ions into TiO_2 at the ZnO/TiO_2 interface at 450°C accompanied with the formation of "voids" in ZnO-core (Kwiatkowski et al., 2017).

Fig. 4.5A, it can be seen that the ZnO nanorods contained numerous small dark spots on its surface after its calcination at 300°C owing to the shrinking phenomenon and formation of a "dented" surface layer of ZnO; meanwhile, the thickness of the external TiO_2 layer had not changed (~ 35 nm) after the additional calcination process. The surface defects of the ZnO/TiO_2 interface revealed a significant improvement of the decolorization rate of methylene blue (MB) and photoassisted decomposition of H_2O under 400 nm monochromatic irradiation.

The photoelectrochemical study conducted by Guerrero-Araque et al. (2017) provided additional information on the charge transfer process across the ZrO_2-TiO_2 heterojunction at different $ZrO_2:TiO_2$ molar ratios (01:99, 05:95, and 10:90), namely, ZT-1, ZT-5, and Z-10, respectively. From the study, the photocatalytic evaluation of phenoxy acetic acid, 2,4-dichlorophenoxyacetic acid, or 4-chlorophenol found that ZT-5 showed the best performance attributed to the surface states that were exhibited in the interface of ZrO_2-TiO_2 heterojunction based on the Mott–Schottky analysis (capacitance measurements). In fact, the intimate contact phase between the crystal plane of TiO_2 of ZrO_2 had clearly portrayed the formation of heterojunction of both semiconductors (see Fig. 4.6B). This intimate contact had produced a defected generation throughout the grain boundary that existed in the crystal structure, thus creating the surface states. These surface states are well known to be present in large numbers, which can lead to the change in position of the band edges. This study also revealed that the illumination response in higher Zr content had reached higher currents and was faster compared to that in pristine TiO_2 and ZrO_2. However, it rapidly decreased over time and with the number of perturbations (see Fig. 4.6A). Based on Fig. 4.6C, the mechanism that was proposed by Mott–Schottky analysis had shown defects in the ZT-5 composite induced energy states, which favored the spatial separation of the photogenerated charge carriers. These energy states formed at the interface might trap the photogenerated electrons, indirectly avoiding their recombination with the photogenerated holes. By considering the ionized states in ZrO_2, the electrons might be transported through them to react with oxygen in the solution. In fact, even in a lower concentration of ZrO_2 incorporation, the increased defect created by the E_{fb} shifting in the Mott-Schottky analysis had caused an upward band bending of the contacted oxides, which proved the generation of energy states at the interface of the oxides. However, it was observed that overgrown ZrO_2 on TiO_2 surfaces has led to less effective charge carrier separation at the interface. Ramírez-Ortega, Meléndez, Acevedo-Peña, González, and Arroyoa (2014) believed that the effect of a higher amount of heterojunction will continue to increase the surface state quantity at the junction interface. However, this increment had led to the larger size of the space charge layer to consequently behave as a barrier in the transportation of the photogenerated electrons within both semiconductors, thus hindering the charge transfer process and indirectly decrement the photocatalytic activity. Hence, the presence of surface states due to grain boundary appearing at the heterojunction interface

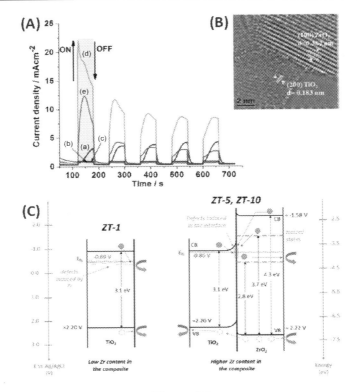

Figure 4.6

(A) Photocurrent response under UV light on—off process at applied voltage of 0.5 V: (a) TiO₂, (b) ZrO₂, (c) ZT-1, (d) ZT-5, and (e) ZT-10. (B) TEM images of the heterojunction in ZT-5 composite. (C) Energy scheme of ZT-x composites and their interaction for spatial separation of photogenerated charge carriers (Guerrero-Araque et al., 2017).

can affect the band bending and modify the transport and separation of charge carriers, thereby changing the photocatalytic activity of the samples (Jafari et al., 2016).

4.3.2 Other semiconductor—semiconductor heterojunction

Presently, numerous types of semiconductors, besides TiO₂, have resulted in a heterojunction for producing visible-light-responsive photocatalysts. Metal oxides (TiO₂, ZnO, Fe₂O₃, Cu₂O, Ag₂O,), metal sulfides (CdS, MoS₂, ZnS), multicomponent oxides (Bi₂WO₆, BiVO₄, Ag₃VO₄, CoTiO₃), metal selenides (MoSe₂, CdSe), metal phosphides (Ni₂P), metal phosphates (Ag₃PO₄), metal halides and oxyhalides (AgBr, BiOBr), and metal-free materials (SiC, g-C₃N₄, and Si) have been used as photocatalysts. These photocatalysts can be sorted out as a wide band gap photocatalyst ($E_g > 3$ eV), such as TiO₂, ZnO, and ZnS, and visible-light-responsive photocatalyst ($E_g \leq 3$ eV), such as Ag₂O,

Bi_2WO_6, Cu_2O, and Ag_3VO_4. Ichimura's group had established the energy band alignment at the heterointerface between Cu_2O and ZnO via the X-ray photoelectron spectroscopy (XPS), revealing a VB maximum at 1.7 eV, with the Cu_2O located higher than that of ZnO. In this perspective, the VB of Cu_2O consisted of the Cu's 3d orbital, which has a higher atomic energy level than O 2p of ZnO at about 4 eV difference. Hence, the construction of type-II heterojunction can be based on the quantum efficiency for photons with energy larger than the band gap of Cu_2O (Ichimura and Song, 2011). Similar to TiO_2-based semiconductor heterojunction, all of these photocatalysts should be coupled with a staggered band alignment to promote the spatial charge separation. For instance, different types of heterojunction photocatalysts, for instance, $CoTiO_3/Ag_3VO_4$ (Wangkawong, Phanichphant, Tantraviwat, & Inceesungvorn, 2015), Fe_2O_3/Bi_2MoO_6 (Shijie, Hu, Zhang, Jiang, & Liu, 2017), and CeO_2/Bi_2MoO_6 (Dai et al., 2018), $NiTiO_3/Ag_3VO_4$ (Inceesungvorn, Teeranunpong, Nunkaew, Suntalelat, & Tantraviwat, 2014), exhibit effective electron−hole separation efficiency, thus widening the light absorption range for an increased mass transfer. Accordingly, the heterojunctions among these photocatalysts have been exploited either in different loadings, such as suitable cocatalysts, or introducing defects (heteroatoms), to promote the surface charge separation and accelerate the surface reaction kinetics.

Zhu, Wang, Ling, and Zhu (2017) stated that the synthesis of $BiPO_4/Bi_2WO_6$ composite photocatalysts can be done via the ultrasonic-calcination method for photocatalytic degradation of organic pollutants. Specifically, the addition of monazite monoclinic $BiPO_4$ nanorods followed 0.5%, 1.0%, 3.0%, 5.0%, 8.0%, and 10.0% to produce $BiPO_4/Bi_2WO_6$ composite photocatalysts. The optimum addition was found to be at 5.0% of $BiPO_4/$ Bi_2WO_6 on the degradation of MB under simulant sunlight irradiation, which was about 25.4 and 3.2 times of pure $BiPO_4$ and Bi_2WO_6, respectively. The proportion in a different loading amount of guest semiconductor to be heterojunctioned to the host semiconductor, as aforementioned, has influenced the heterojunction interface area. Qiu and co-workers fabricated a p−n heterojunction $Bi_2O_2CO_3/BiOBr$ with different $Bi_2O_2CO_3/BiOBr$ ratios, and found that the 1:8 ratio of $Bi_2O_2CO_3/BiOBr$ exhibited the highest photodegradation rate (Qiu et al., 2017). The different amount loadings of BiOBr had revealed different light absorption properties. As shown in Fig. 4.7A, all of the heterojunction samples had shifted the absorption edge from 390 nm to about 440 nm. The efficient visible light absorption ability of $Bi_2O_2CO_3/BiOBr$ was able to generate sufficient photoinduced electron−hole pairs under visible light irradiation, as revealed by the photocurrent response (see Fig. 4.7B).

Besides that, the heterojunction exhibited an interphase in the semiconductor−semiconductor heterojunction, which was basically verified through the morphology and phase evaluation. Structurally, a heterojunction contains individual crystalline phases of both semiconductors to merge with each other through the grain boundary (Li, Chen, Shi, & Li, 2016). The grain

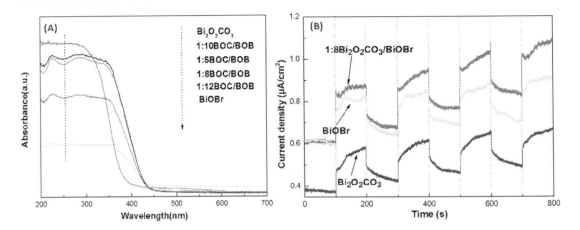

Figure 4.7

(A) Diffuse reflectance spectra (Drs) of the Bi_2O2CO_3, BiOBr, and BOC/BOB series of composite photocatalysts. (B) Transient photocurrent response for sample $Bi_2O_2CO_3$, BiOBr, and 1:8BOC/BOB composite under visible-light irradiation (Zhu, Wang, Ling, & Zhu, 2017).

boundary in the interface lattice will then be matched to promote the migration of photogenerated carriers between the semiconductor—semiconductor interfaces (Wang et al., 2018). Majhi, Bhoi, Samal, and Mishra (2018) constructed a visible-light-driven $CuS-Bi_2O_2CO_3$ heterojunction for chlorpyrifos degradation, where a large interfacial region was observed between the two phases, thus promoting the microscopic contact. The merging of (011) plane of $Bi_2O_2CO_3$ and (006) plane of CuS had led to the heterojunction formation, which was expected to improve the moment of excitons along the grain boundary region. Similarly, a novel CdS/Cu_7S_4 heterostructure nanocrystal that was recently synthesized by Lian et al. (2018) revealed an efficient hot electron injection and long-lived charge separation at the plasmonic p—n heterojunction for photocatalytic H_2 evolution. The formation of CdS phase on disk-shaped Cu_7S_4 had resulted in a dimer structure that led to a well-contact heterostructure, as clearly seen in Fig. 4.8E. The interface lattice-matching had resulted in a small mismatch of 5% in the heterointerface between the triclinic roxbyite Cu_7S_4 and wurtzite CdS phases. The XRD (Fig. 4.8C), TEM-EDX mapping (Fig. 4.8D) and fast Fourier transform (FFT) patterns (Fig. 4.8F and G) supported the well inter-merged structure of the heterojunction.

Through the formation of these heterojunction systems, the hybridized nanocrystal had affected not only the crystal lattice structure but also the nature of the chemical bond as well, especially at the interface. Huang et al. (2018) constructed a $Bi_2O_2CO_3/ZnFe_2O_4$ heterojunction for the photocatalytic removal of nitrogen oxide under visible light irradiation. The heterostructure exhibited superior photocatalytic nitric oxide removal that was mainly attributed to the boosted separation efficiency of the photogenerated

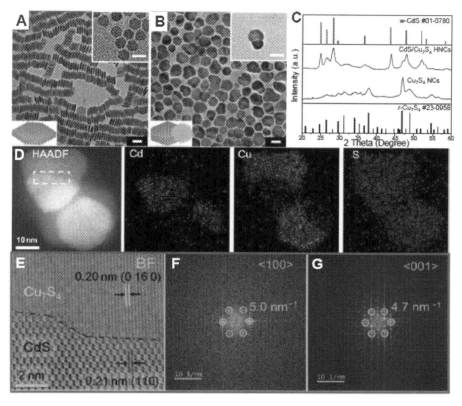

Figure 4.8
TEM images of (A) disk-shaped Cu_7S_4 nanodisk. (B) CdS/Cu_7S_4 heterojunction nanocomposite. (C) XRD patterns of Cu_7S_4 and CdS/Cu_7S_4 heterojunction nanocomposites. (D) STEM-EDS elemental mapping images of CdS/Cu_7S_4 heterojunction nanocomposite. (E) STEM image of the heterointerface of a single CdS/Cu_7S_4 heterojunction nanocomposite generated by a dashed white rectangle of images in (D). (F, G) FFT patterns of the Cu7S4 phase from <100> direction at the upper region of (E) and the CdS phase from <001> direction at the lower region of (E), respectively (Majhi, Bhoi, Samal, & Mishra, 2018).

electron/hole pairs due to the existence of an internal electric field at the $Bi_2O_2CO_3$/$ZnFe_2O_4$ contact interface (see Fig. 4.9B). The charge transfer in the intimate interface of $Bi_2O_2CO_3$ and $ZnFe_2O_4$ phases was clarified by the density functional theory calculation to evaluate the electron density difference (EDD) of this heterojunction. As shown in Fig. 4.9E, the obtained EDD validated the exhibition of charge transfer in the $Bi_2O_2CO_3$/$ZnFe_2O_4$ interface at about 3.07 electrons when transferred from $Bi_2O_2CO_3$ to $ZnFe_2O_4$. This reduction of electron concentration on its surface could be due to the electron transfer activity that had elevated the binding energy in C, Bi, and O peaks, as observed in the XPS.

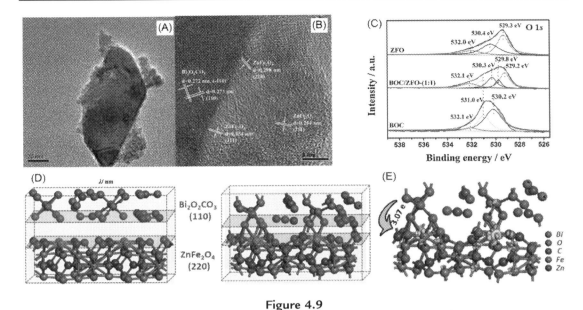

Figure 4.9
(A) TEM and (B) HRTEM of $Bi_2O_2CO_3/ZnFe_2O_4$; (C) constructed and (D) optimized crystal structures of the $Bi_2O_2CO_3/ZnFe_2O_4$ interface. (E) Calculated EDD plot for the $Bi_2O_2CO_3/ZnFe_2O_4$ interface at an isovalue of 0.1 e/$Å^3$, with blue and yellow regions representing electron density accumulation and depletion, respectively (Huang et al., 2018).

Besides that, an ideal lattice-matching pair in the two materials requires to have an identical lattice to reduce the mismatch value (Peng et al., 2017; Wang et al., 2010). A difference in lattice spacing between two semiconductors is likely to induce defect formation, which can trap the photogenerated carriers and prevent the diffusion of electrons and holes (Li et al., 2015). The lattice-matching of hexagonal SiC and hexagonal CdS heterostructure exhibited an H_2 evolution rate of four times greater than the heterostructure of hexagonal SiC and cubic CdS, even with similar band gaps and electropotentials of both cubic and hexagonal CdS. The lattice mismatch can be in any rotation due to the square superlattice possessed by the cubic CdS instead of the diamond-shaped in the hexagonal phase. This makes the cubic CdS particle susceptible to nucleate by itself rather than on the hexagonal SiC surface, hence, making it responsible for the ineffectiveness of carrier transfer in the SiC/CdS heterointerface. The lattice mismatch between hexagonal SiC and cubic CdS heterojunction (interface) would trap the photogenerated electrons and holes, hence, preventing their diffusion.

Besides that, the effect of heterogeneous growth in two semiconductors is a substantial factor in determining the crystal growth and promoting the degree of contact on the heterojunction interface. Otherwise, the mismatched growth will result in defects and, in turn, increase the surface energy and lower the efficiency in charge separation (Peng et al., 2017).

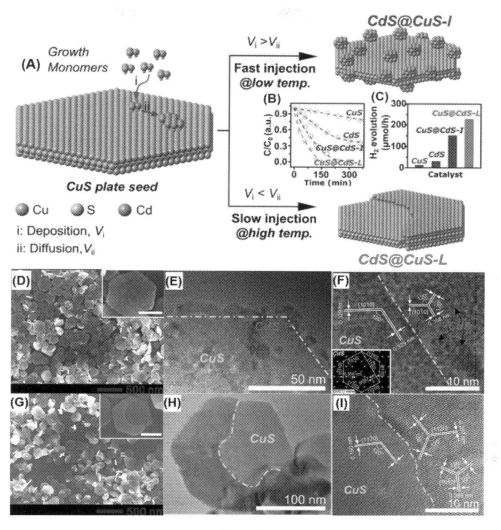

Figure 4.10

(A—C) Fast and low synthesizing CuS@CdS core—shell heterojunctions with controllable interfaces for efficient solar H$_2$ production and organic pollutant degradation. TEM images of CuS@CdS core—shell heterojunctions at (D, E, F) fast injection at lower temperature and (G, H, I) slower injection at high temperature (Vispute, Talyansky, Choopun, Sharma, & Venkatesan, 1998).

Li et al. (2018b) reported the seed-mediated synthesis of CuS@CdS core—shell heterojunction via the manipulation of injection rate of Cd^{2+} precursor over CuS nanoplates (as seeds), which revealed two distinctive growth modes: the island and the layer-by-layer modes. A high injection rate with low reaction temperature of CdS deposition monomers (see Fig. 4.10A) on the surface of the CuS nanoplate had produced an island growth because of the lattice mismatch between CuS and CdS. When the injection rate of Cd^{2+} was reduced and the reaction temperature was increased, a layer-by-layer growth had facilitated the surface diffusion of these monomers. The transmission electron microscopy (TEM) images of both samples found that the crystallinity of slow injection rate of Cd^{2+} at higher temperature gave better crystallinity distribution of CdS over the CuS surface. The lattice mismatch between CuS and CdS had induced an interfacial strain, therefore affecting the electronic band level engineering (Sun, Tian, Zhou, Zhang, & Li, 2019). Meanwhile, good crystallinity of the sample is produced by the slow injection rate of Cd^{2+}, resulting in an improved photocatalytic performance toward dye degradation and H_2 production from water under visible-light irradiation. The results suggested that the crystallinity of CdS formation and small lattice mismatch in heterojunction structure formation had provided excellent charge separation and transfer abilities. Basically, "poor" seeds or mismatched structures on the heterostructure interfaces might lead to phase separations, thus reducing the ability of the photogenerated charge separation (Rashmi, Sivakumar, & Pala, 2019). Hence, well-defined, layer-by-layer growth with favorable crystal nuclei and useful defects is needed in heterostructures (Vispute, Talyansky, Choopun, Sharma, & Venkatesan, 1998). Low interfacial defects with chemically strong bonds across the interface with large contact areas (i.e., interface with small lattice mismatch) are favored for the removal of the diffusion barrier in the interface. Eventually, the route of heterojunction crystallinity growth reveals another perspective in controlling the heterostructure architecture of semiconductors toward shape-dependent physicochemical properties for enhanced photocatalytic activity.

However, some researchers have clarified that forming lattice defect (lattice stain), such as oxygen vacancy, would be great in high visible-light-induced photocatalytic activity (Ruzimuradov, Hojamberdiev, Fasel, & Riedel, 2017). They play a critical role in enhancing photogenerated carrier transfer in the photocatalytic mechanism. As an example, Liu et al. had synthesized the CdS$-$BiOCl photocatalyst via the hydrothermal method (Liu et al., 2014). The results demonstrated that the visible light absorbance at the edge of the CdS$-$BiOCl^{-1} (untreated) had extended to 500 nm, compared to that in pristine BiOCl, which was negligible in the visible range. This can be attributed to the heterostructure formed between the lower band gap of CdS (2.4 eV) and BiOCl (3.5 eV), while the white color was changed to a Khaki brown color. However, when the photocatalyst was treated with UV light irradiation (CdS$-$BiOCl^{-2}), it created a new energy level that lowered the CB of BiOCl via the formation of oxygen vacancies (defects) to give a slight absorbance edge. The oxygen vacancies induced by the low bond energy (337.2 ± 12 kJ/mol) and long bond length (2.318 Å) in Bi$-$O bond turned the BiOCl into the black in color. As shown in inset Fig. 4.11A, the untreated CdS$-$BiOCl^{-1} was excited by CdS (process ①) when the light was irradiated because the untreated CdS$-$BiOCl^{-1} still contained oxygen vacancies; henceforth, process ② would come into existence but with less efficiency. However, the presence of oxygen vacancies in treated CdS$-$BiOCl^{-2} had lowered the CB, thereby

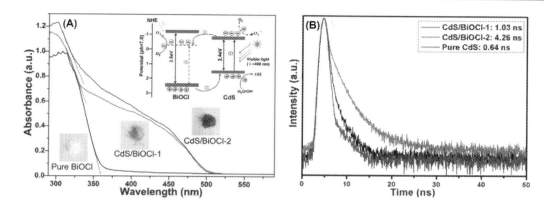

Figure 4.11

UV—vis diffuse reflectance spectra of pure BiOCl, CdS—BiOCl^{-1} and CdS—BiOCl^{-2}, inset: band structure model and photoreaction process on CdS—BiOCl composites under visible light irradiation (Liu et al., 2014).

triggering the electron to accumulate at the low-lying oxygen vacancy in the CB of CdS and the holes to accumulate at the VB of the BiOCl as process ②. This situation had endowed an efficient electron—hole separation, whereas these oxygen vacancies had trapped the excited electrons from BiOCl, thus producing less efficiency. The photocatalytic mechanism that supported the decays of PL transition centered at 470 nm (the state of oxygen vacancy) demonstrated that the treated sample (CdS—BiOCl^{-2}) had shown an elongated PL lifetime of 4.26 ns, compared to the untreated one (CdS—BiOCl^{-1}) at 1.03 ns. This has significantly contributed to a lower recombination rate of the electron—hole pairs in the oxygen vacancies.

Similarly, Ling et al. (2016) reported that the heterojunction structure of the narrow band gap PdO/LaCoO$_3$ had produced oxygen vacancies in the interspace of CB and VB. The purpose of the defect levels as a "step" that supports electron transit has corresponded to the narrow band gap of PdO/LaCoO$_3$. Fundamentally, the imperfect structure at heterostructure grain boundaries leads to a large lattice mismatch between both semiconductors (Zhao, Pan, & Li, 2017). The presence of oxygen vacancy on the heterojunction formation plays a part in decreasing the amount of oxygen lattice, resulting in more oxygen vacancies available. Recently, Lei's group reported that oxygen vacancies that are rich in ultrathin In$_2$O$_3$ porous sheets exhibited a much higher visible-light photocurrent than the oxygen-vacancy-poor case and bulk samples due to the narrowing of the band gap, therefore improving the carrier separation efficiency (Lei et al., 2014).

However, oxygen vacancies or defects are still being debated on how it affects photocatalytic activities. While some studies suggest the role of these defects as recombination centers for electron—hole pairs, others describe the improved properties of the defective samples (Zhao, Pan, & Li, 2017). This disagreement is due to the large number of parameters that describe oxygen vacancies, all of which are hard to determine. Thus their concentration, charge, spatial distribution, presence or absence of other species in their surrounding can all determine the way the oxygen vacancies impact the photocatalytic properties of a heterojunction photocatalyst (Li et al., 2013).

Another important matter in heterojunction semiconductors is in improving the photocatalytic activity over hierarchical photocatalysts. Morphology architecture in semiconductor heterojunction in both surface junctions and texture engineering of the photocatalyst proved to be effective for enhancing light harvesting, and large specific surface area and a number of surface-active sites (Li, Luo, Wang, & Gong, 2018). There are numerous typical hierarchical heterostructures such as core−shell, 3D flower-shaped microspheres, and branched structures have been formed with different nanosized building blocks, including nanorods and nanosheets among others (Li et al., 2016c). For example, Shijie, Hu, Zhang, Jiang, and Liu (2017) demonstrated that the Bi_2MoO_6 nanosheet and Fe_2O_3 NPs can be prepared via a simple solvothermal precipitation−calcination method. The flower-like morphology grown by Bi_2MoO_6 nanosheet through the deposition of Fe_2O_3 NPs on the surface of Bi_2MoO_6 nanosheets had increased the surface area and pore size distribution of the photocatalyst (see Fig. 4.12A and B). The compactly interfacial contact between Bi_2MoO_6 and Fe_2O_3 in flower-like structure was able to increase the number of light interactions and enhance visible light absorption efficacy (see Fig. 4.12C−F). Meanwhile, the photocatalytic degradation of RhB was found to be 4.8 times higher than that of Bi_2MoO_6. Recently, Hou et al. (2019) demonstrated the in situ growth of Bi_2S_3 nanorods from $Bi_4O_5I_2$ flowers, which revealed a strong heterojunction and intimate interfaces between the two semiconductors. The $Bi_4O_5I_2/Bi_2S_3$ heterojunctions exhibited outstanding photocatalytic activities and stability toward the photocatalytic reduction of Cr (VI) under visible light irradiation ($\lambda > 420$ nm). Similarly, Zheng & Zhang, 2018 reported an inspired marigold flower-like hierarchical of $CoO@MnCo_2O_4$ heterostructure (see Fig. 4.13), which displayed a remarkable improvement of the photocatalytic activity toward hexavalent chromium Cr(VI). The decoration of CoO NPs on the nano-leaf of marigold flower-like $MnCo_2O_4$ to form 3D hierarchical architecture $CoO@MnCo_2O_4$ can act as carrier sinks, harvesting more catalytic sites and facilitating the mass transfer of reactants that can indirectly increase the light absorption characteristic.

Moreover, there has been extensive research on the ability of ZnO to grow in numerous morphologies to attain different types of morphologies and orientations of ZnO superstructures (Desai, Vyas, Saratale, & Sartale, 2019). The main reason behind the growth of such morphologies is due to the wurtzite (hexagonal) crystal structure of ZnO, which can generate a large set of different superstructures (Chen, Zhao, Liu, & Yang, 2015; Pirhashemi, Habibi-Yangjeh, & Pouran, 2018). Recently, Long, Wang, Fu, & Liu, 2019 heterojunction the wrinkle-like $ZnFe_2O_4$ nanosheet over the growth of ZnO nanotube

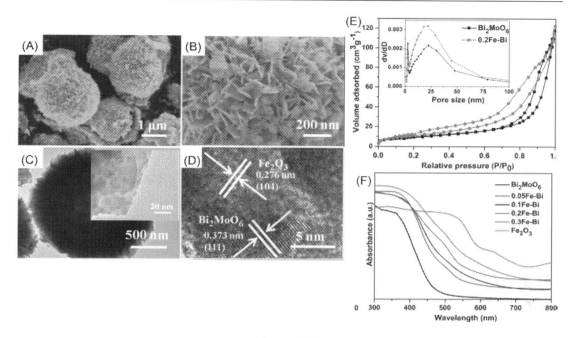

Figure 4.12

SEM (A, B) and TEM (C, D) images of 0.2 Bi_2MoO_6-Fe_2O_3. (E) N_2 adsorption−desorption isotherms of Bi_2MoO_6 and 0.2Fe-Bi. The inset is the pore-size distribution. (F) UV−Vis diffuse reflectance spectra (Drs) of the samples (Shijie, Hu, Zhang, Jiang, & Liu, 2017).

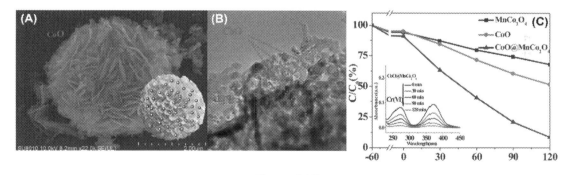

Figure 4.13

(A, B) TEM image of and HRTEM image of $CoO@MnCo_2O_4$. (C) Photocatalytic degradation of tetracycline of different samples under visible-light irradiation (Zheng & Zhang, 2018).

prepared through the three-step strategy. The completely covered ZnFe2O4 nanosheets over the ZnO nanotube arrays demonstrated that the $ZnO/ZnFe_2O_4$ nanotube array had excelled in photoelectrochemical (PEC) water splitting to produce hydrogen due to the outstanding visible-light harvesting ability and large surface area of the wrinkle-like $ZnFe_2O_4$

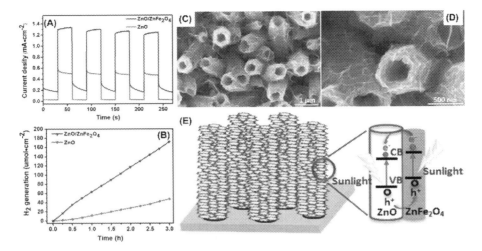

Figure 4.14

(A) Photo-current density of ZnO and hierarchical $ZnO/ZnFe_2O_4$ nanotube arrays at 0 V versus Ag/AgCl. (B) PEC water splitting for H_2 production of the samples at 0.6 V versus Ag/AgCl. (C, D) SEM images of hierarchical $ZnO/ZnFe_2O_4$ nanotube arrays taken at low and high magnifications. (E) Schematic and band alignment of the hierarchical $ZnO/ZnFe_2O_4$ nanorod arrays architecture (Long, Wang, Fu, & Liu, 2019).

nanosheets (see Fig. 4.14). The hierarchical $ZnO/ZnFe_2O_4$ nanotube arrays achieved a drastically enhanced photocurrent density, which was twofold higher than pristine ZnO nanotube arrays. Well-patterned heterostructures of the robust photocatalyst had significantly facilitated the electron–hole separation effectiveness between the ZnO and the $ZnFe_2O_4$. Besides that, the hierarchical nanostructures can potentially enhance light absorption, the abundant active adsorption sites, and photocatalytic reaction sites to improve the homogeneity on the distribution of active sites of the prepared photocatalysts (Wang & He, 2017). More recently, Chen et al. (2019) showed that the branch-like $SnO_2@ZnO$ heterojunction photocatalyst was successfully fabricated via a simple two-step hydrothermal process. The growth of SnO_2 by precipitation on the surface of ZnO that grew into a nanowire structure on the six nonpolar surfaces of ZnO nanorods clearly showed that the branch-like hierarchical exhibited an excellent photodegradation of the RhB (see Fig. 4.15). It is strongly believed that a properly designed hierarchical heterojunction photocatalyst structure can enhance light-scattering effects, which is a key role in enhancing the light-harvesting efficiency while suppressing the electron–hole recombination and optimizing the redox ability of the system. Obviously, the intrinsic interconnected hierarchically heterojunction photocatalyst had also improved the molecular diffusion/transport kinetics and the surface area for a large number of active sites, which can accelerate the surface reaction kinetics.

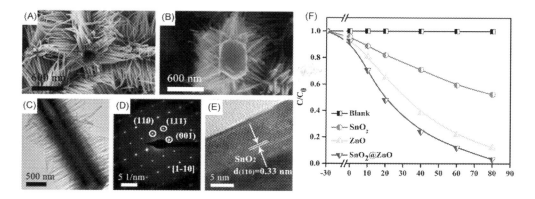

Figure 4.15
(A, B) SEM and (C) TEM patterns of $SnO_2@ZnO$. (D) SAED patterns and (E) HRTEM image of SnO_2 nanowires. (F) The photocatalytic activity of ZnO, SnO_2, and $SnO_2@ZnO$ in 10 mg/L RhB dye (Chen et al., 2019).

4.4 Semiconductor—metal heterojunction

There are several metals, such as metal oxides, transition metals, noble metals, and metal sulfides that are categorized as potential photocatalysts, where they show great performances in photocatalysis due to their heterostructures. In the case of heterojunction photocatalysis using metal-semiconductor as a photocatalyst, the determination of rectifying properties of the interface depends on the energy band structures between two materials. By thermal equilibrium, the Fermi levels of both materials will align to each other due to the charge transfer. After all, the energy bands would be bending when the generation of a built-in electric field within the semiconductor occurs. However, when an insurable energy barrier is formed from band bending to their carriers (either electron in the CB or hole in the VB), this would form a Schottky junction only when the interface is rectified, as opposed to ohmic contact. Meanwhile, the carrier transport mechanism at the Schottky junction is controlled by the probability of most carriers to prevent the barrier height. Under the thermionic emission theory, the mechanism would produce energy difference from the Fermi level of metal to the conduction/VB edge of an n/p-type semiconductor for the electron (hole). In addition, the current density flowing through the interface can be calculated (Sze & Ng, 2006).

According to the mechanism in Fig. 4.16, this study reported that semiconductor—metal junctions, which is a Schottky barrier, had been generated when metal was in contact with a semiconductor. Thus an electron flow from the materials had generated different Fermi-level alignments, for example, Pt/TiO_2. To provide a basis for this mechanism, the utilization of metal must be able to: (1) act as an electron trap to receive photoelectrons from the semiconductor after excitation and (2) improve the charge carrier separation and

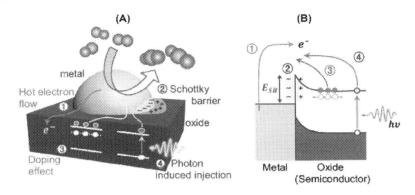

Figure 4.16
(A) Electronic configuration and (B) energy band diagram of metal—oxide catalysts under chemical reactions illustrate the mechanism between both materials to build space-charge separation region (Schottky junction) (Kim et al., 2015).

reduce recombination, as charge cannot flow in the opposite direction (unlike in an ohmic contact) (Moniz, Shevlin, Martin, Guo, & Tang, 2015). The process can be explained accordingly: (1) the flow of fast and energetic charge carriers; (2) formation of a Schottky barrier; (3) effect of dopants and oxygen vacancies; and (4) manipulation of hot carriers generated by photon absorption (Kim, Lee, & Park, 2015). The flow of electron starts from the material that with higher Fermi energy level and move to the material that with lower Fermi energy level. Then, the Fermi energy levels would be aligned together. The basic principles based on the n-type semiconductor and metal were that if a metal is higher than that of the n-type semiconductor (such as TiO_2), the electrons will flow from the semiconductor into the metal to adjust the Fermi energy levels (Wang et al., 2014)

Rapid photogenerated charge separation is influenced by the heterostructure of photocatalysts, combined with the merits of different compounds, including light absorption, charge separation, and charge transfer between different kinds of semiconductors (Chang et al., 2016). The following advantages highlighted the development of semiconductor—metal nanoheterojunctions, which could enhance the photocatalytic performances (Moniz, Shevlin, Martin, Guo, & Tang, 2015; Yang, Wang, Han, & Li, 2013). This is due to the advantages of optical and electrical characteristics between the semiconductor and the metal from their photonic and plasmon—exciton coupling effects. The advantages include: (1) enlarging the cross-sections of the absorption process (Li et al., 2013a); (2) increasing the effective average optical path length (Cushing & Wu, 2013; Thomann et al., 2011); (3) producing an intense electromagnetic near-field (E-field) (Linic, Christopher, & Ingram, 2011); and (4) transferring the energy stored in the local plasmonic field from the metal nanostructures to the semiconductor nanomaterials.

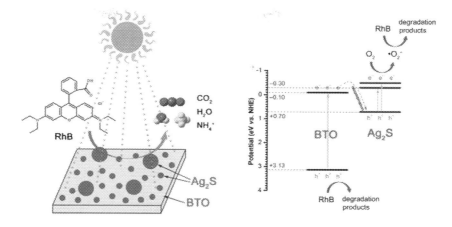

Figure 4.17
$Ag_2S/Bi_4Ti_3O_{12}$ heterojunction mechanism (Zhao, Yang, Li, Cui, & Liu, 2019).

The weaknesses of semiconductor—metal heterojunction are low light utilization, improper band position, fast recombination of charge carriers, and photocorrosion. Thus the investigation on strategies to close the gaps and design more efficient photocatalysts needs to be expedited (Zhan, Sun, & Han, 2019). Electrons can stay longer in the metal NPs upon charge transfer through the junction interfaces, such as Au NPs/TiO_2 and Ag NPs/TiO_2 (Murdoch et al., 2011). By using the Fermi-level equilibration mechanism, noble metal particles will have the capability of having a simultaneous transfer of excited electrons to the semiconductive CB and noble metal particles (Xu et al., 2015). Electrons from Ag metal have a unique formation that can "leave" the metal and transfer to the semiconductor that makes the local electric field in the semiconductor—metal composite system, thus encouraging the electron—hole pair separation. Therefore Ag/$AgGaO_2$ with sol—gel preparation could degrade about 95% of MB within 180 min. In another report, the degradation of tetracycline hydrochloride could achieve 90% within 60 min by using SnO_2 (50%)/ZnO nanocomposites. The process of the electrostatic field at the edges can occur once the negatively charged SnO_2 is transported to the ZnO through stirring and in advance of Fermi echelons adjustments (Lwin, Zhan, Song, Jia, & Zhou, 2019). The interface of the Ag_2S/BTO Z-scheme heterojunction could transport the photogenerated electrons in the CB of BTO to Ag_2S and recombine with the photogenerated holes in the VB of Ag_2S (see Fig. 4.17). Hence, 15 wt.% Ag_2S/$Bi_4Ti_3O_{12}$ will suppress the electron/hole pair recombination in BTO and Ag_2S and degrade 98% of RhB within 120 min (Zhao, Yang, Li, Cui, & Liu, 2019).

4.5 Semiconductor—carbon group heterojunction

To increase the performance of the photocatalytic activity, there are many forms of carbon materials that can be used as a combination with a semiconductor heterojunction photocatalyst, such as amorphous carbon, graphitic-like carbon, activated carbon, carbon nanotubes (CNTs), and graphene (Li & Wu, 2015). The semiconductor–carbon can effectively trap the generated electrons to suppress the recombination of electron–hole pairs (Etacheri, Michlit, Seery, Hinder, & Pillai, 2013; Low, Yu, Jaroniec, Wageh, & Al-Ghamdi, 2017; Wang, 2014; Wen, 2015). In the area of photocatalysis application, the carbonaceous material has gained much interest due to their structures and compositions owing to their intriguing properties and good tunability (Ge, Zhang, Heo, & Park, 2019). One of the best properties of carbonaceous material is good conductivity that is able to boost the visible light absorption capacity, electron transportation, and separation efficiency (El-Sheikh et al., 2017; Shao et al., 2015; Zhao, Wang, Yang, Tang, & Yang, 2012).

According to Jianlong, Zhang, and Park (2019), the limitation of this combination had been discovered recently, where their interface contact between the carbon phase and the semiconductor region is not really close to each other, causing the weakening of bonding strength and electron transportability. Also, it is difficult to achieve the nanostructure of carbons and maintain their performance after recycling due to their powdered morphology in the macroscale with the formation of secondary pollution. Even though 3D carbonaceous photocatalysts had been developed in terms of their mechanical strength and bonding, they are still unstable. Other than that, due to their poor photocatalytic activity, carbonaceous material can only act as a substrate and cocatalyst in the composite system, as compared to the semiconductor itself. Fig. 4.18 illustrates the mechanism of semiconductor photocatalysis that is influenced by the band model, to which their simultaneous reactions should be at least at multiple reactions (equal rates) through the oxidation of photogenerated holes and reduction of photogenerated electrons. The overall process can be

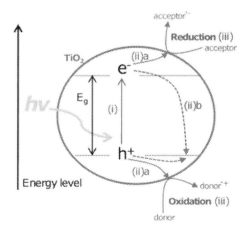

Figure 4.18

Schematic illustration demonstrating the photocatalytic process of semiconductor photocatalysts.

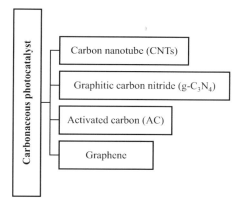

Figure 4.19
Category of carbonaceous photocatalysts.

explained in detail, as follows: (1) photon absorption and electron—hole pair generation; (2) charge separation and migration; (3) surface reaction sites; or (4) recombination sites; and (5) surface chemical reaction at active sites. All of the semiconductor—carbon materials can be classified into the main groups based on their properties for the degradation of organic pollutants: (1) CNTs; (2) graphitic carbon nitrides (g-C_3N_4); (3) activated carbon; and (4) graphene, as shown in Fig. 4.19.

4.5.1 Carbon nanotubes

Recently, carbon nanomaterials that have their nanomaterial properties and fibrous structures are CNTs and carbon nanofibers. Their structural designs and characteristics demonstrate their ability to achieve a large surface area with high aspect ratio, acquire extraordinary electronic properties, have good mechanical and physicochemical properties to be formed as an unique composite photocatalyst system. They are allotropes of carbon with a nanostructure that can have a length-to-diameter ratio greater than 1,000,000 (Saifuddin, Raziah, & Junizah, 2013). CNTs like single-walled or multi-walled carbon nanotubes being as a scaffold supporting example toward TiO_2 and other semiconductors because their properties enhance the charge transport channels, make broader the specific surface area and decrease the potential of charge recombination rate (Ding, Kim, & Erlebacher, 2004; Li et al., 2017).

Scaffold supporting materials became as a booster to modify the electron selectivity of contact properties and influenced the performance and their impedance spectra (Anaya et al., 2017). CNTs are composed of rolling up a graphene sheet that looks like hollow cylinders where an extended planar hexagonal lattice of purely sp^2-bonded carbons. Therefore, the structure of CNTs would become favorable to have high carrier mobility,

Table 4.1: Carbon nanotube (CNT) heterojunction application.

Photocatalyst	Sample	Method	Efficiency	Ref
$TiO_2 - x/CNT$	MO	Solvothermal	MO: 99.6% removal H_2 production: 242.9 μmol/h/g	Zhang et al. (2017)
$CNT/g\text{-}C_3N_4/BWO$	Tetracycline hydrochloride	Hydrothermal	87.65% removal: 90 min	Jiang et al. (2018)
Double Walled CNTs/N-PdTiO$_2$	Eosin yellow	Sol-gel	2.0% DWCNT/N, Pd TiO$_2$ Dye adsorption: 9.93% Degradation (after 180 min): 99.87%	Kuvarega and Mamba (2016)
$Cu_2O/CNTs$	Phenol	—	85.8% removal: 60 min	Yang, Chu, Wang, Wu, and Luo (2016)
CNTs-TiO$_2$@AgI	RhB	Solvothermal	100% removal: 15 min	Yang et al. (2017)
$g\text{-}C_3N_4/TiO_2/CNT$	MB	Hydrothermal	98% removal: $\sim$90 min	Chaudhary, Vankar, and Khare (2017)

CNT, carbon nanotube; MB, methylene blue; MO, methyl orange; RhB, Rhodamine B.

tunable conductivity, narrow band-gap, desirable optical properties, good stability and excellent flexibility (Hu, Hou, Liu, & Cheng, 2019). Herein, we will provide the examples of CNTs as one of the combination with metal photocatalyst. Table 4.1 illustrates the performances of CNTs in heterojunction applications, such as $TiO_2 - x/CNT$, CNT/CN/ BWO, double-walled carbon nanotube/N-PdTiO$_2$, Cu_2O/CNTs, CNTs-TiO$_2$@AgI, and $g\text{-}C_3N_4$/TiO$_2$/CNT.

As a representative work, $TiO_2 - x/CNT$ has been generated as a photocatalyst to enhance the production of hydrogen by using the solvothermal method that completely decolorized methyl orange (MO) as well as can produce a huge amount of hydrogen. CNT and TI^{3+} combination gave a great impact to improve the conductivity based on the rate capability of the electrode. The structure of 3D urchin-like was successfully improved the electrochemical properties by decreasing charge transfer resistance, reduced charge recombination and creating more abundant active sites to enhance light trapping and scattering ability. Other than work, Jiang et al. (2018) used CNT/CN/BWO for the removal of tetracycline hydrochloride for only 1.5 hours. By introducing CNT in their studies, the strong light absorption ability of CNT would increase simultaneously with the visible-light adsorption intensity. Double-walled CNTs/N-PdTiO$_2$ as mentioned in Fig. 4.20 could reduce the formation of agglomeration and generated synergistic effects on the photocatalytic performances. Because of their high electrical conductivity, CNTs are capable to transfer electron to their channels effectively. Hence, they could act as supercapacitors by exhibiting high electron capture ability and high electron storage capacity (Kongkanand & Kamat, 2007).

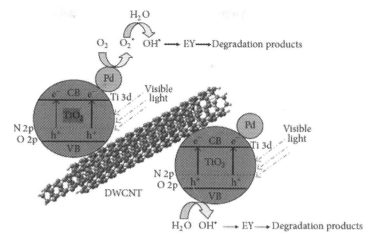

Figure 4.20
Double Walled CNTs/N-PdTiO$_2$ (Kuvarega & Mamba, 2016).

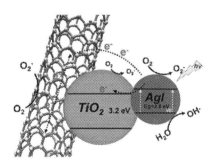

Figure 4.21
Mechanism degradation of RhB by CNTs-TiO$_2$@AgI (Yang et al., 2017).

Yang, Chu, Wang, Wu, and Luo (2016) claimed that CNTs had successfully improved the electronic conductivity of the composite material, as it was able to degrade phenol in 1 h. Nowadays, apart from metals, a variety of newly developed semiconductors have also been coupled with CNTs, such as CNTs-TiO$_2$@AgI. They demonstrated that the exhibited CNTs had enhanced the absorption intensity by expanding the light absorption range, as well as being good electron acceptors that are able to supply another electron and accept the excited electrons from the AgI particles (Fig. 4.21). Moreover, the final composite of g-C$_3$N$_4$/TiO$_2$/CNT and the surface-modified CNTs as an electron acceptor relied on a simple sonication and stirring process, which was easy to operate, energy saving, and able to boost the performance of the photocatalytic process. Consequently, CNTs can be widely utilized as an electron acceptor for all kinds of semiconductors and it can also easily attain superior performance.

4.5.2 Graphitic carbon nitride

The application of g-C_3N_4 (graphitic carbon nitride) as photocatalyst as a semiconductor with great performance consists of carbon and nitrogen elements, which are highly abundant in the earth (Ren, Zeng, & Ong, 2019). It has a unique electronic band structure, medium band gap (~ 2.7 eV), and excellent chemical and thermal stability (Long et al., 2017; Naseri, Samadi, Pourjavadi, Moshfegh, & Ramakrishna, 2017). Hence, it has been used numerously in nanoscience research for multifunctional applications, including remediation of environmental pollutants, water splitting, reduction of carbon dioxide, nitrogen fixation, hydrogen peroxide production, organic synthesis, and bacteria disinfection (Mamba & Mishra, 2016).

To date, there are many researchers seeking to construct a semiconductor heterojunction interface between two dissimilar semiconductors to enhance the intimate contact interface formed by the heterojunction process. This is because the weakness of g-C_3N_4, such as low charge separation and visible-light absorption, could be improved by the formation of the interface. Also, the formation of this interface can simultaneously make up the well-defined electronic structure and band gap configuration while increasing the photocatalytic activity (Jang, Kim, & Lee, 2012; Wang et al., 2013, 2014). There are many examples of the combination of bare g-C_3N_4 with other metals to elevate their performances by contact interface through heterojunction principle, such as metal (metal oxides, transition metal, noble metal, metal chalcogenides, metal phosphides, metal carbides), carbonaceous materials, or other semiconductors (Du, Yi, Wang, Deng, & Wang, 2019; Kong, Chen, Ong, Zhao, & Li, 2019; W. Ong, Putri, Tan, Chai, & Yong, 2016; W. Ong, Tan, Chai, & Yong, 2015; W.J. Ong et al., 2017; D. Zeng, 2018a; M.Y. Zeng, 2018b). Table 4.2 illustrates the performances of g-C_3N_4 heterojunction applications, such as g-C_3N_4 with MoS_2, C-TiO_2, $SrTiO_3$, and CuS.

Table 4.2: Graphitic carbon nitride heterojunction application.

Photocatalyst	Sample	Method	Efficiency	Ref
MoS_2/g-C_3N_4	RhB MO	Facile impregnation and calcination	MO (95%) = 180 min RhB (76%) = 20 min	Li et al. (2016)
C-TiO_2/g-C_3N_4	MO	gC_3N_4: thermal polycondensation (C-TiO_2): hydrothermal	27:8 [C-TiO_2/g-C_3N_4(0.08)] = 98.6% after 1 h	Lu et al. (2017)
g-C_3N_4/ $SrTiO_3$	MB	Sonication mixing	(UV–Vis) K: 0.0220/ min (Visible) K: 0.0071/ min	Konstas, Konstantinou, Petrakis, and Albanis (2018)
g-C_3N_4/CuS	RhB MB	—	MB, 98.1%: 120 min RhB, 96.8%: 120 min	Cai et al. (2017)

MB, methylene blue; *MO*, methyl orange; *RhB*, Rhodamine B.

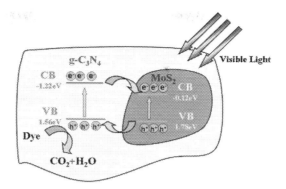

Figure 4.22
Mechanism of the 2D MoS_2/g-C_3N_4 composites (Li et al., 2016).

A number of fascinating reviews revealed that the ability of g-C_3N_4 that was prepared using different methods, such as hydrothermal, sol−gel, and solvothermal. Hao et al. investigated the degradation of RhB that would facilitate the charge transfer between MoS_2 and g-C_3N_4 nanosheets due to the ultrathin thickness that allows to shorten the diffusion distance and time photogenerated carriers. Hence, the interfacial photocatalytic redox reaction would occur. Fig. 4.22 illustrates the mechanism under visible light irradiation for both g-C_3N_4 and MoS_2 that were excited to generate electrons and holes. However, this study shows the holes in the VB of the g-C_3N_4 that acted as the main reactive species to undergo an efficient.

Lu et al. employed C-TiO_2/g-C_3N_4 nanocomposite with chemically bonded interface through C-Ti and N-Ti bonds. Both bonds have formed effective heterojunction by decreasing the recombination of electron−hole pairs and enhancing the separation of the photogenerated carriers. In this study, $\cdot O_2^-$ was much stronger than $\cdot OH$ under visible light. Konstats et al. investigated the Z-scheme mechanism of $SrTiO_3$/g-C_3N_4 with different ratios of $SrTiO_3$ contents toward MB under simulated solar light irradiation. It was observed that the photocatalytic activity increased only at a g-C_3N_4 loading of 20%. This results in the efficient generation of both the hydroxyl and superoxide radicals that were responsible for MB degradation. The initial increase in the CN content had significant effects toward the degradation efficiency due to the charge transfer in the materials' interface, causing a greater response in the visible light region. However, the unlimited g-C_3N_4 content would interrupt the heterointerfaces in the composites, which will then decrease their charge transfer (Wu et al., 2018). In a similar study, Cai et al. observed that g-C_3N_4/CuS heterojunctions have a highly stable structure based on the TEM image results even after five recycling reactions. In addition, $\cdot O_2^-$ and h^+ acted as the main reactive species to oxidize RhB and MO directly, which included the adding of scavengers.

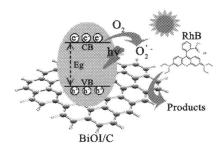

Figure 4.23
Mechanism of BiOI/activated carbon (Hou et al., 2017).

4.5.3 Activated carbon

This work discusses some of the representative studies. Activated carbon became a precursor that aided the pollutant molecules near the active site of TiO_2 for an efficient photodegeneration process. The performance of titanium, polyethylene glycol, and activated carbon would be combined together to degrade MO efficiently (Bagheri, Mohd Khir, A.T., & Abd Hamid, 2016). Other than that, the study (Hou et al., 2017) that investigated BiOI/activated carbon at about 95% RhB removal within 120 min proved that the presence of Bi−C bond acted as a bridge to rapidly transmit the electron from BiOI to carbon in the heterojunction mechanism. Fig. 4.23 shows that the BiOI nanosheets were able to strengthen the surface of the carbon, while simultaneously enhancing their hierarchical structure interfacial interaction to provide high specific surface area. Hence, with the increase in exposed reaction sites, the diffusion pathway of the pollutants was shortened. TiO_2-$MnTiO_3$/HACFs (hollow activated carbon fibers) acted as photocatalysts that could efficiently degrade MB for only 60 min. Activated carbon has become a contributor to the electron charge separation mechanism, as well as to prevent TiO_2 and $MnTiO_3$ NPs from undergoing corrosion, as shown in Fig. 4.24. Moreover, the generation of hydroxyl radicals by the interaction of electrons under visible light can be increased with the presence of carboxyl groups in the graphene of activated carbon fiber matrix (Li, Ng, Kwong, & Chiu, 2017).

4.5.4 Graphene oxide

The semiconductor−graphene mechanism of heterojunction can be divided into Schottky junctions, direct Z-scheme, indirect Z-mechanism, and type-II heterojunction (van der Walls, in plane and multicomponent heterojunction). The application of graphene in heterojunction mechanism can be with its derivatives, such as graphene oxide (GO), reduced graphene oxide (rGO), and others. Based on a study by Putri, Ong, Chang, and Chai (2015), the conduction and VBs of pristine undoped graphene are located at

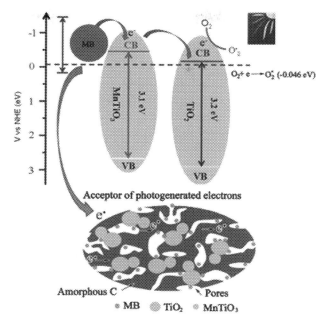

Figure 4.24
Mechanism of TiO_2 and $MnTiO_3$ (Li, Ng, Kwong, & Chiu, 2017).

antibonding π^* orbital and bonding π orbital, respectively. Both bands would degenerate and reach the Brillouin zone, which will create a zero-band gap semimetallic material. Hence, the graphene-based semiconductor is dependent on the concentration of the dopants. Recently, GO was seen to have oxygen-containing functionalities and edge defects on their aromatic scaffold, which would improve the graphene structure to make it an intriguing nanomaterial for visible-light-driven photocatalysis.

Clearly, the selection of graphene as photocatalyst exhibits many advantages, such as being inexpensive, having tunable band structures, high conductivity, superior electron mobility, extremely high specific surface area, photoelectron mediator and acceptor, enhancing the adsorption capacity, tuning the light absorption range and intensity, photothermal effect, macromolecular photosensitizer, and a unique structure for building nanoscale architecture of composites (Han, Zhang, & Xu, 2016; Wang et al., 2014; Xie, Kretschmer, & Wang, 2015). As a result, graphene-based heterojunctions will increase photocatalytic performances. For example, Ahmadi, Dorraji, Rasoulifard, and Amani-Ghadim (2019) reported that the $SrTiO_3$/GO can achieve 100% removal of RhB within 7 h. More importantly, it has also been pointed that rGO became a photosensitizer for the wide band gap of $SrTiO_3$ (STO) semiconductors. According to the nature of graphene, rGO would be photoexcited whereby it was generating electron and hole pairs. Then, it would be moved at the CB of STO and removed an electron from the VB of the STO. Hence, the remaining

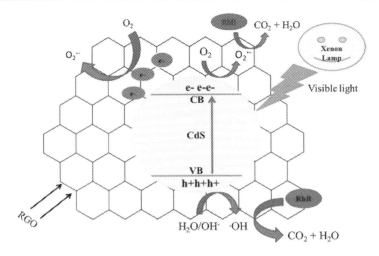

Figure 4.25
CdS—RGO heterojunction (Wei, Ou, Guan, Peng, & Zheng, 2019).

holes on VB of STO react with water molecules and generated hydroxyl radical for the degradation of RhB as targeted pollutants. Also, CdS—RGO heterojunctions were prepared by a hydrothermal process using an ice crystal template method combined with a freeze drying-calcination process (Wei, Ou, Guan, Peng, & Zheng, 2019), which could achieve a degradation efficiency of RhB (97.2%) and acid chrome blue K (65.7%) within 1 h. The presence of RGO could assist the transfer of photoexcited electrons from the CB of CdS to the RGO sheets, as shown in Fig. 4.25. Thus their rapid recombination of photogenerated electron—hole pair and the lifetime of the photogenerated electron pairs could avert and be extended.

4.6 Multicomponent heterojunction

In the last 5 years, multicomponent heterojunction has attracted much attention from researchers due to its ability to widen the light-response range of the photocatalyst and accelerate the separation of electron—hole. These multicomponent heterojunctions function to engineer a migration pathway for an effective photogenerated carrier separation. Basically, the pathway of charge carrier migration that exists as a third heterojunction is as follows: (1) electron collector (sink); (2) tune the CB or VB in the heterojunction system; and (3) bridge both semiconductors. For instance, the carbon or nitrogen doped in heterojunction photocatalyst will lower the VB resulted in the narrowing of the band gap of the heterojunction samples, which is responsible for the effective absorption of visible light. Wang et al. (2018) had narrowed the band gap of $NaTaO_3$ by partial replacement of O^{2-} by N^{3-} in the TaO_6 octahedron, which significantly boosted the photocatalytic performance of

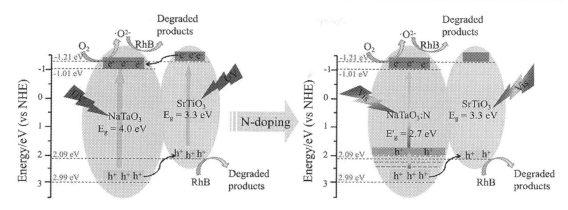

Figure 4.26
Schematic diagram for the band structure and expected charge separation of the STO/NTO and STO/NTON heterojunctions under UV and visible-light irradiation (Wang et al., 2018).

the nitrogen-doped $SrTiO_3$/$NaTaO_3$ (STO/NTON) heterojunction under visible light (see Fig. 4.26). Similarly, (Mohamed et al., 2018) did not only used urea for g-C_3N_4 as precursor but was also responsible for the in situ nitrogen and carbon doping in the mixed lattice structure of titania. The photocatalytic activity of g-C_3N_4@C, N-codoped anatase/rutile demonstrated excellent photodegradation of MO under visible light irradiation, which is accredited to superior interfacial electron—hole separation efficiency.

Apart from that, multiple heterojunction contributed in an effective photogenerated carrier by well aligning the straddling band gap structures that promote the flow charge carriers via the heterojunction while extending light absorbance (Liu et al., 2015). For example, Wang et al. (2016) had increased the visible light absorption capacity by dispersing a narrow band gap of $FeWO_4$ NPs on the surface of $ZnWO_4$/ZnO nanorods while constructing a well-aligned straddling band structure. The heterojunction of $FeWO_4$ over $ZnWO_4$/ZnO photocatalyst displayed higher photocatalytic activity than that of $FeWO_4$ and $ZnWO_4$/ZnO under visible light irradiation. This result had further confirmed that the $FeWO_4$@$ZnWO_4$/ZnO possessed a more efficient separation of the photogenerated electron—hole pairs and provided fast transfer of photoinduced carriers in the ternary system.

Besides that, the presence of a third heterojunction can act as an electron collector and electron transporter, which can extend the lifetime of the photogenerated charge carriers, thus improving the photocatalytic activity (Chang, Wei, & Kuo, 2018). Principally, the combination of heterojunction photocatalyst with noble metal NPs, such as Ag, Au, and Pt, will significantly improve the photocatalytic activity through its capability to accumulate the photogenerated electron, thus limiting the electron—hole recombination (Sun and Li, 2004). Sang, Kuai, Chen, Fang, and Geng (2014) had synthesized $AgVO_3$@AgBr nanobelt heterostructures through the hydrothermal process, an anion exchange reaction. The ternary heterojunction was formed by a light-induced reduction for producing well-dispersed AgBr@Ag NPs on the surface of the AgVO$_3$ nanobelts. The presence of Ag led to electron-rich Ag in the CB, resulting in a decreased recombination opportunity for the electron−hole pairs (Ramírez-Ortega, Meléndez, Acevedo-Peña, González, & Arroyoa, 2014). In particular, the excellent conductivity of Ag NPs greatly promotes the interfacial charge-transfer kinetics among Ag, AgBr, and AgVO$_3$. Also, the presence of visible photoactive Ag in the heterojunction photocatalyst will improve the visible light photocatalytic activity (Kandula & Jeevanandam, 2015). Petronella et al. (2017) conducted a photocatalytic degradation of nalidixic acid on TiO$_2$/Fe$_x$O$_y$/Ag-based nanocrystalline heterostructure, which resulted in a remarkable photocatalytic efficiency of 1.9 times faster than that using commercial TiO$_2$ P25. Such improvement has been attributed to the presence of Ag NPs that enhance visible light photoactivity due to their peculiar plasmonic properties (Ge et al., 2016; Wang et al., 2011). To be specific, the visible-light-induced photoactivation of the Ag heterostructure has two possible mechanisms, the generation of "hot electrons" and "near-field" effect, both of which can occur and result in an increased number of electrons and holes present for the catalytic processes in the juxtaposition of the metal NPs in the heterostructures (Besteiro, Kong, Wang, Gregory, & Govorov, 2017).

Instead of using Ag, the use of Pt as a cocatalyst has been regarded as a popular strategy to improve the photocatalytic performance (Li et al., 2015b). Chen et al. (2016) has integrated the synergetic strategy into well-defined Cu$_{1.94}$S-Zn$_x$Cd$_{1−x}$S heteronanorods ($0 \leq x \leq 1$) to boost the photocatalytic hydrogen evolution rate. High-quality multiple heterojunction can be observed throughout the HR-TEM, whereby high crystalline hetero-interface in the Cu$_{1.94}$S−Zn$_{0.23}$Cd$_{0.77}$S heteronanorods can be obviously seen in Fig. 4.27A. The elemental mapping images of Cu$_{1.94}$S−Zn$_{0.23}$Cd$_{0.77}$S heteronanorods can be visualized in Fig. 4.27B, whereby the Cu (green) is concentrated at the "head," Zn (blue) and Cd (red) are limited to the stem part, and S (yellow) is homogeneously distributed throughout the entire nanorod. However, once the Pt NPs (5 wt.%) were in situ photodeposited over the heteronanorods of Cu$_{1.94}$S-Zn$_x$Cd$_{1−x}$S, a highly improved photoactivity then reached a maximum H$_2$ production rate (13,533 μmol/h/g), with an apparent quantum efficiency of 26.4% at 420 nm (see in Fig. 4.27D). The further deposited Pt cocatalyst has enhanced the additional spatial charge separation of Cu$_{1.94}$S and Zn$_x$Cd$_{1−x}$S by acting as electron collectors to carry out the water-splitting reaction. These heteronanorods exploited the synergetic integration of photocatalytic constituents into a functional heterostructure while opening up opportunities for promoting the overall photocatalytic performance by rational hierarchical nanostructures.

The multicomponent heterojunction also has potential applications in other heterojunction photocatalysts with controllable morphologies and components. As aforementioned, the effect of hierarchical nanostructure greatly affects the light absorption capability to provide a large active adsorption site and photocatalytic reaction. This situation brings an innovative

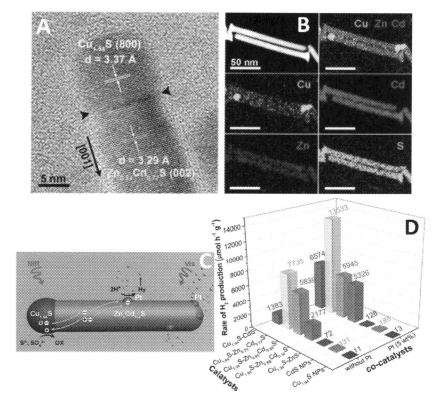

Figure 4.27
(A) HRTEM and (B) STEM-EDX elemental mapping images of $Cu_{1.94}S-Zn_{0.23}Cd_{0.77}S$ heteronanorods. (C) Schematic illustration of the Pt-decorated $Cu_{1.94}S-Zn_xCd_{1-x}S$ heteronanorods. (D) The photocatalytic hydrogen production activity of different samples under visible-light irradiation ($\lambda > 420$ nm) (Chen et al., 2016).

perspective for the multiple-component heterojunction in coupling with the different classes of hierarchal structure. Consequently, this would be a great way of creating a new heterojunction architectural design.

In this regard, another recent work from Sun's group (Sun et al., 2018) is particularly noteworthy. They have selectively constructed a hierarchical hollow black $TiO_2/MoS_2/CdS$ tandem heterojunction photocatalyst that allows for broad-spectrum absorption and boosts the lifetime of the photogenerated electron–hole (see Fig. 4.28A and B). The use of MoS_2 nanosheets in the study is not only cost effective but also acts as a bridge to link two light-harvesting semiconductors in a tandem heterojunction of the CdS NPs and black TiO_2 spheres on both sides efficiently. Subsequently, the photocatalytic hydrogen rate of the black $TiO_2/MoS_2/CdS$ tandem heterojunction resulted in 179 μmol/h per 20 mg photocatalyst under visible-light irradiation, which is nearly three times higher than that in

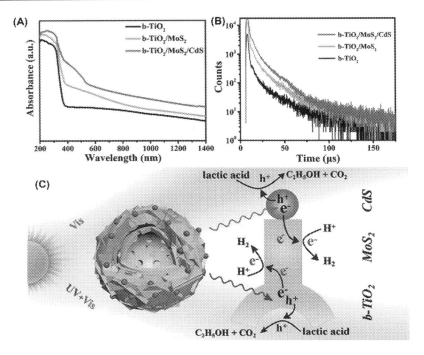

Figure 4.28

(A) Ultraviolet—visible absorption spectra. (B) Time-resolved fluorescence spectra. (C) Schematics of the $TiO_2/MoS_2/CdS$ tandem heterojunctions used for solar-driven water splitting (Sun et al., 2018).

black TiO_2/MoS_2 heterojunctions (57.2 μmol/h). More significantly, the stability of the CdS NPs in black $TiO_2/MoS_2/CdS$ tandem heterojunction was significantly improved compared to that in MoS_2/CdS due to the formation of tandem heterojunctions and the strong UV-absorbing effect of black TiO_2.

Another hierarchical pine-branch-like in ternary heterostructure catalysts was designed and successfully synthesized by Cao, Pan, and Ji (2019) recently. The pine-branch-like CuO/CuS/ZnO heterostructure displayed an excellent photocatalytic activity in the degradation of RhB solution, with 2 times and 4.6 times higher than CuO/CuS and CuO, respectively. The excellent photocatalytic performance was mainly attributed to the unique heterostructure, which effectively enhanced the absorption of sunlight and reduced the charge-carrier recombination. Similar leaf-like hierarchical structure by Lin et al. (2015), as shown in Fig. 4.29A, revealed the insertion of the quantum dot of Ag_3PO_4 on the $InVO_4/BiVO_4$ surfaces, whereby the photocatalytic degradation activity was enhanced. The improvement in photocatalytic performance is associated with the extended absorption in the visible light region and higher surface area that promoted the generation of higher photogenerated electron—hole pairs. This multiple heterojunction also revealed an effective

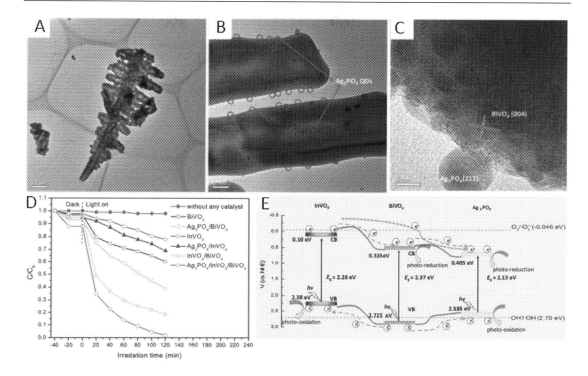

Figure 4.29
TEM (A, B) and HRTEM images (C) of as-synthesized $Ag_3PO_4/InVO_4/BiVO_4$. (D)
Photodegradation efficiencies of RhB as a function of irradiation time for different photocatalysts.
(E) Schematic diagram of the separation and transfer of photogenerated charges in the
heterojunction under visible light irradiation (Lin et al., 2015).

separation of photogenerated carriers at the $Ag_3PO_4/InVO_4/BiVO_4$ interfaces, as seen in
Fig. 4.29E. Hence, the development of multicomponent heterojunction photocatalysts with
hierarchal architectures is expected to provide new insights and meaningful information for
high-efficiency visible-light-driven photocatalysis. Therefore a deeper understanding of the
photocatalysis mechanism over these multicomponent cocatalysts might be helpful for the
development of highly efficient photocatalysts for photocatalytic activity. Meanwhile,
instead of achieving the highest photocatalytic activity, research on the loading methods,
amount, composition, location, and size of the cocatalysts should be highly considered and
optimized.

4.7 Conclusion and challenges

This chapter highlights the heterojunction photocatalyst for four typical categories, which
are the semiconductor—semiconductor, semiconductor—metal, semiconductor—carbon

group, and multicomponent heterojunction. This chapter also discusses the characteristics of each semiconductor combination, the gaps in the current understanding of their mechanism interest in design based on the fundamental principles to achieve a better overall performance of degradation, and the mineralization of the targeted pollutants. It is proven that the study of heterojunction photocatalysts is gaining momentum in this research field. Although extensive research has been conducted on heterojunction photocatalysis at a laboratory scale, the key concepts of its design and construction are of significant influence to increase the photocatalytic performance to developing better semiconductor photocatalysts in the future. However, the applications of heterojunction photocatalysts also have certain limitations that need to be addressed. The application of semiconductor−semiconductor (S−S) presented in this chapter confirms the suitability of their properties to enhance the photocatalytic performance. Apart from that, this chapter briefly reviews the semiconductor−metal heterojunction and some applications were presented to prove the efficiency of this heterojunction process in the degradation of dyes and antibiotics. The semiconductor−carbon group was also included in this work, focusing on their many uses, such as CNTs, graphitic carbon nitrides (g-C_3N_4), activated carbon, and graphene, for pollutant degradation through their unique characteristics. In addition, multicomponent heterojunction processes are known to be very effective in the removal of many hazardous contaminants. Therefore further research for the investigation of heterojunction mechanism towards the degradation of the real wastewater is required to better comprehend the process applications.

References

Adams, D. M., Brus, L., Chidsey, C. E. D., Creager, S., Creutz, C., Kagan, C. R., & Zhu, X (2003). Charge transfer on the nanoscale: Current status. *The Journal of Physical Chemistry. B, 107*(28), 6668−6697.

Ahmadi, M., Dorraji, M. S Sayed, Rasoulifard, M. H., & Amani-Ghadim, A. R. (2019). The effective role of reduced-graphene oxide in visible light photocatalytic activity of wide band gap $SrTiO_3$ semiconductor. *Separation and Purification Technology, 228*, 115771.

Anaya, M., Zhang, W., Hames, B. C., Li., Y, Fabregat-Santiago, F., Calvo, M. E., & Mora-Sero, I. (2017). Electron injection and scaffold effects in perovskite solar cells. *Journal of Materials Chemistry C, 5*(3), 634−644.

Bagheri, S., Mohd Khir, Z. A., A.T., Yousefi, & Abd Hamid, S. B. (2016). Photocatalytic performance of activated carbon-supported mesoporous titanium dioxide. *Desalination and Water Treatment, 57*(23), 10859−10865.

Bai, Y., Mora-Sero, I., Angelis, F. D., Bisquert, J., & Wang, P. (2014). Titanium dioxide nanomaterials for photovoltaic applications. *Chemical Reviews, 114*(19), 10095−10130.

Bak, T., Nowotny, J., Sucher, N. J., & Wachsman, E. (2011). Effect of crystal imperfections on reactivity and photoreactivity of TiO_2 (rutile) with oxygen, water, and bacteria. *The Journal of Physical Chemistry C, 115*(32), 15711−15738.

Bak, T., Chu, D., Francis, A. R., Li, W., & Nowotny, J. (2014). Concentration of electrons at grain boundaries in TiO2 (rutile): Impact on charge transport and reactivity. *Catalysis Today, 224*, 200−208.

Besteiro, L. V., Kong, X. T., Wang, Z., Gregory, H., & Govorov, A. O. (2017). Understanding hot-electron generation and plasmon relaxation in metal nanocrystals: Quantum and classical mechanisms. *Acs Photonics, 4*(11), 2759−2781.

Cai, Z., Zhou, Y., Ma, S., Li, S., Yang, H., Zhao, S., & Wu, W. (2017). Enhanced visible light photocatalytic performance of g-C_3N_4/CuS pn heterojunctions for degradation of organic dyes. *Journal of Photochemistry and Photobiology A: Chemistry, 348*, 168−178.

Cao, F., Pan, Z., & Ji, X. (2019). Enhanced photocatalytic activity of a pine-branch-like ternary CuO/CuS/ZnO heterostructure under visible light irradiation. *New Journal of Chemistry, 43*, 11342−11347.

Cao, S., Shen, B., Tong, T., Fu, J., & Yu, J. (2018). 2D/2D heterojunction of ultrathin MXene/Bi2WO6 nanosheets for improved photocatalytic CO_2 reduction. *Advanced Functional Materials, 28*(21), 1800136.

Chang, C.-J., Wei, Y.-H., & Kuo, W.-S. (2018). Free-standing CuS-ZnS decorated carbon nanotube films as immobilized photocatalysts for hydrogen production. *International Journal of Hydrogen Energy.*

Chang, X., Li, Z., Zhai, X., Sun, S., Gu, D., Dong, L., & Zhu, Y. (2016). Efficient synthesis of sunlightdriven ZnO-based heterogeneous photocatalysts. *Materials & Design, 98*, 324−332.

Chaudhary, D., Vankar, V., & Khare, N. (2017). Noble metal-free g-C_3N_4/TiO_2/CNT ternary nanocomposite with enhanced photocatalytic performance under visible-light irradiation via multi-step charge transfer process. *Solar Energy, 158*, 132−139.

Chen, S., Liu, F., Xu, M., Yan, J., Zhang, F., Zhao, W., . . . Liu, C. (2019). *First-principles calculations and experimental investigation on SnO_2@ZnO heterojunction photocatalyst with enhanced photocatalytic performance. Journal of Colloid and Interface Science, 553*, 613−621.

Chen, Y., Zhao, H., Liu, B., & Yang, H. (2015). Charge separation between wurtzite ZnO polar {0 0 1} surfaces and their enhanced photocatalytic activity. *Applied Catalysis B: Environmental, 163*, 189−197.

Chen, Y., Zhao, S., Wang, X., Peng, Q., Lin, R., Wang, Y., . . . Li, Y. (2016). Synergetic integration of Cu1.94S−$Zn_xCd_{1-x}S$ heteronanorods for enhanced visible-light-driven photocatalytic hydrogen production. *Journal of the American Chemical Society.*

Cushing, N. Q., & Wu, S. K. (2013). Plasmon-enhanced solar energy harvesting. *The Electrochemical Society Interface, 22*, 63−67.

Dai, W., Hu, X., Wang, T., Xiong, W., Luo, X., & Zou, J. (2018). Hierarchical CeO_2/Bi_2MoO_6 heterostructured nanocomposites for photoreduction of CO_2 into hydrocarbons under visible light irradiation. *Applied Surface Science, 434*, 481−491.

Desai, M. A., Vyas, A. N., Saratale, G. D., & Sartale, S. D. (2019). Zinc oxide superstructures: Recent synthesis approaches and application for hydrogen production via photoelectrochemical water splitting. *International Journal of Hydrogen Energy, 44*(4), 2091−2127.

Ding, J., Kim, Y., & Erlebacher, Y. J. (2004). Nanoporous gold leaf: "Ancient technology" advanced material. *Advanced Materials, 166*, 1897−1900.

Du, X., Yi, X., Wang, P., Deng, J., & Wang, C. (2019). Enhanced photocatalytic Cr (VI) reduction and diclofenac sodium degradation under simulated sunlight irradiation over. *Chinese Journal of Catalysis, 40* (1), 70−79.

El-Sheikh, S. M., Khedr, T. M., Hakki, A., Ismail, A. A., Badawy, W. A., & Bahnemann, D. W. (2017). Visible light activated carbon and co-doped mesoporous TiO_2 as efficient photocatalyst for degradation of ibuprofen. *Separation and Purification Technology, 173*, 258−268.

Emeline, A., Kuznetsov, V. N., Ryabchuk, V. K., & Serpone, N. (2013). *Heterogeneous photocatalysis: Basic approaches and terminology. New and future developments in catalysis: Solar photocatalysis* (pp. 1−47). Elsevier.

Etacheri, V., Michlit, G., Seery, M. K., Hinder, S. J., & Pillai, S. C. (2013). A highly effcient $TiO_{2-x}C_x$ nanoheterojunction photocatalyst for visible light induced antibacterial applications. *ACS Applied Materials & Interfaces*, 1663−1672.

Ge, J., Zhang, Y., Heo, Y., & Park, S. (2019). Advanced design and synthesis of composite photocatalysts for the remediation of wastewater: A review. *Catalysts, 9*(2), 122.

Ge, M., Cao, C., Li, S. H., Tang, Y. X., Qi, N., Huang, J., & Lai, Y. K. (2016). *In situ* plasmonic Ag nanoparticle anchored TiO_2 nanotube arrays as visible-light-driven photocatalysts for enhanced water splitting. *Nanoscale, 8*(9), 5226−5234.

Guerrero-Araque, D., Ramírez-Ortega, D., Acevedo-Peña, P., Tzompantzi, F., Calderón, H. A., & Gómez, R. (2017). Interfacial charge-transfer process across ZrO_2-TiO_2 heterojunction and its impact on photocatalytic activity. *Journal of Photochemistry and Photobiology A: Chemistry*, *335*, 276−286.

Han, C., Zhang, N., & Xu, Y.-J. (2016). Structural diversity of graphene materials and their multifarious roles in heterogeneous photocatalysis. *Nano Today*, *11*(3), 351−372.

Hao, R., Wang, G., Jiang, C., Tang, H., & Xu, Q. (2017). *In situ* hydrothermal synthesis of g-C_3N_4/TiO_2 heterojunction photocatalysts with high specific surface area for Rhodamine B degradation. *Applied Surface Science*, *411*, 400−410.

Hou, D., Tang, F., Ma, B., Deng, M., Qiao, X.-q, Liu, Y.-L., & Li, D.-S. (2019). $Bi_4O_5I_2$ flower/Bi_2S_3 nanorod heterojunctions for significantly enhanced photocatalytic performance. *CrystEngComm*, *21*, 4158−4168.

Hou, J., Jiang, K., Sheng, M., Wei, R., Wu, X., Idrees, F., & Cao, C. (2017). Micro and nano hierachical structures of BiOI/activated carbon for efficient visible-light-photocatalytic reactions. *Scientific Reports*, *7*(1), 11665.

Hu, X., Hou, P., Liu, C., & Cheng, H. (2019). Carbon nanotube/silicon heterojunctions for photovoltaic applications. *Nano Materials Science*, *1*(3), 156−172.

Huang, Y., Zhu, D., Zhang, Q., Zhang, Y., Cao, J.-j, Shen, Z., . . . Lee, S. C. (2018). *Synthesis of a $Bi_2O_2CO_3$/ $ZnFe_2O_4$ heterojunction with enhanced photocatalytic activity for visible light irradiation-induced NO removal*. *Applied Catalysis B: Environmental*, *234*, 70−78.

Ichimura, M., & Song, Y. (2011). Band alignment at the Cu_2O/ZnO heterojunction. *Japanese Journal of Applied Physics*, *50*(5R), 051002.

Inceesungvorn, B., Teeranunpong, T., Nunkaew, J., Suntalelat, S., & Tantraviwat, D. (2014). Novel $NiTiO_3$/ Ag_3VO_4 composite with enhanced photocatalytic performance under visible light. *Catalysis Communications*, *54*, 35−38.

Jafari, T., Moharreri, J., Amin, S. S., Miao, R., Song, W., & Suib, S. L. (2016). Photocatalytic water splitting— The untamed dream: A review of recent advances. *Molecules*, *21*(7), 900.

Jang, J. S., Kim, H. G., & Lee, J. S. (2012). Heterojunction semiconductors: A strategy to develop efficient photocatalytic materials for visible light water splitting. *Catalysis Today*, *185*(1), 270−277.

Jiang, D., Ma, W., Xiao, P., Shao, L., Li, D., & Chen, M. (2018). Enhanced photocatalytic activity of graphitic carbon nitride/carbon nanotube/Bi_2WO_6 ternary Z-scheme heterojunction with carbon nanotube as efficient electron mediator. *Journal of Colloid and Interface Science*, *512*, 693−700.

Jianlong, G., Zhang, Y., & Park, S.-J. (2019). Recent advances in carbonaceous photocatalysts with enhanced photocatalytic performances: A mini review. *Materials*, *12*(01916).

Kandula, S., & Jeevanandam, P. (2015). Sun-light-driven photocatalytic activity by ZnO/Ag heteronanostructures synthesized via a facile thermal decomposition approach. *RSC Advances*, *2015*(5), 76150−76159.

Kato, S., Hirano, Y., Iwata, M., Sano, T., Takeuchi, K., & Matsuzawa, S. (2005). Photocatalytic degradation of gaseous sulphur compounds by silver-deposited titanium dioxide. *Applied Catalysis B: Environmental*, *57*(2), 109−115.

Khan, M. M., Pradhan, D., & Sohn, Y. (2017). *Nanocomposites for visible light-induced photocatalysis*. Springer.

Kim, S. M., Lee, H., & Park, J. Y. (2015). Charge transport in metal−oxide interfaces: Genesis and detection of hot electron flow and its role in heterogeneous catalysis. *Catalysis Letters*, *145*(1), 299−308.

Kohtani, S., Kawashima, A., & Miyabe, H. (2017). Reactivity of trapped and accumulated electrons in titanium dioxide photocatalysis. *Catalysts*, *7*(10), 303.

Kong, Z., Chen, X., Ong, W.-J., Zhao, X., & Li, N. (2019). Atomic-level insight into the mechanism of 0D/2D black phosphorus quantum dot/graphitic carbon nitride (BPQD/GCN) metal-free heterojunction for photocatalysis. *Applied Surface Science*, *463*, 1148−1153.

Kongkanand, A., & Kamat, P. V. (2007). Electron storage in single wall carbon nanotubes. Fermi level equilibration in semiconductor−SWCNT suspensions. *ACS Nano*, *1*(1), 13−21.

Konstas, P.-S., Konstantinou, I., Petrakis, D., & Albanis, T. (2018). Synthesis, characterization of g-C_3N_4/$SrTiO_3$ heterojunctions and photocatalytic activity for organic pollutants degradation. *Catalysts*, *8*, 554.

Kuvarega, A. T., & Mamba, B. B. (2016). Double walled carbon nanotube/TiO2 nanocomposites for photocatalytic dye degradation. *Journal of Nanomaterials, 2016*, 12.

Kwiatkowski, M., Chassagnon, R., Heintz, O., Geoffroy, N., Skompska, M., & Bezverkhyy, I. (2017). Improvement of photocatalytic and photoelectrochemical activity of ZnO/TiO$_2$ core/shell system through additional calcination: Insight into the mechanism. *Applied Catalysis B: Environmental, 204*, 200−208.

Lee, K. M., Lai, C. W., Ngai, K. S., & Juan, J. C. (2016). Recent developments of zinc oxide based photocatalyst in water treatment technology: A review. *Water Research, 88*, 428−448.

Lei, F., Sun, Y., Liu, K., Gao, S., Liang, L., Pan, B., & Xie, Y. (2014). Oxygen vacancies confined in ultrathin indium oxide porous sheets for promoted visible-light water splitting. *Journal of the American Chemical Society, 136*(19), 6826−6829.

Li, C., et al. (2018a). Surface, bulk, and interface: Rational design of hematite architecture toward efficient photo-electrochemical water splitting. *Advanced Materials, 30*(30), 1707502.

Li, C., Luo, Z., Wang, T., & Gong, J. (2018). Surface, Bulk, and Interface: Rational Design of Hematite Architecture toward Efficient Photo-Electrochemical Water Splitting. *Advance Material, 30*(20).

Li, J., Liu, E., Ma, Y., Hu, X., Wan, J., Sun, L., & Fan, J. (2016). Synthesis of MoS$_2$/g-C$_3$N$_4$ nanosheets as 2D heterojunction photocatalysts with enhanced visible light activity. *Applied Surface Science, 364*, 694−702.

Li, J., Yin, H., Li, X., Okunishi, E., Shen, Y. L., He, J., & Ding, Y. (2017). Surface evolution of a Pt−Pd−Au electrocatalyst for stable oxygen reduction. *Nature Energy, 2*(17111).

Li, J., Ng, D. H. L., Kwong, F. L., & Chiu, K. L. (2017). Hierarchically porous TiO$_2$-MnTiO$_3$/hollow activated carbon fibers heterojunction photocatalysts with synergistic adsorption-photocatalytic performance under visible light. *Journal of Porous Material, 24*, 1047−1059.

Li, J., et al. (2016a). Synthesis of MoS$_2$/g-C$_3$N$_4$ nanosheets as 2D heterojunction photocatalysts with enhanced visible light activity. *Applied Surface Science, 364*, 694−702.

Li, J., et al. (2017a). Surface evolution of a Pt-Pd-Au electrocatalyst for stable oxygen reduction. *Nature Energy, 2*, 17111.

Li, J., et al. (2017b). Hierarchically porous TiO$_2$-MnTiO$_3$/hollow activated carbon fibers heterojunction photocatalysts with synergistic adsorption-photocatalytic performance under visible light. *Journal of Porous Materials, 24*(4), 1047−1059.

Li, J., Cushing, S. K., Zheng, P., Meng, F., Chu, D., & Wu, N. (2013a). Plasmon-induced photonic and energy-transfer enhancement of solar water splitting by a hematite nanorod array. *Nature Communications, 4*, 2651.

Li, J., & Wu, N. (2015). Semiconductor-based photocatalysts and photoelectrochemical cells for solar fuel generation: A review. *Catalysis Science & Technology, 5*.

Li, M., Chen, C., Shi, Y., & Li, L. (2016). Heterostructures based on two-dimensional layered materials and their potential applications. *Materials Today, 19*(6), 322−335.

Li, N., Du, K., Liu, G., Xie, Y., Zhou, G, Zhu, J., & Cheng, H. (2013). Effects of oxygen vacancies on the electrochemical performance of tin oxide†. *Journal of Materials Chemistry A, 5*.

Li, P., Zhou, Y., Zhao, Z., Xu, Q., Wang, X., Xiao, M., & Zou, Z. (2015). Hexahedron Prism-Anchored Octahedronal CeO$_2$: Crystal Facet-Based Homojunction Promoting Efficient Solar Fuel Synthesis. *J. Am Chem. Soc, 137*(30), 9547−9550.

Lin, X., Guo, X., Shi, W., Guo, F., Che, G., Zhai, H., & Wang, Q. (2015). Ag$_3$PO$_4$ quantum dots loaded on the surface of leaf-like InVO$_4$/BiVO$_4$ heterojunction with enhanced photocatalytic activity. *Catalysis Communications, 71*, 21−27.

Lin, X., Guo, X., Shi, W., Guo, F., Che, G., Zhai, H., & Wang, Q. (2015). Ag$_3$PO$_4$ quantum dots loaded on the surface of leaf-like InVO$_4$/BiVO$_4$ heterojunction with enhanced photocatalytic activity. *Catalyst Communication, 71*, 21−27.

Ling, F., Anthony, O. C., Xiong, Q., Luo, M., Pan, X., Jia, P., & Li, Q. (2016). PdO/LaCoO$_3$ heterojunction photocatalysts for highly hydrogen production from formaldehyde aqueous solution under visible light. *International Journal of Hydrogen Energy, 41*(14), 6115−6122.

Li, M.-Y., et al. (2016b). Heterostructures based on two-dimensional layered materials and their potential applications. *Materials Today, 19*(6), 322−335.

Li, N., et al. (2013b). Effects of oxygen vacancies on the electrochemical performance of tin oxide. *Journal of Materials Chemistry A, 1*(5), 1536−1539.

Li, N., Fu, W., Chen, C., Liu, M., Xue, F., Shen, Q., & Zhou, J. (2018b). Controlling the core-shell structure of CuS@CdS heterojunction via seeded growth with tunable photocatalytic activity. *ACS Sustainable Chemistry & Engineering, 6*(11), 15867−15875.

Li, P., et al. (2015a). Hexahedron prism-anchored octahedronal CeO$_2$: Crystal facet-based homojunction promoting efficient solar fuel synthesis. *Journal of the American Chemical Society, 137*(30), 9547−9550.

Li, X., Yu, J., Low, J., Fang, Y., Xiao, J., & Chen, X. (2015b). Engineering heterogeneous semiconductors for solar water splitting. *Journal of Materials Chemistry A, 3*, 2485−2534.

Li, X., Yu, J., & Jaroniec, M. (2016c). Hierarchical photocatalysts. *Chemical Society Reviews, 45*, 2603−2636.

Lian, Z., Sakamoto, M., Vequizo, J. J. M., Kumara Ranasinghe, C. S., Yamakata, A., Nagai, T., & Teranishi, T. (2018). Plasmonic p−n junction for infrared light to chemical energy conversion. *Journal of the American Chemical Society, 141*(6), 2446−2450.

Lin, X., Guo, X., Shi, W., Guo, F., Che, G., Zhai, H., . . . Wang, Q. (2015). Ag$_3$PO$_4$ quantum dots loaded on the surface of leaf-like InVO$_4$/BiVO$_4$ heterojunction with enhanced photocatalytic activity. *Catalysis Communications, 71*, 21−27.

Ling, F., Anthony, O. C., Xiong, Q., Luo, M., Pan, X., Jia, L., . . . Li, Q. (2016). PdO/LaCoO$_3$ heterojunction photocatalysts for highly hydrogen production from formaldehyde aqueous solution under visible light. *International Journal of Hydrogen Energy, 41*, 6115−6122.

Linic, D. B., Christopher, S., & Ingram, P. (2011). Plasmonic-metal nanostructures for efficient conversion of solar to chemical energy. *Nature Materials, 10*, 911−921.

Liu, B., Xu, W., Sun, T., Chen, M., Tian, L., & Wang, J. (2014). Efficient visible light photocatalytic activity of CdS on (001) facets exposed to BiOCl. *New Journal of Chemistry, 38*(6), 2273−2277.

Liu, D., Yao, W., Wang, J., Liu, Y., Zhang, M., & Zhu, Y. (2015). Enhanced visible light photocatalytic performance of a novel heterostructured Bi$_4$O$_5$Br$_2$/Bi$_{24}$O$_{31}$Br$_{10}$/Bi$_2$SiO$_5$ photocatalyst. *Applied Catalysis B: Environmental, 172−173*, 100−107.

Long, J., Wang, W., Fu, S., & Liu, L. (2019). Hierarchical architectures of wrinkle-like ZnFe$_2$O$_4$ nanosheet-enwrapped ZnO nanotube arrays for remarkably photoelectrochemical water splitting to produce hydrogen. *Journal of Colloid and Interface Science, 536*, 408−413.

Long, B., Zheng, Y., Lin, L., Alamry, K. A., Asiri, A. M., & Wang, X. (2017). Cubic mesoporous carbon nitride polymers with large cage-type pores for visible light photocatalysis. *Journal of Materials Chemistry A, 5*(31), 16179−16188.

Low, J., Yu, J., Jaroniec, M., Wageh, S., & Al-Ghamdi, A. (2017). Heterojunction photocatalysts. *Advanced Materials, 29*(20), 1601−1694.

Lu, Z., Zeng, L., Song, W., Qin, Z., Zeng, D., & Xie, C. (2017). In situ synthesis of C-TiO2/g-C3N4 heterojunction nanocomposite as highly visible light active photocatalyst originated from effective interfacial charge transfer. *Applied Catalysis B: Environmental, 202*, 489−499.

Lwin, H. M., Zhan, W., Song, S., Jia, F., & Zhou, J. (2019). Visible-light photocatalytic degradation pathway of tetracycline hydrochloride with cubic structured ZnO/SnO2 heterojunction nanocatalyst. *Chemical Physics Letters, 736*, 136806.

Majhi, D., Bhoi, Y. P., Samal, P. K., & Mishra, B. G. (2018). Morphology controlled synthesis and photocatalytic study of novel CuS-Bi$_2$O$_2$CO$_3$ heterojunction system for chlorpyrifos degradation under visible light illumination. *Applied Surface Science, 455*, 891−902.

Mamba, G., & Mishra, A. K. (2016). Graphitic carbon nitride (g-C$_3$N$_4$) nanocomposites: A new and exciting generation of visible light driven photocatalysts for environmental pollution remediation. *Applied Catalysis B: Environmental, 198*, 347−377.

Mohameda, M. A., Jafar, J., Zain, M. F. M., Minggu, L. J., Kassima, M. B., Rosmi, M. S., . . . Othman, M. H. D. (2018). In-depth understanding of core-shell nanoarchitecture evolution of g-C$_3$N$_4$@C, N co-doped anatase/rutile: Efficient charge separation and enhanced visible-light photocatalytic performance. *Applied Surface Science, 436*, 302−318.

Mohamed, M. A., Jaafar, J., Zain, M. F. M., Minggu, L. N., Kassim, M. B., Rosmi, M. S., Othman, M. H. D., et al. (2018). In-depth understanding of core-shell nanoarchitecture evolution of g-C$_3$N$_4$@C, N co-doped anatase/rutile: Efficient charge separation and enhanced visible-light photocatalytic performance. *Applied Surface Science Volume*, 302−318.

Moniz, S. J. A., Shevlin, S. A., Martin, D. J., Guo, Z. X., & Tang, J. (2015). Visible-light driven heterojunction photocatalysts for water splitting—A critical review. *Energy & Environmental Science*, 8(3), 731−759.

Murdoch, M., Waterhouse, G. I. N., Nadeem, M. A., Metson, J. B., Keane, M. A., & Idriss, H. (2011). The effect of gold loading and particle size on photocatalytic hydrogen production from ethanol over Au/TiO2 nanopartices. *Nature Chemistry*, 3, 489−492.

Narayanan, H., & Viswanathan, B. (2019). *Hydrogen from photo-electrocatalytic water splitting. Solar hydrogen production* (pp. 419−486). Elsevier.

Naseri, S., Samadi, M., Pourjavadi, A., Moshfegh, A. Z., & Ramakrishna, S. (2017). Graphitic carbon nitride (gC$_3$N$_4$)-based photocatalysts for solar hydrogen generation: Recent advances and future development directions. *Journal of Mateials Chemistry A*, 5(45), 23406−23433.

Nowotny, J. (2008). Titanium dioxide-based semiconductors for solar-driven environmentally friendly applications: Impact of point defects on performance. *Energy & Environmental Science*, 1(5), 565−572.

Nowotny, J., Alim, M. A., Bak, T., Idris, M. A., Lonescu, M., Prince, K., & Sigmund, W. (2015). Defect chemistry and defect engineering of TiO$_2$-based semiconductors for solar energy conversion. *Chemical Society Reviews*, 44(23), 8424−8442.

Ong, W. J., Putri, L. K., Tan, Y. C., Tan, L. L., Li, N., Ng, Y. H., & Chai, S. P. (2017). Unravelling charge carrier dynamics in protonated g-C$_3$N$_4$ interfaced with carbon nanodots as co-catalysts toward enhanced photocatalytic CO$_2$ reduction: A combined experimental and first-principles DFT study. *Nano Research, 10* (5), 1673−1696.

Ong, W., Tan, L.-L., Chai, S.-P., & Yong, S.-T. (2015). Heterojunction engineering of graphitic carbon nitride (g-C$_3$N$_4$) via Pt loading with improved daylight-induced photocatalytic reduction of carbon dioxide to methane. *Dalton Transactions*, 1249−1257.

Ong, W., Putri, L. K., Tan, L.-L., Chai, S.-P., & Yong, S-T. (2016). Heterostructured AgX/g-C$_3$N$_4$ (X = Cl and Br) nanocomposites via a sonication-assisted deposition-precipitation approach: Emerging role of halide ions in the synergistic photocatalytic reduction of carbon dioxide. *Applied Catalysis B: Environmental, 180*, 530−543.

Peng, Y., Han, G., Wang, D., Wang, K., Guo, Z., Yang, J., & Yuan, W. (2017). Improved H$_2$ evolution under visible light in heterostructured SiC/CdS photocatalyst: Effect of lattice match. *International Journal of Hydrogen Energy*, 42, 14409−14417.

Petronella, F., Truppi, A., Sibillano, T., Giannini, C., Striccoli, M., Comparelli, R., & Curri, M. L. (2017). Multifunctional TiO$_2$/FexOy/Ag based nanocrystalline heterostructuresfor photocatalytic degradation of a recalcitrant pollutant. *Catalysis Today*, 284, 100−106.

Pirhashemi, M., Habibi-Yangjeh, A., & Pouran, S. R. (2018). Review on the criteria anticipated for the fabrication of highly efficient ZnO-based visible-light-driven photocatalysts. *Journal of Industrial and Engineering Chemistry*, 62, 1−25.

Putri, L. K., Ong, W. J., Chang, W. S., & Chai, S.-P. (2015). Heteroatom doped graphene in photocatalysis: A review. *Applied Surface Science*, 358, 2−14.

Qiu, F., Li, W., Wang, F., Li, H., Liu, X., & Ren, C. (2017). Preparation of novel pn heterojunction Bi$_2$O$_2$CO$_3$/BiOBr photocatalysts with enhanced visible light photocatalytic activity. *Colloids and Surfaces A: Physicochemical and Engineering Aspects*, 517, 25−32.

Ramírez-Ortega, D., Meléndez, A. M., Acevedo-Peña, P., González, I., & Arroyoa, R. (2014). Semiconducting properties of ZnO/TiO$_2$ composites by electrochemical measurements and their relationship with photocatalytic activity. *Electrochimica Acta*, 140, 541−549.

Rashmi, K., Sivakumar, S., & Pala, R. G. S. (2019). Coherency and lattice misfit strain critically constrains electron-hole separation in isomaterial and heteromaterial type-II heterostructures. *The Journal of Physical Chemistry C*.

Rawool, S. A., Pai, M. R., Benerjee, A. M., Arya, A., Ningthoujam, R. S., Tewari, R., & Bharadwaj, S. R. (2018). p-n Heterojunctions in NiO:TiO$_2$ composites with type-II band alignment assisting sunlight driven photocatalytic H2 generation. *Applied Catalysis B: Environmental, 221*, 443–458.

Ren, Y., Zeng, D., & Ong, W. J. (2019). Interfacial engineering of graphitic carbon nitride (g-C3N4)-based metal sulfide heterojunction photocatalysts for energy conversion: A review. *Chinese Journal of Catalysis, 40*(3), 289–319.

Rosman, N., et al. (2018). Photocatalytic degradation of phenol over visible light active ZnO/Ag2CO3/Ag2O nanocomposites heterojunction. *Journal of Photochemistry and Photobiology A: Chemistry, 364*, 602–612.

Rosman, N., Wan Salleh, W. N., Mohamed, M. A., Harun, Z., Ismail, A. F., & Aziz, F. (2020). Constructing a compact heterojunction structure of Ag$_2$CO$_3$/Ag$_2$O in-situ intermediate phase transformation decorated on ZnO with superior photocatalytic degradation of ibuprofen. *Separation and Purification Technology, 251*, 117391.

Ruzimuradov, O., Hojamberdiev, M., Fasel, C., & Riedel, R. (2017). Fabrication of lanthanum and nitrogen-co-doped SrTiO$_3$–TiO$_2$ heterostructured macroporous monolithic materials for photocatalytic degradation of organic dyes under visible light. *Journal of Alloys and Compounds, 699*, 144–150.

Saifuddin, N., Raziah, A. Z., & Junizah, A. R. (2013). Carbon nanotubes: A review on structure and their interaction with proteins. *Journal of Chemistry, 2013*, 18.

Sang, Y., Kuai, L., Chen, C., Fang, Z., & Geng, B. (2014). Fabrication of a visible-light-driven plasmonic photocatalyst of AgVO3@AgBr@Ag nanobelt heterostructures. *ACS Applied Materials & Interfaces, 6*, 5061–5068.

Shao, Y., Cao, C., Chen, S., He, M., Fang, J., Chen, J., . . . Li, D. (2015). Investigation of nitrogen doped and carbon species decorated TiO$_2$ with enhanced visible light photocatalytic activity by using chitosan. *Applied Catalysis B: Environmental, 179*, 344–351.

Sharma, S., Ruparelia, J., & Patel, M. L. (2011). A general review on advanced oxidation processes for waste water treatment. In: *Nirma University International Conference, Ahmedabad, Gujarat*.

Shen, R., Jiang, C., Xiang, Q., Xie, j., & Li, X. (2019). Surface and interface engineering of hierarchical photocatalysts. *Applied Surface Science, 471*, 43–87.

Shijie, L., Hu, S., Zhang, J., Jiang, W., & Liu, J. (2017). Facile synthesis of Fe$_2$O$_3$ nanoparticles anchored on Bi$_2$MoO$_6$ microflowers with improved visible light photocatalytic activity. *Journal of Colloid and Interface Science, 497*, 93–101.

Sun, B., Zhou, W., Li, H., Ren, L., Qiao, P., Li, W., & Fu, H. (2018). Synthesis of particulate hierarchical tandem heterojunctions toward optimized photocatalytic hydrogen production. *Advanced Materials*, 1804282.

Sun, H., Tian, Z., Zhou, G., Zhang, J., & Li, P. (2019). Exploring the effects of crystal facet in Bi$_2$WO$_6$/BiOCl heterostructures on photocatalytic properties: A first-principles theoretical study. *Applied Surface Science, 469*, 125–134.

Sun, X., & Li, Y. (2004). Colloidal carbon spheres and their core/shell structures with noble-metal nanoparticles. *Angewandte Chemie International Edition, 43*(5), 597–601.

Sze, S. M., & Ng, K. K. (2006). *Physics of semiconductor devices*. John Wiley & Sons.

Thomann, M. L., Pinaud, B. A., Chen, Z., Clemens, B. M., Jaramillo, T. F., & Brongersma, M. L. (2011). Plasmon enhanced solar-to-fuel energy conversion. *Nano Letters, 11*, 3440–3446.

Vispute, R., Talyansky, V., Choopun, S., Sharma, R. P., & Venkatesan, T. (1998). Heteroepitaxy of ZnO on GaN and its implications for fabrication of hybrid optoelectronic devices. *Applied Physics Letters, 73*(3), 348–350.

Wang, F., Li, W., Gu, S., Li, H., Liu, X., & Wang, M. (2016). Fabrication of FeWO$_4$@ZnWO$_4$/ZnO heterojunction photocatalyst: Synergistic effect of ZnWO$_4$/ZnO and FeWO$_4$@ZnWO$_4$/ZnO heterojunction structure on the enhancement of visible-light photocatalytic activity. *ACS Sustainable Chemistry & Engineering, 4*, 6288–6298.

Wang, H., Zhang, L., Chen, Z., Hu, J., Li, S., Wang, Z., . . . Wang, X. (2014). Semiconductor heterojunction photocatalysts: Design, construction, and photocatalytic performances. *Chemical Society Reviews, 43*(15), 5234–5244.

Wang, H. P., & He, J. H. (2017). Toward highly efficient nanostructured solar cells using concurrent electrical and optical design. *Advanced Energy Materials*, 7(23), 1602385.

Wang, K., Chen, J. J., Zeng, Z. M., Tarr, J., Zhou, W. L., Zhang, Y., . . . Mascarenhas, A. (2010). Synthesis and photovoltaic effect of vertically aligned ZnO/ZnS core/shell nanowire arrays. *Applied Physics Letters*, 96 (12), 123105.

Wang, L., Yan, X., Xu, C., Xiao, Z., Yang, L., Zhang, B., & Wang, Q. (2011). Photocatalytical reduction of disulphide bonds in peptides on Ag-loaded nano-TiO_2 for subsequent derivatization and determination. *Analyst*, 136(18), 3602−3604.

Wang, S., Xu, X., Luo, H., Cao, C., Song, X., Zhao, J., . . . Tang, C. (2018). Novel $SrTiO_3$/$NaTaO_3$ and visible-light-driven $SrTiO_3$/$NaTaO_3$:N nano-heterojunctions with high interface-lattice matching for efficient photocatalytic removal of organic dye. *RSC Advances*, 8(34), 19279−19288.

Wang, Y., Wang, Q., Zhan, X., Wang, F., Safdar, M., & He, J. (2013). Visible light driven type II heterostructures and their enhanced photocatalysis properties: A review. *Nanoscale*, 5(18) 8326−8339.

Wang, Y., Lu, N., Luo, M., Fan, L., Zhao, K., Qu, J., . . . Yuan, X. (2019). Enhancement mechanism of fiddlehead-shaped TiO_2-$BiVO_4$ type II heterojunction in SPEC towards RhB degradation and detoxification. *Applied Surface Science*, 463, 234−243.

Wangkawong, K., Phanichphant, S., Tantraviwat, D., & Inceesungvorn, B. (2015). $CoTiO_3$/Ag_3VO_4 composite: A study on the role of $CoTiO_3$ and the active species in the photocatalytic degradation of methylene blue. *Journal of Colloid and Interface Science*, 454, 210−215.

Wei, X.-N., Ou, C.-L., Guan, X.-X., Peng, Z.-K., & Zheng, X.-C. (2019). Facile assembly of CdS-reduced graphene oxide heterojunction with enhanced elimination performance for organic pollutants in wastewater. *Applied Surface Science*, 469, 666−673.

Wen, J. Q., Li, X., Liu, W., Fang, Y., Xie, J., & Xu, Y. (2015). Photocatalysis fundamentals and surface modification of TiO_2 nanomaterials. *Chinese Journal of Catalysis*, 2049−2070.

Wu, Y., Wang, H., Tu, W., Liu, Y., Tan, Y. Z., Yuan, X., & Chew, J.-W. (2018). Quasi-polymeric construction of stable perovskite-type $LaFeO_{3/g}$-C_3N_4 heterostructured photocatalyst for improved Z-scheme photocatalytic activity via solid pn heterojunction interfacial effect. *Journal of Hazardous Materials*, 347, 412−422.

Xie, X., Kretschmer, K., & Wang, G. (2015). Advances in graphene-based semiconductor photocatalysts for solar energy conversion: Fundamentals and materials engineering. *Nanoscale*, 7(32) 13278−13292.

Xu, D., Yang, S., Jin, Y., Chen, M., Fan, W., Luo, B., & Shi, W. (2015). Ag-decorated $ATaO_3$ (A = K, Na) nanocube plasmonic photocatalysts with enhanced photocatalytic water-splitting properties. *Langmuir*, 31 (35), 9694−9699.

Yang, J., Wang, D., Han, H., & Li, C. (2013). Roles of cocatalysts in photocatalysis and photoelectrocatalysis. *Accounts of Chemical Research*, 46(8), 1900−1909.

Yang, L., Chu, D., Wang, L., Wu, X., & Luo, J. (2016). Synthesis and photocatalytic activity of chrysanthemum-like Cu_2O/Carbon Nanotubes nanocomposites. *Ceramics International*, 42(2) 2502−2509.

Yang, L., An, Y., Dai, B., Guo, X., Liu, Z., & Peng, B. (2017). Fabrication of carbon nanotube-loaded TiO_2@AgI and its excellent performance in visible-light photocatalysis. *Korean Journal of Chemical Engineering*, 34(2), 476−483.

Yang, Z., Xiang, H., Xin, T., Tao, Y., Xiangli, L., Libin, Y, & Suchong, W. (2016). Fabrication of BiOBr nanosheets@TiO_2 nanobelts p−n junctionphotocatalysts for enhanced visible-light activity. *Applied Surface Science*, 365, 209−217. Available from https://doi.org/10.1016/j.apsusc.2015.12.249.

Zeng, D., et al. (2018a). Toward noble-metal-free visible-light-driven photocatalytic hydrogen evolution: Monodisperse sub −15 nm Ni2 P nanoparticles anchored on porous g-C3N4 nanosheets to engineer 0D-2D heterojunction interfaces. *Applied Catalysis B: Environmental*, 221, 47−55.

Zeng, M. Y., et al. (2018b). Co2P nanorods as an efficient cocatalyst decorated porous g-C_3N_4 nanosheets for photocatalytic hydrogen production under visible light irradiation. *Particle & Particle Systems Characterization, 35*(1), 1700251.

Zhan, W., Sun, L., & Han, X. (2019). Recent progress on engineering highly efficient porous semiconductor photocatalysts derived from metal-organic frameworks. *Nano-Micro Letters, 11*, 1.

Zhang, X., et al. (2017a). Enhanced visible-light-driven photocatalytic performance of Ag/$AgGaO_2$ metal semiconductor heterostructures. *Journal of Alloys and Compounds, 701*, 16−22.

Zhang, Y., et al. (2017b). 3D urchin-like black TiO_2 − x/carbon nanotube heterostructures as efficient visible-light-driven photocatalysts. *RSC Advances, 7*(1), 453−460.

Zhao, H., Pan, F., & Li, Y. (2017). A review on the effects of TiO_2 surface point defects on CO_2 photoreduction with H_2O. *Journal of Materiomics, 3*(1), 17−32.

Zhao, W. R., Wang, Y., Yang, Y., Tang, J., & Yang, Y. (2012). Carbon spheres supported visible-light-driven CuO-$BiVO_4$ heterojunction: Preparation, characterization, and photocatalytic properties. *Applied Catalysis B: Environmental, 115*, 90−99.

Zhao, X., Yang, H., Li, R., Cui, Z., & Liu, X. (2019). Synthesis of heterojunction photocatalysts composed of Ag_2S quantum dots combined with $Bi_4Ti_3O_{12}$ nanosheets for the degradation of dyes. *Environmental Science and Pollution Research, 26*(6), 5524−5538.

Zheng, H., Li, Y., Liu, H., Yin, X., & Li, Y. (2011). Construction of heterostructure materials toward functionality. *Chemical Society Reviews, 40*(9), 4506−4524.

Zheng, J., & Zhang, L. (2018). Incorporation of CoO nanoparticles in 3D marigold flower-like hierarchical architecture $MnCo_2O_4$ for highly boosting solar light photo-oxidation and reduction ability. *Applied Catalysis B: Environmental, 237*, 1−8.

Zhu, Y., Wang, Y., Ling, Q., & Zhu, Y. (2017). Enhancement of full-spectrum photocatalytic activity over $BiPO_4$/Bi_2WO_6 composites. *Applied Catalysis B: Environmental, 200*, 222−229.

Zhang, Y., Xing, Z., Zou, J., Li, Z., Wu, X., Shen, L., . . . Zhou, W. (2017). 3D urchin-like black TiO_{2-x}/carbon nanotube heterostructures as efficient visible-light-driven photocatalysts. *Royal Society of Chemistry, 7*, 453−460.

Zhang, X., Tang, A., Jia, Y., Wang, Y., Wang, H., & Zhang, S. (2017). Enhanced visible-light-driven photocatalytic performance of Ag/$AgGaO_2$ metal semiconductor heterostructures. *Journal of Alloys and Compounds Volume, 701*, 16−22.

Zeng, D., Xu, W., Ong, W., Xu, J., Ren, H., Chen, Y., . . . Peng, D. (2018). Toward noble-metal-free visible-light-driven photocatalytic hydrogen evolution: Monodisperse sub−15 nm Ni_2P nanoparticles anchored on porous g-C_3N_4 nanosheets to engineer 0D-2D heterojunction interfaces. *Applied Catalysis B: Environmental, 221*, 47−55.

CHAPTER 5

Defective photocatalysts

Y.N. Teja[1], R. Mithun Prakash[1], Adhigan Murali[2] and Mohan Sakar[1]

[1]*Centre for Nano and Material Sciences, Jain University, Bangalore, India* [2]*School for Advanced Research in Polymers (SARP)-ARSTPS, Central Institute of Plastics Engineering & Technology (CIPET), Chennai, India*

5.1 Introduction

Photocatalysis is one of the well-established and versatile processes to potentially address the energy and environmental issues through its applications, such as hydrogen production, CO_2 reduction, pollutant degradation, and NH_3 production (Byrnea, Subramanian, & Pillai, 2018). The fundamental mechanism of the photocatalysis is to employ a photoactive material to produce redox species with appropriate energy to split the water molecules, to convert CO_2 into hydrocarbon fuels, and to degrade various organic pollutants under suitable light sources (Nahar, Zain, Kadhum, Hasan, & Hasan, 2017; Reddy et al., 2016; Takata & Domen, 2019). The subtler insights into this process essentially reveal that the band edge position in the photocatalytic material is the key to engineer them for such versatile processes. Accordingly, the band gap engineering in photocatalysts has been done in many ways, which broadly include doping, composite formation, cocatalyst loading, and size and shape modifications (Chiu et al., 2019; Ge, Zhang, Heo, & Park, 2019; Qi et al., 2020; Tong et al., 2018; Wu et al., 2018). The process of band gap modifications through these techniques varies from each other. For instance, the doping process forms new energy states underneath the conduction band (CB) of the photocatalyst and leads to the reduction in the band gap energy and/or shifting of valence band (VB)/CB levels (Qi et al., 2020). In composites, the formation of p−n junctions alters the band edge positions through band bending or direct charge transfer process (Ge et al., 2019). The cocatalyst or metal loading tunes the band interactions by aligning their potentials and provides more catalytic sites and trapping centers, (Wu et al., 2018) while the control over the particle size and shape tunes the band edge positions through their complex dimension dependent influences on the energy structures (Chiu et al., 2019; Tong et al., 2018). In this direction, defects in the photocatalytic materials play an important role in tuning their band energy structures by controlling conduction and valence bands through cationic and anionic defects (Bai, Zhang, Gao, & Xiong, 2018). Besides, defects offer additional features and functions to the photocatalyst by facilitating charge separation and transportations. Compared to the bulk, defects are found to often form more in nanoscale materials and show excellent properties (Xiong, Di, Xia, Zhu, & Li, 2018).

Nanoscale materials fascinate the researchers by their every possible property, such as size, shape, morphology, and stoichiometry-dependent characteristics, which are completely unaware in the bulk state of materials. The manipulation of materials at the nanoscale influences their physical and chemical properties at their atomic and molecular level. This is one of the essential reasons for the emergence of physical and chemical defects in the nanomaterials and these are conventionally known as crystal defects and atomic vacancy defects, respectively (Zhang, Afzal, Pan, Zhang, & Zou, 2019). Defects fundamentally influence the chemical compositions of the nanomaterials, namely cationic and anionic defects (Hui, Zhang, Ni, & Irvine, 2017; Yu et al., 2020). The mechanism of defects in the chemical structures involves modifying the properties of nanomaterials employing influencing the *electronic environment* of the materials. For instance, the control over electronic movements, transitions, and spins substantially influences the conductive, optical and magnetic properties of the materials, respectively. Therefore the cationic and anionic defects in terms of their vacancies create charge inconsistencies and off-stoichiometry in the material and modify their *electronic environment*, thereby altering their properties (Ran et al., 2020). Consequently, the defects in photocatalytic materials play important roles in many ways concerning their band structures, charge transportations, and other features as well (Zhang, Gao, & Xiong, 2019). In this context, this chapter discusses the defect structures in photocatalytic materials in terms of their origins and mechanisms, and then, it describes the synthesis of defective photocatalytic materials. Then, it discusses the effect of defective photocatalytic materials on various applications of photocatalysis. Finally, it gives a conclusion and outlook on the future directions in the development of defective materials for photocatalytic applications.

5.2 Types, origins, and mechanisms of defects in photocatalytic materials

Defects in photocatalytic materials can be broadly classified into three categories, which are (1) cationic defects (mostly metal) and (2) anionic defects (nonmetal) and (3) dual-ion defects (anion−cation) (Liu, Wei, & Shangguan, 2019). Considering the anionic defects, for instance, the oxygen vacancies (OVs) are well-established defective structure and often found to be promising in the photocatalysis. In contrast, cationic defects giving improved p-type conductivity to the material due to holes' movement. Cationic defects lead to the mutual shifting of valance band maximum (VBM) and conduction band minimum (CBM) toward each other and lead to the formation of new intermediate energy states that facilitate the smooth charge transferability (Bai et al., 2018).

5.2.1 Anionic defects

Anionic defects in photocatalytic materials can be achieved with or without foreign elements called "dopants" by either pre- or posttreatment of the materials (Chen, Liu, Yu, & Mao, 2011).

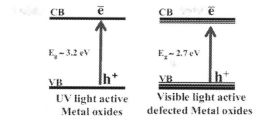

Figure 5.1
Defect-induced alteration in band structures for the conversion of UV-light active metal oxide into visible light active material (Khan et al., 2015).

Anionic defects [typically O (Xu et al., 2020; Yu et al., 2020), N (Xie et al., 2020; Yang, Chu, Jia, Yao, & Liu, 2020), S (Peng, Liu, Zheng, & Fu, 2019; Sun et al., 2020), C (Li et al., 2020; Shen et al., 2019), and halogen (Zhou et al., 2020)] are broadly investigated because of the easy formation and control over the band structure by governing the electronic arrangement and furnishing of a greater number of active centers for the enhanced of photocatalytic reactions (Liu et al., 2019; Zhang et al., 2019). Creating the OVs in metal oxide photocatalysts leads to a reduction in band gap energy as shown in Fig. 5.1 (Khan, Adil, & Al-Mayouf, 2015). Furthermore, it creates the charge-trapping centers on the surface that effectively captures the excited charge carriers and diminishes the charge recombination. This in turn enhances the lifetime of the excitons generated during the photocatalytic reactions. However, the influences of defects in the crystalline lattice toward photocatalytic reactions are still vague that essentially makes the researchers curious for further investigations to understand the mechanism behind the defect engineering (Huo et al., 2019). Engineered oxygen defects adjacent to VB can entice the holes to encourage charge separation and migration (Hu et al., 2020) and also influence the catalytic activity (Shi et al., 2020). The OVs on the surface improve the efficiency further by act as a host for dopants (Hu, Qian, Lin, Ding, & Cui, 2020).

Besides OVs, the nitrogen vacancies (NVs) facilitate a mid-band formation in photocatalysts and narrow down the band gap structure (Liu et al., 2020). N vacancies are often being an influential tool to increase the efficiency of the layered materials, such as g-C_3N_4. The N defects cause the distorted electronic arrangements that lead to fragmented excitons, such as electrons and holes to contribute actively in reaction (Zhou et al., 2020). g-C_3N_4 is an eminent material where both the carbon and NVs can facilitate the reactivity and selectivity toward photocatalytic applications. The carbon vacancies lead to stronger oxidizing capacities (Liu et al., 2020). Similarly, sulfur vacancies in materials have been explored for their efficiency in photocatalysis mainly because of their ability to act as an S^{2-} adsorption active sites toward consumption of holes (Du et al., 2019; Huang, Fang, Yu, Lü, & Cao, 2020). The lattice disorder in sulfur-based materials leads to the absence or dislocation of sulfur defects (Yu, Zhang, & Yakobson, 2015). Though the sulfur vacancies can positively influence, it also leads to a negative impact as these defects form at the interstitial positions (Li, Cheng, Wu, Xu, & Wang, 2019). Recently, defect engineering over halogen-based materials, such as oxyhalides, is getting potential attention due to the synergistic effect of p orbital of halides (X) ions (X: F, Cl, Br, and I) and O $2p$ orbital leading to the reduced band gap in the system. The active sites of halogens can entrap the electrons; thus the recombination will be reduced (Khazaee et al., 2020). Besides the salient redox potentials, the fluorine vacancies facilitate the increased charge transfer rates (He et al., 2020). Defects can be created by the doping of halogen anions, such as Cl and Br. This leads to the formation of OVs on the materials that provide superior photocatalytic activity (Zhang, Zhai, Yu, Su, & Fan, 2020). Furthermore, iodine defects/vacancies concurrently uplift the VBM position of the photocatalyst (e.g., BiOI) and thus the oxidation stability of holes gets improved. The iodine deficiency gives the proficiency to the surface atoms for successive stable photocatalytic reactions (Wang, Zhou, Yin, Zhang, & Liu, 2019; Wei et al., 2019).

5.2.2 Cationic defects

As aforementioned, cationic defects offer an improved p-type conductivity to the materials due to hole movement via a mutual shifting of VBM and CBM; thereby, it leads to the band gap reduction without forming any intermediate energy states (Zhang et al., 2019). But sometimes a new intermediate energy level can also form due to the cationic doping (Neto et al., 2020). The increased hole concentration due to cationic vacancies makes the materials as shallow acceptors instead of donors. The acceptor and donor energy level formation for the photocatalytic water reduction and oxidation is shown in Fig. 5.2 (Reza Gholipour, Dinh, Béland, & Do, 2015).

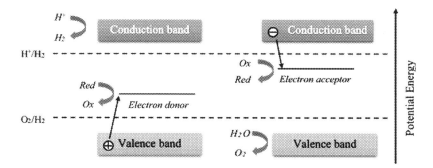

Figure 5.2
Schematic diagrams showing the role of electron donor/acceptor-induced energy levels in the reduction/oxidation reaction toward water splitting (Reza Gholipour et al., 2015).

Cationic defect formation in a material exhibits a range of positive impacts, such as (1) improved electrical conductivity, (2) improved electron mobility for the enhanced interfacial charge transfer that leads to the low recombination rate and high photocatalytic activity, and (3) formation of a greater number of active sites on the surface of the photocatalyst (Bak, Nowotny, Sheppard, & Nowotny, 2008). Meanwhile, sometimes cationic defects can also be ineffective as compared to the anion defects since there is a substantial concentration to overcome the large energy necessity for defect formation and low stability of cations, which can be eventually transformed into defect-free metal oxides (Liu et al., 2019). In these circumstances, numerous approaches have been made for successful defect engineering over cations. For instance, Mohajernia et al. (2020) have proved the presence of two optimized Ti^{3+} defects, which is on the surface and embedded in interstitial position leads to the sudden enhancement over the photocatalytic H_2 generation. In addition, Guo et al. (2020) have analyzed the optimum defect formation in α-Fe_2O_3 that leads to the defects in two facets possibly as large size surface cation defects and smaller size bulk cation defects through different concentrations, where they can develop excellent electron–hole separation process. Furthermore, cation vacancies can extend the absorption range due to the superior number of active sites that concurrently boosts the charge transfer efficiency (He et al., 2019). The improved number of active sites formed via cationic defects also concurrently produces the OVs in the material (Gogoi, Namdeo, Golder, & Peela, 2020). Bellardita et al. (2020) observed the permanent formation of Ti^{3+} active species with OVs and they induced a distorted arrangement in surface charges, which led to the low recombination of charges with enhanced photocatalytic activity.

5.2.3 Dual ionic defects

Developing an ideal photocatalyst consisting of both anion and cation defects has been recognized as a promising strategy rather than a mono (either anion or cation) defective material. The collective influence of cations and anions facilitates the dynamic internal electric field and causes the enhanced separation of charges and thus results in efficient photocatalytic activity (Zhang et al., 2019). Generally, the coexistence of both cation and anion can plausibly form a p–n homojunction having an efficient interfacial charge transfer due to the internal electric field (Liu et al., 2019). Xie, Li, Chen, Jiang, and Zhao (2020) have shown the p–n junction formed between the defective and pristine TiO_2, where it showed the enhanced efficiency in photocatalytic activity due to the band bending across the junction. The formation of p–n junction across the material can able to establish a lower band gap level for smoother energy transportation across the junction, which extends the overall absorption range up to the visible light region (Farooq et al., 2019). Furthermore, Taraka, Gautam, Jain, Bojja, and Pal (2019) have observed that the controlled formation of p-type CuO/n-type ZnO p–n heterojunction showing the improved photocatalytic CO_2 conversion efficiency of ~3855 μmol/g of methanol, which is greater than the pristine materials. Recently, Li group has synthesized a p–n homojunction of rutile (3.02 eV) and anatase TiO_2 (3.20 eV) via hydrogen treatment during calcination. Along with the interfacial charge transfer layer across the junction, the Ti^{3+} defects formed during hydrogen treatment were the reason behind the observed synergistically improved photocatalytic activity (Li et al., 2019).

5.3 Mechanism of defects in photocatalysis

As aforementioned, defect engineering can be one of the most promising strategies that can be positively exploited to overcome the shortfalls of traditional photocatalysts. Different defects offer different pathways toward the enhancement of photocatalytic performance depending on the type and density of the created defects. There are three key roles that are being played by the defects in photocatalysts, which ultimately enhance their photocatalytic performances: (1) enhancing the light absorption ability of photocatalysts, (2) enhancing the transport rate and separation of charge carriers, and (3) enhancing the surface reactions on the photocatalysts.

Light absorption is prime as well as the most crucial step in the photocatalytic process as an ideal photocatalyst is expected to take full advantage of sunlight. Most of the photocatalysts possess a large band gap that axes their ability to be utilized under visible light irradiation. Defects in photocatalysts help to overcome this drawback considerably by narrowing the band gap and extending its light absorption toward visible light (i.e., $420 < \lambda < 780$ nm). The narrowing of band gap is done by the introduction of mid-gap states by defects either below the CB or above the VB, which ultimately results in the reduction of band gap energy of a photocatalyst as shown in Fig. 5.3. These mid-gaps are nothing but new/altered electronic states that are raised by the defects within the band gap structure of a photocatalyst. These mid-gaps essentially form because the bonding at the defect sites is weaker as compared to that of CB and VB states, hence the splitting between bonding and antibonding orbitals is reduced at the defect sites; thereby, it gives rise to the formation of mid-gap states, which

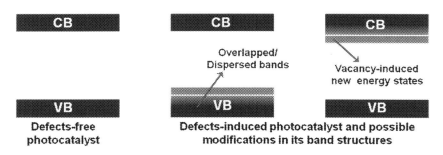

Figure 5.3

General mechanism of defects in alerting the band structure of photocatalysts.

eventually narrows down the band gap leading to the visible light absorption. The position or dispersion of mid-gaps can be tailored by controlling the density of defects and defects' types, such as OV and NV. For example, OVs in TiO_2 resulted in the mid-gaps below the CB edge, (Zhang et al., 2018) while the NVs in g-C_3N4 resulted in the mid-gaps more toward the top of the VB edge (Tu et al., 2017). In addition, the narrowing of band gap occurs when the vacancy-induced mid-gaps overlap with the VB or CB of the photocatalyst, which leads to the downshift of CBM or upshift of VBM. On the other hand, when the defects are created by dopants, the narrowing of band gap is highly influenced by the type of dopants used. For instance, the nonmetal anion dopants result in the formation of mid-gaps on the top of VB or they tend to directly extend the VBM in contrast to metal cation dopants, which tend to form impurity states (of the doped metals) below the CB lowering the CBM or above the VB promoting the VBM or opting for the simultaneous actions.

These mid-gaps often act as "trap states" for the photogenerated charge carriers as the undesired recombination of electron—hole pairs reduces the photocatalytic performance greatly. Mid-gaps can also act as recombination centers depending on the concentration of defects. For instance, bulk vacancies are capable of trapping holes through electrostatic interaction and these holes may act as the recombination centers by recombining with excited electrons. On the other hand, the holes trapped by the surface vacancies react with the adsorbed electron donors before recombining with the excited electrons, thus achieving the charge carrier's separation. The defects can also help in transporting the charge carriers to the surface of the material through a built-in field. The vacancies often create mid-gaps above the VB and below the CB resulting in the formation of type-1 heterojunction between surfaces and bulk, which led to the migration of charge carriers toward the surface of photocatalyst, thus enhancing the photocatalytic properties.

Surface defects are thermodynamically unstable, which essentially means that they highly favor the adsorption and activation of reactant molecules. The defects, such as point defects, can often participate directly in the activation of surface reactants along with the charge carriers that are also localized at the surface defect sites toward the activation of adsorbed molecules. Also, the selectivity of the end-product is possible by surface defects as they offer diverse adsorption and dissociation routes resulting in the formation of various reaction lanes that make it possible for the selectivity of end-product, such as in the case of CO_2 reduction to produce a specific product. Therefore defects must be carefully and systematically designed to fulfill the aforementioned steps one after the other or simultaneously as they are correlated and complementary to each other in devising better photocatalytic performances.

5.4 Synthesis of defective photocatalysts

Among various steps involved in the defect engineering, the synthesis methods of defective material are the most stringent step. The photocatalytic mechanism of the defective materials highly depends on the relationship between the defects created and their photocatalytic performance and this is possible only when the materials with well-defined defects are synthesized. Accordingly, various synthetic methods have been proposed and developed, which can be majorly divided into two categories: introducing defects through (1) pretreatment synthesis and (2) posttreatment synthesis. The former involves the introduction of defects into the preformed host material, while the latter involves the creation of defects during the synthesis of the materials. In the pretreatment synthesis approach, the formation of defects is independent of the growth kinetics of materials. Different kinds of defects, such as vacancies, lattice disorders, voids, and dopants, can be incorporated using the pretreatment synthesis method. In the second approach, the formation of defects will occur during the formation of photocatalytic material, which is the most widely used method, where the bulk defects are the most readily produced defects through this process. These defects typically include dislocations and boundary defects. Accordingly, in this section, we have discussed extensively the strategies that are significantly used for the fabrication of defective photocatalytic materials.

5.4.1 Thermal treatment method

Thermal treatment is one of the conventional and key methods used for the synthesis of defect-rich photocatalytic materials. Thermal treatment is carried out under inert atmospheres, such as Ar, (Chen et al., 2020; Gao et al., 2020) H_2, (Chatzitakis & Sartori, 2019; Liu et al., 2019; Naldoni et al., 2012; Yin et al., 2019) N_2 (Irani et al., 2020; RamezaniSani, Rajabi, & Mohseni, 2020), and NH_3 (Cong et al., 2020; Hara et al., 2003), and the temperature is controlled as desired based on the type of defect in this method. Vacancies are formed when the atoms escape from the lattice due to the bond cleavage caused due to the high-temperature treatment. Especially, OVs are created in the materials through this method as it is well established that an oxygen-depleted environment under elevated temperatures will lead to the rich defective materials with OVs. Thermal treatment has been exploited to create OVs in various photocatalytic materials, such as TiO_2 (Chen et al., 2011; Han, Su, Liu, Ma, & An, 2018; Katal et al., 2019; Li et al., 2019), ZnO (Wang et al., 2012; Wang, Chen, Xiang, & Komarneni, 2018), and WO_3 (Liu et al., 2018; Meng et al., 2018). From the pioneering work of Chen et al. (2011) a high-temperature hydrogen treatment is the most widely used approach for creating the OVs in various metal oxide systems. In their work, they conceptually showed the disordered surface layered nanophase TiO_2 through hydrogenation to improve the solar light absorption. In the hydrogenation process, the first step involves the physical interaction of TiO_2 with hydrogen and followed by, in the second step, the electron transfer occurs from the adsorbed hydrogen to the oxygen atoms in TiO_2 lattice. Finally, the OVs are formed when the lattice oxygen atoms are abstracted from surface of TiO_2, where O atoms leave with H atoms to form H_2O leading to the formation of OVs. The further increase in temperature will lead to the transfer of electrons from H atoms to Ti^{4+} of TiO_2 and Ti^{3+} defects are formed. When the temperature is increased above 500°C, there is a chance for the formation of interstitial Ti defects in the TiO_2 matrix, which occurs due to the transfer of electrons from the OV states to Ti^{4+} leading to a decrease in the intensity of OVs and an increase in Ti^{3+} until the temperature reaches to 600°C. Anatase TiO_2 is more favorable to the formation of OVs than the Ti interstitials as compared to rutile phase TiO_2 that largely favors the formation of Ti interstitials (Morgan & Watson, 2010).

The hydrogenation process also leads to the change of the white color of TiO_2 into black, which greatly extends the light absorption region and boosts the photocatalytic activity extensively. Using thermal treatment, Katal et al. (2019) triggered OVs in TiO_2 by posttreatment under vacuum and hydrogen atmospheres at different temperatures. TiO_2 was initially heated in a tube furnace under the flow of pure hydrogen gas at 400°C for 4 hours and on the other side; another set of the sample was prepared by treating in a vacuum at 400°C for 4 hours. The presence of OVs was confirmed by electron paramagnetic resonance spectroscopy analysis. Furthermore, the observed color change in the samples from white to dark gray after the treatment under hydrogen and vacuum, respectively, further confirmed the formation of OVs. As a result, an enhanced photocatalytic efficacy was also observed, which attributed to the OVs that as electron trapping sites. Gong et al. synthesized WO_3 single crystal nanosheets with rich-OVs via posttreatment under vacuum/H_2 atmosphere, where the yellowish WO_3 powder was turned into olive drab after posttreatment and the enhancement of photocatalytic activity was attributed to the induced plasmon resonance by OVs, which effectively promoted the light-harvesting ability in WO_3 nanosheets (Yan et al., 2015). In another example, Huang et al. employed hydrogenation to synthesize $BiFeO_3$ with surface OVs by heating the pristine $BiFeO_3$ under the H_2 atmosphere with high pressure for 8 hours at various temperatures (Chen et al., 2017).

Other than OVs, hydrogenation can also be utilized for the formation of NVs. Liu et al. synthesized homogeneous self-modification graphitic carbon nitride (g-C_3N_4) with NVs via the hydrogenation process (Niu, Yin, Yang, Liu, & Cheng, 2015). First, g-C_3N_4 was prepared in the conventional way of calcinating the dicyandiamide in air and then the as-prepared g-C_3N_4 was subjected to hydrogenation for 2 hours at various temperatures, but the optimized temperature was found to be between 520°C and 540°C to obtain g-C_3N_4 with self-modification NVs without the destruction of the original layered structure, where the H_2 treatment reduced the lattice N into NH_3 to form NVs. A narrowing of band gap from 2.78 to 2.03 eV was also observed in the material, which significantly boosted the photocatalytic activity. Again in the case of OVs, apart from hydrogenations, other inert gas environments have also been used for the creation of OVs. Chen et al. prepared two predominantly exposed facets {101} and {010} anatase TiO_2 and created the OVs by calcination at 500°C under the argon (Ar) atmosphere and air (Li et al., 2019). It was observed that calcination under the Ar atmosphere showed more favorable conditions to form OVs than the air atmosphere for both facets and the major reason being the dissociation of surface oxygen atoms and removal of surface hydroxyl groups during interaction with Ar. Chen et al. (2020) introduced surface OVs by subjecting the ultrathin 2D ZnO nanosheets to a mixed atmosphere of 25% hydrogen and 75% Ar in volume, which led to an improvement in visible light absorption of target photocatalyst thus enhancing the photocatalytic activity as well. Similar to H_2 annealing, treatment under N_2 and NH_3 also leads to the formation of OVs. One such example was described by Fukushi, Sasaki, Hirabayashi, and Kitano (2017), where the WO_3 nanoparticles were treated under the flow of 10% H_2 and 90% N_2 atmosphere to induce OVs at a temperature range of 800–900°C. RamezaniSani et al. (2020) synthesized the nitrogen-doped multi-phase TiO_2 nanowires, in which the hydrothermally synthesized TiO_2 nanowires were annealed in the N_2 atmosphere for 3 hours at a temperature between 500°C and 700°C. In another study, Cong et al. (2020) prepared an N-doped ceria catalyst by the NH_3-annealing method. The N-doped CeO_2 with two different morphologies (such as coral-like and flower-like) were obtained, where the N-doped coral-like CeO_2 was obtained by annealing the flower-like CeO_2 under NH_3 atmosphere at 900°C for 2 hours. The rich OV phase was formed due to the doping of N atoms that caused the charge imbalance leading to the unsaturation of chemical bonds in CeO_2, thereby leading to the formation of more OVs.

Graphitic carbon nitride (g-C_3N_4) is one of the most explored photocatalysts in defect engineering as it is considered that the introduction of defects into the polymeric network of g-C_3N_4 is relatively simple as compared to other photocatalysts. Wang et al. (2020) introduced the NVs in g-C_3N_4 by annealing the urea and oxalyl dihydrazide (ODH) together in the air atmosphere at 550°C for 4 hours after mechanically grinding them. In this process, ODH was used as a controlling agent to produce NVs in g-C_3N_4 with increased crystallinity. Another example of introducing NVs into g-C_3N_4 involved the synthesis of g-C_3N_4 nano-fragments under the N_2 atmosphere at a temperature of 650°C for various time ranges from 2 to 8 hours (Liang et al., 2020). The introduction of NVs into tri-s-triazine units of g-C_3N_4 formed a mid-energy level and reduced the band gap to enhance the light-harvesting capacity by reducing the CB potential. Ultrathin g-C_3N_4 with enriched surface carbon vacancies was synthesized by Zhang et al. (2019) in the two-step calcination process. In the first step, 2.4 g of bulk g-C_3N_4 was heated for 100 minutes at 500°C and was kept for 2 hours in a muffle furnace and in the second step, 1.2 g of the obtained sample was calcined again under 530°C in a muffle furnace for 2 hours after which the final product of porous white powder with surface carbon vacancies was obtained. Holey g-C_3N_4 nanosheets with carbon vacancies were synthesized by Liang, Li, Huang, Kang, and Yang (2015) via the thermal treatment of bulk g-C_3N_4under NH_3 atmosphere, where the bulk g-C_3N_4 was exfoliated into nanosheets with abundant in-plane holes along with rich carbon vacancies in a single step. In contrast to conventional thermal treatment under N_2, H_2, and NH_3 atmosphere to produce vacancies, Li, Ho, Lv, Zhu, and Lee (2017) induced carbon vacancies Ing-C_3N_4 nanosheets under CO_2 atmosphere where the sodium bicarbonate was used as the source to produce carbon atmosphere.

5.4.2 Chemical reduction method

Chemical reduction is another method that is widely used for the synthesis of defective photocatalysts. By using reducing agents, such as $NaBH_4$ (Tan et al., 2014; Xue et al., 2019), KBH_4 (Zhang et al., 2005), and N_2H_4 (Fageria, Gangopadhyay, & Pande, 2014). Solid-state defects can be introduced in photocatalysts. Zhang, Gao, and Chen (2019) synthesized nitrogen defective g-C_3N_4 using solid-state chemical reduction, where the pristine g-C_3N_4 along with $NaBH_4$ was grounded together and was heated in a tubular furnace at a temperature of $350°C-500°C$ in a vacuum and then treated at $80°C$. The optimum temperature for the formation of nitrogen defects rich g-C_3N_4 was found to be $400°C$, which showed the highest photocatalytic activity as compared to other temperatures. In another example, Liu et al. (2016) synthesized oxygen-deficient blue TiO_2 nanocrystals with coexposed (101) and (001) facets using the chemical reduction method. In this process, first, the TiO_2 with exposed facets was prepared using the hydrothermal method and then the as-prepared samples along with $NaBH_4$ in a ratio of 5:1 were ground together and then calcined in a tube furnace in an Ar atmosphere at $300°C$ during which the $NaBH_4$ is decomposed and reduced Ti^{4+} to Ti^{3+}, thereby producing the OVs. Another example of using $NaBH_4$ as a reducing agent involved the synthesis of colored $TiO_2@TiO_{2-x}$, where OVs were created by calcination of TiO_2 with $NaBH_4$ at various temperatures ranging from room temperature to $500°C$ (Tan et al., 2014). In this method, it is observed that by tuning the reaction time and temperature, a series of samples from light blue to black can be prepared.

Other than the above-mentioned reducing agents, the reducing solvents, such as ethanol, methanol, or ethanolene glycol, have also been used to introduce defects in photocatalysts via in situ chemical reduction process (Gao et al., 2017; Jiang, Zhang, Li, He, & Yin, 2013; Li et al., 2019; Wu et al., 2019; Zhao et al., 2019). For instance, Li et al. (2019) synthesized defective $Zn_2In_2S_5$ nanosheets with rearranged sulfur vacancies and sulfur interstitials without introducing any external elements with the employment of ethanol reducing solvent and gained a 13-fold increase in the H_2 photocatalytic evolution. One more such utilization of reducing solvent involved the synthesis of vanadium-rich vacancy o-$BiVO_4$ layers, where ethylene glycol was used as the reducing agent (Gao et al., 2017). g-C_3N_4 with NVs was synthesized by Wu et al. (2019) via one step $N_2H_4 \cdot H_2O$-assisted thermal polymerization method. The introduction of NVs was supposed to be due to the decomposition of $N_2H_4 \cdot H_2O$ as it is already well known to be a reductant leading to form reductive species, such as H_2 and NH_3, thereby forming the NVs. OVs were introduced into copper-doped coconfined TiO_2 nanosheets using the ethanol reduction strategy by Zhao et al. (2019), which boosted the photocatalytic N_2 fixation drastically. As one more example, BiOCl with OVs was synthesized using the reductive ethylene glycol, which reacted with the oxygen-terminated (001) surface of BiOCl leading to the formation of OVs (Jiang et al., 2013).

5.4.3 Force-induced methods

Force-induced methods, such as plasma etching, liquid-ammonia-assisted lithiation strategy, ultraviolet irradiation, and ball milling, have also been documented for the synthesis of defective photocatalysts. Plasma etching is a powerful method to introduce intrinsic defects in photocatalysts. Li et al. (2016) synthesized sulfur vacancies in basal planes of monolayer MoS_2 using mild Ar plasma. In this study, they observed that the number of sulfur vacancies can be controlled by tuning the exposure time of MoS_2 in Ar plasma and the vacancies helped in tuning the band structure of MoS_2, resulting in enhanced H_2 evolution. In another example, Xu et al. (2016) synthesized the plasma-engraved Co_3O_4 nanosheets with OVs and high surface area for the oxygen evolution reaction, where the pristine Co_3O_4 nanosheets deposited on TiO_2 substrate were exposed to Ar plasma etching. Zhang et al. (2018) synthesized sulfur vacancies on the Zn facet in the monolayered $ZnIn_2S_4$-utilizing lithiation-chemistry approach where MoS_2 quantum dots were positioned to form a heterostructure for photocatalytic H_2 production. Similarly, the ball milling method was used to destroy the structure and create defects in the target materials. One of the examples of this method involved the synthesis of defect-rich $BiPO_4$ rods where bismuth vacancies and OVs were created in $BiPO_4$ rods (Zhu, Ling, Liu, Wang, & Zhu, 2016). This method is also employed majorly for the synthesis of surface defects and by controlling the ball milling time and power, through which it possible to control the concentration of defects but not the type of defects. Zhao et al. (2013) created OVs in BiOCl single-crystalline nanosheets on the exposed facets of {001} and {010} by subjecting the target surface under UV-light irradiation. Interestingly, Ye et al. employed the UV-light irradiation method to create OVs in BiOCl nanosheets and concluded that the reason for the creation of vacancies was due to the low bond energy and long bond length of Bi−O bond in BiOCl, which made it easy to break the Bi-O bonds under high intense UV-light irradiation leading to the formation of OVs on surface of BiOCl nanosheets (Ye, Zan, Tian, Peng, & Zhang, 2011; Ye et al., 2012). However, this method is not employable for various oxides as most of the oxide materials possess strong bond energy, which makes it difficult to break the metal−oxygen bonds to create vacancies.

5.4.4 Other methods

Apart from the above discussed methods, other strategies, such as ultrasonication, vacuum activation, atomic layer deposition, hydrothermal, and postsynthetic etching methods, have also been used for the creation of defective photocatalysts. One of such examples involves the utilization of three methods one after the other in a row to improve the concentration of defects. By following this concept, Yu et al. (2020) synthesized TiO_2 with a certain concentration of OVs by simple hydrothermal method and then the as-prepared sample was subjected to ultrasonication that further increased the OVs, which eventually enhanced the photocatalytic

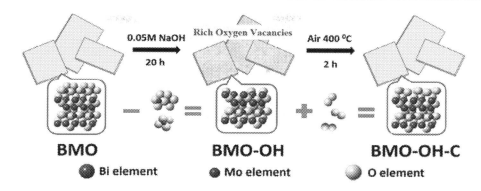

Figure 5.4
Schematic illustration of the synthesis process of these Bi_2MoO_6 nanosheet samples (Chen et al., 2018).

performance drastically. Similarly, the hydrothermally synthesized Bi_2MoO_6 nanosheets were subjected to a postsynthetic etching process using NaOH at different concentrations to create varying numbers of OVs in Bi_2MoO_6 (BMO) nanosheets as shown in Fig. 5.4. The formation of OVs was proposed as follows. Upon the addition of NaOH solution, it partially removed the perovskite-like layer of $(MoO_4)^-$ in BMO (Chen, Yang, Gao, Sun, & Li, 2018). This eventually led to the emergence of charge imbalances in the system, and to reestablish a new electron-neutrality equilibrium in the system, the OVs were created to loosen the $(Bi_2O_2)^{2+}$ layers to neutralize these emerged charge imbalances (BMO-OH). However, upon further annealing of this material (BMO-OH) at 400°C, the formed OVs were found to be reduced to form relatively defect-reduced material BMO-OH-C as shown in Fig. 5.4, through which the OV-defect concentrations were tuned.

The method, such as atomic layer deposition, was used to produce surface OVs on TiO_2 photoelectrodes (Li et al., 2018). In the phase transformation via fast heating process, initially, the precursors of the target sample will be heated in advance at high temperatures and then subjected to calcination at required temperatures. This eventually leads to huge temperature differences driving the exorbitant changes in the phase of the material and finally leading to vacancies in the system. For instance, In_2O_3 with OVs was synthesized using this method, where the $In(OH)_3$ was subjected to fast heating at 400°C for 3 minutes followed by decomposition paved way for the formation of In_2O_3 with porous structure along with OVs (Lei et al., 2014).

5.5 Effect of defective photocatalysts in their applications

5.5.1 Pollutant degradations

Organic pollutants and harmful dyes from various industries are one of the major sources of water pollution. Therefore degradation or conversion of these pollutants into nontoxic form is a must and they require immediate action as well. In this direction, defect-rich photocatalysts can be effectively utilized toward such applications.

Defective $BiFeO_3$ (BFO) with surface OVs was synthesized by Chen et al. through the hydrogenation method under high pressure where the increasing hydrogenation temperature resulted in an increase in OVs and was employed for the degradation of methyl orange (MO) (Chen et al., 2017). The pristine BFO was subjected to hydrogenation at various temperatures and the final samples were denoted as HB-T−P−t, where T, P, and t were the reaction temperature (T = 120, 150, 180, 200°C), pressure (P = 2.0 MPa), and hydrogenation time (t = 8 hours), respectively. The presence of OVs showed a significant influence on the optical properties of BFO as shown in Fig. 5.5A. From UV−vis diffuse reflectance spectra (Drs) (as shown in Fig. 5.5A), it can be observed that the light absorption band of BFO is expanded after the hydrogenation and it can also be seen that the color of BFO is transformed from brownish yellow to brownish black with increasing hydrogenation temperature, which also indicated the further confirmation of light absorption in the visible light region. Accordingly, from the Tauc plot shown in Fig. 5.5B, it can clearly be seen that the hydrogenation has

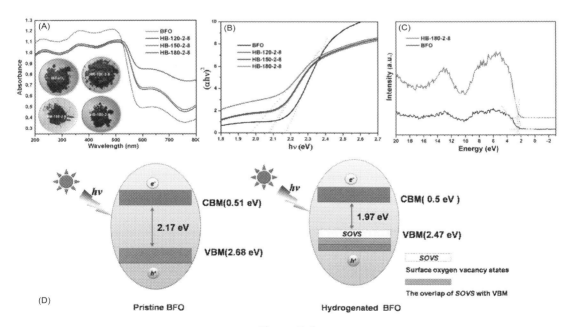

Figure 5.5

(A) UV−Vis diffuse reflectance spectra of pristine BFO and hydrogenated samples (Insets are photographs of pristine BFO and hydrogenated powders), (B) Tauc plots to determine the band gap energy of the samples. (C) Photocatalytic degradation efficiency of MO by different photocatalysts under visible light irradiation. (D) Illustration of energy bands for the pristine BFO and hydrogenated BFO with oxygen vacancies (Chen et al., 2017). *BFO*, Defective $BiFeO_3$; *MO*, methyl orange.

narrowed the band gap from 2.17 eV of pristine BFO to 1.97 eV of hydrogenated BFO (HB-180$-$2$-$8). Along with the alteration in band structure, the surface OVs also acted as trapping centers for photogenerated electrons, which helped the separation of photogenerated electron$-$hole pairs and thereby suppressed the recombination of electron$-$hole pairs as well as produced the predominant active species (hydroxyl radicals, •OH) with H_2O on the hydrogenated BFO surface to effectively degraded target pollutant MO. The photocatalytic degradation performance of BFO in visible light irradiation before and after hydrogenation is shown in Fig. 5.5C, from which it can be clearly seen that the hydrogenation has enhanced the photocatalytic ability drastically that has degraded MO dye almost completely in 240 minutes under visible light irradiation.

In another study, the synthesis of novel carbon and defect comodified carbon nitride (C_xCN) with porous structure through calcination process was developed for the degradation of Bisphenol A (BPA) under visible light irradiation (Wu et al., 2020). Fig. 5.6A shows the UV$-$vis Drs of C_xCN, in which a redshift in all C_xCN samples was observed that represented the presence of carbon and defects modified the band structure of g-C_3N_4. Accordingly, from Fig. 5.6B, the band gap energy of pristine g-C_3N_4 and $C_{1.0}CN$ was calculated to be 2.69 and 2.57 eV, respectively, which further confirmed the modification of the band structure of g-C_3N_4. The photocatalytic performance of C_xCN samples is shown in Fig. 5.6C, where the $C_{1.0}CN$ completely degraded the BPA in 90 minutes of irradiation.

The observed defect-mediated photocatalytic mechanism was proposed as follows. The holes from the defective $C_{1.0}CN$ were essentially possessed stronger oxidation capability for the easier degradation of BPA as the VB of $C_{1.0}CN$ is higher than that of g-C_3N_4, which can be seen from the band energy diagram as well as the band gap of $C_{1.0}CN$ is also less than that of g-C_3N_4 as shown in Fig. 5.7A. This promoted the production of electron$-$hole pairs under visible light irradiation and also the recombination rate of electron$-$hole pairs

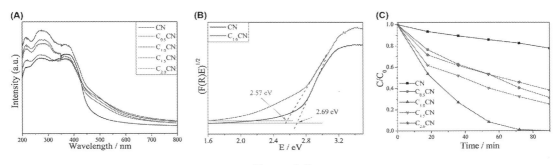

Figure 5.6
(A) UV$-$vis diffuse reflectance spectra of pristine g-C_3N_4 and C_xCN. (B) Tauc plots to determine the band gaps for each sample. (C) Photocatalytic degradation of BPA by g-C_3N_4 and $C_{1.0}CN$ samples under visible light irradiation (Wu et al., 2020).

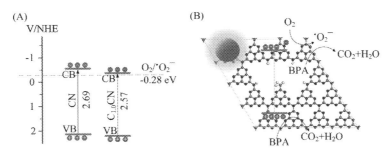

Figure 5.7

(A) The band energy of g-C_3N_4 and $C_{1.0}$CN. (B) Photocatalytic mechanism of $C_{1.0}$CN for degradation of BPA solution under visible light irradiation (Wu et al., 2020).

Table 5.1: Photocatalytic degradation efficiency of various defective photocatalysts.

Photocatalyst	Defect type	Organic pollutant/dye	Degradation time (min)	Efficiency of degradation (%)	Reference
$BiFeO_3$	Oxygen vacancies	MO	240	100	Chen et al. (2017)
C_xg-C_3N_4	Carbon defects	Bisphenol A	90	100	Wu et al. (2020)
ZnO	Oxygen vacancies	MB	50	99	Zhang et al. (2018)
BiOI	Oxygen vacancies	MB	60	99	Zhang et al. (2018)
α-Fe_2O_3/d-C_3N_4	Carbon defects	TOC/RhB/MB/MO	400	97.1/99.1/97.4/94.0	Wang et al. (2020)
ZnO hexagonal plates	Oxygen vacancies	Tetracycline	80	99.9	Xu et al. (2020)
ZnO nanostructures	Oxygen vacancies	RhB	70	97.75	Nandi and Das (2019)
BiO_{2-x}/$Bi_2O_{2.75}$ Z-Scheme	Oxygen vacancies	RhB	240	90.34	Wang et al. (2019)
B-TiO_2	Oxygen vacancies	Phenol	6 h	55	Han et al. (2018)
ZnO	Oxygen vacancies	2,4-Dichloropheno	180	88	Wang et al. (2012)
Bi_2MoO_6/g-C_3N_4 2D/2D heterostructure	Oxygen vacancies	Ciprofloxacin	120	~90	Li et al. (2020)

MO, methyl orange.

of $C_{1.0}$CN is also lower that collectively resulted in the production of more superoxide radicals ($\cdot O_2^-$). Furthermore, these generated superoxide radicals ($\cdot O_2^-$) and holes in defective-$C_{1.0}$CN effectively degraded the BPA into CO_2 and H_2O as shown in Fig. 5.7B. Similarly, various defective photocatalyst systems, their defect types, and degradation efficiencies are given in Table 5.1.

5.5.2 Water splitting

Water splitting is one of the most promising methods to produce clean and sustainable energy. Even though it is an uphill reaction facing various challenges, recently, variety of defective photocatalysts have been successfully demonstrated for the overall water splitting to produce H_2 and O_2 energy and a few of such photocatalysts and their applications toward hydrogen evolution reaction and oxygen evolution reaction have been discussed in this section.

The holey g-C_3N_4 nanosheets (HGCNs) with carbon vacancies with abundant in-plane holes were synthesized by decomposing the as-prepared bulk g-C_3N_4 nanosheets (BGCN) in NH_3 environment. The optimized temperature for the synthesis of the sample with rich carbon vacancies along with a good concentration of holes was observed to be in the range of 490°C–520°C (Liang et al., 2015). The photocatalytic hydrogen evolution of HGCN was found to be 20-fold enhanced as compared to BGCN. This increase in photocatalytic performance was attributed mainly to two reasons. First, the in-plane holes decrease the van der Waals interactions by providing more boundaries as well as providing new active edges and cross-plane diffusion channels, where they significantly increased the transfer and diffusion of photogenerated electron–hole pairs thus enhancing the photocatalytic performance. The second reason was that the carbon vacancies altered the band structure of HGCN and made the light absorption band extended toward near-infrared (Fig. 5.8A and B). Furthermore, the carbon vacancies reduced the recombination of excited charge carriers improved the photocatalytic activity remarkably under visible light. The photocatalytic evolution of hydrogen of HGCN was found to be 82.9 µmol/hour as shown in Fig. 5.8C, which was around 20-fold higher than that of BGCN (4.2 µmol/hour) under visible light irradiation for 3 hour.

Tungsten oxide (WO_3) with enhanced OVs has been demonstrated as one of the suitable photocatalysts for OERs. The defective-WO_3 was synthesized using the solvothermal method by varying the volume of HCl, denoted as W1, W2, W3, W5, where the numbers

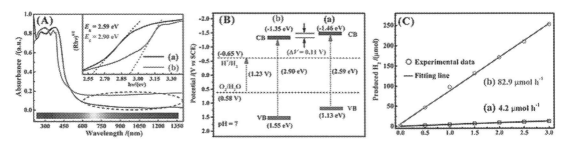

Figure 5.8
(A) Diffuse reflectance spectra (Drs) and Tauc plots, (B) band gap structure, and (C) photocatalytic evolution of hydrogen (a-BGCN, b-HGCN) (Liang et al., 2015).

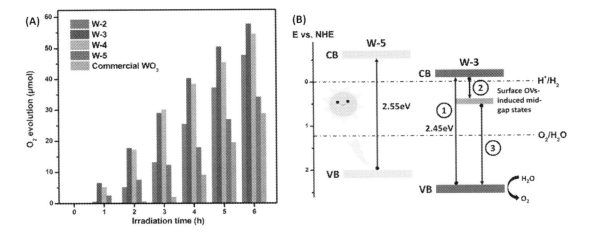

Figure 5.9
(A) Photocatalytic O_2 evolution of all samples under the illumination of simulated solar light in 6 h. (B) Illustration of proposed O_2 evolution reaction scheme for W-5 and W-3 (Chew et al., 2019).

represented the volume of HCl used (Wang et al., 2020). During the synthesis process, the concentration of OVs on the surface of WO_3 was tuned by adjusting the amount of HCl, where the highest amount of OVs was created when the volume of HCl was 3 mL. As a result, the highest O_2 evolution of 57.6 μmol in 6 hours was obtained on W3, while the lowest O_2 evolution was obtained on W5 as shown in Fig. 5.9A. The reduced band gap energy and the enhanced photocatalytic efficiency of this defective system were attributed to the formation of surface OVs-induced mid-energy states underneath the CB of W3 as shown in Fig. 5.9B.

It should be noted in W3 that the surface defects essentially act as trapping centers, in which the excited electrons are trapped and thus prevented the recombination of charge carriers, leading to the enhancement of photocatalytic activity (Fig. 5.9B). Accordingly, in the case of W3, the surface OVs induced the additional mid-gap states that offered additional energy states for the effective transfer of photoexcited electrons rather than directly recombining with holes in VB. Hence, the surface defects on W3 were concluded to be the primary reason for the observed improved photocatalytic performance by promoting the charge carrier separation in the system. Similarly, a list of defective photocatalysts and their water-splitting reaction efficiencies are listed out in Table 5.2.

5.5.3 CO_2 reduction

Photocatalytic CO_2 conversion is a greener approach to convert CO_2 into useful carbon-containing fuels and accordingly, the defect-rich photocatalysts have been successfully employed for such applications as well. In this direction, Z-scheme

Table 5.2: Photocatalytic H_2 and O_2 evolution by various defect-engineered photocatalysts.

Photocatalyst	Defect type	Type of reaction	Yield (μmol/g per h)	Reference
Holey C/g-C_3N_4	Carbon vacancies	HER	82.9 μmol/h	Liang et al. (2015)
g-C_3N_4 (urea)	Carbon vacancies	HER	830.94	Jing et al. (2020)
g-C_3N_4 (melamine)			556.79	
sc-TiO_2-N_2	Oxygen vacancies	HER	196.0	Weng, Zeng, Zhang, Dong, and Wu (2016)
TiO_2/Cu_2O	Oxygen vacancies	HER	32.6 mmol	Wei et al. (2019)
TiO_2	Atomic defects	HER	4.543 mmol/g per h	Li et al. (2020)
g-C_3N_4	Nitrogen vacancies	HER	3166	Liang et al. (2020)
g-C_3N_4	Nitrogen vacancies	HER	5833.1	Wang et al. (2020)
ZnO	Oxygen vacancies	HER	1665	Bak and Kim (2018)
MoS_2/$ZnIn_2S_4$	Sulfur vacancies	HER	6.884 mmol/g per h	Zhang et al. (2018)
WO_3	Oxygen vacancies	OER	57.6 μmol	Chew et al. (2019)
BiOI	Oxygen vacancies	OER	199.26	Ji et al. (2019)
CeO_2	Oxygen vacancies	OER	137.26	Zhang et al. (2018)
BiOCl (010)	Oxygen vacancies	OER	100	Li et al. (2016)
BiOCl (001)			32	
CeO_2	Oxygen vacancies	OER	353	Zhao et al. (2015)
WO_3	Oxygen vacancies	OER	376	Liu et al. (2013)
WO_3	Oxygen vacancies	OER	120 μmol/m^2 per h	Wang et al. (2018)
WO_3	Oxygen vacancies	OER	500 μmol/m^2 per h	Meng et al. (2018)
Mg-TiO_2	Oxygen vacancies	OWS	H_2—850 O_2—425	Gao et al. (2017)
Graphene/$SrTiO_3$	Lattice defects	OWS	H_2—153 mmol/g per h O_2—74 mmol/g per h	Mateo, García-Mulero, Albero, and García (2019)

HER, hydrogen evolution reaction; *OER*, oxygen evolution reaction; *VB*, valence band; *TOC*, total organic carbon; *RhB*, rhodamine B; *MB*, methylene blue; *OWS*, overall water splitting.

photocatalyst is one of the strategies developed for the effective charge separation for the enhancement of photocatalytic activity. Huang et al. reported CO_2 reduction using a defect-engineered Z-scheme-based WO_{3-x}/g-C_3N_4 system, in which the OV-mediated WO_3 nanorods were synthesized by vacuum thermal treatment and the Z-scheme system by chemisorption process (Huang, Long, Ruan, & Zeng, 2019).

The photocatalytic CO_2 reduction mechanism of the synthesized system is shown in Fig. 5.10A, where it was predicted in two ways, one being the traditional heterojunction route and the other one being the Z-scheme. If the heterojunction pathway is taken into consideration then the holes from the VB of WO_3 will be transferred to the HOMO level of g-C_3N_4, while the electrons from the LUMO of g-C_3N_4 will be transferred to CB of WO_3.

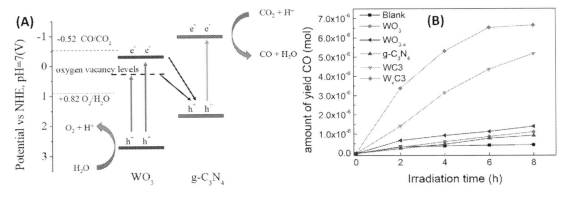

Figure 5.10
(A) Illustration of proposed photocatalytic CO_2 reduction by Z-scheme WO_{3-x}/g-C_3N_4.
(B) Photocatalytic CO_2 reduction efficiency of all the synthesized samples (Huang et al., 2019).

But the CB minimum of WO_3 (-0.37 eV, vs NHE) is found to be less negative than the reduction potential of CO/CO_2 (-0.52 eV, vs NHE), which concludes that the electrons in the CB of WO_3 are not capable of reducing CO_2 to CO. Hence taking this into consideration and analyzing the results of CO yield after reduction CO_2, it is expected that the photocatalytic mechanism was performed by forming a Z-scheme of WO_{3-x}/g-C_3N_4. The OVs present in the WO_{3-x} nanorods introduced the additional energy levels just below the CB and the electrons excited from the VB will be transferred to the OV energy levels. The OV energy bands are near to the HOMO levels of g-C_3N_4 and hence the electrons in the energy levels of OVs will combine with the holes in the HOMO of g-C_3N_4 thus increasing the separation of excited charge carriers in a Z-scheme pathway. The electrons that are left in the LUMO of g-C_3N_4 will react with the surface absorbed CO_2 and H^+ to produce CO and H_2O, while the holes present in the LUMO of WO_{3-x} will oxidize H_2O. The photocatalytic CO_2 reduction results are presented in Fig. 5.10B, where it can be clearly seen that the yield of CO (6.64 μmol) by WO_{3-x}/g-C_3N_4 is the highest. The other similar defective-photocatalytic systems employed for the CO_2 reduction reactions are given in Table 5.3.

5.5.4 N_2 fixation for NH_3 production

Recently, NH_3 production through photocatalytic N_2 fixation has become a trending research due to the demand for NH_3 and the need for an environmentally friendly and a cost-effective approach to produce NH_3.

In this direction, bismuth oxyhalides have been demonstrated for N_2 fixation and studies have proved that the defect engineering in these oxyhalides can enhance their photocatalytic performance on a greater scale. One such example is the synthesis of defect-engineered Fe-doped BiOCl nanosheets (Zhang et al., 2019). BiOCl nanosheets with OVs were

Table 5.3: Photocatalytic CO$_2$ reduction by various defect-engineered photocatalysts.

Photocatalyst	Defect type	Final product	Yield (μmol/g per h)	Reference
WO$_{3-x}$/g-C$_3$N$_4$	Oxygen vacancies	CO	6.64 μmol	Huang et al. (2019)
BiOBr	Oxygen vacancies	CO	87.4	Wu et al. (2018)
Cs$_4$PbBr$_6$/rGO	Oxygen vacancies	CO	34 μmol/g	Wang et al. (2020)
Bi$_{12}$O$_{17}$Cl$_2$	Oxygen vacancies	CO	194.5 μmol/g	Di et al. (2018)
TiO$_{2-x}$/g-C$_3$N$_4$	Oxygen vacancies	CO	77.8	Shi et al. (2019)
g-C$_3$N$_4$	Twin defects	CO	4.3	Lang et al. (2017)
		CH$_4$	0.45	
Bi$_2$MoO$_6$	Oxygen vacancies	CO	3.62	Di et al. (2019)
g-C$_3$N$_4$	Nitrogen vacancies	CO	5.93 μmol/g	Wang, Fu, and Zheng (2019)
		CH$_4$	3.19 μmol/g	
ZnIn$_2$S$_4$	Zinc vacancies	CO	33.2	Jiao et al. (2017)
TiO$_2$	Oxygen vacancies	CH$_4$	80.35	Sorcar, Hwang, Grimes, and In (2017)

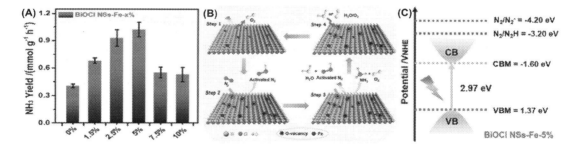

Figure 5.11
(A) Photocatalytic NH$_3$ production rate of BiOCl NSs-Fe-x% in the first 1 h. (B) Schematic illustration of the photocatalytic N$_2$ fixation model. (C) Electronic energy-level diagram for the BiOCl NSs-Fe5% (Zhang et al., 2019).

synthesized through the hydrothermal process by doping various quantities of Fe (BiOCl NSs-Fe-x %). The final samples were used for N$_2$ fixation under full-spectrum irradiation and the photocatalytic NH$_3$ production rate of BiOCl NSs-Fe-x% is shown in Fig. 5.11A. Among the samples, BiOCl NSs-Fe-5% showed the highest photocatalytic performance with the NH$_3$ yield rate of 1.022 mmol/g per hour, which is 2.53 times higher than that of pristine BiOCl nanosheets. The photocatalytic mechanism and the band energy diagram are shown in Fig. 5.11B and C.

It is highlighted that even though a strong reduction ability of excited electrons was exhibited by BiOCl NSs-Fe-5% with a negative CB (-1.60 eV) position, it still lacks the capability to directly reduce N$_2$. Accordingly, the mechanism of N$_2$ fixation was explained

Table 5.4: Photocatalytic N_2 fixation by various defect-engineered photocatalysts.

Photocatalyst	Defect type	NH$_3$ yield (μmol/h per g)	Reference
BiOCl	Oxygen vacancies	1.022	Zhang et al. (2019)
BOBr	Oxygen vacancies	104.2	Li, Shang, Ai, and Zhang (2015)
Bi$_5$O$_7$Br	Oxygen vacancies	1.38	Wang et al. (2017)
Bi$_3$O$_4$Br	Bismuth and oxygen vacancies	50.8	Di et al. (2019)
Bi$_5$O$_7$I	Oxygen vacancies	162.48	Lan et al. (2020)
Bi$_2$O$_2$CO$_3$	Oxygen vacancies	957 μmol/L	Zhang et al. (2016a)
TiO$_2$	Oxygen vacancies	78.9	Zhao et al. (2019)
In$_2$S$_3$	Sulfur vacancies	52.49	He et al. (2019)
g-C$_3$N$_4$/ZnMoCdS	Sulfur vacancies	102.9	Zhang et al. (2016b)
Zn$_{0.1}$Sn$_{0.1}$Cd$_{0.8}$S	Sulfur vacancies	139.7	Hu, Chen, Li, Zhao, and Mao (2016)
g-C$_3$N$_4$	Nitrogen vacancies	435.28	Liang et al. (2020)
g-C$_3$N$_4$	Nitrogen vacancies	1240	Dong, Ho, and Wang (2015)
g-C$_3$N$_4$	Carbon vacancies	5.99 mM/h per g$_{cat}$	Cao et al. (2018)

in four steps. In the first step, under the irradiation along with excited charge carriers, some O atoms from the surface of the photocatalyst (due to the OVs) were introduced on the surface. The second step involved the chemisorption and activation of N_2 on the OVs followed by the third step, in which the injected excited photogenerated electrons into the antibonding π-orbitals of activated N_2 weakened the N−N triple bonds and resulted in the reduction of N_2 and the formation of NH_3. In the final step, the OVs were again reloaded by the vicinal O atoms from H_2O or O_2. Table 5.4 shows the list of defect-engineered photocatalysts for the effective N_2 fixation to produce NH_3.

5.5.5 Other applications

In addition to the above discussed applications, defective photocatalysts have been employed for the other photocatalytic application, such as heavy metal reduction and microbial disinfections. Yao et al. used the lanthanum-doped bismuth titanate nanosheets ($Bi_{3.25}La_{0.75}Ti_3O_{12-x}$/BLTO$_{12-x}$) for the simultaneous methyl orange degradation and Cr (VI) reduction. The creation of OVs into BLTO nanosheets was carried out by employing a facile vacuum-deoxidized treatment, which resulted in abundant surface OVs (Yao, Chen, Li, & Shi, 2020). The photocatalytic mechanism and the respective energy band diagrams are given in Fig. 5.12A and B.

It was found from the obtained UV-Drs spectra that there was a red shift in the absorption spectra after the introduction of OVs. Also, the band gap of BLTO and BLTO$_{12-x}$ was estimated to be 2.86 and 2.56 eV, respectively (Fig. 5.12B) and the reason for the decrease

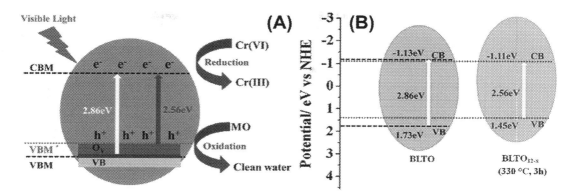

Figure 5.12
(A) Schematic illustration of the proposed photocatalytic mechanism. (B) Electronic energy-level diagram of BLTO before and after introduction of OVs (Yao et al., 2020).

Table 5.5: Photocatalytic heavy metal reduction by various defect-engineered photocatalysts.

Photocatalyst	Defect type	Reduction rate	Reference
$Bi_{3.25}La_{0.75}Ti_3O_{12-x}$	Oxygen vacancies	Cr (VI): ~99% in 3 h	Yao et al. (2020)
(P, Mo)-g-C_3N_x	Oxygen vacancies	Cr (VI): 95% in 2 h	Chen et al. (2019)
Zr-TiO_2	Oxygen vacancies	Cr (VI): 98.9% in 75 min	Wang et al. (2021)
TiO_2/g-C_3N_4	Nitrogen vacancies	TC-HCL: 79.9 in 2 h	Wang, Rao, Wang, Shi, and Zhang (2020)
		Cr (VI): 89.5% in 75 min	
$BiOI_{1-x}$	Iodine vacancies	Gas phase Hg: 68.89% in 30 min	Guan et al. (2020)
CeO_2/$BiOIO_3$	Oxygen vacancies	Gas phase Hg: 86.53% in 30 min	Xiao et al. (2020)

in band gap was ascribed to the presence of abundant surface OVs. The band gap narrowing essentially led to the rise in VBM position, which shifted from 1.73 to 1.45 eV after the introduction of OVs as these OVs form the impurity energy levels just above the VBM position. This rise in VBM led to the expansion of VB width that contributed in the better transport and separation of photogenerated charge carriers. As a result, the advantages due to the decreased recombination rate and the emergence of visible light responsive band gap synergistically enhanced the overall photocatalytic activity toward the simultaneous reduction and degradation of pollutants. The defective photocatalysts employed for heavy metal reduction are listed in Table 5.5.

As another application, the oxygen defect-engineered ZnO was used for antibacterial performance under solar light irradiation (Singh et al., 2019). The oxygen-defective ZnO took around 4.5 hours for the complete killing of Staphylococcus aureus (*S. aureus*) bacteria under sunlight irradiation, while it took around 20 hours for the oxygen-defective ZnO to kill the bacteria without the sunlight irradiation. It was clear that even though the OVs were not effective under nonirradiation conditions, it became highly reactive under light exposure. This clearly indicated that the existence of OVs largely supported the photocatalytic activities toward killing the bacteria. The observed excellent photocatalytic performance was attributed to the presence of oxygen defects, which resulted in narrowing the band gap energy and facilitated the rapid generation of high reactive oxidative species. Similarly, the other photocatalytic materials, such as the porous g-C_3N_4 ultrathin nanosheets with nitrogen defects and WO_3-carbon hybrid nanosheets with oxygen defects, have also been demonstrated for their ability toward photocatalytic disinfection applications (Liu et al., 2020; Wang et al., 2018). The porous g-C_3N_4 ultrathin nanosheets with nitrogen defects were synthesized using thermal polymerization of freeze-dried HCl-pretreated urea and were used for the disinfection of Escherichia coli (*E. coli*) and *S. aureus* bacteria. When compared with pristine g-C_3N_4, NV-induced g-C_3N_4 showed better photocatalytic activity, which performed the disinfection of bacteria within 120 minutes, whereas the pristine sample showed very least performance of disinfection. The factors influencing the enhanced photocatalytic performance were considered to be enhanced light absorption ability due to the reduced band gap energy, more exposed active sites, enhanced charge separation, and well-inhibited recombination of the charge carriers, which collectively facilitated by the nitrogen defects in g-C_3N_4 (Liu et al., 2020). Similarly, the defective WO_3-carbon hybrid nanosheets synthesized through the hydrothermal method showed enhanced photocatalytic disinfection-abilities toward killing almost both *E. coli* and *S. aureus* bacteria within 45 minutes of irradiation time. The reason for the enhancement was attributed to the OVs and carbon matrix, which boosted the photocatalytic performance by triggering a faster transfer and a higher efficient separation of electron—hole pairs, thereby enhancing the production of reductive oxygen species to kill the bacteria (Wang et al., 2018).

5.6 Summary and outlook

Apparently, the emergence of nanoscale materials spotlighted the concept of defects in materials for their potential applications in various fields. In this direction, defects in photocatalytic (PC) materials were also found to play interesting roles in altering or enhancing their properties. It is known that the semiconductor materials are largely used as photocatalysts, where their band edge positions govern their photocatalytic properties. Defects have gained importance in PC materials as they fundamentally help to shift the position of band edges or the dispersion of energy bands. Technically, considering the constituents of a semiconductor PC, the defects namely anionic and cationic defects can be created or they can be created at the sites responsible for the formation VB/HOMO and CB/LUMO so that the band edge structures can be effectively controlled. In this context, the defects of these mentioned kinds have been discussed in this chapter along with their origin, formation mechanism, synthesis strategies and various photocatalytic applications and their mechanism toward enhancing their properties. Despite a considerable amount of works have been done, the origin of defect-induced properties is not clearly defined yet. Accordingly, the computational design of defective materials and their systemic synthesis methods should also be devised. Apart from the influence over the band structures, catalytic sites, charge-trapping characteristics, the other role of defects, such as the defect-driven growth of materials, and phase modifications should be explored toward augmenting the understanding of defective photocatalytic materials and their applications.

Acknowledgment

MS gratefully acknowledges the Department of Science and Technology, Government of India, for the funding support through the DST-INSPIRE Faculty Award [DST/INSPIRE/04/2016/002227, 14 February 2017].

References

Bai, S., Zhang, N., Gao, C., & Xiong, Y. (2018). Defect engineering in photocatalytic materials. *Nano Energy*, *53*, 296−336.

Bak, D., & Kim, J. H. (2018). Oxidation driven ZnS Core-ZnO shell photocatalysts under controlled oxygen atmosphere for improved photocatalytic solar water splitting. *Journal of Power Sources*, *389*, 70−76.

Bak, T., Nowotny, M. K., Sheppard, L. R., & Nowotny, J. (2008). Effect of prolonged oxidation on semiconducting properties of titanium dioxide. *The Journal of Physical Chemistry C*, *112*, 13248−13257.

Bellardita, M., Garlisi, C., Ozer, L. Y., Venezia, A. M., Sá, J., Mamedov, F., . . . Palmisano, G. (2020). Highly stable defective TiO_{2-x} with tuned exposed facets induced by fluorine: Impact of surface and bulk properties on selective UV/visible alcohol photo-oxidation. *Applied Surface Science*, *510*, 145419.

Byrnea, C., Subramanian, G., & Pillai, S. C. (2018). Recent advances in photocatalysis for environmental applications. *Journal of Environmental chemical Engineering*, *6*, 3531−3555.

Cao, S., Fan, B., Feng, Y., Chena, H., Jiang, F., & Wang, X. (2018). Sulfur-doped $g-C_3N_4$ nanosheets with carbon vacancies: General synthesis and improved activity for simulated solar-light photocatalytic nitrogen fixation. *Chemical Engineering Journal*, *353*, 147−156.

Chatzitakis, A., & Sartori, S. (2019). Recent advances in the use of black TiO_2 for production of hydrogen and other solar fuels. *ChemPhysChem*, *20*, 1272−1281.

Chen, D., Liu, J., Jia, Z., Fang, J., Yang, F., Tang, Y., . . . Fang, Z. Q. (2019). Efficient visible-light-driven hydrogen evolution and Cr(VI) reduction over porous P and Mo co-doped $g-C_3N_4$ with feeble N vacancies photocatalyst. *Journal of Hazardous Materials*, *361*, 294−304.

Chen, D., Niu, F., Qin, L., Wang, S., Zhang, N., & Huang, Y. (2017). Defective $BiFeO_3$ with surface oxygen vacancies: Facile synthesis and mechanism insight into photocatalytic performance. *Solar Energy Materials and Solar Cells*, *171*, 24−32.

Chen, Q., Wang, H., Luan, Q., Duan, R., Cao, X., Fang, Y., . . . Hu, X. (2020). Synergetic effects of defects and acid sites of 2D-ZnO photocatalysts on the photocatalytic performance. *Journal of Hazardous Materials*, *385*, 121527.

Chen, X., Liu, L., Yu, P. Y., & Mao, S. S. (2011). Increasing solar absorption for photocatalysis with black hydrogenated titanium dioxide nanocrystals. *Science*, *331*, 746−750.

Chen, Y., Yang, W., Gao, S., Sun, C., & Li, Q. (2018). Synthesis of Bi_2MoO_6 nanosheets with rich oxygen vacancies by post-synthesis etching treatment for enhanced photocatalytic performance. *ACS Applied Nano Materials*, *1*, 3565−3578.

Chew, Y. H., Tang, J. Y., Tan, L. J., Choi, B. W. J., Tan, L. L., & Chai, S. P. (2019). Engineering surface oxygen defects on tungstate oxide to boost photocatalytic oxygen evolution from water splitting. *Chemical Communications*, *55*, 6265−6268.

Chiu, Y. H., Naghadeh, S. B., Lindley, S. A., Lai, T. H., Kuo, M. Y., Chang, K. D., ... Hsu, Y. J. (2019). Yolk-shell nanostructures as an emerging photocatalyst paradigm for solar hydrogen generation. *Nano Energy, 62*, 289–298.

Cong, Q., Chen, L., Wang, X., Ma, H., Zhao, J., Li, S., ... Li, W. (2020). Promotional effect of nitrogen-doping on a ceria unary oxide catalyst with rich oxygen vacancies for selective catalytic reduction of NO with NH_3. *Chemical Engineering Journal, 379*, 122302.

Di, J., Xia, J., Chisholm, M. F., Zhong, J., Chen, C., Cao, X., ... Dai, S. (2019). Defect-tailoring mediated electron–hole separation in single-unit-cell Bi3O4Br nanosheets for boosting photocatalytic hydrogen evolution and nitrogen fixation. *Advanced Materials, 31*, 1807576.

Di, J., Zhao, X., Lian, C., Ji, M., Xia, J., Xiong, J., ... Liu, Z. (2019). Atomically-thin Bi_2MoO_6 nanosheets with vacancy pairs for improved photocatalytic CO_2 reduction. *Nano Energy, 61*, 54–59.

Di, J., Zhu, C., Ji, M., Duan, M., Long, R., Yan, C., ... Liu, Z. (2018). Defect-rich $Bi_{12}O_{17}Cl_2$ nanotubes self-accelerating charge separation for boosting photocatalytic CO_2 reduction. *Angewandte Chemie International Edition, 57*, 14847–14851.

Dong, G., Ho, W., & Wang, C. (2015). Selective photocatalytic N_2 fixation dependent on the g-C_3N_4 induced by nitrogen vacancies. *Journal of Materials Chemistry A, 3*, 23435–23441.

Du, C., Zhang, Q., Lin, Z., Yan, B., Xia, C., & Yang, G. (2019). Half-unit-cell $ZnIn_2S_4$ monolayer with sulfur vacancies for photocatalytic hydrogen evolution. *Applied Catalysis B: Environmental, 248*, 193–201.

Fageria, P., Gangopadhyay, S., & Pande, S. (2014). Synthesis of ZnO/Au and ZnO/Ag nanoparticles and their photocatalytic application using UV and visible light. *RSC Advances, 4*, 24962–24972.

Farooq, M. H., Aslam, I., Shuaib, A., Anam, H. S., Rizwan, M., & Kanwal, Q. (2019). Band gap engineering for improved photocatalytic performance of CuS/TiO_2 composites under solar light irradiation. *Bulletin of the Chemical Society of Ethiopia, 33*, 561–571.

Fukushi, D., Sasaki, A., Hirabayashi, H., & Kitano, M. (2017). Effect of oxygen vacancy in tungsten oxide on the photocatalytic activity for decomposition of organic materials in the gas phase. *Microelectronics Reliability, 79*, 1–4.

Gao, J., Shen, Q., Guan, R., Xue, J., Liu, X., Jia, H., ... Wu, Y. (2020). Oxygen vacancy self-doped black TiO_2 nanotube arrays by aluminothermic reduction for photocatalytic CO_2 reduction under visible light illumination. *Journal of CO_2 Utilization, 35*, 205–215.

Gao, L., Li, Y., Ren, J., Wang, S., Wang, R., Fu, G., & Hu, Y. (2017). Passivation of defect states in anatase TiO_2 hollow spheres with Mg doping: Realizing efficient photocatalytic overall water splitting. *Applied Catalysis B: Environmental, 202*, 127–133.

Gao, S., Gu, B., Jiao, X., Sun, Y., Zu, X., Yang, F., ... Xie, Y. (2017). Highly efficient and exceptionally durable CO_2 photoreduction to methanol over freestanding defective single-unit-cell bismuth vanadate layers. *Journal of the American Chemical Society, 139*, 3438–3445.

Ge, J., Zhang, Y., Heo, Y. J., & Park, S. J. (2019). Advanced design and synthesis of composite photocatalysts for the remediation of wastewater: A review. *Catalysts, 9*, 122.

Gogoi, D., Namdeo, A., Golder, A. K., & Peela, N. R. (2020). Ag-doped TiO_2 photocatalysts with effective charge transfer for highly efficient hydrogen production through water splitting. *International Journal of Hydrogen Energy, 45*, 2729–2744.

Guan, Y., Wu, J., Man, X., Liu, Q., Qi, Y., He, P., & Qi, X. (2020). Rational fabrication of flower-like $BiOI_{1-x}$ photocatalyst by modulating efficient iodine vacancies for mercury removal and DFT study. *Chemical Engineering Journal, 396*, 125234.

Guo, S., Hu, Z., Zhen, M., Gu, B., Shen, B., & Dong, F. (2020). Insights for optimum cation defects in photocatalysis: A case study of hematite nanostructures. *Applied Catalysis B: Environmental, 264*, 118506.

Han, L., Su, B., Liu, G., Ma, Z., & An, X. (2018). Synthesis of oxygen vacancy-rich black TiO_2 nanoparticles and the visible light photocatalytic performance. *Molecular Catalysis, 456*, 96–101.

Hara, M., Hitoki, G., Takata, T., Kondo, J. N., Kobayashi, H., & Domen, K. (2003). TaON and Ta_3N_5 as new visible light driven photocatalysts. *Catalysis Today, 78*, 555–560.

He, C., Li, X., Chen, X., Ma, S., Yan, X., Zhang, Y., . . . Yao, C. (2020). Palygorskite supported rare earth fluoride for photocatalytic nitrogen fixation under full spectrum. *Applied Clay Science, 184*, 105398.

He, Y., Rao, H., Song, K., Li, J., Yu, Y., Lou, Y., . . . Feng, S. (2019). 3D hierarchical $ZnIn_2S_4$ nanosheets with rich Zn vacancies boosting photocatalytic CO_2 reduction. *Advanced Functional Materials, 29*, 1905153.

He, Z., Wang, Y., Dong, X., Zheng, N., Ma, H., & Zhang, X. (2019). Indium sulfide nanotubes with sulfur vacancies as an efficient photocatalyst for nitrogen fixation. *RSC Advances, 9*, 21646−21652.

Hu, H., Qian, D., Lin, P., Ding, Z., & Cui, C. (2020). Oxygen vacancies mediated in-situ growth of noble-metal (Ag, Au, Pt) nanoparticles on 3D TiO_2 hierarchical spheres for efficient photocatalytic hydrogen evolution from water splitting. *International Journal of Hydrogen Energy, 45*, 629−639.

Hu, J., Li, J., Cui, J., An, W., Liu, L., Liang, Y., & Cui, W. (2020). Surface oxygen vacancies enriched FeOOH/ Bi_2MoO_6 photocatalysis-fenton synergy degradation of organic pollutants. *Journal of Hazardous Materials, 384*, 121399.

Hu, S., Chen, X., Li, Q., Zhao, Y., & Mao, W. (2016). Effect of sulfur vacancies on the nitrogen photofixation performance of ternary metal sulfide photocatalysts. *Catalysis Science & Technology, 6*, 5884−5890.

Huang, H. B., Fang, Z. B., Yu, K., Lü, J., & Cao, R. (2020). Visible-light-driven photocatalytic H_2 evolution over CdZnS nanocrystal solid solutions: Interplay of twin structures, sulfur vacancies and sacrificial agents. *Journal of Materials Chemistry A, 8*, 3882−3891.

Huang, S., Long, Y., Ruan, S., & Zeng, Y. J. (2019). Enhanced photocatalytic CO_2 reduction in defect-engineered Z-scheme $WO_{3−x}$/g-C_3N_4 heterostructures. *CS Omega, 4*, 15593−15599.

Hui, J., Zhang, G., Ni, C., & Irvine, J. T. S. (2017). Promoting photocatalytic H_2 evolution by tuning cation deficiency in La and Cr co-doped $SrTiO_3$. *Chemical Communications, 53*, 10038−10041.

Huo, W. C., Li, J. Y., Liu, M., Liu, X. Y., Zhang, Y. X., & Dong, F. (2019). Synthesis of Bi_2WO_6 with gradient oxygen vacancies for highly photocatalytic NO oxidation and mechanism study. *Chemical Engineering Journal, 361*, 129−138.

Irani, R., Ahmet, I. Y., Jang, J. W., Berglund, S. P., Plate, P., Höhn, C., . . . Lardhi, S. (2020). Nature of nitrogen incorporation in $BiVO_4$ photoanodes through chemical and physical methods. *Solar RRL, 4*, 1900290.

Ji, M., Chen, R., Di, J., Liu, Y., Li, K., Chen, Z., . . . Li, H. (2019). Oxygen vacancies modulated Bi-rich bismuth oxyiodide microspheres with tunable valence band position to boost the photocatalytic activity. *Journal of Colloid and Interface Science, 553*, 612−620.

Jiang, J., Zhang, L., Li, H., He, W., & Yin, J. J. (2013). Self-doping and surface plasmon modification induced visible light photocatalysis of BiOCl. *Nanoscale, 5*, 10573−10581.

Jiao, X., Chen, Z., Li, X., Sun, Y., Gao, S., Yan, W., . . . Xie, Y. (2017). Defect-mediated electron-hole separation in one-unit-cell $ZnIn_2S_4$ layers for boosted solar-driven CO_2 reduction. *Journal of the American Chemical Society, 139*, 7586−7594.

Jing, L., Wang, D., Xu, Y., Xie, M., Yan, J., He, M., . . . Li, H. (2020). Porous defective carbon nitride obtained by a universal method for photocatalytic hydrogen production from water splitting. *Journal of Colloid and Interface Science, 566*, 171−182.

Katal, R., Eshkalak, S. K., Masudy-panah, S., Kosari, M., Saeedikhani, M., Zarinejad, M., & Ramakrishna, S. (2019). Evaluation of solar-driven photocatalytic activity of thermal treated TiO_2 under various atmospheres. *Nanomaterials, 9*, 163.

Khan, M. M., Adil, S. F., & Al-Mayouf, A. (2015). *Metal oxides as photocatalysts*. Elsevier.

Khazaee, Z., Khavar, A. H. C., Mahjoub, A. R., Motaee, A., Srivastava, V., & Sillanpää, M. (2020). Template-confined growth of X-Bi_2MoO_6 (X: F, Cl, Br, I) nanoplates with open surfaces for photocatalytic oxidation; experimental and DFT insights of the halogen doping. *Solar Energy, 196*, 567−581.

Lan, M., Zheng, N., Dong, X., Hua, C., Ma, H., & Zhang, X. (2020). Bismuth-rich bismuth oxyiodide microspheres with abundant oxygen vacancies as an efficient photocatalyst for nitrogen fixation. *Dalton Transactions, 49*(49), 9123−9129.

Lang, Q., Hu, W., Zhou, P., Huang, T., Zhong, S., Yang, L., . . . Bai, S. (2017). Twin defects engineered Pd cocatalyst on C_3N_4 nanosheets for enhanced photocatalytic performance in CO_2 reduction reaction. *Nanotechnology, 28*, 484003.

Lei, F., Sun, Y., Liu, K., Gao, S., Liang, L., Pan, B., & Xie, Y. (2014). Oxygen vacancies confined in ultrathin indium oxide porous sheets for promoted visible-light water splitting. *Journal of the American Chemical Society, 136*, 6826−6829.

Li, B., Lai, C., Qin, L., Chu, C., Zhang, M., Liu, S., . . . Chen, L. (2020). Anchoring single-unit-cell defect-rich bismuth molybdate layers on ultrathin carbon nitride nanosheet with boosted charge transfer for efficient photocatalytic ciprofloxacin degradation. *Journal of Colloid and Interface Science, 560*, 701−713.

Li, C., Wang, T., Zhao, Zj, Yang, W., Li, J., Li, A., . . . Gong, J. (2018). Promoted fixation of molecular nitrogen with surface oxygen vacancies on plasmon-enhanced TiO_2 photoelectrodes. *Angewandte Chemie International Edition, 57*, 5278−5282.

Li, H., Shang, J., Ai, Z., & Zhang, L. (2015). Efficient visible light nitrogen fixation with BiOBr nanosheets of oxygen vacancies on the exposed {001} facets. *Journal of the American Chemical Society, 137*, 6393−6399.

Li, H., Shang, J., Zhu, H., Yang, Z., Ai, Z., & Zhang, L. (2016). Oxygen vacancy structure associated photocatalytic water oxidation of BiOCl. *ACS Catalysis, 6*, 8276−8285.

Li, H., Sun, B., Yang, F., Wang, Z., Xu, Y., Tian, G., . . . Zhou, W. (2019). Homojunction and defect synergy-mediated electron−hole separation for solar-driven mesoporous rutile/anatase TiO_2 microsphere photocatalysts. *RSC Advances, 9*, 7870−7877.

Li, H., Tsai, C., Koh, A. L., Cai, L. L., Contryman, A. W., Fragapane, A. H., . . . Zheng, X. L. (2016). Activating and optimizing MoS_2 basal planes for hydrogen evolution through the formation of strained sulphur vacancies. *Nature Materials, 15*, 48.

Li, H., Wu, S., Hood, Z. D., Sun, J., Hu, B., Liang, C., . . . Jiang, B. (2020). Atomic defects in ultra-thin mesoporous TiO_2 enhance photocatalytic hydrogen evolution from water splitting. *Applied Surface Science, 513*, 145723.

Li, T., Shen, Z., Shu, Y., Li, X., Jiang, C., & Chen, W. (2019). Facet-dependent evolution of surface defects in anatase TiO_2 by thermal treatment: Implications for environmental applications of photocatalysis. *Environmental Science: Nano, 6*, 1740−1753.

Li, X., Cheng, Y., Wu, Q., Xu, J., & Wang, Y. (2019). Synergistic effect of the rearranged sulfur vacancies and sulfur interstitials for 13-fold enhanced photocatalytic H_2 production over defective $Zn_2In_2S_5$ nanosheets. *Applied Catalysis B: Environmental, 240*, 270−276.

Li, Y., Gu, M., Shi, T., Cui, W., Zhang, X., Dong, F., . . . Lv, K. (2020). Carbon vacancy in C_3N_4 nanotube: Electronic structure, photocatalysis mechanism and highly enhanced activity. *Applied Catalysis B: Environmental, 262*, 118281.

Li, Y., Ho, W., Lv, K., Zhu, B., & Lee, S. C. (2017). Carbon vacancy-induced enhancement of the visible light-driven photocatalytic oxidation of NO over g-C_3N_4 nanosheets. *Applied Surface Science, 30*, 380−389.

Liang, C., Niu, H. Y., Guo, H., Niu, C. G., Huang, D. W., Yang, Y., . . . Feng, H. P. (2020). Insight into photocatalytic nitrogen fixation on graphitic carbon nitride: Defect-dopant strategy of nitrogen defect and boron dopant. *Chemical Engineering Journal, 396*, 125395.

Liang, L., Shi, L., Wang, F., Wang, H., Yan, P., Cong, Y., . . . Qi, W. (2020). g-C3N4 nano-fragments as highly efficient hydrogen evolution photocatalysts: Boosting effect of nitrogen vacancy. *Applied Catalysis A, General, 599*, 117618.

Liang, Q., Li, Z., Huang, Z. H., Kang, F., & Yang, Q. H. (2015). Holey graphitic carbon nitride nanosheets with carbon vacancies for highly improved photocatalytic hydrogen production. *Advanced Functional Materials, 25*, 6885−6892.

Liu, G., Han, J., Zhou, X., Huang, L., Zhang, F., Wang, X., . . . Li, C. (2013). Enhancement of visible-light-driven O_2 evolution from water oxidation on WO3 treated with hydrogen. *Journal of Catalysis, 307*, 148−152.

Liu, H., Ma, S., Shao, L., Liu, H., Gao, Q., Li, B., . . . Zhou, J. (2020). Defective engineering in graphitic carbon nitride nanosheet for efficient photocatalytic pathogenic bacteria disinfection. *Applied Catalysis B: Environmental, 261*, 118201.

Liu, J., Liu, B., Ren, Y., Yuan, Y., Zhao, H., Yang, H., & Liu, S. F. (2019). Hydrogenated nanotubes/nanowires assembled from TiO_2 nanoflakes with exposed {111} facets: Excellent photo-catalytic CO_2 reduction activity and charge separation mechanism between (111) and (111) polar surfaces. *Journal of Materials Chemistry A, 7,* 14761−14775.

Liu, J., Wei, Z., & Shangguan, W. (2019). Defects engineering in photocatalytic water splitting materials. *ChemCatChem, 11,* 6177−6189.

Liu, L., Jiang, Y., Zhao, H., Chen, J., Cheng, J., Yang, K., & Li, Y. (2016). Engineering co-exposed {001} and {101} facets in oxygen-deficient TiO_2 nanocrystals for enhanced CO_2 photoreduction under visible light. *ACS Catalysis, 6,* 1097−1108.

Liu, M., Zhang, D., Han, J., Liu, C., Ding, Y., Wang, Z., & Wang, A. (2020). Adsorption enhanced photocatalytic degradation sulfadiazine antibiotic using porous carbon nitride nanosheets with carbon vacancies. *Chemical Engineering Journal, 382,* 123017.

Liu, Q., Wang, F., Lin, H., Xie, Y., Tong, N., Lin, J., . . . Wang, X. (2018). Surface oxygen vacancy and defect engineering of WO3 for improved visible light photocatalytic performance. *Catal. Sci. Technol, 8,* 4399−4406.

Liu, X., Jing, B., Lun, G., Wang, Y., Wang, X., Fang, C., . . . Li, C. H. (2020). Integrating nitrogen vacancies into crystalline graphitic carbon nitride for enhanced photocatalytic hydrogen production. *Chemical Communications, 56,* 3129−3182.

Mateo, D., García-Mulero, A., Albero, J., & García, H. (2019). N-doped defective graphene decorated by strontium titanate as efficient photocatalyst for overall water splitting. *Applied Catalysis B: Environmental, 252,* 111−119.

Meng, J., Lin, Q., Chen, T., Wei, X., Li, J., & Zhang, Z. (2018). Oxygen vacancy regulation on tungsten oxides with specific exposed facets for enhanced visible-light-driven photocatalytic oxidation. *Nanoscale, 10,* 2908−2915.

Mohajernia, S., Andryskova, P., Zoppellaro, G., Hejazi, S., Kment, S., Zboril, R., . . . Schmuki, P. (2020). Influence of Ti^{3+} defect-type on heterogeneous photocatalytic H_2 evolution activity of TiO_2. *Journal of Materials Chemistry A, 12,* 38211−38221.

Morgan, B. J., & Watson, G. W. (2010). Intrinsic n-type defect formation in TiO_2: A comparison of rutile and anatase from GGA +U calculations. *The Journal of Physical Chemistry C, 114,* 2321−2328.

Nahar, S., Zain, M. F. M., Kadhum, A. A. H., Hasan, H. A., & Hasan, Md. R. (2017). Advances in photocatalytic CO_2 reduction with water: A review. *Materials, 10,* 629.

Naldoni, A., Allieta, M., Santangelo, S., Marelli, M., Fabbri, F., Cappelli, S., . . . Dal Santo, V. (2012). Effect of nature and location of defects on bandgap narrowing in black TiO_2 nanoparticles. *Journal of the American Chemical Society, 134,* 7600−7603.

Nandi, P., & Das, D. (2019). Photocatalytic degradation of rhodamine-B dye by stable ZnO nanostructures with different calcination temperature induced defects. *Applied Surface Science, 465,* 546−556.

Neto, N. A., Nascimento, L., Correa, M., Bohn, F., Bomio, M., & Motta, F. (2020). Characterization and photocatalytic application of Ce^{4+}, Co^{2+}, Mn^{2+} and Ni^{2+} doped Fe_3O_4 magnetic nanoparticles obtained by the co-precipitation method. *Materials Chemistry and Physics, 242,* 122489.

Niu, P., Yin, L. C., Yang, Y. Q., Liu, G., & Cheng, H. M. (2015). Increasing the visible light absorption of graphitic carbon nitride (Melon) photocatalysts by homogeneous self-modification with nitrogen vacancies. *Advanced Materials, 26,* 8046−8052.

Peng, H., Liu, D., Zheng, X., & Fu, X. (2019). N-doped carbon-coated ZnS with sulfur-vacancy defect for enhanced photocatalytic activity in the visible light region. *Nanomaterials, 12,* 1657.

Qi, K., Xing, X., Zada, A., Li, M., Wang, Q., Liue, S. Y., . . . Wang, G. (2020). Transition metal doped ZnO nanoparticles with enhanced photocatalytic and antibacterial performances: Experimental and DFT studies. *Ceramics International, 46,* 1494−1502.

RamezaniSani, S., Rajabi, M., & Mohseni, F. (2020). Influence of nitrogen doping on visible light photocatalytic activity of TiO_2 nanowires with anatase-rutile junction. *Chemical Physics Letters, 744,* 137217.

Ran, L., Hou, J., Cao, S., Li, Z., Zhang, Y., Wu, Y., . . . Sun, L. (2020). Defect engineering of photocatalysts for solar energy conversion. *Solar RRL, 4,* 1900487.

Reddy, P. A. K., Reddy, P. V. L., Kwon, E., Kim, K. H., Akter, T., & Kalagara, S. (2016). Recent advances in photocatalytic treatment of pollutants in aqueous media. *Environment International, 91,* 94−103.

Reza Gholipour, M., Dinh, C. T., Béland, F., & Do, T. O. (2015). Nanocomposite heterojunctions as sunlight-driven photocatalysts for hydrogen production from water splitting. *Nanoscale, 7*, 8187−8208.

Shen, M., Zhang, L., Wang, M., Tian, J., Jin, X., Guo, L., . . . Shi, J. (2019). Carbon-vacancy modified graphitic carbon nitride: Enhanced CO_2 photocatalytic reduction performance and mechanism probing. *Journal of Materials Chemistry A, 7*, 1556−1563.

Shi, H., Long, S., Hu, S., Hou, J., Ni, W., Song, C., . . . Guo, X. (2019). Interfacial charge transfer in 0D/2D defect-rich heterostructures for efficient solar-driven CO_2 reduction. *Applied Catalysis B: Environmental, 245*, 760−769.

Shi, J., Chang, Y., Tang, Y., Wang, X., Wang, X., Zhang, X., & Cao, J. (2020). Hydrogenated $LaFeO_3$ with oxygen vacancies for enhanced visible light photocatalytic performance. *Ceramics International, 46*, 5315−5322.

Singh, J., Juneja, S., Palsaniya, S., Manna, A. K., Sonia, R. K., & Bhattacharya, J. (2019). Evidence of oxygen defects mediated enhanced photocatalytic and antibacterial performance of ZnO nanorods. *Colloids and Surfaces B: Biointerfaces, 184*, 110541.

Sorcar, S., Hwang, Y., Grimes, C. A., & In, S. I. (2017). Highly enhanced and stable activity of defect-induced titania nanoparticles for solar light-driven CO_2 reduction into CH_4. *Materials Today, 20*, 507−514.

Sun, B., Liang, Z., Qian, Y., Xu, X., Han, Y., & Tian, J. (2020). Sulfur vacancy-rich O-doped 1T-MoS_2 nanosheets for exceptional photocatalytic nitrogen fixation over CdS. *ACS Applied Materials & Interfaces, 12*, 7257.

Takata, T., & Domen, K. (2019). Particulate photocatalysts for water splitting. Recent advances and future prospects. *ACS Energy Letters, 4*, 542−549, 4.

Tan, H., Zhao, Z., Niu, M., Mao, C., Cao, D., Cheng, D., . . . Sun, Z. (2014). A facile and versatile method for preparation of colored TiO_2 with enhanced solar-driven photocatalytic activity. *Nanoscale, 6*, 10216−10223.

Taraka, T. P. Y., Gautam, A., Jain, S. L., Bojja, S., & Pal, U. (2019). Controlled addition of Cu/Zn in hierarchical CuO/ZnO pn heterojunction photocatalyst for high photoreduction of CO_2 to MeOH. *Journal of CO_2 Utilization, 31*, 207−214.

Tong, H., Zhou, Y., Chang, G., Li, P., Zhu, R., & He, Y. (2018). Anatase TiO_2 single crystals with dominant {001} facets: Synthesis, shape-control mechanism and photocatalytic activity. *Applied Surface Science, 444*, 267−275.

Tu, W., Xu, Y., Wang, J., Zhang, B., Zhou, T., Yin, S., . . . Xu, R. (2017). Investigating the role of tunable nitrogen vacancies in graphitic carbon nitride nanosheets for efficient visible-light-driven H_2 evolution and CO_2 reduction. *ACS Sustainable Chemistry & Engineering, 5*, 7260.

Wang, J., Chen, R., Xiang, L., & Komarneni, S. (2018). Synthesis, properties and applications of ZnO nanomaterials with oxygen vacancies: A review. *Ceramics International, 44*, 7357−7377.

Wang, J., Wang, Z., Huang, B., Ma, Y., Liu, Y., Qin, X., . . . Dai, Y. (2012). Oxygen vacancy induced band-gap narrowing and enhanced visible light photocatalytic activity of ZnO. *ACS Applied Materials & Interfaces, 4*, 4024−4030.

Wang, K., Fu, J., & Zheng, Y. (2019). Insights into photocatalytic CO_2 reduction on C_3N_4: Strategy of simultaneous B, K co-doping and enhancementby N vacancies. *Applied Catalysis B: Environmental, 254*, 270−282.

Wang, M., Tan, G., Zhang, D., Li, B., Lv, L., Wang, Y., . . . Liu, Y. (2019). Defect-mediated Z-scheme BiO_{2-x}/$Bi_2O_{2.75}$ photocatalyst for full spectrum solar-driven organic dyes degradation. *Applied Catalysis B: Environmental, 254*, 98−112.

Wang, R., Zhang, W., Zhu, W., Yan, L., Li, S., Chen, K., . . . Wang, J. (2018). Enhanced visible-light-driven photocatalytic sterilization of tungsten trioxide by surface-engineering oxygen vacancy and carbon matrix. *Chemical Engineering Journal, 348*, 292−300.

Wang, S., Hai, X., Ding, X., Chang, K., Xiang, Y., Meng, X., . . . Ye, J. (2017). Light-switchable oxygen vacancies in ultrafine Bi_5O_7Br nanotubes for boosting solar-driven nitrogen fixation in pure water. *Advanced Materials, 29*, 1701774.

Wang, S., Teng, Z., Xu, Y., Yuan, M., Zhong, Y., Liu, S., . . . Ohno, T. (2020). Defect as the essential factor in engineering carbon-nitride-based visible-light-driven Z-scheme photocatalyst. *Applied Catalysis B: Environmental, 260*, 118145.

Wang, S., Zhang, Z., Huo, W., Zhang, X., Fang, F., Xie, Z., & Jiang, J. (2021). Single-crystal-like black Zr-TiO_2 nanotube array film: An efficient photocata- lyst for fast reduction of Cr(VI). *Chemical Engineering Journal, 403*, 126331.

Wang, X., Li, K., He, J., Yang, J., Dong, F., Mai, W., & Zhu, M. (2020). Defect in reduced graphene oxide tailored selectivity of photocatalytic CO_2 reduction on Cs4PbBr6 pervoskite hole-in-microdisk structure. *Nano Energy, 78*, 105388.

Wang, X., Zhou, C., Yin, L., Zhang, R., & Liu, G. (2019). Iodine-deficient BiOI nanosheets with lowered valence band maximum to enable visible light photocatalytic activity. *ACS Sustainable Chemistry & Engineering, 7*, 7900−7907.

Wang, Y., Cai, J., Wu, M., Chen, J., Zhao, W., Tian, Y., . . . Li, X. (2018). Rational construction of oxygen vacancies onto tungsten trioxide to improve visible light photocatalytic water oxidation reaction. *Applied Catalysis B: Environmental, 239*, 398−407.

Wang, Y., Rao, L., Wang, P., Guo, Y., Shi, Z., Guo, X., & Zhang, L. (2020). Synthesis of nitrogen vacancies g-C_3N_4 with increased crystallinity under the controlling of oxalyl dihydrazide: Visible-light-driven photocatalytic activity. *Applied Surface Science, 505*, 144576.

Wang, Y., Rao, L., Wang, P., Shi, Z., & Zhang, L. (2020). Photocatalytic activity of N-TiO_2/O-doped N vacancy g-C_3N_4 and the intermediates toxicity evaluation under tetracycline hydrochloride and Cr(VI) coexistence environment. *Applied Catalysis B: Environmental, 262*, 118308.

Wei, T., Zhu, Y. N., An, X., Liu, L., Cao, X., Liu, H., & Qu, J. (2019). Defect modulation of Z-scheme TiO_2/Cu_2O photocatalysts for durable water splitting. *ACS Catalysis, 9*, 8346−8354.

Wei, Y., Su, H., Zhang, Y., Zheng, L., Pan, Y., Su, C., . . . Long, M. (2019). Efficient peroxodisulfate activation by iodine vacancy rich bismuth oxyiodide: A vacancy induced mechanism. *Chemical Engineering Journal, 375*, 121971.

Weng, X., Zeng, Q., Zhang, Y., Dong, F., & Wu, Z. (2016). A Facile Approach for the Syntheses of Ultrafine TiO_2 Nano-crystallites with Defects and C Heterojunction for Photocatalytic Water Splitting. *ACS Sustainable Chemistry & Engineering, 4*, 4314−4320.

Wu, J., Li, N., Fang, H. B., Li, X., Zheng, Y. Z., & Tao, X. (2019). Nitrogen vacancies modified graphitic carbon nitride: Scalable and one-step fabrication with efficient visible-light-driven hydrogen evolution. *Chemical Engineering Journal, 358*, 20−29.

Wu, J., Li, X., Shi, W., Ling, P., Sun, Y., Jiao, X., . . . Xie, Y. (2018). Efficient visible-light-driven CO_2 reduction realized by defect-mediated BiOBr atomic layers. *Angewandte Chemie International Edition, 130*, 8855−8859.

Wu, M., He, X., Jing, B., Wang, T., Wang, C., Qin, Y., . . . An, T. (2020). Novel carbon and defects co-modified g-C_3N_4 for highly efficient photocatalytic degradation of bisphenol A under visible light. *Journal of Hazardous Materials, 384*, 121323.

Wu, M. H., Li, L., Liu, N., Wang, D. J., Xue, Y. C., & Tang, L. (2018). Molybdenum disulfide (MoS_2) as a co-catalyst for photocatalytic degradation of organic contaminants: A review. *Process Safety and Environment Protection, 118*, 40−52.

Xiao, Y., Tan, S., Wang, D., Wu, J., Jia, T., Liu, Q., . . . Zhou, M. (2020). CeO_2 /$BiOIO_3$ heterojunction with oxygen vacancies and Ce^{4+}/Ce^{3+} redox centers synergistically enhanced photocatalytic removal heavy metal. *Applied Surface Science, 530*, 147116.

Xie, H., Li, N., Chen, X., Jiang, J., & Zhao, X. (2020). Surface oxygen vacancies promoted photodegradation of benzene on TiO_2 film. *Applied Surface Science, 511*, 145597.

Xie, Y., Li, Y., Huang, Z., Zhang, J., Jia, X., Wang, X. S., & Ye, J. (2020). Two types of cooperative nitrogen vacancies in polymeric carbon nitride for efficient solar-driven H_2O_2 evolution. *Applied Catalysis B: Environmental, 265*, 118581.

Xiong, J., Di, J., Xia, J., Zhu, W., & Li, H. (2018). Surface defect engineering in 2D nanomaterials for photocatalysis. *Advanced Functional Materials, 28*, 1801983.

Xu, L., Jiang, Q., Xiao, Z., Li, X., Huo, J., Wang, S., & Dai, L. (2016). Plasma-engraved Co_3O_4 nanosheets with oxygen vacancies and high surface area for the oxygen evolution reaction. *Angewandte Chemie International Edition, 128*, 5363−5367.

Xu, Y., Li, H., Sun, B., Qiao, P., Ren, L., Tian, G., . . . Zhou, W. (2020). Surface oxygen vacancy defect-promoted electron-hole separation for porous defective ZnO hexagonal plates and enhanced solar-driven photocatalytic performance. *Chemical Engineering Journal, 379,* 122295.

Xue, C., Li, D., Li, Y., Li, N., Zhang, F., Wang, Y., . . . Hu, S. (2019). 3D-carbon dots decorated black TiO_2 nanotube Array@Ti foam with enhanced photothermal and photocatalytic activities. *Ceramics International, 45,* 17512−17520.

Yan, J., Wang, T., Wu, G., Dai, W., Guan, N., Li, L., & Gong, J. (2015). Tungsten oxide single crystal nanosheets for enhanced multichannel solar light harvesting. *Advanced Materials, 27,* 1580−1586.

Yang, Z., Chu, D., Jia, G., Yao, M., & Liu, B. (2020). Significantly narrowed bandgap and enhanced charge separation in porous, nitrogen-vacancy red g-C_3N_4 for visible light photocatalytic H_2 production. *Applied Surface Science, 504,* 144407.

Yao, L., Chen, Z., Li, J., & Shi, C. (2020). Creation of oxygen vacancies to activate lanthanum-doped bismuth titanate nanosheets for efficient synchronous photocatalytic removal of Cr(VI) and methyl orange. *Journal of Molecular Liquids, 314,* 113613.

Ye, L., Deng, K., Xu, F., Tian, L., Peng, T., & Zan, L. (2012). Increasing visible-light absorption for photocatalysis with black BiOCl. *Physical Chemistry Chemical Physics, 14,* 82−85.

Ye, L., Zan, L., Tian, L., Peng, T., & Zhang, J. (2011). The {001} facets-dependent high photoactivity of BiOCl nanosheets. *Chemical Communications, 47,* 6951−6953.

Yin, J., Xing, Z., Kuang, J., Li, Z., Zhu, Q., & Zhou, W. (2019). Dual oxygen vacancy defects-mediated efficient electron-hole separation via surface engineering of Ag/Bi_2MoO_6 nanosheets/TiO_2nanobelts ternary heterostructures. *Journal of Industrial and Engineering Chemistry, 78,* 155−163.

Yu, F., Wang, C., Ma, H., Song, M., Li, D., Li, Y., . . . Liu, Y. (2020). Revisiting Pt/TiO_2 photocatalyst in the thermally assistedphotocatalytic reduction of CO_2. *Nanoscale, 12,* 7000−7010.

Yu, Y., Yao, B., He, Y., Cao, B., Ma, W., & Chang, L. (2020). Oxygen defect-rich In-doped ZnO nanostructure for enhanced visible light photocatalytic activity. *Materials Chemistry and Physics, 244,* 122672.

Yu, Z. G., Zhang, Y. W., & Yakobson, B. I. (2015). An anomalous formation pathway for dislocation-sulfur vacancy complexes in polycrystalline monolayer MoS_2. *Nano Letters, 15,* 6855−6861.

Zhang, F., Jin, R., Chen, J., Shao, C., Gao, W., Li, L., & Guan, N. (2005). High photocatalytic activity and selectivity for nitrogen in nitrate reduction on Ag/TiO_2 catalyst with fine silver clusters. *Journal of Catalysis, 232,* 424−431.

Zhang, H., Cai, J., Wang, Y., Wu, M., Meng, M., Tian, Y., . . . Gong, J. (2018). Insights into the effects of surface/bulk defects on photocatalytic hydrogen evolution over TiO_2 with exposed {001}. *Applied Catalysis B: Environmental, Facets, 220,* 126.

Zhang, J., Chen, Z., Qiu, Y., Li, M., Yang, H., Huang, Y., & Chen, J. (2018). Preparing BiOI photocatalyst for degradation of methyl blue in wastewater. *Inorganic Chemistry Communications, 98,* 58−61.

Zhang, N., Gao, C., & Xiong, Y. (2019). Defect engineering: A versatile tool for tuning the activation of key molecules in photocatalytic reactions. *Journal of Energy Chemistry, 37,* 43−57.

Zhang, N., Li, L., Shao, Q., Zhu, T., Huang, X., & Xiao, X. (2019). Fe-doped BiOCl nanosheets with light-switchable oxygen vacancies for photocatalytic nitrogen fixation. *ACS Applied Energy Materials, 2,* 8394−8398.

Zhang, Q., Hu, S., Fan, Z., Liu, D., Zhao, Y., Ma, H., & Li, F. (2016a). Bismuth subcarbonate with designer defects for broad spectrum photocatalytic nitrogen fixation. *Dalton Transactions, 45,* 3497−3505.

Zhang, Q., Hu, S., Fan, Z., Liu, D., Zhao, Y., Ma, H., & Li, F. (2016b). Preparation of g-C_3N_4/ZnMoCdS hybrid heterojunction catalyst with outstanding nitrogen photofixation performance under visible light via hydrothermal post-treatment. *Dalton Transactions, 45,* 3497−3505.

Zhang, Q., Xu, M., You, B., Zhang, Q., Yuan, H., & Ostrikov, K. (2018). Oxygen vacancy-mediated ZnO nanoparticle photocatalyst for degradation of methylene blue. *Applied Sciences, 8,* 353.

Zhang, S., Liu, X., Liu, C., Luo, S., Wang, L., Cai, T., . . . Liu, Y. (2018). MoS_2 quantum dots growth induced by S vacancy in ZnIn2S4 monolayer: Atomic-level heterostructure for photocatalytic hydrogen production. *ACS Nano, 12,* 751−758.

Zhang, Y., Di, J., Ding, P., Zhao, J., Gu, K., Chen, X., . . . Li, H. (2019). Ultrathin g-C$_3$N$_4$ with enriched surface carbon vacancies enables highly efficient photocatalytic nitrogen fixation. *Journal of Colloid and Interface Science*, *553*, 530−539.

Zhang, Y., Gao, J., & Chen, Z. (2019). A solid-state chemical reduction approach to synthesize graphitic carbon nitride with tunable nitrogen defects for efficient visible-light photocatalytic hydrogen evolution. *Journal of Colloid and Interface Science*, *535*, 331−340.

Zhang, Y., Zhai, Y., Yu, Y., Su, Z., & Fan, X. (2020). The halogen atoms induced different oxygen vacancies in RbNa$_2$B$_6$O$_{10}$X (X = Cl, Br) for the enhanced photo-dechlorination properties. *Applied Surface Science*, *504*, 144498.

Zhang, Y. C., Afzal, N., Pan, L., Zhang, X., & Zou, J. J. (2019). Structure-activity relationship of defective metal-based photocatalysts for water splitting: Experimental and theoretical perspectives. *Advanced Science*, *6*, 1900053.

Zhang, Y. C., Li, Z., Zhang, L., Pan, L., Zhang, X., Wang, L., . . . Zou, J.-J. (2018). Role of oxygen vacancies in photocatalytic water oxidation on ceria oxide: Experiment and DFT studies. *Applied Catalysis B: Environmental*, *224*, 101−108.

Zhao, K., Qi, J., Yin, H., Wang, Z., Zhao, S., Ma, X., . . . Tang, Z. (2015). Efficient water oxidation under visible light by tuning surface defects on ceria nanorods. *Journal of Materials Chemistry A*, *3*, 20465.

Zhao, K., Zhang, L., Wang, J., Li, Q., He, W., & Yin, J. J. (2013). Surface structure-dependent molecular oxygen activation of BiOCl single-crystalline nanosheets. *Journal of the American Chemical Society*, *135*, 15750−15753.

Zhao, Y., Zhao, Y., Shi, R., Wang, B., Waterhouse, G. I. N., Wu, L. Z., . . . Zhang, T. (2019). Tuning oxygen vacancies in ultrathin TiO$_2$ nanosheets to boost photocatalytic nitrogen fixation up to 700 nm. *Advanced Materials*, *31*, 1806482.

Zhou, Y., Li, R., Tao, L., Li, R., Wang, X., & Ning, P. (2020). Solvents mediated-synthesis of 3D-BiOX (X = Cl, Br, I) microspheres for photocatalytic removal of gaseous Hg0 from the zinc smelting flue gas. *Fuel*, *268*, 117211.

Zhou, Z., Li, K., Deng, W., Li, J., Yan, Y., Li, Y., . . . Wang, T. (2020). Nitrogen vacancy mediated exciton dissociation in carbon nitride nanosheets: Enhanced hydroxyl radicals generation for efficient photocatalytic degradation of organic pollutants. *Journal of Hazardous Materials*, *387*, 122023.

Zhu, Y., Ling, Q., Liu, Y., Wang, H., & Zhu, Y. (2016). Photocatalytic performance of BiPO$_4$ nanorods adjusted via defects. *Applied Catalysis B: Environmental*, *187*, 204−211.

CHAPTER 6

Artificial Z-scheme-based photocatalysts: design strategies and approaches

KrishnaRao Neerugatti Eswar and Jaeyeong Heo

Department of Materials Engineering, Optoelectronics Convergence Research Center, Chonnam National University, Gwangju, Republic of Korea

6.1 Introduction

Heterogeneous photocatalysis is found to be one of the promising solutions for a sustainable environment with clean water and energy (Loddo et al., 2018). The revolution in this field has begun with the discovery of water splitting with TiO_2 by Fujishima and Honda (1972). The mechanism initially involves the separation of charged species in a semiconductor upon irradiating light with an appropriate wavelength, and furthermore, these charge carriers generate reactive radical species that participate in the chemical transformation. Such chemical transformation predominantly helps in water decontamination, air purification, and clean energy production (Fox & Dulay, 1993; Fujishima et al., 2007; Ibhadon & Fitzpatrick, 2013) as shown in Fig. 6.1.

For a photocatalytic reaction to be successful, its efficiency relies on a few critical parameters as follows: first, the generation of charge carriers upon light irradiation, where the term quantum efficiency is coined to determine the ratio between incident photons and generated charge carriers; second, availability of the generated electrons and holes for the catalytic reactions; third, active participation of the generated charge carriers to perform redox reactions; and finally, charge-carrier recombination happening at the surface or in the bulk of the photocatalyst (Linsebigler et al., 1995). The positions and time scales from generation to recombination for a typical photocatalyst are shown in Fig. 6.2. In particular, the recombination of the generated charge carriers poses a major challenge to the efficiency of a photocatalyst (Rothenberger et al., 1985).

In addition, visible light responsiveness has become a necessary requirement to utilize the full spectrum of solar radiation (Banerjee et al., 2014; Lang et al., 2014). Therefore impressive charge-carrier separation and visible spectrum performances are the most crucial aspects that decide the practical application of a photocatalyst. Comprehensive efforts have been made to

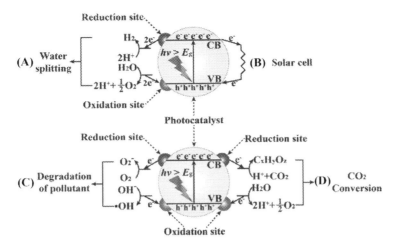

Figure 6.1

Various applications of a semiconductor. Water splitting (A), solar cell (B), water purification (C), and other organic conversion reactions (D). Source: *Reprinted with permission from Zhou, P., et al. (2014). All-solid-state Z-scheme photocatalytic systems.* Advanced Materials, *26(29), 4920–4935. Copyright 2014 John Wiley and Sons.*

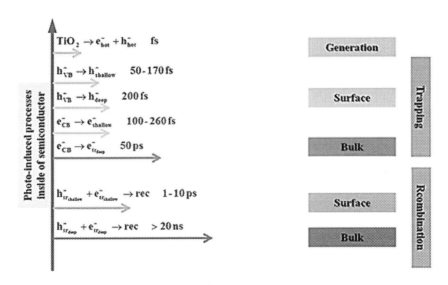

Figure 6.2

Time scales of photoinduced reactions in a photocatalyst (TiO_2). Source: *Reprinted with permission from Schneider, J., et al. (2014). Understanding TiO2 photocatalysis: Mechanisms and materials.* Chemical Reviews, *114(19), 9919–9986. Copyright 2014 American Chemical Society.*

address these crucial parameters to improve the operable viability of a photocatalyst. Incorporation of impurities/dopants, addition of external electron acceptors, surface modification of the photocatalyst, and formation of semiconducting heterojunction composites (Fujishima et al., 2008; Hashimoto et al., 2005; Wu & Wark, 2016) are some of the strategies to overcome the limitations of a photocatalyst. Among these strategies, semiconductor heterojunctions are currently trending to develop efficient photocatalytic systems where both the separation of charge species and engineering of bandgap are achieved simultaneously. Heterojunctions offer synergistic retrieval of the properties of each component in a photocatalytic composite. In particular, heterojunctions following the Z-scheme mechanism have been evolving for the past several decades to address the drawbacks in conventional photocatalysts. Heterojunctions that follow a Z-type charge migration can be denoted as Z-scheme photocatalysts where the excited electrons having low reducing potential from photocatalyst II (PC-II) recombine with the holes of photocatalyst I (PC-I). Meanwhile, long-living electrons of PC-I and holes of PC-II having high reducing and oxidizing abilities, respectively, will be available for the redox reactions. This differs from the conventional heterojunctions by facilitating the electrostatic attraction of charge carriers from electron-rich PC-II to hole-rich PC-I. Z-scheme photocatalysts not only address the charge separation in the heterojunctions but also avoid the electrostatic repulsion between electron—electron or hole—hole pairs. Therefore to have a thorough understanding, it is highly essential to summarize the basic principles or mechanistic aspects, recent discoveries, and the progress in the fabrication of artificial Z-scheme photocatalysts. This book chapter initially describes the importance of heterojunctions and their classifications. Furthermore, a brief overview of the history of Z-scheme mechanisms and designs, implications in photocatalysis, and their advances is discussed. It also includes an updated overview of the recent developments in Z-scheme photocatalysts and the suitable features that are needed for practical implementation are discussed concisely.

6.2 Types of heterojunctions

The fundamental mechanism of a photocatalyst reiterates that the recombination of the photogenerated charge carriers is deleterious. To achieve higher photocatalytic efficiency, it is required to separate the charge carriers and must be transported to the surface or interface. Besides, to understand the mechanism of a Z-scheme photocatalyst, it is essential to have a preliminary knowledge of the semiconductor heterojunctions. The design and fabrication of photocatalysts having semiconductors coupled with metal, noble metal (NM), and other semiconductors lead to the formation of active heterojunctions. Heterojunction semiconductors can be classified based on their components and design. Based on components, it can be categorized into (1) semiconductor—semiconductor (S—S)-, (2) semiconductor—metal (S—M)-, and (3) semiconductor—carbon (S—C)-based compounds and (4) multicomponent heterojunctions.

6.2.1 Semiconductor—semiconductor (S-S) based heterojunctions

S—S-based heterojunctions can further be divided into two categories: (1) p—n and (2) non-p—n heterojunction systems. p—n heterojunction photocatalysts possess the most effective design for a better charge collection and separation process. Diffusion of the charge carriers at the interface across the semiconductors creates a space-charge region to form a p—n junction as shown in Fig. 6.3A. Such diffusion continues to occur until the Fermi level equilibrium is reached. This causes development of built-in potential that facilitates the migration of electrons and holes in the opposite direction. Due to such potential, the photogenerated electrons migrate to the conduction band of n-type semiconductor and the holes to the valence band of p-type semiconductor (Hagfeldt & Graetzel, 1995).

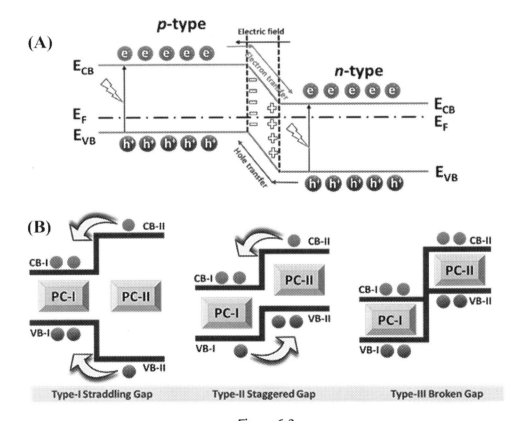

Figure 6.3
Different types of semiconductor—semiconductor (S—S) heterojunctions. p—n heterojunction (A) and non-p—n heterojunctions (B). PC, CB, and VB stand for photocatalyst, conduction band, and valence band respectively.

p—n junction photocatalysts found to be productive in charge separation and transfer, maintaining a long lifetime of electrons/holes and further enabling localized redox reactions. Non-p—n heterojunctions can be classified broadly into four categories, such as type-I, -II, and -III and Z-scheme. Among them, only type-II heterojunction systems are the most favorable for photocatalytic reactions whereas type-I and -III suffer from high recombination due to the spatial separation of excited charge carriers and nonoverlapping bandgaps, respectively, as shown in Fig. 6.3B. Basically, in the type-II mechanism, upon excitation, electrons from the semiconductor-I (PC-I) having a high conduction band (CB) position moves to semiconductor-II (PC-II) at a relatively lower CB position whereas the holes move in the reverse direction. Charge distribution and built-in potential between the semiconductors in a staggered-band photocatalyst depend on their work function difference. Shifu et al. developed NiO/TiO_2 p—n junction photocatalyst using sol—gel method for the reduction of $Cr_2O_7^-$ and oxidation of rhodamine B. In addition, they also studied the effect of heat treatment on the photocatalytic activity of p—n junction photocatalyst (Shifu et al., 2008). It was clearly established that the photocatalytic activity using p—n junction photocatalyst was significantly high compared to the individual components. Over the years, versatile p—n junction photocatalysts, such as TiO_2 and SnO_2/Bi_2WO_6/CdS/ZnO/Cu_2O (Chen et al., 2008; Shang et al., 2009; Shifu et al., 2009; Yu et al., 2003; Wang et al., 2009), Bi_2O_3-Bi_2WO_6 (H. Wang, et al., 2013), $BiVO_4$ and MnO_x/WO_3/FeOOH-NiOOH/AgI/CoO_x (Hong et al., 2011; Kim & Choi, 2014; Li et al., 2014; Petala et al., 2019; Zhao et al., 2019), Ta_3N_5-IrO_2 (D. Wang et al., 2013), WO_3-Ag_2CO_3 (Gao et al., 2019), and $FeVO_4$-BiOCl (Eshaq et al., 2020), have been developed using different approaches and tested for various reactions, such as photodegradation of dyes, organic compounds, and water splitting.

6.2.2 Semiconductor—metal (S-M)based heterojunctions

S—M heterojunction photocatalysts rely on the formation of a Schottky barrier at the interface between semiconductor and metal (mostly NM) where migration of electrons takes place based on their Fermi levels leaving the holes behind (Dutta et al., 2015; Ingram et al., 2011; Nosaka et al., 1984; Wu, 2018). NMs like Au, Ag, Pt, Pd, Ru, Ir, and Rh have been deposited using different fabrication processes on semiconductors to enhance the overall efficiency by allowing effective electron transfer across the interfaces. Such photocatalysts have been widely tested for both energy- and environmental-related photocatalytic reactions. In addition, localized surface plasmon resonance contributed by the metals improves light absorption and quantum efficiency of the photocatalyst.

6.2.3 Semiconductor—carbon (S-C)based heterojunctions

Carbon and carbon-based semiconducting compounds are utilized in forming heterojunctions with other metal oxide/sulfide semiconductors. Carbons, such as activated carbon, carbon nanotubes, graphene, and reduced graphene oxide, are predominantly used for their high surface area. Activated carbon with metal oxide like TiO_2 (Tanguay et al., 1989) increases the overall surface area of the photocatalysts, thereby providing more sites for adsorption and catalysis. Besides surface area, carbon nanotubes also possess excellent metallic conductivity allowing the formation of Schottky barrier heterojunction with the semiconductor (Hoffmann et al., 1995; Woan et al., 2009). Similar to carbon nanotubes, incorporating graphene with the semiconductor can also provide excellent charge transfer properties. In addition, several studies show that particle nucleation on planar carbon structures like graphene, graphene oxide, and reduced graphene oxide grants a thorough dispersion of semiconducting particles on their surfaces. Although graphene oxide is basically nonconductive, reduction of graphene oxide to reduce graphene oxide and further fabrication of S−C heterojunctions have been found to be useful for photocatalytic reactions. Electrons transferred from the semiconductor to carbon structures are usually scavenged by the dissolved oxygen present in the system, thereby contributing to charge-carrier separation and reducing the recombination as well (Du et al., 2011; Gao et al., 2011; Huang et al., 2012). Furthermore, graphitic carbon nitride (g-C_3N_4) (Zheng et al., 2012), a recent carbon-based semiconductor, is widely being used for the formation of S−C heterojunctions with other semiconductors (Akhundi et al., 2019). g-C_3N_4 is similar to graphene oxide with carbon and nitrogen instead of oxygen atoms forming interconnected tri-s-triazenes. Although it is considered as an intrinsic metal-free semiconductor, it is appropriate to include it under this section for a better understanding of the categorization of heterojunctions (Cheng et al., 2015; Darkwah & Oswald, 2019; Dong & Cheng, 2015).

6.2.4 Multicomponent heterojunctions

Multicomponent heterojunctions are fabricated primarily for two reasons. (1) to facilitate a "spatial-integrated" charge transfer system and (2) to bestow both UV and visible light responsiveness for the photocatalytic system. Such systems usually consist of binary, ternary, or quaternary components. Each individual photocatalyst has its characteristic bandgap in either UV or visible region. To allow better charge transfer and also to prevent recombination, the UV and visible light photocatalysts are spatially integrated with a metal heterojunction. At such junctions, the photogenerated electrons and holes from different components migrate toward the metal junction, which further gets stored or undergoes recombination. Meanwhile, a "vectorial" electron transfer occurs from one semiconductor to other through the metal junction (Schottky barrier) based on their Fermi level. Meanwhile, the electrons with better reduction potential and holes with high oxidation potential from each semiconductor participate in the photocatalytic reactions (Marschall, 2014; Wang et al., 2014).

6.3 History of Z-scheme reactions

Although the direct Z-scheme photocatalysts are a recent innovation, the history of Z-scheme reactions goes past several decades. Artificial Z-scheme photocatalysis mimics the natural photosynthesis process consisting of two semiconductors meant for oxidation and reduction. Z-scheme charge transfer reactions initially originated with the liquid phase further evolving to solid state and currently reached a direct Z-scheme process as shown in Fig. 6.4.

Hill and Bendall (1960) proposed a working hypothesis on a spontaneous electron transfer mechanism between CO_2 and water molecules in cytochrome components of plants. Such a complex process has inspired to the development of several simple photocatalytic systems using heterojunctions and mediators (Grätzel, 2001).

$$\text{acceptor} + e^- (\text{CB of SC-II}) \rightarrow \text{donor} \tag{6.1}$$

$$\text{donor} + h^+ (\text{VB of SC-I}) \rightarrow \text{acceptor} \tag{6.2}$$

Initially, a Z-scheme catalyst was proposed to perform charge transfer reactions in the liquid phase (Bard, 1979) stimulated using a shuttle redox mediator (first-generation Z-scheme), such as Fe^{3+}/Fe^{2+}, IO_3^-/I^-, and NO_3^-/NO_2^-, between two different photocatalysts as given in Eqs. (6.1) and (6.2) (Abe et al., 2005). Even complex molecules, such as $[Co(phen)_3]^{3+}/[Co(phen)_3]^{2+}$ and $[Co(bpy)_3]^{3+}/[Co(bpy)_3]^{2+}$, were also utilized as redox mediators by Sasaki et al. (2013). The crucial step in such a reaction is to couple charge carriers of weak potential with the help of mediators. These redox ion mediators behave as acceptor and donor, thereby freeing the charged species with high potential for reduction and oxidation reactions, respectively, as shown in Fig. 6.5A. Despite their

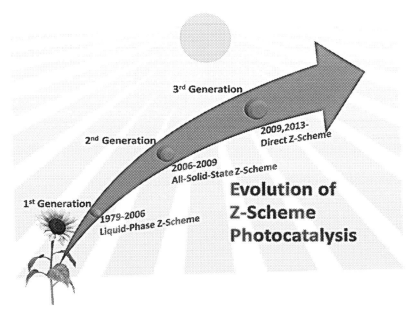

Figure 6.4

Schematic representation of the evolution of Z-scheme photocatalytic transfer reactions.

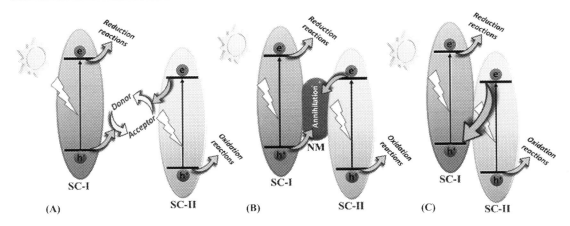

Figure 6.5

Schematic representation of liquid phase Z-scheme—first generation (A), all-solid-state Z-scheme—second generation (B), and direct Z-scheme—third generation (C) photocatalysts. SC and NM stand for semiconductor and noble metal, respectively.

advantages, first-generation photocatalysts possess several physiochemical limitations especially their unstable nature. For instance, the ratio of redox ion pairs constantly varies as the Z-scheme reaction proceeds, resulting in poor charge transferability overall. The ionic species in the photocatalytic system tend to consume the strong redox-driven photogenerated electrons and holes with thermodynamic feasible back reactions. Furthermore, a decrease in charge transfer rates due to diffusion-limited ion pairs, being sensitive to pH and light, hinders the rate of Z-scheme reactions. Besides these disadvantages, first-generation Z-scheme photocatalysts can work only for liquid-phase reactions. Meanwhile, an "all-solid-state (ASS)" (second generation) was developed to counter the problems faced with liquid-phase reactions with an insoluble electron mediator like an NM nanoparticle (Tada et al., 2006).

Introducing the NM between the semiconductors creates interfaces where the electrons from SC-II are easily transferred, thereby reducing the distance of migration as shown in Fig. 6.5B (Veldurthi et al., 2018a, 2018b, 2018c). NM as electron mediator possesses better work function, strong light absorption, nonphotocorrosive property, and inertness in the reaction medium. Furthermore, they also impart plasmonic property to the photocatalyst. In addition, it facilitates a low resistance Ohmic contact where the photogenerated electrons from SC-II recombine with holes from SC-I which leaves strong reductive/oxidative electrons and holes respectively for redox reactions. Nevertheless, NM as an electron mediator does carry certain drawbacks. NM suppresses the light absorption for the semiconductors due to their strong absorbing nature. Besides, they are highly expensive to construct Z-scheme photocatalysts on a large scale. Considering these limitations, X. Wang et al. (2009) and Yu et al. (2013) proposed and developed a direct Z-scheme photocatalyst.

Direct Z-scheme photocatalysts (third generation) imbibe the benefits of both first- and second-generation Z-scheme photocatalysts, that is, enhanced charge-carrier separation with tunable redox potential regardless of introducing any liquid/solid phase redox mediators. In the direct Z-scheme photocatalysts, electrons from SC-II combine with the holes from SC-I remaining the strong reducing electrons and oxidizing holes of SC-I and SC-II respectively as shown in Fig. 6.5C. Such a simplified process of direct charge transfer based on the band edges of the semiconductor components provides enough degree of freedom to engineer the Z-scheme photocatalysts for specific reactions. Moreover, constructing a direct Z-scheme photocatalyst is highly economical.

6.4 Confirmation of direct Z-scheme mechanism

6.4.1 Experimental

Since direct Z-scheme is an enhanced version of type-II band configuration, it is necessary to assure that the photocatalytic charge transfer process follows a direct Z-scheme path. Several techniques are currently being followed to confirm the electron/hole migration path and each method has a unique approach to identify the same. For instance, photocatalytic reactions like hydrogen evolution are a "self-confirming" reaction for direct Z-scheme mechanism where only the photogenerated electrons with appropriate reducing potential can perform the water-splitting reaction. Possibilities of conventional type-II heterojunctions (Veldurthi et al., 2018a, 2018b, 2018c) can be ruled out if the electrons from one semiconductor migrate to another having a low CB. Such migration cannot perform reduction reactions (as shown in Table 6.1), such as water splitting or electron-rich CO_2 reduction to methane. Similarly, targeted the photodeposition of NMs on an "electron-rich" reduction region of one of the semiconductor components would signify the potential Z-scheme transfer path (Xu et al., 2014). Furthermore, the photocatalytic reaction in the presence of scavengers (charge trapping molecules) would result in understanding the fate of reactive radical species (for e^-, h^+, $O_2^{-\bullet}$ and $OH^{-\bullet}$) to verify the Z-scheme mechanism (Meng et al., 2015). On the other hand, the binding energies of

Table 6.1: Various reduction reactions and their appropriate reducing potentials.

Redox potential[a] (V)	Reduction reaction	
	Reactants	**Products**
-0.24	$CO_2 + 8H^+ + 8e^-$	$CH_4 + 2H_2O$
-0.38	$CO_2 + 6H^+ + 6e^-$	$CH_3OH + H_2O$
-0.41	$2H^+ + 2e^-$	H_2
-0.48	$CO_2 + 4H^+ + 4e^-$	$HCHO + H_2O$
-0.53	$CO_2 + 2H^+ + 2e^-$	$CO + H_2O$
-0.61	$CO_2 + 2H^+ + 2e^-$	$HCOOH$

[a]Reduction potential versus Normal Hydrogen Electrode (NHE) at pH 7.

elements in photocatalytic Z-scheme components would vary upon undergoing photocatalytic reactions. Such variation can be identified and analyzed using X-ray photoelectron spectroscopy (Aguirre et al., 2017; Di et al., 2017).

6.4.2 Theoretical

Density functional theory (DFT)-based theoretical "first-principles simulation" of the component semiconductors in a Z-scheme system may offer much deeper insights on charge-carrier dynamics (Low et al., 2017). The prerequisite for such calculation is the bandgap of the component semiconductors obtained using the first-principle studies by DFT. Furthermore, a parabolic fitting of the conduction and valence band maxima would give the effective mass of the photogenerated charge carriers using Eqs. (6.3) and (6.4), respectively.

$$m^* = \hbar^2 \left(d^2E/dk^2 \right)^{-1} \tag{6.3}$$

$$v = \frac{\hbar k}{m^*} \tag{6.4}$$

$$D = \frac{m_h^*}{m_e^*} = \left| \left(d^2E/dk^2 \right)_{\text{CB}} / \left(d^2E/dk^2 \right)_{\text{VB}} \right| \tag{6.5}$$

m^*, $\hbar$, d^2E/dk^2, k, and v represents the effective mass of the charge carrier, Planck's constant, second-order differentiated term of quadratic fit of $E(k)$ for individual band edge, wave vector, and charge-carrier transfer rates respectively. The rates of recombination can be found out easily using Eq. (6.5) where D represents the relative ratio of the effective masses of the charge carriers. A higher value of D signifies a lower recombination rate of the photogenerated electron−hole pairs (Yu et al., 2015; Zhang et al., 2012). In addition to effective mass calculations, simulation of the internal electric field by considering the surface and the interface-formation energy would give further insights on the performance of a direct Z-scheme photocatalyst. The band offsets between the semiconductor components are very critical that describes the interface and performance. First principles and "potential-line-up method" together aid in assessing the band offsets. The efficient interface formation also depends on the high-active sites of the photocatalysts. For instance, among all other surfaces, (100) of TiO_2 tends to be highly active for photocatalysis due to the efficient distribution of Ti_{5c} atoms on the surface. Liu et al. (2016) have calculated band offsets by constructing the interface using a monolayer of (001) g-C_3N_4 nanosheet and (100) TiO_2. The difference in the three-dimensional charge density was computed to realize the charge transfer and separation dynamics. Fig. 6.6A and B shows the charge redistribution happening predominantly at the interface of g-C_3N_4 and TiO_2. Electrons have migrated from g-C_3N_4 to TiO_2 across the interface leaving holes behind which can be

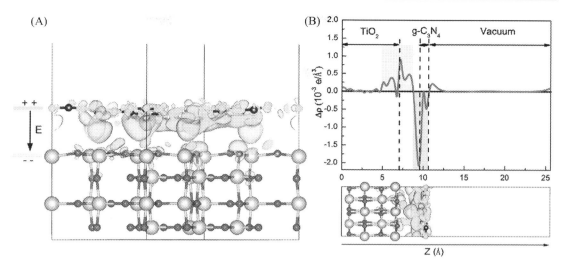

Figure 6.6

Charge density difference (A) and planar-averaged density difference for g-C$_3$N$_4$/TiO$_2$. The yellow-colored region denotes the electron accumulation whereas cyan represents depletion. Source: *Reprinted with permission from Liu, J., et al. (2016). A new understanding of the photocatalytic mechanism of the direct Z-scheme g-C3N4/TiO2 heterostructure. Physical Chemistry Chemical Physics, 18(45), 31175–31183. Copyright 2016 The Royal Society of Chemistry.*

distinctly seen from the accumulation and depletion regions of the planar-averaged charge density calculation.

6.5 Synthesis and various design strategies for fabricating Z-scheme photocatalysts

As discussed earlier, a firm contact is essential between the semiconductor components to construct a high-active, efficient direct Z-scheme photocatalyst. Therefore the design and synthesis strategies play a crucial role in deciding the interface formation. Usually, one or more synthesis methods are followed for developing a direct Z-scheme photocatalyst. Versatile synthesis procedures for fabricating direct Z-scheme photocatalysts are tabulated in Table 6.2. It is well-known that the size and configuration of a nanomaterial determine its physiochemical properties. In addition, growth confinement, site/surface-specific growth, morphological variations, etc., as well contribute to the fate of interfaces and heterojunctions formation by altering the electronic band structures. A direct Z-scheme photocatalyst predominantly consists of two semiconductors meant for oxidation and reduction reactions. Individual photocatalysts may perform better for a specific reaction. However, when two or more photocatalytic components are embedded together, the geometry of construction becomes decisive affecting the efficiency of the photocatalyst.

Table 6.2: Different synthesis strategies, critical governing parameters, advantages, and limitations for developing direct Z-scheme photocatalyst.

Synthesis	Process	Critical parameter	Advantages	Limitations
Mechanical	Physical agitation of the individual semiconductor components	External physical force, such as milling, sonication, and grinding	Quick process	Poor interfaces, agglomeration
Solid state	Calcination of semiconductor precursors at high temperatures and further annealing	Temperature	Fast, easy	Low surface area. Mostly followed for oxides and carbon compounds
Precipitation	Heterogeneous nucleation of one of the semiconductors on a base/substrate semiconductor	Concentration of precursors, pH, temperature, stirring	Simple and reliable	Formation of clusters of particles at high concentration
Hydro/solvo-thermal	Temperature and pressure-controlled synthesis using an autoclave	Temperature, pressure, solvent, time	High crystalline, narrow size distribution, morphology control	Mostly a two-step process
Ion exchange	Exchange of ions by simultaneously creating an interface	Time, temperature, solvent chemical potential, inert atmosphere	Janus, uniform size, excellent interface, precise control	Low yield, challenging to scale up
Electrospinning	A polymeric solution of precursors jetted using voltage onto a substrate to get nanofibers	Viscosity, dispersibility, voltage, radius of the needle orifice	Ultrafine 1 D uniform fibers, can be scaled up easily	Annealed nanofibers may become fragile
Self-assembly/self-organized	Spontaneous, free energy-assisted ordering of components	Electrostatic attraction, Van der Waals forces, temperature, pH, surface tension	Janus, excellent interface, 0D/1D/2D over 2D heterojunctions	Self-organizing is random

6.5.1 Importance of geometrical configurations for direct Z-scheme photocatalysts

Several studies have been performed to realize the effect of geometrical configurations to narrow the charge transfer of photocatalytic systems toward the Z-scheme path. Such configurations can be broadly classified into three main categories: (1) surface-embedded geometry (Fig. 6.7A), (2) Janus-type geometry (Fig. 6.7B), and (3) core–shell configuration (Fig. 6.7C). Each type of geometry has its own benefits not only for an overall photocatalytic efficiency but also for specific reactions as well.

6.5.1.1 Surface-embedded geometry

In the surface-embedded configuration, a small-size photocatalyst is embedded (grown) on the surface of a large-size component for an overall Z-scheme system as shown in Fig. 6.8. The base photocatalyst can be simply a nanoparticle or nanomaterial with any morphology as well. The surface area of the base photocatalyst decides the nucleation and growth of the succeeding photocatalyst. Developing a direct Z-scheme photocatalyst with this configuration is much easier compared to other methods (Wang et al., 2016) and, however,

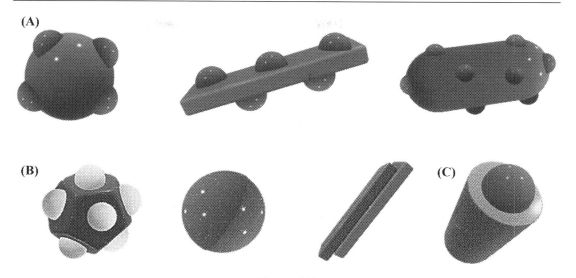

Figure 6.7

Various geometrical configurations for fabricating direct Z-scheme photocatalyst. Surface-embedded configuration (A), Janus-type geometry (B), and core—shell configuration (C).

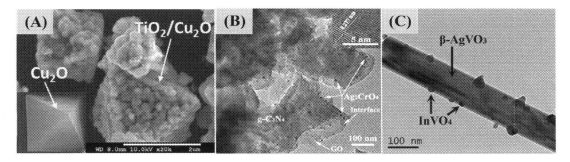

Figure 6.8

Direct Z-scheme photocatalysts. TiO_2 onto pyramid morphology of Cu_2O (A), Ag_2CrO_4 nanoparticles on g-C_3N_4 nanosheet (B), and $InVO_4$ on β-$AgVO_3$ (C) via cation exchange. Source: *Reprinted with permission from Aguirre, M. E., et al. (2017). Cu_2O/TiO_2 heterostructures for CO_2 reduction through a direct Z-scheme: Protecting Cu_2O from photocorrosion.* Applied Catalysis B: Environmental, 217, 485—493. Copyright 2017 Elsevier B.V. Xu, D., et al. (2018). *Ag_2CrO_4/g-C_3N_4/graphene oxide ternary nanocomposite Z-scheme photocatalyst with enhanced CO_2 reduction activity.* Applied Catalysis B: Environmental, 231, 368—380. Copyright 2018 Elsevier B.V. Yang, J., et al. (2019). *$InVO_4$/β-$AgVO_3$ nanocomposite as a direct Z-scheme photocatalyst toward efficient and selective visible-light-driven CO_2 reduction.* ACS Applied Materials & Interfaces, 11, 35, 32025—32037. Copyright 2019 American Chemical Society.

accompanied by certain limitations. Since nucleation and growth are preferred over a large surface area, clustering or agglomeration of the second photocatalyst is likely to happen as seen in Fig. 6.8A. However, better dispersion of nanoparticles can be achieved if the base component has a large/planar surface as shown in Fig. 6.8B. Non-site-specific nucleation and growth result in varied charge-carrier dynamics and may confine the light absorption. On the other hand, photodeposition of the second component onto the base photocatalyst may bring more control to the nucleation process and could be site specific as well. While the base photocatalyst is excited, the charge carrier is transported to the surface where the reactants of the second photocatalytic component participate in the reaction to get deposited on high-energy sites, thereby creating a firm contact/interface (Jiang et al., 2018).

Besides direct Z-scheme, ASS Z-scheme photocatalysts can also be subcategorized into two categories, such as photocatalytic "composite particles" and "particulate sheets" that lie under similar geometrical configurations as shown in Fig. 6.9A and B (Xia et al., 2019). Composite particles include closely integrated SC-I, SC-II, and NM to enable a smooth migration of the photogenerated charge carriers and quenching/annihilation at the NM junction. On the other hand, particulate sheet photocatalysts consist of immobilized SC-I and SC-II onto a conductive thin film. Such design eases the resistance for charge transfer and enhances the overall redox reaction capacity.

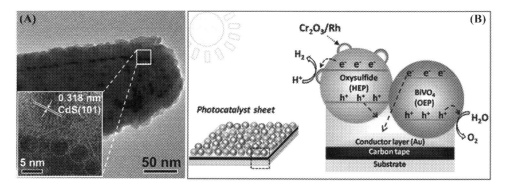

Figure 6.9
All-solid-state Z-scheme composite (A) and particulate sheet (B) photocatalysts. Composite photocatalyst consisting core—shell structure having ZnO and CdS with Au nanoparticles embedded in between. Particulate photocatalyst having two different semiconductors immobilized onto Au layer. Source: *(A) Reprinted with permission from Zhang, N., et al., (2016). Vertically aligned ZnO—Au@CdS core—shell nanorod arrays as an all-solid-state vectorial Z-scheme system for photocatalytic application.* Journal of Materials Chemistry A, 4(48), 18804—18814. *Copyright 2016 Royal Society of Chemistry. (B) Reprinted with permission from Sun, S., et al. (2018). Efficient redox-mediator-free Z-scheme water splitting employing oxysulfide photocatalysts under visible light.* ACS Catalysis, 8(3), 1690—1696. *Copyright 2018 American Chemical Society.*

6.5.1.2 Janus-type configuration

Individual photocatalytic components with approximately similar sizes synthesized in situ exhibiting distinct properties as required for direct Z-scheme photocatalysts. Although such configuration provides highly efficient photocatalysts, synthesizing Janus particles is very challenging. Masking of surfaces, selective nucleation to grow particles with different properties, self-assembly, and phase separation are some of the routes followed so far to grow Janus particles (Jiang et al., 2010; Li et al., 2011). In addition to having a distinct chemical recipe, scaling up of synthesis process on a large scale is not quite certain. Yet, their interesting properties have attracted much attention for fabricating direct Z-scheme photocatalysts. Yuan et al. (2017) have developed γ-MnS and Cu_7S_4 direct Z-scheme Janus heterojunctions through cation-exchange (Fig. 6.10A—A-5) for overall water-splitting reactions. Such particles with firm interfaces enabled efficient charge transfer and performed site-specific redox half-reactions on the surfaces of semiconductor components. A similar type of synthesis method was followed by Yang et al. to grow $InVO_4$ particles on β-$AgVO_3$ nanorods as seen in Fig. 6.8C. Likewise, Li et al. (2019) have developed $ZnIn_2S_4$/WO_3 Janus heterostructures in the form of planar configuration by liquid exfoliation and lithiation, respectively, as depicted in Fig. 6.10B and C. These planar type heterostructures allow for large interface area, thereby enabling better charge transport properties and reducing the charge migration distance across the interface to surface as well.

6.5.1.3 Core—shell-type architecture

Z-scheme photocatalysts that are differentiated with definite boundaries as inner (core) and outer (shell) semiconductors belong to core—shell architecture as shown in Fig. 6.10D and E. The photocatalytic efficiency of a core—shell Z-scheme catalyst primarily depends on the interactions of core—shell components, geometrical configuration arising due to reactivity of surfaces during the growth process. Such geometry can be synthesized using a wide variety of methods. However, the hydrothermal method is mostly preferred to obtain uniform growth. Zhang et al. have developed a core—shell WO_3/SnS_2 direct Z-scheme heterojunction using the two-step hydrothermal method where the core (WO_3) is synthesized initially and further reacted with shell precursors to form core—shell geometry (Zhang et al., 2019). Although core—shell architecture facilitates an excellent interface for heterojunction, it still possesses few crucial drawbacks. The core semiconductor is not directly exposed to the reaction environment. This leads to the accumulation of photogenerated charge carriers, which further restricts the charge migration between the core—shell components. Furthermore, thick layers of shell components limit the light absorption by the core material. However, researchers have found a way to mitigate these problems by developing porous core—shell configuration. Xu et al. have developed TiO_2/NiS core—shell direct Z-scheme nanofibers by combining electrospinning and hydrothermal synthesis procedures, resulting in TiO_2 core decorated with few layers of NiS shell nanoplates (F. Xu et al., 2018).

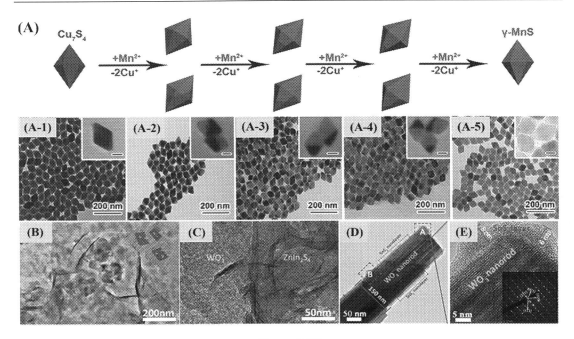

Figure 6.10

Fabrication of direct Z-scheme photocatalysts with Janus-type geometry. Synthesis of γ-MnS and Cu₇S₄ using the cation-exchange method with precise control on the reaction time. Schematic representation of cation exchange reaction (A), initial Cu7S4 nanocrystals (A-1), growth of γ-MnS on Cu₇S₄ after 0.5 h (A-2), 1 h (A-3), and 1.5 h (A-4), and complete transformation to γ-MnS (A-5). Planar nanosheets of ZnIn₂S₄/WO₃ Janus heterostructures (B and C). Core—shell architecture of WO₃/SnS₂ direct Z-scheme photocatalyst (D and E). Source: (A (A, A-1 to A-5)): *Noble-Metal-Free Janus-like Structures by Cation Exchange for Z-Scheme Photocatalytic Water Splitting under Broadband Light Irradiation. Angewandte Chemie 129(15): 4270—4274 6.10 (B and C): Two-dimensional Janus heterostructures for superior Z-scheme photocatalytic water splitting. Nano Energy 59: 537—544 6.10 (D and E): Construction of core-shell structured WO3@SnS2 hetero-junction as a direct Z-scheme photo-catalyst. Journal of Colloid and Interface Science 554: 229—238. Copyright 2019 Elsevier.*

6.6 Future perspectives in developing direct Z-scheme photocatalytic systems

The three basic types of photocatalytic applications are water purification, water-splitting reactions, and carbon dioxide conversion. Each reaction has its own electrochemical potential that needs to be overcome by the flat band potential of the photocatalyst for the reaction to happen. Most of the metal oxides with some exceptions have their valence bands located appropriately for oxidative reactions. On the other hand, the metal sulfides have their conduction band positions suitably located near the reduction potential region. Several researchers have combined these materials together to achieve better photocatalytic efficiency. With respect to nanomaterials, metal oxides, metal sulfides, mixed metal oxides/sulfides (M-O or M-S), other compounds, such as metal halides (M-X, X = F, Cl, Br, I), phosphides (M-P), phosphates (M-PO_4), vanadates (M-VO_x), molybdates (M-MoO_x), and tungstates (M-WO_x), have been developed for various photocatalytic reactions. Recently, semiconductor integrated with carbon compounds like graphene, and especially graphitic carbon nitride (g-C_3N_4), is paid more attention. Synthesizing graphitic carbon nitride is much simpler than other semiconductors. Besides, planar nanosheet-like structures of g-C_3N_4 enable better interface formation with component semiconductors. With the ease of making planar structures, metal sulfides form good interfacial heterojunctions with both carbonaceous and noncarbonaceous planar structures like g-C_3N_4 and boron nitride for making direct Z-scheme photocatalyst. Among other non-conventional Z-scheme photocatalysts, layered double hydroxides (LDHs) and metal-organic frameworks (MOFs) are given more importance owing to their excellent properties. LDHs are basically complex-layered compounds with a divalent and trivalent cation intercalated by charge-balancing anions. LDHs can be empirically expressed as $[M_{1-x}^{2+} M_x^{3+} (OH)_2] (A^{n-})^{x/n} \cdot mH_2O$ (Yin & Tang, 2016) where M represents metal and A denotes NO_3, CO_3, Cl ions respectively. LDHs possess semiconductor properties comprising plenty of surface hydroxyl groups, tunable ionic species, in addition to their anisotropic nature, to become promising photocatalytic components. Hybrid interfaces with other planar or particulate semiconductors would impart the extended wavelength absorption and absolves from the ill effects of poor charge ability of LDHs. However, there is only a little research established for this new photocatalytic composite to develop an overall direct Z-scheme photocatalyst. In addition, a recent upsurge of a new class of highly porous, three-dimensional compounds called MOFs competes with LDHs realizing their potential in photocatalysis. MOFs are compounds with metal ions (clusters) complexed with organic ligand molecules to form a three-dimensional porous architecture. Their porous structures with large specific surface areas and active metal sites help in providing better sorption properties. These complex organic ligands also enable "proton-coupled electron transfer"; "linker to cluster" charge transfer for redox reactions at metal sites makes them suitable candidates for photocatalytic reactions (Furukawa et al., 2013). Despite being widely employed in catalytic reactions, only a few works are available in employing MOFs for direct Z-scheme photocatalysis.

As described earlier, a direct Z-scheme photocatalyst predominantly consists of two or more components that can be considered as the best only when it possesses both reduction and oxidation components for respective redox reactions. There are several research articles published on synthesizing direct Z-scheme-based photocatalysts for water decontamination or pollutant degradation. However, such reactions are thermodynamically favorable spontaneous reactions. Therefore solving thermodynamic uphill reactions for managing the energy and environmental concerns must be given high priority. Likewise, most of the direct Z-scheme photocatalysts developed so far were solely utilized for hydrogen evolution reactions, that is, utilizing the complete photocatalytic system for one-half reaction. This has led to a rapid increase in the number of publications with more focus on the combination of material composites but without any significant progress for the rapidly growing environmental requirements. An important fact often overseen is the challenge in the simultaneous balancing of the thermodynamics and kinetics of these two reactions. To do so, thorough knowledge is necessary to understand the oxidation and reduction sites and further site-specific reactions as well. First-principle studies, DFT calculations could help in designing highly active direct Z-scheme photocatalysts. However, engineering such compounds with precise band positions is an uphill task. Furthermore, utilizing sacrificial reagents for HER (hydrogen evolution reaction) and OER (oxygen evolution reaction) never resembles the real-life reaction environment. Some of these challenges in photocatalytic experiments can be overcome by employing photoelectrocatalysis. Photoelectrocatalysis is a much more viable solution for green energy where reproducible photoelectrodes take part in the photochemical reaction. However, those reactions should preferably be an overall water-splitting reaction with better "solar-to-hydrogen" conversion efficiency. In addition to water-splitting reactions, the conversion of carbon dioxide to value-added products must be addressed with finesse by developing highly active thin films of direct Z-scheme catalysts. As it is well-known, unlike pollutant transformation or water-splitting reactions, gas-phase reactions are surface limited since light cannot penetrate through the bulk of the catalytic bed. Thus photocatalytic or photoelectrocatalytic thin films with high surface area, multiple adsorption sites, and sites with greater affinity may give better conversion yield. Therefore for a better Z-scheme photocatalyst, the component semiconductors should be able to absorb light across a wide wavelength. It must be able to facilitate excellent charge separation. To achieve that, a firm and highly dispersed contact is essential where multiple heterojunctions act as an individual reaction site. On the other hand, a huge difference between the flat band potential and the electrochemical potential of a desired reaction causes a drop in photovoltage necessary for driving the reaction. In such cases, a supplementary deposition of a thin layer of semiconductor of appropriate bandgap with respect to the bandgap of the base material may induce band bending that leads to an increase in the open-circuit voltage upon light irradiation. Separation of reduction and oxidation sites with Janus-type architecture of an overall catalyst will drastically improve the efficiency of the reaction. Thus to build an inexpensive yet efficient direct Z-scheme photocatalytic system, the choice of semiconductors, design and band engineering of the components, tunable geometry, built-in electric field-driven charge transfer, and excellent interfacial properties may result in superior photocatalytic ability. Yet, there is still a huge scope to address not only the crucial drawbacks but also the availability of ample space in developing an efficient artificial direct Z-scheme photocatalyst at par with the natural Z-scheme process.

References

Abe, R., Sayama, K., & Sugihara, H. (2005). Development of New Photocatalytic Water Splitting into H2 and O2 using Two Different Semiconductor Photocatalysts and a Shuttle Redox Mediator IO3-/I. *The Journal of Physical Chemistry B, 109*(33), 16052−16061.

Aguirre, M. E., Zhou, R., Eugene, A. J., Guzman, M. I., & Grela, M. A. (2017). Cu2O/TiO2 heterostructures for CO2 reduction through a direct Z-scheme: Protecting Cu2O from photocorrosion. *Applied Catalysis B: Environmental, 217*, 485−493.

Akhundi, A., Habibi-Yangjeh, A., Abitorabi, M., & Rahim Pouran, S. (2019). Review on photocatalytic conversion of carbon dioxide to value-added compounds and renewable fuels by graphitic carbon nitride-based photocatalysts. *Catalysis Reviews, 61*(4), 595−628.

Banerjee, S., Pillai, S. C., Falaras, P., O'Shea, K. E., Byrne, J. A., & Dionysiou, D. D. (2014). New Insights into the Mechanism of Visible Light Photocatalysis", The. *Journal of Physical Chemistry Letters, 5*(15), 2543−2554.

Bard, A. J. (1979). Photoelectrochemistry and heterogeneous photo-catalysis at semiconductors. *Journal of Photochemistry, 10*(1), 59−75.

Chen, S., Zhao, W., Liu, W., & Zhang, S. (2008). Preparation, characterization and activity evaluation of p−n junction photocatalyst p-ZnO/n-TiO2. *Applied Surface Science, 255*(5), 2478−2484, Part 1.

Cheng, H., Hou, J., Takeda, O., Guo, X.-M., & Zhu, H. (2015). A unique Z-scheme 2D/2D nanosheet heterojunction design to harness charge transfer for photocatalysis. *Journal of Materials Chemistry A, 3* (20), 11006−11013.

Darkwah, W. K., & Oswald, K. A. (2019). Photocatalytic Applications of Heterostructure Graphitic Carbon Nitride: Pollutant Degradation, Hydrogen Gas Production (water splitting), and CO2 Reduction. *Nanoscale Research Letters, 14*(1), 234.

Di, T., Zhu, B., Cheng, B., Yu, J., & Xu, J. (2017). A direct Z-scheme g-C3N4/SnS2 photocatalyst with superior visible-light CO2 reduction performance. *Journal of Catalysis, 352*, 532−541.

Dong, X., & Cheng, F. (2015). Recent development in exfoliated two-dimensional g-C3N4 nanosheets for photocatalytic applications. *Journal of Materials Chemistry A, 3*(47), 23642−23652.

Du, J., Lai, X., Yang, N., Zhai, J., Kisailus, D., Su, F., Wang, D., & Jiang, L. (2011). Hierarchically Ordered Macro − Mesoporous TiO2 − Graphene Composite Films: Improved Mass Transfer, Reduced Charge Recombination, and Their Enhanced Photocatalytic Activities. *Acs Nano, 5*(1), 590−596.

Dutta, S. K., Mehetor, S. K., & Pradhan, N. (2015). Metal Semiconductor Heterostructures for Photocatalytic Conversion of Light Energy", The. *Journal of Physical Chemistry Letters, 6*(6), 936−944.

Eshaq, G., Wang, S., Sun, H., & Sillanpää, M. (2020). Core/shell FeVO4@BiOCl heterojunction as a durable heterogeneous Fenton catalyst for the efficient sonophotocatalytic degradation of p-nitrophenol. *Separation and Purification Technology, 231*, 115915.

Fox, M. A., & Dulay, M. T. (1993). Heterogeneous photocatalysis. *Chemical Reviews, 93*(1), 341−357.

Fujishima, A., & Honda, K. (1972). Electrochemical photolysis of water at a semiconductor electrode. *Nature, 238*(5358), 37.

Fujishima, A., Zhang, X., & Tryk, D. A. (2007). Heterogeneous photocatalysis: From water photolysis to applications in environmental cleanup. *International Journal of Hydrogen Energy, 32*(14), 2664−2672.

Fujishima, A., Zhang, X., & Tryk, D. A. (2008). TiO2 photocatalysis and related surface phenomena. *Surface Science Reports, 63*(12), 515−582.

Furukawa, H., Cordova, K. E., O'Keeffe, M., & Yaghi, O. M. (2013). The Chemistry and Applications of Metal-Organic Frameworks. *Science, 341*(6149), 1230444.

Gao, E., Wang, W., Shang, M., & Xu, J. (2011). Synthesis and enhanced photocatalytic performance of graphene-Bi2WO6 composite. *Physical Chemistry Chemical Physics, 13*(7), 2887−2893.

Gao, M., You, L., Guo, L., & Li, T. (2019). Fabrication of a novel polyhedron-like WO3/Ag2CO3 p-n junction photocatalyst with highly enhanced photocatalytic activity. *Journal of Photochemistry and Photobiology A: Chemistry, 374*, 206−217.

Grätzel, M. (2001). Photoelectrochemical cells. *Nature, 414*(6861), 338−344.

Hagfeldt, A., & Graetzel, M. (1995). Light-Induced Redox Reactions in Nanocrystalline Systems. *Chemical Reviews, 95*(1), 49−68.

Hashimoto, K., Irie, H., & Fujishima, A. (2005). TiO2Photocatalysis: A Historical Overview and Future Prospects. *Japanese Journal of Applied Physics, 44*(12), 8269−8285.

Hill, R., & Bendall, F. A. Y. (1960). Function of the Two Cytochrome Components in Chloroplasts: A Working Hypothesis. *Nature, 186*(4719), 136−137.

Hoffmann, M. R., Martin, S. T., Choi, W., & Bahnemann, D. W. (1995). Environmental Applications of Semiconductor Photocatalysis. *Chemical Reviews, 95*(1), 69−96.

Hong, S. J., Lee, S., Jang, J. S., & Lee, J. S. (2011). Heterojunction BiVO4/WO3 electrodes for enhanced photoactivity of water oxidation. *Energy & Environmental Science, 4*(5), 1781−1787.

Huang, X., Qi, X., Boey, F., & Zhang, H. (2012). Graphene-based composites. *Chemical Society Reviews, 41*(2), 666−686.

Ibhadon, A. O., & Fitzpatrick, P. (2013). Heterogeneous Photocatalysis: Recent Advances and Applications. *Catalysts, 3*(1), 189−218.

Ingram, D. B., Christopher, P., Bauer, J. L., & Linic, S. (2011). Predictive Model for the Design of Plasmonic Metal/Semiconductor Composite Photocatalysts", Acs. *Catalysis, 1*(10), 1441−1447.

Jiang, S., Chen, Q., Tripathy, M., Luijten, E., Schweizer, K. S., & Granick, S. (2010). Janus Particle Synthesis and Assembly. *Advanced Materials, 22*(10), 1060−1071.

Jiang, W., Zong, X., An, L., Hua, S., Miao, X., Luan, S., Wen, Y., Tao, F. F., & Sun, Z. (2018). Consciously Constructing Heterojunction or Direct Z-Scheme Photocatalysts by Regulating Electron Flow Direction. *Acs Catalysis, 8*(3), 2209−2217.

Kim, T. W., & Choi, K.-S. (2014). Nanoporous BiVO<sub>4</sub> Photoanodes with Dual-Layer Oxygen Evolution Catalysts for Solar Water Splitting. *Science, 343*(6174), 990.

Lang, X., Chen, X., & Zhao, J. (2014). Heterogeneous visible light photocatalysis for selective organic transformations. *Chemical Society Reviews, 43*(1), 473−486.

Li, F., Josephson, D. P., & Stein, A. (2011). Colloidal Assembly: The Road from Particles to Colloidal Molecules and Crystals. *Angewandte Chemie International Edition, 50*(2), 360−388.

Li, R., Han, H., Zhang, F., Wang, D., & Li, C. (2014). Highly efficient photocatalysts constructed by rational assembly of dual-cocatalysts separately on different facets of BiVO4. *Energy & Environmental Science, 7*(4), 1369−1376.

Li, Z., Hou, J., Zhang, B., Cao, S., Wu, Y., Gao, Z., Nie, X., & Sun, L. (2019). Two-dimensional Janus heterostructures for superior Z-scheme photocatalytic water splitting. *Nano Energy, 59*, 537−544.

Linsebigler, A. L., Lu, G. Q., & Yates, J. T. (1995). Photocatalysis on Tio2 Surfaces - Principles, Mechanisms, and Selected Results. *Chemical Reviews, 95*(3), 735−758.

Liu, J., Cheng, B., & Yu, J. (2016). A new understanding of the photocatalytic mechanism of the direct Z-scheme g-C3N4/TiO2 heterostructure. *Physical Chemistry Chemical Physics, 18*(45), 31175−31183.

Loddo, V., Bellardita, M., Camera-Roda, G., Parrino, F., & Palmisano, L. (2018). Chapter 1 - Heterogeneous Photocatalysis: A Promising Advanced Oxidation Process. In A. Basile, S. Mozia, & R. Molinari (Eds.), *Current Trends and Future Developments on (Bio-) Membranes* (pp. 1−43). Elsevier.

Low, J., Jiang, C., Cheng, B., Wageh, S., Al-Ghamdi, A. A., & Yu, J. (2017). A Review of Direct Z-Scheme Photocatalysts. *Small Methods, 1*(5), 1700080.

Marschall, R. (2014). Semiconductor Composites: Strategies for Enhancing Charge Carrier Separation to Improve Photocatalytic Activity. *Advanced Functional Materials, 24*(17), 2421−2440.

Meng, S., Ning, X., Zhang, T., Chen, S.-F., & Fu, X. (2015). What is the transfer mechanism of photogenerated carriers for the nanocomposite photocatalyst Ag3PO4/g-C3N4, band−band transfer or a direct Z-scheme?". *Physical Chemistry Chemical Physics, 17*(17), 11577−11585.

Nosaka, Y., Norimatsu, K., & Miyama, H. (1984). The function of metals in metal-compounded semiconductor photocatalysts. *Chemical Physics Letters, 106*(1), 128−131.

Petala, A., Noe, A., Frontistis, Z., Drivas, C., Kennou, S., Mantzavinos, D., & Kondarides, D. I. (2019). Synthesis and characterization of CoOx/BiVO4 photocatalysts for the degradation of propyl paraben. *Journal of Hazardous materials, 372*, 52−60.

Rothenberger, G., Moser, J., Graetzel, M., Serpone, N., & Sharma, D. K. (1985). Charge carrier trapping and recombination dynamics in small semiconductor particles. *Journal of the American Chemical Society, 107* (26), 8054−8059.

Sasaki, Y., Kato, H., & Kudo, A. (2013). [Co(bpy)3]3 + /2 + and [Co(phen)3]3 + /2 + Electron Mediators for Overall Water Splitting under Sunlight Irradiation Using Z-Scheme Photocatalyst System. *Journal of the American Chemical Society, 135*(14), 5441−5449.

Schneider, J., Matsuoka, M., Takeuchi, M., Zhang, J., Horiuchi, Y., Anpo, M., & Bahnemann, D. W. (2014). Understanding TiO2 Photocatalysis: Mechanisms and Materials. *Chemical Reviews, 114*(19), 9919−9986.

Shang, M., Wang, W., Zhang, L., Sun, S., Wang, L., & Zhou, L. (2009). 3D Bi2WO6/TiO2 Hierarchical Heterostructure: Controllable Synthesis and Enhanced Visible Photocatalytic Degradation Performances. *The Journal of Physical Chemistry C, 113*(33), 14727−14731.

Shifu, C., Sujuan, Z., Wei, L., & Wei, Z. (2008). Preparation and activity evaluation of p−n junction photocatalyst NiO/TiO2. *Journal of Hazardous materials, 155*(1), 320−326.

Shifu, C., Sujuan, Z., Wei, L., & Wei, Z. (2009). Study on the Photocatalytic Activity of p–n Junction Photocatalyst Cu2O/TiO2. *Journal of Nanoscience and Nanotechnology, 9*(7), 4397−4403.

Sun, S., Hisatomi, T., Wang, Q., Chen, S., Ma, G., Liu, J., Nandy, S., Minegishi, T., Katayama, M., & Domen, K. (2018). Efficient Redox-Mediator-Free Z-Scheme Water Splitting Employing Oxysulfide Photocatalysts under Visible Light. *Acs Catalysis, 8*(3), 1690−1696.

Tada, H., Mitsui, T., Kiyonaga, T., Akita, T., & Tanaka, K. (2006). All-solid-state Z-scheme in CdS−Au−TiO2 three-component nanojunction system. *Nature Materials, 5*(10), 782−786.

Tanguay, J. F., Suib, S. L., & Coughlin, R. W. (1989). Dichloromethane photodegradation using titanium catalysts. *Journal of Catalysis, 117*(2), 335−347.

Veldurthi, N. K., Eswar, N. K., Singh, S. A., & Madras, G. (2018). Cocatalyst free Z-schematic enhanced H2 evolution over LaVO4/BiVO4 composite photocatalyst using Ag as an electron mediator. *Applied Catalysis B: Environmental, 220*, 512−523.

Veldurthi, N. K., Eswar, N. K., Singh, S. A., & Madras, G. (2018). Heterojunction ZnWO4/ZnFe2O4 composites with concerted effects and integrated properties for enhanced photocatalytic hydrogen evolution. *Catalysis Science & Technology, 8*(4), 1083−1093.

Veldurthi, N. K., KrishnaRao Eswar, N., Singh, S. A., & Madras, G. (2018). Cooperative effect between BaTiO3 and CaFe2O4 in a cocatalyst-free heterojunction composite for improved photochemical H2 generation. *International Journal of Hydrogen Energy, 43*(51), 22929−22941.

Wang, C., Shao, C., Zhang, X., & Liu, Y. (2009). SnO2 Nanostructures-TiO2 Nanofibers Heterostructures: Controlled Fabrication and High Photocatalytic Properties. *Inorganic Chemistry, 48*(15), 7261−7268.

Wang, D., Hisatomi, T., Takata, T., Pan, C., Katayama, M., Kubota, J., & Domen, K. (2013a). Core/Shell Photocatalyst with Spatially Separated Co-Catalysts for Efficient Reduction and Oxidation of Water. *Angewandte Chemie International Edition, 52*(43), 11252−11256.

Wang, H., Zhang, L., Chen, Z., Hu, J., Li, S., Wang, Z., Liu, J., & Wang, X. (2014). Semiconductor heterojunction photocatalysts: design, construction, and photocatalytic performances. *Chemical Society Reviews, 43*(15), 5234−5244.

Wang, H., Li, S., Zhang, L., Chen, Z., Hu, J., Zou, R., Xu, K., Song, G., Zhao, H., Yang, J., & Liu, J. (2013b). Surface decoration of Bi2WO6 superstructures with Bi2O3 nanoparticles: an efficient method to improve visible-light-driven photocatalytic activity. *Crystengcomm, 15*(44), 9011−9019.

Wang, T., Quan, W., Jiang, D., Chen, L., Li, D., Meng, S., & Chen, M. (2016). Synthesis of redox-mediator-free direct Z-scheme AgI/WO3 nanocomposite photocatalysts for the degradation of tetracycline with enhanced photocatalytic activity. *Chemical Engineering Journal, 300*, 280−290.

Wang, X., Liu, G., Chen, Z.-G., Li, F., Wang, L., Lu, G. Q., & Cheng, H.-M. (2009). Enhanced photocatalytic hydrogen evolution by prolonging the lifetime of carriers in ZnO/CdS heterostructures. *Chemical Communications* (23), 3452−3454.

Woan, K., Pyrgiotakis, G., & Sigmund, W. (2009). Photocatalytic Carbon-Nanotube−TiO2 Composites. *Advanced Materials, 21*(21), 2233−2239.

Wu, N. (2018). Plasmonic metal−semiconductor photocatalysts and photoelectrochemical cells: a review. *Nanoscale, 10*(6), 2679−2696.

Wu, Y., & Wark, M. (2016). *CHAPTER 6 New Concepts in Photocatalysis. Photocatalysis: Fundamentals and Perspectives* (pp. 129−161). The Royal Society of Chemistry.

Xia, X., Song, M., Wang, H., Zhang, X., Sui, N., Zhang, Q., Colvin, V. L., & Yu, W. W. (2019). Latest progress in constructing solid-state Z scheme photocatalysts for water splitting. *Nanoscale, 11*(23), 11071−11082.

Xu, D., Cheng, B., Wang, W., Jiang, C., & Yu, J. (2018). Ag2CrO4/g-C3N4/graphene oxide ternary nanocomposite Z-scheme photocatalyst with enhanced CO2 reduction activity. *Applied Catalysis B: Environmental, 231*, 368−380.

Xu, F., Zhang, L., Cheng, B., & Yu, J. (2018). Direct Z-Scheme TiO2/NiS Core−Shell Hybrid Nanofibers with Enhanced Photocatalytic H2-Production Activity. *Acs Sustainable Chemistry & Engineering, 6*(9), 12291−12298.

Xu, F., Xiao, W., Cheng, B., & Yu, J. (2014). Direct Z-scheme anatase/rutile bi-phase nanocomposite TiO2 nanofiber photocatalyst with enhanced photocatalytic H2-production activity. *International Journal of Hydrogen Energy, 39*(28), 15394−15402.

Yang, J., Hao, J., Xu, S., Wang, Q., Dai, J., Zhang, A., & Pang, X. (2019). InVO4/β-AgVO3 Nanocomposite as a Direct Z-Scheme Photocatalyst toward Efficient and Selective Visible-Light-Driven CO2 Reduction. *ACS Applied Materials & Interfaces, 11*(35), 32025−32037.

Yin, H., & Tang, Z. (2016). Ultrathin two-dimensional layered metal hydroxides: an emerging platform for advanced catalysis, energy conversion and storage. *Chemical Society Reviews, 45*(18), 4873−4891.

Yu, J. C., Wu, L., Lin, J., Li, P., & Li, Q. (2003). Microemulsion-mediated solvothermal synthesis of nanosized CdS-sensitized TiO2 crystalline photocatalyst. *Chemical Communications, 13*, 1552−1553.

Yu, J., Wang, S., Low, J., & Xiao, W. (2013). Enhanced photocatalytic performance of direct Z-scheme g-C3N4−TiO2 photocatalysts for the decomposition of formaldehyde in air. *Physical Chemistry Chemical Physics, 15*(39), 16883−16890.

Yu, W., Xu, D., & Peng, T. (2015). Enhanced photocatalytic activity of g-C3N4 for selective CO2 reduction to CH3OH via facile coupling of ZnO: a direct Z-scheme mechanism. *Journal of Materials Chemistry A, 3*(39), 19936−19947.

Yuan, Q., Liu, D., Zhang, N., Ye, W., Ju, H., Shi, L., Long, R., Zhu, J., & Xiong, Y. (2017). Noble-Metal-Free Janus-like Structures by Cation Exchange for Z-Scheme Photocatalytic Water Splitting under Broadband Light Irradiation. *Angewandte Chemie, 129*(15), 4270−4274.

Zhang, H., Liu, L., & Zhou, Z. (2012). Towards better photocatalysts: first-principles studies of the alloying effects on the photocatalytic activities of bismuth oxyhalides under visible light. *Physical Chemistry Chemical Physics, 14*(3), 1286−1292.

Zhang, N., Xie, S., Weng, B., & Xu, Y.-J. (2016). Vertically aligned ZnO−Au@CdS core−shell nanorod arrays as an all-solid-state vectorial Z-scheme system for photocatalytic application. *Journal of Materials Chemistry A, 4*(48), 18804−18814.

Zhang, X., Zhang, R., Niu, S., Zheng, J., & Guo, C. (2019). Construction of core-shell structured WO3@SnS2 hetero-junction as a direct Z-scheme photo-catalyst. *Journal of Colloid and Interface Science, 554*, 229−238.

Zhao, W., Li, J., Dai, B., Cheng, Z., Xu, J., Ma, K., Zhang, L., Sheng, N., Mao, G., Wu, H., Wei, K., & Leung, D. Y. C. (2019). Simultaneous removal of tetracycline and Cr(VI) by a novel three-dimensional AgI/BiVO4 p-n junction photocatalyst and insight into the photocatalytic mechanism. *Chemical Engineering Journal, 369*, 716−725.

Zheng, Y., Liu, J., Liang, J., Jaroniec, M., & Qiao, S. Z. (2012). Graphitic carbon nitride materials: controllable synthesis and applications in fuel cells and photocatalysis. *Energy & Environmental Science, 5*(5), 6717−6731.

Zhou, P., Yu, J., & Jaroniec, M. (2014). All-Solid-State Z-Scheme Photocatalytic Systems. *Advanced Materials, 26*(29), 4920−4935.

CHAPTER 7

Plasmonic photocatalysis: an extraordinary way to harvest visible light

Saikumar Manchala[1,2], Vijayakumar Elayappan[3], Hai-Gun Lee[3] and Vishnu Shanker[1,2]

[1]*Department of Chemistry, National Institute of Technology, Warangal, India,* [2]*Centre for Advanced Materials, National Institute of Technology, Warangal, India,* [3]*Department of Materials science and Engineering, Korea University, Seoul, Republic of South Korea*

7.1 Introduction

As an exponential growth in the world population and extension of industrial activities continue to build up, there is a led to exploit new green technologies in the sustainable growth of human society for environmental pollution and alternative clean energy supplies. Since, the first report by Fujishima and Honda on water splitting, the semiconductor-mediated photocatalysis has very much attracted great attention (Manchala, Nagappagari, Muthukonda Venkatakrishnan, & Shanker, 2018; Tonda, Kumar, Gawli, Bhardwaj, & Ogale, 2017). Photocatalysis is an emerging pollution-free technology in the interdisciplinary area of sustainable nano-science and technology, in which sunlight is used as an energy source. It is a nontoxic and cost-effective method and offers a variety of applications in the areas of wastewater treatment, soil-pollution treatment, air-pollution treatment in the terms of CO_2 reduction, self-cleaning, clean energy production, organic synthesis, bacterial elimination, etc. (Kumar, Karthikeyan, & Lee, 2018; Manchala et al., 2019; Tian et al., 2014).

Unfortunately, the performance of the presently developed photocatalysts under the visible light is still limited caused by some exciting issues like slight light-response range, surface structure, and high recombination rate. In previous years, remarkable efforts have been made for improving photocatalysis under visible-light irradiation. In this context, superficial deposition of noble-metal nanoparticles (NPs) (Pt, Au, Ag, Pd, Ru, and Cu) on the surface of semiconductors is considered as an excellent method to improve the photocatalytic performance under visible-light irradiation. Emergent exploitation of various noble-metal NPs in photocatalysis has been received much attention due to its response to visible light [localized surface plasmonic resonance (LSPR)] since the first report of Mie in 1908. Plasmonic photocatalyst usually refers to the combination of noble-metal NPs with semiconductors. The major advantages of the LSPR effect of noble-metal NPs in plasmonic photocatalysts promote the charge migration, act as an electron trapping agent, and then reduce excitons' recombination effect. The noble-metal NPs in the plasmonic photocatalyst could also improve the light-absorption capacity of semiconductors through a plasmonic enhancement effect. Due to these advantages, researchers have been developed various potential plasmonic photocatalysts by different synthesis approaches and utilized them in several applications (Gupta, Mohapatra, & Bahadur, 2017; Li, Yu, Jaroniec, & Chen, 2019; Manchala, Nagappagari, Venkatakrishnan, & Shanker, 2019; Tian et al., 2014; Tonda, Kumar, & Shanker, 2016).

7.1.1 Localized surface plasmon resonance

LSPR is a basic and unique optical phenomenon exhibited by the noble-metal NPs when illuminated with light. It displays the strong and coherent collective oscillation or excitations of electrons of the noble-metal NPs by adjusting the electromagnetic field of the incident light, as visualized in Fig. 7.1A (Zhang, Chen, Liu, & Tsai, 2013). LSPR largely improves the light absorption of the wide and small band gap (TiO_2 and g-C_3N_4) semiconductor

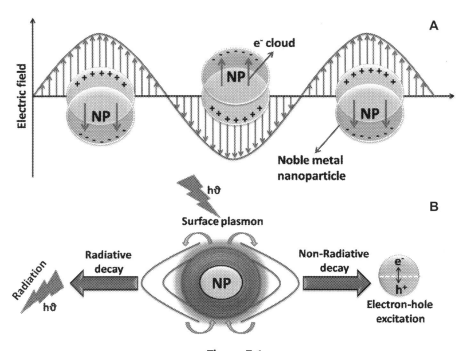

Figure 7.1
Illustration of the localized surface plasmonic resonance evolution in the noble-metal nanoparticles concerning light source (A) and localized surface plasmonic resonance decay processes (B).

photocatalytic systems to visible and near-infrared (NIR) regions (Cai et al., 2019; Tonda et al., 2016). It already established that the magnitude of absorption for an optical cross-section of noble-metal NPs is "5" times higher when compared to that of particular dye-sensitizer molecules (Jain, Lee, El-Sayed, & El-Sayed, 2006). More importantly, the LSPR effect depends on the nature, morphology, size, aggregation, and composition of the noble-metal NPs, electronic interactions between the stabilizing agents and NPs, and dielectric characteristics of the surrounding environment (Zhang et al., 2018). The controlling of LSPR is the main object to analyze the performance of photocatalytic reactions, photonic and optical devices. Therefore the development of photocatalytic heterojunctions with noble-metal NPs is a suitable way to harvest the visible light and improves the efficiency of solar energy conversion. As displayed in Fig. 7.1B, the surface plasmons can decay either radiatively or nonradiatively to transfer the energy, which subsequently leads to a quick generation of hot excitons. LSPR effect can be used in surface- and tip-enhanced Raman spectroscopy (SERS and TERS, respectively) technique for the sensitive identification of short lifetime organic intermediates to explain the reaction mechanism (Wang, Wang, & Mu, 2019).

7.1.1.1 Typical advantages of localized surface plasmonic resonance in plasmonic photocatalysis

The unique characteristic feature of plasmonic photocatalysts is the LSPR of the noble-metal NPs concerning the electric field of the incident light. This renders various major advantages to the plasmonic photocatalysts, as represented in Fig. 7.2 (Leong et al., 2018; Zhang et al., 2013, 2018). First, the noble-metal NPs, including Au, Ag, Pt, and Pd, are strong visible-light absorbers in the total solar range. They provide visible-light absorption to the wide band gap semiconductors, such as TiO_2 and ZnO, and improve the visible light—absorption property of narrow band gap semiconductors like g-C_3N_4. Commonly, noble-metal NPs form Fermi level

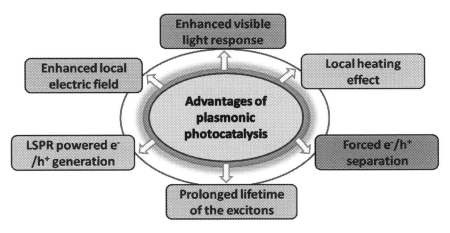

Figure 7.2
Typical advantages of the plasmonic photocatalysis.

(Ef) in between the conduction band (CB) and the valence band of a semiconductor, and the deposited NPs act as light-scattering and -trapping centers. It can allow the transfer of electrons between the noble-metal NPs and semiconductors. In this way, the light can be largely accumulated on the semiconductor; thus light-absorption property can be enhanced. Furthermore, LSPR can speed up the rapid electron transfer between the noble-metal NPs and semiconductors. And it is proved that the lifetime of photo-generated electrons is higher in noble-metal/semiconductor heterojunction compared to that of the bare semiconductor. Besides, the plasmonic NPs can decay inelastically and release energy as heat through plasmon—phonon interaction. Thus the temperature of the surrounding medium can also increase considerably which intern increases the rate of photocatalytic reaction according to the Arrhenius equation. The LSPR also supplies the large and strong local electromagnetic field, which helps in the excitation of more excitons and polarization of the nonpolar molecular species for superior adsorption to enhance the rate of redox reaction and mass transfer.

7.1.2 Basics of plasmonic photocatalysis

The formation of notable Schottky junction between the noble-metal NPs with semiconductors is classified as plasmonic photocatalyst. The word "plasmonic" is directly linked to the distinctive features of the LSPR and induced effects between the noble-metal NPs and a semiconductor. Generally, when plasmonic nanostructure is exposed to light energy it oscillates quickly and the absorbed energy can be transformed into a large amount of plasmonic energy. But individually plasmon cannot attain a high quantum yield of photocatalytic efficiency due to a short lifetime of excitons. On the other side, when plasmon combines with semiconductor, it is very easy to attain high quantum yield of photocatalytic efficiency, in which plasmonic structure acts as a light antenna and improves the light-harvesting power of semiconductor to visible light. In addition, the excitation of more number of excitons enhances the efficiency of the redox reactions. The typical qualities or effects of plasmons are described in Fig. 7.2. These qualities may affect the important chemical and physical properties of a semiconductor like light absorption, charge migration, charge separation, recombination, and charge injection into the redox or active species (Leong et al., 2018; Zhang et al., 2018).

7.1.2.1 Proposed principles/mechanisms for energy transfer between noble-metal nanoparticles and semiconductor in plasmonic photocatalysis

There are several reports that explain the influence of plasmons on the light absorption and charge separation of the semiconductors, but few reports available on the charge migration and recombination in the semiconductors. Majorly, "3" types of mechanisms have been proposed to demonstrate the energy transfer from plasmon to the semiconductor: (1) light scattering/trapping, (2) plasmon-induced resonance energy transfer (PIRET), and (3) hot-electron injection (Huang et al., 2019; Kawamura & Matsuda, 2019; Wu, 2018).

7.1.2.1.1 Light scattering/trapping

Generally, the LSPR band of light-induced noble-metal NPs arises from the total energy contribution of light absorption and scattering. And mostly in the large plasmonic NPs (>50 nm) scattering is the predominant phenomenon. When large plasmonic NPs are integrated with the semiconductor NPs to make plasmonic photocatalyst, the irradiated light can be scattered elastically by these large plasmonic noble-metal NPs and deliver the photons into the bulk or semiconductor or other metal NPs, which can eventually enhance the photon flux of the other counterparts; and this phenomenon is called as light trapping (Fig. 7.3A). Therefore the light-scattering effect of noble-metal NPs can increase the light

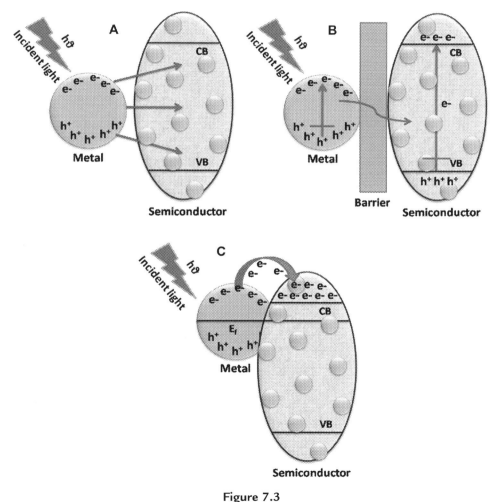

Figure 7.3
Basic principles/mechanisms (A–C) involved in plasmonic photocatalysis for energy transfer between noble-metal nanoparticles and semiconductors.

absorption and charge separation of other counterparts especially semiconductor NPs and thus leads to trigger the performance of photocatalytic reactions. In this type of plasmonic energy transfer, the distance between noble-metal NPs and semiconductor NPs is 100 nm or lesser, but the direct junction is not required. The light-scattering/trapping effect of plasmonic noble-metal NPs depends upon the shape, size, aggregation, composition, and dielectric characteristics of the surrounding environment.

7.1.2.1.2 Plasmon-induced resonance energy transfer

The incident light excites the noble-metal NPs and generates excitons, that is leads to the development of strong dipoles. As displayed in Fig. 7.3B, the plasmonic energy can be transferred from plasmonic noble-metal NPs to semiconductor nonradiatively through a dipole—dipole near-field interconnection and generates the excitons in the semiconductor. It involves a blue-shift energy transfer approach. The presence of a 25-nm insulating barrier between plasmonic noble-metal NPs and semiconductors is also not an issue to take place in this mechanism. Besides, the efficiency of the PIRET mechanism is mainly influenced by "2" factors: first, the distance between noble-metal NPs and their counterparts like semiconductors and, second, an overall spectral contribution of the absorption bands of LSPR and semiconductor.

7.1.2.1.3 Hot-electron injection

It also involves a nonradiative plasmonic energy transfer from noble-metal NPs by a generation of hot energy excitons through Landau damping with both intra- and interband electronic transitions. When the generated hot electrons have higher energy ($1-4$ eV) as compared to Schottky barrier energy (1 eV), they can directly get injected into the CB of the semiconductor as displayed in Fig. 7.3C. After transfer, the electrons in the CB of semiconductors do not lost their energy and still can be considered/act as hot-electrons. Also, in noble-metal/semiconductor heterojunctions, such as Au/TiO_2, the hot electrons generated and injected into the TiO_2 exhibit "2" times of excited-state lifetime compared to that of electrons photo-excited by the interband transition in TiO_2 through UV excitation (DuChene et al., 2014). In other words, the hot electrons migrated from plasmonic noble-metal NP to semiconductor can have a low recombination rate and higher thermodynamic driving force for a photocatalytic reaction than that of the electrons photo-excited directly within a semiconductor. But, for maximum efficiency of hot-electron injection, there should be close contact between the noble-metal NPs and semiconductors. But in the case of metal—metal/semiconductor heterojunction, the hot-electron transfer is an inefficient process because of the high recombination rate and extreme back electron transfer. After the failure of Fowler's theory for small noble-metal NPs (Fowler, 1931), Govorov's (Govorov, Zhang, Demir, & Gun'ko, 2014; Zhang & Govorov, 2014) and Kumarasinghe's (Kumarasinghe, Premaratne, Bao, & Agrawal, 2015; Kumarasinghe, Premaratne, Gunapala, & Agrawal, 2016) groups have proposed a single-electron model to predict hot-electron generation from noble-metal NPs and transfer to the semiconductor quantitatively. According to these theories, the hot-electron generation and injection highly depend on the morphology and size of noble-metal NPs as well as the presence of "hot spots".

These "3" mechanisms could enhance the photo-conversion efficiency of a photocatalytic reaction by adjusting light absorption, charge migration, charge separation, and charge recombination processes in the semiconductor. And especially, the band gap of the semiconductor is crucial to determine the nature of the energy transfer between the noble-metal NPs and semiconductors. If the band gap of a semiconductor is >3.1 eV and it is coupled with noble-metal NPs, the hot-electron transfer is the major phenomenon observed under visible or NIR light. This provides a little enhancement in the photo-conversion efficiency. And when the band gap of a semiconductor is <2.5 eV and if it is coupled with noble-metal NPs, then there is a possiblity for the emergence of "3" types of energy transfer mechanisms in the system which would take place under visible or NIR light irradiation. This leads to a larger enhancement in the photo-conversion efficiency (Cushing, Bristow, & Wu, 2015).

Apart, many works reported majorly hot-electron injection mechanisms as energy transfer for improved plasmonic photocatalysis in various applications, including dye degradation, CO_2 reduction, H_2O splitting, and organic transformations. The transfer time of hot electrons, lifetime, and the existence of hot electrons in the CB of a semiconductor can be studied thoroughly by femtosecond (fs) transient absorption spectroscopy, ultrafast dynamics spectroscopy, and X-ray absorption fine structure spectroscopy. The hot electrons can be moved from plasmonic noble-metal NPs to semiconductors through either indirect electron transfer (IET) or direct electron transfer (DET). Otherwise, there is a possibility to get back hot electrons to the noble-metal NPs and relax as shown in Fig. 7.4C (Zhang et al., 2018).

7.1.2.1.3.1 Indirect electron transfer Conventional IET can occur via a "2" step mechanism on the time range of picoseconds (ps). Initially, the photo-induced hot excitons can generate within the noble-metal NPs. Then, the generated hot electrons are injected into the CB of the semiconductor as displayed in Fig. 7.4A.

7.1.2.1.3.2 Direct electron transfer Recent studies revealed that the electron transfer can proceed on the order of fs, not on the order of ps. It is less than that of IET time. These studies evidenced that electron transfer is occurring other than IET. Then, Long and Prezhdo (2014) reported that the electron transfer can also occur via DET with a probability of ~50% along with IET. DET proceeds through a single-step mechanism. Here, the hot electrons directly transferred to CB of the semiconductor and leaves the holes in the noble-metal NPs as represented in Fig. 7.4B. And it is believed that the DET is more efficient than IET.

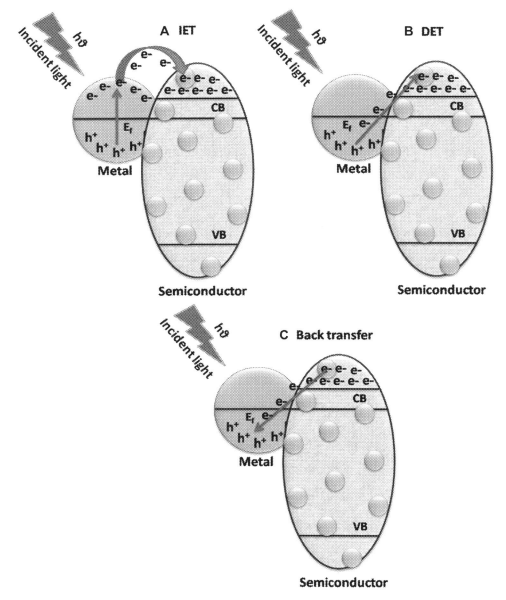

Figure 7.4
Various types of processes (A—C) involved in the hot-electron transfer between the noble-metal
nanoparticles and semiconductors of plasmonic photocatalysis.

7.2 Synthesis methods for noble-metal nanoparticles

As represented in Fig. 7.5, noble-metal NPs can be synthesized by employing various
methods, such as photo-deposition, impregnation, chemical reduction, micro-emulsion,

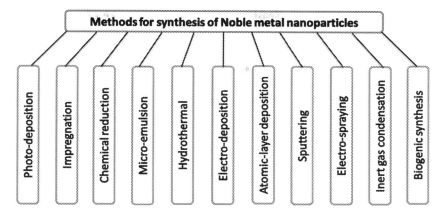

Figure 7.5
Major synthesis approaches for noble-metal nanoparticles.

hydrothermal, electro-deposition, atomic-layer deposition (ALD), sputtering, electro-spraying, inert-gas condensation, and biogenic synthesis based on the suitability for an application (Bueno-Alejo et al., 2020; Cheng et al., 2017; Dhand et al., 2015; Kavinkumar et al., 2019; Kumar et al., 2011; Maicu et al., 2014; Manchala et al., 2019; Pelletier & Thiébaut, 2016; Shen et al., 2018; Soleimani Zohr Shiri, Henderson, & Mucalo, 2019; Tahir, Tahir, Saidina Amin, & Alias, 2016; Yang, Guo, & Li, 2017; Zhou, Liu, Yu, & Fan, 2012).

7.2.1 Photo-deposition

Photo-deposition has widely employed method for the synthesis of noble-metal NPs, since the first report of Clark and Vondjidis in 1965 on photo-deposition of "Ag" NPs on the surface of TiO_2 and the report of Kraeutler and Bard in 1978, where "Pt" NPs well-deposited on the surface of the anatase TiO_2, by irradiating the mixture consists of anatase TiO_2 powder, H_2PtCl_6, HCl, sodium carbonate, and acetic acid. Generally, photo-deposition involves the mechanism of light-induced electrochemistry and the reductive photo-deposition of a metal (M) is represented in the following equation:

$$M^{n+} + n^{e^-} + hJ \rightarrow (s) \tag{7.1}$$

In photo-deposition the noble-metal NPs, such as Pt, Au, Ag, and Ru, are synthesized on the surface of the semiconductors in the presence of light. Several reports have been already demonstrated for the synthesis of plasmonic photocatalysts by this route. Many scientists have been tried to optimize the ratio of noble-metal NPs with the semiconductor material. Especially, some people are developed the bimetallic or trimetallic plasmonic photocatalysts, in which two or three different noble-metal NPs can deposit simultaneously or one by one.

7.2.2 Impregnation

Impregnation is also an often-used method for the synthesis of noble-metal NPs on the surface of the semiconductors, where the metal precursor-containing solution (aqueous or organic) mixed with the solid catalyst. Then, the catalyst can be dried for depositing noble-metal NPs on the surface of a solid support. The maximum percentage of the loading of noble-metal NPs is limited by the solubility of the metal precursor in the solution.

7.2.3 Chemical reduction

Chemical reduction is a common method for the deposition of noble-metal NPs on the semiconductor's surface. If polyethylene glycol (PEG) was used as the reducing agent, then it was called a polyol synthesis. In the polyol method, PEG also acts as a stabilizing agent. Apart, several chemical-reducing agents are developed for the fabrication of noble-metal NPs on semiconductor materials, such as $NaBH_4$, $NaOH$, hydrazine, and hydrazine hydrate. But the usage of these chemicals leads to hazardous for both human health and the environment. Thus there is a need to develop alternative methods.

7.2.4 Micro-emulsion technique

Micro-emulsion is also a widely used method for the synthesis of noble-metal NPs. It can consist of macroscopically mono-dispersed, thermally stable, homogenous, and optically transparent spherical bubbles (usually 600—8000 nm) often made up of surfactant, aqueous phase (commonly water) and nonaqueous phase (commonly hydrocarbon liquid or oil). It can be divided into micelle and reverse micelle types based on the surfactant. Reverse micelle route may serve as the best method for the synthesis of noble-metal NPs.

7.2.5 Hydrothermal technique

Hydrothermal synthesis employs a chemical solution containing two or more phases [solid, liquid (aqueous and nonaqueous), and gas] to synthesize noble-metal NPs under controlled conditions of temperature and pressure. This method offers several advantages, including morphology, controlled size, high reaction rate, inexpensive, and less time. Instead of water, any solvent is used for the synthesis; it is considered as a solvothermal process.

7.2.6 Electro-deposition

Electro-deposition simply refers to electrochemical deposition/reduction, a prominent method to coat or deposit noble-metal NPs from metal salts on the surface of a substrate dipped in an electrolyte by applying an electric current. Generally, it referred to as electroplating. Particularly, in metallurgy metals were extracted from their metal ores by using this method and the obtained metallic deposit was crystalline, that is why this method is also called as electro-crystallization. This method has a lot of advantages like size control and uniformity of deposited NPs, single step, low cost, and room temperature and can employ water as a solvent to deposit noble-metal NPs on semiconductors and vice versa.

7.2.7 Atomic-layer deposition

ALD was also a powerful method to synthesize uniform, precise, and atomic size thin-film layers of noble-metal NPs than that of physical- and chemical-vapor deposition (CVD and PVD) methods by varying compositions of precursors. It is also similar to CVD method, where precursors of noble-metal NPs are in the vapor phase can be deposited sequentially and continuously on a substrate.

7.2.8 Sputtering

Sputtering is also a type of vapor deposition method on a substrate from a solid surface. Here, high-velocity ions or electrons can be bombarded with a solid target of interest and form nanoclusters and atoms and deposited on a substrate under inert atmosphere and high pressure.

7.2.9 Electro-spraying

The electro-spraying method uses the electro-mechanical tool in which the homogeneous dispersion of polymer and metal salt produce charged droplets by the help of a syringe and at the capillary tip of the syringe a high voltage can be applied. Finally, the solvent is dry out and NPs or nanofibers can be obtained as the product.

7.2.10 Inert-gas condensation

Inert-gas condensation was a basic method for the synthesis of noble-metal NPs with the help of liquid N_2-cooled substrate in the presence of inert gas (He and Ar). Here, the vaporized materials carried through the inert gas and condensed onto the substrate connected with the liquid N_2.

7.2.11 Biogenic synthesis

Apart from the above synthesis methods, biogenic synthesis of noble-metal NPs was an emerging field in nowadays due to its low cost, simple synthesis, and eco-friendly nature. Various plant extracts, microorganisms, and bio-molecules (ascorbic acid, sodium citrate, glucose, sucrose, and fructose) based methods have been reported for the biogenic synthesis of noble-metal NPs.

7.3 Applications of plasmonic photocatalysis

Nowadays, harvesting energy from solar resources has been a more significant issue due to the depletion of nonrenewable energy sources. And also, the energy harvest from solar light is considered cleaner and more cost-effective. From the last decade, plasmonic nanomaterials are considered as more effective photocatalyst, which can be applied in various photocatalytic energy-harvesting fields, such as wastewater treatment, H_2 production by water splitting, CO_2 reduction, and production of fine chemicals. Mainly plasmonic photocatalysis has "2" prominent properties, LSPR and formation of Schottky barrier/junction, which help in the more visible light—absorption capacity and enhance the ability of photo-generated excitons separation (Zhang et al., 2013). Therefore the development of plasmonic photocatalysts is highly desirable as it can function efficiently under sunlight in various energy-harvesting applications (Fig. 7.6).

7.3.1 Wastewater treatment

Increasing water demand and water shortage issues around the globe can be managed with the advancement of powerful and effective techniques of water purification. Recently, a lot of research has been explored to purify and enhance the quality of water from wastewater by plasmonic photocatalysis techniques. Water contamination happens mainly from multiple sources, such as dyes, BTEX (benzene, toluene, ethylbenzene, and xylene), petroleum derivatives, and organic solvents, and compounds like trichloroethylene, phenol, chloro-phenol, nitrophenol, pesticides, antibiotics, pharmaceutical drugs, etc. The decomposition of organic compounds can be driven by the energy of photons in photocatalytic processes that converts light into chemical energy utilized for desirable surface reactions. Among the developed photocatalysts, the plasmonic photocatalysts are

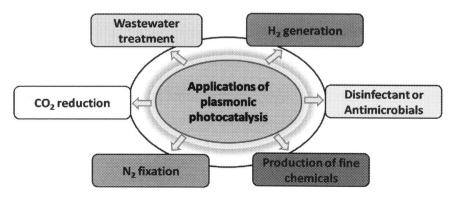

Figure 7.6
Various applications of plasmonic photocatalysis.

significant in which the LSPR property of noble-metal NPs has the tendency to improve the performance of photocatalysts.

Gomez, Sebastian, Arruebo, Santamaria, and Cronin (2014) decorated the 5-nm "Au" nanospheres on silica NPs and studied the photocatalytic activity under visible-light irradiation. Their results showed effective decomposition of methyl orange (MO) dye with the Au-coated silica, rather than without plasmonic "Au" nanospheres. Besides, they verified the definite decomposition of trichloroethylene by visible-light irradiation, which worked with excellent stability for long-term application. Guan et al. (2020) prepared "Pt" plasmonic material, which was embedded into $TiO_2/H_3PW_{12}O_{40}$ nanopore and they studied the photocatalytic degradation and detoxification of various types of chloro-phenols as pollutants in water. They observed that these chloro-phenols were effectively removed through the plasmonic photocatalytic activity. Moreover, their study revealed an outstanding recycling performance, which can be cost-effective with the minimum risk of secondary pollution. Chatterjee, Kar, Wulferding, Lemmens, and Pal (2020) developed flower-like morphology of BiO_2 microspheres decorated with plasmonic "Au" NPs and used them for dual detoxification of organic and inorganic water pollutants. This plasmonic hybrid photocatalyst exhibited a superior photocatalytic activity effectively toward the detoxification pollutants of organic (MO) as well as inorganic (Cr(VI)) water pollutants under visible and red light ($\lambda > 600$ nm) irradiation. Irani and Amoli-Diva (2020) prepared a plasmonic Ag-doped $ZnO@Fe_3O_4$/MWCNT nanocomposite associated with membrane filtration to efficiently remove the amoxicillin (AMOX) using a solar-simulated "Xe" lamp irradiation. This work revealed an efficient removal of toxic chemicals with outstanding performance and provided cost-effective water purification reactors to generate nontoxic clean water.

Yang et al. (2020) demonstrated imidacloprid photocatalytic degradation by plasmon resonance and oxygen vacancy defects in chlorine-doped $Bi-BiO_{2-x}$. Tun, Wang, Khaing, Wu, and Zhang (2020) fabricated plasmonic functionalized Ag-loaded Bi_2O_3/montmorillonite nanocomposite photocatalysts, which showed good potent in the removal of tetracycline (TC) antibiotics and Rhodamine-B (RhB) organic dye in wide pH range. Wan et al. (2020) synthesized plasmon-coupled 2D $Au/Bi_2WO_6-MoS_2$ heterojunction for superior photocatalytic detoxification of Cr(VI) and TC. Interestingly, these results showed 1.90 times enhanced photocatalytic detoxification activity when compared with the absence of plasmonic NPs. Shen et al. (2018) performed a useful degradation of phenol and 4-nitrophenol (4-NP) by plasmonic "Ag" modified Bi_2MoO_6 photocatalysts with surface oxygen vacancies. Under visible-light irradiation, the optimized sample exhibited the highest photocatalytic activity for the degradation of both phenol and 4-NP. This extraordinary photo-degradation performance of organic substances is attributed to the synergistic effect between the surface oxygen vacancies, metallic "Ag" NPs, and Bi_2MoO_6. Apart, many plasmonic photocatalysts have been synthesized as shown in Table 7.1 for the removal of various organic and inorganic pollutants from water.

Table 7.1: Various plasmonic photocatalysts reported for wastewater treatment and H_2 generation.

S. no.	Plasmonic photocatalyst	Synthesis method	Application description	Reference
1	Au/SnO_2 quantum dots	One-pot solvothermal	Methylene blue degradation under 100 W visible light source ($\lambda > 400$ nm)	Bathula, Koutavarapu, Shim, and Yoo (2020)
2	Bi_2WO_6/Ag	Photo-reduction	RhB and malachite green degradation under 300 W tungsten—halogen lamp	Kavinkumar et al. (2019)
3	Au/ZnO, Ag/ZnO, Pd/ZnO	Borohydride reduction	Congo red degradation under 100 W UV light ($\lambda = 360$ nm)	Güy and Özacar (2016)
4	Ga_2O_3 (labeled as M-P and MO-C; M = Ag, Pt, and Pd)	Photocatalytic and chemical deposition	Ciprofloxacin, Ronidazole, and RhB under 25 W mercury lamp ($\lambda = 245$ nm)	Li et al. (2019)
5	Ag/Bi_3TaO_7	Photo-reduction	TC degradation under 250 W "Xe" lamp ($\lambda > 420$ nm)	Luo et al. (2015)
6	$Ag/WO_3/SBA-15$	Hydrothermal	Atrazine (Herbicide) degradation under 450 W "Xe" arc lamp ($\lambda > 400$ nm)	Gondal et al. (2016)
7	Ag/graphene codoped TiO_2	Surfactant-free solvothermal	Paraoxon (pesticide) degradation under 570 W "Xe" lamp	Keihan, Hosseinzadeh, Farhadian, Kooshki, and Hosseinzadeh (2016)
8	Ag/comodified Bi_2WO_6	Impregnation-calcination	Phenol and 4-NP degradation under 400 W halogen lamp	Shen et al. (2018)
9	Ag/TiO_2	Photo-reduction	AMOX and 2,4-dichlorophenol degradation under 500 W tungsten—halogen lamp	Leong, Gan, Ibrahim, and Saravanan (2014)
10	$g-C_3N_4/Ag/LaFeO_3$	Photo-deposition and hydrothermal	RhB and phenol degradation under 400 W gold-halide lamp	Gao, Shang, Liu, and Nie (2019)
11	Au NPs@MoS_2	Hydrothermal	H_2 production under "Xe" lamp ($\lambda \sim 420$ nm)	Yang et al. (2017)
12	Au-loaded WO_3/TiO_2	Sol—gel-assisted photo-deposition	H_2 production under "Xe" lamp ($\lambda \sim 420$ nm)	Tahir et al. (2020)
13	CdS core-Au@SiO_2	Hydrothermal	H_2 production under "Xe" lamp	Xu et al. (2018)
14	Au/TiO_2 nanosheets	Light reduction followed by urea reduction method	H_2 production under "PLS LED 100" lamp ($\lambda \sim 365$ nm)	Cheng et al. (2019)
15	$g-C_3N_4$ quantum dots/Au/highly branched TiO_2 fibers	Vapor deposition method	H_2 production under "Xe" lamp ($\lambda = 420$ nm)	Nasir, Yang, Ayub, Wang, and Yan (2020)
16	$Au/Al-SrTiO_3$	Chemical reduction	H_2 production under 150 W "Xe" lamp	Saadetnejad and Yıldırım (2018)
17	$Au/LaFeO_3$	Biogenic reduction	H_2 production under 500 W "Xe" lamp	Huang et al. (2019)

RhB, Rhodamine-B.

7.3.2 H_2 generation

The depletion of nonrenewable energy sources globally emphasizes the need to find alternative renewable energy sources. In this context, H_2 is considered a very clean fuel and is a good alternative source of renewable energy. Nowadays, production of H_2 fuel via the conversion of solar energy through photocatalytic water splitting is proving to have good efficacy and is eco-friendly. Recently, H_2 production through plasmonic photocatalytic materials is much more energetic and is of great interest than traditional photocatalysis. This may be due to the plasmonic-metal-based photocatalysts possessing certain hallmark features like strong light absorption in the visible region, production of hot energetic charge carriers, and more catalytically active sites. Very recently, Jiang's group prepared "Au" nanorod-decorated 2D-CoFe-based metal−organic framework nanosheets, which are utilized for H_2 evolution under light illumination. The authors ascribed that, the catalytic activity increased more than "4" times under illumination than the dark condition, which is attributed to the LSPR effect of the plasmonic material (Wang et al., 2019). A pioneer work of plasmon-assisted H_2 generation is conducted by Van Dao et al. using Pt-functionalized Au@CeO_2 core−shell catalyst under visible light. The inference drawn by the authors shows that the higher amount of H_2 evolution is due to the LSPR effect, which provides efficient light-harvesting and charge carrier conversion of the plasmonic photocatalyst (Van Dao et al., 2020).

Moreover, plasmonic "Au" NPs loaded on 1D-branched TiO_2/g-C_3N_4 quantumdot photocatalyst were prepared by Nasir et al. and they performed H_2 evolution. The prepared photocatalyst seems to exhibit good improvement and recorded 2.22 mmol/g^m/h^m of H_2 evolution with 19.5% quantum efficiency at 420 nm of visible light. The enhancement is mainly due to the type-II heterostructure, which remarkably promotes the photo-generated excitons. Besides, the LSPR effect of plasmonic "Au" greatly improves the photocatalytic activity with effective reduction of charge recombination and fast electron movement (Nasir et al., 2020). Chen et al. was able to synthesized Au-capped CdS−PbS hetero-octahedron photocatalytic material to generate 400.6 μmol h^m/g^m of H_2, which is around 6.33 times higher than the pure CdS−PbS and bare "Au" nanocap. This higher percentage of H_2 evolution by Au-capped CdS−PbS photocatalyst is because "Au" plasmonic antenna facilitates superior optical absorption to generate more excitons (Chen et al., 2020). Likewise, recently some plasmonic photocatalysts, such as TiO_2/g-C_3N_4 heterojunction coupled with plasmon "Ag" NPs, Au/Pd-TiO_2, plasmonic Au/TiO_2 nanosheets with small "Au" NPs, Au/Ta_2O_5, and bimetallic Ag/Au-decorated grapheme, are developed prominently to boost the yield of H_2 production (Cai et al., 2019; Cheng et al., 2019; Fudo, Tanaka, & Kominami, 2019; Manchala et al., 2019; Yang et al., 2020). Besides, several plasmonic photocatalysts have been fabricated for H_2 generation as represented in Table 7.1.

7.3.3 CO_2 reduction

The drastically increasing amount of CO_2 emissions is the global environmental catastrophe, and researchers are exploring for pronounced techniques to find a valuable remedy to this vexing problem. In this way, CO_2 is using a feedstock precursor to synthesize more value-added chemical products and small chain hydrocarbon fuels through photocatalytic CO_2 reduction. Moreover, CO_2 released from automobile exhaust and industrial production can also recyclable through photocatalytic CO_2 reduction. Recently, CO_2 photocatalytic reduction through plasmon-mediated photocatalysts is gaining excessive attention due to their rapid catalytic activity and overall performance. Very recently, Xie et al. developed tunable MoO_{3-x}-TiO_2 LSPR photocatalyst, which greatly enhanced the CO_2 photo-reduction under visible light. The MoO_{3-x}-TiO_2 plasmonic photocatalyst generated a significantly higher amount of CO and CH_4 than bare MoO_{3-x}, which is due to its superior visible-light absorption. Furthermore, the prepared MoO_{3-x}-TiO_2 revealed long-term stability with excellent performance. The CO and CH_4 quantities were linearly increased within 12 hour duration, and for every 12 hours, the reactor was degassed and CO_2 was re-introduced (Xie et al., 2020). Zhao et al. successfully fabricated the plasmonic Au-decorated TiO_2 photocatalyst with an ultrathin Al_2O_3 interlayer and reduce the CO_2 into CO and CH_4. The LSPR induced hot-electron injection improves the charge separation, which is the main reason for the observed high photocatalytic reduction of CO_2 molecules. Moreover, the Al_2O_3 interlayer present in the photocatalyst served as the passivation layer to suppress exciton recombination in the TiO_2. These aforementioned properties of the $Au/Al_2O_3/TiO_2$ photocatalyst effectively reduced the CO_2 into low carbon chain products (Zhao, Zheng, Feng, & Li, 2018). Liao et al. prepared first perovskite-based plasmonic $CsPbBr_3$-Au photocatalyst for CO_2 reduction (Liao et al., 2020). In this work, plasmonic "Au" NPs anchored on the perovskite $CsPbBr_3$ through a mercapto-propanoic organic molecule linker. The authors state that photo-generated electrons are produced under visible-light irradiation (>420 nm) from $CsPbBr_3$ and then injected into "Au" NPs, which efficiently suppressed the excitons recombination and thus greatly enhance the photocatalytic activity. High rate of CO_2 photo-reduction into was achieved using Au-TiO_2 plasmonic photonic crystal, which was prepared by Zeng et al. The authors stated that the photocatalytic reduction of CO_2 proceeds through two different mechanistic pathways. Pathway 1 (implicates hot electrons generated through plasmon damping of "Au" NPs and transferred to the TiO_2) can occur under simulated sunlight and the final product is CH_4 (302 μmol/g$_{catalyst}$/h) and pathway 2 (electron−hole pairs produced interband transitions in TiO_2 followed by the Z-scheme mechanism and transferred to "Au" NPs) can occur under UV light and produces HCHO and CO (420 and 323 μmol/g$_{catalyst}$/h) as final products (Zeng et al., 2020).

Furthermore, Cai et al. developed an Ag-CeO_2 photocatalyst by hydrothermal method and the synthesized plasmonic photocatalyst effectively reduces the CO_2 into CH_4

(99.98 μmol/g) and CH_3OH (34.96 μmol/g). The enhanced performance is attributed to the LSPR effect of "Ag" NPs, which can effectively boost the incident light absorption to facilitate the excitons' separation and transport efficiency (Cai et al., 2018). Also, a variety of plasmonic photocatalysts have been reported for the CO_2 reduction as displayed in Table 7.2.

7.3.4 Production of fine chemicals

The major environmental pollution issues are generated from traditional chemical industries and they seek alternative technologies with clean and low pollution for the production of chemicals. Among various clean energy sources, sunlight offers endless solar energy, which provides a clean and green way to produce fine chemicals via photocatalytic oxidation, reduction, and cross-coupling reactions. But the conventional photocatalytic reactions are suffering from low quantum efficiency due to its intrinsic nature. Hence, the researchers focus on the development of a new type of photocatalysts utilizing visible light more effectively for the different kinds of organic reactions. Recently, it has been revealed that the aforementioned properties are more appropriate for plasmonic nanomaterials because they can absorb a wide range of visible light and perform a variety of chemical reactions. Very recently, Wang et al. have developed plasmonic Au−Cu@LDH photocatalyst and used it for the preparation of methyl benzoate through oxidative esterification of benzyl alcohol with methanol under visible-light irradiation. Due to the LSPR effect of fused plasmonic "Au" with "Cu" NPs, the ability to activate the O_2 molecules to O2⁻ radicals enhanced which indirectly stimulate the cleavage of the C−H bond of benzyl alcohol. The authors conclude that oxidative esterification of aromatic alcohols through this plasmonic photocatalyst is simple, cost-effective, and environmentally benign (Wang, Wang, Wang, Fan, & Zheng, 2020).

Then, Cui et al. used the Au/CeO_2 plasmonic photocatalyst to selectively convert various benzyl alcohols to their corresponding benzaldehydes by oxidation reaction under visible-light illumination. This states that upon introduction "Au" NPs on CeO_2 effectively improve the visible-light absorbance efficiency as well as provides rich surface catalytic active sites. Besides, Au/CeO_2 also effectively converts other alcohols to corresponding aldehydes or ketones selectively under visible-light irradiation (Cui et al., 2018). Moreover, Au/CeO_2 plasmonic photocatalyst is found to be efficient in the reduction of ketones to alcohols, deoxygenation of epoxides to alkenes, reduction of nitro-aromatics to azo-compounds, and hydrogenation of azo-benzene to hydrazo-benzene. This excellent reduction reaction rate is improved mainly due to the LSPR effect of "Au" NPs under visible-light irradiation (Ke et al., 2013). It has also been reported that the Ag−Pd nanocages employed for the reduction of styrene to ethylbenzene in the presence of H_2 under visible-light irradiation. During the reaction, "Pd" provides active sites for hydrogenation reaction whereas "Ag" offers plasmonic properties to convert light into heat (Zhao, Long, Liu, Luo, & Xiong, 2015). Also, a wide variety of heterogeneous plasmonic photocatalysts have been developed for the production of fine chemicals as shown in Table 7.2.

Table 7.2: Various plasmonic photocatalysts reported for CO_2 reduction and production of fine chemicals.

S. no.	Plasmonic photocatalyst	Synthesis method	Application description	Reference
1	Ag-Bi/BiVO$_4$	Galvanic replacement reaction	CO_2 reduction to CO under "Xe" lamp ($\lambda > 420$ nm)	Duan, Zhao, Wei, and Chen (2020)
2	Cu/m-LaVO$_4$ hollow microspheres	Hydrothermal	CO_2 reduction to CH$_4$ under "Xe" lamp	Duan, Feng, and Chen (2020)
3	Au-g-C$_3$N$_4$	NaBH$_4$ reduction method	CO_2 reduction to CO and CH$_4$ under "Xe" lamp	Li, Gao, Xiong, Liao, and Shih (2018)
4	Ag/TiO$_2$	Biogenic synthesis	CO_2 reduction to CO and CH· under "Xe" lamp ($\lambda < 420$ nm)	Cheng et al. (2017)
5	Ag-SrTiO$_3$	Photo-deposition	CO_2 reduction to CO under "Xe" lamp ($\lambda = 420$ nm)	Shao et al. (2018)
6	Ag-MWCNT@TiO$_2$	Calcination followed by stirring	CO_2 reduction to CH$_4$, CO, and C$_2$H$_4$ under 15 W visible-light source	Gui, Wong, Chai, and Mohamed (2015)
7	Au/ZnO	Photo-deposition	CO_2 reduction to CH$_4$, CO, and C$_2$H$_6$ under "Xe" lamp	Zhao et al. (2019)
8	Au/TiO$_2$ nanotubes	Electrochemical deposition method	CO_2 reduction to CH$_4$ under "Xe" lamp	Khatun, Abd Aziz, Sim, and Monir (2019)
9	Cu-loaded In$_2$O$_3$/TiO$_2$	Solid-state chemical method	CO_2 reduction to CH$_4$ and CH$_3$OH under "Xe" lamp	Tahir et al. (2016)
10	Au-ZnO	Chemical reduction	Oxidation of n-hexane under UV-LED	Bueno-Alejo et al. (2020)
11	Cu/Cu@UiO-66	Advanced double solvents approach	Partial oxidation of aromatic alcohols under "Xe" lamp ($\lambda \geq 400$ nm)	Xiao et al. (2019)
12	Ag/β-Ag$_2$WO$_4$/BiVO$_4$	Photo-deposition	Oxidation of alcohols under tungsten lamp ($\lambda = 550$ nm)	Askari, Mohebbi, and Do (2018)
13	Ordered mesoporous Au/TiO$_2$	Chemical reduction	Oxidation of benzylamine under halogen-tungsten lamp ($\lambda > 420$ nm)	Yang and Mou (2018)
14	Ag/SnNb$_2$O$_6$Au/SnNb$_2$O$_6$	Chemical reduction	Reduction of p-nitrophenol under "Xe" lamp ($\lambda > 420$ nm)	Wu et al. (2019)
15	Au/CdMoO$_4$	Chemical reduction	Oxidation of benzylic alcohol under "Xe" lamp ($\lambda = 420$ nm)	Bi et al. (2015)
16	Au/TiO$_2$	Impregnation and reduction	Aerobic oxidation of benzyl alcohol under "Xe" lamp ($\lambda > 420$ nm)	Verma, Mori, Kuwahara, Cho, and Yamashita (2020)
17	Au-Ag/ZnO	Cophoto-deposition method	Oxidation of ethylene under "Xe" lamp	Zhai et al. (2020)

7.3.5 Disinfection and antimicrobial activity

To avoid millions of death due to disinfections by harmful microorganisms, several disinfectants are used, such as detergents, alcohols, and chlorine compounds. But the problem is that these materials are environmentally benign and also ineffective for the long term. Although UV radiation is used effectively, the limitations of direct usage of UV radiation are only in medical and technical purposes. Hence, the urgent need for controlling the disinfections caused by microorganisms becomes imperative. In the past decades, TiO_2-based photocatalytic materials have been widely used to inactivate or kill the various bacteria, viruses, fungi, and some harmful microorganisms, but it has some limits/limitations. Furthermore, to improve the photocatalytic activity, that is, antimicrobial activity toward the unwanted living factors, recently employing plasmonic materials for water disinfection has gained enormous momentum. In most of the antimicrobial applications, Ag-based TiO_2 photocatalytic materials have shown superior photocatalytic activity than other materials. Some examples of recently developed materials are silver-halide-supported TiO_2 photocatalyst for antibacterial organisms, Ag-TiO_2 nanocomposites for *Escherichia coli* and *Staphylococcus aureus*, and AgCl@Ag@TiO_2 sandwich-structured photocatalyst for gram-negative bacteria *E. coli* K12 (Tian et al., 2014). These experiments have been conducted both in the presence and absence of light irradiation. In the presence of light irradiation, the catalyst exhibits complete inactivation of *E. coli* colonies than the absence of light irradiation (killing response only 6%). This confirmed the antimicrobial efficacy due to the presence of plasmonic "Ag" NPs, which can destroy the cell wall of antimicrobial bacteria and inactivate the function by interacting with "S" and "P" present in the DNA. Furthermore, the antimicrobial activity of the "Ag" particle is size and shape dependent. Small-sized "Ag" NPs possess a larger surface area, which can be beneficial for a high percentage of interaction with the cell membrane than larger particles. The excitons are created when light is irradiated to the photocatalyst, due to the LSPR effect of "Ag" NPs. With the aid of these excitons, reactions can undergo further with the dissolved oxygen and water to form reactive radical species, such as $O2^{-}\cdot$ and $\cdot OH$. These aforementioned properties of "Ag" NPs will act as a good biocidal agent that could directly enhance the antimicrobial activity when they are coupled with TiO_2 photocatalysts (Dong et al., 2019). Recently, some Ag-coupled photocatalytic materials, such as Ag-loaded WO_3 nano-plates, Ag@CeO composite, ZnO-CdS-Ag ternary nanocomposite, graphene sheet-grafted plasmonic Ag/AgX, Ag-embedded ZnO nanocomposite, TiO_2 nanotubes/Ti plates modified by Ag-benzene, and Ag-3D-ordered mesoporous CeO_2 (Ayodhya, Veerabhadram, & Veerabhadram, 2019; Faraji, Mohaghegh, & Abedini, 2018; Gupta et al., 2016, 2017; Xia et al., 2016; Zhou, Ma, & Zhan, 2018; Zhu, Liu, Yu, Zhou, & Yan, 2016), were reported to be good plasmonic enhanced photocatalysts. Besides, different types of plasmonic photocatalysts are reported to inactivate or kill the several microorganisms, which are mentioned in Table 7.3.

Table 7.3: Various plasmonic photocatalysts reported for antimicrobial activity and N_2 fixation.

S. no	Plasmonic photocatalyst	Synthesis method	Application description	Reference
1	ZnO-Se	Biogenic reduction	Antibacterial activity against *Staphylococcus aureus* under 200 W tungsten bulb	Ahmad et al. (2020)
2	ZnO-CdS-Ag	Impregnation	Antibacterial activity against *Escherichia coli* (MG1655) under 125 W mercury vapor lamp and 400 W metal-halide lamp	Gupta, Eswar, Modak, and Madras (2017)
3	Ag/WO$_3$	photo-reduction	Antibacterial activity against *E. coli* (ATCC 8739) and *Bacillus subtilis* under visible light	Zhu, Liu, Yu, Zhou, and Yan (2016)
4	Ag-3D-ordered mesoporous CeO$_2$	Impregnation and photo-reduction	Antibacterial activity against *E. coli* under 300 W "Xe" arc lamp with a UV cutoff ($\lambda < 420$ nm)	Zhou, Ma, and Zhan (2018)
5	Multilayer ultrathin Ag-δ-Bi$_2$O$_3$	photo-reduction	N_2 fixation under 400 W "Xe" lamp with a cutoff ($\lambda < 400$ nm)	Gao, Shang, Liu, and Fu (2019)
6	Au/TiO$_2$	Chemical reduction	N_2 fixation under UV—vis or visible light ($\lambda > 400$ nm)	Bu et al. (2019)

7.3.6 N_2 fixation

In modern agriculture and the chemical industry, the fixation of atmospheric N_2 into NH_3 is a crucial process. NH_3 is a fertilizer need for the plant's growth. One-third of the whole atmospheric earth is surrounded by molecular N_2 but it cannot be utilized nutritionally by the living organisms particularly plants because of the strong N≡N bond (need 941 kJ/mol activation energy to cleave the bond) in molecular N_2. Recently, solar-induced NH_3 synthesis has received enormous interest and it is considered a clean and economical approach. During fixation, the conventional photocatalyst yields poor results compared to plasmonic photocatalysts. The plasmonic metal NPs have been combined as cocatalysts with semiconductors to enrich the light absorption and provide hot electrons for N_2 activation. Yang et al. reported Au/TiO$_2$ plasmonic photocatalyst and under NIR illumination, the hot electrons created in "Au" NPs are directly injecting into the CB of TiO$_2$, which further enters through the antibond orbital of N≡N to form NH_4^+ and then further yields NH_3 with a quantum efficiency of 0.82% at 550 nm (Yang et al., 2018). Likewise, plasmon-assisted N_2 fixation (114.3 μmol/g^m/h^m of NH_3) is achieved over Au/CeO$_2$ (Jia et al., 2019). Xiong et al. reported plasmonic AuRu$_{0.31}$ photocatalysts with an Au-rich core as a light absorber and achieved an NH_3 production rate of 101.4 μmol/g^m/h^m (Hu et al., 2019). Wang et al. fabricated plasmon-enhanced Ag/black-silicon as nitrogen fixation photocatalyst to reach a high Faraday efficiency of 55.05% with an NH_3 yield of 2.87 μmol/h/cm^2. The higher yield is due to the substantial increment in photo-generated excitons' separation efficiency (Wang et al., 2020).

Likewise, plasmonic Au-loaded g-C_3N_4 nanosheets are developed by Wu et al. and they achieved a maximum NH_3 yield of 184 and 93 $\mu mol/g^m$ under full-light and visible-light irradiation, respectively (Wu et al., 2020). Few plasmonic photocatalysts also developed for the N_2 fixation as displayed in Table 7.3.

7.4 Comparison of photocatalytic performances for monometallic, bimetallic, and trimetallic plasmonic photocatalysts

Especially plasmonic photocatalysts can address "3" major challenges, including visible-light absorption, low recombination rate, and performance of the photocatalyst. Generally, monometallic, bimetallic, and trimetallic plasmonic photocatalysts comprise single, two, and three types of noble-metal NPs. And these bimetallic and trimetallic plasmonic photocatalysts can be in core–shell and alloy nanostructures. But bimetallic and trimetallic plasmonic photocatalysts may yield high efficiency to that of monometallic plasmonic photocatalysts due to synergistic effect between different types of noble-metal NPs employed during the

Table 7.4: Comparison of photocatalytic performances for monometallic, bimetallic, and trimetallic plasmonic photocatalysts.

S. no.	Plasmonic photocatalyst	Synthesis method	Application description	Comparison of efficiency	Reference
1	Au-TiO$_2$Pt-TiO$_2$Pt/Au-TiO$_2$	Chemical reduction	Phenol degradation under 1000 W "Xe" lamp ($\lambda > 420$ nm)	0.4 $\mu mol/dm^3/min$6.3 $\mu mol/dm^3/$min6.1 $\mu mol/dm^3/min$ (degradation rates)	Gołą biewska et al. (2017)
2	Pt-TiO$_2$Pd-TiO$_2$Pt/Pd-TiO$_2$	Sol–gel method	Phenol degradation under 1000 W "Xe" lamp ($\lambda > 420$ nm)	6.20 $\mu mol/dm^3/min$1.76 $\mu mol/dm^3/$min6.84 $\mu mol/dm^3/min$	Zielińska-Jurek and Hupka (2014)
3	Au/Pd/Pt-NaTaO$_3$Au/Pd/Pt-TiO$_2$ (Rutile)	Surface immobilization of core–shell NPs followed by green reduction	H$_2$ generation under 250 W "Xe" lamp	41.6 $\mu mol/min$213.8 $\mu mol/min$ (H$_2$ generation rates)	Malankowska et al. (2018)
4	Ag/graphemeAu/graphemeAg/Au-graphene	Green reduction	H$_2$ generation under sunlight	20 $\mu mol/h/g$28 $\mu mol/h/g$40 $\mu mol/h/g$	Manchala et al. (2019)
5	Pt/g-C$_3$N$_4$Au/g-C$_3$N$_4$Au/Pt/g-C$_3$N$_4$	Photo-deposition	Tetracycline hydrochloride degradation under 500 W "Xe" lamp ($\lambda = 420$ nm)	0.1809 min^{-1}0.2495 min^{-1}0.4286 min^{-1}	Xue, Ma, Zhou, Zhang, and He (2015)

synthesis. Several researchers already synthesized different types of monometallic, bimetallic, and trimetallic plasmonic photocatalysts and compared their photocatalytic performances for different applications as displayed in Table 7.4. Lee et al. (2019) prepared ZnO-supported Au/Pd bimetallic nanocomposites for plasmon-improved photocatalytic activity for methylene blue degradation under visible-light irradiation. Compared with plasmonic monometallic NPs, bimetallic NPs, such as Pd/Pt, Pt/Au, Au/Pd, Pt/Ag, and Au/Ag, showed enhanced catalytic, magnetic, electrocatalytic, electronic, and optical properties due to the synergistic effects.

7.5 Principle of plasmonic photocatalysis

Since decade, many researchers have been developed various types of potential plasmonic photocatalysts in several applications as mentioned above. In common, a semiconductor photocatalytic cycle comprises "3" steps. But in plasmonic photocatalysis, it takes "4" steps. It includes light absorption, excitons generation, and separation, and finally, the photo-induced excitons can play a part in redox (oxidation or reduction) reactions with adsorbed species. Also, electron transfer between the noble-metal NPs and semiconductors is an additional step as displayed in Fig. 7.7. In this way, the lifetime of photo-induced excitons is increased. As a result, incident light energy can be employed to direct advantageous chemical reactions (Fan & Leung, 2016).

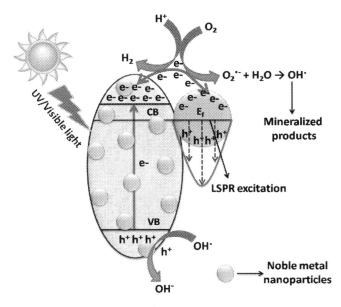

Figure 7.7
General principle involved in the plasmonic photocatalysis.

7.6 Conclusions and future perspective

Plasmonic photocatalysis or usage of noble-metal NPs in photocatalysis was considered to be one of the best technologies to improve the performance of the photocatalyst in the visible region. This chapter describes the fundamentals and different types of energy transfer mechanisms associated with the plasmonic photocatalysis in detail. Then, advantages of loading of noble-metal NPs on semiconductors were explained. Besides, it demonstrated various synthesis methods for the deposition or synthesis of noble-metal NPs onto the semiconductor. Finally, it explained the several applications of plasmonic photocatalysis, including wastewater treatment, H_2 generation, CO_2 reduction, production of fine chemicals, and antimicrobial applications. And a general mechanism of plasmonic photocatalysis has been discussed. Efficiency is still not up to the mark, more research should be dedicated toward the surface morphology and optimization. Also, the reports on the synthesis of bimetallic and trimetallic NPs have been steadily increasing for different applications to address the limitations of monometallic NPs by providing more catalytic active sites and additional plasmon decay path for hot-electron generation. LSPR effect or plasmonics not only utilized in the development of potential photocatalysts but also can be used in the fabrication of optical devices, photonic devices, solar cells, fuel cells, biosensors, and photo-electrochemical devices. Besides, the LSPR effect of noble-metal NPs can also be applied in SERS and TERS for the sensitive structural identification of various organic molecules and their intermediates, such as radicals and radical ions.

Moreover, a lot of investigation has been conducted for hot-electron generation and its transfer. In most of the cases, the photo-induced charge vacancies, that is, hot holes are scavenged by different scavenging agents. Therefore in the future, there is a need to conduct serious investigations on the role of hot holes in plasmonic photocatalysis. And compared to Pt, Au, and Ag plasmonic metals, Cu, Ni, Pd, and Al are cheaply available. Thus more research should be conducted on these metals. At the last, we hope that the topics of this chapter will facilitate to open a unique path for the development of advanced, new, and potential plasmonic photocatalysts that can utilize hot excitons fully for enhancing desirable reactions.

References

Ahmad, A., Ullah, S., Ahmad, W., Yuan, Q., Taj, R., Khan, A. U., . . . Khan, U. A. (2020). Zinc oxide-selenium heterojunction composite: Synthesis, characterization and photo-induced antibacterial activity under visible light irradiation. *Journal of Photochemistry and Photobiology B: Biology*, *203*, 111743. Available from https://doi.org/10.1016/j.jphotobiol.2019.111743.

Askari, P., Mohebbi, S., & Do, T. (2018). High performance plasmonic activation of Ag on β-Ag2WO4/BiVO4 as nanophotocatalyst for oxidation of alcohols by incident visible light. *Journal of Photochemistry and Photobiology A: Chemistry*, *367*, 56−65. Available from https://doi.org/10.1016/j.jphotochem.2018.080.013.

Ayodhya, D., Veerabhadram, G., & Veerabhadram, G. (2019). Green synthesis of garlic extract stabilized Ag@CeO composites for photocatalytic and sonocatalytic degradation of mixed dyes and antimicrobial studies. *Journal of Molecular Structure*.

Bathula, B., Koutavarapu, R., Shim, J., & Yoo, K. (2020). Facile one-pot synthesis of gold/tin oxide quantum dots for visible light catalytic degradation of methylene blue: Optimization of plasmonic effect. *Journal of Alloys and Compounds*, *812*, 152081. Available from https://doi.org/10.1016/j.jallcom.2019.152081.

Bi, J., Zhou, Z., Chen, M., Liang, S., He, Y., Zhang, Z., & Wu, L. (2015). Plasmonic Au/CdMoO4 photocatalyst: Influence of surface plasmon resonance for selective photocatalytic oxidation of benzylic alcohol. *Applied Surface Science*, *349*, 292−298. Available from https://doi.org/10.1016/j.apsusc.2015.040.213.

Bu, T.-A., Hao, Y.-C., Gao, W.-Y., Su, X., Chen, L.-W., Zhang, N., & Yin, A.-X. (2019). Promoting photocatalytic nitrogen fixation with alkali metal cations and plasmonic nanocrystals. *Nanoscale*, *11*, 10072−10079. Available from https://doi.org/10.1039/C9NR02502B.

Bueno-Alejo, C. J., Graus, J., Arenal, R., Lafuente, M., Bottega-Pergher, B., & Hueso, J. L. (2020). Anisotropic Au-ZnO photocatalyst for the visible-light expanded oxidation of *n*-hexane. *Catalysis Today*. Available from https://doi.org/10.1016/j.cattod.2020.030.063.

Cai, W., Shi, Y., Zhao, Y., Chen, M., Zhong, Q., & Bu, Y. (2018). The solvent-driven formation of multi-morphological Ag-CeO2 plasmonic photocatalysts with enhanced visible-light photocatalytic reduction of CO2. *RSC Advances*, *8*, 40731−40739. Available from https://doi.org/10.1039/c8ra08938h.

Cai, X., Chen, Q., Wang, R., Wang, A., Wang, J., Zhong, S., . . . Bai, S. (2019). Integration of plasmonic metal and cocatalyst: an efficient strategy for boosting the visible and broad-spectrum photocatalytic H2 evolution. *Advanced Materials Interfaces*, *6*, 1−10. Available from https://doi.org/10.1002/admi.201900775.

Chatterjee, A., Kar, P., Wulferding, D., Lemmens, P., & Pal, S. K. (2020). Flower-like BiOI microspheres decorated with plasmonic gold nanoparticles for dual detoxification of organic and inorganic water pollutants. *ACS Applied Nano Materials*, *3*, 2733−2744. Available from https://doi.org/10.1021/acsanm.0c00090.

Chen, K., Ding, S.-J., Ma, S., Wang, W., Liang, S., Zhou, L., & Wang, Q.-Q. (2020). Enhancing photocatalytic activity of Au-capped CdS−PbS heterooctahedron by morphology control. *The Journal of Physical Chemistry C*, *124*, 7938−7945. Available from https://doi.org/10.1021/acs.jpcc.0c00349.

Cheng, L., Zhang, D., Liao, Y., Li, F., Zhang, H., & Xiang, Q. (2019). Constructing functionalized plasmonic gold/titanium dioxide nanosheets with small gold nanoparticles for efficient photocatalytic hydrogen evolution. *Journal of Colloid and Interface Science*, *555*, 94−103. Available from https://doi.org/10.1016/j.jcis.2019.07.060.

Cheng, X., Dong, P., Huang, Z., Zhang, Y., Chen, Y., Nie, X., & Zhang, X. (2017). Green synthesis of plasmonic Ag nanoparticles anchored TiO2 nanorod arrays using cold plasma for visible-light-driven photocatalytic reduction of CO2. *Journal of CO2 Utilization*, *20*, 200−207. Available from https://doi.org/10.1016/j.jcou.2017.040.009.

Cui, Z., Wang, W., Zhao, C., Chen, C., Han, M., Wang, G., . . . Zhao, H. (2018). Spontaneous redox approach to the self-assembly synthesis of Au/CeO2 plasmonic photocatalysts with rich oxygen vacancies for selective photocatalytic conversion of alcohols. *ACS Applied Materials & Interfaces*, *10*, 31394−31403. Available from https://doi.org/10.1021/acsami.8b10705.

Cushing, S. K., Bristow, A. D., & Wu, N. (2015). Theoretical maximum efficiency of solar energy conversion in plasmonic metal−semiconductor heterojunctions. *Physical Chemistry Chemical Physics*, *17*, 30013−30022. Available from https://doi.org/10.1039/C5CP04512F.

Dhand, C., Dwivedi, N., Loh, X. J., Jie Ying, A. N., Verma, N. K., Beuerman, R. W., . . . Ramakrishna, S. (2015). Methods and strategies for the synthesis of diverse nanoparticles and their applications: A comprehensive overview. *RSC Advances*, *5*, 105003−105037. Available from https://doi.org/10.1039/C5RA19388E.

Dong, P., Yang, F., Cheng, X., Huang, Z., Nie, X., Xiao, Y., & Zhang, X. (2019). Plasmon enhanced photocatalytic and antimicrobial activities of Ag-TiO2 nanocomposites under visible light irradiation prepared by DBD cold plasma treatment. *Materials Science and Engineering C*, *96*, 197−204. Available from https://doi.org/10.1016/j.msec.2018.110.005.

DuChene, J. S., Sweeny, B. C., Johnston-Peck, A. C., Su, D., Stach, E. A., & Wei, W. D. (2014). Prolonged hot electron dynamics in plasmonic-metal/semiconductor heterostructures with implications for solar photocatalysis. *Angewandte Chemie International Edition*, *53*, 7887−7891. Available from https://doi.org/10.1002/anie.201404259.

Duan, Z., Feng, X., & Chen, L. (2020). Cu/m-LaVO4 hollow composite microspheres for photocatalytic CO2 reduction. *Journal of Solid State Chemistry*, *286*, 121298. Available from https://doi.org/10.1016/j.jssc.2020.121298.

Duan, Z., Zhao, X., Wei, C., & Chen, L. (2020). Ag-Bi/BiVO4 chain-like hollow microstructures with enhanced photocatalytic activity for CO2 conversion. *Applied Catalysis A: General*, *594*, 117459. Available from https://doi.org/10.1016/j.apcata.2020.117459.

Fan, W., & Leung, M. (2016). Recent development of plasmonic resonance-based photocatalysis and photovoltaics for solar utilization. *Molecules (Basel, Switzerland)*, *21*, 180. Available from https://doi.org/10.3390/molecules21020180.

Faraji, M., Mohaghegh, N., & Abedini, A. (2018). TiO2 nanotubes/Ti plates modified by silver-benzene with enhanced photocatalytic antibacterial properties. *New Journal of Chemistry*, *42*, 2058−2066. Available from https://doi.org/10.1039/c7nj03554c.

Fowler, R. H. (1931). The analysis of photoelectric sensitivity curves for clean metals at various temperatures. *Physical Review*, *38*, 45−56. Available from https://doi.org/10.1103/PhysRev.38.45.

Fudo, E., Tanaka, A., & Kominami, H. (2019). Cocatalyst-free plasmonic H2 production over Au/Ta2O5 under irradiation of visible light. *Chemistry Letters*, *48*, 939−942. Available from https://doi.org/10.1246/cl.190379.

Gao, X., Shang, Y., Liu, L., & Nie, W. (2019). A plasmonic Z-scheme three-component photocatalyst g-C3N4/Ag/LaFeO3 with enhanced visible-light photocatalytic activities. *Optical Materials*, *88*, 229−237. Available from https://doi.org/10.1016/j.optmat.2018.110.030.

Gao, X., Shang, Y., Liu, L., & Fu, F. (2019). Multilayer ultrathin Ag-δ-Bi2O3 with ultrafast charge transformation for enhanced photocatalytic nitrogen fixation. *Journal of Colloid and Interface Science*, *533*, 649−657. Available from https://doi.org/10.1016/j.jcis.2018.080.091.

Gomez, L., Sebastian, V., Arruebo, M., Santamaria, J., & Cronin, S. B. (2014). Plasmon-enhanced photocatalytic water purification. *Physical Chemistry Chemical Physics*, *16*, 15111. Available from https://doi.org/10.1039/c4cp00229f.

Gondal, M. A., Suliman, M. A., Dastageer, M. A., Chuah, G.-K., Basheer, C., Yang, D., & Suwaiyan, A. (2016). Visible light photocatalytic degradation of herbicide (Atrazine) using surface plasmon resonance induced in mesoporous Ag-WO3/SBA-15 composite. *Journal of Molecular Catalysis A: Chemical*, *425*, 208−216. Available from https://doi.org/10.1016/j.molcata.2016.100.015.

Govorov, A. O., Zhang, H., Demir, H. V., & Gun'ko, Y. K. (2014). Photogeneration of hot plasmonic electrons with metal nanocrystals: Quantum description and potential applications. *Nano Today*, *9*, 85−101. Available from https://doi.org/10.1016/j.nantod.2014.020.006.

Gołąbiewska, A., Lisowski, W., Jarek, M., Nowaczyk, G., Michalska, M., Jurga, S., & Zaleska-Medynska, A. (2017). The effect of metals content on the photocatalytic activity of TiO2 modified by Pt/Au bimetallic nanoparticles prepared by sol-gel method. *Molecular Catalysis*, *442*, 154−163. Available from https://doi.org/10.1016/j.mcat.2017.090.004.

Guan, J., Li, L., Chen, C., Lu, P., Yan, Y., Wang, Z., . . . Huo, M. (2020). Enhanced solar-photocatalytic activity for the simultaneous degradation and detoxification of multiple chlorophenols by embedding plasmonic Pt into TiO2/H3PW12O40 nanopore. *Applied Surface Science*, *513*, 145833. Available from https://doi.org/10.1016/j.apsusc.2020.145833.

Gui, M. M., Wong, W. M. P., Chai, S. P., & Mohamed, A. R. (2015). One-pot synthesis of Ag-MWCNT@TiO2 core-shell nanocomposites for photocatalytic reduction of CO2 with water under visible light irradiation. *Chemical Engineering Journal*, *278*, 272−278. Available from https://doi.org/10.1016/j.cej.2014.090.022.

Gupta, J., Mohapatra, J., & Bahadur, D. (2017). Visible light driven mesoporous Ag-embedded ZnO nanocomposites: reactive oxygen species enhanced photocatalysis, bacterial inhibition and photodynamic therapy. *Dalton Transactions, 46*, 685−696. Available from https://doi.org/10.1039/c6dt03713e.

Gupta, R., Eswar, N. K., Modak, J. M., Eswar, N. K., Modak, J. M., & Madras, G. (2016). Effect of morphology of zinc oxide in ZnO-CdS-Ag ternary nanocomposite towards photocatalytic inactivation of E. coli under UV and visible light. *Chemical Engineering Journal.*

Gupta, R., Eswar, N. K., Modak, J. M., & Madras, G. (2017). Effect of morphology of zinc oxide in ZnO-CdS-Ag ternary nanocomposite towards photocatalytic inactivation of E. coli under UV and visible light. *Chemical Engineering Journal, 307*, 966−980. Available from https://doi.org/10.1016/j.cej.2016.080.142.

Güy, N., & Özacar, M. (2016). The influence of noble metals on photocatalytic activity of ZnO for Congo red degradation. *International Journal of Hydrogen Energy, 41*, 20100−20112. Available from https://doi.org/10.1016/j.ijhydene.2016.070.063.

Hu, C., Chen, X., Jin, J., Han, Y., Chen, S., Ju, H., . . . Xiong, Y. (2019). Surface plasmon enabling nitrogen fixation in pure water through a dissociative mechanism under mild conditions. *Journal of the American Chemical Society, 141*, 7807−7814. Available from https://doi.org/10.1021/jacs.9b01375.

Huang, H. J., Wu, J. C.-S., Chiang, H.-P., Chou Chau, Y.-F., Lin, Y.-S., Wang, Y. H., & Chen, P.-J. (2019). Review of experimental setups for plasmonic photocatalytic reactions. *Catalysts, 10*, 46. Available from https://doi.org/10.3390/catal10010046.

Huang, Y., Liu, J., Cao, D., Liu, Z., Ren, K., Liu, K., . . . Wang, Z. (2019). Separation of hot electrons and holes in Au/LaFeO3 to boost the photocatalytic activities both for water reduction and oxidation. *International Journal of Hydrogen Energy, 44*, 13242−13252. Available from https://doi.org/10.1016/j.ijhydene.2019.030.182.

Irani, E., & Amoli-Diva, M. (2020). Hybrid adsorption−photocatalysis properties of quaternary magneto-plasmonic ZnO/MWCNTs nanocomposite for applying synergistic photocatalytic removal and membrane filtration in industrial wastewater treatment. *Journal of Photochemistry and Photobiology A: Chemistry, 391*, 112359. Available from https://doi.org/10.1016/j.jphotochem.2020.112359.

Jain, P. K., Lee, K. S., El-Sayed, I. H., & El-Sayed, M. A. (2006). Calculated absorption and scattering properties of gold nanoparticles of different size, shape, and composition: applications in biological imaging and biomedicine. *The Journal of Physical Chemistry. B, 110*, 7238−7248. Available from https://doi.org/10.1021/jp057170o.

Jia, H., Du, A., Zhang, H., Yang, J., Jiang, R., Wang, J., & Zhang, C. Y. (2019). Site-selective growth of crystalline ceria with oxygen vacancies on gold nanocrystals for near-infrared nitrogen photofixation. *Journal of the American Chemical Society, 141*, 5083−5086. Available from https://doi.org/10.1021/jacs.8b13062.

Kavinkumar, V., Verma, A., Masilamani, S., Kumar, S., Jothivenkatachalam, K., & Fu, Y.-P. (2019). Investigation of the structural, optical and crystallographic properties of Bi2WO6/Ag plasmonic hybrids and their photocatalytic and electron transfer characteristics. *Dalton Transactions, 48*, 10235−10250. Available from https://doi.org/10.1039/C9DT01807G.

Kawamura, G., & Matsuda, A. (2019). Synthesis of plasmonic photocatalysts for water splitting. *Catalysts, 9*, 982. Available from https://doi.org/10.3390/catal9120982.

Ke, X., Zhang, X., Zhao, J., Sarina, S., Barry, J., & Zhu, H. (2013). Selective reductions using visible light photocatalysts of supported gold nanoparticles. *Green Chemistry: An International Journal and Green Chemistry Resource, 15*, 236−244. Available from https://doi.org/10.1039/c2gc36542a.

Keihan, A. H., Hosseinzadeh, R., Farhadian, M., Kooshki, H., & Hosseinzadeh, G. (2016). Solvothermal preparation of Ag nanoparticle and graphene co-loaded TiO2 for the photocatalytic degradation of paraoxon pesticide under visible light irradiation. *RSC Advances, 6*, 83673−83687. Available from https://doi.org/10.1039/C6RA19478H.

Khatun, F., Abd Aziz, A., Sim, L. C., & Monir, M. U. (2019). Plasmonic enhanced Au decorated TiO2 nanotube arrays as a visible light active catalyst towards photocatalytic CO2 conversion to CH4. *Journal of Environmental Chemical Engineering, 7*, 103233. Available from https://doi.org/10.1016/j.jece.2019.103233.

Kumar, M. K., Krishnamoorthy, S., Tan, L. K., Chiam, S. Y., Tripathy, S., & Gao, H. (2011). Field effects in plasmonic photocatalyst by precise SiO2 thickness control using atomic layer deposition. *ACS Catalysis*, *1*, 300−308. Available from https://doi.org/10.1021/cs100117v.

Kumar, S., Karthikeyan, S., & Lee, A. (2018). g-C3N4-based nanomaterials for visible light-driven photocatalysis. *Catalysts*, *8*, 74. Available from https://doi.org/10.3390/catal8020074.

Kumarasinghe, C. S., Premaratne, M., Bao, Q., & Agrawal, G. P. (2015). Theoretical analysis of hot electron dynamics in nanorods. *Scientific Reports*, *5*, 12140. Available from https://doi.org/10.1038/srep12140.

Kumarasinghe, C. S., Premaratne, M., Gunapala, S. D., & Agrawal, G. P. (2016). Theoretical analysis of hot electron injection from metallic nanotubes into a semiconductor interface. *Physical Chemistry Chemical Physics*, *18*, 18227−18236. Available from https://doi.org/10.1039/C6CP03043B.

Lee, S. J., Jung, H. J., Koutavarapu, R., Lee, S. H., Arumugam, M., Kim, J. H., & Choi, M. Y. (2019). ZnO supported Au/Pd bimetallic nanocomposites for plasmon improved photocatalytic activity for methylene blue degradation under visible light irradiation. *Applied Surface Science*, *496*, 143665. Available from https://doi.org/10.1016/j.apsusc.2019.143665.

Leong, K. H., Aziz, A. A., Sim, L. C., Saravanan, P., Jang, M., & Bahnemann, D. (2018). Mechanistic insights into plasmonic photocatalysts in utilizing visible light. *Beilstein Journal of Nanotechnology*, *9*, 628−648. Available from https://doi.org/10.3762/bjnano.9.59.

Leong, K. H., Gan, B. L., Ibrahim, S., & Saravanan, P. (2014). Synthesis of surface plasmon resonance (SPR) triggered Ag/TiO2 photocatalyst for degradation of endocrine disturbing compounds. *Applied Surface Science*, *319*, 128−135. Available from https://doi.org/10.1016/j.apsusc.2014.060.153.

Li, H., Gao, Y., Xiong, Z., Liao, C., & Shih, K. (2018). Enhanced selective photocatalytic reduction of CO2 to CH4 over plasmonic Au modified g-C3N4 photocatalyst under UV−vis light irradiation. *Applied Surface Science*, *439*, 552−559. Available from https://doi.org/10.1016/j.apsusc.2018.01.071.

Li, M., Yu, Z., Hou, Y., Liu, Q., Qian, L., Lian, C., ... Yang, X. (2019). Charge trapping and transfer mechanisms of noble metals and metal oxides deposited Ga2O3 toward typical contaminant degradation. *Chemical Engineering Journal*, *370*, 1119−1127. Available from https://doi.org/10.1016/j.cej.2019.030.291.

Li, X., Yu, J., Jaroniec, M., & Chen, X. (2019). Cocatalysts for selective photoreduction of CO2 into solar fuels. *Chemical Reviews*, *119*, 3962−4179. Available from https://doi.org/10.1021/acs.chemrev.8b00400.

Liao, J., Cai, Y., Li, J., Jiang, Y., Wang, X., Chen, H.-Y., & Kuang, D.-B. (2020). Plasmonic CsPbBr3-Au nanocomposite for excitation wavelength dependent photocatalytic CO2 reduction. *Journal of Energy Chemistry*. Available from https://doi.org/10.1016/j.jechem.2020.040.017.

Long, R., & Prezhdo, O. V. (2014). Instantaneous generation of charge-separated state on TiO2 surface sensitized with plasmonic nanoparticles. *Journal of the American Chemical Society*, *136*, 4343−4354. Available from https://doi.org/10.1021/ja5001592.

Luo, B., Xu, D., Li, D., Wu, G., Wu, M., Shi, W., & Chen, M. (2015). Fabrication of a Ag/Bi3TaO7 plasmonic photocatalyst with enhanced photocatalytic activity for degradation of tetracycline. *ACS Applied Materials & Interfaces*, *7*, 17061−17069. Available from https://doi.org/10.1021/acsami.5b03535.

Maicu, M., Schmittgens, R., Hecker, D., Glöß, D., Frach, P., & Gerlach, G. (2014). Synthesis and deposition of metal nanoparticles by gas condensation process. *Journal of Vacuum Science & Technology A: Vacuum, Surfaces and Films*, *32*, 02B113. Available from https://doi.org/10.1116/1.4859260.

Malankowska, A., Kobylański, M. P., Mikolajczyk, A., Cavdar, O., Nowaczyk, G., Jarek, M., ... Zaleska-Medynska, A. (2018). TiO2 and NaTaO3 decorated by trimetallic Au/Pd/Pt core−shell nanoparticles as efficient photocatalysts: experimental and computational studies. *ACS Sustainable Chemistry & Engineering*, *6*, 16665−16682. Available from https://doi.org/10.1021/acssuschemeng.8b03919.

Manchala, S., Nagappagari, L. R., Muthukonda Venkatakrishnan, S., & Shanker, V. (2018). Facile synthesis of noble-metal free polygonal Zn2TiO4 nanostructures for highly efficient photocatalytic hydrogen evolution under solar light irradiation. *International Journal of Hydrogen Energy*, *43*, 13145−13157. Available from https://doi.org/10.1016/j.ijhydene.2018.050.035.

Manchala, S., Nagappagari, L. R., Venkatakrishnan, S. M., & Shanker, V. (2019). Solar-light harvesting bimetallic Ag/Au decorated graphene plasmonic system with efficient photoelectrochemical performance for the enhanced water reduction process. *ACS Applied Nano Materials*, 2, 4782−4792. Available from https://doi.org/10.1021/acsanm.9b00684.

Manchala, S., Tandava, V. S. R. K., Nagappagari, L. R., Muthukonda Venkatakrishnan, S., Jampaiah, D., Sabri, Y. M., . . . Shanker, V. (2019). Fabrication of a novel ZnIn2S4/g-C3N4/graphene ternary nanocomposite with enhanced charge separation for efficient photocatalytic H2 evolution under solar light illumination. *Photochemical & Photobiological Sciences: Official Journal of the European Photochemistry Association and the European Society for Photobiology*, 18, 2952−2964. Available from https://doi.org/10.1039/C9PP00234K.

Nasir, M. S., Yang, G., Ayub, I., Wang, S., & Yan, W. (2020). In situ decoration of g-C3N4 quantum dots on 1D branched TiO2 loaded with plasmonic Au nanoparticles and improved the photocatalytic hydrogen evolution activity. *Applied Surface Science*, 519, 146208. Available from https://doi.org/10.1016/j.apsusc.2020.146208.

Pelletier, F., & Thiébaut, B. (2016). Improvement of noble metal based photocatalysts by spray pyrolysis processes. *Johnson Matthey Technology Review*, 60, 39−54. Available from https://doi.org/10.1595/205651315X689829.

Saadetnejad, D., & Yıldırım, R. (2018). Photocatalytic hydrogen production by water splitting over Au/Al-SrTiO3. *International Journal of Hydrogen Energy*, 43, 1116−1122. Available from https://doi.org/10.1016/j.ijhydene.2017.100.154.

Shao, K., Wang, Y., Iqbal, M., Lin, L., Wang, K., Zhang, X., . . . He, T. (2018). Modification of Ag nanoparticles on the surface of SrTiO3 particles and resultant influence on photoreduction of CO2. *Applied Surface Science*, 434, 717−724. Available from https://doi.org/10.1016/j.apsusc.2017.110.004.

Shen, H., Xue, W., Fu, F., Sun, J., Zhen, Y., Wang, D., . . . Tang, J. (2018). Efficient degradation of phenol and 4-nitrophenol by surface oxygen vacancies and plasmonic silver co-modified Bi2MoO6 photocatalysts. *Chemistry—A European Journal*, 24, 18463−18478. Available from https://doi.org/10.1002/chem.201804267.

Soleimani Zohr Shiri, M., Henderson, W., & Mucalo, M. R. (2019). A review of the lesser-studied microemulsion-based synthesis methodologies used for preparing nanoparticle systems of the noble metals, Os, Re, Ir and Rh. *Materials (Basel)*, 12, 1896. Available from https://doi.org/10.3390/ma12121896.

Tahir, M., Tahir, B., Saidina Amin, N. A., & Alias, H. (2016). Selective photocatalytic reduction of CO2 by H2O/H2 to CH4 and CH3OH over Cu-promoted In2O3/TiO2 nanocatalyst. *Applied Surface Science*, 389, 46−55. Available from https://doi.org/10.1016/j.apsusc.2016.060.155.

Tahir, M., Siraj, M., Tahir, B., Umer, M., Alias, H., & Othman, N. (2020). Au-NPs embedded Z−scheme WO3/TiO2 nanocomposite for plasmon-assisted photocatalytic glycerol-water reforming towards enhanced H2 evolution. *Applied Surface Science*, 503, 144344. Available from https://doi.org/10.1016/j.apsusc.2019.144344.

Tian, B., Dong, R., Zhang, J., Bao, S., Yang, F., & Zhang, J. (2014). Sandwich-structured AgCl at Ag at TiO2 with excellent visible-light photocatalytic activity for organic pollutant degradation and E. coli K12 inactivation. *Applied Catalysis B: Environmental*, 158−159, 76−84. Available from https://doi.org/10.1016/j.apcatb.2014.040.008.

Tonda, S., Kumar, S., & Shanker, V. (2016). Surface plasmon resonance-induced photocatalysis by Au nanoparticles decorated mesoporous g-C3N4 nanosheets under direct sunlight irradiation. *Materials Research Bulletin*, 75, 51−58. Available from https://doi.org/10.1016/j.materresbull.2015.11.011.

Tonda, S., Kumar, S., Gawli, Y., Bhardwaj, M., & Ogale, S. (2017). g-C3N4(2D)/CdS(1D)/rGO(2D) dual-interface nano-composite for excellent and stable visible light photocatalytic hydrogen generation. *International Journal of Hydrogen Energy*, 42, 5971−5984. Available from https://doi.org/10.1016/j.ijhydene.2016.11.065.

Tun, P. P., Wang, J., Khaing, T. T., Wu, X., & Zhang, G. (2020). Fabrication of functionalized plasmonic Ag loaded Bi2O3/montmorillonite nanocomposites for efficient photocatalytic removal of antibiotics and organic dyes. *Journal of Alloys and Compounds*, *818*, 152836. Available from https://doi.org/10.1016/j.jallcom.2019.152836.

Van Dao, D., Nguyen, T. T. D., Le, T. D., Kim, S.-H., Yang, J.-K., Lee, I.-H., & Yu, Y.-T. (2020). Plasmonically driven photocatalytic hydrogen evolution activity of a Pt-functionalized Au@CeO2 core−shell catalyst under visible light. *Journal of Materials Chemistry A*, 7687−7694. Available from https://doi.org/10.1039/d0ta00811g.

Verma, P., Mori, K., Kuwahara, Y., Cho, S. J., & Yamashita, H. (2020). Synthesis of plasmonic gold nanoparticles supported on morphology-controlled TiO2 for aerobic alcohol oxidation. *Catalysis Today*, *352*, 255−261. Available from https://doi.org/10.1016/j.cattod.2019.100.014.

Wan, J., Zhang, Y., Wang, R., Liu, L., Liu, E., Fan, J., & Fu, F. (2020). Effective charge kinetics steering in surface plasmons coupled two-dimensional chemical Au/Bi2WO6-MoS2 heterojunction for superior photocatalytic detoxification performance. *Journal of Hazardous Materials*, *384*. Available from https://doi.org/10.1016/j.jhazmat.2019.121484.

Wang, B., Yao, L., Xu, G., Zhang, X., Wang, D., Shu, X., . . . Wu, Y.-C. (2020). Highly efficient photoelectrochemical synthesis of ammonia using plasmon-enhanced black silicon under ambient conditions. *ACS Applied Materials & Interfaces*. Available from https://doi.org/10.1021/acsami.0c00881.

Wang, J., Wang, X., & Mu, X. (2019). Plasmonic photocatalysts monitored by tip-enhanced raman spectroscopy. *Catalysts*, *9*, 109. Available from https://doi.org/10.3390/catal9020109.

Wang, S. S., Jiao, L., Qian, Y., Hu, W. C., Xu, G. Y., Wang, C., & Jiang, H. L. (2019). Boosting electrocatalytic hydrogen evolution over metal−organic frameworks by plasmon-induced hot-electron injection. *Angewandte Chemie International Edition*, *58*, 10713−10717. Available from https://doi.org/10.1002/anie.201906134.

Wang, X., Wang, R., Wang, J., Fan, C., & Zheng, Z. (2020). The synergistic role of the support surface and Au-Cu alloys in a plasmonic Au-Cu@LDH photocatalyst for the oxidative esterification of benzyl alcohol with methanol. *Physical Chemistry Chemical Physics*, *22*, 1655−1664. Available from https://doi.org/10.1039/c9cp05992j.

Wu, J., Wang, J., Wang, T., Sun, L., Du, Y., Li, Y., & Li, H. (2019). Photocatalytic reduction of *p*-nitrophenol over plasmonic M (M = Ag, Au)/SnNb2O6 nanosheets. *Applied Surface Science*, *466*, 342−351. Available from https://doi.org/10.1016/j.apsusc.2018.090.222.

Wu, N. (2018). Plasmonic metal−semiconductor photocatalysts and photoelectrochemical cells: a review. *Nanoscale*, *10*, 2679−2696. Available from https://doi.org/10.1039/C7NR08487K.

Wu, S., Chen, Z., Liu, K., Yue, W., Wang, L., & Zhang, J. (2020). Chemisorption-induced and plasmon-promoted photofixation of nitrogen on Au-loaded carbon nitride nanosheets. *ChemSusChem*, 1−8. Available from https://doi.org/10.1002/cssc.202000818.

Xia, D., An, T., Li, G., Wang, W., Zhao, H., & Wong, P. K. (2016). Synergistic photocatalytic inactivation mechanisms of bacteria by graphene sheets grafted plasmonic AgAgX (X = Cl, Br, I) composite photocatalyst under visible light irradiation. *Water Research*, *99*, 149−161. Available from https://doi.org/10.1016/j.watres.2016.040.055.

Xiao, L., Zhang, Q., Chen, P., Chen, L., Ding, F., Tang, J., . . . Yin, S.-F. (2019). Copper-mediated metal-organic framework as efficient photocatalyst for the partial oxidation of aromatic alcohols under visible-light irradiation: Synergism of plasmonic effect and schottky junction. *Applied Catalysis B: Environmental*, *248*, 380−387. Available from https://doi.org/10.1016/j.apcatb.2019.020.012.

Xie, S., Zhang, H., Liu, G., Wu, X., Lin, J., Zhang, Q., & Wang, Y. (2020). Tunable localized surface plasmon resonances in MoO3 − x-TiO2 nanocomposites with enhanced catalytic activity for CO2 photoreduction under visible light. *Chinese Journal of Catalysis*, *41*, 1125−1131. Available from https://doi.org/10.1016/S1872-2067(20)63566-5.

Xu, J., Yang, W.-M., Huang, S.-J., Yin, H., Zhang, H., Radjenovic, P., . . . Li, J.-F. (2018). CdS core-Au plasmonic satellites nanostructure enhanced photocatalytic hydrogen evolution reaction. *Nano Energy*, *49*, 363−371. Available from https://doi.org/10.1016/j.nanoen.2018.040.048.

Xue, J., Ma, S., Zhou, Y., Zhang, Z., & He, M. (2015). Facile photochemical synthesis of Au/Pt/g-C3N4 with plasmon-enhanced photocatalytic activity for antibiotic degradation. *ACS Applied Materials & Interfaces*, *7*, 9630−9637. Available from https://doi.org/10.1021/acsami.5b01212.

Yang, B., Ma, Z., Li, Q., Liu, X., Liu, Z., Yang, W., . . . Jia, X. (2020). Regulation of surface plasmon resonance and oxygen vacancy defects in chlorine doped Bi-BiO2-X for imidacloprid photocatalytic degradation. *New Journal of Chemistry, 44*, 1090−1096. Available from https://doi.org/10.1039/c9nj04936c.

Yang, J., & Mou, C. (2018). Ordered mesoporous Au/TiO2 nanospheres for solvent-free visible-light-driven plasmonic oxidative coupling reactions of amines. *Applied Catalysis B: Environmental, 231*, 283−291. Available from https://doi.org/10.1016/j.apcatb.2018.020.054.

Yang, J., Guo, Y., Jiang, R., Qin, F., Zhang, H., Lu, W., . . . Yu, J. C. (2018). High-efficiency "working-in-tandem" nitrogen photofixation achieved by assembling plasmonic gold nanocrystals on ultrathin titania nanosheets. *Journal of the American Chemical Society, 140*, 8497−8508. Available from https://doi.org/10.1021/jacs.8b03537.

Yang, L., Guo, S., & Li, X. (2017). Au nanoparticles@MoS2 core-shell structures with moderate MoS 2 coverage for efficient photocatalytic water splitting. *Journal of Alloys and Compounds, 706*, 82−88. Available from https://doi.org/10.1016/j.jallcom.2017.020.240.

Yang, Y., Lu, C., Ren, J., Li, X., Ma, Y., Huang, W., & Zhao, X. (2020). Enhanced photocatalytic hydrogen evolution over TiO2/g-C3N4 2D heterojunction coupled with plasmon Ag nanoparticles. *Ceramics International, 46*, 5725−5732. Available from https://doi.org/10.1016/j.ceramint.2019.110.021.

Zeng, S., Vahidzadeh, E., VanEssen, C. G., Kar, P., Kisslinger, R., Goswami, A., . . . Shankar, K. (2020). Optical control of selectivity of high rate CO2 photoreduction via interband- or hot electron Z-scheme reaction pathways in Au-TiO2 plasmonic photonic crystal photocatalyst. *Applied Catalysis B: Environmental, 267*, 118644. Available from https://doi.org/10.1016/j.apcatb.2020.118644.

Zhai, H., Liu, X., Wang, Z., Liu, Y., Zheng, Z., Qin, X., . . . Huang, B. (2020). ZnO nanorod decorated by Au-Ag alloy with greatly increased activity for photocatalytic ethylene oxidation. Chinese Journal of Catalysis, 41, 1613-1621. Available from https://doi.org/S1872-2067(19)63473-X

Zhang, H., & Govorov, A. O. (2014). Optical generation of hot plasmonic carriers in metal nanocrystals: the effects of shape and field enhancement. *The Journal of Physical Chemistry C, 118*, 7606−7614. Available from https://doi.org/10.1021/jp500009k.

Zhang, X., Chen, Y. L., Liu, R.-S., & Tsai, D. P. (2013). Plasmonic photocatalysis. *Reports on Progress in Physics, 76*, 046401. Available from https://doi.org/10.1088/0034-4885/76/4/046401.

Zhang, Y., He, S., Guo, W., Hu, Y., Huang, J., Mulcahy, J. R., & Wei, W. D. (2018). Surface-plasmon-driven hot electron photochemistry. *Chemical Reviews, 118*, 2927−2954. Available from https://doi.org/10.1021/acs.chemrev.7b00430.

Zhao, H., Zheng, X., Feng, X., & Li, Y. (2018). CO2 reduction by plasmonic Au nanoparticle-decorated TiO2 photocatalyst with an ultrathin Al2O3 interlayer. *The Journal of Physical Chemistry C, 122*, 18949−18956. Available from https://doi.org/10.1021/acs.jpcc.8b04239.

Zhao, J., Liu, B., Meng, L., He, S., Yuan, R., Hou, Y., . . . Long, J. (2019). Plasmonic control of solar-driven CO2 conversion at the metal/ZnO interfaces. *Applied Catalysis B: Environmental, 256*, 117823. Available from https://doi.org/10.1016/j.apcatb.2019.117823.

Zhao, X., Long, R., Liu, D., Luo, B., & Xiong, Y. (2015). Pd-Ag alloy nanocages: Integration of Ag plasmonic properties with Pd active sites for light-driven catalytic hydrogenation. *Journal of Materials Chemistry A, 3*, 9390−9394. Available from https://doi.org/10.1039/c5ta00777a.

Zhou, Q., Ma, S., & Zhan, S. (2018). Superior photocatalytic disinfection effect of Ag-3D ordered mesoporous CeO2 under visible light. *Applied Catalysis B: Environmental, 224*, 27−37. Available from https://doi.org/10.1016/j.apcatb.2017.100.032.

Zhou, X., Liu, G., Yu, J., & Fan, W. (2012). Surface plasmon resonance-mediated photocatalysis by noble metal-based composites under visible light. *Journal of Materials Chemistry, 22*, 21337. Available from https://doi.org/10.1039/c2jm31902k.

Zhu, W., Liu, J., Yu, S., Zhou, Y., & Yan, X. (2016). Ag loaded WO3 nanoplates for efficient photocatalytic degradation of sulfanilamide and their bactericidal effect under visible light irradiation. *Journal of Hazardous Materials, 318*, 407−416. Available from https://doi.org/10.1016/j.jhazmat.2016.060.066.

Zielińska-Jurek, A., & Hupka, J. (2014). Preparation and characterization of Pt/Pd-modified titanium dioxide nanoparticles for visible light irradiation. *Catalysis Today, 230*, 181−187. Available from https://doi.org/10.1016/j.cattod.2013.090.045.

CHAPTER 8

Cocatalyst-integrated photocatalysts for solar-driven hydrogen and oxygen production

Chinh Chien Nguyen[1,2], Thi Hong Chuong Nguyen[1,2] and Minh Tuan Nguyen Dinh[3]

[1]*Institute of Research and Development, Duy Tan University, Da Nang, Viet Nam,* [2]*Faculty of Environmental and Chemical Engineering, Duy Tan University, Da Nang, Viet Nam,* [3]*The University of Da Nang, University of Science and Technology, Faculty of Chemical Engineering, Da Nang, Viet Nam*

8.1 Introduction

The last decades have witnessed a momentum toward the running out of fossil fuels, which can lead to the end of the oil, gas and coal resource era. Moreover, the increases in their combustion products in the global atmosphere have caused numerous issues for climate and environmental problems, which are CO_2 emission and environmental pollution (i.e., NO_x, SO_x, VOCs) (Lenferna, 2018; Zhao et al., 2019). These obstacles motivate the 21st homo sapiens to forage alternative solutions to address the increasing energy demands and protect the global environment.

Exploiting solar energy has been considered as a paradigm-shifting in energy utilization because its potential (i.e., plenty, safe, and renewable) can satisfy the current and future energy demand of humans (Pacesila, Burcea, & Colesca, 2016; Lewis, 2016). Thus the earth receives nearly million exajoules (1 EJ $= 10^{18}$ J) of solar energy in which 5×10^4 EJ is feasibly harvestable, which can fulfill the energy demand of the entire world (Kabir, Kumar, Kumar, Adelodun, & Kim, 2018). Furthermore, harvested energy from the sun can be employed to take place chemical reactions to produce fuels, which are storable, transportable, and usable, while minimizing the environmental impacts. However, current technological approaches are found to struggle to perform this idea on a large scale owing to a limited solar-to-fuel conversion of utilized materials leading to extremely high production costs.

In 1972, Fujishima first reported the use of TiO_2 as a photoanode material in a photoelectrochemical system to produce H_2 from water (Fujishima, 1972). This extraordinary study opened a new avenue to harness solar energy and has inspired

Photocatalytic Systems by Design.
DOI: https://doi.org/10.1016/B978-0-12-820532-7.00004-7

numerous researchers who have devoted an enormous effort to probe scalable photocatalysts for solar fuel production over the past decades. It can be said that semiconductor-based photocatalysts are one of the mainstreams in the present research trends of science because it possesses many superior properties, which are full of promise for solar-driven fuel generations. Indeed, most photocatalysts can be simply synthesized via simple calcination of their precursors in the air (i.e., TiO_2, WO_3, g-C_3N_4, Peroskites). Furthermore, the photocatalytic reactors can be feasibly designed and enlarged for wide-scale applications. A photocatalytic reactor producing similar solar energy conversion efficiencies as a solar panel could significantly reduce the cost for fuel production by a factor of 3 (Miseki & Sayama, 2019). Therefore the fabrication of high-performance materials, which can reach such a desired target, is the priority in this research field.

The fundamental of photocatalysis could be described by the band gap model, as shown in Fig. 8.1A. Readers are recommended to see Chapter 1, Principles and Mechanisms of Photocatalysis, for further information. The valence band (VB) and the conduction band (CB) are separated by a region of forbidden energy, which is a band gap. The energy of incident light, which is either equal or higher than the band gap value of the semiconductor, excites an electron from the VB to the CB and leaves a hole (h^+) in the VB. The photoexcited electron takes place reduction reactions with electron acceptors (donated as A). The CB potential should be more negative than that of the $A + e \rightarrow A^-$ reaction to promote the reduction reaction (Wang et al., 2014; Nguyen, Vu, & Do, 2015). The photogenerated electron can be employed for many essentially photocatalytic applications, such as hydrogen production, CO_2 reduction, and nitrogen fixation (Wang, Li, & Domen, 2019; Vu, Kaliaguine, & Do, 2019; Ithisuphalap et al., 2019). In the VB, the photogenerated holes possessing a more positive potential than that

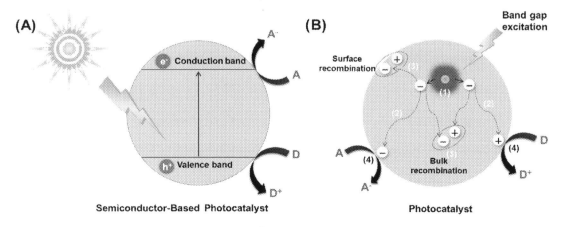

Figure 8.1
(A) Schematic illustration of the fundamental of semiconductor-based photocatalysts. (B) The principle of photocatalysis: (1) band gap excitation, (2) charge migration, (3) charge recombination, and (4) reductive and oxidative reactions (Nguyen et al., 2015).

of the $D + h^+ \rightarrow D^+$ reaction can also carry out an oxidative reaction. This hole can be utilized to generate oxygen gas and decompose pollutants (Yu et al., 2019; Irfan et al., 2019).

The mechanism of photocatalysis occurs in the four-step process, as shown in Fig. 8.1B. Electron—hole pairs generated under light illumination (step 1) migrate to the surface of the semiconductor (step 2). Electron and hole could recombine because of the tendency to maintain charge balance (step 3). Thus the electron—hole recombination in bulks takes place in several ps after the excitation event, which is much faster than charge transportation requiring several hundred ps (Li, Cai, Shang, Yu, & Zhang, 2016). The free photogenerated charges, then, participate in the surface reactions with the electron acceptor and electron donor (step 4) (Nguyen et al., 2015). In this context, steps 3 and 4 hold a critical function that notably affects the photocatalytic performance. Therefore over the past three decades have been witnessed numerous efforts aiming to address this confronting issue.

Employing cocatalysts has offered a brilliant approach to enhance photocatalytic performance as they intensify the charge separation. The presence of reduction and/or oxidation cocatalyst(s) on a semiconductor can selectively attract either photogenerated electrons or holes (Chen, Takata, & Domen, 2017). Therefore the promoted charge separation results in a remarkable enhancement in photocatalytic activities. Moreover, the cocatalysts also provide the proper active sites, which favor the redox reactions. For example, the active sites facilitating the hydrogen reduction reaction can be inactive for the CO_2 reduction reaction owing to the limited interaction between the surface of a catalyst and CO_2 molecules. Hence, cocatalysts firmly impact the selectivity of products, especially in the multiproduct photocatalysis, such as the decontamination of organic compounds or reduction of CO_2 (Kuehnel, Orchard, Dalle, & Reisner, 2017).

The aim of this chapter is to provide the state-of-the-art development of photocatalysts-integrated cocatalysts. The concept and role of cocatalysts will be stated in detail. Moreover, the readers will find the latest development of various cocatalysts tailored for solar light-driven hydrogen and oxygen productions, which have been attracted the increasing attention as the auspicious solutions to address the overexploitation of fossil fuels.

8.2 Fundamental of cocatalysts

As proof-of-concept, cocatalysts are deposited on a semiconductor-based photocatalyst to trap the as-photogenerated charges. The cocatalyst can be a metal, metal oxide, or complex molecules. Fig. 8.2 exhibits the migration of electron—hole pairs in a cocatalysts/photocatalyst system. Electrons migrate to the reduction cocatalysts while the presence of an oxidation cocatalyst attracts the holes (Nguyen et al., 2015; Yang, Wang, Han, & Li, 2013; Meng, Zhang, Cheng, & Yu, 2019). The behavior of electrons and holes toward either metal or metal oxide-based cocatalysts can be explained by the polarization model. The presence of

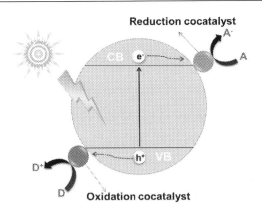

Figure 8.2
The migration of a photogenerated electron—hole pair toward the reduction and oxidation cocatalysts, respectively (Nguyen et al., 2015).

cocatalysts induces a redistribution of charges at the interface leading to generate an internal electric field as a driving force promoting the charge separation under the irradiation condition (Bai et al., 2015; Zhu et al., 2017; Bai, Jiang, Zhang, & Xiong, 2015).

The Schottky junction formed at the interfacial metal—semiconductor junction is widely accepted to explain the formation of the built-in electric field. Theoretically, both semiconductor and metal contain a work function, which is defined as "the minimum energy needed to remove an electron from the E_f to the vacuum level" (Bai et al., 2015). The work functions of noble metals are located between n- and p-type semiconductors, which mean that the Fermi levels (E_f) of the noble metals are lower than those of n-type semiconductors but higher than those of p-type semiconductors. The contact between a cocatalyst (i.e., noble metals) and host photocatalyst (i.e., n-type semiconductors) leads to redistribute charge carriers at the interface where electrons diffuse to the metal leaving holes on the semiconductor and produce an internal electric field. The electron accumulation causes the depletion of free electrons near the metal—semiconductor interface compared to the bulk, which induces upward band bending and therefore forms a Schottky barrier (Bai et al., 2015; Víctor, 2013). The successive generation of electrons under light irradiation tends to break this barrier because photogenerated electrons tend to localize in the positive region in bulk under the influence of the electric field. The high accumulation of electrons in the bulk prompts to recovery of the E_{fs} to the initial position. However, the noble metal, in this case, will show its influence through the ability to accept the photogenerated electrons owing to its lower Fermi level. As a result, a flow of photoexcited electrons transfers from the host photocatalyst to cocatalyst leading to improve the charge separation of electron—hole pairs. Noted that the E_{fm} is hardly changed because of the high density of free electrons in metal whereas the E_{fs} is more flexible due to its level depends on the concentration of charge carriers. The Schottky barrier also plays a crucial role by prohibiting the backward movement of electrons (Bai et al., 2015).

The oxidation cocatalysts are utilized to attract the photogenerated holes. The mechanism of the attraction can be described by the p—n junction. The Fermi level of an n-semiconductor photocatalyst is considerably higher than that of a metal oxide cocatalyst, which is a p-semiconductor. The junction between p- and n-semiconductors results in the accumulation of negative and positive charges on the cocatalyst and photocatalyst sides, respectively. This redistribution of charges forms an internal electric field at the space charge region. Under the illumination, photoexcited electrons tend to accumulate in photocatalysts while holes migrate to the oxidation cocatalyst under the assistance of the electric field. The partial separation of photogenerated electron—hole pairs breaks the E_f equilibrium. However, the E_f equilibrium will be recovered once charges take place during the redox reactions. Thus this configuration greatly facilitates charge separation (Bai et al., 2015; Cao, Wang, Fu, & Chen, 2017; Meng et al., 2019).

In addition to dealing with electron—hole pair separation, the cocatalysts are served as the active sites performing the surface chemical reactions and impeding the back reactions. The solar conversion efficiency and selectivity of reactions intensely depend on the employed cocatalysts. Thus the behavior of active sites toward the adsorption of reactant molecules, activation, intermediates, and desorption of products drives the outcome (Xu & Xu, 2015; Chang, Wang, & Gong, 2016). Particularly, a weak cocatalyst-reactant affinity reduces the overall efficiency because fewer molecules can reach the surface at the same time. Also, the performance of materials could be decreased owing to the tight adsorption capacity of the cocatalyst toward the intermediates that can block the reactants approaching the active sites (Rawalekar & Mokari, 2013). After the reaction, the newly formed products need to desorb and diffuse to the medium rather than stay on the surface and perform the back reactions. Indeed, a rock—solid interaction of cocatalyst with the products will impact the catalytic performance negatively because adsorbed products will block the active sites hindering the reactants while the presence of as-formed products on the surface opens a new opportunity for back and/or side reactions (Marszewski, Cao, Yu, & Jaroniec, 2015). Therefore the selection and preparation of cocatalysts are taken into account as the crucial step in the fabrication of an entire photocatalytic system. The following parts aim at providing the cutting-edge research of cocatalysts that can be helpful for the selection of proper cocatalyst(s), which is beneficial for the desired applications.

8.3 Cocatalysts for sun-light driven hydrogen production and overall water splitting

Solar-driven hydrogen has emerged as the most attractive solar fuel because it offers high energy density and zero pollutant emission during combustion (Mazloomi & Gomes, 2012). Numerous efforts have been devoted to develop high-performance photocatalysts for solar to hydrogen production as well as overall water splitting over the past decades. Thus the efficiency of the overall water splitting photocatalytic system requires a 10% of solar to hydrogen conversion along with a lifetime of 10 years, an annual depreciation rate of 4%, and an allowable cost of 102 USD m^{-2} to make the technology to be competitive (Mazloomi & Gomes, 2012; Hisatomi & Domen, 2019). Therefore the fabrication of high-end materials occupies the central role. Cocatalysts that can boost the solar conversion efficiency possess the essential position in this pathway.

As proof-of-concept, excellent electrocatalysts, which exhibit high H$_2$ reduction reaction efficiencies, are the golden option for the usage of cocatalysts because these materials produce the appropriate adsorption energy of H on the catalyst surface. The Tafel slope is utilized to predict the proper materials toward the photocatalytic hydrogen production reaction, as shown in Fig. 8.3. The volcano curve indicates that the position, where the ΔG_{H^*} ($\Delta G_{H^*} \to 0$) and i$_0$ values are close to the peak, implies the outperformance of the catalyst (Eftekhari, 2017; Jiao, Zheng, Jaroniec, & Qiao, 2015; Wang & Domen, 2020; Zheng et al., 2014). Therefore noble metals (i.e., Pt, Rh) are the appropriate candidates to accelerate the hydrogen production reaction. Various designs of cocatalysts are found to be favorable for hydrogen production. In this section, five attractive groups of materials being metal, sulfur, MXene, graphene, phosphor, and core/shell structure-based cocatalysts are stated and discussed in detail.

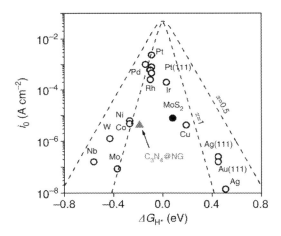

Figure 8.3
The relation between the HER j$_0$ and hydrogen adsorption-free energy for the surfaces of various metals, alloys, and nonmetallic materials. Source: *Reproduced by permission from Zheng, Y., Jiao, Y., Zhu, Y., Li, L. H., Han, Y., Chen, Y. . . . Qiao, S. Z. (2014). Hydrogen evolution by a metal-free electrocatalyst.* Nature Communications, *5, 3783.*

8.3.1 Metal-based cocatalysts

8.3.1.1 Single metal atom cocatalysts

The utilization of single metal atom cocatalysts is an outstanding route to enhance the photocatalytic efficiency because low-coordinated metal atoms function as forceful active sites while minimizing material loading (Lang & Bernhardt, 2012). Thus the number of active sites in the single metal atom catalysts can be orders of magnitude higher than those of heterogeneous catalysts, which own a bigger size leading to the small fraction of exposed metal sites to the reactants (Wang, Li, & Zhang, 2018). Therefore employing single atom cocatalysts has been offered an essentially novel avenue in the context of photocatalyst development as it not only advances the performance but also minimizes the metal cocatalyst usage.

In 2016, Li et al. reported the preparation of Pt single atom on graphitic carbon nitride (g-C_3N_4) for the photocatalytic hydrogen production (Li et al., 2016). Isolated single Pt atoms were successfully prepared via a facile reaction between g-C_3N_4 and H_2PtCl_6 followed by an annealing step at mild temperature. High-angle annular dark-field scanning transmission electron microscopy (STEM) confirms the uniform distribution of Pt atoms on the g-C_3N_4 surface where the size of 99.4% Pt is less than 0.2 nm indicating the existence of isolated single atoms. Furthermore, Pt single atoms impact the structure of g-C_3N_4 by forming Pt$-$N bonds. These features result in a remarkable improvement compared to the conventional material (g-C_3N_4 + Pt particles) for the photocatalytic hydrogen production. This superior activity can be originated from the increasing of charge separation induced by Pt single atoms. The utilization of single atom-based Pt is also employed for various photocatalysts, such as TiO_2 and metal-organic framework (MOF) (Al-TCPP), which also greatly boosted the photocatalytic hydrogen production performance (Xing et al., 2014; Fang et al., 2018; DeRita et al., 2019).

Nonnoble metal single-atom cocatalysts have also received great attention for the solar-driven hydrogen evolution as they are earth-abundant elements and offer a low production cost and low toxicity (Ran, Zhang, Yu, Jaroniec, & Qiao, 2014). Moreover, transition metal cocatalyst (i.e., Ni, Co, Cu) showed the capability to encourage charge separation leading to enhance the amount of evolved hydrogen (Xu & Xu, 2015). Very recently, Lee et al. reported the single atom of copper on TiO_2 hollow photocatalyst. The presence of copper atoms on the TiO_2 support can be observed by TEM and STEM EDS elemental mapping, as provided in Fig. 8.4A$-$C. Impressively, the study has shown the behavior of Cu single atoms toward the suppression of the electron$-$hole recombination issue. Thus the photogenerated electrons accumulated in the Cu single atoms changes their valence state, which induces the activation of the TiO_2 adjacent environment into the active state. These superiorities result in an unambiguous improvement in charge separation and photocatalytic

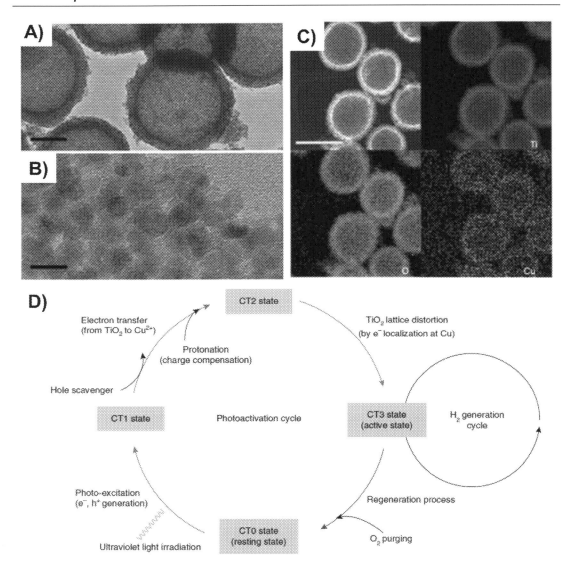

Figure 8.4

(A and B) TEM images of Cu/TiO_2 at low and high magnification. Scale bar, 100 nm (A), 10 nm (B). (C) STEM EDS elemental mapping of Cu/TiO_2. Scale bar, 250 nm. (D) Illustration of the photoactivation cycle of Cu/TiO_2. *STEM*, Scanning transmission electron microscopy. Source: *Reprinted by permission from Lee, B.-H., Park, S., Kim, M., Sinha, A. K., Lee, S. C., Jung, E., . . . Cho, S.-P. (2019). Reversible and cooperative photoactivation of single-atom Cu/TiO2 photocatalysts.* Nature Materials, 18, *620.*

performance. In contrast, the photogenerated electrons in TiO_2 standing alone are localized in its CB. This indicates the crucial role of isolated Cu atoms, which significantly impacts the structural properties of the material and catalytic performance (Lee et al., 2019).

In addition to Cu, single atoms of Co and Ni have also revealed a strong influence on Co-nitrogen-doped graphene (NG)-CdS and Ni-NG-CdS photocatalysts, respectively. The presence of these cocatalysts was demonstrated to remarkably prompt the charge separation leading to enhance the photocatalytic hydrogen production activity (Zhao, Yao, Huang, Wu, & Xu, 2017; Zhao et al., 2018).

It can be realized that the single-atom cocatalyst has revealed a novel solution toward the development of high-end materials. There is plenty of space for researchers in this field. However, the most confronting issue needed to be concerned is the aggregation of single atoms owing to the high surface energy of atom size. Moreover, methods employing to prepare the single atom cocatalysts with high dispersion and durability are also highly needed.

8.3.1.2 Bimetallic cocatalysts

The use of bimetallic cocatalysts has offered an up-and-coming approach to enhance light absorption and address the charge recombination. Bimetallic nanoparticles formed by two different metals can exhibit outperformance compared to monometallic counterparts owing to the outstanding electronic properties obtained by the synergistic effect. In addition, in the case of noble metal cocatalysts, employing bimetallic cocatalysts can reduce the noble metal loading without lowering the activity (Zhang, Yang, Wang, Dou, & Liu, 2018; Aronovitch et al., 2015).

Wang et al. investigated the role of AuPt bimetallic as a cocatalyst on TiO_2 photocatalyst. The alloy PtAu remarkably improves the photocatalytic activity toward hydrogen production. Thus the amount of evolved hydrogen was around 10 times higher than that of the monometallic counterparts. It turns out that the stronger interaction between TiO_2 and PtAu bimetallic, as well as a higher electron population on the $PtAu/TiO_2$ photocatalyst, induces an increase in the electron—hole separation boosting the photocatalytic performance (Wang et al., 2015). In another investigation, Pd was employed to form the PdAu bimetallic cocatalyst on the TiO_2 support. The bimetallic structure of Pd_{shell}-Au_{core} affects the kinetics of electron migration. Particularly, the fermi level (E_F) of bimetallic below the CB minimum of TiO_2 is beneficial for the electron transfer. Moreover, Pd_{shell}-Au_{core} nanoparticles contain unoccupied d-orbital states that can prolong the photogenerated lifetime of electrons in photocatalytic reactions. Also, the bimetallic structure possesses the capability to promote the storage and release of photoexcited electrons for catalysis. Consequently, the produced hydrogen over the $PdAu/TiO_2$ nanocomposite exhibited a great improvement compared to their monometallic cocatalysts (Su et al., 2014). This investigation thus proves that bimetallic structure-derived electronic properties can mitigate the electron—hole recombination and provide the proper active sites for the catalysis reactions.

Transition metal bimetallic also attracts the increasing attention owing to the features that could not find in the monometallic counterparts. For example, Muñoz-Batista, recently, introduced the CuNi/TiO$_2$ photocatalysts in which the presence of Ni strongly impacts the behavior of Cu. Thus the study shows that the nickel functions the crucial role as follows: (1) shape the phase contact among reduced and oxidized copper phases and (2) control the electronic and structural properties. These outstanding modifications were found to be the main reasons for the improvement of the produced hydrogen (Muñoz-Batista, Motta Meira, Colón, Kubacka, & Fernández-García, 2018). In another investigation, AgCu bimetallic is also found to be a promising cocatalyst to promote the hydrogen evolution reaction. Indeed, Zhu reported the deposition of AgCu bimetallic nanoparticles on well-ordered graphitic carbon nitride nanotubes (C$_3$N$_4$ NTs). Ag-Cu bimetallic NPs obtained through the reduction reaction method enhance the light absorption. Moreover, the existence of Ag causes intimate contact with the support and contributes to the presence of the more metallic copper state. Such advantages accelerate the charge separation leading to a sharp improvement compared to their monometallic counterparts and Pt/C$_3$N$_4$ NTs (Zhu, Marianov, Xu, Lang, & Jiang, 2018).

It can be said that bimetallic cocatalysts display unambiguous modifications at the interfacial contact, which is beneficial to suppress the electron—hole recombination and accelerate the redox reactions. Further investigation is highly needed to obtain an in-depth understanding of electron—hole separation and charge transfer manners. In addition, the exploration of new types of bimetallic cocatalysts in the context of composition and particle size also functions an essential role in this pathway to discover the most suitable cocatalyst.

8.3.2 Sulfur based cocatalysts

8.3.2.1 MoS$_2$

Over the past decades, two-dimensional layer materials have emerged as a rose-colored candidate to deal with the electron—hole recombination and active site issues. Thus 2D materials possess superior conductivity and electron mobility properties, which are favorable for charge transportation and separation. Also, large surface area of this type of material offers an abundance of active sites facilitating the redox reactions (Xiang, Yu, & Jaroniec, 2012; Low, Cao, Yu, & Wageh, 2014). This section provides a typical 2D layer cocatalyst that has been demonstrated the ability to enhance the hydrogen photocatalytic performance, which is MoS$_2$.

Both computation and experimental results prove that MoS$_2$ plates contain the potential active sites for the hydrogen reduction reactions, which is originated from the edge sites of

MoS_2. Thus the conductivity of the edge of MoS_2 layers is meaningfully higher than that of its basal plane (Jaramillo et al., 2007; Li et al., 2011). Therefore employing MoS_2 layers in the preparation of photocatalyst composite has been identified as a prominent alternative approach to replace the usage of noble metals. Thus coupling MoS_2 nanolayers to CdS was found to promote the solar-to-hydrogen conversion. The existence of MoS_2 on CdS remarkably boosts the rate of hydrogen production owing to the intimate contact between two materials, which is beneficial for electron transportation, and the excellent H_2 activation property of MoS_2 (Zong et al., 2008). Similarly, MoS_2 nanolayers also boost the photoactivity of various semiconductors (i.e., CdSe, $Zn_{0.4}Ca_{0.6}In_2S_4$ nanosheets, $ZnIn_2S_4$, etc.) (Frame & Osterloh, 2010; Shi, Fujitsuka, & Majima, 2018; Zhang et al., 2017).

The most essential issue is to engineer the proper contact between the MoS_2 and host photocatalysts to maximize the electron transfer channel. Motivated by this challenge, He et al. engineered the MoS_2/TiO_2 edge-on heterostructures, in which the edge of MoS_2 selectively contacts on the TiO_2 nanorods (NRs), through the chemical vapor deposition (CVD) strategy associated with the sulfurization process. Scanning electron microscopy (SEM) and high-resolution transmission electron microscopy (HR-TEM) images reveal the morphology and the existence of MoS_2 on TiO_2. The high-resolution annular dark-field scanning transmission electron microscopy (ADF-STEM) and electron energy loss spectrum (EELS) analyses of TiO_2/MoS_2 results indicate the high quality of obtained MoS_2 and a smooth interface between the edges of MoS_2 and the anatase TiO_2. Importantly, the EELS of the interface proves that the MoS_2 nanoplates form the edge-on heterojunction rather than the coat on the TiO_2 surface. It turns out that this configuration greatly enhances the interfacial conductivity and governs the transportation pathway of photoexcited along the basal planes leading to enhance electron—hole separation efficiency. The photoactivity therefore increases remarkably, compared to the bare TiO_2 and MoS_2/TiO_2 B, in which MoS_2 nanoplates connect through the basal plane (He et al., 2016). The extraordinary investigation proves the behavior of MoS_2 layer structure toward the photogenerated charges and provides a new approach to engineer MoS_2-based cocatalysts for the hydrogen production reactions. In another study, Bian et al. vertically constructed MoS_2 on a g-C_3N_4 few layers in an attempt to prepare an energetic photocatalyst for hydrogen evolution. Employing the solid-state reaction, the MoS_2/g-C_3N_4 heterojunction was successfully obtained and MoS_2 was found to grow on the few-layer g-C_3N_4 vertically. The vertical structure of MoS_2 on g-C_3N_4 was confirmed by the results from the transmission electron microscopy (TEM) and Raman analysis. It is realized that the interfacial sites of the vertical MoS_2/C_3N_4 act as electron-accumulation regions, which is confirmed by the density functional theory (DFT) calculations, facilitating the electron—hole separation (Bian et al., 2018). In this case, it can manner that the intimate contact formed between MoS_2 and the few-layer g-C_3N_4 on the edge of MoS_2 is the main reason for the charge separation enhancement.

Two major issues confronting the development of MoS_2 as a cocatalyst for solar-driven hydrogen production is originated from synthesis methods and designing of facial contact. First, the methods, which can produce MoS_2 with high quality, green, and large-scale production of the monolayer with abundant edge sites, are highly needed (Wu et al., 2019). The novel approach to advance the properties of MoS_2 like using MoS_2 quantum dots, changing morphology, or doping strategy is also prominent to improve the cocatalyst function of MoS_2. Second, it is predicted that MoS_2 continues to receive the increasing attention of researchers in the next years. The rational design of the face contact between MoS_2 and photocatalysts will attract more concern to maximize the electron transportation and reaction efficiency.

8.3.2.2 NiS_x cocatalyst

NiS_x materials are also a promising option as a cocatalyst for the photocatalytic hydrogen production. The NiS_x cocatalyst has shown the capability to boost the photocatalytic production of H_2. Thus the evolved hydrogen on NiS/CdS, which was found to be higher than that of Pt/CdS, indicates its ability (Zhang, Wang, Wang, Zhong, & Xu, 2010). Thus the Ni^{2+} species connecting to the surface of the host photocatalyst (CdSe/CdS) particles via dangling bonds facilitate the photogenerated electron transportation from CdSe/CdS to Ni^{2+} sites. Therefore the separated electrons on Ni^{2+} sites promote the hydrogen production reaction (Li et al., 2013; Ji, Lv, Chen, Huang, & Zhang, 2018).

In 2016, Liu et al. reported the preparation of a high-end $Cd_{0.5}Zn_{0.5}S$ (CZS) nanotwinned unanchored the NiS_x cocatalyst for the extraordinary photocatalytic performance for hydrogen production. The study introduced a novel route, collision-contact, to accelerate the photoexcited electron transfer while impeding the back reactions. Fig. 8.5A depicts that the Ni−H bond is established between the adsorbed water and NiS_x clusters, which collide with CZS NRs to extract the photogenerated electrons for the H_2O reduction reaction to produce H_2. It is noted that the water reduction on the NiSx cocatalyst taking place after the collision thus overcomes the back reaction issue leading to an increase in the hydrogen production rate. Fig. 8.5B displays the band alignment and the photocatalytic process occurring in the material whereas Fig. 8.5C exhibits the photocatalytic activity of the samples with and without Ni suggesting the essential role of NiS_x toward the production of hydrogen. It was found that the internal quantum efficiency (Fig. 8.5D) reached approximately 100% at 425 nm implying the role of the unanchored NiS_x cluster cocatalyst (Liu et al., 2016). In another investigation, NiS nanoparticles also promoted the photocatalytic hydrogen production of g-C_3N_4 notably. It was proved that the existence of NiS greatly enhances the charge separation. Moreover, H-NiS bonds formed on the surface of the cocatalyst are beneficial for H_2 evolution. Consequently, the evolved hydrogen was 16.4 mmol g^{-1} h^{-1} with 0.76 wt.% NiS loading, which is extremely higher than that of pure g-C_3N_4 (Zhao et al., 2018). This obtained activity suggests the superior ability of NiS_x cocatalyst for the solar-driven hydrogen production. Furthermore, the NiS cocatalyst was also very active when coupled with various semiconductors like $Zn_xCd_{1-x}S$ and TiO_2 (Zhao et al., 2018; Zhang, Tian, Chen, & Zhang, 2012). Indeed, Zhao provided a deep understanding of the features of NiS species in $Zn_xCd_{1-x}S$/NiS system fabricated from the

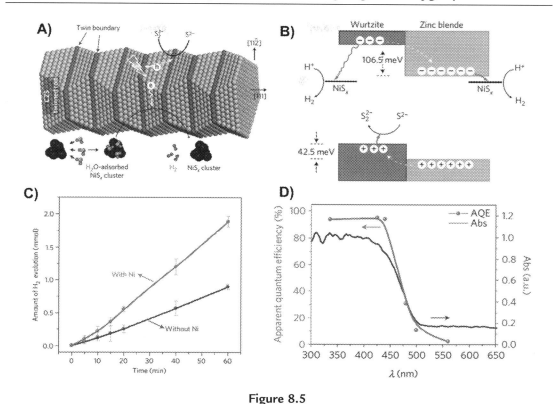

Figure 8.5

(A) Schematic illustration for the improved H2 production of NiS$_x$ in which NiS$_x$ cluster undergoes the collision to prompt the electron transfer. (B) Band alignment and photocatalytic process in CZS and (C) produced hydrogen of the CZS film photocatalysts with and without Ni.
Source: Reprinted by permission from Liu, M., Chen, Y., Su, J., Shi, J., Wang, X., & Guo, L. (2016). Photocatalytic hydrogen production using twinned nanocrystals and an unanchored NiSx co-catalyst. Nature Energy, 1, 16151.

MOF. NiS cocatalyst lowers the barrier for the water dissociation into *H and *OH. Therefore more surface hydrogen is supplied for the reduction reaction (Zhao et al., 2018).

The mentioned examples suggest that NiS$_x$ cocatalyst is the attractive material for the preparation of highly active photocatalytic hydrogen production systems in which the g-C$_3$N$_4$ and sulfide-based photocatalysts are commonly selected as the host semiconductors. Moreover, anchoring NiS$_x$ on the other materials is also required to discover proper configurations possessing excellent activity and stability.

8.3.3 MXene-based cocatalysts

MXene, two-dimensional transition metal carbides, and nitrides have emerged as the hot topic for fuel generation and energy storage research fields owing to their outstanding properties. MXenes are formed from MAX phases, where M, A, and X represent an early transition metal, the main group IIIA or IVA element, and carbon and/or nitrogen atom, respectively. The chemical stoichiometry of MAX phases is $M_{n+1}AX_n$ ($n = 1, 2,$ or 3). The "A" containing bonds is more chemically active than M−X bonds. Thus M−X bond is governed by covalent/metallic/ionic bonds whereas the M−A bond is metallic. Therefore the "A" is possibly removed from the MAX phase through an etching step to generate a new 2D material MX, which is known MXene (Naguib, Mochalin, Barsoum, & Gogotsi, 2014). The first MXene, that is, Ti_3C_2, prepared from Ti_3AlC_2 via the etching step with HF is reported by Naguib et al. (2011). The utilization of HF turns out as the key step to remove Al from the precursor phase. Thus this etchant has been found to be the potential to prepare numerous various MXenes (i.e., Mo_2C, Zr_3C_2). Moreover, the investigation of alternative etchants has attracted attention. NH_2HF_2, the mixture of HCl (H_2SO_4) and LiF (NaF, KF, CsF, CaF_2), and tetrabutylammonium fluoride have been proposed to prepare many kinds of MXenes. Recently, CVD offers a new method to synthesize MXenes with a free fluorine-containing etchant (Xu et al., 2015).

MXenes possess both metallic conductivity of transition metal carbide and the hydrophilic nature originated from their hydroxyl or oxygen terminated surfaces. MXene-based materials have been explored as the promising cocatalyst for the photocatalytic hydrogen production because they contain many advantages for an outstanding cocatalyst, such as fast interfacial charge transfer, electron shuttling, proper Gibbs-free energy for H_2 evolution, hydrophilic property causing the sufficient contact with water, and durability (Guo, Zhou, Zhu, & Sun, 2016; Liu, Xiao, & Goddard, 2016; Seh et al., 2016). Thus employing density functional theory calculations, Gao et al. demonstrated that a mixture of O*- and OH*-terminated 2D MXenes are the stable states results in the materials owing to metallic properties and excellent electrical conductivity. Therefore O*/OH*-terminated MXenes are beneficial for charge transfer. In addition, the study showed that the surface oxygen atoms offer suitable interaction strength between H* and 2D MXenes. H_2 molecules feasibly release from the surface after the reaction that is favorable for the hydrogen evolution reaction. Therefore coupling MXenes to a proper photocatalyst can produce a high-quality system for solar-driven hydrogen generation (Gao, O'Mullane, & Du, 2016).

In 2017, Ran reported the first employment of MXene, which is Ti_3C_2 as a cocatalyst, on CdS photocatalyst for the enhanced visible-light photocatalytic hydrogen production. The Ti_3C_2/CdS system fabricated by the hydrothermal method witnessed a remarkable improvement in light absorption, charge separation, transfer, and surface catalytic reactions. Particularly, Fig. 8.6A displays the HR-TEM image indicating the intimate Ti_3C_2/CdS contact and confirming the successfully synthesized method. The obtained Ti_3C_2/CdS showed extremely high photocatalytic activity for hydrogen production. The amount of produced hydrogen, Fig. 8.6B, was higher than that of CdS using well-known cocatalysts (i.e., Pt, NiS, Ni, MoS_2). This comparison suggests the extraordinary capability of Ti_3C_2

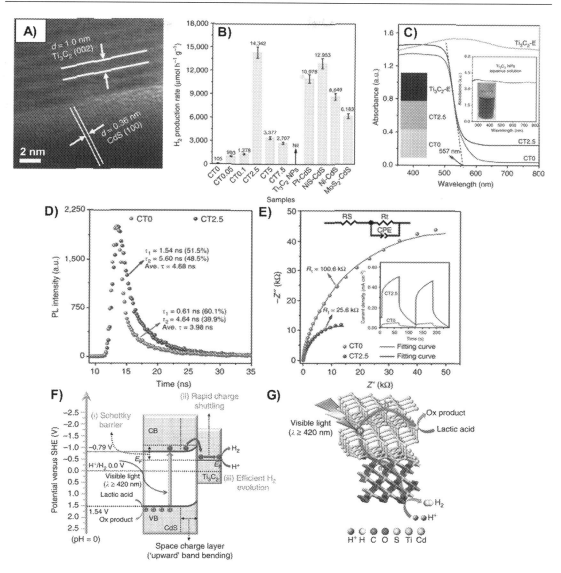

Figure 8.6

(A) HR-TEM image. (B) The hydroproduction of Ti_3C_2/CdS compared to different cocatalysts. (C) The light absorption capacity. (D and E) Time-resolved photoluminescence. (F) Electron transfer. (G) Schematic illustration of electron transportation between CdS and MXenes. *Source: Reprinted by permission from Ran, J., Gao, G., Li, F.-T., Ma, T.-Y., Du, A., & Qiao, S.-Z. (2017), Ti3C2 MXene co-catalyst on metal sulfide photo-absorbers for enhanced visible-light photocatalytic hydrogen production.* Nature Communications, 8, 13907.

MXene as a cocatalyst. The ultraviolet-visible diffuse reflectance spectra, as shown in Fig. 8.6C, displays the sturdy light-harvesting capability of Ti_3C_2/CdS (denoted as CT2.5) from 350 to 800 nm owing to the color of Ti_3C_2 NPs.

Fig. 8.6D−E exhibits the charge separation and transfer efficiency of Ti_3C_2/CdS obtained from time-resolved and steady-state photoluminescence spectra, electrochemical impedance spectra, and transient photocurrent response, respectively. It can be seen that the presence of MXene restricts the electron−hole combination and prolongs the lifetime of charge carriers (Fig. 8.6D). Moreover, this configuration also facilitates interfacial charge transfer, resulting in the lower interfacial resistance and higher photocurrent compared to bare CdS, as shown in Fig. 8.6E. These outstanding properties result in the superior photocatalytic activity for hydrogen production under visible light suggesting the essential role of Ti_3C_2 as the cocatalysts. Indeed, it turns out that the produced hydrogen was found to be higher than that of various cocatalysts (i.e., Pt, Ni, and NiS) (Ran, Zhu, & Qiao, 2017). A similar manner has been successfully applied to many kinds of photocatalytic systems, such as Ti_3C_2/TiO_2 and Ti_3C_2/C_3N_4 (Li, Deng, Tian, Liang, & Cui, 2018; An et al., 2018; Su et al., 2019).

The mentioned analysis indicates the potential of MXene-based cocatalysts for solar-driven hydrogen production. Looking forward, 2D MXene-based nanomaterials are burgeoning and a hot spot research topic. Tremendous efforts in both computation and experiments are needed to probe the appropriate solution to prepare high-end MXene cocatalysts for practical applications in renewable energy.

8.3.4 Graphene-based cocatalysts

2D graphene nanosheet has emerged as a good cocatalyst for photocatalytic applications as its unique nature can motivate the charge carrier separation and transportation. The early studies showed that graphene oxide (GO) possesses excellent electron conductivity derived from π-conjugated structure and large surface area. Therefore it is an excellent platform for photocatalyst dispersion, which offers more light harvesting sites and effective charge separation. Consequently, the photocatalytic activity can enhance significantly (Li et al., 2011). It turns out that the GO nanosheets can undergo a reduction by the photoexcited electrons to form the reduced GO (rGO) (Mukherji, Seger, Lu, & Wang, 2011). Therefore integrating graphene to second cocatalysts provides a promising solution in which rGO-second cocatalyst configuration (i.e., Pt, Ni, MoS_2) simultaneously mitigates the charge recombination and active sites. The rGO is an electron transport bridge while the second cocatalyst with a high dispersion is the chemical reaction sites, respectively. These features can generate outperformance for hydrogen evolution compared to the other cocatalyst system. Xiang et al. demonstrated the synergetic effect of graphene and MoS_2 toward the enhancement of H_2 production of TiO_2 nanoparticles, as shown in Fig. 8.7. This nanocomposite was fabricated via the hydrothermal process. TEM and HR-TEM

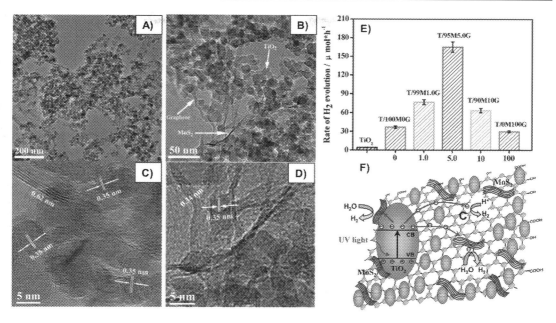

Figure 8.7

TEM (A and B) and HR-TEM (C and D) images of MoS_2/graphene/TiO_2. (E) The photocatalytic activity toward hydrogen of different samples. (F) The schematic illustration of electron transfer and redox reactions. Source: *Reprinted with permission from Xiang, Q., Yu, J., & Jaroniec, M. (2012). Synergetic effect of MoS2 and graphene as cocatalysts for enhanced photocatalytic H2 production activity of TiO2 nanoparticles.* Journal of the American Chemical Society, 134, 6575—6578.

(Fig. 8.7A−D) images indicate the presence of MoS_2 and TiO_2 on graphene with good dispersion. The coexistence of graphene and MoS_2 greatly contributes to improving the interfacial charge transportation and provides numerous active sites for the chemical reaction. These outstanding features intensified the photocatalytic hydrogen production, as shown in Fig. 8.7E. It was proposed that photogenerated electrons migrate to MoS_2 through the graphene sheets as a bridge, as shown in Fig. 8.7F (Xiang, Yu, & Jaroniec, 2012).

Similarly, Yuan incorporated MoS_2-graphene composite cocatalyst on $ZnIn_2S_4$. This system shows not only broad visible light absorption originated from the narrow band gap structure of $ZnIn_2S_4$ (i.e., 2.39 eV) but also enhanced charge transportation and reaction sites archived from MoS_2-graphene cocatalyst. In this case, photoexcited electrons transfer through a transport bridge, which is a graphene to reach the active sites (MoS_2) to perform the photoreactions. The produced hydrogen of the optimum $ZnIn_2S_4$/graphene/MoS_2 was found to be 22.8 times higher than that of pure $ZnIn_2S_4$. Interestingly, this photocatalytic activity is also higher than Pt/$ZnIn_2S_4$ system suggesting a solution in lieu of noble metal cocatalysts (Yuan et al., 2016). The same manner is also employed for the configuration of WS$_2$/graphene/CdS (Xiang, Cheng, & Lang, 2016).

It turns out that the usage of graphene-based cocatalysts, no doubt, is an attractive approach because of the outstanding properties of graphene. Therefore the development of graphene-based cocatalyst still holds an important category. Several issues require further investigations as follow: (1) the preparation of high-quality graphene has faced many challenges in the context of large scale and electronic features, (2) coupling reduced graphene to either semiconductor or cocatalysts is also a crucial concern as the limitation of functional groups of graphene leads to poor interfacial contact with the counterpart semiconductor that could hinder the charge transportation, and (3) in-depth understanding of electron transfer within the graphene-photocatalyst composite is also required the extensive investigation both experimental evidence and theoretical calculation because the current knowledge in this issue is still ambiguous (Li, Yu, Wageh, Al-Ghamdi, & Xie, 2016).

8.3.5 P-based cocatalysts

The usage of phosphorous-based cocatalysts for photocatalysis has become the hot spot research field recently as they have the potential to improve photocatalytic hydrogen production. Accordingly, in this section, two main phosphorous-containing nanomaterials, i.e. phosphide, phosphorene have been discussed in detail. In addition, the role of phosphates in the reaction medium assisting to suppress the electron–hole recombination and facilitate the chemical reactions is also stated (Hu, Shen, & Yu, 2017).

8.3.5.1 Transition metal phosphide

In 2013, Popczun et al. reported that the first preparation of Ni$_2$P nanoparticles as an electrocatalyst for hydrogen production reaction opened a new avenue for the use of transition metal phosphide as the cocatalysts for photocatalytic hydrogen evolution reaction (Popczun et al., 2013). Numerous theoretical and experimental studies have demonstrated that transition metal phosphides (i.e., Ni$_2$P, Co$_2$P, MoP) own the features containing the capability to address the electron–hole recombination issues and provide active sites accelerating the hydrogen production reactions (Liu & Rodriguez, 2005; Cao, Wang, Fu, & Chen, 2017). Therefore integrating metal phosphide into photocatalysts provides a great occasion to improve the photocatalytic activity. Thus Sun et al. investigated the combination of Ni$_2$P and CdS NRs toward the photocatalytic hydrogen production under visible light. This nanocomposite was fabricated via the facile solvothermal method. The size of Ni$_2$P was found to be a few nanometers (Fig. 8.8A and B). The intimate interfacial contact between Ni$_2$P and CdS NRs is an essential factor for the steady improvement in the context of charge separation and transportation. Thus it can be seen from Fig. 8.8C that time-resolved photoluminescence spectra suggest a shorter lifetime of the Ni$_2$P/CdS NR

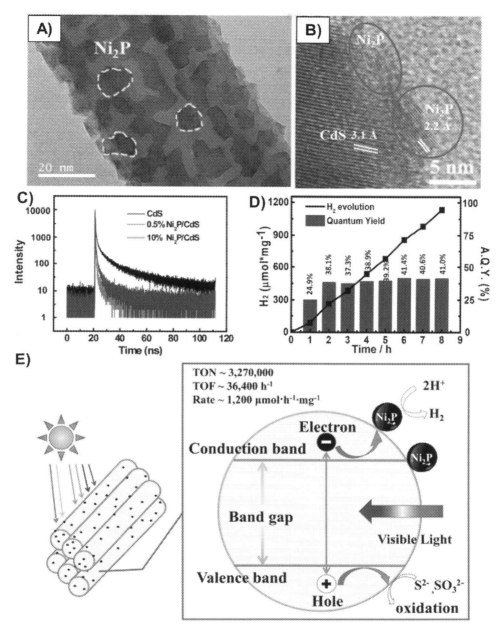

Figure 8.8

(A and B) TEM and HR-TEM images of Ni_2P/CdS, respectively; (C) time-resolved photoluminescence spectra; (D) the stability; and (E) the schematic illustration of the enhanced photocatalytic performance. Source: *Reprinted by permission from Sun, Z., Zheng, H., Li, J., & Du, E. (2015). Extraordinarily efficient photocatalytic hydrogen evolution in water using semiconductor nanorods integrated with crystalline Ni2P cocatalysts.* Science & Environmental Science, 8, 2668–2676.

samples owing to the robust charge separation and transportation from CdS to Ni_2P cocatalysts. Consequently, Ni_2P/CdS nanocomposite exhibited superior activity and stability for hydrogen production under visible light, as depicted in Fig. 8.8D. This configuration is among the most compelling free-noble metal photocatalysts. This impressive study thus has inspired many metal phosphide/photocatalyst nanocomposite systems for the flowing years (Sun, Zheng, Li, Du, & Science, 2015).

In another study, Ni_2P nanoparticles loaded on g-C_3N_4 function as a superior cocatalyst for efficient photocatalytic hydrogen generation. The presence of Ni_2P is beneficial for the enhanced charge transportation at the N_2P/C_3N_4 heterojunction and accelerates the chemical reactions. The archived N_2P/C_3N_4 nanocomposite boosts the produced hydrogen, which is 60 times higher than that of bare C_3N_4 after 24 working hours (Indra et al., 2017). In addition to CdS and g-C_3N_4, Ni_2P nanoparticles have been found to be a forcible cocatalyst for various photocatalytic systems like $Cd_{0.5}Zn_{0.5}S$, $Cd_{0.9}Zn_{0.1}S$/g-C3N4 (Qin, Xue, Chen, Shen, & Guo, 2017; Shao et al., 2018).

The employment of the other transition metal phosphides also attracts attention. Many efforts have been devoted to seeking new cocatalyst-based transition metal phosphides in which Fe_xP has emerged as an attractive candidate. Thus Zhao et al. prepared g-C_3N_4/Fe_xP composite as an efficient photocatalyst. Thus Fe_xP cocatalyst embraced on C_3N_4 intensified the generated hydrogen by a factor of 277 compared to the pristine g-C_3N_4. The robust enhancement is originated from intense charge separation and transportation caused by the presence of Fe_xP. Thus it was found that dual proton adsorption sites induced by the adjacent Fe and P atoms in Fe_xP synergistically accelerate the H_2 production from water (Zhao, Wang, Dong, & Jiang, 2017). A similar manner is also demonstrated with CoP/g-C_3N_4 photocatalysts suggesting that transition metal phosphides are very promising cocatalysts for solar-driven hydrogen production (Li, Du, Wang, & Xu, 2017).

The coming years will witness intensive efforts in the development of metal sulfide cocatalysts. Moreover, mechanism investigation and morphology optimization are also essential issues, which are required more attention to obtain an in-depth understanding of the behavior of this type of cocatalyst.

8.3.5.2 Phosphorene

Phosphorene, a monolayer of black phosphorous, has emerged as the potential metal-free cocatalysts owing to its outstanding features of large surface area, excellent charge mobility, and tunable band structure (Woomer et al., 2015; Carvalho et al., 2016). The DFT calculations suggested that the few-layer phosphorene nanosheet (FPS) possesses the forceful reduction and oxidation abilities, limits electron−hole recombination, and produces abundant active sites and well-established electronic properties as integrated with photocatalysts (Ran, Zhu, & Qiao, 2017). Based on such calculations, phosphorene incorporated on CdS was fabricated for the enhanced photocatalytic hydrogen production. Thus this configuration, which is prepared by a mechanical mixing route, resulted in the intimate interfacial contact between CdS and phosphorene, built up the channel that is beneficial for electron transfer, and provides plenty of active centers for the chemical reactions. Consequently, FPS/CdS nanocomposite exhibited an excellent activity for hydrogen production. The amount of produced hydrogen was much higher than that of other cocatalysts and an impressive apparent quantum yield (AQY) of 34.7% at 420 nm, which is one of the most energetic photocatalytic systems (Ran, Zhu, & Qiao, 2017). In this circumstance, phosphorene is the cocatalyst attracting the photogenerated electrons and performing the photoproduction of hydrogen. Phosphorene also facilitates photoactivity when coupling with $Zn_xCd_{1-x}S$ and g-C_3N_4 (Ran et al., 2018; Ran, Wang, Zhu, & Qiao, 2017). Therefore looking forward to the future, it can be said that phosphorene is one the most potential cocatalysts. However, the double efforts require looking into the novel and feasible routes to prepare high-quality phosphorene. In addition, the long-term stability of phosphorene is a crucial challenge due to its air-sensitive property. The coupling to high-performance semiconductors is another consideration to maximize the charge separation and not to shield incident light within the photocatalytic system.

8.3.5.3 Phosphate-containing medium

It is well known that some phosphates function an essential role in the phosphorylation, in which they can assist to transfer electron and pump protons for the reaction taking place under the light. Motivated by this idea, Liu et al. mimicked phosphate-involving natural photosynthesis to boost the photocatalytic H_2 production of Pt/g-C_3N_4 nanosheets by adding KH_2PO_4 to TEOA-containing normal reaction solution. The authors discovered that the optimized amount of KH_2PO_4 functions as the electron carrier and mediator, which assist the electron consumption and supply the proton source. Moreover, the addition of KH_2PO_4 is also beneficial for the hole removal to oxidize the sacrificial agent (TEOA). Both processes thus suppress the electron−hole recombination and accelerate the reduction reaction. The evolved hydrogen, in this circumstance, was found to be remarkably higher than that of bulk Pt-C_3N_4 and Pt-C_3N_4 nanosheets. The AQY is approximately 26.1% at 420 ± 14.5 nm, which is one of the highest AQY values for C_3N_4-based photocatalysts (Liu et al., 2015).

8.3.6 Metal−metal oxide core−shell cocatalysts

In this section, the unique design of core−shell-based cocatalysts will be discussed in detail for the photocatalytic hydrogen production and overall water splitting. Thus the fabrication of core−shell structure has attracted considerable attention as this fascinating configuration possesses the capability to simultaneously address electron−hole recombination, reactant adsorption, and suppress the back reactions. It means that this type of cocatalyst is very promising for photocatalytic hydrogen generation and overall water splitting. Herein, we introduce two different core–shell cocatalysts, which have outperformed conventional materials.

The early metal/metal oxide core–shell cocatalysts have been employed by Professor Kazunari Domen's group since 2006 for photocatalytic overall water splitting (Maeda et al., 2006). It has been demonstrated that Rh/Cr_2O_3 (core/shell) prepared the photodeposition method on $(Ga_{1-x}Zn_x)(N_{1-x}O_x)$ solid solution enhanced the performance of evolved H_2 and O_2 in the overall water splitting reaction. It turns out that the Rh core accelerates the separation of electron–hole pairs generated in $(Ga_{1-x}Zn_x)(N_{1-x}O_x)$ photocatalyst. Photogenerated electrons transport to the Cr_2O_3 shell where the hydrogen production reactions are carried out while hindering the formation of water from H_2 and O_2. As a result, the generated products are steadily enhanced, as shown in Fig. 8.9 (Yoshida et al., 2009). In the same manner, Rh/Cr_2O_3 core–shell nanoparticles have been found as an efficient cocatalyst for various semiconductors like barium tantalate mixed oxide composites, GaN Nanowire Arrays, and $KTaO_3/Ta_3N_5$ (Soldat, Marschall, & Wark, 2014; Wang et al., 2011; Wang et al., 2018). It can be realized that the Rh/Cr_2O_3 core–shell offers a compelling pathway to simultaneously improve the charge separation,

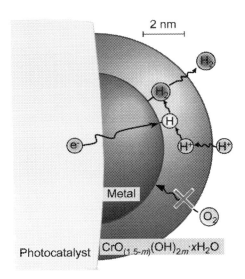

Figure 8.9

Schematic illustration of the improvement of photoactivity via Rh/Cr_2O_3 core–shell cocatalyst configuration. Source: *Reprinted by permission from Yoshida, M., Takanabe, K., Maeda, K., Ishikawa, A., Kubota, J., Sakata, Y., . . . Domen, K. (2009). Role and function of noble-metal/Cr-layer core/shell structure cocatalysts for photocatalytic overall water splitting studied by model electrodes. The Journal of Physical Chemistry C, 113, 10151–10157.*

provide active sites, and retard the back reactions. Therefore this configuration is one of the most weighty candidates for the selection of cocatalysts for photocatalytic overall water splitting.

The Ni/NiO core—shell structure has received great attention for the solar water splitting application when integrated on appropriate semiconductors. Thus the early reports have shown that the Ni/NiO core—shell structure capably restricts the charge recombination (Tsai, Chen, Liu, Asakura, & Chan, 2011; Domen, Kudo, Onishi, Kosugi, & Kuroda, 1986). However, this cocatalyst undergoes deactivation raising the concern about the stability of the material (Hondow, Chou, Sader, Brydson, & Douthwaite, 2010). Recently, Han et al. explained the photocatalytic enhancement derived from Ni/NiO$_x$ core—shell. The obvious modifications in the composition and structure of the Ni@NiO$_x$ core—shell particles on the host photocatalyst, SrTiO$_3$, could be observed. The Ni/NiO$_x$ cocatalyst undergoes the structural rearrangement to form Ni/NiO$_x$/NiOOH. The continued illumination leads to loss of the metallic Ni and further facilities for the formation of NiOOH (Han, Kreuger, Mei, & Mul, 2017). Therefore it can be said that the loss of metallic Ni is the original reason for the deactivation of Ni/NiO$_x$ cocatalyst. This investigation thus provides important evidence toward the behavior of Ni/NiO$_x$ core—shell cocatalysts, which could assist the researcher to design the proper configuration.

8.4 Oxidation cocatalysts in the multiple component systems

Oxidation cocatalysts in multiple component systems have attracted the increasing attention owing to their ability, which attracts the photogenerated holes leading to improve charge separation. Moreover, the selection of oxidation cocatalysts plays an essential role in photocatalytic systems, which are designed for either water oxidation processes or overall water splitting. The volcano plot for the OER, reveals that RuO$_2$, IrO$_2$, and CoO$_x$ are the most active oxidative cocatalysts, which are a good agreement with the experimental data obtained by various studies over the past decades (Man et al., 2011; Wang & Domen, 2020). The following parts will be discussed the advantages of these materials in detail.

8.4.1 RuO$_2$ cocatalysts

The RuO$_2$ nanoparticles contain the ability to attract photogenerated holes to produce oxygen when integrated on CdS material. Moreover, the existence of RuO$_2$ also stabilizes the CdS from the corrosion by photoinduced holes (Kalyanasundaram, Borgarello, Duonghong, & Grätzel, 1981). Teramura proved that the presence of RuO$_2$ on the $(Ga_{1-x}Zn_x)(N_{1-x}O_x)$ photocatalyst significantly encouraged the overall water-splitting performance under visible light (Teramura et al., 2005). Thus RuO$_2$ species have been found to be an effective oxidation cocatalyst for the generation of oxygen over various photocatalytic systems, especially a Z-scheme photocatalytic system for overall water splitting (Maeda, Abe, & Domen, 2011; Nakada et al., 2017). Recently Wang presented the scalable particulate photocatalytic sheets composed by La- and Rh-codoped $SrTiO_3/Au/Mo$-doped $BiVO_4$ Z-scheme system in which Ru and RuO_x function the reductive and oxidative sites, respectively. Interestingly, this configuration which exhibited an AQY of 30% at 419 nm and an STH of 1.1% in overall pure water splitting reveals as the most robust photocatalytic system for overall water splitting (Wang et al., 2016).

8.4.2 CoO_x cocatalysts

In 2015, Li et al. reported that the direct evidence obtained by surface photovoltage spectroscopy proves the capability of CoO_x nanoparticles to prolong the lifetime of photogenerated holes. The hole collection effect can be assigned to the heterojunction between TiO_2 and CoO_x tuning the charge transfer direction. Thus the work function difference between TiO_2 and CoO_x at the interface causes a strong interfacial electric field, which assists the transportation of holes from TiO_2 to CoO_x to facilitate the oxidation reaction (Li, Hou, Zhang, Chen, & Lin, 2015).

In another investigation, Ivanova et al. proved that the presence of CoO nanoparticles on the La-doped $NaTaO_3$ photocatalyst is beneficial for the O_2 evolution reaction under the presence of $AgNO_3$ as an electron sacrificial agent. Thus CoO nanoparticles have an affinity toward the photogenerated holes that assist in prolonging the lifetime of the holes. In addition, CoO cocatalyst offers active sites and decreases the activation energy prompting the O_2 generation. Moreover, the existence of two cocatalysts, that is, RuO_2 and CoO are beneficial for the simultaneous formation of H_2 and O_2 in pure water (Ivanova, Kandiel, Cho, Choi, & Bahnemann, 2018). The analogous conclusion is also drawn from the work with CoO_x/Ta_3N_5 photocatalysts in which the roles of CoO_x are to capture the holes and reduce the defects. Together, these outstanding impacts of CoO_x significantly prolong the lifetime of both photogenerated electrons and holes (Vequizo, Hojamberdiev, Teshima, & Yamakata, 2018). Therefore, it turns out that CoO_x nanoparticles decorated photocatalysts advance the charge separation and provide the active sites in many nanocomposite photocatalytic systems for water splitting and hydrogen production applications, such as $CoO_x/BiVO_4/GRO/$metal sulfide and CoO_x/CdS (Iwase et al., 2016; Liu et al., 2018).

8.4.3 IrO_x cocatalysts

The use of IrO_x cocatalysts has shown a positive impact on the photocatalytic performance of the host materials because IrO_x nanoparticles function as the hole collector improving the charge separation. Thus Wang reported the investigation of core/shell photocatalyst in which Pt and IrO_2 nanoparticles are located inside and outside of a Ta_3N_5 hollow sphere. It turns out that two spatially separated cocatalysts drive the photogenerated carriers in different directions leading to enhance photocatalytic water splitting firmly. In this context, IrO_2 decorated on the Ta_3N_5 wall thus attracts the photogenerated hole while the electrons migrate to Pt nanoparticles. The noteworthy improvement in charge separation thus boosts the photocatalytic performance in comparison to the case of a single cocatalyst (Wang et al., 2013). Moreover, IrO_2 nanoparticles also positively contribute to strengthening charge separation in the $Pt/SiC/IrO_2$ photocatalyst. Thus the presence of IrO_2 and Pt accelerates the charge separation, which boosts the photocatalytic performance by a factor of 130 in comparison to the bare SiC (Wang, Wang, Wang, Guo, & Yuan, 2017). It can be said that IrO_x-based oxidation cocatalysts are the interesting material that requires more attention both theoretical and experimental aspects to generate the proper configurations for excellent photocatalytic performance.

8.5 Conclusion

The solar-driven fuel production has remained the holy grail in material science in the attempts to resolve energy crises and environmental issues. The last decades have been witnessed a huge effort in the development of solar-driven hydrogen production, oxygen evolution as well as overall water splitting employing photocatalysts. Using cocatalysts has attracted extensive attention as the cocatalysts can not only mitigate the electron—hole recombination, which is one of the bottleneck problems resulting in poor catalytic performance, but also offer active sites for the chemical reactions. This chapter provides the cutting-edge development of reduction and oxidation cocatalysts for solar-driven hydrogen and oxygen evolution. The pros and cons of each type of cocatalyst have also mentioned that assist to engineer proper cocatalysts for each host material. Moreover, the loading of cocatalysts is also needed to be considered as the under or overloading drives the photocatalytic performance negatively (Ran et al., 2014).

The future may observe the innovative exploration of novel multicomponent cocatalysts, such as dual cocatalysts or nanocomposite cocatalysts. The investigations of the interfacial charge kinetics and active sites in both theoretical and experimental aspects are highly required to obtain an in-depth understanding of the nature of each cocatalyst. Finally, the preparation of novel cocatalysts is also highly needed. To this end, there is still large space for the researcher in this field and more efforts devoted to this topic will accelerate the development of material toward the desired destination.

References

An, X., Wang, W., Wang, J., Duan, H., Shi, J., & Yu, X. (2018). The synergetic effects of Ti3C2 MXene and Pt as co-catalysts for highly efficient photocatalytic hydrogen evolution over gC3N4. *Physical Chemistry Chemical Physics*, *20*, 11405—11411.

Aronovitch, E., Kalisman, P., Mangel, S., Houben, L., Amirav, L., & Bar-Sadan, M. (2015). Designing bimetallic co-catalysts: A party of two. *The Journal of Physical Chemistry Letters, 6,* 3760–3764.

Bai, S., Yang, L., Wang, C., Lin, Y., Lu, J., Jiang, J., & Xiong, Y. (2015). Boosting photocatalytic water splitting: Interfacial charge polarization in atomically controlled core−shell cocatalysts. *Angewandte Chemie International Edition, 54,* 14810–14814.

Bai, S., Jiang, J., Zhang, Q., & Xiong, Y. (2015). Steering charge kinetics in photocatalysis: Intersection of materials syntheses, characterization techniques and theoretical simulations. *Chemical Society Reviews, 44,* 2893–2939.

Bian, H., Ji, Y., Yan, J., Li, P., Li, L., Li, Y., & Liu, S. (2018). In situ synthesis of few-layered g-C3N4 with vertically aligned MoS$_2$ loading for boosting solar-to-hydrogen generation. *Small (Weinheim an der Bergstrasse, Germany), 14,* 1703003.

Cao, S., Wang, C. J., Fu, W. F., & Chen, Y. (2017). Metal phosphides as co-catalysts for photocatalytic and photoelectrocatalytic water splitting. *ChemSusChem, 10,* 4306–4323.

Carvalho, A., Wang, M., Zhu, X., Rodin, A. S., Su, H., & Neto, A. H. C. (2016). Phosphorene: From theory to applications. *Nature Reviews Materials, 1,* 16061.

Chang, X., Wang, T., & Gong, J. (2016). CO$_2$ photo-reduction: Insights into CO$_2$ activation and reaction on surfaces of photocatalysts. *Energy & Environmental Science, 9,* 2177–2196.

Chen, S., Takata, T., & Domen, K. (2017). Particulate photocatalysts for overall water splitting. *Nature Reviews Materials, 2,* 17050.

DeRita, L., Resasco, J., Dai, S., Boubnov, A., Thang, H. V., Hoffman, A. S., . . . Pacchioni, G. (2019). Structural evolution of atomically dispersed Pt catalysts dictates reactivity. *Nature Materials, 1.*

Domen, K., Kudo, A., Onishi, T., Kosugi, N., & Kuroda, H. (1986). Photocatalytic decomposition of water into hydrogen and oxygen over nickel (II) oxide-strontium titanate (SrTiO$_3$) powder. 1. *Structure of the Catalysts, 90,* 292–295.

Eftekhari, A. (2017). Electrocatalysts for hydrogen evolution reaction. *International Journal of Hydrogen Energy, 42,* 11053–11077.

Fang, X., Shang, Q., Wang, Y., Jiao, L., Yao, T., Li, Y., . . . Jiang, H. L. (2018). Single Pt atoms confined into a metal−organic framework for efficient photocatalysis. *Advanced Materials, 30,* 1705112.

Frame, F. A., & Osterloh, F. E. (2010). CdSe-MoS$_2$: A quantum size-confined photocatalyst for hydrogen evolution from water under visible light. *The Journal of Physical Chemistry C, 114,* 10628–10633.

Fujishima, A. (1972). Electrochemical photolysis of water at a semiconductor electrode. *Nature, 238,* 37–38.

Gao, G., O'Mullane, A. P., & Du, A. (2016). 2D MXenes: A new family of promising catalysts for the hydrogen evolution reaction. *ACS Catalysis, 7,* 494–500.

Guo, Z., Zhou, J., Zhu, L., & Sun, Z. (2016). MXene: A promising photocatalyst for water splitting. *Journal of Materials Chemistry A, 4,* 11446–11452.

Han, K., Kreuger, T., Mei, B., & Mul, G. (2017). Transient behavior of Ni@NiO$_x$ functionalized SrTiO$_3$ in overall water splitting. *ACS Catalysis, 7,* 1610–1614.

He, H., Lin, J., Fu, W., Wang, X., Wang, H., Zeng, Q., . . . Tay, B. K. (2016). MoS$_2$/TiO$_2$ edge-on heterostructure for efficient photocatalytic hydrogen evolution. *Advanced Energy Materials, 6,* 1600464.

Hisatomi, T., & Domen, K. (2019). Reaction systems for solar hydrogen production via water splitting with particulate semiconductor photocatalysts. *Nature Catalysis, 1.*

Hondow, N., Chou, Y.-H., Sader, K., Brydson, R., & Douthwaite, R. E. (2010). v: The role of ion migration and alloy formation on the stability of core shell cocatalysts for photoinduced water splitting. *The Journal of Physical Chemistry C, 114,* 22758–22762.

Hu, Z., Shen, Z., & Yu, J. C. (2017). Phosphorus containing materials for photocatalytic hydrogen evolution. *Green Chemistry, 19,* 588–613.

Indra, A., Acharjya, A., Menezes, P. W., Merschjann, C., Hollmann, D., Schwarze, M., . . . Driess, M. (2017). Boosting visible-light-driven photocatalytic hydrogen evolution with an integrated nickel phosphide−carbon nitride system. *Angewandte Chemie International Edition, 56,* 1653–1657.

Irfan, M., Sevim, M., Kocak, Y., Balci, M., Metin, Ö., & Ozensoy, E. (2019). Enhanced photocatalytic NO_x oxidation and storage under visible-light irradiation by anchoring Fe_3O_4 nanoparticles on mesoporous graphitic carbon nitride (mpg-C_3N_4). *Applied Catalysis B: Environmental, 249*, 126−137.

Ithisuphalap, K., Zhang, H., Guo, L., Yang, Q., Yang, H., & Wu, G. (2019). Photocatalysis and photoelectrocatalysis methods of nitrogen reduction for sustainable ammonia synthesis. *Small Methods, 3*, 1800352.

Ivanova, I., Kandiel, T. A., Cho, Y.-J., Choi, W., & Bahnemann, D. (2018). Mechanisms of photocatalytic molecular hydrogen and molecular oxygen evolution over La-doped $NaTaO_3$ particles: Effect of different cocatalysts and their specific activity. *ACS Catalysis, 8*, 2313−2325.

Iwase, A., Yoshino, S., Takayama, T., Ng, Y. H., Amal, R., & Kudo, A. (2016). Water splitting and CO_2 reduction under visible light irradiation using Z-scheme systems consisting of metal sulfides, CoO_x-loaded $BiVO_4$, and a reduced graphene oxide electron mediator. *Journal of the American Chemical Society, 138*, 10260−10264.

Jaramillo, T. F., Jørgensen, K. P., Bonde, J., Nielsen, J. H., Horch, S., & Chorkendorff, I. (2007). Identification of active edge sites for electrochemical H_2 evolution from MoS_2 nanocatalysts. *Science (New York, N.Y.), 317*, 100−102.

Ji, L., Lv, C., Chen, Z., Huang, Z., & Zhang, C. (2018). Nickel-based (photo) electrocatalysts for hydrogen production. *Advanced Materials, 30*, 1705653.

Jiao, Y., Zheng, Y., Jaroniec, M., & Qiao, S. Z. (2015). Design of electrocatalysts for oxygen-and hydrogen-involving energy conversion reactions. *Chemical Society Reviews, 44*, 2060−2086.

Kabir, E., Kumar, P., Kumar, S., Adelodun, A. A., & Kim, K.-H. (2018). Solar energy: Potential and future prospects. *Renewable and Sustainable Energy Reviews, 82*, 894−900.

Kalyanasundaram, K., Borgarello, E., Duonghong, D., & Grätzel, M. (1981). Cleavage of water by visible-light irradiation of colloidal CdS solutions; inhibition of photocorrosion by RuO_2. *Angewandte Chemie International Edition, 20*, 987−988.

Kuehnel, M. F., Orchard, K. L., Dalle, K. E., & Reisner, E. (2017). Selective photocatalytic CO_2 reduction in water through anchoring of a molecular Ni catalyst on CdS nanocrystals. *Journal of the American Chemical Society, 139*, 7217−7223.

Lang, S. M., & Bernhardt, T. M. (2012). Gas phase metal cluster model systems for heterogeneous catalysis. *Physical Chemistry Chemical Physics, 14*, 9255−9269.

Lee, B.-H., Park, S., Kim, M., Sinha, A. K., Lee, S. C., Jung, E., . . . Cho, S.-P. (2019). Reversible and cooperative photoactivation of single-atom Cu/TiO_2 photocatalysts. *Nature Materials, 18*, 620.

Lenferna, G. A. (2018). Can we equitably manage the end of the fossil fuel era? *Energy Research & Social Science, 35*, 217−223.

Lewis, N. S. (2016). Research opportunities to advance solar energy utilization. *Science (New York, N.Y.), 351*, aad1920.

Li, C., Du, Y., Wang, D., Yin, S., Tu, W., Chen, Z., & Xu, R. (2017). Unique PCoN surface bonding states constructed on g-C3N4 nanosheets for drastically enhanced photocatalytic activity of H2 evolution. *Advanced Functional Materials, 27*, 1604328.

Li, J., Cai, L., Shang, J., Yu, Y., & Zhang, L. (2016). Giant enhancement of internal electric field boosting bulk charge separation for photocatalysis. *Advanced Materials, 28*, 4059−4064.

Li, Q., Guo, B., Yu, J., Ran, J., Zhang, B., Yan, H., & Gong, J. R. (2011). Highly efficient visible-light-driven photocatalytic hydrogen production of CdS-cluster-decorated graphene nanosheets. *Journal of the American Chemical Society, 133*, 10878−10884.

Li, S., Hou, L., Zhang, L., Chen, L., Lin, Y., Wang, D., & Xie, T. (2015). Direct evidence of the efficient hole collection process of the CoO_x cocatalyst for photocatalytic reactions: A surface photovoltage study. *Journal of Materials Chemistry A, 3*, 17820−17826.

Li, X., Bi, W., Zhang, L., Tao, S., Chu, W., Zhang, Q., . . . Xie, Y. (2016). Single-atom Pt as co-catalyst for enhanced photocatalytic H_2 evolution. *Advanced Materials, 28*, 2427−2431.

Li, X., Yu, J., Wageh, S., Al-Ghamdi, A. A., & Xie, J. J. (2016). Graphene in photocatalysis: A review. *Small, 12*, 6640−6696.

Li, Y., Wang, H., Xie, L., Liang, Y., Hong, G., & Dai, H. (2011). MoS2 nanoparticles grown on graphene: An advanced catalyst for the hydrogen evolution reaction. *Journal of the American Chemical Society, 133*, 7296–7299.

Li, Y., Deng, X., Tian, J., Liang, Z., & Cui, H. (2018). Ti_3C_2 MXene-derived Ti_3C_2/TiO_2 nanoflowers for noble-metal-free photocatalytic overall water splitting. *Applied Materials Today, 13*, 217–227.

Li, Z. J., Wang, J. J., Li, X. B., Fan, X. B., Meng, Q. Y., Feng, K., ... Wu, L. Z. (2013). *An exceptional artificial photocatalyst, Nih-CdSe/CdS core/shell hybrid, made in situ from cdse quantum dots and nickel salts for efficient hydrogen evolution,* . *Advanced Materials* (25, pp. 6613–6618).

Liu, G., Wang, T., Zhang, H., Meng, X., Hao, D., Chang, K., ... Ye, J. (2015). Nature-inspired environmental "phosphorylation" boosts photocatalytic H_2 production over carbon nitride nanosheets under visible-light irradiation. *Angewandte Chemie International Edition, 54*, 13561–13565.

Liu, M., Chen, Y., Su, J., Shi, J., Wang, X., & Guo, L. (2016). Photocatalytic hydrogen production using twinned nanocrystals and an unanchored NiS_x co-catalyst. *Nature Energy, 1*, 16151.

Liu, P., & Rodriguez, J. A. (2005). Catalysts for hydrogen evolution from the [NiFe] Hydrogenase to the Ni2P (001) surface: The importance of ensemble effect. *Journal of the American Chemical Society, 127*, 14871–14878.

Liu, Y., Xiao, H., & Goddard, W. A., III (2016). Schottky-barrier-free contacts with two-dimensional semiconductors by surface-engineered MXenes. *Journal of American Chemical Society, 138*, 15853–15856.

Liu, Y., Ding, S., Shi, Y., Liu, X., Wu, Z., Jiang, Q., ... Hu, J. (2018). Construction of CdS/CoO$_x$ core-shell nanorods for efficient photocatalytic H_2 evolution. *Applied Catalysis B: Environmental, 234*, 109–116.

Low, J., Cao, S., Yu, J., & Wageh, S. (2014). Two-dimensional layered composite photocatalysts. *Chemical communications, 50*, 10768–10777.

Maeda, K., Teramura, K., Lu, D., Saito, N., Inoue, Y., & Domen, K. (2006). Noble-metal/Cr$_2$O$_3$ core/shell nanoparticles as a cocatalyst for photocatalytic overall water splitting. *Angewandte Chemie International Edition, 45*, 7806–7809.

Maeda, K., Abe, R., & Domen, K. (2011). Role and function of ruthenium species as promoters with TaON-based photocatalysts for oxygen evolution in two-step water splitting under visible light. *The Journal of Physical Chemistry C, 115*, 3057–3064.

Man, I. C., Su, H. Y., Calle-Vallejo, F., Hansen, H. A., Martínez, J. I., Inoglu, N. G., ... Rossmeisl, J. (2011). Universality in oxygen evolution electrocatalysis on oxide surfaces. *ChemCatChem, 3*, 1159–1165.

Marszewski, M., Cao, S., Yu, J., & Jaroniec, M. (2015). Semiconductor-based photocatalytic CO$_2$ conversion. *Materials Horizons, 2*, 261–278.

Mazloomi, K., & Gomes, C. (2012). Hydrogen as an energy carrier: Prospects and challenges. *Renewable and Sustainable Energy Reviews, 16*, 3024–3033.

Meng, A., Zhang, L., Cheng, B., & Yu, J. (2019). Dual cocatalysts in TiO$_2$ photocatalysis. *Advanced Materials*, 1807660.

Miseki, Y., & Sayama, K. (2019). Photocatalytic water splitting for solar hydrogen production using the carbonate effect and the Z-scheme reaction. *Advanced Energy Materials, 9*, 1801294.

Mukherji, A., Seger, B., Lu, G. Q., & Wang, L. (2011). Nitrogen doped Sr$_2$Ta$_2$O$_7$ coupled with graphene sheets as photocatalysts for increased photocatalytic hydrogen production. *ACS Nano, 5*, 3483–3492.

Muñoz-Batista, M. J., Motta Meira, D., Colón, G., Kubacka, A., & Fernández-García, M. (2018). Phase-contact engineering in mono- and bimetallic Cu-Ni Co-catalysts for hydrogen photocatalytic materials. *Angewandte Chemie International Edition, 57*, 1199–1203.

Naguib, M., Kurtoglu, M., Presser, V., Lu, J., Niu, J., Heon, M., ... Barsoum, M. W. (2011). Two-dimensional nanocrystals produced by exfoliation of Ti$_3$AlC$_2$. *Advanced Materials, 23*, 4248–4253.

Naguib, M., Mochalin, V. N., Barsoum, M. W., & Gogotsi, Y. (2014). 25th anniversary article: MXenes: A new family of two-dimensional materials. *Advanced Materials, 26*, 992–1005.

Nakada, A., Nishioka, S., Vequizo, J. J. M., Muraoka, K., Kanazawa, T., Yamakata, A., ... Maeda, K. (2017). Solar-driven Z-scheme water splitting using tantalum/nitrogen co-doped rutile titania nanorod as an oxygen evolution photocatalyst. *Journal of Materials Chemistry A, 5*, 11710–11719.

Nguyen, C. C., Vu, N. N., & Do, T.-O. (2015). Recent advances in the development of sunlight-driven hollow structure photocatalysts and their applications. *Journal of Materials Chemistry A, 3,* 18345−18359.

Pacesila, M., Burcea, S. G., & Colesca, S. E. (2016). Analysis of renewable energies in European Union. *Renewable and Sustainable Energy Reviews, 56,* 156−170.

Popczun, E. J., McKone, J. R., Read, C. G., Biacchi, A. J., Wiltrout, A. M., Lewis, N. S., & Schaak, R. E. (2013). Nanostructured nickel phosphide as an electrocatalyst for the hydrogen evolution reaction. *Journal of American Chemical Society, 135,* 9267−9270.

Qin, Z., Xue, F., Chen, Y., Shen, S., & Guo, L. (2017). Spatial charge separation of one-dimensional Ni_2P-$Cd_{0.9}Zn_{0.1}S$/g-C_3N_4 heterostructure for high-quantum-yield photocatalytic hydrogen production. *Applied Catalysis B: Environmental, 217,* 551−559.

Ran, J., Zhang, J., Yu, J., Jaroniec, M., & Qiao, S. Z. (2014). Earth-abundant cocatalysts for semiconductor-based photocatalytic water splitting. *Chemical Society Reviews, 43,* 7787−7812.

Ran, J., Zhu, B., & Qiao, S.-Z. (2017). Phosphorene co-catalyst advancing highly efficient visible-light photocatalytic hydrogen production. *Angewandte Chemie International Edition, 56,* 10373−10377.

Ran, J., Wang, X., Zhu, B., & Qiao, S.-Z. (2017). Strongly interactive 0D/2D hetero-structure of a $Zn_xCd_{1-x}S$ nano-particle decorated phosphorene nano-sheet for enhanced visible-light photocatalytic H 2 production. *Chemical Communications, 53,* 9882−9885.

Ran, J., Guo, W., Wang, H., Zhu, B., Yu, J., & Qiao, S. Z. (2018). Metal-free 2D/2D phosphorene/g-C_3N_4 Van der Waals heterojunction for highly enhanced visible-light photocatalytic H_2 production. *Chemical Communications, 30,* 1800128.

Ran, J., Zhu, B., & Qiao, S.-Z (2017). Phosphorene co-catalyst advancing highly efficient visible-light photocatalytic hydrogen production. *Angewandte Chemie International Edition, 56,* 10373−10377.

Rawalekar, S., & Mokari, T. (2013). Rational design of hybrid nanostructures for advanced photocatalysis. *Advanced Energy Materials, 3,* 12−27.

Seh, Z. W., Fredrickson, K. D., Anasori, B., Kibsgaard, J., Strickler, A. L., Lukatskaya, M. R., . . . Vojvodic, A. (2016). Two-dimensional molybdenum carbide (MXene) as an efficient electrocatalyst for hydrogen evolution. *ACS Energy Letters, 1,* 589−594.

Shao, Z., He, Y., Zeng, T., Yang, Y., Pu, X., Ge, B., & Dou, J. (2018). Compounds, highly efficient photocatalytic H_2 evolution using the Ni_2P-$Zn_{0.5}Cd_{0.5}S$ photocatalyst under visible light irradiation. *Journal of Alloys and Compounds, 769,* 889−897.

Shi, X., Fujitsuka, M., & Majima, T. (2018). Electron transfer dynamics of quaternary sulfur semiconductor/MoS_2 layer-on-layer for efficient visible-light H_2 evolution. *Applied Catalysis B: Environmental, 235,* 9−16.

Soldat, J., Marschall, R., & Wark, M. (2014). Improved overall water splitting with barium tantalate mixed oxide composites. *Chemical Science, 5,* 3746−3752.

Su, R., Tiruvalam, R., Logsdail, A. J., He, Q., Downing, C. A., Jensen, M. T., . . . Bechstein, R. (2014). Designer titania-supported Au−Pd nanoparticles for efficient photocatalytic hydrogen production. *ACS Nano, 8,* 3490−3497.

Su, T., Hood, Z. D., Naguib, M., Bai, L., Luo, S., Rouleau, C. M., . . . Wu, Z. (2019). Monolayer $Ti_3C_2T_x$ as an effective Co-catalyst for enhanced photocatalytic hydrogen production over TiO_2. *ACS Applied Energy Materials, 2,* 4640−4651.

Sun, Z., Zheng, H., Li, J., Du, P. J. E., & Science, E. (2015). Extraordinarily efficient photocatalytic hydrogen evolution in water using semiconductor nanorods integrated with crystalline Ni_2P cocatalysts. *Science & Environmental Science, 8,* 2668−2676.

Teramura, K., Maeda, K., Saito, T., Takata, T., Saito, N., Inoue, Y., & Domen, K. (2005). Characterization of ruthenium oxide nanocluster as a cocatalyst with $(Ga_{1-x}Zn_x)(N_{1-x}O_x)$ for photocatalytic overall water splitting. *The Journal of Physical Chemistry B, 109,* 21915−21921.

Tsai, C.-W., Chen, H. M., Liu, R.-S., Asakura, K., & Chan, T.-S. (2011). Ni@NiO core−shell structure-modified nitrogen-doped $InTaO_4$ for solar-driven highly efficient CO_2 reduction to methanol. *The Journal of Physical Chemistry C, 115,* 10180−10186.

Vequizo, J. J. M., Hojamberdiev, M., Teshima, K., & Yamakata, A. (2018). Role of CoO_x cocatalyst on Ta_3N_5 photocatalysts studied by transient visible to mid-infrared absorption spectroscopy. *Journal of Photochemistry and Photobiology A: Chemistry, 358*, 315–319.

Víctor, A. (2013). *The role of co-catalysts: Interaction and synergies with semiconductors, design of advanced photocatalytic materials for energy and environmental applications* (pp. 195–216). Springer.

Vu, N. N., Kaliaguine, S., & Do, T. O. (2019). Critical aspects and recent advances in structural engineering of photocatalysts for sunlight-driven photocatalytic reduction of CO_2 into fuels. *Advanced Functional Materials*, 1901825.

Wang, A., Li, J., & Zhang, T. (2018). Heterogeneous single-atom catalysis. *Nature Reviews Chemistry, 2*, 65.

Wang, D., Pierre, A., Kibria, M. G., Cui, K., Han, X., Bevan, K. H., ... Mi, Z. (2011). Wafer-level photocatalytic water splitting on GaN nanowire arrays grown by molecular beam epitaxy. *Nano Letters, 11*, 2353–2357.

Wang, D., Hisatomi, T., Takata, T., Pan, C., Katayama, M., Kubota, J., & Domen, K. (2013). Core/shell photocatalyst with spatially separated co-catalysts for efficient reduction and oxidation of water. *Angewandte Chemie International Edition, 52*, 11252–11256.

Wang, D., Wang, W., Wang, Q., Guo, Z., & Yuan, W. (2017). Spatial separation of Pt and IrO_2 cocatalysts on SiC surface for enhanced photocatalysis. *Materials Letters, 201*, 114–117.

Wang, F., Jiang, Y., Lawes, D. J., Ball, G. E., Zhou, C., Liu, Z., & Amal, R. (2015). Analysis of the promoted activity and molecular mechanism of hydrogen production over fine Au−Pt alloyed TiO_2 photocatalysts. *ACS Catalysis, 5*, 3924–3931.

Wang, H., Zhang, L., Chen, Z., Hu, J., Li, S., Wang, Z., ... Wang, X. (2014). Semiconductor heterojunction photocatalysts: Design, construction, and photocatalytic performances. *Chemical Society Reviews, 43*, 5234–5244.

Wang, Q., & Domen, K. (2020). Particulate photocatalysts for light-driven water splitting: Mechanisms, challenges, and design strategies. *Chemical Reviews, 120*, 919–985.

Wang, Q., Hisatomi, T., Jia, Q., Tokudome, H., Zhong, M., Wang, C., ... Shibata, N. (2016). Scalable water splitting on particulate photocatalyst sheets with a solar-to-hydrogen energy conversion efficiency exceeding 1%. *Nature Materials, 15*, 611.

Wang, Z., Inoue, Y., Hisatomi, T., Ishikawa, R., Wang, Q., Takata, T., ... Domen, K. (2018). Overall water splitting by Ta_3N_5 nanorod single crystals grown on the edges of $KTaO_3$ particles. *Nature Catalysis, 1*, 756.

Wang, Z., Li, C., & Domen, K. (2019). Recent developments in heterogeneous photocatalysts for solar-driven overall water splitting. *Chemical Society Reviews, 48*, 2109–2125.

Woomer, A. H., Farnsworth, T. W., Hu, J., Wells, R. A., Donley, C. L., & Warren, S. C. (2015). Phosphorene: Synthesis, scale-up, and quantitative optical spectroscopy. *ACS Nano, 9*, 8869–8884.

Wu, C., Zhang, J., Tong, X., Yu, P., Xu, J. Y., Wu, J., ... Chueh, Y. L. (2019). A critical review on enhancement of photocatalytic hydrogen production by molybdenum disulfide: From growth to interfacial activities. *Small (Weinheim an der Bergstrasse, Germany), 15*, 1900578.

Xiang, Q., Yu, J., & Jaroniec, M. (2012). Synergetic effect of MoS_2 and graphene as cocatalysts for enhanced photocatalytic H_2 production activity of TiO_2 nanoparticles. *Journal of the American Chemical Society, 134*, 6575–6578.

Xiang, Q., Yu, J., & Jaroniec, M. (2012). Graphene-based semiconductor photocatalysts. *Chemical Society Reviews, 41*, 782–796.

Xiang, Q., Cheng, F., & Lang, D. (2016). Hierarchical layered WS_2/graphene-modified CdS nanorods for efficient photocatalytic hydrogen evolution. *ChemSusChem, 9*, 996–1002.

Xing, J., Chen, J. F., Li, Y. H., Yuan, W. T., Zhou, Y., Zheng, L. R., ... Zhao, H. J. (2014). Stable isolated metal atoms as active sites for photocatalytic hydrogen evolution. *Chemistry−A European Journal, 20*, 2138–2144.

Xu, C., Wang, L., Liu, Z., Chen, L., Guo, J., Kang, N., ... Ren, W. (2015). Large-area high-quality 2D ultrathin Mo2C superconducting crystals. *Nature Materials, 14*, 1135.

Xu, Y., & Xu, R. (2015). Nickel-based cocatalysts for photocatalytic hydrogen production. *Applied Surface Science, 351*, 779–793.

Yang, J., Wang, D., Han, H., & Li, C. (2013). Roles of cocatalysts in photocatalysis and photoelectrocatalysis. *Accounts of Chemical Research, 46*, 1900−1909.

Yoshida, M., Takanabe, K., Maeda, K., Ishikawa, A., Kubota, J., Sakata, Y., . . . Domen, K. (2009). Role and function of noble-metal/Cr-layer core/shell structure cocatalysts for photocatalytic overall water splitting studied by model electrodes. *The Journal of Physical Chemistry C, 113*, 10151−10157.

Yu, H., Jiang, L., Wang, H., Huang, B., Yuan, X., Huang, J., . . . Zeng, G. (2019). Modulation of Bi_2MoO_6-based materials for photocatalytic water splitting and environmental application: A critical review. *Small (Weinheim an der Bergstrasse, Germany)*, 1901008.

Yuan, Y.-J., Tu, J.-R., Ye, Z.-J., Chen, D.-Q., Hu, B., Huang, Y.-W., . . . Zou, Z.-G. (2016). MoS_2-graphene/$ZnIn_2S_4$ hierarchical microarchitectures with an electron transport bridge between light-harvesting semiconductor and cocatalyst: A highly efficient photocatalyst for solar hydrogen generation. *Applied Catalysis B: Environmental, 188*, 13−22.

Zhang, B. W., Yang, H. L., Wang, Y. X., Dou, S. X., & Liu, H. K. (2018). A comprehensive review on controlling surface composition of Pt-based bimetallic electrocatalysts. *Advanced Energy Materials, 8*, 1703597.

Zhang, L., Tian, B., Chen, F., & Zhang, J. (2012). Nickel sulfide as co-catalyst on nanostructured TiO_2 for photocatalytic hydrogen evolution. *International Journal of Hydrogen Energy, 37*, 17060−17067.

Zhang, S., Liu, X., Liu, C., Luo, S., Wang, L., Cai, T., . . . Pei, Y. (2017). MoS_2 quantum dot growth induced by S vacancies in a $ZnIn_2S_4$ monolayer: Atomic-level heterostructure for photocatalytic hydrogen production. *ACS Nano, 12*, 751−758.

Zhang, W., Wang, Y., Wang, Z., Zhong, Z., & Xu, R. (2010). Highly efficient and noble metal-free NiS/CdS photocatalysts for H_2 evolution from lactic acid sacrificial solution under visible light. *Chemical Communications, 46*, 7631−7633.

Zhao, H., Wang, J., Dong, Y., & Jiang, P. (2017). Noble-metal-free iron phosphide cocatalyst loaded graphitic carbon nitride as an efficient and robust photocatalyst for hydrogen evolution under visible light irradiation. *ACS Sustainable Chemistry & Engineering, 5*, 8053−8060.

Zhao, H., Zhang, H., Cui, G., Dong, Y., Wang, G., Jiang, P., . . . Zhao, N. (2018). A photochemical synthesis route to typical transition metal sulfides as highly efficient cocatalyst for hydrogen evolution: From the case of NiS/g-C_3N_4. *Applied Catalysis B: Environmental, 225*, 284−290.

Zhao, Q., Yao, W., Huang, C., Wu, Q., & Xu, Q. (2017). Effective and durable Co single atomic cocatalysts for photocatalytic hydrogen production. *ACS Applied Materials & Interfaces, 9*, 42734−42741.

Zhao, Q., Sun, J., Li, S., Huang, C., Yao, W., Chen, W., . . . Xu, Q. (2018). Single nickel atoms anchored on nitrogen-doped graphene as a highly active cocatalyst for photocatalytic H_2 evolution. *ACS Catalysis, 8*, 11863−11874.

Zhao, X., Feng, J., Liu, J., Shi, W., Yang, G., Wang, G. C., & Cheng, P. (2018). An efficient, visible-light-driven, hydrogen evolution catalyst $NiS/Zn_xCd_{1-x}S$ nanocrystal derived from a metal−organic framework. *Angewandte Chemie International, 130*, 9938−9942.

Y. Zhao, W. Gao, S. Li, G.R. Williams, A.H. Mahadi, D. Ma, Solar- versus thermal-driven catalysis for energy conversion, Joule, (2019).

Zheng, Y., Jiao, Y., Zhu, Y., Li, L. H., Han, Y., Chen, Y., . . . Qiao, S. Z. (2014). Hydrogen evolution by a metal-free electrocatalyst. *Nature Communications, 5*, 3783.

Zhu, J., Pang, S., Dittrich, T., Gao, Y., Nie, W., Cui, J., . . . Li, C. (2017). Visualizing the nano cocatalyst aligned electric fields on single photocatalyst particles. *Nano Letters, 17*, 6735−6741.

Zhu, Y., Marianov, A., Xu, H., Lang, C., & Jiang, Y. (2018). Bimetallic Ag−Cu supported on graphitic carbon nitride nanotubes for improved visible-light photocatalytic hydrogen production. *ACS Applied Materials & Interfaces, 10*, 9468−9477.

Zong, X., Yan, H., Wu, G., Ma, G., Wen, F., Wang, L., & Li, C. (2008). Enhancement of photocatalytic H_2 evolution on CdS by loading MoS_2 as cocatalyst under visible light irradiation. *Journal of the American Chemical Society, 130*, 7176−7177.

CHAPTER 9

Materials and features of ferroelectric photocatalysts: the case of multiferroic BiFeO₃

U. Bharagav[1], N. Ramesh Reddy[1], K. Pratap[2], K.K. Cheralathan[3], M.V. Shankar[1], P.K. Ojha[4] and M. Mamatha Kumari[1]

[1]Nanocatalysis and Solar Fuels Research Laboratory, Department of Materials Science & Nanotechnology, Yogi Vemana University, Kadapa, India, [2]Centre for Advanced Studies in Electronics Science and Technology (CASEST), School of Physics, University of Hyderabad, Hyderabad, India, [3]Department of Chemistry, School of Advanced Sciences, Vellore Institute of Technology, Vellore, India, [4]Naval Materials Research Laboratory (NMRL), Ambernath, India

9.1 Introduction

Ferroelectrics are a class of materials which exhibits reversible polarization on application of an electric field. The first ferroelectric material was discovered 100 years ago, that is, Rochelle salt which exhibits the sudden electric polarization (Si et al., 2019). Ferroelectrics are famous for their extensive properties such as narrow bandgap values, spontaneous electric polarization, superior magnetic properties, and they had several applications in capacitors, storage memories, wave guides, optical memory display, displacement transducers, etc. The list of ferroelectrics contains titanates ($BaTiO_3$, $PbTiO_3$, and $SnTiO_3$) (Alammar, Hamm, Wark, & Mudring, 2015), niobates ($LiNbO_3$, $KNbO_3$, $NaNbO_3$, and $AgNbO_3$) (Zlotnik, Tobaldi, Seabra, Labrincha, & Vilarinho, 2016), tantalates ($LiTaO_3$, $KTaO_3$, $NaTaO_3$, and $AgTaO_3$) (Yogamalar, Kalpana, Senthil, & Chithambararaj, 2018), and perovskites containing iron, that is, $BiFeO_3$ and $LaFeO_3$.

In the field photocatalysis, narrow bandgap with suitable energy band potential and stability of the catalyst plays a crucial role. $BiFeO_3$ is gaining more and more attention because of its visible-active bandgap for photocatalytic applications, this is the major reason behind choosing perovskite materials as catalysts (Yogamalar et al., 2018). However, it also has some limitations as other photocatalysts do, but majority of the characteristic properties were preferable for making an efficient catalytic material (Lin et al., 2014). $BiFeO_3$ in short BFO is one of the promising candidates for the photocatalytic applications (Yao, Wenchao

Photocatalytic Systems by Design.
DOI: https://doi.org/10.1016/B978-0-12-820532-7.00003-5

Liu, Chiwah Leung, & Ploss, 2011). It is eco-friendly in nature, and because of its absorbance in visible-light region, it can be activated by sunlight, thus utilizing more solar energy (Wrzesinska, Khort, Bobowska, Busiakiewicz, & Wypych-Puszkarz, 2019). BFO is not only ferroelectric it is also said to have piezoelectric properties, and it is the only single-phase multiferroic material that can exhibit both high ferroelectric curie temperature at temperatures near 1103K and high antiferromagnetic Neel temperature at 643K (You et al., 2019). The bandgap of BFO is in between 2.6 and 3.0 eV. $BiFeO_3$ is an ABO_3 perovskite material with rhombohedral structure.

One of the major constraints to BFO in many applications is its leakage current density. To restrain the leakage current density, several efforts were made and some of them were successful in reducing the leakage current. To utilize BFO in photocatalytic fields, it has been doped with several materials for enhancing various parameters. Some dopants offer high surface area, whereas some other offer higher charge carrier life time. In this manner, many of the dopants were loaded on BFO for enhancing its properties.

9.2 Preparation techniques used for the synthesis of BFO

Before digging it deeper, synthesis criteria are briefing here. Several methods were used for the synthesis of BFO and comprehensively named herewith a simple classification. BFO synthesis mechanisms can be mainly classified into two types: one from the Oxide precursors and another is via wet chemical routes. Solid-state reaction of single oxides at suitable temperatures is the majorly reported arena for the synthesis of perovskite materials. It includes preparation methods such as rapid liquid sintering and mechanical activation, but as per the recent reports, it was found that wet chemical routes are getting impeccable attention. Preparation methods mentioned in the classification were traditional methods for the preparation of single crystalline BFO. Solid-state reaction method is much reported method than the others, whereas sol–gel method takes the second place in the list. Every method is of with its own benefits and defects. Hydrothermal is the one of the finest methods to prepare the novel nanostructures, which is not confined to the particle size criteria (Feng & Li, 2017). In sol–gel method, dissolution into solvents to form homogeneous mixture takes place in first step, followed by adding other components. After gel formation, drying will be done at the end (Rao, Mukherjee, & Reddy, 2017). Self-propagating high-temperature synthesis and the thermal explosion are the two important modes of combustion methods (Yeh, 2016). In sonochemical method the ultrasound waves originated to form extreme transient conditions are induced at higher temperatures (Xu, Zeiger, & Suslick, 2013). Nitrates and other salts were utilised as precursors in the co-precipitation process to produce metal nanoparticles and metal oxide nanoparticles, which were then calcined at higher temperatures. (Haron, Wisitsoraat, & Wongnawa, 2014). Agglomeration is one of the major constraints that takes place due to enlarged grain size

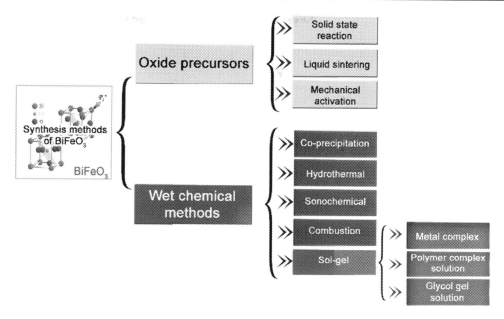

Figure 9.1
Classification of methods used for the preparation of BiFeO₃.

during the synthesis procedure. Evolution of wet chemical methods fills the lack in synthesis criteria, but fine tuning of some parameters is still necessary during synthesis to achieve a bigger difference in the outcome. Highly reported methods next to the sol—gel in the class of wet chemical methods are hydrothermal, sonochemical, co-precipitation, and combustion. Fig. 9.1 classifies information on preparative methods of BiFeO₃.

9.3 Photocatalytic mechanisms in BiFeO₃

In broad definition the steps involve in photocatalysis mechanism are as follows: (1) generating charge carriers on giving light as an energy source; (2) separation of charge carriers to avoid recombination; (3) transfer of generated electrons and holes onto the surface of photocatalyst; and (4) utilizing the evolved charge carriers for redox reactions.

To execute all the mentioned steps successfully and efficiently the photocatalysts should have some major properties such as suitable bandgap and band edge potentials, high surface area, stability, and electron trapping sites.

In the case of BFO the bandgap of BFO is visible active, that is, it is suitable for harvesting solar light. But while coming to the band edge potentials for the reduction potential of BFO, it is not suitable for photocatalytic water splitting to generate H_2. However, it is capable of capturing major portion of sunlight with its bandgap and high surface area. This is one of the causes for limited applications of BFO in the field of H_2 generation via photocatalysis, that is, pristine BFO is not suitable for photocatalytic water splitting to generate hydrogen in alone condition as BFO did not have the suitable conduction band edge potential to liberate H_2. But still on fine tuning its band edge potentials, BFO is reporting with several combinations for photocatalytic hydrogen evolution. Au-sensitized BFO is also reported for H_2 evolution in which SPR (surface plasmon resonance) effect also takes part and promotes the efficiency (Bera, Ghosh, Shyamal, Bhattacharya, & Basu, 2019). The plausible reaction mechanism involved in Au-sensitized BFO is shown in Fig. 9.2A-B.

Basith et al. reported RGO-BFO combination for hydrogen production in which after forming the heterojunction the band edge positions were also shifted (Anjum, Lamia, Arafat, Mahboob, & Basith, 2018). From the results the amount of H_2 generated is found to be higher than the pristine Degussa P25. The above two reports gave an idea about how BFO is fine tuning for photocatalytic water-splitting application. For photocatalytic dye degradation, several combinations are reported. But the basic overview is depicted in a simple manner in Fig. 9.3.

Ponraj et al. reported BFO for photocatalytic dye degradation. The mechanism involved is depicted in Fig. 9.3. In degradation, superoxide radical and hydroxy radical play a cruel role, which further react with the dye and degrade it effectively (Ponraj, Vinitha, & Joseph, 2019).

9.4 Bandgap engineering in multiferroic BiFeO₃

Whatever may be the application based on photocatalysis the key reaction involved is redox reaction. Oxidation, which takes place at the valence band, and reduction, which takes place at

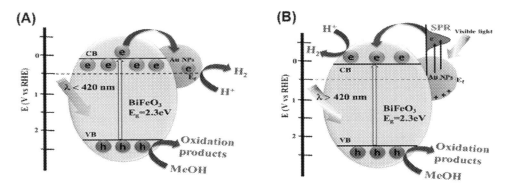

Figure 9.2

(A and B) Photocatalytic H_2 production mechanism in Au-sensitized BFO at different incident intensities. Source: *Reprinted with permission from Bera, S., Ghosh, S., Shyamal, S., Bhattacharya, C., & Basu, R. N. (2019). Photocatalytic hydrogen generation using gold decorated BiFeO₃ heterostructures as an efficient catalyst under visible light irradiation.* Solar Energy Materials and Solar Cells, 194, 195–206. *https://doi.org/10.1016/j.solmat.2019.01.042, copyright (2019), Elsevier.*

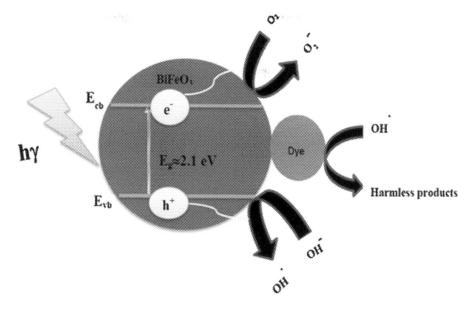

Figure 9.3

Photocatalytic dye degradation mechanism in BFO. Source: *Reprinted with permission from Ponraj, C., Vinitha, G., & Daniel, J. (2020). Visible light photocatalytic activity of Mn-doped BiFeO3 nanoparticles, International Journal of Green Energy, 17(1), 71—83. https://doi.org/10.1080/15435075.2019.1688158, copyright (2020), Elsevier.*

the conduction band, are completely associated with the bandgap of the respective catalyst. Hence, in order to synthesize an efficient photocatalyst, fine tuning of band edge potential, that is, bandgap is must. Here, in the case of BFO, varying of the bandgap depends on the synthesis parameters; from the earlier reports the optical bandgap values of BFO are in between 2.2 and 2.8 eV. There are several factors which can be fine tuned to engineer the bandgap of BFO.

The bandgap can be tuned by several pathways. One of the simplest paths is controlling the processing temperature. Literature proves that synthesis temperature plays a crucial role in the bandgap outcome of BFO. Palai et al. (2008) reported how the bandgap value decreases with the increase in the temperature, as shown in Fig. 9.4.

Another pathway is synthesis of different semiconductor combinations, that is, nanocomposites. By the addition of another semiconductor material that has suitable reduction potential the utilization of BFO will get improved for several applications. Fan, Chen, Tang, Ni, and Lu (2015) reported degradation of Rhodamine B dye via photocatalysis using the combination of graphitic carbon nitride-associated BFO. $BiVO_4$-combined BFO is prepared via hydrothermal method and, as reported by Ting et al., heterojunction is formed between 2.0 eV bandgap $BiVO_4$ and 2.3 eV bandgap BFO. The generated charge carriers were successfully utilized for degradation of Rh B dye (Fan, Chen, & Tang, 2016). TiO_2-coupled BFO, proposed by

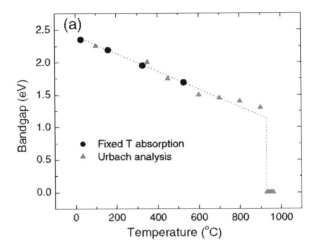

Figure 9.4

Effect of temperature on bandgap of BFO. Source: *Reprinted with permission from Palai, R., Katiyar, R. S., Schmid, H., Tissot, P., Clark, S. J., Robertson, J., ..., Scott, J. F. (2008). β Phase and γ-β metal-insulator transition in multiferroic BiFeO₃. Physical Review B—Condensed Matter and Materials Physics, 77 (1), 1—11. https://doi.org/10.1103/PhysRevB.77.014110, copyright (2008), American Physical Society.*

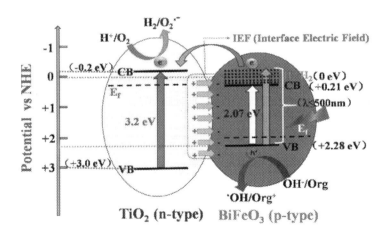

Figure 9.5

Plausible reaction mechanism for degradation and photocatalytic hydrogen production using TiO₂-BFO composite. Source: *Reprinted with permission from Humayun, M., Zada, A., Li, Z., Xie, M., Zhang, X., Qu, Y., Raziq, F., Jing, L. (2016). Enhanced visible-light activities of porous BiFeO3 by coupling with nanocrystalline TiO2 and mechanism, Applied Catalysis B: Environmental,180, 219—226. https://doi.org/10.1016/j.apcatb.2015.06.035., copyright (2016), Elsevier.*

Humayun et al. (2016), is used for both phenol degradation and H_2 production activity. Mechanism proposed in the case of TiO_2-BFO is based on the generation of internal electric field between p- and n-type semiconductor, as shown in Fig. 9.5.

Apart from them, one of the best paths for fine tuning the bandgap is doping with metal ions. While doping with metal ions that are having less or more ionic radius than the iron in BFO, the size of the entire material is predominantly affected. By this way including the band edge potentials, structural, optical, and morphological properties were also strongly influenced and can be altered as per the requirement. On this basis, the effect of nonradioactive transition metals on BFO was covered in Section 9.4.

9.5 Effect of d-block dopants on BiFeO₃

9.5.1 Effect of scandium

Scandium is the first d-block element, which has the ionic radius of 0.885 Å in its Sc^{3+} state (Shi, Liu, Hao, & Hu, 2011; Wang et al., 2015). On the basis of literature survey, it was found that replacement of Fe^{3+} ion in the lattice is observed in BFO by the addition of dopants (Dutta, Mandal, Naik, Lawes, & Tyagi, 2013; Titus, Balakumar, Sakar, Das, & Srinivasu, 2017). Dutta et al. reported scandium-doped BFO, in which slight change in bond length of Fe−O and bond angle of Fe−O−Fe is observed. This change is attributed to occupied Sc ions at Fe sites of BFO. In BFO, Fe^{3+} (0.738 Å) state has smaller ionic radius than that of Sc (0.885 Å), this change in ionic radius is responsible for the change in bond parameters (Dutta et al., 2013). Wang, Li et al. (2019) stated that when Sc dopant is entering into the Fe site of BFO, Fe−O bond lengths are elongated while Bi−O bond lengths are shortened, which is good in agreement with the statement of Dutta and coworkers. From the morphological analysis, it was found that on doping with scandium the particles of BFO are destabilized, hence couldn't find applications in photocatalysis. The morphological analysis confirms that by the addition of Sc, the nanocubes morphology is getting influenced and destabilized. In Sc-doped BFO, very few nanocubes were found; majority portion is changed to faceted morphology, which conveys that Sc doping is not worthy for enhancing the photocatalytic behavior of BFO.

9.5.2 Effect of titanium

Anjum et al. reported the bulk and nanocomposites of titanium-doped BFO. Doping effect of Ti is studied and the influence of Ti on bandgap of BFO is investigated, as shown in Fig. 9.6A-B. BFTO samples represent 10% TiO_2-doped BFO (Anjum et al., 2018). Change in absorption edge is observed in DR−UV−visible spectrum; the red shift is clearly observed and the corresponding wavelength displacements are taken place from 730 to 620 nm in bulk materials and 620 to 605 nm nanocomposites. From Kubelka−Munk plots the bandgap depletion is observed. The narrow down of bandgap energy is also attributed to the change in particles size from the morphological information of BFO and BFTO. It was found that Ti-doped BFO shows a 90-nm particle size, whereas BFTO is about 100 nm in range. Rong et al. reported that the addition of Ti at increased quantities leads to slightly distorted hexagonal structure by Ti

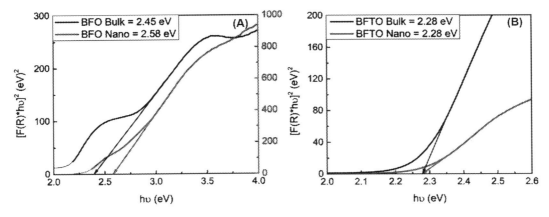

Figure 9.6

(A and B) Bandgap evolution of bulk and nano-undoped and Ti-doped BiFeO₃. Source: *Reprinted with permission from Anjum, N., Lamia, S. N. E., Yeashir Arafat, Md., Mahboob, M., & Basith, M. A. (2018). Photocatalytic properties of Ti-doped BiFeO₃ bulk and nanoparticles for solar hydrogen fuel generation. AIP Conference Proceedings, 1980. https://doi.org/10.1063/1.5044372, copyright (2018), American Institute of Physics.*

substitution. The optical properties of Ti-doped BFO were also studied and the change in bandgap was observed. The bandgaps of Ti (0.05)−BFO and Ti (0.1)−BFO are estimated to be about 1.97 and 2.05 eV, respectively. The decrease in bandgap is due to the presence of Ti at various proportions, the energy states of Ti 3d will affect the valence band and conduction band positions of the composite. This way the bandgap of the BFO is further tuned as per the required application. Ti-based semiconducting materials are prominent for the photocatalysis and here the addition of Ti also makes a considerable impact on the optical and structural properties of BFO (Rong et al., 2016).

9.5.3 Effect of chromium

Chromium doping influences the morphology of BiFeO₃. Du et al. reported the difference between pure and Cr-doped BFO structure. Though the structure did not change from Rhombohedral phase, there is a strong change in morphology. Pure BFO is in spheroidal shape, whereas by the addition of Cr in higher dosages the morphology changes to octahedral shape. The roughness and the size of the Cr-doped BFO are decreased. Change in bond length is also observed, especially increment in the Fe−O bond length is clearly observed from 2.12 to 2.22 Å on increasing the Cr content. The complete compositions of Cr doping and in detail bond length changes are shown in Fig. 9.7 (Du, Cheng, Dou, Shahbazi, & Wang, 2010). The percentage of roughness on the surface of the catalyst influences the electron trapping behavior of the catalyst. So, the addition of Cr doping in minor quantities will be applauded instead of going for higher ones (Sinha et al., 2019).

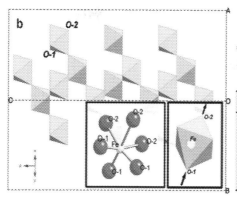

Calculated lattice parameters of $BiFe_{1-x}Cr_xO_3$ ($0 \leq x \leq 0.1$) by refinement of XRD data.

Samples	Fe–O (2) (Å)
$BiFeO_3$	2.12(7)
$BiFe_{0.975}Cr_{0.025}O_3$	2.11(9)
$BiFe_{0.95}Cr_{0.05}O_3$	2.12(8)
$BiFe_{0.925}Cr_{0.075}O_3$	2.15(3)
$BiFe_{0.9}Cr_{0.1}O_3$	2.22(4)

Figure 9.7

Fe–O bond lengths of undoped BiFeO₃ and Cr-doped BiFeO₃. Source: *Reprinted with permission from Du, Y., Cheng, Z. X., Dou, S. X., Shahbazi, M., & Wang, X. L. (2010). Enhancement of magnetization and dielectric properties of chromium-doped BiFeO₃ with tunable morphologies.* Thin Solid Films, 518 (24 Suppl.), 8–11. https://doi.org/10.1016/j.tsf.2010.03.118, copyright (2010), Elsevier.

9.5.4 Effect of manganese

Manganese is doped on BFO via different methods in a countable number of reports (Chung, Lin, Wu, & Ming, 2006; Ponraj et al., 2019). From the literature, it was found that noticeable change has taken place in bandgap of BFO on doping Mn to it (Wang et al., 2017). Recently, Yang, Ma, Xu, Deng, and Wang (2019) investigated the effect of Mn on BFO and reported that crystallite size increases with Mn doping from $x = 0$ to 0.08 in $BiFe_{1-x}Mn_xO_3$. The bandgap varies in such a way that by the addition of 8% of Mn the bandgap of BFO is depleted to 1.92 from 2.12 eV. The bandgap values of 0%, 4%, 8%, and 12% Mn-added BFO are reported as 2.12, 2.0, 1.92, 1.97 eV, respectively. These results remain as an evident for improved light responsive behavior of BFO with Mn doping. The absorbance of the composites is depicted in Fig. 9.8. The improved absorption and tuned bandgap are attributed to the reduced oxygen vacancy density, improved grain size, and marvelous stability of the perovskite structure (Liu et al., 2014). These possible changes could lead to efficient usage of BFO photocatalytic hydrogen production.

9.5.5 Effect of cobalt

The effect of cobalt as dopant is predominant towards BFO, since its addition, the particle size is highly varied. Barrionuevo et al. found the formation of impurity phase. From the XRD analysis the formation of impurity phase is confirmed with the evident peak at 28.5 degrees, which corresponds to $BiFe_{0.97}Co_{0.03}O_3$. This impurity phase causes the structural distortion and forms a rhombohedral distorted perovskite structure (Barrionuevo, Singh, & Tomar, 2011). Khajonrit et al. stated that the addition of higher cobalt helps in preventing the formation of $Bi_2Fe_4O_9$ and

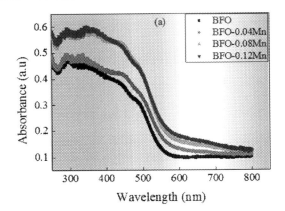

Figure 9.8

Change in absorption edge of undoped BiFeO$_3$ and Mn-doped BiFeO$_3$. Source: *Reprinted with permission from Yang, S., Ma, G., Xu, L., Deng, C., & Wang, X. (2019). Improved ferroelectric properties and band-gap tuning in BiFeO$_3$ films: Via substitution of Mn. RSC Advances, 9, 29238–29245. https:// doi.org/10.1039/c9ra05914h, copyright (2019), Royal Society of Chemistry.*

leads to the formation of spinel CoFe$_2$O$_4$. The particle sizes of the Co-doped BFO is studied using the XRD and BET also. The results conclude that both the characterizations are well in agreement with each other, and on increasing the concentration of cobalt in composites the particle size is decreased (Khajonrit, Wongpratat, Kidkhunthod, Pinitsoontorn, & Maensiri, 2018). This decrease in particle size is further studied using the XANES data. This data confirms that the involvement of Co^{2+} and Co^{3+} at iron sites of BFO is the reason for reduced particle size. The phenomenon takes place here is higher atomic radius element like Fe^{3+} (0.78 Å) is replaced by lower ionic radius element Co^{2+} (0.65 Å), Co^{3+} (0.545 Å). Rhaman et al. studied the optical properties of Co-doped BFO. From the DRS−UV result, a reduced bandgap is observed by the addition of cobalt. The bandgap of the composite is taken down to 1.6 eV from the pristine material bandgap values of 2.1 eV. This change in bandgap is attributed to varied particle size, change in bond length of Fe−O and by the addition of Co cation dopant (Rhaman et al., 2019). Contrarily, Sinha et al. reported an increase in the bandgap from 2.07 to 2.43 eV on increasing the cobalt-doping concentration from 0% to 7%. This increase in the bandgap is due to the decrease in particle size and incorporation of Co^{3+} ions in the lattice (Kumar Sinha, 2020). The decreased particle size favors the synthesis of nanocomposites with higher surface area, and tunable bandgap facilitates better light absorption. Hence, it could be possible to utilize Co-doped BFO for better photocatalytic hydrogen production (Fig. 9.9A-B).

9.5.6 Effect of nickel

Doping Ni results in the addition of a new phase in crystal structure which is said to have nickel content that leads to the increment in magnetic moment. It was reported that the alone BFO

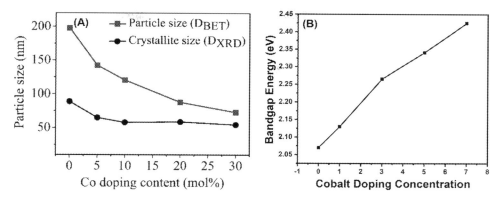

Figure 9.9

Influence of Co-doping on (A) particle size of BiFeO₃ and (B) bandgap of BFO. Source: *Reprinted with permission from (A) Khajonrit, J., Wongpratat, U., Kidkhunthod, P., Pinitsoontorn, S., Maensiri, S. (2018). Effects of Co doping on magnetic and electrochemical properties of BiFeO3 nanoparticles, Journal of Magnetism and Magnetic Materials, 449, 423–434, https://doi.org/10.1016/j.jmmm.2017.10.092., copyright (2017), Elsevier; (B) Sinha, A.K. (2021). Band gap energy calculation of Cobalt doped Bismuth Ferrite Nanoparticles, Materials Today: Proceedings, 42(2), 1519–1521, https://doi.org/10.1016/j. matpr.2021.01.912., copyright (2021), Elsevier.*

material synthesized via sol−gel route is in inhomogeneous state, whereas by loading appropriate quantity of Ni the homogeneity can be attained (Wang & Qi, 2012). Wang et al. studied one of the major obstacles of BFO, that is, leakage current density, which restricts the usage of BFO for several real time applications. As per the data obtained on calculating the current density values, 2 wt.% loading of Ni is consistent and shows lower leakage (Wang, Han, Xu, & Yang, 2019). On the basis of XRD results, it was found that Ni-dopant levels from 0 to 2 wt.% will increase the grain size, and above 2 wt.%, the grain size will decrease, which shows that the crystallization of BFO films becomes better with the increase of Ni-doped content from 0 to 2 wt.%. On looking further, it was also confirmed that on increasing the Ni concentration the intensity is lowering gradually; this is attributed to the occupation of Fe^{2+} position by Ni^{2+} in the lattice. The reason for this could be the large ion radius of Ni ion in 2 + state, that is, 0.069 nm than the Fe ion in 3 + state 0.0645 nm (Wang, Yang, Zhang, & Fan, 2018). Wang, Han et al. (2019) gave the information about grain size with different Ni-doping amounts calculated from SEM and XRD. By the addition of Ni in appropriate quantity, the homogeneity will increase and grain size will decrease; both these lead to improved efficiency in photocatalysis (Fig. 9.10A-B).

9.5.7 Effect of copper

Samran et al. studied the crystallography of the Cu-doped BFO in which the planes (1 1 0) and (−1 1 0) of Cu-BFO composite were seems to be slightly shifted. This shift is due to the change in ionic radius by the involvement of Cu in the crystal structure. Cu radius, that is, Fe^{3+} (0.64 Å)

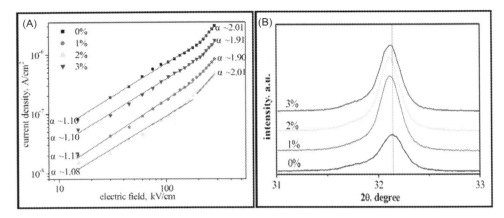

Figure 9.10

(A) Leakage current densities of Ni-doped BiFeO₃; (B) XRD peaks of undoped BiFeO₃ and Ni-doped BiFeO₃. *Source: Reprinted with permission from Wang, L., Yang, S., Zhang, F., & Fan, S. (2018). Perfection of leakage and ferroelectric properties of Ni-doped BiFeO₃ thin films.* Micro & Nano Letters, 13, 502−505. https://doi.org/10.1049/mnl.2017.0554, copyright (2018), IET Digital Library.

is substituted with Cu^{2+} (0.73 Å) ions. The major advantage of adding Cu to the BFO is the change in the surface area; there is a huge change in surface area of the Cu-doped BFO. BFO alone has a 3.83 m²/g surface area, whereas the composite with 1 mol.% Cu-doped BFO has 10.25 m²/g. Such an increment in area is predominant and promising for extensive implementation of photocatalytic applications with the higher efficiencies. Addition of Cu shows its identity in a small change in bandgap variation as well, that is, reduced bandgap of about 0.02 eV is observed from 2.14 to 2.12 eV. PL study strengthens the other characterization results and Cu-doped BFO shows much lower intensity values, the depletion is symbol of reduced recombination rate of charge carriers. From the aforementioned data, Cu addition is highly favorable for efficient utilization BFO for photocatalytic applications (Samran, Krongkitsiri, & Chaiwichian, 2018).

9.5.8 Effect of zinc

Zinc as a dopant leads to the formation of dense structures of BFO. As BFO alone condition forms like uneven surface material with wide pores, the addition of Zn leads to uniformity and increased grain size as well. Yang et al. reported that with the implementation of 2% Zn in BFO the current density values decrease that leads to reduced leakage current of BFO. Zn-added composite shows a higher crystallinity; from the XRD data shown below a small shift is observed—magnified XRD, that is, the intensity is shifted to smaller angles on increasing the concentration of Zn. This is due to the Zn ionic radius in its 2 + state (0.074 nm) which is higher than the iron ionic radius in 3 + state value (0.0645 nm) (Yang et al., 2018). Notonegoro et al. reported another procedure for depleting the current leakage of BFO on doping Zn to it, that is, increasing the bandgap of the BFO.

However, it prevents the current leaks and the efficiency of photocatalysis is reduced (Notonegoro, Soegijono, Rafsanjani, Mudzakir, & Raksa, 2019).

9.5.9 Effect of yttrium

Yttrium-doped BFO is not studied much. Narayan, Alemu, MacHeli, Thakurdesai, and Rao (2009) reported Y-doped TiO_2 in which Y^{3+} has shown predominant impact on particle size and also takes a portion in the depletion of charge carrier recombination. Haydous et al. recently studied about the influence of yttrium thickness on the strain and structural properties of the BFO. From the XRD data of 22, 31, 61, and 75 nm yttrium-doped BFO thin films, FWHM values and strain percentage were discussed. From the results, it was found that strain percentage decreases with the increase in the thickness of Y content (Haydous et al., 2018). On substituting Y in BFO structure, the BFO major peaks shift to lower angle. The reason behind this structural phase transition is reported as the difference in the radius of the doped Y^{3+} ions. Structural changes are majorly observed on choosing Y as dopant towards BFO. In most of the cases other factors such as optical properties, particle size, and bandgap do not vary much. This could be the reason for less usage of Y as dopant for BFO in the field of photocatalysis (Maleki, 2019).

9.5.10 Effect of zirconium

Preparation of oxygen-deficient BFO using Zr as a dopant is reported by Wang et al. It was reported that Fe^{3+} is replaced with Zr^{4+} ions in BFO nanoparticles. Structural deformation is not taking place, but while coming to the bandgap, there is a slight extension from 1.9 to 2.2 eV. Zr enlarges the bandgap slightly, whereas some other dopants take part in reducing it. PL studies about the different percentage of Zr-doped BFO conveys that there is a chance of restraining the recombination on doping with Zr, it happens at a particular concentration. From the reports of Wang et al., 2% Zr-doped BFO shows charge carriers with longer life time. An important factor that impacts the photocatalytic behavior is surface area. Effect of Zr as dopant increases the surface area, which is shown in Fig. 9.11A-B (Wang et al., 2017). Hence, Zr doping can enhance one of the essential parameters for photocatalysis.

9.5.11 Effect niobium

Nb-doped BFO is studied scarcely. From the reports, it was found that Nb-doped BFO shows enhanced electric properties. Major changes observed in the structure of the material was changed to pseudo-cubic structure and the grain size decreased (Azough et al., 2010; Chung et al., 2006; Yan, Xing, Islam, Li, & Lu, 2012). Reports by Yan et al. and Cheng et al. also mention the same case of depletion in grain size by the increment of Nb concentration. Yan et al. also remarked that the addition of Nb causes a shift in interplanar spacing and an increase in bond length, which is due to the strain generated by the iron

(A)

Samples	Mean particle size (nm)*	BET surface area (m²/g)
BFO	319	19.6
Zr1%-BFO	148	24.9
Zr2%-BFO	133	26.2
Zr3%-BFO	129	27.3

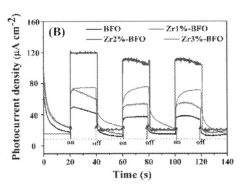

Figure 9.11
(A) influence of Zr doping on surface area of BiFeO₃. (B) Influence of Zr doping on photocurrent density of BiFeO₃. Source: *Reprinted with permission from Wang, F., Chen, D., Zhang, N., Wang, S., Qin, L., Sun, X., & Huang, Y. (2017). Oxygen vacancies induced by zirconium doping in bismuth ferrite nanoparticles for enhanced photocatalytic performance. Journal of Colloid and Interface Science, 508, 237–247. https://doi.org/10.1016/j.jcis.2017.08.056, copyright (2017), Elsevier.*

ions, which have a smaller ionic radius than Nb. Yan et al. (2012) Since it facilitates less grain size, it was not recommended as others for photocatalytic applications.

9.5.12 Effect of molybdenum

The ionic radius of Mo^{6+} (0.41 Å) is much smaller than the Fe^{3+}; hence, a change in crystal structure is reported by the addition of Mo (Ohno, Tanigawa, Fujihara, Izumi, & Matsumura, 1999). It is clearly observed that several new peaks are formed after the addition Mo. The additional peaks confirm the monoclinic phase loading of Mo having space group $C^{2/c}$. Zhang et al. reported Mo-doped BFO for water oxidation reactions. From the absorption and bandgap analysis, it was evident that by the addition of Mo (1 0 4) plane of BFO is further enhanced, as shown in Fig. 9.12A. Bandgap energy is much influenced, that is, decrease in bandgap from 2.3 to 1.8 eV is observed by the addition of Mo in increasing order from 0 to 5 wt.%, which can be seen from Fig. 9.12B (Zhang et al., 2020). Apart from the all-other dopants, it is found that MO addition significantly increases the grain size; however, it also forms dense structures, both of these changes can be observed in FE SEM images given in Fig. 9.12B (Murtaza, Salmani, Ali, & Khan, 2018). On the basis of literature, it was found that Mo doping was not preferred which gave increased grain size.

9.5.13 Effect of ruthenium

Yan et al. reported the influence of ruthenium as dopant on BFO. From this report, it was found that Ru influences the structural properties, morphology, and limiting leakage current.

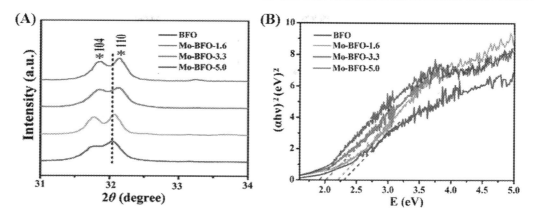

Figure 9.12
(A) XRD plots of undoped and Mo-doped $BiFeO_3$. (B) Bandgap of undoped and Mo-doped
$BiFeO_3$. *Source: Reprinted with permission from Zhang, H., Wang, L., Guo, C., Ning, J., Zhong, Y., & Hu,
Y. (2020). Enhanced photoactivity and photostability for visible-light-driven water oxidation over BiFeO₃ porous
nanotubes by modification of Mo doping and carbon nanocoating.* ChemNanoMat, 6 (9), 1325−1331.
https://doi.org/10.1002/cnma.202000140, copyright (2020), Wiley.

Structurally rhombohedral phase was changed to orthorhombic phase; a little amount of
lattice reduction is observed in Ru-doped BFO. As mentioned earlier, the replacement of Fe
atom is the major cause for the distortion in structure; here, Ru ionic radius is smaller and
hence the structural parameters were restrained. In morphology the dopant made the prime
catalyst surface smooth which is confirmed from the cross-sectional images taken using FE
SEM; an increase in size is also observed in Ru-doped BFO than the pristine material (Yan,
Zhu, Lai, & Lu, 2010). The leakage current is also limited by the Ru, which is due to the
depletion of oxygen vacancies on adding the high-valence metal ion such as Ru; hence, the
leakage current density is reduced. So, for most of the applications-based BFO, Ru doping
is favorable, but a slight limitation is there in morphological aspect which adversely
impacts the photocatalytic hydrogen generation.

9.5.14 Effect of palladium

As a noble metal, Pd is doped on different kinds of materials to enhance the properties and
applications of the source material. Lam et al. reported Pd-doped BFO composite. From the
characterization results, it was found that Pd did not affect the crystallography, but it was
involved in the extension of absorption edge to visible region (Jaffari, Lam, & Sin, 2019). In
the case of well-reported photocatalysts such as CdS, TiO_2, and other materials also, Pd plays a
crucial role in enhancing their absorption to visible region to make them efficient. Here, in the
case of BFO also the absorption edge is increased. While coming to the bandgap, the bandgap
of the Pd-doped BFO came down to 2.03 from 2.33 eV as shown in the inset of Fig. 9.13A,

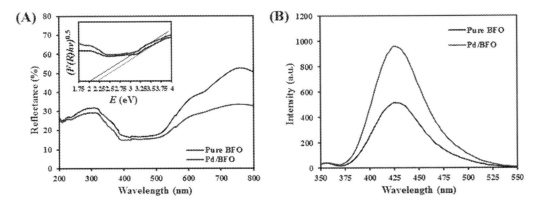

Figure 9.13

(A) Reflectance of pristine and Pd-doped BiFeO₃, influence on bandgap of BiFeO₃. (B) Charge carrier recombination in undoped and Pd-doped BiFeO₃. Source: *Reprinted with permission from Lam, S. M., Jaffari, Z. H., & Sin, J. C. (2018). Hydrothermal synthesis of coral-like palladium-doped BiFeO₃ nanocomposites with enhanced photocatalytic and magnetic properties.* Materials Letters, 224, 1—4. *https:// doi.org/10.1016/j.matlet.2018.04.058, copyright (2018), Elsevier.*

which is a considerable advantage for photocatalytic applications (Lam, Jaffari, & Sin, 2018). Another foremost salient factor in photocatalysis is recombination, Pd influence on charge carrier recombination of BFO is studied using PL data. The photoluminescence data was given in Fig. 9.13B. From this, it was clear that role of Pd as a dopant is exceptional in the case of BFO as well. However, it is not so economical but utilizing its properties on doping in minor quantities will lead to tremendous development in the entire field and efficiency will also get enhanced much.

9.5.15 Effect of silver

Ag is another noble metal in the list, which is studied in most of the combinations earlier. Ag-doped BFO predicts how the bond strength of Bi—O is getting weak after doping with Ag. Recently, Bagwaiya et al. reported the Ag-doped BFO as a gas-sensing material. From the study, it was found that the addition of Ag will replace the Bi ion in the lattice of BFO, which further leads to decrease in lattice distortion because of the differences in the ionic radius (Bagwaiya et al., 2018). This change could be understood from the XRD data. Xu et al. (2014) reported that addition of Ag leads to defective formation of energies that are negative under oxygen-rich conditions. It means that Ag can be doped directly to the BFO without any external energy consumption, and from the experimental studies, it was confirmed that addition of Ag will increase the conductivity of BFO (Tang et al., 2018). Ag doping is preferred in photocatalysis because of SPR effect. It also provides better surface-active sites.

9.5.16 Effect of cadmium

Cd-doped BFO is a rare combination, which was first reported in 2009. Bellaki and his coworker synthesized and studied its magnetic properties. Nevertheless, it is also a d-block element but Cd did not affect the structure of BFO. From the XRD peaks, it is confirmed that the structure is not distorted to any other phase even after adding in different proportions (Bellakki & Manivannan, 2010). The same context of not influencing the structure of the prime catalyst has also been recently discussed; in the case of ZnO nanorods even after doping CdS, it remains in its original phase (Shahmoradi et al., 2019). A small change is observed in bond lengths—Fe$-$O slightly enlarged from 1.949 to 1.961 Å. The presence of Cd is further confirmed from the micro-EDX. Cd is a divalent metal and its ionic radius is nearer to calcium; it increases the magnetic properties of the material. It was confirmed from the reports that the magnetic moment increases on doping Cd to the ferroelectric materials. This increment is due to the structural disorder and variation in bond angle of Fe$-$O$-$Fe which got increased from 115.106 to 156.172 degrees. Andronic, Enesca, Vladuta, and Duta (2009) reported that addition of Cd increases the active sites on catalyst surface, and from the study of Cd-doped TiO$_2$, it was found that Cd doped using cadmium acetate gave a better photocatalytic performance than cadmium chloride.

9.5.17 Effect of lanthanum

Fine nanoparticles of La-doped BFO were synthesized by Ahmed et al.; the particles observed are good in dispersion, no agglomeration is observed in entire TEM micrograph attached. Synthesis is done on introducing La^{3+} with different contents, that is, $x = 015$ and $x = 0.35$. From the results, it was found that La^{3+} with higher content gave a less particle size of about 9 nm, whereas with the lower content, the resulted particle size is slightly enlarged to 11 nm. XRD data conveys that a small depletion is present in lattice parameters on doping La to it (Ahmed, Dhahri, El-Dek, & Ayoub, 2013). This is attributed to the decrease in molecular weight and ionic radius. Meng et al. and Ting et al. stated that lanthanum doping will lessen the formation of impurity phases of BFO. A slight change attained by the BFO on adding La dopant is confirmed by DR$-$UV$-$visible spectra. Fig. 9.14A shows the change in light absorption and bandgap values as well. BFO with 15% lanthanum shows bandgap of 2.06 eV, whereas pristine BFO shows 2.10 eV. This change in bandgap is due to the La^{3+} impurity level formed below the conduction band to enhance the light absorption. From this study of La-doped BFO, Meng and his coworkers mentioned the charge recombination behavior also (Lin et al., 2014; Meng et al., 2016). The PL analysis of undoped BFO and 15% La-doped BFO is shown in Fig. 9.14B. From the results, it was found that La is effective in depleting the charge carrier recombination that favors the photocatalytic applications predominantly (Meng et al., 2016).

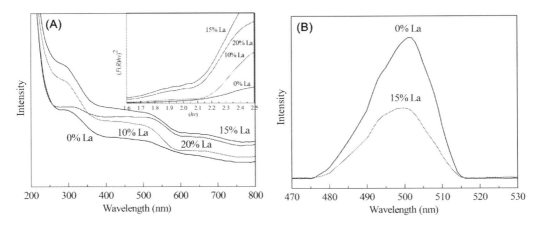

Figure 9.14
Influence of La doping on (A) bandgap and (B) charge carrier recombination of BiFeO₃. Source: *Reprinted with permission from Meng, W., Hu, R., Yang, J., Du, Y., Li, J., & Wang, H. (2016). Influence of lanthanum-doping on photocatalytic properties of BiFeO₃ for phenol degradation.* Cuihua Xuebao/Chinese Journal of Catalysis, *37 (8), 1283−1292. https://doi.org/10.1016/S1872-2067(16)62449-X, copyright (2016), Elsevier.*

9.5.18 Effect of tantalum

Jun et al. reported that the addition of Ta leads to its substitution in the lattice of BFO which causes structural distortion in oxygen position. This distortion leads to the formation of octahedral phase in the structure (Xu et al., 2009). While coming to grain size, from the literature, it was found Ta restrains the formation of large grains, that is, the grain size is predominantly decreasing in the case where particularly the Ta-doped BFO is synthesized via solid-state reaction method. The grain size reduction is observed at greater rates when compared with pristine material; the same kind of phenomena also occurred in BaTiO₃. This could be due to the formation of more resistive grain boundary (Jun et al., 2007). Islam et al. (2018) found less leakage current in BFO by adding 15% Ta to it which is confirmed by $I-V$ characteristics where the reduction in oxygen vacancy is clearly observed. So, in photovoltaic and photocatalytic applications, Ta loading on BFO is preferable, which leads to decreased grain size and higher surface area.

9.5.19 Effect of tungsten

Tungsten-doped BFO nanoparticles and nanostructures gain much changes in their structural and morphological properties. Raghavan et al. stated that morphology of the BFO is changed by introducing tungsten. From the morphological data, they found change in structure density. W-doped BFO shows denser morphology with greatly reduced pores and distribution of grains is also homogeneous. The effects may vary depending on the synthesis and other parameters (Raghavan, Kim, Choi, Kim, & Kim, 2015). Bernardo, Jardiel, Peiteado, and Caballero (2016) reported that by adding tungsten via solid-state reaction method, an intermediate multiphase structure is observed, which is due to the rich content or iron presence and stoichiometry, but as per the report of Waghmare et al., the addition of tungsten did not lead to any change in the structure and remains as distorted rhombohedral phase. An important parameter changed by the addition of W is enhanced surface area. The nanoscale phenomenon of surface-to-volume ratio has increased here. Precisely, undoped and doped BFO have the surface area of 5.0386 and 8.0814 m^2/g, respectively. This enhancement is highly beneficial for photocatalytic applications (Waghmare et al., 2018).

9.5.20 Effect of platinum

Pt is a noble metal that is loaded widely on the surface of the many photocatalysts to increase the photocatalytic active sites. Here in this case also, Pt is impregnated on BFO. The structural analysis clears that the addition of Pt did not show any adverse impact on the rhombohedral structure of BFO (Tsurumaki, Yamada, & Sawa, 2012). Optical properties of the BFO is strongly influenced by the Pt doping. The absorption edge is extended from 580 to 700 nm, this extension is attributed to the SPR effect. With the increment of Pt loading, the SPR effect was found to be increased. From UV−DR spectra shown in Fig. 9.15, the bandgap of the Pt-loaded BFO catalyst is calculated. The catalyst pristine BFO, 0.5, 1.0, and 1.5 wt.% Pt-loaded BFO, shows the bandgap values of 2.16, 2.14, 2.12, and 1.98 eV. This change in bandgap also supports the SPR effect of Pt-loaded BFO. The depletion of bandgap is also due to the lower energy levels produced by Pt as a result of heterojunction formation and Schottky barrier at the interface of Pt and BFO (Niu et al., 2015).

9.5.21 Effect of gold

Au is a noble metal that can exhibit the SPR effect (Shuai, Zhou, Brger, Helm, & Schmidt, 2011). Here the doping Au leads to considerable changes in BFO (Fig. 9.16A-B). The crystallographic phases of BFO were not changed, that is, Au did not lead to any phase distortion. From the SEM and TEM analysis reported by Li et al., it was found that the density of the nanoflakes is influenced. By the addition of Au, the number of BFO nanoflakes formed increased with the increase of Au doping. The size of the grains and other morphological properties were not influenced much, while the optical properties were influenced and DRS−UV plots give the picture of how the Au loading is impacted on BFO. In general, the BFO is active in UV and visible regions at around 580 nm. A greater enhancement in absorption edge which is about 725 nm is observed by the addition of Au at different proportions which is attributed the SPR effect. The bandgap energies are

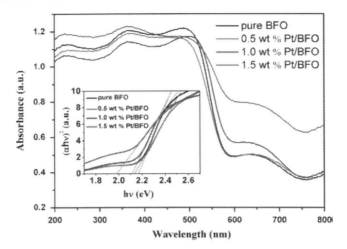

Figure 9.15

Influence of Pt doping on bandgap of BiFeO₃. Source: *Reprinted with permission from Niu, F., Chen, D., Qin, L., Gao, T., Zhang, N., Wang, S., . . . Huang, Y. (2015). Synthesis of Pt/BiFeO<inf>3</Inf> heterostructured photocatalysts for highly efficient visible-light photocatalytic performances.* Solar Energy Materials and Solar Cells, *143, 386–396. https://doi.org/10.1016/j.solmat.2015.07.008, copyright (2015), Elsevier.*

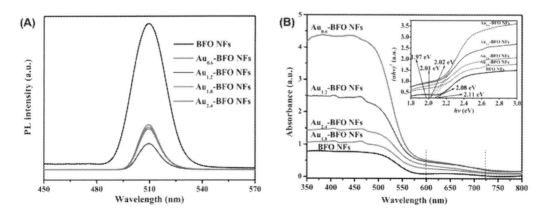

Figure 9.16

Influence of Au doping on (A) charge carrier recombination, (B) bandgap of BiFeO₃. Source: *Reprinted with permission from Li, Y., Li, J., Chen, L., Sun, H., Zhang, H., Guo, H. and Feng, L. (2019) In situ Synthesis of Au-Induced Hierarchical Nanofibers/Nanoflakes Structured BiFeO3 Homojunction Photocatalyst with Enhanced Photocatalytic Activity.* Front. Chem. 6, 649. *https://doi.org/10.3389/fchem.2018.00649, copyright (2019), Frontiers in Chemistry.*

Table 9.1: Effect of individual dopants on bandgap of BFO.

Name of the dopant	Bandgap (eV)		Reference
	Pristine BFO	Doped BFO	
Ti	2.58	2.28	Rong et al. (2016)
Mn	2.12	1.97	Yang et al. (2019)
Co	2.1	1.6	Rhaman et al. (2019)
Cu	2.14	2.12	Samran et al. (2018)
Zr	1.9	2.2	Wang et al. (2017)
Pd	2.33	2.03	Lam et al. (2018)
La	2.10	2.06	Meng et al. (2016)
Pt	2.16	1.98	Niu et al. (2015)
Au	2.11	1.97	Y. Li et al. (2019)

calculated to be 2.11, 2.08, 2.02, 2.01, and 1.97 eV for $x = 0$, 0.6, 1.2, 1.8, 2.4 wt.%, (Fig. 9.16B) respectively (Li et al., 2019). Slightly restrained bandgap values are observed by the incremental addition of Au (Li et al., 2017). This varied bandgap achievement that favors the tuning of band edge positions of BFO is attributed to the Schottky junction formation at the interfacial position of BFO and Au. The efficacy of a photocatalyst is largely determined by the rate of recombination. Accordingly, the charge carrier shuttling and reduced recombination were validated by the PL spectra results of Au-doped BFO, as shown in Fig. 9.16A, where 1.2 wt.% Au-doped BFO shows very less recombination that favors for the photocatalytic applications extensively.

Herein, the influence of most of the d-block elements is covered, but still some elements such as V, Hf, Re, Os, and Ir were not studied, since the reported literature is very limited and not relevant to the individual dopants criteria. Table 9.1 gives information about bandgap enhancement and reduction in BFO due to dopants.

9.6 Summary

The effect of nonradioactive d-block dopants on optical, structural, and morphological properties of BFO perovskite was discussed. Parameters that influence the photocatalytic activity from the aspect of catalyst were covered. Most of the transition elements lead to distortion in structural phases of BFO and tend to change grain size by replacing the Fe position in BFO. Improved surface properties by the addition of dopants and change in the porosity and roughness due to the dopant's presence are studied. SPR effect of noble metals and bigger and smaller ionic radius effect of dopants and their corresponding consequences were comprehended on the basis of literature. Optical properties and tunable bandgap values were discussed, and the rate of charge carrier recombination is also studied. The entire chapter could be helpful in further studies in understanding dopant effects on BiFeO₃ and in designing of successful BiFeO₃-based photocatalyst.

Acknowledgment

The financial support received from Naval Research Board (NRB), DRDO, New Delhi, India (Project. No. NRB-391/MAT/17-18) is gratefully acknowledged.

References

Ahmed, M. A., Dhahri, E., El-Dek, S. I., & Ayoub, M. S. (2013). Size confinement and magnetization improvement by La^{3+} doping in $BiFeO_3$ quantum dots. *Solid State Sciences, 20*, 23−28. Available from https://doi.org/10.1016/j.solidstatesciences.2013.02.023.

Alammar, T., Hamm, I., Wark, M., & Mudring, A. V. (2015). Low-temperature route to metal titanate perovskite nanoparticles for photocatalytic applications. *Applied Catalysis B: Environmental, 178*, 20−28. Available from https://doi.org/10.1016/j.apcatb.2014.11.010.

Andronic, L., Enesca, A., Vladuta, C., & Duta, A. (2009). Photocatalytic activity of cadmium doped TiO_2 films for photocatalytic degradation of dyes. *Chemical Engineering Journal, 152*(1), 64−71. Available from https://doi.org/10.1016/j.cej.2009.03.031.

Anjum, N., Lamia, S. N. E., Arafat, M. Y., Mahboob, M., & Basith, M. A. (2018). Photocatalytic properties of Ti-doped $BiFeO_3$ bulk and nanoparticles for solar hydrogen fuel generation. *AIP Conference Proceedings, 1980*. Available from https://doi.org/10.1063/1.5044372.

Azough, F., Freer, R., Thrall, M., Cernik, R., Tuna, F., & Collison, D. (2010). Microstructure and properties of Co-, Ni-, Zn-, Nb- and W-modified multiferroic $BiFeO_3$ ceramics. *Journal of the European Ceramic Society, 30*(3), 727−736. Available from https://doi.org/10.1016/j.jeurceramsoc.2009.09.016.

Bagwaiya, T., Khade, P., Reshi, H. A., Bhattacharya, S., Shelke, V., Kaur, M., . . . Gadkari, S. C. (2018). Investigation on gas sensing properties of Ag doped $BiFeO_3$. *AIP Conference Proceedings, 1942*, 1−5. Available from https://doi.org/10.1063/1.5028910.

Barrionuevo, D., Singh, S. P., & Tomar, M. S. (2011). Multiferroic properties of Co doped $BiFeO_3$ and some composite films. *Integrated Ferroelectrics, 124*(1), 41−47. Available from https://doi.org/10.1080/10584587.2011.573716.

Bellakki, M. B., & Manivannan, V. (2010). Synthesis, and study of magnetic properties, of Bi1-x Cd XFeO₃. *Journal of Materials Science, 45*(4), 1137−1142. Available from https://doi.org/10.1007/s10853-009-4065-8.

Bera, S., Ghosh, S., Shyamal, S., Bhattacharya, C., & Basu, R. N. (2019). Photocatalytic hydrogen generation using gold decorated $BiFeO_3$ heterostructures as an efficient catalyst under visible light irradiation. *Solar Energy Materials and Solar Cells, 194*(December 2018), 195−206. Available from https://doi.org/10.1016/j.solmat.2019.01.042.

Bernardo, M. S., Jardiel, T., Peiteado, M., & Caballero, A. C. (2016). Metastable nature of donor-doped $BiFeO_3$ obtained by mechanochemical synthesis. *Journal of the Ceramic Society of Japan, 124*(1), 92−97. Available from https://doi.org/10.2109/jcersj2.15191.

Chung, F., Lin, C., Wu, J. P., & Ming, E. J. (2006). Influence of Mn and Nb dopants on electric properties of chemical-solution-deposited $BiFeO_3$ films. *Applied Physics Letters, 88*(24), 3−6. Available from https://doi.org/10.1063/1.2214138.

Du, Y., Cheng, Z. X., Dou, S. X., Shahbazi, M., & Wang, X. L. (2010). Enhancement of magnetization and dielectric properties of chromium-doped $BiFeO_3$ with tunable morphologies. *Thin Solid Films, 518*(24 Suppl.), 8−11. Available from https://doi.org/10.1016/j.tsf.2010.03.118.

Dutta, D. P., Mandal, B. P., Naik, R., Lawes, G., & Tyagi, A. K. (2013). Magnetic, ferroelectric, and magnetocapacitive properties of sonochemically synthesized Sc-doped $BiFeO_3$ nanoparticles. *The Journal of Physical Chemistry C, 117*, 2382−2389.

Fan, T., Chen, C., & Tang, Z. (2016). Hydrothermal synthesis of novel $BiFeO_3/BiVO_4$ heterojunctions with enhanced photocatalytic activities under visible light irradiation. *RSC Advances, 6*(12), 9994−10000. Available from https://doi.org/10.1039/c5ra26500b.

Fan, T., Chen, C., Tang, Z., Ni, Y., & Lu, C. (2015). Synthesis and characterization of G-C$_3$N$_4$/BiFeO$_3$ composites with an enhanced visible light photocatalytic activity. *Materials Science in Semiconductor Processing*, *40*, 439−445. Available from https://doi.org/10.1016/j.mssp.2015.06.054.

Feng, S. H., & Li, G. H. (2017). *Hydrothermal and solvothermal syntheses. Modern inorganic synthetic chemistry* (2nd ed.). Elsevier. Available from https://doi.org/10.1016/B978-0-444-63591-4.00004-5.

Haron, W., Wisitsoraat, A., & Wongnawa, S. (2014). Comparison of nanocrystalline LaMO$_3$ (M = Co, Al) perovskite oxide prepared by Co-precipitation method. *International Journal of Chemical Engineering and Applications*, *5*(2), 123−126. Available from https://doi.org/10.7763/ijcea.2014.v5.364.

Haydous, F., Scarisoreanu, N. D., Birjega, R., Ion, V., Lippert, T., & Dum, N. (2018). Rolling dopant and strain in Y-doped BiFeO$_3$ epitaxial thin films for photoelectrochemical water splitting. *Scientific Reports*, *8*, 15826. Available from https://doi.org/10.1038/s41598-018-34010-9.

Humayun, M., Zada, A., Li, Z., Xie, M., Zhang, X., Qu, Y., . . . Jing, L. (2016). Enhanced visible-light activities of porous BiFeO$_3$ by coupling with nanocrystalline TiO$_2$ and mechanism.". *Applied Catalysis B: Environmental*, *180*, 219−226. Available from https://doi.org/10.1016/j.apcatb.2015.06.035.

Islam, M. R., Islam, M. S., Zubair, M. A., Usama, H. M., Azam, M. S., & Sharif, A. (2018). Evidence of superparamagnetism and improved electrical properties in Ba and Ta Co-doped BiFeO$_3$ ceramics.". *Journal of Alloys and Compounds*, *735*, 2584−2596. Available from https://doi.org/10.1016/j.jallcom.2017.11.323.

Jaffari, Z. H., Lam, S. M., & Sin, J. C. (2019). Photocatalytic degradation of organic pollutants using magnetic Pd-doped BiFeO$_3$ composites under visible light irradiation. *AIP Conference Proceedings*, *2157*. (September). Available from https://doi.org/10.1063/1.5126576.

Jun, Y. K., Lee, S. B., Kim, M., Hong, S. H., Kim, J. W., & Kim, K. H. (2007). Dielectric and magnetic properties in Ta-substituted BiFeO$_3$ ceramics. *Journal of Materials Research*, *22*(12), 3397−3403. Available from https://doi.org/10.1557/jmr.2007.0421.

Khajonrit, J., Wongpratat, U., Kidkhunthod, P., Pinitsoontorn, S., & Maensiri, S. (2018). Effects of Co doping on magnetic and electrochemical properties of BiFeO$_3$ nanoparticles. *Journal of Magnetism and Magnetic Materials*, *449*, 423−434. Available from https://doi.org/10.1016/j.jmmm.2017.10.092.

Kumar Sinha, A. (2020). Band gap energy calculation of cobalt doped bismuth ferrite nanoparticles. *Materials Today: Proceedings*, *42*, 1519−1521. Available from https://doi.org/10.1016/j.matpr.2021.01.912.

Lam, S. M., Jaffari, Z. H., & Sin, J. C. (2018). Hydrothermal synthesis of coral-like palladium-doped BiFeO$_3$ nanocomposites with enhanced photocatalytic and magnetic properties. *Materials Letters*, *224*, 1−4. Available from https://doi.org/10.1016/j.matlet.2018.04.058.

Li, F. Z., Zheng, H. W., Zhu, M. S., Zhang, X. A., Yuan, G. L., Xie, Z. S., . . . Zhang, W. F. (2017). Photovoltaic enhancement by Au surface-plasmon effect for La doped BiFeO$_3$ films. *Journal of Materials Chemistry C*, *5*(40), 10615−10623. Available from https://doi.org/10.1039/c7tc03371k.

Li, Y. 'an, Li, J., Chen, L., Sun, H., Zhang, H., Guo, H., & Feng, L. (2019). In situ synthesis of Au-induced hierarchical nanofibers/nanoflakes structured BiFeO$_3$ homojunction photocatalyst with enhanced photocatalytic activity. *Frontiers in Chemistry*, *7*. (JAN). Available from https://doi.org/10.3389/fchem.2018.00649.

Lin, P. T., Li, X., Zhang, L., Yin, J. H., Cheng, X. W., Wang, Z. H., . . . Wu, G. H. (2014). La-doped BiFeO$_3$: Synthesis and multiferroic property study. *Chinese Physics B*, *23*(4), 0−6. Available from https://doi.org/10.1088/1674-1056/23/4/047701.

Liu, W., Tan, G., Dong, G., Yan, X., Ye, W., Ren, H., & Xia, A. (2014). Structure transition and multiferroic properties of Mn-doped BiFeO$_3$ thin films. *Journal of Materials Science: Materials in Electronics*, *25*(2), 723−729. Available from https://doi.org/10.1007/s10854-013-1636-x.

Maleki, H. (2019). Characterization and photocatalytic activity of Y-doped BiFeO$_3$ ceramics prepared by solid-state reaction method. *Advanced Powder Technology*, *30*(11), 2832−2840. Available from https://doi.org/10.1016/j.apt.2019.08.031.

Meng, W., Hu, R., Yang, J., Du, Y., Li, J., & Wang, H. (2016). Influence of lanthanum-doping on photocatalytic properties of BiFeO$_3$ for phenol degradation. *Cuihua Xuebao/Chinese Journal of Catalysis*, *37*(8), 1283−1292. Available from https://doi.org/10.1016/S1872-2067(16)62449-X.

Murtaza, T., Salmani, I. A., Ali, J., & Khan, M. S. (2018). Effect of Mo doping at the B site on structural and electrical properties of multiferroic $BiFeO_3$. *Journal of Superconductivity and Novel Magnetism, 31*(6), 1955−1959. Available from https://doi.org/10.1007/s10948-017-4443-4.

Narayan, H., Alemu, H., MacHeli, L., Thakurdesai, M., & Rao, T. K. G. (2009). Synthesis and characterization of Y^{3+}-doped TiO_2 nanocomposites for photocatalytic applications. *Nanotechnology, 20*(25). Available from https://doi.org/10.1088/0957-4484/20/25/255601.

Niu, F., Chen, D., Qin, L., Gao, T., Zhang, N., Wang, S., . . . Huang, Y. (2015). Synthesis of $Pt/BiFeO_3$ heterostructured photocatalysts for highly efficient visible-light photocatalytic performances. *Solar Energy Materials and Solar Cells, 143*, 386−396. Available from https://doi.org/10.1016/j.solmat.2015.07.008.

Notonegoro, H. A., Soegijono, B., Rafsanjani, R. A., Mudzakir, I., & Raksa, T. Y. (2019). Structural and optical studies of zinc doped $BiFeO_3$ on the Bi-site. *IOP Conference Series: Materials Science and Engineering, 599*(1). Available from https://doi.org/10.1088/1757-899X/599/1/012018.

Ohno, T., Tanigawa, F., Fujihara, K., Izumi, S., & Matsumura, M. (1999). Photocatalytic oxidation of water by visible light using ruthenium-doped titanium dioxide powder. *Journal of Photochemistry and Photobiology A: Chemistry, 127*(1−3), 107−110. Available from https://doi.org/10.1016/S1010-6030 (99)00128-8.

Palai, R., Katiyar, R. S., Schmid, H., Tissot, P., Clark, S. J., Robertson, J., . . . Scott, J. F. (2008). β phase and γ-β metal-insulator transition in multiferroic $BiFeO_3$. *Physical Review B - Condensed Matter and Materials Physics, 77*(1), 1−11. Available from https://doi.org/10.1103/PhysRevB.77.014110.

Ponraj, C., Vinitha, G. D., & Joseph. (2019). Visible light photocatalytic activity of Mn-doped $BiFeO_3$ nanoparticles. *International Journal of Green Energy, 00*(00), 1−13. Available from https://doi.org/ 10.1080/15435075.2019.1688158.

Raghavan, C. M., Kim, J. W., Choi, J. Y., Kim, J. W., & Kim, S. S. (2015). Effects of donor W^{6+}-ion doping on the microstructural and multiferroic properties of *Aurivillius* $Bi_7Fe_3Ti_3O_{21}$ thin film. *Applied Surface Science, 346*, 201−206. Available from https://doi.org/10.1016/j.apsusc.2015.04.020.

Rao, B. G., Mukherjee, D., & Reddy, B. M. (2017). *Nanostructures for novel therapy novel approaches for preparation of nanoparticles. Nanostructures for novel therapy.* Elsevier Inc, https://doi.org/10.1016/B978-0-323-46142-9/00001-3.

Rhaman, M. M., Matin, M. A., Hossain, M. N., Mozahid, F. A., Hakim, M. A., & Islam, M. F. (2019). Bandgap engineering of cobalt-doped bismuth ferrite nanoparticles for photovoltaic applications.". *Bulletin of Materials Science, 42*(4), 1−6. Available from https://doi.org/10.1007/s12034-019-1871-8.

Rong, N., Chu, M., Tang, Y., Zhang, C., Cui, X., He, H., . . . Xiao, P. (2016). Improved photoelectrocatalytic properties of Ti-doped $BiFeO_3$ films for water oxidation. *Journal of Materials Science, 51*(12), 5712−5723. Available from https://doi.org/10.1007/s10853-016-9873-z.

Samran, B., Krongkitsiri, P., & Chaiwichian, S. (2018). Effect of copper dopants on visible-light-driven photocatalytic activity of $BiFeO_3$ photocatalysts. *Modern Environmental Science and Engineering, 4*(3), 234−243. Available from https://doi.org/10.15341/mese(2333-2581)/03.04.2018/006.

Shahmoradi, B., Farahani, F., Kohzadi, S., Maleki, A., Pordel, M., Zandsalimi, Y., Gong, Y., et al. (2019). Application of cadmium-doped ZnO for the solar photocatalytic degradation of phenol. *Water Science and Technology, 79*(2), 375−385. Available from https://doi.org/10.2166/wst.2019.061.

Shi, C., Liu, X., Hao, Y., & Hu, Z. (2011). Structural, magnetic and dielectric properties of Sc modified $(1 − y)$ $BiFeO_3$-$YBaTiO_3$ ceramics. *Solid State Sciences, 13*(10), 1885−1888. Available from https://doi.org/ 10.1016/j.solidstatesciences.2011.08.001.

Shuai, Y., Zhou, S., Brger, D., Helm, M., & Schmidt, H. (2011). Nonvolatile bipolar resistive switching in Au/ $BiFeO_3$/Pt. *Journal of Applied Physics, 109*(12), 1−5. Available from https://doi.org/10.1063/1.3601113.

Si, M., Saha, A. K., Gao, S., Qiu, G., Qin, J., Duan, Y., Jian, J., et al. (2019). A ferroelectric semiconductor field-effect transistor. *Nature Electronics, 2*(12), 580−586. Available from https://doi.org/10.1038/s41928-019-0338-7.

Sinha, A. K., Bhushan, B., Jagannath., Sharma, R. K., Sen, S., Mandal, B. P., Meena, S. S., et al. (2019). Enhanced dielectric, magnetic and optical properties of Cr-doped $BiFeO_3$ multiferroic nanoparticles synthesized by Sol-gel route. *Results in Physics*, *13*(April), 102299. Available from https://doi.org/10.1016/j.rinp.2019.102299.

Tang, J., Wang, R., Liu, M., Zhang, Z., Song, Y., Xue, S., . . . Dionysiou, D. D. (2018). Construction of novel Z-scheme Ag/FeTiO$_3$ /Ag/BiFeO$_3$ photocatalyst with enhanced visible-light-driven photocatalytic performance for degradation of norfloxacin. *Chemical Engineering Journal*, *351*, 1056−1066. Available from https://doi.org/10.1016/j.cej.2018.06.171.

Titus, S., Balakumar, S., Sakar, M., Das, J., & Srinivasu, V. V. (2017). Electron spin resonance studies of Bi1-XScxFeO$_3$ nanoparticulates: Observation of an enhanced spin canting over a large temperature range. *Solid State Communications*, *268*, 61−63. Available from https://doi.org/10.1016/j.ssc.2017.08.006.

Tsurumaki, A., Yamada, H., & Sawa, A. (2012). Impact of Bi deficiencies on ferroelectric resistive switching characteristics observed at P-type schottky-like Pt/Bi 1-ΔFeO$_3$ interfaces. *Advanced Functional Materials*, *22*(5), 1040−1047. Available from https://doi.org/10.1002/adfm.201102883.

Waghmare, S. D., Jadhav, V. V., Shaikh, S. F., Mane, R. S., Rhee, J. H., & O'Dwyer, C. (2018). Sprayed tungsten-doped and undoped bismuth ferrite nanostructured films for reducing and oxidizing gas sensor applications. *Sensors and Actuators, A: Physical*, *271*, 37−43. Available from https://doi.org/10.1016/j.sna.2018.01.008.

Wang, C. A., Pang, H. Z., Zhang, A. H., Lu, X. B., Gao, X. S., Zeng, M., & Liu, J. M. (2015). Enhanced ferroelectric polarization and magnetization in BiFe1-XScxO$_3$ ceramics. *Materials Research Bulletin*, *70*, 595−599. Available from https://doi.org/10.1016/j.materresbull.2015.05.027.

Wang, F., Chen, D., Zhang, N., Wang, S., Qin, L., Sun, X., & Huang, Y. (2017). Oxygen vacancies induced by zirconium doping in bismuth ferrite nanoparticles for enhanced photocatalytic performance. *Journal of Colloid and Interface Science*, *508*, 237−247. Available from https://doi.org/10.1016/j.jcis.2017.08.056.

Wang, H. U. A., Han, D., Xu, J., & Yang, L. (2019). Influence of Ni doping on the structural, ferroelectric, magnetic and optical properties of Bi$_{0.85}$Nd$_{0.15}$Fe$_{1-x}$Ni$_x$O$_3$ thin films. *Bulletin of Materials Science*, *43*, 163.

Wang, L., Yang, S., Zhang, F., & Fan, S. (2018). Perfection of leakage and ferroelectric properties of Ni-doped BiFeO$_3$ thin films. *Micro & Nano Letters*, *13*, 502−525. Available from https://doi.org/10.1049/mnl.2017.0554.

Wang, N., Li, Y., Wang, F. L., Zhou, S. D., Zhu, L., Wang, Y. G., & Chen, F. G. (2019). Structure, magnetic and ferroelectric properties of Sm and Sc doped BiFeO$_3$ polycrystalline ceramics. *Journal of Alloys and Compounds*, *789*, 894−903. Available from https://doi.org/10.1016/j.jallcom.2019.03.132.

Wang, Y.-H., & Qi, X. (2012). The effects of nickel substitution on bismuth ferrite. *Procedia Engineering*, *36*, 455−461. Available from https://doi.org/10.1016/j.proeng.2012.03.066.

Wrzesinska, A., Khort, A., Bobowska, I., Busiakiewicz, A., & Wypych-Puszkarz, A. (2019). Influence of the La^{3+}, Eu^{3+}, and Er^{3+} doping on structural, optical, and electrical properties of BiFeO$_3$ nanoparticles synthesized by microwave-assisted solution combustion method. *Journal of Nanomaterials*, *2019*. Available from https://doi.org/10.1155/2019/5394325.

Xu, H., Zeiger, B. W., & Suslick, K. S. (2013). Sonochemical synthesis of nanomaterials. *Chemical Society Reviews*, *42*(7), 2555−2567. Available from https://doi.org/10.1039/c2cs35282f.

Xu, J., Huan, H., Ke, D. C., Jia, W., Wang., & Zhou, Y. (2009). Microwave-dielectric and magnetic properties of Ta-doped BiFeO$_3$ nanopowders. *Philosophical Magazine Letters*, *89*(11), 701−710. Available from https://doi.org/10.1080/09500830903267066.

Xu, Q., Sobhan, M., Anariba, F., Ho, J. W. C., Chen, Z., & Wu, P. (2014). Transition metal-doped BiFeO$_3$ nanofibers: Forecasting the conductivity limit. *Physical Chemistry Chemical Physics*, *16*(42), 23089−23095. Available from https://doi.org/10.1039/c4cp03045a.

Yan, F., Xing, G. Z., Islam, M., Li, S., & Lu, L. (2012). Orientation-dependent surface potential behavior in Nb-doped BiFeO$_3$. *Applied Physics Letters*, *100*(17), 2010−2015. Available from https://doi.org/10.1063/1.4705405.

Yan, F., Zhu, T. J., Lai, M. O., & Lu, L. (2010). Enhanced multiferroic properties and domain structure of Ru-doped BiFeO$_3$ thin films. *Scripta Materialia*, *63*(7), 780−783. Available from https://doi.org/10.1016/j.scriptamat.2010.06.013.

Yang, S., Ma, G., Xu, L., Deng, C., & Wang, X. (2019). Improved ferroelectric properties and band-gap tuning in BiFeO₃ films: Via substitution of Mn.". *RSC Advances*, *9*(50), 29238–29245. Available from https://doi.org/10.1039/c9ra05914h.

Yang, S., Zhang, F., Xie, X., Sun, H., Zhang, L., & Fan, S. (2018). Enhanced leakage and ferroelectric properties of Zn-doped BiFeO₃ thin films grown by sol-gel method. *Journal of Alloys and Compounds*, *734*, 243–249. Available from https://doi.org/10.1016/j.jallcom.2017.11.014.

Yao, Y., Wenchao Liu, Y. C., Chiwah Leung, C. M., & Ploss, B. (2011). Studies of rare-Earth-doped BiFeO₃ ceramics.". *International Journal of Applied Ceramic Technology*, *8*(5), 1246–1253. Available from https://doi.org/10.1111/j.1744-7402.2010.02577.x.

Yeh, C.-L. (2016). Combustion synthesis: Principles and applications. *Reference Module in Materials Science and Materials Engineering*, 1–8, no. February 2015. Available from https://doi.org/10.1016/b978-0-12-803581-8.03743-7.

Yogamalar, N. R., Kalpana, S., Senthil, V., & Chithambararaj, A. (2018). *Ferroelectrics for photocatalysis. Multifunctional photocatalytic materials for energy.* Elsevier Ltd. Available from https://doi.org/10.1016/B978-0-08-101977-1.00014-4.

You, H., Wu, Z., Zhang, L., Ying, Y., Liu, Y., Fei, L., Chen, X., et al. (2019). Harvesting the vibration energy of BiFeO₃ nanosheets for hydrogen evolution. *Angewandte Chemie*, *131*(34), 11905–11910. Available from https://doi.org/10.1002/ange.201906181.

Zhang, H., Wang, L., Guo, C., Ning, J., Zhong, Y., & Hu, Y. (2020). Enhanced photoactivity and photostability for visible-light-driven water oxidation over BiFeO₃ porous nanotubes by modification of Mo doping and carbon nanocoating. *ChemNanoMat*, *6*(9), 1325–1331. Available from https://doi.org/10.1002/cnma.202000140.

Zlotnik, S., Tobaldi, D. M., Seabra, P., Labrincha, J. A., & Vilarinho, P. M. (2016). Alkali niobate and tantalate perovskites as alternative photocatalysts. *Chemphyschem: A European Journal of Chemical Physics and Physical Chemistry*, *17*(21), 3570–3575. Available from https://doi.org/10.1002/cphc.201600476.

CHAPTER 10

Metal organic framework-based photocatalysts for hydrogen production

Lakshmana Reddy Nagappagari[1], Kiyoung Lee[1] and Shankar Muthukonda Venkatakrishnan[2]

[1]School of Nano & Materials Science and Engineering, Kyungpook National University, Sangju, South Korea, [2]Nano Catalysis and Solar Fuels Research Laboratory, Department of Materials Science & Nanotechnology, Yogi Vemana University, Kadapa, India

10.1 Introduction of metal-organic frameworks

In the recent years, the organic photochemistry has become a mature science and has been achieving a remarkable properties or chemical reactions upon light absorption (Noh & Jung, 2016). The discovery of porous materials in 1990s namely porous coordination polymers (PCPs) or metal-organic frameworks (MOFs), which resulted from the coordination of organic and inorganic materials, has grabbed significant interest among the researchers. MOFs consist of extremely large surface areas ($1000-10,000$ m^2/g) due to ultrahigh porosity and the size of their pores ranges from the micro- to mesoporous regime (Stock & Biswas, 2012; Zhu & Xu, 2014). As depicted in Fig. 10.1, the structures of MOFs are composed of organic ligands (or linkers) and the metal clusters can serve as connecters (Cao, 2016). Due to a high degree of variability of the inorganic and organic components, there is a large library of esthetically pleasing structures that have vast applications, such as clean energy storage (methane and hydrogen), photocatalysis (Zhang & Lin, 2014), CO_2 capture, organic transformations (Chughtai, Ahmad, Younus, Laypkov, & Verpoort, 2015), and various separation processes (Li, Kuppler, & Zhou, 2009).

10.1.1 Historical developments

MOFs represent a special group of compounds that arise through the linking of metal ions by coordinate bonds to either organic or inorganic ligands. These polymers comprise various structures like one-, two-, or three-dimensional networks. Some of the examples of—with networks of Cu, Zn, Ag, and Cd—the late transition metals are represented in Fig. 10.2, and these coordination polymer structures could be determined only with the advantage of single-crystal X-ray diffraction. MOFs are not new since it was coined in

Photocatalytic Systems by Design.
DOI: https://doi.org/10.1016/B978-0-12-820532-7.00007-2

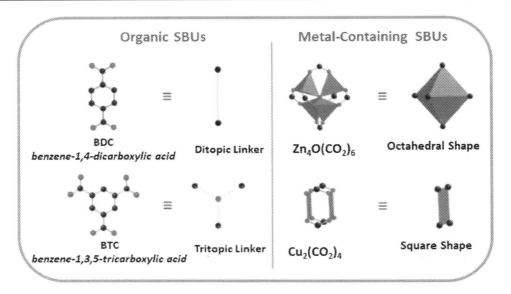

Figure 10.1
The structure of metal-organic frameworks, composed of organic ligands connected by metal clusters. Source: *Reprinted with permission from Cao, W. (2016). Semiconductor photocatalysis: Materials, mechanisms and applications. IntechOpen. Copyright IntechOpen 2016.*

early 1916, but it was not possible to verify the nature of the MOFs without XRD. At present, the term MOF is used to denote porous coordination polymers, and also named as coordination polymer (Keggin & Miles, 1936). The name MOF is basically applied to solid compounds that encompass tightly bonding, linking units of a highly crystalline and well-defined geometric composition, and that can be modified through posttreatment. Moreover, there are several key factors that can be used to explain the chemistry of these compounds.

10.1.2 Properties of metal organic frameworks

MOFs are intriguing class of crystalline porous materials composed of rigid bi- or multipodal organic linkers and metal ions or clusters have well-defined pore structures. Their expose unique chemical properties, interestingly the structures and surface properties of MOFs, are tunable (Dhakshinamoorthy, Alvaro, & Garcia, 2012; Dhakshinamoorthy, Asiri, & García, 2016). If the metal-containing units are linked, yield architecturally robust crystalline MOF structures with a high porosity of greater than 50% of the MOF crystal volume and the surface area of such MOFs typically range from 1000 to 10,000 m^2/g, which is much higher than those of traditional porous materials, such as zeolites and mesoporous carbons. MOFs have the largest pore aperture (98 Å) and lowest density (0.13 g/cm^3) (Zhang & Lin, 2014; Zhu & Xu, 2014). The chemical and thermal stability of

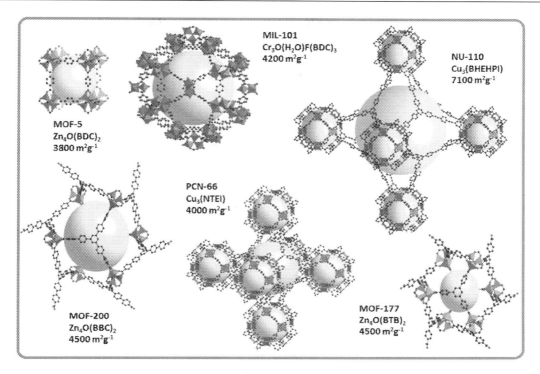

Figure 10.2

Crystal structures of selected metal-organic frameworks (MOFs) with high BET surface area and chemical formula for each MOF. The chemical formula and BET surface areas are provided for each MOF. Source: *Reprinted with permission from Cao, W. (2016). Semiconductor photocatalysis: Materials, mechanisms and applications. IntechOpen, Copyright IntechOpen 2016.*

MOFs has made them amenable to postsynthetic covalent organic and metal-complex functionalization. MOF structures are agreeable to the expansion and incorporation of multiple functional groups within their interiors. Careful selection of MOF constituents can yield crystals of ultrahigh porosity and high thermal and chemical stability. The highly porous nature of MOFs has made them to be successfully used in wide range of applications, including gas storage, separations, and heterogenous catalysis (Dhakshinamoorthy, Opanasenko, Čejka, & Garcia, 2013; Dias & Petit, 2015; Li et al., 2016). Moreover, the proof of permanent porosity of MOFs was obtained by the measurement of nitrogen and carbon dioxide isotherms on layered zinc terephthalate MOF. The MOF-5, 2200 m^2/cm^3, is the very best reported among them (for comparison, the value for NU-110 is 1600 m^2/cm^3). The topology of these isoreticular MOFs is naturally represented with a three-letter code, pcu, called as primitive cubic net. One of the smallest isoreticular structures of MOF-5 is Zn4O(fumarate)$_3$. MOFs are entirely composed of strong bonds (e.g., C−C, C−H, C−O, and M−O); hence, they show high thermal stability ranging from 250°C to 500°C (Cao, 2016). Some of the important properties of MOFs were depicted in Fig. 10.3 for easy understanding of the readers.

10.2 Synthesis strategies of metal organic frameworks

Identification of effective synthesis methods of MOFs is an important task for various applications especially for photocatalytic water splitting. Many methods have been discovered for the preparation of various types of MOFs from laboratory scale to industrial scale. But still some challenges remain unclear. Here, we discuss most successful and important methods of MOF preparation. The timeline of the important synthetic approaches patented for the synthesis of MOFs is represented in Fig. 10.4. Here, we briefly discuss the energy-efficient processes with less synthesis times that facilitate the scale-up process and the continuous operation, such as in electrochemical (Mueller et al., 2006), sol—gel (Hay & Raval, 2001), and mechanochemistry (Kouznetsov et al., 2016) approaches and the more recent routes, the spray dryer, and flow chemistry (Rubio-Martinez et al., 2016; Shekhah et al., 2007).

10.2.1 Electrochemical synthesis

Electrochemical method has become one of the efficient methods for MOF synthesis. The BASF company first patented the synthesis procedure of MOFs in 2005 (Mueller et al., 2006). The synthesis procedure consisted of immersing a copper plate in a solution containing the organic linker, 1,3,5-benzenetricarboxylic acid (BTC), and an electrolyte.

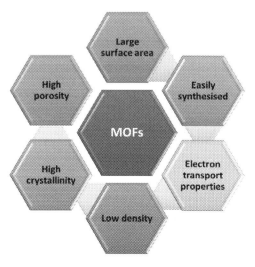

Figure 10.3
Common properties of the metal organic frameworks.

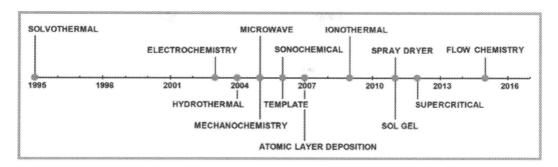

Figure 10.4

Timeline of the most common synthetic approaches patented for the synthesis of metal organic frameworks. Source: *Reprinted with permission from Rubio-Martinez, M., Avci-Camur, C., Thornton, A. W., Imaz, I., Maspoch, D., & Hill, M. R. (2017). New synthetic routes towards MOF production at scale, Chemical Society Reviews, 46, 3453–3480. doi:10.1039/C7CS00109F, Copyright Royal Society of Chemistry 2017.*

The copper plate (acts as electrode) was used as the source of Cu(II) ions. If required amount of current or voltage was applied, the Cu(II) ions were released from the copper electrode to the solution and reacted with the dissolved linker. The other advantages of this method are to run the synthesis of MOFs in a continuous way without hurdles. It also allows to work under milder conditions than typical solvothermal or microwave (MW) syntheses, reducing the reaction time. The schematic illustration of the electrosynthesis of MOFs by anodic dissolution is shown in Fig. 10.5.

10.2.2 Microwave synthesis

MW synthesis is a well-known method in organic chemistry. In recent years, it is familiar for the synthesis of inorganic nanomaterials like zeolites and MOFs (Bilecka & Niederberger, 2010; de la Hoz, Díaz-Ortiz, & Moreno, 2005; Gharibeh, Tompsett, Yngvesson, & Conner, 2009; Khan & Jhung, 2015). This method is based on the interaction of electromagnetic waves with any material containing mobile electric charges, such as polar molecules in a solvent or conducting ions in a solid. In MW synthesis, the faster heating can be expected due to direct interaction of the irradiation with the reactants, resulting in more efficient contrary to classical solvothermal methods, where thermal energy is transferred from the heat source to the solution through the reaction vessel. In 2005 Jhung, Lee, and Chang (2005) reported the water-based synthesis of the chromium trimesate MIL-100 MOF in the presence of hydrofluoric acid. The synthesis was performed at 220°C in an MW oven for 1, 2, or 4 h with the reaction mixture in a sealed Teflon autoclave. Similarly, Ni and Masel (2006) reported MW synthesis of IRMOF-1, IRMOF-2, and IRMOF-3 microcrystals with a relatively uniform size and identical cubic

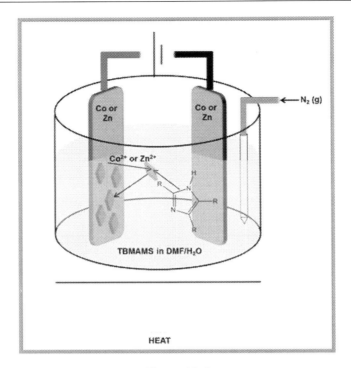

Figure 10.5
The electrosynthesis of metal-organic frameworks (MOFs) by anodic dissolution. Schematic
illustration showing the anodic dissolution cell for synthesis of ZIF MOF. Source: *Reprinted with
permission from Worrall, S. D., Mann, H., Rogers, A., Bissett, M. A., Attfield, M. P., Dryfe, R. A. W.
(2016). Electrochemical deposition of zeolitic imidazolate framework electrode coatings for supercapacitor
electrodes.* Electrochimica Acta, 197, 228–240. https://doi.org/10.1016/j.electacta.2016.02.145.
Copyright Elsevier 2016.

morphology in less than 2 min. Recently, the one-pot MW synthesis of MCM-41/Cu-based
MOF composite was reported by Tari, Tadjarodi, Tamnanloo, and Fatemi (2016) and the
schematic representation of the synthesis procedure is depicted in Fig. 10.6.

10.2.3 Mechanochemical synthesis

Mechanosynthesis is the most important technique in metallurgy and mineral processing
but in the last few decades it has expanded rapidly into many areas of chemistry, such
as inorganic chemistry, catalysis, and pharmaceutical synthesis (Friščić, 2012; Garay,
Pichon, & James, 2007). The main concept behind this synthesis method is to promote
chemical reactions by milling or grinding solids with only minimal amounts of solvents
(Stolle, Szuppa, Leonhardt, & Ondruschka, 2011). In this method, the conventional

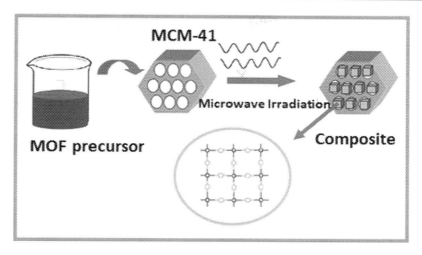

Figure 10.6

The synthesis of MCM-41@Cu(BDC) composite by one-pot microwave-assisted method. *Source: Reprinted with permission from Tari, N. E., Tadjarodi, A., Tamnanloo, J., Fatemi, S. (2016). One pot microwave synthesis of MCM-41/Cu based MOF composite with improved CO2 adsorption and selectivity.* Microporous and Mesoporous Materials, 231, 154–162. https://doi.org/10.1016/j. micromeso.2016.05.027. Copyright Elsevier 2016.

solvothermal MOF reactors are substituted by a mortar and pestle or in a mechanical process by automated ball mills. In addition to the solvent-free conditions, this technique leads to a faster and more efficient synthesis of MOFs obtaining quantitative yields and allows to use MOF precursors with low solubility, such as oxides, hydroxides, and carbonates. The only limitation lies in upscaling mechanosynthesis; it is essentially a batch processing technique with a relatively low rate of production. James group (Pichon, Lazuen-Garay, & James, 2006) first reported the solvent-free mechanochemical (SFM) method, briefly milling a dry mixture of copper acetate and isonicotinic acid (Hina) powder for 10 min, resulting in the formation of copper(II) isonicotinate MOF with acetic acid and water molecules blocked in the pores (Pichon et al., 2006). The important advantage of SFM method involves that the synthesized MOFs via SFM method provides only water as a by-product, allowing the complete elimination of the unreacted products during the purification stage. The mechanochemical synthesis of a yttrium-based metal organic framework reported by Singh, Hardi, and Balema (2013) was shown in Fig. 10.7. Recently, Xu and coworkers reported the synthesis of MIL-101(Cr) without the addition of solvent and hydrofluoric acid (Leng, Sun, Li, Sun, & Xu, 2016). The chromium salt and the terephthalic acid were ground for 30 min at room temperature and then transferred into an autoclave at 220°C for 4 h, yielding a material with a Brunauer-Emmett-Teller (BET) surface area of 3517 m^2/g.

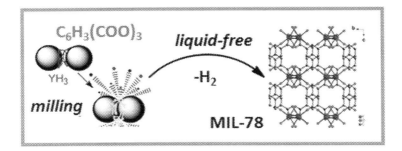

Figure 10.7
Schematic representation of mechanochemical synthesis of an yttrium-based metal organic
framework. Source: *Reprinted with permission from Singh, N. K., Hardi, M., Balema, V. P. (2013).*
Mechanochemical synthesis of an yttrium based metal—organic framework. Chemical Communications, 49,
972—974. doi:10.1039/C2CC36325A, Copyright Royal Society of Chemistry 2013.

10.2.4 Spray-drying synthesis

Spray-drying (SD) process is also a well-studied method in industry for decades. The basic
idea behind this method is the production of dispersed powder from a liquid or slurry that is
rapidly evaporated with a hot gas. The development of the SD method evolved over a
period from 1870s through early 1900s. Maspoch et al. in 2013 expanded this concept to
the synthesis of supramolecular materials, particularly MOFs (Carné-Sánchez, Imaz, Cano-
Sarabia, & Maspoch, 2013). The main principle of the process was based on the fast drying
of atomized microdroplets of a solution that contains the MOF precursors (Fig. 10.8A and B).
The process starts with the atomization of a solution of the MOF precursors into a spray of
microdroplets, which was followed by simultaneous injection of one or more solutions, at a
certain feed rate, and compressed air or nitrogen gas, at another flow rate. Various types of
MOFs were synthesized using the SD process as reported in the literature. To mention few
HKUST-1 (also known as Cu-BTC or Basolitet C300) can be synthesized by SD a solution of
$Cu(NO_3)_20.2 \cdot 5H_2O$ and trimesic acid (H3BTC) (3:2 molar ratio) in DMF, ethanol, and water
(1:1:1) with a feed rate of 4.5 mL/min, a flow rate of 336 mL/min, and an inlet temperature
of 180°C (Carné-Sánchez et al., 2013; Garzón-Tovar et al., 2016). Also, other MOFs such as
MIL-88A, ZIF-8, Fe-BTC/MIL-100, UiO-66, UiO-66-NH$_2$, and Fe-BTC/MIL-100 have also
been successfully prepared by this method for many catalytic applications (Carné-Sánchez
et al., 2013; Garcia Marquez et al., 2013; Garzón-Tovar et al., 2016; Garzón-Tovar,
Rodríguez-Hermida, Imaz, & Maspoch, 2017).

10.3 Metal-organic frameworks as photocatalysts

The ever-increasing environmental pollution and increasing demand for clean energy have
let the development of new technologies for future energy crisis and environmental

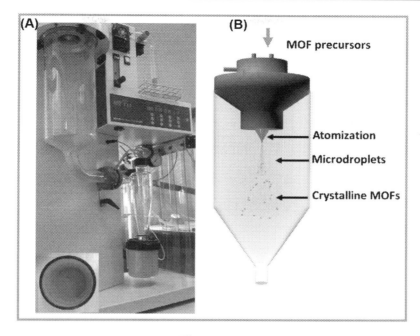

Figure 10.8

Spray-drying method for the production of metal-organic frameworks. (A) Photograph of the spray dryer while it is used to fabricate HKUST-1. (B) Schematic illustration of the spray-drying synthesis of metal-organic frameworks. Source: *Reprinted with permission from Carné-Sánchez, A., Imaz, I., Cano-Sarabia, M., & Maspoch, D. (2013). A spray-drying strategy for synthesis of nanoscale metal—organic frameworks and their assembly into hollow superstructures.* Nature Chemistry, 5, 203—211. *doi:10.1038/nchem.1569, Copyright Nature 2013.*

remediation. The solar photocatalytic water splitting has become one of the hot topic of research for clean energy production (Kim & Choi, 2010; Kudo & Miseki, 2009; Li, 2017). In this connection, many semiconducting photocatalytic materials have been developed for solar fuel generation through the water-splitting process. The metal, metal oxide, nonmetals, composite-based materials, core—shell-type materials, and graphitic carbon nitride-based composite photocatalysts have been extensively investigated (He et al., 2017; Lakshmana Reddy et al., 2018; Ran, Zhang, Yu, Jaroniec, & Qiao, 2014; Zhou et al., 2016). Recently, MOF-based materials have attracted more attention in photocatalysis due to their high surface area, high crystallinity, ease of synthesis, high porosity, framework flexibility, and chemical stability (Alshammari, Jiang, & Cordova, 2016; Chughtai et al., 2015; Lee et al., 2009; Sun, Yu, Zhang, Wei, & Yang, 2016; Zhang & Lin, 2014). The important features of MOFs are that they can act as (1) integrated photosensitizers and catalytic components in a single material by immobilizing the active sites on metal nodes, organic linkers, or encapsulated guest molecules inside the pores. The limitless choices of metal nodes and organic linkers in MOFs offer to improve the use of the visible spectrum of sunlight. (2) The high porosity of MOFs allows fast transport and diffusion of substrates and products from catalytic sites. The well-defined crystalline nature of MOFs provides a unique platform to examine the energy transfer mechanism of the photocatalytic process, which is difficult to study in other photocatalytic systems. (3) Unlike homogeneous photocatalysts, MOFs can be easily separated from the reaction systems and can be reused multiple times. Therefore it will extend the lifetime of the photocatalysts and minimized the waste and contamination. Due to the semiconducting properties of the MOFs, they can be directly used as photocatalysts, as cocatalysts, and as photo-sensitizers. The key merits of the MOFs as catalysts for photocatalytic H_2 evolution are explained by Zhu, Zou, and Xu (2018) and shown in Fig. 10.9.

Let us discuss each role separately in the following sections. First focus on the photocatalytic role of the MOFs. In 2007, Garcia et al. reported the experimental proof for the behavior of MOF-5 as a semiconductor and it can be used as photocatalyst (Alvaro, Carbonell, Ferrer, i Xamena, & Garcia, 2007). The Zn_4O clusters of MOF-5 can be considered as semiconductor dots that are isolated and distributed regularly in the

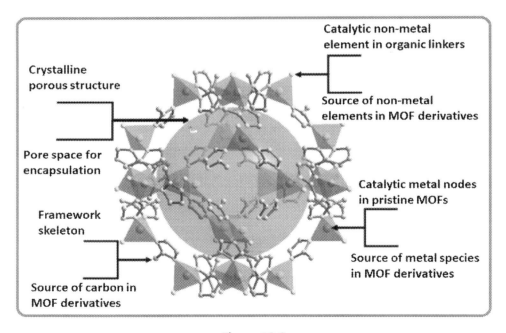

Figure 10.9
The key merits of metal-organic frameworks as catalysts for HER (ZIF-67 is used here as an example, green ball: pore space; purple: Co; yellow: nitrogen; gray: carbon). Source: *Reprinted with permission from Zhu, B., Zou, R., & Xu, Q. (2018). Metal—organic framework based catalysts for hydrogen evolution. Advanced Energy Materials, 8, 1801193. doi:10.1002/aenm.201801193, Copyright Wiley-VH 2018.*

framework. A charge-separated state of MOF-5 was observed and the band gap was estimated to be 3.4 eV. After that, a variety of MOFs have been studied for photocatalysis, such as UiO-66(Zr) (Cavka et al., 2008) and MIL-125(Ti) (Dan-Hardi et al., 2009). In 2009 Mori and coworkers (Kataoka et al., 2009) employed Ru-MOFs ($[Ru_2(p\text{-}BDC)_2]_n$, $Ru_2(CH_3COO)_4BF_4$, and $[Ru_2^{II,III}(p\text{-}BDC)_2BF_4]_n$) as H_2 production catalysts (HPCs), which showed catalytic activity for the photochemical reduction of water into H_2 molecules in the presence of light-harvesting systems with photosensitizer $Ru(bpy)_3^{2+}$ (bpy = 2,2′-bipyridine), electron relay MV^{2+} (methyl viologen), and sacrificial donors EDTA-2Na under visible light irradiation as shown in Fig. 10.10. Some other MOFs like Rh-MOFs ($[Rh_2(p\text{-}BDC)_2]_n$ and $[Rh_2(C_6H_5COO)_4]$) were used as H_2 production catalysts for photochemical reduction of water under visible light irradiation (Kataoka et al., 2010). In 2008 Cavka et al. reported two highly water-stable Zr-based MOFs [UiO-66: $Zr_6O_4(OH)_4(BDC)_{12}$ (Cavka et al., 2008), BDC: 1,4-benzenedicarboxylate, where UiO = University of Oslo, and NH_2-UiO-66: $Zr_6O_4(OH)_4(ATA)_{12}$, where ATA = 2-aminoterephthalate] as photocatalysts for photocatalytic H_2 evolution from methanol and water/methanol mixture (Gomes Silva, Luz, i Xamena, Corma, & García, 2010). UiO-66 and NH_2-UiO-66 are isoreticular MOFs with $Zr_6(O)_4(OH)_4(CO_2)_{12}$ secondary building units, and they both feature high crystallinities. Many other semiconductor@MOF heterostructures like ZnO@ZIF-8 (Wang et al., 2016), $Cu_3(BTC)_2@TiO_2$ (Li et al., 2014), $BiVO4@MIL\text{-}101$ (Xu et al., 2015), $UiO\text{-}66@g\text{-}C_3N_4$ (Wang et al., 2015), $MoS_2@UiO\text{-}66@CdS$ (Shen et al., 2015), and $Cd_{0.2}Zn_{0.8}S@UiO\text{-}66\text{-}NH_2$ (Su, Zhang, Liu, & Wang, 2017) have been synthesized and used for photocatalytic H_2 generation applications. Therefore MOFs have become one of the promising porous materials for solar fuel generations through photocatalysis.

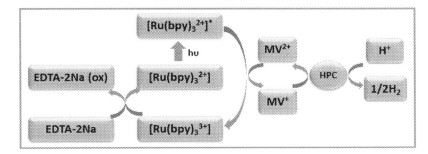

Figure 10.10

The reaction scheme of photochemical hydrogen production from water using Ru-MOFs as H_2 production catalyst in the presence of $Ru(bpy)_3^{2+}$, MV^{2+}, and EDTA-2Na. Source: *Reprinted with permission from Li, S.-L., & Xu, Q. (2013). Metal—organic frameworks as platforms for clean energy.* Energy & Environmental Science, 6, 1656—1683. *doi:10.1039/C3EE40507A, Copyright Royal Society of Chemistry 2013.*

10.3.1 Metal organic frameworks as cocatalysts

In addition to the catalytic role of MOFs as discussed, they can act as cocatalysts as well for H_2 generation. Many factors exist for the MOFs as cocatalysts for photocatalysis, because MOFs feature the combined catalytic characteristics of transition metal ions (e.g., Co, Zn, Mn) and functional organic ligands (e.g., imidazolate entities) together with high thermal and chemical stabilities (Hayashi, Côté, Furukawa, O'Keeffe, & Yaghi, 2007; Park et al., 2006; Zhang, Zhang, Lin, & Chen, 2012). In 2009 Mori and his group reported the first example of MOFs as a cocatalyst for the photochemical reduction of water under visible-light irradiation (Kataoka et al., 2009). In their photocatalytic system, the porous MOF $[Ru_2(p\text{-BDC})_2]_n(p\text{-BDC} = p\text{-benzenedicarboxylate})$ is the cocatalyst, while Ru $(bpy)_3^{2+}$ serves as a photosensitizer, together with MV^{2+} and EDTA-2Na acting as an electron relay and a sacrificial reductant, respectively. Ott et al. recently synthesized a photocatalytic MOF UiO-66-[FeFe](dcbdt)(CO)$_6$ via the incorporation of a molecular proton reduction catalyst [FeFe](dcbdt)(CO)$_6$ (1,dcbdt = 1,4-dicarboxylbenzene-2,3-dithiolate) into the framework of UiO-66 by the postsynthetic exchange strategy (Pullen, Fei, Orthaber, Cohen, & Ott, 2013). Recently, Liu, Zhang, and Ao (2018) reported heterostructured $ZnIn_2S_4@NH_2\text{-MIL-125(Ti)}$ nanocomposites for visible-light-driven H_2 production. Here, the optimal content of $NH_2\text{-MIL-125(Ti)}$ about 40 wt.% showed the photocatalytic H_2 production rate of 2204.2 μmol/h/g (with an apparent quantum

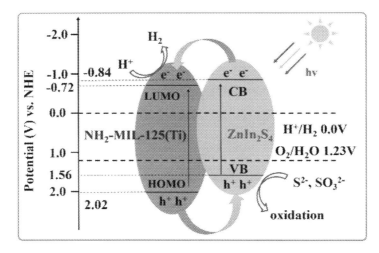

Figure 10.11
The proposed charge transfer mechanism of the $ZnIn_2S_4@NH_2\text{-MIL-125(Ti)}$ composite for photocatalytic H_2 production. Source: *Reprinted with permission from Liu, H., Zhang, J., & Ao, D. (2018). Construction of heterostructured ZnIn2S4@NH2-MIL-125(Ti) nanocomposites for visible-light-driven H2 production.* Applied Catalysis B: Environmental, 221, 433–442. https://doi.org/10.1016/j.apcatb.2017.09.043. Copyright Elsevier 2018.

efficiency of 4.3% at 420 nm), which was 6.5 times higher than that of pure $ZnIn_2S_4$. The NH_2-MIL-125(Ti) can act as cocatalyst for the reduction of H^+ ions into H_2. The photocatalytic reaction mechanism is displayed in Fig. 10.11.

10.3.2 Metal organic frameworks as sensitizers

Besides the cocatalytic role of MOF, they can also used as photosensitizers to catalyze the photo-reaction. Due to intrinsic channels and/or cavities as well as the decorative organic ligands, MOFs are readily functionalized with catalytic active sites and/or photoredox components during synthesis or through postsynthetic modifications, which renders them attractive hosts for heterogeneous photocatalysis (Cohen, 2012; Li, Kozlowski, Doud, Blakely, & Rosi, 2013; Nasalevich, Goesten, Savenije, Kapteijn, & Gascon, 2013; O'Keeffe & Yaghi, 2012). Wang, deKrafft, and Lin (2012) studied Pt nanoparticle@photoactive MOF assemblies for H_2 evolution by visible light with trimethylamine (TEA) as a sacrificial reductant. Authors claimed that the high catalytic activity of these assemblies was attributed to the vicinity of the iridium-based photosensitizer and the Pt cocatalyst that favored

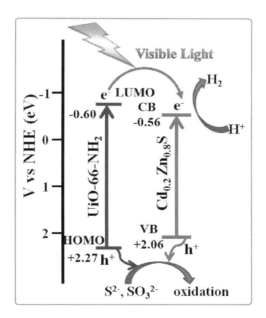

Figure 10.12

Schematic diagram illustrating the sensitizing role of UiO-66-NH$_2$ and photogenerated electrons and holes' transfer in the Cd$_{0.2}$Zn$_{0.8}$S@UiO-66-NH$_2$ composite for photocatalytic H$_2$ production under visible light irradiation. Source: *Reprinted with permission from Su, Y., Zhang, Z., Liu, H., Wang, Y. (2017). Cd0.2Zn0.8S@UiO-66-NH2 nanocomposites as efficient and stable visible-light-driven photocatalyst for H2 evolution and CO2 reduction.* Applied Catalysis B: Environmental, 200, *448−457. https://doi.org/10.1016/j.apcatb.2016.07.032. Elsevier 2017.*

Table 10.1: Summary of photocatalytic reactions on H_2 evolution using various metal organic framework (MOF)-based catalysts.

Photocatalyst	Light source	Sacrificial agent	H_2 production (μmol/h/g)	Reference
Pt@MOF-1 or Pt@MOF-2	450 W Xe-lamp visible	TEA	3400	Wang et al. (2012)
H_2TCPP[AlOH]$_2$(DMF$_3$-(H$_2$O)$_2$)	300 W Xe lamp visible	EDTA	200	Fateeva et al. (2012)
{[Ln$_2$Cu$_5$(OH)$_2$(pydc)$_6$(H$_2$O)$_8$] · I$_8$}(Ln: Tb)	500 W Hg lamp UV	CH$_3$OH	2105	Hu et al. (2013)
MOF-253-Pt	300 W Xe lamp visible	TEOA	3000	Zhou et al. (2013)
Pt/NH$_2$-UiO-66	200 W Xe lamp visible	CH$_3$OH	2800	Gomes Silva et al. (2010)
NH$_2$-UiO-66(Zr/Ti)-120−16 days	300 W Xe lamp visible	CH$_3$CN and TEOA	3500	Sun, Liu, Qiu, Zhang, and Li (2015)
RhB/UiO-66(Zr)-100	300 W Xe lamp visible	TEOA	33.9	He et al. (2014)
Pt-UiO-66−30	300 W Xe lamp visible	CH$_3$OH	37	Yuan et al. (2015)
Pt/Ti-MOF-Ru(tpy)$_2$	500 W Xe lamp visible	TEOA	5.1	Toyao et al. (2014)
Pt/[Cu (en)$_2$]$_4$[PNb$_{12}$O$_{40}$(VO)$_6$] · (OH)$_5$ · 8H$_2$O	125 W Hg lamp UV	CH$_3$OH	44.3	Shen et al. (2014)
Fe$_2$O$_3$@TiO$_2$/Pt	Xe lamp	TEA	0.8	Dekrafft, Wang, and Lin (2012)
Pt@CdS/MIL-101(Cr)	300 W Xe lamp visible	Lactic acid	150	He et al. (2013)
UiO-66/CdS/rGO	300 W Xe lamp visible	Na$_2$S, Na$_2$SO$_3$	105	Lin et al. (2014)
Pt/NH$_2$-MIL-125(Ti)	Xe lamp	TEOA	33	Horiuchi et al. (2012)
Pt(1.5)/NH$_2$-MIL-125(Ti)	CH$_3$OH	TEOA	15.5	Toyao et al. (2013)
Pt/MIL-125(Ti)	Xe lamp	TEOA	38.6	Shen, Luo, Huang, Feng, and Wu (2015)
Co@NH$_2$-MIL-125(Ti)	500 W Xe lamp	TEA	37	Nasalevich et al. (2015)
(Cu$_3$(BTC)$_2$(H$_2$O)$_3$)@ZnO/GO	300 W Xe lamp visible	CH$_3$OH	129	Zhang et al. (2018)
Cu-I-bipy	300 W Xe lamp visible	TEOA	7090	Qin, Wang, and Wang (2017)
NH$_2$-MIL-125(Ti)/0.75CN/ Ni$_{15.8}$Pd$_{4.1}$	300 W Xe lamp visible	TEOA	7840	Choi et al. (2017)
(AlOH)$_2$H$_2$TCPP@Pt	300 W Xe lamp visible	TEOA	129	Zhang et al. (2016)

Bipy, 4,4′-bipyridine; *EDTA*, ethylenediminetetraacetic acid; *PMOF*, porphyrin metal-organic framework; *TEA*, triethylamine; *TEOA*, triethanolamine; *tpy*, terpyridine.

electron transfer between the two components. Recently, Su et al. (2017) reported $Cd_{0.2}Zn_{0.8}S@UiO-66-NH_2$ composite photocatalyst with the H_2 evolution rate of 5846.5 μmol/h/g. Under visible-light irradiation, UiO-66-NH_2 and $Cd_{0.2}Zn_{0.8}S$ were excited to generate electron—hole pairs. Since the LUMO potential of UiO-66-NH_2(-0.60 eV vs NHE) is more negative than the CB potential of $Cd_{0.2}Zn_{0.8}S$ (-0.56 eV vs NHE), the photoinduced electrons on the LUMO of UiO-66-NH_2 can directly transfer to the CB of $Cd_{0.2}Zn_{0.8}S$, which could inhibit the recombination of photoinduced electron—hole pairs as shown in Fig. 10.12. Some other MOFs were studied as photocatalytic hosts for H_2 evolution as reported in the literature (Nasalevich et al., 2015; Sasan, Lin, Mao, & Feng, 2014). Therefore the different roles of MOFs in heterogeneous photocatalysis have been systematically explored and in addition the recent literature of MOFs on solar photocatalytic H_2 generation is listed in Table 10.1.

10.4 Challenges and future prospects

MOFs exhibited promising future as new class of photocatalytic materials that attracted extensive research interest, whereas they face significant challenges for widespread application. To tackle these issues, great research efforts should be taken care on

1. Band gap and band positions should be carefully altered through tuning the metal or metal cluster nodes as well as target-directed design of the ligands.
2. Incorporating inorganic semiconductors, metals, polymer semiconductors, and surface modification may enrich the performance of MOFs photocatalysts.
3. Developing and applying spectral and photoelectrochemical characterization techniques.
4. In particular, transient characterizations, for precisely exploring the exact photocatalytic and photoelectrochemical catalysis mechanism.
5. Demonstrating the potentials of MOF-based photocatalysts at pilot scale.

Acknowledgment

The authors acknowledge the National Research Foundation of Korea Grant funded by the Korean Government (NRF-2019R111A3A01041454).

References

Alshammari, A., Jiang, Z., & Cordova, K. E. (2016). Metal organic frameworks as emerging photocatalysts. In W. Cao (Ed.), *Semiconductor Photocatalysis*. Rijeka: IntechOpen. Available from http://doi.org/10.5772/63489.

Alvaro, M., Carbonell, E., Ferrer, B., i Xamena, F. X., & Garcia, H. (2007). Semiconductor behavior of a metal-organic framework (MOF). *Chemistry—A European Journal*, *13*, 5106—5112. Available from https://doi.org/10.1002/chem.200601003.

Bilecka, I., & Niederberger, M. (2010). Microwave chemistry for inorganic nanomaterials synthesis. *Nanoscale*, *2*, 1358−1374. Available from https://doi.org/10.1039/B9NR00377K.

Cao, W. (2016). Semiconductor photocatalysis: Materials, mechanisms and applications. IntechOpen.

Carné-Sánchez, A., Imaz, I., Cano-Sarabia, M., & Maspoch, D. (2013). A spray-drying strategy for synthesis of nanoscale metal−organic frameworks and their assembly into hollow superstructures. *Nature Chemistry*, *5*, 203−211. Available from https://doi.org/10.1038/nchem.1569.

Cavka, J. H., Jakobsen, S., Olsbye, U., Guillou, N., Lamberti, C., Bordiga, S., & Lillerud, K. P. (2008). A new zirconium inorganic building brick forming metal organic frameworks with exceptional stability. *Journal of the American Chemical Society*, *130*, 13850−13851. Available from https://doi.org/10.1021/ja8057953.

Choi, K. M., Kim, D., Rungtaweevoranit, B., Trickett, C. A., Barmanbek, J. T. D., Alshammari, A. S., & Yaghi, O. M. (2017). Plasmon-enhanced photocatalytic CO_2 conversion within metal−organic frameworks under visible light. *Journal of the American Chemical Society*, *139*, 356−362. Available from https://doi.org/10.1021/jacs.6b11027.

Chughtai, A. H., Ahmad, N., Younus, H. A., Laypkov, A., & Verpoort, F. (2015). Metal−organic frameworks: Versatile heterogeneous catalysts for efficient catalytic organic transformations. *Chemical Society Reviews*, *44*, 6804−6849. Available from https://doi.org/10.1039/C4CS00395K.

Cohen, S. M. (2012). Postsynthetic methods for the functionalization of metal−organic frameworks. *Chemical Reviews*, *112*, 970−1000. Available from https://doi.org/10.1021/cr200179u.

Dan-Hardi, M., Serre, C., Frot, T., Rozes, L., Maurin, G., Sanchez, C., & Ferey, G. (2009). A new photoactive crystalline highly porous titanium(IV) dicarboxylate. *Journal of the American Chemical Society*, *131*, 10857−10859. Available from https://doi.org/10.1021/ja903726m.

de la Hoz, A., Díaz-Ortiz, Á., & Moreno, A. (2005). Microwaves in organic synthesis. Thermal and non-thermal microwave effects. *Chemical Society Reviews*, *34*, 164−178. Available from https://doi.org/10.1039/B411438H.

Dekrafft, K. E., Wang, C., & Lin, W. (2012). Metal-organic framework templated synthesis of Fe2O3/TiO2 nanocomposite for hydrogen production. *Advanced Materials*, *24*, 2014−2018. Available from https://doi.org/10.1002/adma.201200330.

Dhakshinamoorthy, A., Alvaro, M., & Garcia, H. (2012). Commercial metal−organic frameworks as heterogeneous catalysts. *Chemical Communications*, *48*, 11275−11288. Available from https://doi.org/10.1039/C2CC34329K.

Dhakshinamoorthy, A., Asiri, A. M., & García, H. (2016). Metal−organic framework (MOF) compounds: Photocatalysts for redox reactions and solar fuel production. *Angewandte Chemie International Edition*, *55*, 5414−5445. Available from https://doi.org/10.1002/anie.201505581.

Dhakshinamoorthy, A., Opanasenko, M., Čejka, J., & Garcia, H. (2013). Metal organic frameworks as heterogeneous catalysts for the production of fine chemicals. *Catalysis Science & Technology*, *3*, 2509−2540. Available from https://doi.org/10.1039/C3CY00350G.

Dias, E. M., & Petit, C. (2015). Towards the use of metal−organic frameworks for water reuse: A review of the recent advances in the field of organic pollutants removal and degradation and the next steps in the field. *Journal of Materials Chemistry A*, *3*, 22484−22506. Available from https://doi.org/10.1039/C5TA05440K.

Fateeva, A., Chater, P. A., Ireland, C. P., Tahir, A. A., Khimyak, Y. Z., Wiper, P. V., & Rosseinsky, M. J. (2012). A water-stable porphyrin-based metal−organic framework active for visible-light photocatalysis. *Angewandte Chemie International Edition*, *51*, 7440−7444. Available from https://doi.org/10.1002/anie.201202471.

Friščić, T. (2012). Supramolecular concepts and new techniques in mechanochemistry: Cocrystals, cages, rotaxanes, open metal−organic frameworks. *Chemical Society Reviews*, *41*, 3493−3510. Available from https://doi.org/10.1039/C2CS15332G.

Garay, A. L., Pichon, A., & James, S. L. (2007). Solvent-free synthesis of metal complexes. *Chemical Society Reviews*, *36*, 846−855. Available from https://doi.org/10.1039/B600363J.

Garcia Marquez, A., Horcajada, P., Grosso, D., Ferey, G., Serre, C., Sanchez, C., & Boissiere, C. (2013). Green scalable aerosol synthesis of porous metal−organic frameworks. *Chemical Communications*, *49*, 3848−3850. Available from https://doi.org/10.1039/C3CC39191D.

Garzón-Tovar, L., Cano-Sarabia, M., Carné-Sánchez, A., Carbonell, C., Imaz, I., & Maspoch, D. (2016). A spray-drying continuous-flow method for simultaneous synthesis and shaping of microspherical high nuclearity MOF beads. *Reaction Chemistry & Engineering, 1*, 533−539. Available from https://doi.org/10.1039/C6RE00065G.

Garzón-Tovar, L., Rodríguez-Hermida, S., Imaz, I., & Maspoch, D. (2017). Spray drying for making covalent chemistry: Postsynthetic modification of metal−organic frameworks. *Journal of the American Chemical Society, 139*, 897−903. Available from https://doi.org/10.1021/jacs.6b11240.

Gharibeh, M., Tompsett, G. A., Yngvesson, K. S., & Conner, W. C. (2009). Microwave synthesis of zeolites: Effect of power delivery. *The Journal of Physical Chemistry B, 113*, 8930−8940. Available from https://doi.org/10.1021/jp900400d.

Gomes Silva, C., Luz, I., i Xamena, F. X., Corma, A., & García, H. (2010). Water stable Zr−benzenedicarboxylate metal−organic frameworks as photocatalysts for hydrogen generation. *Chemistry—A European Journal, 16*, 11133−11138. Available from https://doi.org/10.1002/chem.200903526.

Hay, J. N., & Raval, H. M. (2001). Synthesis of organic − inorganic hybrids via the non-hydrolytic sol − gel process. *Chemistry of Materials, 13*, 3396−3403. Available from https://doi.org/10.1021/cm011024n.

Hayashi, H., Côté, A. P., Furukawa, H., O'Keeffe, M., & Yaghi, O. M. (2007). Zeolite A imidazolate frameworks. *Nature Materials, 6*, 501−506. Available from https://doi.org/10.1038/nmat1927.

He, J., Wang, J., Chen, Y., Zhang, J., Duan, D., Wang, Y., & Yan, Z. (2014). A dye-sensitized Pt@UiO-66(Zr) metal-organic framework for visible-light photocatalytic hydrogen production. *Chemical Communications, 50*, 7063−7066. Available from https://doi.org/10.1039/C4CC01086H.

He, J., Yan, Z., Wang, J., Xie, J., Jiang, L., Shi, Y., & Sun, Y. (2013). Significantly enhanced photocatalytic hydrogen evolution under visible light over CdS embedded on metal−organic frameworks. *Chemical Communications, 49*, 6761−6763. Available from https://doi.org/10.1039/C3CC43218A.

He, K., Xie, J., Yang, Z., Shen, R., Fang, Y., Ma, S., & Li, X. (2017). Earth-abundant WC nanoparticles as an active noble-metal-free co-catalyst for the highly boosted photocatalytic H2 production over g-C3N4 nanosheets under visible light. *Catalysis Science & Technology, 7*, 1193−1202. Available from https://doi.org/10.1039/C7CY00029D.

Horiuchi, Y., Toyao, T., Saito, M., Mochizuki, K., Iwata, M., Higashimura, H., & Matsuoka, M. (2012). Visible-light-promoted photocatalytic hydrogen production by using an amino-functionalized Ti(IV) metal−organic framework. *The Journal of Physical Chemitry C, 116*, 20848−20853. Available from https://doi.org/10.1021/jp3046005.

Hu, X.-L., Sun, C.-Y., Qin, C., Wang, X.-L., Wang, H.-N., Zhou, E.-L., & Su, Z-M. (2013). Iodine-templated assembly of unprecedented 3d−4f metal−organic frameworks as photocatalysts for hydrogen generation. *Chemical Communications, 49*, 3564−3566. Available from https://doi.org/10.1039/C3CC39173F.

Jhung, Sung-Hwa, Lee, Jin-Ho, & Chang, J.-S. (2005). Microwave synthesis of a nanoporous hybrid material, chromium trimesate. *Bulletin of the Korean Chemical Society, 26*, 880−881. Available from https://doi.org/10.5012/BKCS.2005.26.6.880.

Kataoka, Y., Sato, K., Miyazaki, Y., Masuda, K., Tanaka, H., Naito, S., & Mori, W. (2009). Photocatalytic hydrogen production from water using porous material [Ru2(p-BDC)2]n. *Energy & Environmental Science, 2*, 397−400. Available from https://doi.org/10.1039/B814539C.

Kataoka, Y., Sato, K., Miyazaki, Y., Suzuki, Y., Tanaka, H., Kitagawa, Y., & Mori, W. (2010). Photocatalytic hydrogen production from water using heterogeneous two-dimensional rhodium coordination polymer [Rh2 (p-BDC)2]n. *Chemistry Letters, 39*, 358−359. Available from https://doi.org/10.1246/cl.2010.358.

Keggin, J. F., & Miles, F. D. (1936). Structures and formulæ of the prussian blues and related compounds. *Nature, 137*, 577−578. Available from https://doi.org/10.1038/137577a0.

Khan, N. A., & Jhung, S. H. (2015). Synthesis of metal-organic frameworks (MOFs) with microwave or ultrasound: Rapid reaction, phase-selectivity, and size reduction. *Coordination Chemistry Reviews, 285*, 11−23. Available from https://doi.org/10.1016/j.ccr.2014.10.008.

Kim, J., & Choi, W. (2010). Hydrogen producing water treatment through solar photocatalysis. *Energy & Environmental Science, 3*, 1042−1045. Available from https://doi.org/10.1039/C003858J.

Kouznetsov, V. V., Merchán-Arenas, D. R., Martínez-Bonilla, C. A., Macías, M. A., Roussel, P., & Gauthier, G. H. (2016). Grinding and milling: Two efficient methodologies in the solvent-free phosphomolybdic acid-catalyzed and mechanochemical synthesis of cis-4-amido-N-yl-2-methyl-tetrahydroquinolines. *Journal of the Brazilian Chemical Society, 27*, 2246–2255.

Kudo, A., & Miseki, Y. (2009). Heterogeneous photocatalyst materials for water splitting. *Chemical Society Reviews, 38*, 253–278. Available from https://doi.org/10.1039/b800489g.

Lakshmana Reddy, N., Navakoteswara Rao, V., Mamatha Kumari, M., Kakarla, R. R., Ravi, P., Sathish, M., & Inamuddin. (2018). Nanostructured semiconducting materials for efficient hydrogen generation. *Environmental Chemistry Letters, 16*, 765–796. Available from https://doi.org/10.1007/s10311-018-0722-y.

Lee, J., Farha, O. K., Roberts, J., Scheidt, K. A., Nguyen, S. T., & Hupp, J. T. (2009). Metal-organic framework materials as catalysts. *Chemical Society Reviews, 38*, 1450–1459. Available from https://doi.org/10.1039/B807080F.

Leng, K., Sun, Y., Li, X., Sun, S., & Xu, W. (2016). Rapid synthesis of metal−organic frameworks MIL-101 (Cr) without the addition of solvent and hydrofluoric acid. *Crystal Growth & Design, 16*, 1168–1171. Available from https://doi.org/10.1021/acs.cgd.5b01696.

Li, J.-R., Kuppler, R. J., & Zhou, H.-C. (2009). Selective gas adsorption and separation in metal−organic frameworks. *Chemical Society Reviews, 38*, 1477–1504. Available from https://doi.org/10.1039/B802426J.

Li, R. (2017). Latest progress in hydrogen production from solar water splitting via photocatalysis, photoelectrochemical, and photovoltaic-photoelectrochemical solutions. *Chinese Journal of Catalysis, 38*, 5–12. Available from https://doi.org/10.1016/S1872-2067(16)62552-4.

Li, R., Hu, J., Deng, M., Wang, H., Wang, X., Hu, Y., & Xiong, Y. (2014). Integration of an inorganic semiconductor with a metal−organic framework: A platform for enhanced gaseous photocatalytic reactions. *Advanced Materials, 26*, 4783–4788. Available from https://doi.org/10.1002/adma.201400428.

Li, S., Yang, K., Tan, C., Huang, X., Huang, W., & Zhang, H. (2016). Preparation and applications of novel composites composed of metal−organic frameworks and two-dimensional materials. *Chemical Communications, 52*, 1555–1562. Available from https://doi.org/10.1039/C5CC09127F.

Li, S.-L., & Xu, Q. (2013). Metal−organic frameworks as platforms for clean energy. *Energy & Environmental Science, 6*, 1656–1683. Available from https://doi.org/10.1039/C3EE40507A.

Li, T., Kozlowski, M. T., Doud, E. A., Blakely, M. N., & Rosi, N. L. (2013). Stepwise ligand exchange for the preparation of a family of mesoporous MOFs. *Journal of the American Chemical Society, 135*, 11688–11691. Available from https://doi.org/10.1021/ja403810k.

Lin, R., Shen, L., Ren, Z., Wu, W., Tan, Y., Fu, H., & Wu, L. (2014). Enhanced photocatalytic hydrogen production activity via dual modification of MOF and reduced graphene oxide on CdS. *Chemical Communications, 50*, 8533–8535. Available from https://doi.org/10.1039/C4CC01776E.

Liu, H., Zhang, J., & Ao, D. (2018). Construction of heterostructured ZnIn2S4@NH2-MIL-125(Ti) nanocomposites for visible-light-driven H2 production. *Applied Catalysis B: Environmental, 221*, 433–442. Available from https://doi.org/10.1016/j.apcatb.2017.09.043.

Mueller, U., Schubert, M., Teich, F., Puetter, H., Schierle-Arndt, K., & Pastré, J. (2006). Metal−organic frameworks—prospective industrial applications. *Journal of Materials Chemistry, 16*, 626–636. Available from https://doi.org/10.1039/B511962F.

Nasalevich, M. A., Becker, R., Ramos-Fernandez, E. V., Castellanos, S., Veber, S. L., Fedin, M. V., & Gascon, J. (2015). Co@NH2-MIL-125(Ti): Cobaloxime-derived metal−organic framework-based composite for light-driven H2 production. *Energy & Environmental Science, 8*, 364–375. Available from https://doi.org/10.1039/C4EE02853H.

Nasalevich, M. A., Goesten, M. G., Savenije, T. J., Kapteijn, F., & Gascon, J. (2013). Enhancing optical absorption of metal−organic frameworks for improved visible light photocatalysis. *Chemical Communications, 49*, 10575–10577. Available from https://doi.org/10.1039/C3CC46398B.

Ni, Z., & Masel, R. I. (2006). Rapid production of metal − organic frameworks via microwave-assisted solvothermal synthesis. *Journal of the American Chemical Society, 128*, 12394–12395. Available from https://doi.org/10.1021/ja0635231.

Noh, T. H., & Jung, O.-S. (2016). Recent advances in various metal−organic channels for photochemistry beyond confined spaces. *Accounts of Chemical Research, 49*, 1835−1843. Available from https://doi.org/10.1021/acs.accounts.6b00291.

O'Keeffe, M., & Yaghi, O. M. (2012). Deconstructing the crystal structures of metal−organic frameworks and related materials into their underlying nets. *Chemical Reviews, 112*, 675−702. Available from https://doi.org/10.1021/cr200205j.

Park, K. S., Ni, Z., Côté, A. P., Choi, J. Y., Huang, R., Uribe-Romo, F. J., & Yaghi, O. M. (2006). Exceptional chemical and thermal stability of zeolitic imidazolate frameworks. *Proceedings of the National Academy of Sciences of the United States of America, 103*, 10186−10191. Available from https://doi.org/10.1073/pnas.0602439103.

Pichon, A., Lazuen-Garay, A., & James, S. L. (2006). Solvent-free synthesis of a microporous metal−organic framework. *CrystEngComm, 8*, 211−214. Available from https://doi.org/10.1039/B513750K.

Pullen, S., Fei, H., Orthaber, A., Cohen, S. M., & Ott, S. (2013). Enhanced photochemical hydrogen production by a molecular diiron catalyst incorporated into a metal−organic framework. *Journal of the American Chemical Society, 135*, 16997−17003. Available from https://doi.org/10.1021/ja407176p.

Qin, J., Wang, S., & Wang, X. (2017). Visible-light reduction CO2 with dodecahedral zeolitic imidazolate framework ZIF-67 as an efficient co-catalyst. *Applied Catalysis B: Environmental, 209*, 476−482. Available from https://doi.org/10.1016/j.apcatb.2017.03.018.

Ran, J., Zhang, J., Yu, J., Jaroniec, M., & Qiao, S. Z. (2014). Earth-abundant cocatalysts for semiconductor-based photocatalytic water splitting. *Chemical Society Reviews, 43*, 7787−7812. Available from https://doi.org/10.1039/C3CS60425J.

Rubio-Martinez, M., Avci-Camur, C., Thornton, A. W., Imaz, I., Maspoch, D., & Hill, M. R. (2017). New synthetic routes towards MOF production at scale. *Chemical Society Reviews, 46*, 3453−3480. Available from https://doi.org/10.1039/C7CS00109F.

Rubio-Martinez, M., Hadley, T. D., Batten, M. P., Constanti-Carey, K., Barton, T., Marley, D., & Hill, M. R. (2016). Scalability of continuous flow production of metal−organic frameworks. *ChemSusChem, 9*, 938−941. Available from https://doi.org/10.1002/cssc.201501684.

Sasan, K., Lin, Q., Mao, C., & Feng, P. (2014). Incorporation of iron hydrogenase active sites into a highly stable metal−organic framework for photocatalytic hydrogen generation. *Chemical Communications, 50*, 10390−10393. Available from https://doi.org/10.1039/C4CC03946G.

Shekhah, O., Wang, H., Kowarik, S., Schreiber, F., Paulus, M., Tolan, M., & Woll, C. (2007). Step-by-step route for the synthesis of metal − organic frameworks. *Journal of the American Chemical Society, 129*, 15118−15119. Available from https://doi.org/10.1021/ja076210u.

Shen, J.-Q., Zhang, Y., Zhang, Z.-M., Li, Y.-G., Gao, Y.-Q., & Wang, E.-B. (2014). Polyoxoniobate-based 3D framework materials with photocatalytic hydrogen evolution activity. *Chemical Communications, 50*, 6017−6019. Available from https://doi.org/10.1039/C3CC49245A.

Shen, L., Luo, M., Huang, L., Feng, P., & Wu, L. (2015). A clean and general strategy to decorate a titanium metal−organic framework with noble-metal nanoparticles for versatile photocatalytic applications. *Inorganic Chemistry, 54*, 1191−1193. Available from https://doi.org/10.1021/ic502609a.

Shen, L., Luo, M., Liu, Y., Liang, R., Jing, F., & Wu, L. (2015). Noble-metal-free MoS2 co-catalyst decorated UiO-66/CdS hybrids for efficient photocatalytic H2 production. *Applied Catalysis B: Environmental, 166−167*, 445−453. Available from https://doi.org/10.1016/j.apcatb.2014.11.056.

Singh, N. K., Hardi, M., & Balema, V. P. (2013). Mechanochemical synthesis of an yttrium based metal−organic framework. *Chemical Communications, 49*, 972−974. Available from https://doi.org/10.1039/C2CC36325A.

Stock, N., & Biswas, S. (2012). Synthesis of metal-organic frameworks (MOFs): Routes to various MOF topologies, morphologies, and composites. *Chemical Reviews, 112*, 933−969. Available from https://doi.org/10.1021/cr200304e.

Stolle, A., Szuppa, T., Leonhardt, S. E. S., & Ondruschka, B. (2011). Ball milling in organic synthesis: Solutions and challenges. *Chemical Society Reviews, 40*, 2317−2329. Available from https://doi.org/10.1039/C0CS00195C.

Su, Y., Zhang, Z., Liu, H., & Wang, Y. (2017). Cd0.2Zn0.8S@UiO-66-NH2 nanocomposites as efficient and stable visible-light-driven photocatalyst for H2 evolution and CO2 reduction. *Applied Catalysis B: Environmental*, *200*, 448−457. Available from https://doi.org/10.1016/j.apcatb.2016.07.032.

Sun, D., Liu, W., Qiu, M., Zhang, Y., & Li, Z. (2015). Introduction of a mediator for enhancing photocatalytic performance via post-synthetic metal exchange in metal−organic frameworks (MOFs). *Chemical Communications*, *51*, 2056−2059. Available from https://doi.org/10.1039/C4CC09407G.

Sun, X., Yu, Q., Zhang, F., Wei, J., & Yang, P. (2016). A dye-like ligand-based metal-organic framework for efficient photocatalytic hydrogen production from aqueous solution. *Catalysis Science & Technology*, *6*, 3840−3844. Available from https://doi.org/10.1039/C5CY01716E.

Tari, N. E., Tadjarodi, A., Tamnanloo, J., & Fatemi, S. (2016). One pot microwave synthesis of MCM-41/Cu based MOF composite with improved CO2 adsorption and selectivity. *Microporous and Mesoporous Materials*, *231*, 154−162. Available from https://doi.org/10.1016/j.micromeso.2016.05.027.

Toyao, T., Saito, M., Dohshi, S., Mochizuki, K., Iwata, M., Higashimura, H., & Matsuoka, M. (2014). Development of a Ru complex-incorporated MOF photocatalyst for hydrogen production under visible-light irradiation. *Chemical Communications*, *50*, 6779−6781. Available from https://doi.org/10.1039/C4CC02397H.

Toyao, T., Saito, M., Horiuchi, Y., Mochizuki, K., Iwata, M., Higashimura, H., & Matsuoka, M. (2013). Efficient hydrogen production and photocatalytic reduction of nitrobenzene over a visible-light-responsive metal-organic framework photocatalyst. *Catalysis Science & Technology*, *3*, 2092−2097. Available from https://doi.org/10.1039/C3CY00211J.

Wang, C., deKrafft, K. E., & Lin, W. (2012). Pt nanoparticles@photoactive metal−organic frameworks: Efficient hydrogen evolution via synergistic photoexcitation and electron injection. *Journal of the American Chemical Society*, *134*, 7211−7214. Available from https://doi.org/10.1021/ja30053Пp.

Wang, R., Gu, L., Zhou, J., Liu, X., Teng, F., Li, C., & Yuan, Y. (2015). Quasi-polymeric metal−organic framework UiO-66/g-C3N4 heterojunctions for enhanced photocatalytic hydrogen evolution under visible light irradiation. *Advanced Materials Interfaces*, *2*, 1500037. Available from https://doi.org/10.1002/admi.201500037.

Wang, X., Liu, J., Leong, S., Lin, X., Wei, J., Kong, B., & Wang, H. (2016). Rapid construction of ZnO@ZIF-8 heterostructures with size-selective photocatalysis properties. *ACS Applied Materials & Interfaces*, *8*, 9080−9087. Available from https://doi.org/10.1021/acsami.6b00028.

Worrall, S. D., Mann, H., Rogers, A., Bissett, M. A., Attfield, M. P., & Dryfe, R. A. W. (2016). Electrochemical deposition of zeolitic imidazolate framework electrode coatings for supercapacitor electrodes. *Electrochimica Acta*, *197*, 228−240. Available from https://doi.org/10.1016/j.electacta.2016.02.145.

Xu, Y., Lv, M., Yang, H., Chen, Q., Liu, X., & Wei, F. (2015). BiVO4/MIL-101 composite having the synergistically enhanced visible light photocatalytic activity. *RSC Advances*, *5*, 43473−43479. Available from https://doi.org/10.1039/C4RA11383G.

Yuan, Y.-P., Yin, L.-S., Cao, S.-W., Xu, G.-S., Li, C.-H., & Xue, C. (2015). Improving photocatalytic hydrogen production of metal−organic framework UiO-66 octahedrons by dye-sensitization. *Applied Catalysis B: Environmental*, *168−169*, 572−576. Available from https://doi.org/10.1016/j.apcatb.2014.11.007.

Zhang, H., Wei, J., Dong, J., Liu, G., Shi, L., An, P., & Ye, J. (2016). Efficient visible-light-driven carbon dioxide reduction by a single-atom implanted metal−organic framework. *Angewandte Chemie International Edition*, *55*, 14310−14314. Available from https://doi.org/10.1002/anie.201608597.

Zhang, J.-P., Zhang, Y.-B., Lin, J.-B., & Chen, X.-M. (2012). Metal azolate frameworks: From crystal engineering to functional materials. *Chemical Reviews*, *112*, 1001−1033. Available from https://doi.org/10.1021/cr200139g.

Zhang, M., Wang, L., Zeng, T., Shang, Q., Zhou, H., Pan, Z., & Cheng, Q. (2018). Two pure MOF-photocatalysts readily prepared for the degradation of methylene blue dye under visible light. *Dalton Transactions*, *47*, 4251−4258. Available from https://doi.org/10.1039/C8DT00156A.

Zhang, T., & Lin, W. (2014). Metal−organic frameworks for artificial photosynthesis and photocatalysis. *Chemical Society Reviews*, *43*, 5982−5993. Available from https://doi.org/10.1039/C4CS00103F.

Zhou, L., Zhang, H., Sun, H., Liu, S., Tade, M. O., Wang, S., & Jin, W. (2016). Recent advances in non-metal modification of graphitic carbon nitride for photocatalysis: A historic review. *Catalysis Science & Technology, 6,* 7002−7023. Available from https://doi.org/10.1039/C6CY01195K.

Zhou, T., Du, Y., Borgna, A., Hong, J., Wang, Y., Han, J., & Xu, R. (2013). Post-synthesis modification of a metal−organic framework to construct a bifunctional photocatalyst for hydrogen production. *Energy & Environmental Science, 6,* 3229−3234. Available from https://doi.org/10.1039/C3EE41548A.

Zhu, B., Zou, R., & Xu, Q. (2018). Metal−organic framework based catalysts for hydrogen evolution. *Advanced Energy Materials, 8,* 1801193. Available from https://doi.org/10.1002/aenm.201801193.

Zhu, Q.-L., & Xu, Q. (2014). Metal−organic framework composites. *Chemical Society Reviews, 43,* 5468−5512. Available from https://doi.org/10.1039/C3CS60472A.

CHAPTER 11

Transition metal chalcogenide—based photocatalysts for small-molecule activation

Afsar Ali[1] and Arnab Dutta[2]

[1]Chemsitry Discipline, IIT Gandhinagar, Gandhinagar, India, [2]Chemistry Department, IIT Bombay, Mumbai, India

11.1 Introduction

The rapid growth of the human population and industrial expansion has instigated environmental pollution and energy crisis on a global scale. Currently, the major focus has been shifted toward the generation of green and sustainable energy resources in place of fossil fuels for the development of environmental-friendly energy infrastructure (Balat & Balat, 2009; Walter et al., 2010). The consumption of fossil fuel in various sectors, including transportation, industry, and household, enhances the emission of greenhouse gases, such as carbon dioxide, carbon monoxide, nitrogen oxide, and sulfur oxide in the atmosphere (Haque, Daeneke, Kalantar, & Ou, 2018). The renewable energy resources like solar, wind, hydropower, and geothermal have emerged as a potential alternative for the adequate supply of sustainable and clean energy. However, the lack of appropriate technological advancements has stalled a worldwide implementation of renewable energy. Sunlight is the predominant and most readily available renewable resource of clean energy, which has a high potential for the production of environmental-friendly energy via photocatalytic (PC) reaction. Although several factors, such as variation in solar irradiation over the different times in a day, season, and geographical position, have severely affected its usage (Rahmanian, Malekfar, & Pumera, 2017; Su, Shao, Qin, Guo, & Wu, 2018).

An efficient strategy for the conversion of sunlight to chemical energy is the key for proper utilization of solar energy. The natural photosynthesis is a perfect example of this approach, which has successfully sustained the life on earth by harvesting solar energy. The presence of a photocatalyst plays a pivotal role in this energy transformation process (Maeda et al., 2006; Yoon, Ischay, & Du, 2010). The photocatalysis can not only aid the generation of green energy but also in the alleviation of hazardous pollutants. Typically, organic dye molecules and metal-based semiconductors are the primary choices for the role of photocatalyst (Hernandez-Alonso, Fresno, Suarez, & Coronado, 2009). In 1972 Fujishima and Honda reported the use of semiconductor TiO_2 electrode for PC water splitting (Inoue, Fujishima, Konishi, & Honda, 1979).

The semiconductors are activated by appropriately energetic electromagnetic irradiation, which is equal or greater than the band gap energy of the material. During this excitation, the electrons are excited from the valence band (VB) to the conduction band (CB) to induce charge separation. The hot electrons generated in the CB can potentially perform a reduction reaction in the presence of appropriate cocatalyst (such as H_2 production from water; Fig. 11.1A; Zhang, Chen, & Bahnemann, 2009). Reduction of CO_2 competes with hydrogen production during this step (Maeda et al., 2006). Photoelectrons can reduce CO_2 into CO, CH_4, or other forms of reduced carbon in the presence of water. Analogously, the hot holes can induce oxidation reactions, such as O_2 evolution from water (Fig. 11.1A). An ideal photocatalyst typically contains two characteristic functional features. First, the band-edge values of the semiconductor ensures that both reduction potential of H_2/H^+ (-4.44 eV at pH = 0) and oxidation potential of H_2O/O_2 (-5.67 eV at pH = 0) are satisfied. Second, the band gap complements the absorption of solar irradiation in the visible region. A fraction of these photogenerated electrons and holes can recombine and lose the energy in the form of heat and (Fig. 11.1B). Besides, the PC oxidation process can take place on the surface of semiconductors by the formation of hydroxyl radical and radical anions. As a result, this is helpful for the degradation of organic compounds (dyes, textile industry waste, pesticide, and other pollutants; Akpan & Hameed, 2009). Holes generated in the VB have lower potential than OH^-/H_2O and react with water molecule or OH^- to produce hydroxyl radical that is strong

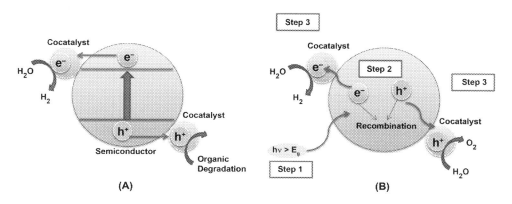

Figure 11.1
(A) Schematic energy diagram for photocatalysts and (B) recombination of electrons and holes (Peng, Li, Zhang, & Zhang, 2017). Source: *Taken with permission from Peng, W., Li, Y., Zhang, F., & Zhang, G. (2017). Roles of two-dimensional transition metal dichalcogenides as cocatalysts in photocatalytic hydrogen evolution and environmental remediation.* Industrial & Engineering Chemistry Research, *56, 4611 – 462.*

oxidizing agents. Meanwhile, electrons in CB are received by electron-accepting species and generate radical anions (Turchi & Ollis, 1990).

Till now, a range of semiconductor reported like TiO_2, ZnO, WO_3, MoO_3, and CdS, which act as photocatalysts for the utilization of solar energy, but these have limited application due to the rapid electron—hole recombination (Fig. 11.1; Liu, Yu, & Zhang, 2013; Perera, Mariano, Nour, Seitz, & Chabal, 2012). The PC activity of semiconductors can be enhanced by the loading of noble metal cocatalysts. However, noble metals are very rare and expensive. Here, the synthesis of cheap and highly active cocatalyst is a difficult task in the field of photocatalyst (Lu, Yu, Ma, Chen, & Zhang, 2016).

Transition metal chalcogenides (TMCs) like WS_2, MoS_2, and CdS_2 have shown excellent activity as cocatalysts for the modification of semiconductors (Bhandavat, David, & Singh, 2012). The properties of TMCs are variable due to these found in different crystalline structures and sequences of their nanosheets. Loading of TMCs as cocatalyst creates more interface with semiconductors that are responsible for enhancing charge separation and electron transport and decrease the activation energy and overpotential in case of PC reaction (water splitting), leading to activity improvements (Zong et al., 2008). Here, TMCs have a special 2D-layered structure that provides a more active site, reduce the mobility, and avoid agglomeration of semiconductor nanoparticles. On the basis of different types, properties of TMCs make more feasible candidates compare to noble metals for the synthesis of photocatalysts (Bai, Wang, Chen, Du, & Xiong, 2014).

11.2 Structural classification of transition metal chalcogenides

TMCs have various applications in the field of photocatalysts, development of sensor, tunneling device, transistor, optoelectronics, and hydrogen evolution. The catalytic activity of TMCs can be enhanced in two ways: the first one by increasing the number of active sites and the second one by modifying chemical composition to reduce ΔG_H (intrinsic; Chianelli et al., 2006). A broad range of TMCs has been synthesized by changing the structure and composition (extrinsic and intrinsic properties) of TMCs. Based on the architectural arrangement, TMCs have been classified in the following categories: (1) binary, (2) ternary, and (3) quaternary structure.

11.2.1 Binary structure of transition metal chalcogenides

Binary TMCs generally known as transition metal dichalcogenide (TMD) contains one type of transition metal and one type of chalcogen in their chemical formula. Here, the band gap of TMCs is $\sim 1-2$ eV, which does not exist in graphene; so TMCs are capable of absorbing the full spectrum of solar irradiation, ranging from the visible to infrared. This feature is crucial for potential applications of TMCs in the field of optoelectronics and photonics.

TMDs exist in three polymorphs: 1T (trigonal), 2H (hexagonal), and 3R (rhombohedral), where the preceding numerical value indicates the number of layers in a unit cell. Based on symmetry, 2H and 1T belong to the point groups d_{6h} and d_{3d}, respectively. TMC assemblies can exhibit metallic, semiconductor, or insulator behaviors depending on the number of d-electrons of the transition metal. D_{3d}- and D_{3h}-based symmetries generally break in two sets $(d_z^2, d_{x^2-y^2}$ and $d_{xy,xz,yz})$ and three sets $(d_z^2, d_{x^2-y^2,xy}$ and $d_{xz,yz})$ of degeneracy, respectively (Fig. 11.2).

MoS_2, WSe_2, and $MoSe_2$ are representative examples of TMDs. Naturally, MoS_2 occurs in the 2H form with trigonal prismatic coordination of the metal atoms, in which $S-Mo-S$

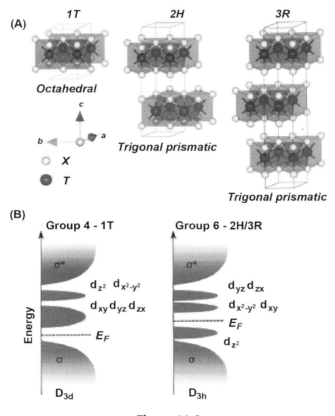

Figure 11.2
(A) Transition metal dichalcogenides (T = W, Mo, X = S, Se, Te) are layered materials. (B) Schematic representation of density state of transition metal group, corresponding to d-orbital splitting due to symmetry (Agnieszka & Thomas, 2015). Source: *Taken with permission from Agnieszka, K., & Thomas, H. (2015). The electronic structure calculations of two-dimensional transition-metal dichalcogenides in the presence of external electric and magnetic fields.* Chemical Society Reviews, 44 *(9),* 2603–2614.

layers have van der Waals interaction with each other to allow spontaneous exfoliation (Kisielowski et al., 2010). The thickness between each layer of MoS_2 has been found to be 0.64 nm and prevails in 2H (two units per unit cell) and 1T (trigonal) phases. Here, 2H polytype is more dominant and stable compared to 1T polytype (Tiwari et al., 2015). On the other hand, a single layer of WSe_2 contains a 0.67-nm thick slab in which $Se-W-Se$ found in a sandwich layer. WSe_2 exhibits 2H crystal and metallic 1T phase, in which 2H phase has two layers per unit cell stack in hexagonal symmetry with trigonal prismatic coordination whereas the metastable 1T metallic phase consists of one unit cell in tetragonal symmetry with octahedral coordination (Cheng, Sun, & Hu, 2016).

11.2.2 Ternary structure of transition metal chalcogenides

Typically, ternary TMCs contain two different combinations of transition metal and chalcogenide. In the first category, two different transition metals and one chalcogenide produce the TMC (e.g., HfGeTf4) while in the other form, one transition metal combines with two different chalcogenides (e.g., MoSSe; Guan, Ni, & Hu, 2018; Sheeba, Israel, & Saravanakumar, 2016). The ratio of transition metal and chalcogens in the TMCs regulates their electronic and optical properties. $HfGeTf_4$ acts as noble layered materials and contains the van der Waals interaction gap in the zigzag pattern to increase the

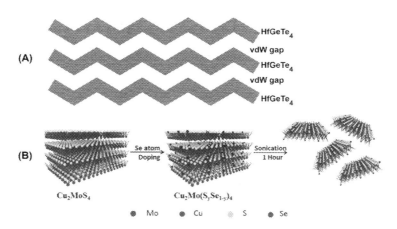

Figure 11.3

(A) Crystal structure of $HfGeTf_4$ containing van der Waals interaction gap (vdW gap) between layers. (B) Layered and crystal structure of quaternary $Cu_2Mo(S_ySe_{1-y})_4$ (Mar & Ibers, 1993; Tiwari, Kim, Kim, Prakash, & Lee, 2016). Source: *Taken with permission from Mar, A., & Ibers, J. A. (1993). The layered ternary germanium tellurides $ZrGeTe_4$, $HfGeTe_4$, and $TiGeTe_6$. Structure, bonding, and physical properties.* Journal of the American Chemical Society, 115 (8), 3227; Tiwari, A. P., Kim, D., Kim, Y., Prakash, O., & Lee, H. (2016). *Highly active and stable layered ternary transition metal chalcogenide for hydrogen evolution reaction.* Nano Energy, 28, 366—372.

surface area compared to TMDs. $HfGeTf_4$ is found in the orthorhombic crystal structure that contains interlayer distance $3.6-4.3$ A$^\circ$, depending on the combination of two atoms (Fig. 11.3).

11.2.3 Quaternary structure of transition metal chalcogenides

Quaternary TMCs contain two different transition metals and two distinct chalcogen atoms in their chemical formula, such as $Cu_2Mo(S_ySe_{1-y})_4$. This Cu-based quaternary TMC is typically synthesized by solution processing methods and it possesses excellent electrochemical properties. In quaternary TMCs, the layers are oriented via van der Waals interactions and covalent bonds (Tiwari et al., 2016). In quaternary TMCs, the chalcogen lattice has distorted cubic close-packed chalcogen atoms and the transition metal occupies 3/4 tetrahedral hole in alternate layers. Quaternary $Mo_{1-x}W_xSe_{2(1-x)}S_2$ has shown higher band gap variability, ranging from 1.60 to 2.03 eV, in comparison to binary TMCs (Pruss, Snyder, & Stacy, 1993).

11.3 Methods for the enhancement of photocatalytic performance of transition metal chalcogenides

Different types of methods have been reported for enhancing the PC properties of TMCs through elemental doping, surface functionalization, and heterojunction.

11.3.1 Elemental doping

Surface properties and band structure of the materials can be altered by elemental doping for enhancing the redox activities and efficient harvesting of visible light. Generally, the dopant has been classified into two types: metallic and nonmetallic (Zhang et al., 2009). Nonmetal elements like N, B, F, P, and C have been reported to act as a dopant for transition metal oxide chalcogenides (TMOCs). In the case of nonmetallic dopant, the band gap of TMOCs is typically narrow due to the overlapping between the orbitals of dopant and oxygen/chalcogenide atoms. As a result, photogenerated electrons excited from the dopant band state instead of VB edge to the CB (Asahi, Morikawa, Ohwaki, Aoki, & Taga, 2001). Liu et al. have synthesized N-doped MoS_2 nanosheets by using sol−gel methods, in which thiourea $[(NH_2)_2CS]$ acts as N dopant source. After doping of N, band energy in N-doped MoS_2 (2.08 eV) was found to be slightly lower than MoS_2 (2.17 eV) nanosheets. N-doped MoS_2 contains PC activities toward photodegradation of Rhodium B (RhB) dye at a rate of 134 μmol/g/h, whereas MoS_2 nanosheets remain inactive (Fig. 11.4; Liu, Liu, Ye, Ma, & Gao, 2016). Similarly, N-doped TiO_2 also demonstrated H_2 production activity with an almost double catalytic rate compared to TiO_2 nanosheets. N-doped TiO_2 enhanced visible light absorption from 350 to 500 nm and increases the number of exposed

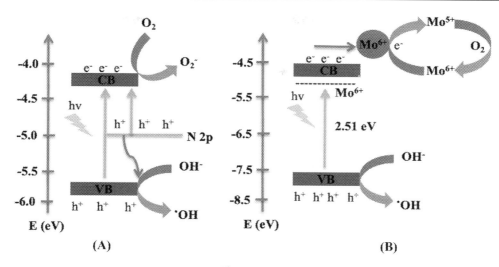

Figure 11.4
(A) Nonmetallic doping (N doped in MoS_2 nanosheets). (B) Metallic doping (Mo doped in WO_3 nanosheets); both are reported for Rhodium B photodegradation (Li, Teng, Zhang, Zhou, & Liu, 2015). *Source: Taken with permission from Li, N., Teng, H., Zhang, L., Zhou, J., & Liu, M. (2015). Synthesis of Mo doped WO_3 nanosheets with enhanced visible-light-drive photocatalytic properties.* RSC Advances, 5 (115), 95394—95400; Lin, Z., Li, J., Zheng, Z., Li, L., Yu, L., Wang, C., & Yang G. (2016). A floating sheet for efficient photocatalytic water splitting. Advanced Energy Materials, 6, 1600510.

facets {0 0 1} in crystal structure, which is highly active catalytic site for TiO_2 nanosheets (Xiang, Yu, Wang, & Jaroniec, 2011; Xu, Ouyang, Li, Kako, & Ye, 2013).

In the case of metallic doping, the ionic radius and valence state of the dopant play a crucial role in the determination of electronic properties and occupied sites. Li et al. synthesized Mo-doped WO_3 nanosheets by using ammonium molybdate as the dopant source. Mo ions distributed inside Wo lattice without disturbing the monoclinic crystal structure because the ionic radius of Mo is closer to W. Doping of WO_3 with Mo generates a donor level near to the CB of WO_3 to increase the absorption intensity of visible light. Here, band gap for Mo-doped WO_3 also decreases from 2.56 to 2.36 eV (Li et al., 2015). Both metallic and nonmetallic dopants helpful for the separation of photogenerated charge barriers and at optimum concentration dopant act as a trap site for one type of charge carrier and for another one allowing to surface for desired chemical redox reaction (Serpone, Lawless, Disdier, & Herrmann, 1994). Oxygen vacancies in TMOs also deviate band gap energy via the formation of localized defect within the band gap. Few examples of 2D TMOCs can accommodate a large number of oxygen/chalcogen vacancies, which increase free charge carrier concentration up to $10^{21}/cm^3$ and transform semiconductor to quasimetallic properties (Alsaif, Chrimes, Daeneke, Balendhran, & Bellisario, 2016).

11.3.2 Surface functionalization

Functionalization of 2D TMOC, with specific organic molecules, can improve PC activity by increasing the adsorption of surface species. Xue et al. functionalized 2D ZnO nanosheets with an amine group to enhance the surface capture of CO_2 molecules for PC molecular CO_2 reduction. CO_2 activation takes place by the formation of the $N-C$ bond between the amine group and CO_2 molecules and produced carbamate as intermediate. The reduction of carbamate into CO and CH_4 carried out by direct interaction with Zn^{2+} to received electrons from ZnO. Therefore the rate for CO_2 reduction into CO and CH_4 is higher to amine-functionalized ZnO nanosheets compared to bare ZnO (Liao, Hu, Gu, & Xue, 2015).

The incorporation of dye molecules on the surface of TMOC is an emerging approach to enhance visible light harvesting. Noble metal-free zinc porphyrin complexes and various dye molecules act as photosensitizers that have been reported for dye-sensitized solar cells and contain the highest conversion efficiency ($>10\%$; Mathew, Yella, Gao, Baker, & Curchodet, 2014). Zinc porphyrin complex is an attractive photosensitizer due to delocalized p-electrons, intense absorption in the visible light region, suitable redox potential for electron injection, and dye regeneration.

Yuan et al. have reported 2D MoS_2 nanosheets heterostructured with TiO_2, ZnO, and incorporated with zinc porphyrin for hydrogen production from water (Fig. 11.5; Yuan, Tu, Ye, Lu, & Ji, 2015). In this system, after visible light illumination, electrons excited from HOMO to LUMO of zinc porphyrin complex. After that, electrons are injected from zinc porphyrin complexes to TMO because dye molecules have relatively negative oxidation potential.

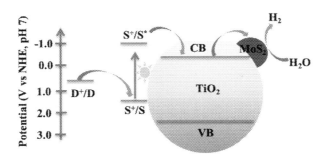

Figure 11.5

The schematic diagram of electronic band structure for Zn porphyrin-based dye-sensitized on the surface of TiO_2/MoS_2 (Yuan, Lu, Ji, Zhong, & Ding, 2015). Source: *Taken with permission from Yuan, Y., Lu, H., Ji, Z., Zhong, J., & Ding, M. (2015). Enhanced visible-light-induced hydrogen evolution from water in a noble metal-free system catalyzed by ZnTCPP-MoS₂/TiO₂ assembly*. Chemical Engineering Journal, 275, 8–16.

After that, an excited electron has been collected by 2D MoS_2 nanosheets which act as cocatalyst for hydrogen evolution reaction (HER) from water. The longevity of this dye-sensitized zinc porphyrin-based system is not promising due to the short lifetime of the dye.

11.3.3 Heterojunction

A correct combination of VB and CB energy levels between the semiconductor and 2D TMOC has an efficient transfer pathway for photogenerated charge carriers from one to another. Generally, two types of alignments are known, in which type II is most popular compared to type I. In case of type-II, the photogenerated electrons move from a higher postive CB edge to the lower, while the holes move from a higher negative VB edge to the lower. Whereas in type I assembly, the photogenerated electrons and holes from CB and VB, respectively relocate to another semiconductor (Fig. 11.6; Jafari et al., 2016). The size difference between the coupled semiconductor and dimensionality plays a vital role in hetero-interfacial contact. Here, defect density and crystallinity can alter the band structure of materials. 2D/2D heterostructure of semiconductor shows better stability coupling heterointerface due to short exciton diffusion length and large contact surface area compared to other 2D/low dimensional hetero-structure, which is essential in the transfer and separation of photoexcited pairs (Zhou, Lou, & Xie, 2013).

2D MoS_2-based heterojunction has excellent electrocatalytic HER performance due to high electron mobility. Graphic carbon nitride (g-C_3N_4) is a popular 2D semiconductor that

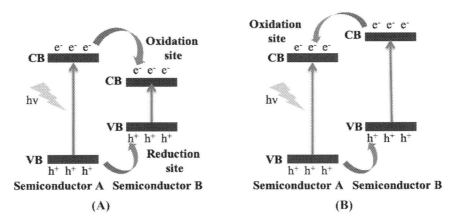

Figure 11.6

Semiconductor heterojunction (A) type I and (B) type II band alignment (Lo, Mirkovic, Chuang, Burda, & Scholes, 2011). Source: *Taken with permission from Lo, S. S., Mirkovic, T., Chuang, C. H., Burda, C., & Scholes, G. D. (2011). Emergent properties resulting from type-II band alignment in semiconductor nano heterostructures.* Advanced Materials, 23,*180—197.*

forms heterojunction with 2D MoS_2 for water splitting as it contains proper CB and VB positions and optimum band gap energy (2.7 eV), suitable for visible light harvesting (Ge, Han, Xiao, & Guo, 2013).

In addition, g-C_3N_4 has edge potential -2.8 eV versus vacuum, which is less negative than MoS_2 (-4.2 V), allowing migration of electron from g-C_3N_4 to MoS_2. On the other side, hole generated from MoS_2 transferred to g-C_3N_4 due to more negative VB potential of MoS_2 (-6.0 V) compared to g-C_3N_4 (-5.5 V); hence, efficient charge separation takes place. Heterojunction of 2D MoS_2 with g-C_3N_4 can improve delocalized conductivity and low charge recombination and enhance the absorption of solar light compared to individual g-C_3N_4 (Mahler, Hoepfner, Liao, & Ozin, 2014). Therefore g-C_3N_4 loaded with 1% 2D MoS_2 has onefold higher HER performance (35.6 μmol/g/h) and 20% more enhancement PC degradation of RhB and methylene blue (MB) compared to g-C_3N_4 without 2D MoS_2 loading (Ge et al., 2013).

2D ZnO is another example of heterojunction with 2D MoS_2, which contains type II band alignment. In this heterojunction, 2D ZnO and 2D MoS_2 have very close potential, -4.4 and -4.2 V, respectively (Awasthi et al., 2016).

In addition, $CuInS_4$ and $ZnInS_4$ investigated, which have narrow band gap (1.5–1.9 eV for $CuInS_4$ and 2.4–2.5 eV for $ZnInS_4$) and suitable band structure for heterojunction with 2D MoS_2. Here, the rate for hydrogen production for 2 wt.% 2D MoS_2 with $CuInS_2$ is two times higher (316 μmol/g/h) compared to $CuInS_2$ and three times higher than platinum (Pt)-loaded $CuInS_2$. On the other hand, 3 wt.% 2D MoS_2/$ZnInS_4$ heterojunction has two times higher rate for hydrogen production compared to Pt-loaded $ZnInS_4$ (Wei et al., 2014).

Within the type II alignment, inclusion of p–n heterojunction is another practical approach for decreasing charge recombination. In this system, p- and n-type combinations produce a space charge layer and enhance the electric field that can increase the probability of electron–hole separation (Ida, Takashiba, Koga, Hagiwara, & Ishihara, 2014). Meng et al. have synthesized p–n heterojunction by using p-2D MoS_2 and n-2D reduced graphene oxide (rGO) to improve the PC activity of MoS_2. In this combination, MoS_2 acts as the dual role of catalytic and photo centers for absorbing solar light (Fig. 11.7; Meng, Li, Cushing, Zhi, & Wu, 2013).

Xing et al. also synthesized p–n heterojunction by using of n-CdS and p-CdTe nanocrystals to improve the charge recombination suppression and enhance absorption of visible light. After that, 2D $Ti_{0.91}O_2$ nanosheets are combined with p–n heterojunction to provide an electron sink and generate catalytic site for hydrogen production. As a result, the hydrogen production rate is 463 μmol/g/h observed in the case of p–n heterojunction with TiO_2 nanoparticles, but TiO_2 nanosheets exhibited no hydrogen production (Xing et al., 2016).

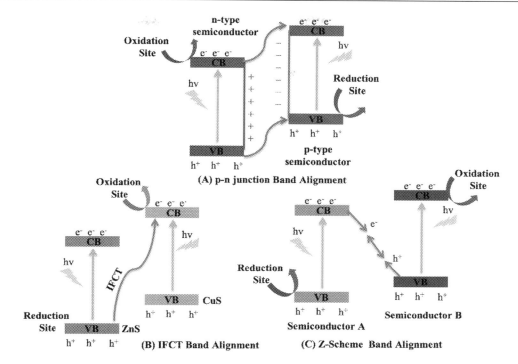

Figure 11.7

Schematic images for (A) p—n junction band alignment, (B) interfacial charge transfer band alignment, and (C) Z-scheme band alignment (Zha, Nadimicherla, & Guo, 2015; Zhang, Yu, Zhang, Li, & Gong, 2011; Zhu et al., 2016). Source: *Taken with permission from Zha, R., Nadimicherla, R., & Guo, X. (2015). Ultraviolet photocatalytic degradation of methyl orange by nanostructured TiO$_2$/ZnO heterojunctions.* Journal of Materials Chemistry A, 3 *(12), 6565—6574, Zhang, J., Yu, J., Zhang, Y., Li, Q., & Gong, J. R. (2011). Visible light photocatalytic H$_2$-production activity of CuS/ZnS porous nanosheets based on photoinduced interfacial charge transfer.* Nano Letters, 11 *(11), 4774—4779, and Zhu, C., Zhang, L., Jiang, B., Zheng, J., Hu, P., Li, S., . . . Wu, W. (2016). Fabrication of Z-scheme Ag$_3$PO$_4$/MoS$_2$ composites with enhanced photocatalytic activity and stability for organic pollutant degradation.* Applied Surface Science, 377, 99—108.

Zhang et al. developed photocatalyst containing 2D ZnS-0D CuS nanocluster heterojunction. Photoelectron from VB of ZnS transferred to CB of CuS cluster due to the interfacial charge transfer (IFCT) mechanism after the irradiation of visible light (Fig. 11.7B). After the partial reduction of CuS to Cu$_2$S , it initiates the effective reduction of H$^+$ to H$_2$. In this case, the potential for CuS/Cu$_2$S is higher than H$^+$/H$_2$O. Hydrogen production rate is 4.2 mmol/g/h, which is eight times larger than Cu-doped Zn nanoparticles (Zhang, Yu, Zhang, Li, & Gong, 2011).

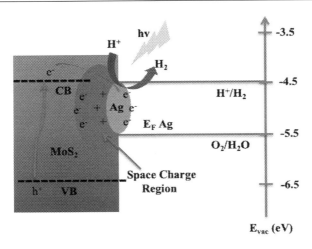

Figure 11.8

Schematic diagram of heterojunction between 2D MoS_2 nanosheet with Ag nanoparticle (Liang et al., 2015). Source: *Taken with permission from Liang, Y., Guo, N., Li, L., Li, R., Ji, G., & Gan, S. (2015). Fabrication of porous 3D flower-like Ag/ZnO heterostructure composites with enhanced photocatalytic performance.* Applied Surface Science, 332, 32–39.

Yu et al. developed 2D ZnO–CdS nanoparticle heterojunction. In this system, the recombination of electrons with holes occurs at the interface of heterojunction. As a result, the retention of photogenerated electrons takes place in CdS with higher CB position and holes in ZnO with a lower VB position. In this mechanism, ZnO acts as a metallic polar surface, which initiates the Z-scheme transfer process (Yu et al., 2013).

Similar, a rectified charge transfer can take place in the Schottky barrier between semiconductor and metal, which depends upon their Fermi level position (Linsebigler, Lu, & Yates, 1995). Generally, in this type of system, Fermi level lift toward more negative value after the irradiation of visible light. After that, energy difference at the semiconductor–metal interface is responsible for transferring electrons from the CB of semiconductor into the metal. The size of metal taken in the form of nanoparticles and weight ratios corresponding to 2D TMOCs are optimized to avoid the surface overloading effect (Liang et al., 2015).

Cheah et al. (2015) have reported heterojunction of Ag/MoS_2 nanocomposite photocatalyst for hydrogen gas evolution. The MoS_2 nanoparticles were decorated with in situ reduced Ag nanoparticles. In this assembly, Ag nanoparticles act as cocatalysts and enhance visible light absorption. Heterojunction generated from 20 wt.% Ag/MoS_2 nanocomposites shows PC hydrogen gas evolution (179.5 μmol H_2/g_{cat}), which is approximately two times higher than MoS_2 nanoparticles (92.0 μmol H_2/g_{cat}; Fig. 11.8; Liang et al., 2015).

11.4 Synthesis methods for transition metal chalcogenide

Generally, synthesis methods are classified into two categories, bottom-up approach, and top-down process. In a top-down approach, nanomaterials are synthesized by the removal of material from bulk solids. Whereas in a bottom-up approach, a range of atomic and molecular precursors are utilized for nanomaterial production either via continuous growth or self-assembly.

11.4.1 Top-down approach

Nanomaterial synthesis takes place by physical or chemical methods: the physical top-down approach, applied by mechanical force or ultrasonic wave to van der Waals solid materials into 2D nanosheets, and chemical top-down approach, applied by ion exchange or by heat. Synthesis methods by top-down approach have been classified in the following methods:

1. Micromechanical exfoliation;
2. Ultrasonic exfoliation;
3. Ion-exchange exfoliation; and
4. Lithium-intercalated exfoliation.

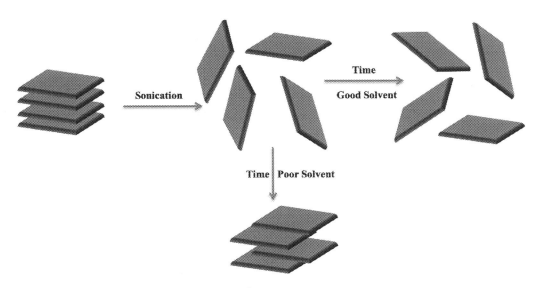

Figure 11.9

The solvent effect on the synthesis of nanosheets by applying ultrasonic exfoliation methods (Wang, Yang, & Xie, 2015). Source: *Taken with permission from Wang, H., Yang, X., & Xie, Y. (2015). Ultrathin black phosphorus nanosheets for efficient singlet oxygen generation.* Journal of the American Chemical Society, 137 (35),11376—11382.

11.4.1.1 Micromechanical exfoliation

This method is useful for obtaining single and few-layer TMOC from bulk solids. Crystal structure and all other properties (stoichiometry ratio and stacking order) remain unchanged after synthesis methods. Here, this method is reliable only for layered van der Waals solids, obtaining small-scale production due to the low yield of desired TMOC. This method is very beneficial for the synthesis of WS_2 and MoS_2 nanosheets (Chhowalla et al., 2013).

11.4.1.2 Ultrasonic exfoliation

Ultrasonic exfoliation is another method for delaminating van der Waals solids into a single few-layer nanosheet and more effective and higher productive compared to mechanical exfoliation. In this case, the time for sonication and suitable of appropriate solvents regulate the exfoliation. The selection of apropriate solvents avoids re-aggregation of exfoliated nanosheets and such solvents can also be used as dispersing media for delaminating the van der Waals solids. This method is also useful for obtaining high-purity single-layer 2D material, which is primarily needed for electronic application.WS_2, CdS, and MoS_2 nanosheets can be synthesized by ultrasonic methods (Fig. 11.9; Coleman, Lotya, & Neill, 2011; Smith, King, Lotya, & Coleman, 2011).

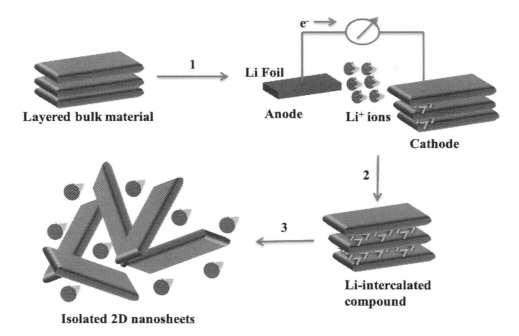

Figure 11.10

Lithium-intercalated exfoliation method for fabrication of 2D MoS_2 nanosheets (Eda, Yamaguchi, & Chhowalla, 2011). Source: *Taken with permission from Eda, G., Yamaguchi, H., & Chhowalla, M. (2011). Photoluminescence from chemically exfoliated MoS_2. Nano Letters, 11 (12), 5111–5116.*

11.4.1.3 Ion-exchange exfoliation

The abovementioned synthetic methods are useful for van der Waals solids, but in the case of ionic solid, either of them is not feasible. The strong ionic bonding in the layer structure is the primary reason for this observation. Ion-exchange exfoliation method has emerged as one of the best alternative methods. Ju et al. synthesized $LiCoO_2$ monolayer by the ion-exchange exfoliation method (Kim, Oh, & Hwang, 2009).

11.4.1.4 Lithium-intercalated exfoliation

Lithium-intercalated exfoliation method has produced single nanosheet, which is not possible by other exfoliating methods. In this method, the yield of single-layer formation is controlled by the formation of Li_xXS_2. Here, this method has 100% single-layer formation to TMC. On the other hand, the reaction has been carried out very carefully due to a high temperature for a long time, preventing the formation of metal nanoparticles and the precipitation of Li_2S. In this case, formations of intercalated ions are responsible for weakening interlayer bond in the basal plane by charge transfer (Fig. 11.10; Zhang, Shin, & Eda, 2013).

Chhowalla et al. synthesized monolayer of MoS_2 by lithium intercalation and exfoliating methods. Here, bulk MoS_2 crystals were taken in hexane solution containing butyllithium

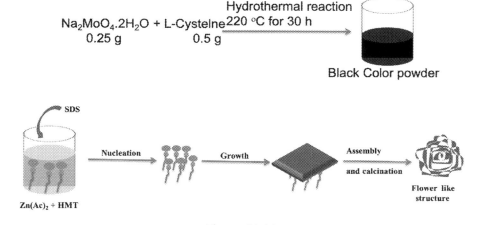

Figure 11.11
Schematic representation of hydrothermal methods for 2D and 3D nanosheets (Das, Mishra, Srivastava, & Kayastha, 2017; Miao, Zhang, Yuan, Jiao, & Zhu, 2016). Source: *Taken with permission from Das, R., Mishra, H., Srivastava, A., & Kayastha, A. M. (2017). Covalent immobilization of b-amylase onto functionalized molybdenum sulfide nanosheets, its kinetics and stability studies: A gateway to boost enzyme application.* Chemical Engineering Journal 328, 215—227; Miao, Y., Zhang, H., Yuan, S., Jiao, Z., & Zhu, X. (2016). Preparation of flower-like ZnO architectures assembled with nanosheets for enhanced photocatalytic activity. Journal of Colloid and Interface Science, 462, 9—18.

under anaerobic atmosphere. The follow-up exfoliation was executed by ultrasonicating Li_xMoS_2 in water for 1 hour (Eda et al., 2011). In this method, the semiconducting 2H phase of MoS_2 transformed into the metallic 1T phase.

11.4.2 Bottom-up approach

The bottom-up strategies are used for the synthesis of nanosheets from atomic and molecular precursors, which are suitable for reaction growth and self-assemble into more complex structures. The following are the classifications of the available bottom-up methods:

1. Hydro/solvothermal method;
2. Template method;
3. Microwave-assisted method; and
4. Chemical vapor deposition (CVD) method.

11.4.2.1 Hydro/solvothermal method

Hydro/solvothermal method is more sophisticated compared to other available synthetic methods due to low temperature ($100-240°C$), reaction time, reactant ratio, etc. MoS_2, TiO_2, ZnO, and Co_3O_4 can be synthesized by this method (Fig. 11.11A and B; Chen & Xue, 2014).

2D MoS_2 nanosheets have been synthesized by using MoO_3 and potassium thiocyanate (KSCN) as the sources of Mo and S, respectively. Besides, molybdate salts and organic acids are used for hydrothermal synthesis. Hydrothermal methods can synthesize three-dimensional nanosheet. Micro flower-shaped ($\{0\ 0\ 1\}$ orientation) 3D ZnO nanosheets have been produced in the presence of citric acid. However, 1D needle-like morphology prevails in the absence of citric acid. This structure was generated due to the etching in $\{001\}$

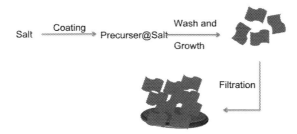

Figure 11.12

Schematic representation of the template method to synthesize the transition metal oxide chalcogenide (Xiao, Song, Lin, Zhou, & Zhan, 2016). Source: *Taken with permission from Xiao, X., Song, H., Lin, S., Zhou, Y., & Zhan, X. (2016). Scalable salt template synthesis of two-dimensional transition metal oxides. Nature Communications, 7, 11296.*

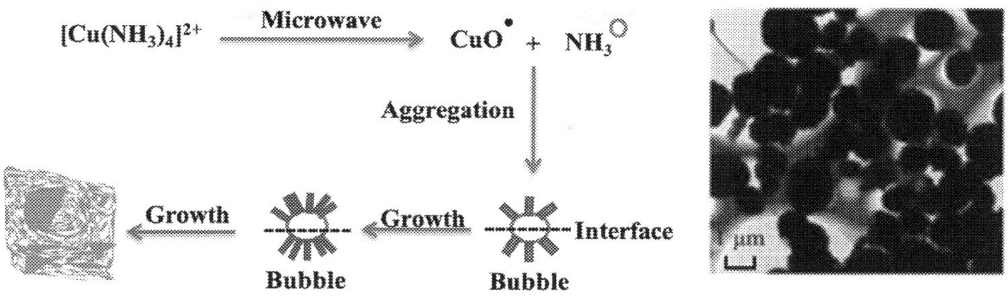

Figure 11.13

Schematic representation for microwave-assisted method (Zhang et al., 2015). Source: *Taken with permission from Zhang, Y., Liu, S., Liu, W., Liang, T., Yang, X., Xu, M., & Chen, H. (2015). Two-dimensional MoS₂-assisted immediate aggregation of poly-3-hexylthiophene with high mobility*. Physical Chemistry Chemical Physics, 17 (41), 27565—27572.

direction, in which KOH participated in the reaction (Mujtaba, Sun, Fang, Ahmad, & Zhu, 2015).

11.4.2.2 Template method

Template method used for the synthesis of nanosheets with confined in a specific direction and template can be removed by high temperature or adjusting pH. All types of nanomaterials like quantum dot (0D), nanowire (1D), 2D nanosheets, and 3D nanosheets can be synthesized, such as $CuInS_2$ and Fe_2O_3 (Fig. 11.12; Bi, Zhou, Ma, & Xie, 2012).

Cheon et al. have synthesized ultrathin ZrS_2 nanodiscs by using the template method in the presence of long-chain oleylamine or oleic acid, which acts as surfactants. In this method, the intermediate is composed of 2D arrays of $ZrCl_4$ and alky amine in which alkyl amine generally acts as the soft colloidal template (Zhu, Cao, & Li, 2014).

11.4.2.3 Microwave-assisted method

This method is more advantageous due to a much shorter time (a few minutes), higher energy efficiency and controlled shape and size of TMC compared to other methods. SnO_2, α-Ni $(OH)_2$, $K_{0.17}MnO_2$, and CuSe can be synthesized by this strategy. For example, α-Ni$(OH)_2$ has been synthesized by using $Ni(NO_3)_2 \cdot 6H_2O$, urea, deionized water, and ethylene glycol as a starting material (Fig. 11.13; Jang, Jeong, & Cheon, 2011; Zhu, Li, & Xie, 2013).

11.4.2.4 Chemical vapor deposition method

CVD method has a tendency to synthesize high-quality TMC layers with a scalable size, large surface area, uniform and controllable thickness, and excellent electronic properties. In this method, the desired product can be deposited on the surface of the substrate by applying high temperature. CVD method is mainly studied for synthesizing thin film

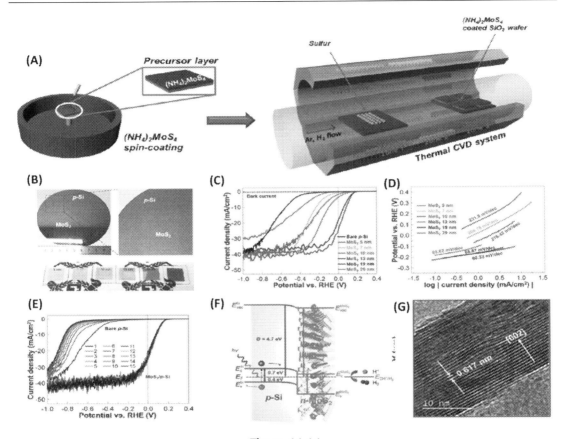

Figure 11.14

(A) Synthesis of MOS₂ thin films by chemical vapor deposition method. (B) Photograph of MoS₂ with different thicknesses on glass substrates. (C) J—V polarization curves for MoS₂ film of different thickness. (D) Tafel slopes of the MoS₂ layers (log(*j*) vs Reversible Hydrogen Electrode (RHE)). (E) Graph to compare stabilities of bare p-Si and MoSi/p-Si photocathodes. (F) Energy band diagram of the photocathode. (G) Transmission Electron Microscopy (TEM) images of MoS₂ (high resolution) (Tan et al., 2014). Source: *Taken with permission from Tan, Y., Liu, P., Chen, L., Cong, W., Ito, Y., Han, J., . . . Hirata, A. (2014). Monolayer MoS₂ films supported by 3D nanoporous metals for high-efficiency electrocatalytic hydrogen production.* Advanced Materials, 26 *(47), 8023—8028.*

coating of extensive range materials like metals, semiconductors, and insulators. CVD method is promising for the formation of high-quality, high-purity 2D nanomaterials with controlled properties. Graphene, MoS₂, and h-BN can be synthesized by CVD method (Kim, Hsu, & Kong, 2012; Li, Cai, An, & Rodney, 2009). Tan et al. (2014) have been reported coating of monolayer MoS₂ on nanoporous gold surface, in resultant catalytic activity of MoS₂ could improve by porous gold substrate (Fig. 11.14).

11.5 Applications

In the recent decade, nanomaterials based on TMC have useful application and explosive development in the field of research and technology due to their unique physical and chemical properties. TMC nanomaterials have promising applications in catalysis, electronics, energy conversion and storage, biomedicine and pollutant degradation, etc. (Acerce, Voiry, & Chhowalla, 2015; Kappera et al., 2014).

11.5.1 Catalysis

Mainly, catalytic activity depends upon surface functionality, chemical composition, and intrinsic electronic structure of the catalyst. Various factors play a pivotal role in the selection of catalysts like tunable surface properties, unique electronic structure, low production cost, and earth abundance. 2D TMC nanosheets tend catalytic applications, including HER, CO_2 reduction, and pollutant degradation (Wang, Zadeh, Kis, Coleman, & Strano, 2012).

11.5.1.1 Hydrogen evolution reaction

Production of H_2 as a clean and renewable energy vector from water is a major challenge in the field of PC chemistry. Pt is the most dominant element used as a catalyst for HER, but this is highly expensive and rarely found in earth's crust. So, researchers are focusing on the synthesis of photocatalyst for HER based on abundant, inexpensive, and novel hybrid materials. TMC- and TMC-based materials act as photocatalyst for HER from water splitting (Walter et al., 2010).

Norskov et al. have calculated free energy of hydrogen bonding ($\Delta G_H{}^*$) to MoS_2 and analyzed while basal plane of MoS_2 is catalytically inert; its sulfide Mo-edges are active for the HER. Also, the $\Delta G_H{}^*$ for sulfide Mo-edges (0.1 V) has been found nearly close to HER electrocatalysts like Pt, nitrogenase, and hydrogenase. MoS_2 supported with graphene also has good catalytic HER activity with an overpotential $0.1-0.2$ V at pH $= 0$ (Hinnemann et al., 2005). On the other hand, Lassalle-Kaiser et al. (2015) found that protonation and reduction of outer S2 site around Mo(III) is the rate-limiting step for HER activity. Some studies also revealed that in the case of pristine MoS_2, Mo-edges is more favorable than S-edges for HER, but doped MoS_2 containing both Mo- and S-edges contribute to HER activity, whereas other sulfur- and selenium-based TMCs from group 6d transition metal have the same trend as MoS_2. The stability of the active site is vital for the electrocatalytic process (Xiao, Liu, et al., 2016). Tsai et al. shown that catalytic activity and stability of the edges and basal plane of various TMC depend on the free energy of hydrogen adsorption on chalcogen atoms. The value of adsorption energy revealed that the robustness of the basal plane is lower for metallic TMC compared to that for semiconducting TMC (Tsai, Chan, Nørskov, & Pedersen, 2015).

11.5.1.1.1 Choice and selection of photocatalyst for water splitting

The selection of semiconductor materials for the PC process depends on the standard reduction potential for the redox process. For the catalytic activity, semiconductor material should match with the oxidation and reduction potentials of HER (from water) or water-splitting reaction (Preethi & Kanmani, 2013). Here, band positions of the catalyst have minimum potential difference 1.23 eV, in which the CB should have relatively negative potential than H^+/H_2 (0 V vs NHE) and VB should have comparatively positive potential than O_2/H_2O (1.23 V; Wu, Jiang, & Roy, 2015). Photocatalyst absorb photons and produce electron and hole pairs in the presence of suitable light source followed by the transfer of photogenerated carriers. For water splitting, photoexcited electron should react with water to generate H^+ reduction reaction to form hydrogen, and holes react with water to form oxygen through the oxidation reaction. The catalytic efficiency of photocatalysis depends on the excited electrons at the water/catalyst interface and minimum recombination of the excitons (Jiang, Moniz, Wang, Zhang, & Tang, 2017; Pinaud et al., 2013). The frequently utilized photocatalysts for water splitting absorb mostly in the visible region of the solar spectrum. A range of metal oxide photocatalysts like ZnO, TiO_2, WO_3, SnO_2, Ta_2O_5, CeO_2, CuO, ZrO_2, Fe_2O_3, and metal chalcogenides (ZnS, CdS, etc.), and their composites [$SrTiO_3$, TiO_2/C_3N_4, TiO_2-ZnO, Pt/TiO_2, etc., contain band gaps in the range of ultraviolet (UV)−visible light; Wang et al., 2016; Zhang, Zhou, et al., 2018]. The rate of hydrogen production from water splitting can be enhanced by tuning the band gap (3.1−1.6 eV) of the semiconductor. TMC has a high surface area and a large number of surface atoms due to the presence of the ultrathin layer (few to monolayer) compared to transition metal oxide. As a result, this is helpful for the reduction of exciton recombination, and it can also affect the redox potential of photocatalyst (Ahmad, Kamarudin, Minggu, & Kassim, 2015). 2D TMCs showcase anisotropic optical and electronic properties and a blue shift in the absorption. Suitable hybrid (heterogeneous nanocomposite) photocatalyst can enhance the activity of water splitting by decreasing the interaction of electron−hole pairs. Hybrid photocatalysts have specific features like harnessing the visible light spectra and reducing the recombination of excitons (Luo, Liu, & Wang, 2016).

11.5.1.1.2 Water-splitting mechanism

The various hydrogen production techniques have been classified into PC, photoelectrochemical (PEC), and photovoltaic−PEC (Jiang et al., 2017). Water-splitting reaction is a combination of HER and oxygen evolution reaction (OER), which proceeds through multiple electron transfer processes. These reactions are feasible in the presence of suitable photocatalyst with sufficient potential to attain overall water splitting (Fig. 11.15; Li, 2017).

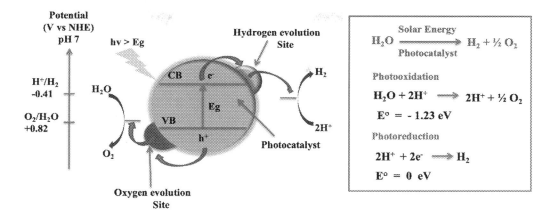

Figure 11.15

The schematic energy diagram for water splitting (Maeda & Domen, 2010). Source: *Taken with permission from Maeda, K., & Domen, K. (2010). Photocatalytic water splitting: Recent progress and future challenges.* The Journal of Physical Chemistry Letters, 1 *(18), 2655—2661.*

A typical PEC cell contains two segments. The first part is photoanode (semiconductor), and the second one is a counter electrode. In PEC cell, electron and hole are generated after irradiation of solar light of suitable frequency. The separation of the charged carriers occurs at the semiconductor—liquid interface. In the case of n-type semiconductor, photoexcited electron travels toward the counter electrode via an outer circuit and reduces water to H_2 (HER) while the hole moves toward the semiconductor surface and instigates water oxidation to produce O_2 (OER), whereas in the case of p-type semiconductor, reaction takes in the opposite direction (Jiang et al., 2017). For highly efficient water splitting, semiconductor's band-edge energies should reciprocate to the potential energy required for HER and OER in aqueous environment. The efficiency of hydrogen production from water mainly depends on properties of photocatalyst, such as surface adsorption capacity, charge utilization, transport and separation, light harvesting, stability, and durability (Preethi & Kanmani, 2013).

11.5.1.1.3 Role of transition metal chalcogenide for water splitting

2D TMC materials possess excellent electronic and optical properties, tunable thickness, and weak interlayer van der Waals forces that make them suitable for advanced device preparation (Wang & Mi, 2017). TMC family has approximately 60 compounds of the general formula MX_2, where X for chalcogen (S, Se, or Te) and M could be Sn, Re, Mo, W, or Pb (Ataca, Şahin, & Ciraci, 2012). TMC materials are classified into different groups based on electrical properties, such as insulator (HfS_2), semiconductor (WS_2, $MoSe_2$, and MoS_2), semimetals (WTe_2 and TcS_2), and pure metallic (VSe_2, NbS_2; Rasmussen & Thygesen, 2015).

Proper band alignment and tuning of band gap can influence the charge separation and catalytic HER rate. TMC semiconductor cocatalysts are used for the formation of heterojunction between two semiconductors that are actively involved in the enhancement of absorption of visible light and the effective separation of photogenerated carriers. Semiconducting TMC, such as MoS_2, WS_2, WSe_2, and $MoSe_2$, possesses direct band gap, and it existed in few to monolayer forms (Flatten et al., 2016). Mukherjee et al. shown that this band gap induces a blue shift with a decrease in size up to 4 nm of MoS_2 nanocrystals (Mukherjee, Maiti, Midya, Das, & Ray, 2015). The heterostructure formation with 2D TMC semiconductor as cocatalyst for the PC-based HER aided in the tuning of band gap toward the higher wavelength region of visible light. Few- to monolayer nanosheets of TMC, such as WS_2, $MoSe_2$, MoS_2, and WS_2, with large surface area thermodynamically favor the water splitting and possess extensive PC HER activity compared to pristine TMCs. CdS semiconductor possesses a narrow band gap (Eg = 2.4 eV), which is suitable for PC-based HER. However, the rate for hydrogen evolution is negatively impacted due to surface recombination and bulk recombination of the carriers (Chava, Do, & Kang, 2018). The catalytic activity of CdS can be improved by the formation of heterojunction with MoS_2 acting as a cocatalyst. In this case, MoS_2 immediately takes the photoexcited electron from the CB of CdS before they recombine with the holes. The holes accumulate at the VB of MoS_2 during this process and react with specific sacrificial agents. Hence, heterojunction of CdS with cocatalyst MoS_2 generates p–n junction, resists the photo corrosion, and enhances hydrogen evolution efficiency (Fig. 11.16; Zong et al., 2008).

Kumar et al. reported ultrathin MoS_2/CdS nanorod containing heterostructured photocatalyst for hydrogen evolution (HER rate of 174 mM/g/h) in the presence of sunlight (Kumar, Hong, Reddy, & Kim, 2016). Yin et al. also reported MoS_2/CdS nanosheets on nanorod hybrid heterostructure for hydrogen evolution under visible light (Yan, Du, & Phillips, 2017). Similarly, WS_2/CdS heterosystem displayed elevated hydrogen evolution rate (14.1 mM/g/h) in comparison to pure CdS (0.015 mM/g/h). Unsaturated active S atoms of MoS_2 at the exposed edge are responsible for this upgraded PC hydrogen evolution (Zhang, Fujitsuka, Du, & Majima, 2018).

Meng et al. reported p-MoS_2/n-rGO photocatalyst with a rate of 0.160 mM/g/h for hydrogen evolution. In this photocatalyst, p–n junction created space charge region near to junction and increased the rate for exciton generation to reduce the charge recombination (Meng et al., 2013).

Yuan and coworkers developed robust visible light active g-C_3N_4/RGO/MoS_2 photocatalyst with a rate of 0.317 mM/g/h. Exposed edge of MoS_2 in nanocomposite acts as the active region for hydrogen evolution and accepts electrons from electron-rich RGO. The graphene layer is crucial in promoting charge transfer between g-C_3N_4 and MoS_2 (Yuan, Yang, et al., 2018). Chava et al. synthesized one-dimensional CdS-Au/MoS_2 heterostructure

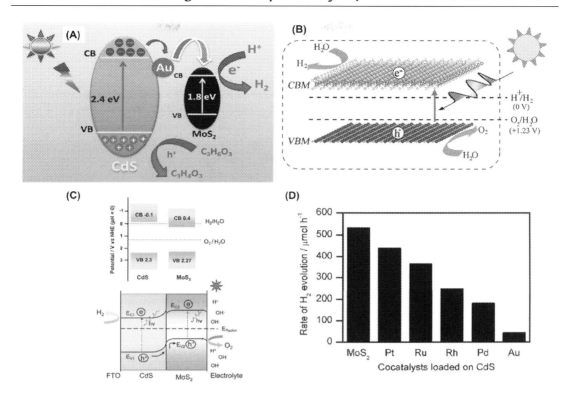

Figure 11.16

(A) Schematic representation of charge separation of CdS-Au/MoS₂ CSHNSs (Chava et al., 2018). (B) van der Waals heterostructure as photocatalyst for water splitting (Liao, Sa, Zhou, Ahuja, & Sun, 2014). (C) MoS₂ form p—n junction with CdS (Liu et al., 2013). (D) Activity of HER for CdS/ different cocatalysts (Zong et al., 2008). *Source: Taken with permission from Chava, R. K., Do, J. Y., & Kang, M. (2018). Smart hybridization of Au coupled CdS nanorods with few layered MoS₂ nanosheets for high performance photocatalytic hydrogen evolution reaction.* ACS Sustainable Chemistry & Engineering, 6 *(5), 6445—6457; Kumar, D. P., Hong, S., Reddy, D. A., & Kim, T. K. (2016). Noble metal-free ultrathin MoS₂ nanosheet-decorated CdS nanorods as an efficient photocatalyst for spectacular hydrogen evolution under solar light irradiation.* Journal of Materials Chemistry A, 4, *18551—18558; Liao, J., Sa, B., Zhou, J., Ahuja, R. & Sun, Z. (2014). Design of high-efficiency visible-light photocatalysts for water splitting: MoS₂/AlN(GaN) heterostructures.* The Journal of Physical Chemistry C, 118 *(31), 17594—17599; Liu, Y., Yu, Y. X., & Zhang, W. D. (2013). MoS₂/CdS heterojunction with high photoelectrochemical activity for H₂ evolution under visible light: The role of MoS₂.* The Journal of Physical Chemistry C, 117 *(25), 12949—12957; Zong, X., Yan, H. J., Wu, G. P., Ma, G. J., Wen, F. Y., Wang, L., & Li, C. (2008). Enhancement of photocatalytic H₂ evolution on CdS by loading MoS₂ as cocatalyst under visible light irradiation.* Journal of the American Chemical Society, 130 *(23), 7176—7177.*

photocatalyst with a rate of 7 mM/g/h. In this case, MoS_2 acts as cocatalyst and robust generator for charge carriers (Chava et al., 2018).

In a similar way, Yuan et al. (2016) also designed 2D MoS_2/TiO_2 photocatalyst for HER with a rate of 2.145 mM/g/h, which is 36.4 times higher than of pristine TiO_2. As per the density functional theory (DFT), the free energy of atomic hydrogen bonding to the MoS_2 is approximately zero (similar to Pt), which makes MoS_2 robust photosensitizer to hydrogen evolution (Hinnemann et al., 2005). Yuan et al. (2017) also developed $MoS_2/$rGO photocatalyst, Zn(II)-5, 10, 15, 20-tetrakis (4-methylpyridyl)porphyrin ($[ZnTMPPyP]^{4+}$) photosensitizer, and triethanolamine (TEOA) as a sacrificial electron donor for HER in visible light, which operates at a rate of 2.560 mM/g/h for hydrogen evolution.

Sun et al. proposed $MoS_2/CdS/TiO_2$, a ternary nanocomposite system, containing 3& w/w MoS_2, that produces H_2 at a rate of 4.146 mM/g/h for HER under visible light. In this case, MoS_2 lies in between CdS and TiO_2, which enhances charge separation and reduced recombination of carriers (Sun, Wang, Fang, Sun, & Yu, 2016). Recently, Yuan et al. also synthesized $CuInS_2/TiO_2/MoS_2$ ternary nanocomposite with the rate of 1.034 mM/g/h under UV-near IR range, in which MoS_2 and $CuInS_2$ have loading of 0.5% and 0.6 mM/g, respectively (Yuan, Fang, et al., 2018).

Zong et al. demonstrated that CdS/WS_2 photocatalyst, with 100:1 loading ratio for WS_2:CdS, produces H_2 at a rate of 0.420 mM/g/h under visible light irradiation (Zong et al., 2011). Akple et al. reported g-C_3N_4/WS_2 photocatalyst with the optimum loading of 0.01 wt.% WS_2 and the rate of 0.101 mM/g/h for HER (Akple et al., 2015). Reddy et al. synthesized $CdS/WS_2/MoS_2$ ternary nanocomposite photocatalyst with the excellent rate of 209.79 mM/g/h for HER under visible light, which is 83 times higher than pure CdS nanorods (Reddy, Park et al., 2017). $MoSe_2$ is another promising candidate that has similar physical and chemical properties to MoS_2 and WS_2. Kou, Guo, Xu, and Yang (2018) reported $MoSe_2/TiO_2$ photocatalyst for HER with the rate of 0.468 mM/g/h under visible light, which is comparatively higher than pure TiO_2 (0.0055 mM/g/h) and $MoSe_2$ (0.070 mM/g/h). Tahir, Nabi, Iqbal, Sagir, and Rafique (2018) reported $MoSe_2/WO_3$-CNT photocatalyst that has 4 wt.% $MoSe_2$ into the WO_3-CNT composite and showed higher rate of 0.1 mM/g/h than pure $MoSe_2$ and WO_3-CNT. Lin et al. (2016) synthesized floating sheet with WSe_2/Cu_2O-based photocatalyst with the rate of 64.85 mM/g/h, which is 13.2 times better compared to the powder form of pure components. The rate of hydrogen evolution depends not only on the composition of photocatalyst but also on the following factors, such as physicochemical reaction on the surface of photocatalyst, stability of photocatalyst, nature of sacrificial reagents, light absorption capability of the photocatalyst, purging flow rate of inert gas, light source, and gas sampling (Majrik, Paszti, & Korecz, 2018; Reddy, Emin, Valant, & Shankar, 2017; Schneider & Bahnemann, 2013).

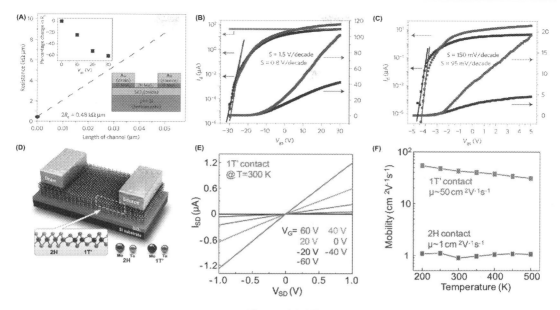

Figure 11.17

(A) Plot of resistance versus length of 2H-MoS$_2$ and 1T-MoS$_2$ with Au electrode deposition (B) bottom-gated and (C) top-gated devices (*blue curve* for devices with Au electrode coated on 1T-MoS$_2$, *black curves* for devices with Au electrode coated on 2H-MoS$_2$, and *red curve* for I$_d$ of the device on the 1T-MoS$_2$ channel with Au electrode; Kappera et al., 2014). (D) Schematic diagram of device with 1T/2H-MoTe$_2$ homojunction. (E) Source-drain current (I$_{SD}$) characteristics of device based on 1T/2H-MoTe$_2$ homojunction at gate voltages (−60 to 60 V). (F) Field-effect mobility as a function of temperature on the 1T/2H-MoTe2 homojunction-based device (Cho et al., 2015). *Source: Taken with permission from Cho, S., Kim, S., Kim, J. H., Zhao, J., Seok, J., Keum, D. H., ... Yang, H. (2015). Device technology: Phase patterning for ohmic homojunction contact in MoTe$_2$. Science, 349 (6248), 625—628; Kappera, R., Voiry, D., Yalcin, S. E., Branch, B., Gupta, G., Mohite, A. D., & Chhowalla, M. (2014). Phase-engineered low-resistance contacts for ultrathin MoS$_2$ transistors. Nature Materials, 13 (12), 1128—1134.*

11.5.1.1.4 Electronics

2D TMC nanosheets are used for the formation of semiconductor channel materials. These particular systems are useful for the fabrication of high-quality electronic devices with high efficiency, due to their modulated mobility, high surface area, and tunable band gap (Koppens et al., 2014). Here, 2D TMC nanosheets have excellent performance for the fabrication of photodiodes, photodetectors, and field-effect transistors (FET; Ganatra & Zhang, 2014). Kappera et al. synthesized metallic 1T phase MoS$_2$ nanosheets, promising candidate for the fabrication of FET. In this synthesis, *n*-butyllithium was used for the partial conversion of mechanically exfoliated 1H-MoS$_2$ nanosheets into metallic 1T phase. The work function of 1T phase and CB energy of the 1H phase are close to each other. So, they are covered by polymethyl methacrylate to avoid any phase change. As a result, the contact resistance of the FET was reduced to $200-300$ μm at a zero gate bias from the high resistance of $700-10000$ μm. On the basis of phase selection, back-gated 1T-MoS$_2$-based FET displayed unique properties like highest on/off ratio larger than 10^6, high drive current of 85 μA/μm, and high mobility of 50 cm^2/V/s. In comparison, 1H-MoS$_2$-based FET with Au electrode has low drive current 30 μA/μm and low mobility of $15-20$ cm^2/V/S. On the other hand, top-gated device based on 1T-MoS$_2$ has excellent saturation current and low subthreshold values of 90 mV/decade. However, analogous top-gated device based on 1H-MoS$_2$ has high subthreshold values of 150 mV/decade and low saturation current. Performance of electronics device can be regulated on the basis of phase selection for TMC (Fig. 11.17A−C; Kappera et al., 2014). Similarly, Cho et al. reported homojunction between 2H and 1T′ MoTe$_2$ that is useful for the fabrication of FET. Laser irradiation technique was employed to create local phase transition of 2H to 1T′ in MoTe$_2$ nanosheets. Homojunction-based field-effect transistor exhibited n-type transport characteristic with high on/off current ratio of 10^6 and device can operate in a high-temperature environment (up to 300°C). In addition, 2H-MoTe$_2$/1T′-MoTe$_2$ contact has fast electron transfer because small energy difference between work function of 1T′-MoTe$_2$ and electron affinity of 2H-MoTe$_2$ (Fig. 11.17D−F; Cho et al., 2015).

11.5.1.1.5 Degradation of organic pollutants

Industrial effluents are a source of water pollution, which are toxic and carcinogenic to our environment. From the last few decades, several water purification methods, such as sonochemical degradation, membrane process, coagulation, precipitation, biodegradation, micellar-enhanced ultrafiltration and advanced oxidation process (AOP), have been utilized for water purification (Rasalingam, Peng, & Koodali, 2014). Among them, AOP is most the promising method for the treatment of contaminant water followed by the formation of reactive oxygen species (ROS). For this purpose, AOP mostly depends on a broad range of semiconductor photocatalyst, such as TMO and TMC materials. For organic degradation, TMC materials are more attractive and reactive worldwide compared to TMO because of the superior optical absorption due to its tunable indirect band gap energy ($1-2$ eV), more active site, and low recombination of carriers (Yang, 2018).

Yu et al. synthesized WS$_2$/WO$_3$/H$_2$O heterostructure photocatalyst to showcase PC activity toward methyl orange (approximately 90%) under visible light irradiation. It has been found that the structural alignment between WS$_2$ and WO$_3$-H$_2$O plays essential roles for the efficient photogenerated charge separation, which is responsible for good catalytic activity (Yu, Prevot, Guijarro, & Sivula, 2015).

Peng et al. reported CdS/MoS$_2$/rGO heterostructure photocatalyst for the reduction of nitroaromatic compounds to aromatic amines under visible light irradiation (Wc, Chen, & Li, 2016). Liu et al. (2016) synthesized α-Fe$_2$O$_3$@N-doped MoS$_2$ heterostructure photocatalyst for the removal of RhB under visible light irradiation, which has 26.4 times

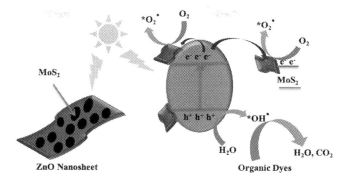

Figure 11.18

Photocatalysis mechanism of ZnO nanosheet/ MoS_2 monolayer (Lin et al., 2014). Source: *Taken with permission from Lin, Z., Li, J., Zheng, Z., Li, L., Yu, L., Wang, C., & Yang, G. (2014). A floating sheet for efficient photocatalytic water splitting. Advanced Energy Materials, 6, 1600510.*

higher degradation rate with respect to MoS_2 nanosheets. Similarly, Yang et al. synthesized ZnO, doped with P, coated by MoS_2 photocatalyst with appreciable degradation efficiency (95%) toward MB and RhB under solar irradiation. Almost 3.4 times better degradation rate was observed here than P25 because light adsorption efficiency has been enhanced by MoS_2 monolayer (Fig. 11.18; Liu, Xie, & Li, 2014).

Haiyang et al. synthesized 1T@2H-MoS_2/Ag nanocomposite photocatalyst by microwave hydrothermal methods for the degradation of MB (rate of 41%) under visible light irradiation. It has been found that PC activity of 1T@2H-MoS_2/Ag is higher compared to 2H-MoS_2 mostly due to the difference in light absorption capacity (Liu, Wu, & Tian, 2018). In another study, Xi et al. reported PC properties of amorphous and crystalline $MoSe_2$ for the degradation of rhodamine B and MB under visible light irradiation. Amorphous $MoSe_2$ has good PC activity in comparison to crystalline $MoSe_2$ due to excellent visible light absorption and narrow band gap (1–2 eV), formation of superoxide radicals, and more unsaturated atoms (Yang, Wu, Liu, & Fan, 2018). Molla et al. synthesized Ag/In/Ni/S nanocomposites with different shapes for the degradation of MB under dark and visible light irradiation. These nanocomposites displayed excellent degradation rate of MB (time taken for complete degradation of dye: under dark 12 minutes and in visible light 2 minutes). In this study, ROS were generated during the experiment (Molla, Sahu, & Hussain, 2015).

Luo et al. also reported Ni-doped/CdS hollow sphere photocatalyst for the degradation of RhB and phenol under visible light irradiation. In addition, 1.2 mol% of Ni-doped CdS has excellent degradation activity toward degradation of organic pollutants because of the low recombination of carriers. Zhang et al. also reported SnS_2/g-C_3N_4 heterojunction photocatalyst

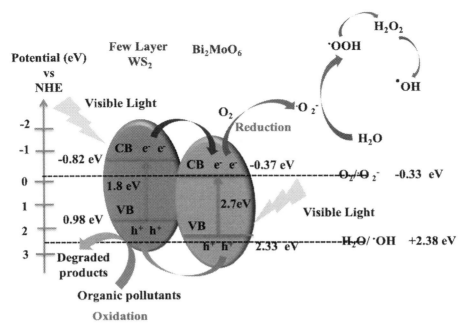

Figure 11.19

The mechanism for catalytic activity of $WS_2/BiMoO_6$ heterocomposite photocatalyst. *Source: Taken with permission from Li, X., Su, M., & Zhu, G. (2018). Fabrication of a novel few-layer WS_2/Bi_2MoO_6 plate-onplate heterojunction structure with enhanced visible-light photocatalytic activity.* Dalton Transactions, 47 (30), 10046−10056.

for PC degradation of RhB and phenols under visible light irradiation, which illustrated excellent PC activity compared to g-C_3N_4 and SnS_2 nanosheets (Luo, Liu, & Hu, 2012).

Xiang et al. reported $WS_2/BiMoO_6$ heterojunction composites for the degradation of organic pollutants, such as RhB, ciprofloxacin, MB, and methimazole. This material has excellent degradation rate due to high surface area for IFCT and low recombination rate (Fig. 11.19; Li, Su, & Zhu, 2018).

11.6 Conclusion

The intrinsic photophysical properties, such as narrow band gap, high visible light absorption, and low recombination energy of nontoxic TMCs, have propelled them for potential PC applications. Variation in the transition metal and chalcogenide components resulted in a library of TMCs with interesting electronic and optical properties. Here, in this chapter, we have detailed the structural classification, synthetic procedures, tunability of photophysical properties, and potential applications of TMCs.

References

Acerce, M., Voiry, D., & Chhowalla, M. (2015). Metallic 1T phase MoS_2 nanosheets as supercapacitor electrode materials. *Nature Nanotechnology, 10*(4), 313–318.

Agnieszka, K., & Thomas, H. (2015). The electronic structure calculations of two-dimensional transition-metal dichalcogenides in the presence of external electric and magnetic fields. *Chemical Society Reviews, 44*(9), 2603–2614.

Ahmad, H., Kamarudin, S. K., Minggu, L. J., & Kassim, M. (2015). Hydrogen from photo-catalytic water splitting process: A review. *Renewable & Sustainable Energy Reviews, 43*, 599–610.

Akpan, U., & Hameed, B. (2009). Parameters affecting the photocatalytic degradation of dyes using TiO_2-based photocatalysts: A review. *Journal of Hazardous Materials, 170*(2–3), 520–529.

Akple, M. S., Low, J., Wageh, S., AlGhamdi, A. A., Yu, J., & Zhang, J. (2015). Shape-dependent photocatalytic hydrogen evolution activity over a Pt nanoparticle coupled g-C_3N_4 photocatalyst. *Applied Surface Science, 358*, 196–203.

Alsaif, M. M., Chrimes, A. F., Daeneke, T., Balendhran, S., & Bellisario, D. O. (2016). High-performance field effect transistors using electronic inks of 2D molybdenum oxide nanoflakes. *Advanced Functional Materials, 26*, 91–100.

Asahi, R., Morikawa, T., Ohwaki, T., Aoki, K., & Taga, Y. (2001). Visible-light photocatalysis in nitrogen-doped titanium oxides. *Science, 293*(5528), 269–271.

Ataca, C., Şahin, H., & Ciraci, S. (2012). Stable, single-layer MX_2 transition-metal oxides and dichalcogenides in a honeycomb-like structure. *The Journal of Physical Chemistry C, 116*(16), 8983–8999.

Awasthi, G. P., Adhikari, S. P., Ko, S., Kim, H. J., Park, C. H., & Kim, C. S. (2016). Facile synthesis of ZnO flowers modified graphene-like MoS_2 sheets for enhanced visible-light-driven photocatalytic activity and antibacterial properties. *Journal of Alloys and Compounds, 682*, 208–215.

Bai, S., Wang, L. M., Chen, X. Y., Du, J. T., & Xiong, Y. J. (2014). Chemically exfoliated metallic MoS_2 nanosheets: A promising supporting co-catalyst for enhancing the photocatalytic performance of TiO_2 nanocrystals. *Nano Research, 8*(1), 175–183.

Balat, M., & Balat, H. (2009). Biogas as a renewable energy source—A review. *Energy Sources, Part A: Recovery, Utilization, and Environmental Effects, 31*(14), 1280–1293.

Bhandavat, R., David, L., & Singh, G. (2012). Synthesis of surface-functionalized WS_2 nanosheets and performance as Li-ion battery anodes. *The Journal of Physical Chemistry Letters, 3*(11), 1523–1530.

Bi, W., Zhou, M., Ma, Z., & Xie, Y. (2012). $CuInSe_2$ ultrathin nanoplatelets: Novel self-sacrificial template-directed synthesis and application for flexible photodetectors. *Chemical Communications, 48*(73), 9162–9164.

Chava, R. K., Do, J. Y., & Kang, M. (2018). Smart hybridization of Au coupled CdS nanorods with few layered MoS_2 nanosheets for high performance photocatalytic hydrogen evolution reaction. *ACS Sustainable Chemistry & Engineering, 6*(5), 6445–6457.

Cheah, A. J., Chiu, W. S., Khiew, P. S., Nakajima, H., Saisopa, T., Songsiriritthigul, P., . . . Hamid, M. A. A. (2015). Facile synthesis of a Ag/MoS_2 nanocomposite photocatalyst for enhanced visible-light-driven hydrogen gas evolution. *Catalysis Science & Technology, 5*(8), 4133–4143.

Chen, Y., & Xue, J. M. (2014). Ultrasmall Fe_3O_4 nanoparticle/MoS_2 nanosheet composites with superior performances for lithium-ion batteries. *Small (Weinheim an der Bergstrasse, Germany), 10*, 1536–1543.

Cheng, P., Sun, K., & Hu, Y. H. (2016). Memristive behavior and ideal memristor of 1T phase MoS_2 nanosheets. *Nano Letters, 16*(1), 572–576.

Chhowalla, M., Shin, H. S., Eda, G., Li, L. J., Loh, K. P., & Zhang, H. (2013). The chemistry of two-dimensional layered transition metal dichalcogenide nanosheets. *Nature Chemistry, 5*(4), 263–275.

Chianelli, R. R., Siadati, M. H., De la Rosa, M. P., Berhault, G., Wilcoxon, J. P., Bearden, R., & Abrams, B. L. (2006). Catalytic properties of single layers of transition metal sulfide catalytic materials. *Catalysis Reviews, 48*, 1–41.

Cho, S., Kim, S., Kim, J. H., Zhao, J., Seok, J., Keum, D. H., . . . Yang, H. (2015). Device technology: Phase patterning for ohmic homojunction contact in $MoTe_2$. *Science, 349*(6248), 625–628.

Coleman, J. N., Lotya, M., & Neill, A. O. (2011). Two-dimensional nanosheets produced by liquid exfoliation of layered materials. *Science, 331*(6017), 568−571.

Das, R., Mishra, H., Srivastava, A., & Kayastha, A. M. (2017). Covalent immobilization of b-amylase onto functionalized molybdenum sulfide nanosheets, its kinetics and stability studies: A gateway to boost enzyme application. *Chemical Engineering Journal, 328*, 215−227.

Eda, G., Yamaguchi, H., & Chhowalla, M. (2011). Photoluminescence from chemically exfoliated MoS_2. *Nano Letters, 11*(12), 5111−5116.

Flatten, L. C., He, Z., Coles, D. M., Trichet, A. A. P., Powell, A. W., Taylor, R. A., ... Smith, J. M. (2016). Room-temperature exciton-polaritons with two-dimensional WS_2. *Scientific Reports, 6*, 33134.

Ganatra, R., & Zhang, Q. (2014). Few-layer MoS_2: A promising layered semiconductor. *ACS Nano, 8*(5), 4074−4099.

Ge, L., Han, C., Xiao, X., & Guo, L. (2013). Synthesis and characterization of composite visible light active photocatalysts MoS_2−g-C_3N_4 with enhanced hydrogen evolution activity. *International Journal of Hydrogen Energy, 38*(17), 6960−6969.

Guan, Z., Ni, S., & Hu, S. (2018). Tunable electronic and optical properties of monolayer and multilayer janus MoSSe as a photocatalyst for solar water splitting: A first-principles study. *The Journal of Physical Chemistry C, 122*(11), 6209−6216.

Haque, F., Daeneke, T., Kalantar, K., & Ou, J. Z. (2018). Two-dimensional transition metal oxide and chalcogenide-based photocatalysts. *Nano-Micro Letters, 10*(2), 23.

Hernandez-Alonso, M. D., Fresno, F., Suarez, S., & Coronado, J. M. (2009). Development of alternative photocatalysts to TiO_2: Challenges and opportunities. *Energy & Environmental Science, 2*(12), 1231−1257.

Hinnemann, B., Moses, P. G., Bonde, J., Jørgensen, K. P., Nielsen, J. H., Horch, S., ... Nørskov, J. K. (2005). Biomimetic hydrogen evolution: MoS_2 nanoparticles as catalyst for hydrogen evolution. *Journal of the American Chemical Society, 127*(15), 5308−5309.

Ida, S., Takashiba, A., Koga, S., Hagiwara, H., & Ishihara, T. (2014). Potential gradient and photocatalytic activity of an ultrathin p−n junction surface prepared with two-dimensional semiconducting nanocrystals. *Journal of the American Chemical Society, 136*(5), 1872−1878.

Inoue, T., Fujishima, A., Konishi, S., & Honda, K. (1979). Photoelectrocatalytic reduction of carbon dioxide in aqueous suspensions of semiconductor powders. *Nature, 277*, 637−638.

Jafari, T., Moharreri, E., Amin, A. S., Miao, R., Song, W., & Suib, S. L. (2016). Photocatalytic water splitting-the untamed dream: A review of recent advances. *Molecules, 21*(7), 900.

Jang, J. T., Jeong, J., & Cheon, J. (2011). Ultrathin zirconium disulfide nanodiscs. *Journal of the American Chemical Society, 133*(20), 7636−7639.

Jiang, C., Moniz, S. J. A., Wang, A., Zhang, T., & Tang, J. (2017). Photoelectrochemical devices for solar water splitting—Materials and challenges. *Chemical Society Reviews, 46*(15), 4645−4660.

Kappera, R., Voiry, D., Yalcin, S. E., Branch, B., Gupta, G., Mohite, A. D., & Chhowalla, M. (2014). Phase-engineered low-resistance contacts for ultrathin MoS_2 transistors. *Nature Materials, 13*(12), 1128−1134.

Kim, K. K., Hsu, A., & Kong, J. (2012). Synthesis of monolayer hexagonal boron nitride on Cu foil using chemical vapour deposition. *Nano Letters, 12*(1), 161−166.

Kim, T. W., Oh, E. J., & Hwang, S. J. (2009). Soft-chemical exfoliation route to layered cobalt oxide monolayers and its application for film deposition and nanoparticle synthesis. *Chemistry—A European Journal, 15*(41), 10752−10761.

Kisielowski, C., Ramasse, Q. M., Hansen, L. P., Brorson, M., Carlsson, A., Molenbroek, A. M., ... Helveg, S. (2010). Imaging MoS_2 nanocatalysts with single-atom sensitivity. *Angewandte Chemie International Edition, 49*(15), 2708−2710.

Koppens, F. H. L., Mueller, T., Avouris, P., Ferrari, A. C., Vitiello, M. S., & Polini, M. (2014). Photodetectors based on graphene, other two-dimensional materials and hybrid systems. *Nature Nanotechnology, 9*(10), 780−793.

Kou, S., Guo, X., Xu, X., & Yang, J. (2018). TiO_2 on $MoSe_2$ nanosheets as an advanced photocatalyst for hydrogen evolution in visible light. *Catalysis Communications, 106*, 60−63.

Kumar, D. P., Hong, S., Reddy, D. A., & Kim, T. K. (2016). Noble metal-free ultrathin MoS_2 nanosheet-decorated CdS nanorods as an efficient photocatalyst for spectacular hydrogen evolution under solar light irradiation. *Journal of Materials Chemistry A, 4*, 18551−18558.

Lassalle-Kaiser, B., Merki, D., Vrubel, H., Gul, S., Yachandra, V. K., Hu, X., & Yano, J. (2015). Evidence from in situ X-ray absorption spectroscopy for the involvement of terminal disulfide in the reduction of protons by an amorphous molybdenum sulfide electrocatalyst. *Journal of the American Chemical Society, 137*(1), 314−321.

Li, N., Teng, H., Zhang, L., Zhou, J., & Liu, M. (2015). Synthesis of Mo doped WO_3 nanosheets with enhanced visible-light-drive photocatalytic properties. *RSC Advances, 5*(115), 95394−95400.

Li, R. (2017). Photocatalytic nitrogen fixation: An attractive approach for artificial photocatalysis. *Journal of Catalysis, 38*(7), 5−12.

Li, X., Cai, W., An, J., & Rodney, S. R. (2009). Large-area synthesis of high-quality and uniform graphene films on copper foils. *Science, 324*(5932), 1312−1314.

Li, X., Su, M., & Zhu, G. (2018). Fabrication of a novel few-layer WS_2/Bi_2MoO_6 plate-onplate heterojunction structure with enhanced visible-light photocatalytic activity. *Dalton Transactions, 47* (30), 10046−10056.

Liang, Y., Guo, N., Li, L., Li, R., Ji, G., & Gan, S. (2015). Fabrication of porous 3D flower-like Ag/ZnO heterostructure composites with enhanced photocatalytic performance. *Applied Surface Science, 332*, 32−39.

Liao, J., Sa, B., Zhou, J., Ahuja, R., & Sun, Z. (2014). Design of high-efficiency visible-light photocatalysts for water splitting: MoS_2/AlN(GaN) heterostructures. *The Journal of Physical Chemistry C, 118*(31), 17594−17599.

Liao, Y., Hu, Z., Gu, Q., & Xue, C. (2015). Amine-functionalized ZnO nanosheets for efficient CO_2 capture and photoreduction. *Molecules, 20*(10), 18847−18855.

Lin, Z., Li, J., Zheng, Z., Li, L., Yu, L., Wang, C., & Yang, G. (2016). A floating sheet for efficient photocatalytic water splitting. *Advanced Energy Materials, 6*, 1600510.

Linsebigler, A. L., Lu, G., & Yates, J. T. (1995). Photocatalysis on TiO_2 surfaces: Principles, mechanisms, and selected results. *Chemical Reviews, 95*(3), 735−758.

Liu, H., Wu, R., & Tian, L. (2018). Synergetic photocatalytic effect between 1 T@2H-MoS_2 and plasmon resonance induced by Ag quantum dots. *Nanotechnology, 29*(28)285402.

Liu, P., Liu, Y., Ye, W., Ma, J., & Gao, D. (2016). Flower-like n-doped MoS_2 for photocatalytic degradation of RhB by visible light irradiation. *Nanotechnology, 27*(22)225403.

Liu, Y., Xie, S., & Li, H. (2014). A highly efficient sunlight driven ZnO nanosheet photocatalyst, synergetic effect of P-doping and MoS_2 atomic layer loading. *ChemCatChem, 6*(9), 2522−2526.

Liu, Y., Yu, Y. X., & Zhang, W. D. (2013). MoS_2/CdS heterojunction with high photoelectrochemical activity for H_2 evolution under visible light: The role of MoS_2. *The Journal of Physical Chemistry C, 117*(25), 12949−12957.

Lo, S. S., Mirkovic, T., Chuang, C. H., Burda, C., & Scholes, G. D. (2011). Emergent properties resulting from type-II band alignment in semiconductor nano heterostructures. *Advanced Materials, 23*, 180−197.

Lu, Q. P., Yu, Y. F., Ma, Q. L., Chen, B., & Zhang, H. (2016). 2D transition-metal-dichalcogenide-nanosheet-based composites for photocatalytic and electrocatalytic hydrogen evolution reactions. *Advanced Materials, 28*(10), 1917−1933.

Luo, B., Liu, G., & Wang, L. (2016). Recent advances in 2D materials for photocatalysis. *Nanoscale, 8*(13), 6904−6920.

Luo, M., Liu, Y., & Hu, J. (2012). One-pot synthesis of CdS and Ni-doped CdS hollow spheres with enhanced photocatalytic activity and durability. *ACS Applied Materials & Interfaces, 4*(3), 1813−1821.

Maeda, K., & Domen, K. (2010). Photocatalytic water splitting: Recent progress and future challenges. *The Journal of Physical Chemistry Letters, 1*(18), 2655−2661.

Maeda, K., Teramura, K., Lu, D., Takata, T., Saito, N., Inoue, Y., & Domen, K. (2006). Photocatalyst releasing hydrogen from water. *Nature, 440*(7082), 295.

Mahler, B., Hoepfner, V., Liao, K., & Ozin, G. A. (2014). Colloidal synthesis of 1T-WS$_2$ and 2H-WS$_2$ nanosheets: Applications for photocatalytic hydrogen evolution. *Journal of the American Chemical Society*, *136*(40), 14121–14127.

Majrik, K., Paszti, Z., & Korecz, L. (2018). Study of PtO$_x$/TiO$_2$ photocatalysts in the photocatalytic reforming of glycerol: The role of co-catalyst formation. *Materials*, *11*, 1927.

Mar, A., & Ibers, J. A. (1993). The layered ternary germanium tellurides ZrGeTe$_4$, HfGeTe$_4$, and TiGeTe$_6$. Structure, bonding, and physical properties. *Journal of the American Chemical Society*, *115*(8), 3227.

Mathew, S., Yella, A., Gao, P., Baker, R. H., & Curchodet, B. F. (2014). Dye-sensitized solar cells with 13% efficiency achieved through the molecular engineering of porphyrin sensitizers. *Nature Chemistry*, *6*(3), 242–247.

Meng, F., Li, J., Cushing, S. K., Zhi, M., & Wu, N. (2013). Solar hydrogen generation by nanoscale p–n junction of p-type molybdenum disulfide/n-type nitrogen-doped reduced graphene oxide. *Journal of the American Chemical Society*, *135*(28), 10286–10289.

Miao, Y., Zhang, H., Yuan, S., Jiao, Z., & Zhu, X. (2016). Preparation of flower-like ZnO architectures assembled with nanosheets for enhanced photocatalytic activity. *Journal of Colloid and Interface Science*, *462*, 9–18.

Molla, A., Sahu, M., & Hussain, S. (2015). Under dark and visible light: Fast degradation of methylene blue in the presence of Ag–In–Ni–S nanocomposites. *Journal of Materials Chemistry A*, *3*(30), 15616–15625.

Mujtaba, J., Sun, H., Fang, F., Ahmad, M., & Zhu, J. (2015). Fine control over the morphology and photocatalytic activity of 3D ZnO hierarchical nanostructures: Capping vs. etching. *RSC Advances*, *5*(69), 56232–56238.

Mukherjee, S., Maiti, R., Midya, A., Das, S., & Ray, S. K. (2015). Tunable direct bandgap optical transitions in MoS$_2$ nanocrystals for photonic devices. *ACS Photonics*, *2*(6), 760–768.

Peng, W., Li, Y., Zhang, F., & Zhang, G. (2017). Roles of two-dimensional transition metal dichalcogenides as cocatalysts in photocatalytic hydrogen evolution and environmental remediation. *Industrial & Engineering Chemistry Research*, *56*, 4611–4626.

Perera, S. D., Mariano, R. G., Nour, Vu. K., Seitz, N., & Chabal, O. (2012). Hydrothermal synthesis of graphene-TiO$_2$ nanotube composites with enhanced photocatalytic activity. *ACS Catalysis*, *2*(6), 949–956.

Pinaud, B. A., Benck, J. D., Seitz, L. C., Forman, A. J., Chen, Z., Deutsch, T. G., ... Jaramillo, T. F. (2013). Technical and economic feasibility of centralized facilities for solar hydrogen production via photocatalysis and photoelectrochemistry. *Energy & Environmental Science*, *6*, 1983–2002.

Preethi, V., & Kanmani, S. (2013). Photocatalytic hydrogen production. *Materials Science in Semiconductor Processing*, *16*(3), 561–575.

Pruss, E. A., Snyder, B. S., & Stacy, A. M. (1993). A new layered ternary sulfide: Formation of Cu$_2$WS$_4$ by reaction of WS and Cu$^+$ ions. *Angewandte Chemie International Edition*, *32*, 256–257.

Rahmanian, E., Malekfar, R., & Pumera, M. (2017). Nanohybrids of two-dimensional transition-metal dichalcogenides and titanium dioxide for photocatalytic applications. *Chemistry—A EuropeanJournal*, *24*(1), 18–31.

Rasalingam, S., Peng, R., & Koodali, R. T. (2014). Removal of hazardous pollutants from wastewaters: Applications of TiO$_2$-SiO$_2$ mixed oxide materials. *Journal of Nanomaterials*, *2014*, 1–42.

Rasmussen, F. A., & Thygesen, K. S. (2015). Computational 2D materials database: Electronic structure of transition-metal dichalcogenides and oxides. *The Journal of Physical Chemistry C*, *119*(23), 13169–13183.

Reddy, D. A., Park, H., Ma, R., Kumar, D. P., Lim, M., & Kim, T. K. (2017). Heterostructured WS$_2$-MoS$_2$ ultrathin nanosheets integrated on CdS nanorods to promote charge separation and migration and improve solar-driven photocatalytic hydrogen evolution. *ChemSusChem*, *10*, 1563–1570.

Reddy, N. L., Emin, S., Valant, M., & Shankar, M. V. (2017). Nanostructured Bi$_2$O$_3$@TiO$_2$ photocatalyst for enhanced hydrogen production. *International Journal of Hydrogen Energy*, *42*(10), 6627–6636.

Schneider, J., & Bahnemann, D. W. (2013). Undesired role of sacrificial reagents in photocatalysis. *The Journal of Physical Chemistry Letters*, *4*, 3479–3483.

Serpone, N., Lawless, D., Disdier, J., & Herrmann, J. M. (1994). Spectroscopic, photoconductivity, and photocatalytic studies of TiO_2 colloids: Naked and with the lattice doped with Cr^{3+}, Fe^{3+}, and V^{5+} cations. *Langmuir*, *10*(3), 643—652.

Sheeba, R., Israel, S., & Saravanakumar, S. (2016). Investigation of the van der Waals epitaxy gap in isostructural semiconducting germanium tellurides: $HfGeTe_4$ and $ZrGeTe_4$. *Chinese Journal of Physics*, *54*(5), 668.

Smith, R. J., King, P. J., Lotya, M., & Coleman, J. N. (2011). Large-scale exfoliation of inorganic layered compounds in aqueous surfactant solutions. *Advanced Materials*, *23*(34), 3944—3948.

Su, T., Shao, Q., Qin, Z., Guo, Z., & Wu, Z. (2018). Role of interfaces in two-dimensional photocatalyst for water splitting. *ACS Catalysis*, *8*(3), 2253—2276.

Sun, M., Wang, Y., Fang, Y., Sun, S., & Yu, Z. (2016). Construction of $MoS_2/CdS/TiO_2$ ternary composites with enhanced photocatalytic activity and stability. *Journal of Alloys and Compounds*, *684*, 335—341.

Tahir, M. B., Nabi, G., Iqbal, T., Sagir, M., & Rafique, M. (2018). Role of $MoSe_2$ on nanostructures WO_3-CNT performance for photocatalytic hydrogen evolution. *Ceramics International*, *44*, 6686—6690.

Tan, Y., Liu, P., Chen, L., Cong, W., Ito, Y., Han, J., . . . Hirata, A. (2014). Monolayer MoS_2 films supported by 3D nanoporous metals for high-efficiency electrocatalytic hydrogen production. *Advanced Materials*, *26*(47), 8023—8028.

Tiwari, A. P., Kim, D., Kim, Y., Prakash, O., & Lee, H. (2016). Highly active and stable layered ternary transition metal chalcogenide for hydrogen evolution reaction. *Nano Energy*, *28*, 366—372.

Tiwari, A. P., Yoo, H., Lee, J., Kim, D., Park, J. H., & Lee, H. (2015). Prevention of sulfur diffusion using MoS_2-intercalated 3D-nanostructured graphite for high-performance lithium-ion batteries. *Nanoscale*, *7*(28), 11928—11933.

Tsai, C., Chan, K., Nørskov, J. K., & Pedersen, F. A. (2015). Theoretical insights into the hydrogen evolution activity of layered transition metal dichalcogenides. *Surface Science*, *640*, 133—140.

Turchi, C. S., & Ollis, D. F. (1990). Photocatalytic degradation of organic water contaminants: Mechanisms involving hydroxyl radical attack. *Journal of Catalysis*, *122*(1), 178—192.

Walter, M. G., Warren, E. L., McKone, J. R., Boettcher, S. W., Mi, Q., Santori, E. A., & Lewis, N. S. (2010). Solar water splitting cells. *Chemical Reviews*, *110*(11), 6446—6473.

Wang, H., Yang, X., & Xie, Y. (2015). Ultrathin black phosphorus nanosheets for efficient singlet oxygen generation. *Journal of the American Chemical Society*, *137*(35), 11376—11382.

Wang, Q. H., Zadeh, K. K., Kis, A., Coleman, J. N., & Strano, M. S. (2012). Electronics and optoelectronics of two-dimensional transition metal dichalcogenides. *Nature Nanotechnology*, *7*(11), 699—712.

Wang, X., Long, R., Liu, D., Yang, D., Wang, C., & Xiong, Y. (2016). Metal—organic framework assisted and in situ synthesis of hollow CdS nanostructures with highly efficient photocatalytic hydrogen evolution. *Nano Energy*, *24*(17), 87—93.

Wang, Z., & Mi, B. (2017). Environmental applications of 2D molybdenum disulfide (MoS2) nanosheets. *Environmental Science & Technology*, *51*(15), 8229—8244.

Wc, P., Chen, Y., & Li, X. (2016). MoS_2/reduced graphene oxide hybrid with CdS nanoparticles as a visible light-driven photocatalyst for the reduction of 4-nitrophenol. *Journal of Hazardous Materials*, *309*, 173—179.

Wei, L., Chen, Y., Lin, Y., Wu, H., Yuan, R., & Li, Z. (2014). MoS_2 as non-noble-metal co-catalyst for photocatalytic hydrogen evolution over hexagonal $ZnIn_2S_4$ under visible light irradiations. *Applied Catalysis B*, *144*, 521—527.

Wu, W., Jiang, C., & Roy, V. A. L. (2015). Recent progress in magnetic iron oxide—semiconductor composite nanomaterials as promising photocatalysts. *Nanoscale*, *7*(1), 38—58.

Xiang, Q., Yu, J., Wang, W., & Jaroniec, M. (2011). Nitrogen self-doped nano-sized TiO_2 sheets with exposed 001 facets for enhanced visible-light photocatalytic activity. *Chemical Communications*, *47*, 6906—6908.

Xiao, W., Liu, P., Zhang, J., Song, W., Feng, Y. P., Gao, D., & Ding, J. (2016). Dual-functional N dopants in edges and basal plane of MoS_2 nanosheets toward efficient and durable hydrogen evolution. *Advanced Energy Materials*, *7*, 1602086.

Xiao, X., Song, H., Lin, S., Zhou, Y., & Zhan, X. (2016). Scalable salt template synthesis of two-dimensional transition metal oxides. *Nature Communications*, *7*, 11296.

Xing, Z., Zong, X., Zhu, Y., Chen, Z., Bai, Y., & Wang, L. (2016). Ananohybrid of CdTe@CdS nanocrystals and titania nanosheets with p−n nanojunctions for improved visible light-driven hydrogen production. *Catalysis Today*, *264*, 229−235.

Xu, H., Ouyang, S., Li, P., Kako, T., & Ye, J. (2013). High-active anatase TiO_2 nanosheets exposed with 95% 100 facets toward efficient H_2 evolution and CO_2 photoreduction. *ACS Applied Materials & Interfaces*, *5*, 1348−1354.

Yan, Z., Du, L., & Phillips, D. L. (2017). Multilayer core−shell MoS_2/CdS nanorods with very high photocatalytic activity for hydrogen production under visible-light excitation and investigation of the photocatalytic mechanism by femtosecond transient absorption spectroscopy. *RSC Advances*, *7*, 55993.

Yang, X. (2018). Amorphous molybdenum selenide as highly efficient photocatalyst for the photodegradation of organic dyes under visible light. *Applied Surface Science*, *457*, 214−220.

Yang, X., Wu, R., Liu, H., & Fan, H. (2018). Amorphous molybdenum selenide as highly efficient photocatalyst for the photodegradation of organic dyes under visible light. *Applied Surface Science*, *457*, 214−220.

Yoon, T. P., Ischay, M. A., & Du, J. (2010). Visible light photocatalysis as a greener approach to photochemical synthesis. *Nature Chemistry*, *2*(7), 527−532.

Yu, X. Y., Prevot, M. S., Guijarro, N., & Sivula, K. (2015). Self-assembled 2D WSe_2 thin films for photoelectrochemical hydrogen production. *Nature Communications*, *6*(147), 7596.

Yu, Z. B., Xie, Y. P., Liu, G., Lu, G. Q., Ma, X. L., & Cheng, H. M. (2013). Self-assembled CdS/Au/ZnO heterostructure induced by surface polar charges for efficient photocatalytic hydrogen evolution. *Journal of Materials Chemistry A*, *1*, 2773−2776.

Yuan, Y., Lu, H., Ji, Z., Zhong, J., & Ding, M. (2015). Enhanced visible-light-induced hydrogen evolution from water in a noble metal-free system catalyzed by ZnTCPP-MoS_2/TiO_2 assembly. *Chemical Engineering Journal*, *275*, 8−16.

Yuan, Y. J., Chen, D., Zhong, J., Yang, L. X., Wang, J. J., Yu, Z. T., & Zou, Z. G. (2017). Construction of a noble-metal-free photocatalytic H_2 evolution system using MoS_2/reduced graphene oxide catalyst and zinc porphyrin photosensitizer. *The Journal of Physical Chemistry C*, *121*(44), 24452−24462.

Yuan, Y. J., Fang, G., Chen, D., Huang, Y., Yang, L. X., Cao, D. P., ... Zou, Z. G. (2018). High light harvesting efficiency $CuInS_2$ quantum dots /TiO_2 /MoS_2 photocatalysts for enhanced visible light photocatalytic H_2 production. *Dalton Transactions (Cambridge, England: 2003)*, *47*(16), 5652−5659.

Yuan, Y. J., Tu, J. R., Ye, Z. J., Lu, H. W., & Ji, Z. G. (2015). Visible light-driven hydrogen production from water in a noble-metal free system catalysed by zinc porphyrin sensitized MoS_2/ZnO. *Dyes and Pigments*, *123*, 285−292.

Yuan, Y. J., Yang, Y., Li, Z., Chen, D., Wu, S., Fang, G., ... Zou, Z. G. (2018). 3D hierarchical g-C_3N_4 architectures assembled by ultrathin self-doped nanosheets: Extremely facile HMTA activation and superior photocatalytic hydrogen evolution. *ACS Applied Energy Materials*, *1*, 1400−1407.

Yuan, Y. J., Ye, Z. J., Lu, H. W., Hu, B., Li, Y. H., Chen, D. Q., ... Zou, Z. G. (2016). Constructing anatase TiO_2 nanosheets with exposed (001) facets/layered MoS_2 two-dimensional nanojunctions for enhanced solar hydrogen generation. *ACS Catalysis*, *6*(2), 532−541.

Zha, R., Nadimicherla, R., & Guo, X. (2015). Ultraviolet photocatalytic degradation of methyl orange by nanostructured TiO_2/ZnO heterojunctions. *Journal of Materials Chemistry A*, *3*(12), 6565−6574.

Zhang, H., Chen, G., & Bahnemann, D. W. (2009). Photoelectrocatalytic materials for environmental applications. *Journal of Materials Chemistry*, *19*, 5089−5121.

Zhang, H., Shin, H. S., & Eda, G. (2013). The chemistry of two-dimensional layered transition metal dichalcogenide nanosheets. *Nature Chemistry*, *5*(4), 263−275.

Zhang, J., Yu, J., Zhang, Y., Li, Q., & Gong, J. R. (2011). Visible light photocatalytic H2-production activity of CuS/ZnS porous nanosheets based on photoinduced interfacial charge transfer. *Nano Letters*, *11*(11), 4774−4779.

Zhang, K., Fujitsuka, M., Du, Y., & Majima, T. (2018). 2D/2D heterostructured CdS/WS$_2$ with efficient charge separation improving H$_2$ evolution under visible light irradiation. *ACS Applied Materials & Interfaces*, *10*(24), 20458–20466.

Zhang, X., Zhou, Y. Z., Wu, D. Y., Liu, X. H., Zhang, R., Liu, H., ... Du, X. W. (2018). Self-powered, visible-blind ultraviolet photodetector based on n-ZnO nanorods/i-MgO/p-GaN structure light-emitting diodes. *Journal of Materials Chemistry A*, *6*, 9057–9063.

Zhang, Y., Liu, S., Liu, W., Liang, T., Yang, X., Xu, M., & Chen, H. (2015). Two-dimensional MoS$_2$-assisted immediate aggregation of poly-3-hexylthiophene with high mobility. *Physical Chemistry Chemical Physics*, *17*(41), 27565–27572.

Zhou, M., Lou, X. W. D., & Xie, Y. (2013). Two-dimensional nanosheets for photoelectrochemical water splitting: Possibilities and opportunities. *Nano Today*, *8*(6), 598–618.

Zhu, C., Zhang, L., Jiang, B., Zheng, J., Hu, P., Li, S., ... Wu, W. (2016). Fabrication of Z-scheme Ag$_3$PO$_4$/MoS$_2$ composites with enhanced photocatalytic activity and stability for organic pollutant degradation. *Applied Surface Science*, *377*, 99–108.

Zhu, J., Li, Q., & Xie, Y. (2013). Ultra-rapid microwave-assisted synthesis of layered ultrathin birnessite K$_{0.17}$MnO$_2$ nanosheets for efficient energy storage. *Journal of Materials Chemistry A*, *1*, 8154–8159.

Zhu, Y., Cao, C., & Li, Y. D. (2014). Ultrathin nickel hydroxide and oxide nanosheets: Synthesis, characterization and excellent supercapacitor performance. *Scientific Reports*, *4*, 5787.

Zong, X., Han, J., Ma, G., Yan, H., Wu, G., & Li, C. (2011). Photocatalytic H$_2$ evolution on CdS loaded with WS$_2$ as cocatalyst under visible light irradiation. *The Journal of Physical Chemistry C*, *115*(24), 12202–12208.

Zong, X., Yan, H. J., Wu, G. P., Ma, G. J., Wen, F. Y., Wang, L., & Li, C. (2008). Enhancement of photocatalytic H$_2$ evolution on CdS by loading MoS$_2$ as cocatalyst under visible light irradiation. *Journal of the American Chemical Society*, *130*(23), 7176–7177.

CHAPTER 12

MXene-based photocatalysts

Yolice P. Moreno Ruiz[1], William Leonardo da Silva[2] and João H. Zimnoch dos Santos[3]

[1]Centro de Tecnologias Estratégicas do Nordeste, Recife, Brazil, [2]Universidade Franciscana, Santa Maria, Brazil, [3]Instituto de Química—Universidade Federal do Rio Grande do Sul, Porto Alegre, Brazil

12.1 Principle of heterogeneous photocatalysis

The photocatalysis process is based on the photocatalyst absorption of an irradiated light to generate an electron—hole pair, followed by its separation and transference to photocatalyst surface and utilization of the excited charges for reduction and oxidation reactions. For this process to happen, the photocatalyst energy band gap should be equal or, at least, smaller than the energy of the absorbed photon to promote the electron transition from the valence band (VB) to the conduction band (CB) (Liu, Tang, et al., 2018; Park, 2017). Thereby, oxidant and reducing sites arise at the surface allowing the redox reactions of donor/acceptor species adsorbed on the semiconductor (Wen, Yin, & Dai, 2014; Malato, Fernández-Ibáñez, Maldonado, Blanco, & Gernjak, 2009). Along with other factors, the efficiency of a photocatalyst relies on the electron—hole pair recombination, which dissipates part of the harvest energy. For photodegradation reactions, oxygen is also of great importance since the formed hydroxyl and superoxide radicals are primary oxidants and photocatalytic process is one of the most efficient ways of obtaining these radicals, as displayed in Fig. 12.1 (Verbruggen, 2015).

The hydroxyl radicals ($\bullet$OH) also play a fundamental role on advanced oxidative processes (AOPs), which are based on the formation of species with high oxidizing power (2.8 V). These radicals are highly effective in the degradation and treatment of recalcitrant organic pollutants (Kanakaraju, Glass, & Oelgemöller, 2018; Miklos et al., 2018). Thus the great advantage of AOPs for treatment is that the organic compounds are destroyed and transferred from one phase to another, as is the case in some conventional treatment processes (Sillanpää, Ncibi, & Matilainen, 2018).

Among the AOPs, heterogeneous photocatalysis stand out as process that involves redox reactions induced by radiation on the surface of semiconductors (the photocatalysts)

Photocatalytic Systems by Design.
DOI: https://doi.org/10.1016/B978-0-12-820532-7.00012-6

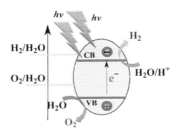

Figure 12.1

Generation of the electron/hole pair from photoexcitation and its water reduction and oxidation reactions. *VB*, valence band; *CB*, conduction band.

(Loddo, Bellardita, Camera-Roda, Parrino, & Palmisano, 2018; Wetchakun, Wetchakun, & Sakulsermsuk, 2019). The semiconductors are characterized by two energy bands: one is low energy (the VB) and another is higher energy (the CB) (Wang, Zhang, & Xie, 2018). Between these two bands is located a band gap zone that corresponds to the minimum energy required for excitation of the electron from the VB to the CB (see Fig. 12.1) (Augugliaro et al., 2012; Bethi, Sonawane, Bhanvase, & Gumfekar, 2016).

The groups of semiconductors used in heterogeneous photocatalysis processes generally involve oxides and sulfides since these groups have properties that favor the separation and mobility of the electronic charges on the surface of the semiconductors (Yin & Hill, 2017; Wang, Zhao, Liu, & Huang, 2018). Moreover, to be used in the processes of heterogeneous photocatalysis, the semiconductors should exhibit some characteristics (Swaminathan, 2018; Zou et al., 2019; Navarro Yerga, Alvarez-Galván, Vaquero, Arenales, & Fierro, 2013), namely:

1. Photoactivity with radiation that comprises a near UV−vis range; thus it is possible to use sunlight as the semiconductor activation source;
2. A redox potential of the CB positive enough to trigger the mineralization of the organic matter;
3. Photo-stability, that is, it must be resistant to anodic or cathodic photocorrosion;
4. Efficiency in the catalysis of the oxygen reduction reaction;
5. Low toxicity, given the possibility of photocorrosion for some semiconductors with extremely toxic ions; and
6. Low cost since this point is determinant as to the application of the semiconductor in the industry or on large scale.

In an attempt to draw a panorama of the studies that have being carried out with the heterogeneous photocatalysis theme in the literature (review articles), Table 12.1 shows the main research trends in photocatalysis.

As illustrated in Table 12.1, the heterogeneous photocatalysis presents a series of potential applications, such as the degradation of certain pollutants (e.g., dyes, drugs, and pesticides),

Table 12.1: Bibliographic reviews of semiconductors for application in heterogeneous photocatalysis.

Photocatalyst	Comments	Reference
TiO_2	Hydro-/solvothermal synthesis for fabrication of TiO_2 nanostructures for indoor air purification, analyzing the influence of preparation parameters (time, temperature, pH, solvent and calcination temperature)	Mamaghani, Haghighat and Lee (2019)
TiO_2 -halloysite nanotubes	Photocatalytic activities to degradation of the small amount of antibiotics, drug components, and pesticides using halloy site-based nanocomposites with TiO_2	Papoulis (2019)
Bismuth-based composite oxide	Photocatalytic hydrogen generation by bismuth-based composite oxides, mainly in the energy band engineering, Z-scheme overall water splitting, and strategies for photocatalytic activity improvement	Fang and Shangguan (2019)
g-C_3N_4	Photocatalytic applications in the O_2 evaluation, H_2 production and CO_2 reduction and photocatalytic degradation of organic dyes in aqueous solution using g-C_3N_4 based ternary composite	Prasad, Tang and Bahadur (2019)
TiO_2 photoanode catalysts	Topics of oxygen evolution catalysts, water reduction catalysts, and a system assembly capable of completing the four-electron water-oxidation reaction for efficient photocatalytic water splitting	Agbe et al. (2019)
Geopolymers	Geopolymers for further activity on additional areas of relevance to wastewater treatment such as photocatalysis, disinfection, and H_2-energy production from wastewater	Rasaki, Bingxue, Guarecuco, Thomas and Minghui (2019)
Hybrids of Ti, Zn, W, Fe, and polymers	Treatment of priority pollutant by the hybrids functional material and the related issues of concerns	Debnath, Gupta and Ghosa (2019)
Bi_2O_3, TiO_2, ZnO, and ZrO_2	Multifarious catalysts including Bi_2O_3, TiO_2, ZnO, and ZrO_2 were used to investigate the impact on photocatalytic degradation of 4-nitrophenol under halogen lamp	Muersha and Pozan Soylu (2018)
Doped and heterojunction-based semiconductor metal sulfide nanostructures (MSNSs)	Enhance photocatalytic degradation of toxic dyes by using various MSNSs (such as ZnS, CdS, CuS, Ag_2S, Bi_2S_3, CoS, FeS, and PbS) to make it a flexible and cost-effective commercial dye treatment technology	Ayodhya and Veerabhadram (2018)
$ZnS/PbSZnS/CdSZnS/Ag_2S$	Photocatalytic activity of ZnS-based dual nano-semiconductors for degradation of the Rhodamine B (cationic dye) and Methyl orange (anionic dye)	Wang, et al. (2018)

(Continued)

Table 12.1: (Continued)

Photocatalyst	Comments	Reference
Ag_3PO_4/Ti_3C_2	Ag_3PO_4/Ti_3C_2 Schottky catalyst was used to photocatalytic degradation of the 2,4-Dinitrophenol and tetracycline hydrochloride	Cai et al. (2018)
$CdS/Ti_3C_2T_x$	$CdS/Ti_3C_2T_x$ sheet-onto-sheet heterostructures were applied to the photoreduction of nitro organics	Xie, Zhang, Tang, Anpo and Xu (2018)
Quasi-core-shell In_2S_3/TiO_2@metallic $Ti_3C_2T_x$ hybrids	$TiO_2/Ti_3C_2T_x$ can be a potentially novel platform for constructing efficient photocatalysts both for wide-ranging applications and unraveling the transfer behavior of photoexcited electrons based on charge transfer channels	Wang, Wu, et al. (2018)
TiO_2/platinum-modified	Photocatalytic oxidation of carboplatin (antineoplastic drug) using TiO_2 photocatalysts (pure and modified by photodeposition and recovery of platinum)	Kitsiou, Zachariadis, Lambropoulou, Tsiplakides and Poulios (2018)
Bi_2S_3/TiO_2—Montmorillonite nanocomposites	Application of Bi_2S_3/TiO_2-montmorillonite for Ketoprofen (KP) degradation under near UV—vis irradiation (NUV-Vis)	Djouadi et al. (2018)
HT/Fe/TiO_2 (TiO_2 modified with hydrotalcite and iron oxide)	Evaluate ADMI color removal from a biologically treated textile mill effluent by heterogeneous photocatalysis with UV—visible irradiation (UV—vis) using a novel catalyst composed of TiO_2 supported on hydrotalcite and doped with iron oxide (HT/Fe/TiO_2)	Arcanjo et al. (2018)
Fe_3O_4/ZnO magnetite core shell	Three chore-shell magnetite ZnO catalysts (Fe_3O_4/ZnO: M1, $Fe_3O_4/SiO_2/ZnO$: M2, Fe_3O_4/SiO_2-APTS/ZnO: M3) were prepared, characterized and tested for their photocatalysis in the degradation of methylene blue dye under sunlight irradiation	Atla et al. (2018)

photocatalytic water purification, and the inactivation of microorganisms. In addition, among the researched literature reviews, TiO_2 was the most studied semiconductor as the photocatalyst, due to its properties, such as: high photocatalytic activity under UV radiation, chemical stability over a wide range of pH (3—13), low cost (US$ 1750—2300/tonne), and low toxicity and insolubility in water (Basha et al., 2010; Marcone, Oliveira, Almeida, Umbuzeiro, & Jardim, 2012; Collivignarelli, Abbà, Carnevale Miino, & Damiani, 2019). However, TiO_2 has limitations under visible light irradiation.

Independent of the nature of the photocatalyst, the photocatalytic process may suffer some interferences, such as (1) the presence of large amounts of oils, greases, and solids, which affect the useful life of their energy sources; (2) the presence of solids on the surface of the reactor wall at large scale, so preventing the passage of the radiation and its contact with the oxidizing agent; (3) concentration of the organic pollutant to be treated; (4) concentration of the photocatalyst; and (5) light intensity of the radiation source. However, due to the way the catalyst is homogenized in the effluent, the contact of the irradiation occurs easily with the catalyst (Chen, Tang, et al., 2019; Finčur et al., 2017; Trawiński & Skibiński, 2019; Yasmina, Mourad, Mohammed, & Khaoula, 2014).

More recently, a new series of alternative advanced materials to conventional semiconductors have been published in the literature with high versatility and photocatalytic activity under visible light irradiation too: the MXene family. MXenes have higher absorbed photon fluxes than typical photovoltaic materials, such as Si, GaAs, poly(3-hexylthiophene), graphene, and even transition metal dichalcogenides, MoS_2 and WS_2, indicating their potential application in photovoltaics too (Wong, Tan, Yang, & Xu, 2018). In the following, the fundamental aspects of the MXenes will be presented.

12.2 MXenes structure and applications for photocatalysis

MXenes are highly efficient cocatalytic systems. As semiconductors, they have electronic functionality and metal conductivity, which is the function of the nature of the transition metal (Cheng, Li, Zhang, & Xiang, 2019). The MXenes are obtained by removing a thin layer "A" from the three-dimensional layered materials of the so-called MAX phases. MAX phases (or structures) are a group of solids with layered hexagonal structures and $P6_3/mmc$

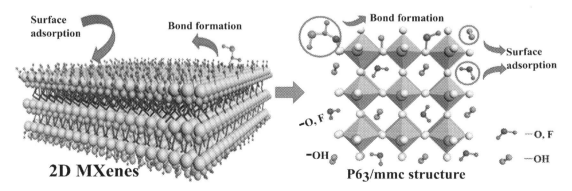

Figure 12.2

Surface adsorption of MXene in photocatalysis and their $P6_3/mmc$ hexagonal structure. *Source: Reprinted with permission from Cheng, L., Li, X., Zhang, H., & Xiang, Q. (2019). Two-dimensional transition metal MXene-based photocatalysts for solar fuel generation.* The Journal of Physical Chemistry Letters, *10(12), 3488–3494, Copyright (2019) American Chemical Society.*

space group symmetry (see Fig. 12.2). Their composition is represented by the chemical formula $M_{n+1}AX_n$, where $n = 1$, 2, or 3, "M" is a transition metal (Sc, Ti, Zr, Hf, V, Nb, Ta, Cr, or Mo), "A" is an element from groups III—VI of the Periodic Table (Al, Ga, In, Tl, Si, Ge, Sn, Pb, P, As, Bi, S, or Te), and "X" means carbon and/or nitrogen. "A" atoms organize themselves as a hexagonal lattice. The M—X bonds are the strongest ones in MAX structures (Barsoum, 2000; Barsoum, 2014). The layer An is very reactive: it can be easily corroded and eliminated from MAX phases by chemical exfoliation, or micro-mechanical stripping method (Cheng et al., 2019), or combination of exfoliation and ultrasound process (Ying et al., 2015). They can also be obtained from direct synthesis via chemical vapor deposition (CVD) (Gogotsi, 2015; Wong et al., 2018). The chemical exfoliation allows to obtain reproducible and photostable MXenes photocatalysts (Cheng et al., 2019). Some examples of MAX phases are M_3AX_2 and M_3X_2 (Khazaei, Mishra, Venkataramanan, Singh, & Yunoki, 2019).

The X layer atoms are surrounded by two layers of transition metals, forming $M_{n+1}X_n$ layers, and every two $M_{n+1}X_n$ layers are interleaved with a layer of A atoms. When MAX phases are treated with strong acid solution (HF or similar chemicals), the "A" atoms (Al or Si) are removed from the MAX materials, leading to multiple layers of $M_{n+1}X_n$, which can be separated into monolayers by sonication. This process generates a distribution of termination group on surface, for example, $-O$, $-F$, and $-OH$ of MXene photocatalysts (Loddo et al., 2018; Wetchakun et al., 2019). For example, some $M_{n+1}AX_n$ phases Ti_2AlC result into Ti_2C MXenes, V_2AlC into V_2C, Nb_2AlC into Nb_2C (Naguib et al., 2013), Zr_3AlC_2 into Zr_3C_2 (Zhou et al., 2016), and Ti_4AlN_3 into Ti_4N_3 (Urbankowski et al., 2016). MXenes transform from a metallic state to a semiconductive state by surface absorption of $-O$, $-F$, and $-OH$ groups, and the band gap can be tailored by surface functionalization (Si, Zhou, & Sun, 2015; Guo, Zhou, Zhu, & Sun, 2016).

The photocatalytic (and electrochemical) performance is affected by the surface chemistry of materials during the photocatalytic process. In the case of MXenes photocatalysts, the

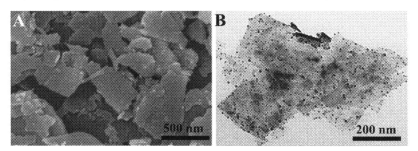

Figure 12.3

(A) Scanning electron microscopy (SEM) image and (B) Transmission electron microscopy (TEM) image of $Ti_3C_2T_x$-10% nanosheets after complete exfoliation. Source: *Reprinted with permission from Ying, Y., Liu, Y., Wang, X., Mao, Y., Cao, W., Hu, P., & Peng, X. (2015). Two-dimensional titanium carbide for efficiently reductive removal of highly toxic chromium(VI) from water.* ACS Applied Materials & Interfaces, 7(3), 1795—1803, Copyright (2015) American Chemical Society.

functional groups ($-O$, $-OH$, and $-F$) dispersed over the surface of MXene layers influence the chemical, electronic, optical properties, carrier mobility, and thermodynamic stability (Guo et al., 2016; Liao, Sa, Zhou, Ahuja, & Sun, 2014), as well as the photocatalytic efficiency (Cheng et al., 2019). MXene semiconductors with $-F$ and $-OH$ surface functional groups, such as $Ti_3C_2F_2$ and $Ti_3C_2(OH)_2$, respectively, have small band gaps and high electronic conductivity turning them in excellent photocatalysts (Wang, Shen, et al., 2015).

Currently, 60 MAX phases are known, but one of the most widely studied and a promising members of this family is Ti_3AlC_2 (Guo et al., 2016; Wang & Zhou, 2010; Tzenov & Barsoum, 2000). Ti_3C_2 2D MXene is obtained from Ti_3AlC_2 by chemical exfoliation using a strong acid solution (Gao et al., 2015), see Fig. 12.3, for example, some etchants are HF (Naguib, Mashtalir, et al., 2012; Naguib, Mochalin, Barsoum, & Gogotsi, 2014), LiF + HCl (Ghidiu, Lukatskaya, Zhao, Gogotsi, & Barsoum, 2014), and NH_4HF_2 to transform into 2D MXenes (Khazaei et al., 2019).

Anasori, Dahlqvist, et al. (2015) synthetized a special type of MAX phases with organized double transition metals, named oMAX, the layers have just one type of transition metal. These are MAX phases, out-of-plane organized double transition metals. For instance, $Mo_2Ti_2AlC_3$ was treated with acid solution resulting in $Mo_2Ti_2C_3$ (Anasori, Xie, et al., 2015). On the other hand, MAX phases are found to be dependent on the nature of the transition metals, which in turn may affect the characteristics of the resulting lattice (Dahlqvist et al., 2017; Lu et al., 2017; Tao et al., 2017; Dahlqvist, Petruhins, Lu, Hultman, & Rosen, 2018). For example, $(Nb_{2/3}Sc_{1/3})_2AlC$ iMAX phases result in $Nb_{1.33}C$ MXene after the chemical-exfoliating process (Halim et al., 2018).

Ghidiu et al. (2016), Halim et al. (2014), and Wang et al. (2017) have done efforts to replace the HF solution by low-toxicity etching agents: $NaBF_4$, HCl, and NH_4HF_2 (Halim et al., 2014); LiF in HCl (Ghidiu et al., 2016); and FeF_3/HCl (Wang et al., 2017) to prepare $Ti_3C_2T_x$ MXenes, introducing new opportunities to improve experimental methodologies. Other important aspects for the preparation of MXenes are: (1) the concentration of etching agents because higher concentration would dissolve the MAX/MXenes materials or produce ternary fluorides and alter the intercalation process (Ying et al., 2015). (2) The effect of the degradation of MXenes over time due to the temperature, which might result in the transformation of MXenes into bulk transition-metal carbides or oxides (Khazaei et al., 2019). (3) The temperature is an important parameter of formation of MXenes. Wong et al. (2018) showed that the formation of $Ti_{2(1-x)}Zr_{2x}CO_2$, $Ti_{2(1-x)}Hf_{2x}CO_2$ alloys is thermodynamically favorable at temperatures beyond 1300 and 1000K, respectively, whereas $Zr_{2(1-x)}Hf_{2x}CO_2$ alloys are already stable at 300K. These temperatures are lower than the typical fabricating temperatures of the parent MAX phases (1600K$-$1900K) (Naguib et al., 2011) and comparable to that of direct MXene synthesis via CVD ($\sim$1400K) (Gogotsi, 2015).

The two-dimensional (2D) photocatalysts, MXenes, have outstanding properties compared to traditional photocatalysts, such as TiO_2, ZnO, Fe_3O_4, and CdS. The main characteristics that identify MXenes materials are (Cheng et al., 2019; Khazaei et al., 2019; Naguib, Mashtalir, et al., 2012; Ran et al., 2017; Wong et al., 2018; Ying et al., 2015):

1. higher electrical conductivity;
2. surface hydrophilic functionalities;
3. good structural stabilities;
4. ordered layered structure;
5. adsorption ability;
6. great charge-carrier transfer;
7. powerful redox reactivity;
8. stable in aqueous solutions;
9. excellent capability to transport electrons;
10. high specific surface area for photocatalytic reactions; and
11. higher concentration of active adsorption sites on surface.

These above properties allow adjusting the surface functionality for different purposes. The potential applications of MXenes are influenced by capacity to tune the termination/functionalization of the surface with different groups (Hu et al., 2013; Naguib, Mashtalir, et al., 2012). MXene-based photocatalysts have been explored for: lithium-ion battery anodes (Naguib, Come, et al., 2012; Ahmed, Anjum, Hedhili, Gogotsi, & Alshareef, 2016), hybrid electrochemical capacitors (Guo et al., 2016; Er, Li, Naguib, Gogotsi, & Shenoy, 2014), electrocatalysts for hydrogen evolution reaction (HER) (She et al., 2016), energy storage (Tan, Jin, Sullivan, Anasori, & Gogotsi, 2017), water purification (Ying et al., 2015), electromagnetic interference shielding (Luo et al., 2014), catalysis (Ran et al., 2017), dye-sensitized solar cells (Dall'Agnese, Dall'Agnese, Anasori, Sugimoto, & Mori, 2018), sensors (dos Santos & Cichero, 2019), and others. MXenes are flexibles and versatiles, so they have also been tested in polymer bulk as polyethylene and polypyrrole too (Shahzad et al., 2018). MXene materials have been highlighted for applications in heterogeneous photocatalysis. Other examples of MXenes for photocatalysis are: $(Ti_{0.5},Nb_{0.5})_2C$, Ti_3CN, $(V_{0.5},Cr_{0.5})_3C_2$, and Ta_4C_3 (Wang, de Godoi, et al., 2015). These catalytic systems applied to the production of hydrogen and degradation of pollutants will be discussed in detail.

12.3 MXenes for water splitting

According to Hydrogen Program Overview (2019) of Energy Department, there are 40 retail hydrogen stations in the United States. Great investment has been applied for high throughput of H_2-fueling technologies for trucks (US\$ 6M), hydrogen production and utilization (US\$ 12M) and H_2@ scale pilot-integrated systems (US\$ 10M) (Satyapal, 2019). The production of hydrogen from H_2O using sunlight has been a clean process and environmentally friendly thanks to the use of semiconductor as photocatalysts (Ran et al., 2017).

Hydrogen production by conventional catalytic process is expensive, and the commercialization is reduced because of the use of noble metals, such as Pt, Ru, Rh, Au, Ag, and Pd as cocatalysts (Ran et al., 2017; Sun et al., 2019; Wang, de Godoi, et al., 2015). For water splitting, platinum is one of the most efficient photocatalysts (Khazaei et al., 2019).

The best water-splitting photocatalysts should exhibit the following characteristics: large difference in effective masses of the holes and electrons, appropriate band positions, and high absorbed photon flux in the solar spectrum (Wong et al., 2018), low hole—electron recombination rate (Singh, Mathew, Zhuang, & Hennig, 2015), and a band gap of larger than 1.23 eV, the minimum value for this reaction to occur (Guo et al., 2016; Wong et al., 2018). Furthermore, the photocatalyst must have a hydrogen reduction potential (H^+/H_2) lower than the CB minimum (CBM, more negative) and the water oxidation potential (H_2O/O_2) should be higher than the VB maximum (VBM, more positive) (Singh et al., 2015). However, the performance of photocatalysts is restricted by undesirable processes, such as hole—electron recombination. Hence, extensive approaches have been explored for reach high activity photocatalytic using MXenes; some of these summarized in Table 12.2.

According to Table 12.2, Ti_3C_2 is the most widely investigated MXene due to its outstanding properties (Sun et al., 2019). Ti_3C_2 has shown a high hole—electron generation rate by efficient absorption of solar energy and spatial separation of the holes and electrons (Wong et al., 2018). The oxidation process of MXene in H_2O solution is described as follows:

$$aTi_2CT_x + H_2O_2 \rightarrow bTiO_2 + (1-b)Ti_2CT_x + CO_y \qquad (12.1)$$

where Ti_2CT_x oxidizes to form TiO_2 and emits CO/CO_2 (Ahmed et al., 2016). $Ti_3C_2T_x$ can be synthetized by thermal treatment from Ti_2AlC and TiC at 1350°C by 2 hours, followed by stirring with an HF solution by 15 hours (Wang et al., 2016), or heating at 1150°C in air (Naguib, Mashtalir, et al., 2014), then treating with H_2O_2 at 25°C (Ahmed et al., 2016) and flowing with CO_2 at 850°C (Zhang, Kim, et al., 2016).

Semiconductors, such as TiO_2, ZnO, and CdS, have also been used for hydrogen production (Behara et al., 2015; Krishnappa et al., 2015; Chowdhury, Gomaa, & Ray, 2015). The combination of these photocatalysts with MXenes has been one of the strategies explored to achieve high catalytic efficiency. Sun et al. (2019). Oxidized Ti_3C_2 MXene in water at 60°C during different times to form TiO_2/Ti_3C_2 on amorphous carbon composites by in situ hydrothermal oxidation in the presence of H_2O and O_2. They determined that oxidation process of Ti_3C_2 to TiO_2 is improved with the increase in H_2O/O_2 treatment time (48 hours) and decreasing oxidation temperature from 473°C (Ti_3C_2) to 370°C (TiO_2/Ti_3C_2). The hydrogen production rate was 33.4 μmol/hour/g. The Ti_3C_2 concentration was an important parameter. In some cases, increasing the quantity of Ti_3C_2 affected the surface area of

Table 12.2: Researches on MXene-based photocatalysts for hydrogen production and other applications under visible light irradiation.

MXenes	Synthesis	Comments	Reference
$TiO_2/Ti_2CT_xTiO_2/Nb_2CT_x$	Hydrothermally process using Ti_3AlC_2 using HF (49%) at stirring for 24 h at 30°C, $TiCl_4$ as titania precursor, and Nb_2CT_x at 95°C by 4 h	Hydrogen energy reaction (HER), 31 μmol/h per g and 45 μmol/h per g of hydrogen production	Wang et al. (2016)
g-$C_3N_4/Pt/GO$	Reduction chemical. Sensibilization with Eosin Y, organic dyes	HER, 3820 μmol/h per g of hydrogem production	Wang, Guan, Li and Yang (2018)
TiO_2/Ti_3C_2	In situ hydrothermal oxidation with H_2O/O_2 at 60ₒC. Sensibilization with EY	HER, 33.4 μmol/h per g hydrogen production	Sun et al. (2019)
CdS/Ti_3C_2	Chemical exfoliation of Ti_3AlC_2 by HF solution. Hydrothermal process for produce CdS on Ti_3C_2 using ethanediamine and Thiourea at 160°C for 48 h	HER, 2407 μmol/h per g hydrogen production. Synthesis of ultrathin Mxene nanosheets	Xiao et al. (2019)
$CdS/MoS_2/Ti_3C_2$	T_3AlC_2 added in HF solution. CdS and MoS_2 by hydrothermal method at 160°C by 24 h	HER, 9679 μmol/h per g hydrogen production. The synergy between Mo_2S and Ti_3C_2	Chen, Wang, Chen, Wang and Ao (2019)
$CdS/Ti_3C_2CdS/NiS/Ti_3C_2$	Immersion of Ti_3AlC_2 in HF solution, and hydrothermal process for produce CdS/Ti_3C_2	HER, 14,342 and 18,560 μmol/h per g, respectively, of hydrogen production	Ran et al. (2017)
Mo_2CT_x and Ti_2CT_x	Exfoliation from Mo_2Ga_2C and Ti_2AlC to prepare Mo_2CT_x and Ti_2CT_x	He used theoretical calculations (DFT) and experimental synthesis for determine HER activity, measured by cyclic voltammetry. Determined the hydrogen adsorption free energy by DFT	She et al. (2016)
Zr_2CO_2 and Hf_2CO_2	Theoretical studies by density functional theory (DFT)	Zr_2CO_2 (1.76 eV) and Hf_2CO_2 (1.79 eV) MXenes had a large optical absorption for HER	Guo et al. (2016)
Zr_2CO_2 and Hf_2CO_2 MXenes versus $Ti_{2(1-x)}Zr_{2x}CO_2$, $Ti_{2(1-x)}Hf_{2x}CO_2$, and, $Zr_{2(1-x)}Hf_{2x}CO_2$	Theoretical studies by DFT	$Ti_{2(1-x)}Zr_{2x}CO_2$ with x = 0.2778, Eg = 1.48 eV, was chosen as the best photocatalyst for HER	Wong et al. (2018)
$T_3C_2T_x$	Exfoliation process of T_3AlC_2 with HF for 10 h at 25°C and treated with H_2O_2 at room temperature	As anode in Li-on battery	Ahmed et al. (2016)

(Continued)

Table 12.2: (Continued)

MXenes	Synthesis	Comments	Reference
$T_3C_2T_x$	Thermal oxidation at 1150°C	As anode Li-on Battery	Naguib et al. (2014)
$T_3C_2T_x$	Flash oxidation treating with CO_2 at 850°C	MXenes films for dye-sensitized solar cell	Dall'Agnese et al. (2018)
$TiO_2@C/g-C_3N_4$	T_3AlC_2 was exfoliating with HF at 25°C by 4 h. Mixed with $C_3H_6N_6$ by ultrasound, calcined at 550°C by 2 h.	The nitrogen reduction reaction, achieving an NH_3 production rate of 250.6 μmol/g per h	Liu et al. (2018)

Figure 12.4

SEM images of (A, A1) Ti_3C_2, (B, B1) Ti_3C_2 treated at 180°C, (E, E1) 0.002 mol, and (G, G1) 0.004 mol of TiO_2/Ti_3C_2. Source: *Reprinted with permission from Gao, Y., Wang, L., Zhou, A., Li, Z., Chen, J., Bala, H., ... Cao, X. (2015). Hydrothermal synthesis of TiO_2/Ti_3C_2 nanocomposites with enhanced photocatalytic activity.* Materials Letters, *150, 62–64, Copyright (2015) Elsevier.*

catalytic system, with decreasing active sites on surface, thus reducing redox catalytic reactions. An adequate ratio of cocatalyst and photocatalyst is, thus very important (Ran et al., 2017). Fig. 12.4 shows images of some examples of these photocatalytic systems, $TiO_2/Ti_3C_2T_x$.

Other photocatalytic systems, such as TiO_2/Ti_2CT_x and TiO_2/Nb_2CT_x (5 wt.% of MXene), which promote the transport of photogenerated electrons from TiO_2 to MXenes, exhibited appreciable activity (31 and 45 μmol/hour/g, respectively) to hydrogen production under irradiation of visible light (Wang et al., 2016). MXene was synthesized by etching Ti_3AlC_2 using HF (49%) at stirring for 24 hours at 30°C.

Ran et al. (2017) used experimental synthesis and theoretical studies, as density functional theory (DFT) to evaluate the efficiency of Ti_3C_2 as cocatalyst, under visible-light irradiation. Different photocatalytic systems were investigated, such as Ti_3C_2/CdS producing high amount of hydrogen of 14,342 μmol/hour/g, NiS−CdS (12,953 μmol/hour/g), Ni−CdS (8649 μmol/hour/g), and MoS_2−CdS (6183 μmol/hour/g). The highest efficiency catalytic was 18,560 μmol/hour/g employing CdS/1 mol.% NiS/2.5 wt.% Ti_3C_2 system, which NiS is a p-type semiconductor, it was simultaneously loaded with Ti_3C_2 nanoparticles on CdS submicrosphere, a n-type semiconductor. $CdS/NiS/Ti_3C_2$ system stimulates a faster charge transfer from CdS to NiS, and CdS to Ti_3C_2, inducing holes and electrons, respectively. Hence, the migration distances of the photogenerated holes and electrons to the reaction interface were reduced, the possibility of hole − electron recombination was decreased, and the highest hydrogen production was obtained. The theoretical calculations showed that O-terminated Ti_3C_2 presented higher positive value of Fermi level, 1.88 V, which indicated an outstanding capacity to get electrons from semiconductor CdS. Thus Ran et al. (2017) recommended increased hydrophilic functionality on Ti_3C_2 surface with −O termination in the course of etching process to favor the density active sites (Ran et al., 2017).

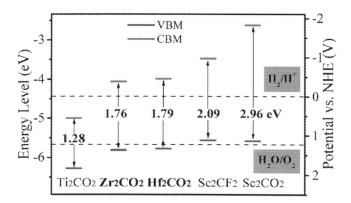

Figure 12.5

The water redox potentials of MXene photocatalysts. Source: *Reprinted with permission from Guo, Z., Zhou, J., Zhu, L. & Sun, Z. (2016). MXene: A promising photocatalyst for water splitting.* Journal of Materials Chemistry A, 4 *(29), 11446−11452, Copyright (2016) The Royal Society of Chemistry—* Journal of Materials Chemistry A.

The optical and electrical properties of MXenes change as a function of the terminal group on the surface. MXenes with a surface exposed to O atoms exhibit better absorption of visible light than the other MXenes with surface termination $-F$ and $-OH$. MXenes with $-OH$ group on their surface can be employed as Schottky barrier free, thus participating as electron (hole) donors (Liu, Xiao, & Goddard, 2016). Ling, Shi, Ouyang, and Wang (2016) and Gao, O'Mullane, and Du (2017) have shown that MXenes with atoms O on their surface and have Gibbs-free energies close to 0 eV for hydrogen adsorption (Gao et al., 2017; Ling et al., 2016). W_2CO_2 (Chaudhari et al., 2017), $ZrCO_2$, and Hf_2CO_2 present the band edges close to water redox potentials, becoming photocatalysts suitable for water splitting as displayed in Fig. 12.5. She et al. (2016) showed that the Mo_2CT_x has high surface activity for HER in comparison to Ti_2CT_x. They used theoretical simulations and observed that it is possible to improve charge transfer to the O$-$ atom of water, incorporating Fe atom that is less electronegative.

According to Fig. 12.5, the band edges of some MXenes, such as Ti_2CO_2 (1.28 eV), Sc_2CF_2 (2.09 eV), and Sc_2CO_2 (2.96 eV), do not surpass the entire water redox potentials (Naguib, Mochalin, et al., 2014). In addition, too small band gaps may not provide enough energy for water splitting, for example, 0.74 eV of $Sc_2C(OH)_2$ and 0.84 eV of Mo_2CF_2 (Guo et al., 2016). To utilize the sunlight efficiently is important that the band gap stands between 1.55 and 3.0 eV (Wong et al., 2018). However, Zr_2CO_2 (1.76 eV) and Hf_2CO_2 (1.79 eV) MXenes have a large optical absorption and smaller bad gap that improve the efficiency in the utilization of solar energy. The process of electron$-$hole pairs' migration and separation in MXenes has been studied by theoretical simulations using DFT (Guo et al., 2016; Wong et al., 2018) and cluster expansion method (Chen, Schmidt, Schneider, & Wolverton, 2011; Herder, Bray, & Schneider, 2015). These systems use models with the uniform adsorption of one or few groups (F, OH, and/or O) at active sites of MXene surface to photocatalytic water splitting (Guo et al., 2016; Wong et al., 2018).

The properties of Zr_2CO_2 and Hf_2CO_2 (Guo et al., 2016), such as the electronic structure and band gap (E_g) of the exchange-correlation functional and the generalized gradient approximations (Perdew & Wang, 1992) of Perdew$-$Burke$-$Ernzerhof (Perdew, Burke, & Wang, 1996), were studied by computational approaches. The simulations showed that the carrier mobility of Zr_2CO_2 and Hf_2CO_2 photocatalysts has anisotropic character (Guo et al., 2016). The transfer and separation mechanism of photogenerated electrons and holes is very fast. The electron mobility in the y direction is 10 times larger than that in the x direction, while for holes, the mobility in the x direction is about 50% larger than that in the y direction. In other words, electrons have a tendency to migrate along the y direction and while holes tending to migrate along the x direction (Guo et al., 2016). The effective electron masses are highly anisotropic and heavier than the effective hole masses that are isotropic, suggesting that the electron presents lower mobilities (Wong et al., 2018). Hence, Zr_2CO_2 and Hf_2CO_2 MXene photocatalysts can efficiently facilitate the migration and separation of electron−hole pairs because they exhibit unexpectedly high and directionally anisotropic carrier mobility (Guo et al., 2016). To improve the absorption of the Zr_2CO_2 and Hf_2CO_2 within the solar spectrum, two approaches can develop surface functionalization and alloying (Tan, Jin, et al., 2017).

Currently, Zr_2CO_2 and Hf_2CO_2 MXenes versus $Ti_{2(1−x)}Zr_{2x}CO_2$, $Ti_{2(1−x)}Hf_{2x}CO_2$, and $Zr_{2(1−x)}Hf_{2x}CO_2$, where $0 < x < 1$, were evaluated over 3 million MXenes (Wong et al., 2018). The results indicated that $Ti_{2(1−x)}Zr_{2x}CO_2$, $Ti_{2(1−x)}Hf_{2x}CO_2$, and $Zr_{2(1−x)}Hf_{2x}CO_2$ MXenes are stable from thermodynamic and mechanical perspectives. The structural analysis displayed that $Ti_{2(1−x)}Zr_{2x}CO_2$ and $Ti_{2(1−x)}Hf_{2x}CO_2$ have a higher degree of structural distortions possessing more positive formation energies than $Zr_{2(1−x)}Hf_{2x}CO_2$ with the least distortions having the less positive formation energies (Wong et al., 2018). The MXene-based photocatalyst had an indirect band gap of 1.25−1.80 eV. $Ti_{2(1−x)}Zr_{2x}CO_2$ with $x = 0.2778$ was selected as the best photocatalyst, with outstanding properties, such as a broad band gap ($E_g = 1.48$ eV) along with properly aligned band positions [$E_{CBM} = −0.02$ eV vs normal hydrogen electrode (NHE); $E_{VBM} = +1.45$ eV vs NHE]; a large difference in effective masses of the holes and electrons to prevent electron − hole recombination; and high absorbed photon flux.

Other MXene photocatalysts, such as T_2C, Sc_2C, and V_2C, have been proposed too as reversible hydrogen storage materials under ambient conditions, for instance, 8.6 wt.%, from theoretical studies by DFT (Hu et al., 2013). The aNANT MXene database (Indian Institute Science) have information about 23,000 MXenes (Khazaei et al., 2019), which the electronic properties have been predicted employing DFT. These theoretical studies expand the number of MXenes structures, as efficient cocatalysts that can be applied to hydrogen production. Next, efficient MXenes for the degradation of organic pollutants will be presented.

12.4 MXenes for environmental pollutants' degradations

MXenes can suffer partial or complete oxidation into transition metal oxides, permitting better catalytic activity (Dall'Agnese et al., 2018). These photocatalysts have characteristics of 2D materials composed of a single layer of atoms, with thickness of only single- or few-atoms thick (typically less than 5 nm) and large lateral size (up to 100 nm or up to a few micrometers), providing for these materials with specific textural characteristics, such as considerable porosity, high specific surface area, and ultimately exposed surface atom, making them desirable for application in the treatment of environmental solutions, such as adsorption and photocatalysis (Deng et al., 2016; Tan, Cao, et al., 2017).

The benefits from the unique atomic structure, the MXene materials, show a series of advantage application in the degradation of pollutants for photocatalysis (Chen, Shen,

Table 12.3: Examples of MXene materials for environmental pollutants' degradations.

MXene	Synthesis	Comment	Reference
$T_3C_2T_x$	Delaminating ultrasonically in HF solution	$Ti_3C_2T_x$ nanosheets with Ti-surfaces covered with hydroxyl and fluorine groups were used for Cr (VI) removal in treated water by simple in situ reduction, of 250 mg/g	Ying et al. (2015)
$TiO_2/Ti_3C_2T_x$	Chemical exfoliation from Ti_3AlC_2. Thermal treatment of T_3C_2 with HCl and $NaBF_4$ at 160°C by 12 h	$Ti_3C_2T_x$ (T_x = OH, O, or F) and its heterostructural derivative TiO_2 facet-decorated ($Ti_3C_2T_x$ MXene [001-T/MX]) were used to study of the decomposition of the antiepileptic carbamazepine drug under UV and visible light	Shahzad et al. (2018)
TiO_2/Ti_3C_2 composites	Hydrothermally process with $TiSO_4$ and Ti_3C_2 at 180°C by 18 h	Photodegradation of metal orange under UV illumination	Gao et al. (2015)
$Ag-C_3N_4/Ti_3C_2$	HF treatment at 60°C, ultrasonic hydrolysis at 150°C, and reduction of $AgNO_3$ on C_3N_4/Ti_3C_2	Degradation of aniline 81.8% within 8 h under visible-light irradiation photocatalytic activity	Ding et al. (2019)
Ag_3PO_4/Ti_3C_2	Ti_3AlC_2 was exfoliated in NaF + HCl solution for 12 h at 60°C. Ag_3PO_4/Ti_3C_2 was generated by electrostatically method	Ag_3PO_4/Ti_3C_2 MXene catalyst was used to evaluate photodegradation of the antibiotics and persistent organic pollutant, such as methyl orange, 2,4-dinitrophenol, tetracycline hydrochloride, thiamphenicol, and chloramphenicol, under visible light irradiation	Cai et al. (2018)
Co_3O_4/Ti_3C_2	One-step heating method, Ti_3AlC_2 was etching at 40°C by 72 h. Co $(NO_3)_2 \cdot 6H_2O$ was mixed with $Ti_3C_2T_x$ by stirring at 60°C to dry sample	Photodegradation of Bisphenol A ($C_{15}H_{16}O_2$), 95% within 7 min in broad pH range (4—10)	Liu et al. (2018)
$Co_3O_4/3D$ nitrogen-doped graphene aerogel (NGA)	Hydrothermal procedure of Co $(NO_3)_2.6H_2O$ and $CO(NH_2)_2$ at 120°C by 16 h. After calcination of $Co_2CO_3(OH)_2$ under 450 °C for 2 h	3D Co_3O_4/NGA composite was used to degradation of the acid orange 7 dye, evaluating the influences of reaction parameters such as catalyst, peroxymonosulfate (PMS) concentration, initial solution pH and anions onto the activity photocatalytic	Yuan et al. (2018)
$Co_3O_4/Ti_3C_2T_x$	Ti_3AlC_2 was eatched by HF at 40°C by 72 h. $Co(NO_3)2 \cdot 6H_2O$ mixed with H_2O at 60°C by 12 h	$Co_3O_4/MXene$ composite was used to activate PMS for bisphenol A degradation	Liu et al. (2018)
$Fe_3O_4/MXene$	Ti_3AlC_2 was added to HF solution at 25°C for 10 h. $FeSO_4$ and $FeCl_3$ were mixed at 80°C for 4 h	$Fe_3O_4/MXene$ composites used for phosphate removal from wastewater	Zhang et al. (2016)

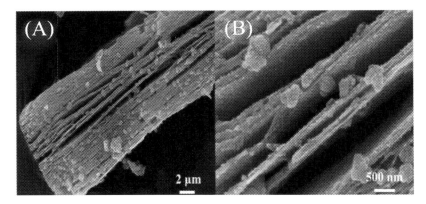

Figure 12.6
SEM images from Ag-C$_3$N$_4$/Ti$_3$C$_2$ nanosheets. Source: *Reprinted with permission from Ding, X., Li, C., Wang, L., Feng, L., Han, D., & Wang, W. (2019). Fabrication of hierarchical g-C3N4/MXene-AgNPs nanocomposites with enhanced photocatalytic performances.* Materials Letters, 247, 174−177, Copyright (2019) Elsevier.

Guo, & Mao, 2010; Li & Wu, 2019). The band gap energy of MXene materials is between 0.92 and 1.75 eV, and the light absorbance is at wavelengths in the range of 300−500 nm (Zhang, Yang, et al., 2016). Thus to increase their applications to visible radiation, surface modifications of these materials can be advantageous to fulfill the requirements for photocatalysis (Tang, Zhou, & Shen, 2012). Table 12.3 shows some of MXene materials that are been used to environmental pollutants' degradations by photocatalysis.

Ti$_3$C$_2$ is an excellent electrical conductor, one of the most popular photocatalysts among MXene materials. Recently, to improve the photodegradation of pollutants, Ti$_3$C$_2$ has been combined with other transition metals (Ag, Fe, Ti, and Co) to take advantage of the properties as the cocatalyst and as a semiconductor. For instance, Ag nanoparticles with g-C$_3$N$_4$ as photocatalyst, and Ti$_3$C$_2$ MXene as cocatalyst, result in Ag-C$_3$N$_4$/Ti$_3$C$_2$, as shown in Fig. 12.6, a catalytic system applied to aniline photodegradation (Ding et al., 2019). At interface Ti$_3$C$_2$ and g-C$_3$N$_4$ sheets, the Schottky barrier appears. The Ag deposition on C$_3$N$_4$/Ti$_3$C$_2$ sheets enhanced optical absorption and decreased band gap energy due to the surface plasmon resonance effect of Ag nanoparticles. The Ag nanoparticles helped the electron's transportation.

Several pollutants, such as methyl orange, 2,4-dinitrophenol, tetracycline hydrochloride, thiamphenicol, and chloramphenicol, were degraded under visible light irradiation. Combining the properties of the Ti$_3$C$_2$ with Ag$_3$PO$_4$ (Cai et al., 2018), a Schottky barrier is created at the interfaces, to render easier carrier separation as well as electrons' transfer. Some specific characteristics of Ag$_3$PO$_4$/Ti$_3$C$_2$ photocatalytic system are: (1) the marked

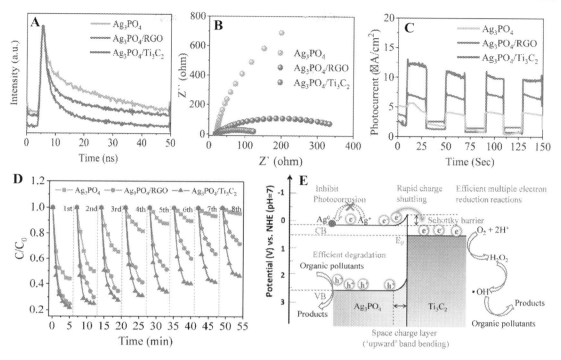

Figure 12.7

Time-resolved photoluminescence decay spectra (A), EIS Nyquist plots (B), and transient photocurrent responses (C) of Ag_3PO_4, Ag_3PO_4/RGO, and Ag_3PO_4/Ti_3C_2. Photodegradation of tetracycline hydrochloride (D) and mechanism of photocatalytic of Ag_3PO_4/Ti_3C_2 (E). Source: *Reprinted with permission from Cai, T., Wang, L., Liu, Y., Zhang, S., Dong, W., Chen, H., ... Luo, S. (2018). Ag3PO4/Ti3C2 MXene interface materials as a Schottky catalyst with enhanced photocatalytic activities and anti-photocorrosion performance.* Applied Catalysis B: Environmental 239, 545—554, *Copyright (2018) Elsevier.*

charge separation; (2) an efficient transfer of electrons that decreased the probability of electrons return from Ti_3C_2 to Ag_3PO_4; and (3) great •OH production obtained by the reactivity of metal Ti sites on surface that developed several electron reduction reactions ($O_2 \rightarrow H_2O_2 \rightarrow$ •OH) (Cai et al., 2018).

The charge transfer process between Ti_3C_2 with Ag_3PO_4 was determined by electrochemical impedance spectroscopy (EIS) and transient photocurrent responses. The radiative lifetimes of Ag_3PO_4/Ti_3C_2 were only 1.35 ns, according to the photoluminescence spectra analysis, which can inhibit the carrier recombination as shown in Fig. 12.7A. The better separation of carriers was verified by EIS Nyquist plots and transient photocurrent responses. Ag_3PO_4/Ti_3C_2 had smallest arc radius in Nyquist plot, that is, lower resistance to electric current flow. Furthermore, the photocurrent responses were the highest compared to others, such as Ag_3PO_4 and Ag_3PO_4/RGO photocatalysts, as shown in Fig. 12.7B and C. The previous results allowed to achieve an excellent degradation activity of some pollutants, such as chloramphenicol, tetracycline hydrochloride, and methyl orange, with a higher rate constant (k) of 0.025, 0.32, and 0.094/min, respectively, using Ag_3PO_4/Ti_3C_2 MXene. This photocatalytic system has lost a little efficiency of 31.6% after eight degradation cycles, indicating an improvement in the anticorrosion process, as displayed in Fig. 12.7D (Cai et al., 2018).

The functionalization and flexibility of terminal groups on Ti_3C_2 surface and incorporation of transition metals, such as Co_3CO_4, TiO_2, and Ag-C_3N_4, have shown a great photocatalytic activity in the degradation of several pollutants, such as bisphenol A (Liu, Luo, et al., 2018), metal orange (Gao et al., 2015), carbamazepine (Shahzad et al., 2018), and aniline (Ding et al., 2019). These catalytic systems showed a great pollutant removal capacity due to their properties. The intimate and strong contact between the host photocatalyst and MXene enhances migration and separation of the photogenerated carrier from host photocatalyst to the MXene under light illumination (Cheng et al., 2019; Wang, Wu, et al., 2018; Cao, Shen, Tong, Fu, & Yu, 2018). According to Tables 12.2 and 12.3, MXene presents a promising prospect for hydrogen production and environmental pollutant removal and extends their photocatalytic performance to other applications under solar light.

12.5 Challenges

As discussed earlier, the application perspectives of MXene materials are broad and computational simulations foresee more than 20,000 MXene-based photocatalysts with outstanding characteristics, but there are still some challenges to overcome, such as:

1. Synthesis or preparation of MXenes with ideal crystal structure with a uniform and pure surface having only one type of −O, −F, or −OH termination group.
2. The synthesis process of MXene materials has to be environmentally friendly, cheaper, safe, reproducible and with controlled physicochemical properties for large-scale preparation because HF solution is the most commonly used etchant (Cheng et al., 2019).
3. Increase the stability, as MXenes can suffer degradation over time due to temperature and can be transformed into metal oxides or carbides (Khazaei et al., 2019; Srivastava, Mishra, Mizuseki, Lee, & Singh, 2016).
4. Fundamental concepts, chemical, optical, and electronic properties, thermodynamic stability, and photocatalytic mechanism have to be studied in detail to develop positive effects in photocatalytic process and expand the applications in the industry.
5. Cost-effective, robust, scarcity, instability, high activity, and insufficient hydrophilic characteristics at large-scale are still issues to be target.

12.6 Conclusion and remarks

As discussed in this chapter, many strategies have been explored in immobilizing MXene semiconductor, either as a catalyst, cocatalyst, and/or catalyst support. Currently, MXenes have been coated with metals, such as Ti, Ag, Co, Fe, Zr, Mo, CdS, and Ni among others, to improve their efficiency in photocatalytic processes and to reduce electron/hole pair recombination. Thus the relationships between the synthesis, resulting physicochemical properties, electronic structures, carrier mobility, stability, and aspects that influence the performance of MXene photocatalysts have to be understood to determine the photocatalytic mechanism and to boost the catalytic efficiency of MXenes. They have shown functional versatility with a surface rich in active catalytic sites, and strategies, such as MXene-based multicomponent systems, should be under future research views.

Acknowledgments

This project was partially financed by the Conselho Nacional de Desenvolvimento Científico e Tecnológico (CNPq—310408/2019-9) and Fundação de Amparo à Pesquisa do Estado do Rio Grande do Sul (FAPERGS—16/2551-0000470-6). Yolice P. Moreno Ruiz is grateful for the grant provided by CNPq—300367/2019-8. William Leonardo da Silva is grateful for the grant provided by Coordenação de Aperfeiçoamento de Pessoal de Nível Superior (CAPES)—Brazil.

References

Agbe, H., Nyankson, E., Raza, N., Dodoo-Arhin, D., Chauhan, A., Osei, G., ... Kim, K.-H. (2019). Recent advances in photoinduced catalysis for water splitting and environmental applications. *Journal of Industrial and Engineering Chemistry*, *72*, 31−49.

Ahmed, B., Anjum, D. H., Hedhili, M. N., Gogotsi, Y., & Alshareef, H. N. (2016). H_2O_2 assisted room temperature oxidation of Ti_2C MXene for Li-ion battery anodes. *Nanoscale*, *8*(14), 7580−7587.

Anasori, B., Dahlqvist, M., Halim, J., Moon, E. J., Lu, J., Hosler, B. C., ... Barsoum, M. W. (2015). Experimental and theoretical characterization of ordered MAX phases Mo_2TiAlC_2 and $Mo_2Ti_2AlC_3$. *Journal of Applied Physics*, *118*(9)094304.

Anasori, B., Xie, Y., Beidaghi, M., Lu, J., Hosler, B. C., Hultman, L., ... Barsoum, M. W. (2015). Two-dimensional, ordered, double transition metals carbides (MXenes). *ACS Nano*, *9*(10), 9507−9516.

Arcanjo, G. S., Mounteer, A. H., Bellato, C. R., Silva, L. M. Md, Brant Dias, S. H., & Silva, P. Rd (2018). Heterogeneous photocatalysis using TiO_2 modified with hydrotalcite and iron oxide under UV−visible irradiation for color and toxicity reduction in secondary textile mill effluent. *Journal of Environmental Management*, *211*, 154−163.

Atla, S. B., Lin, W. R., Chien, T. C., Tseng, M. J., Shu, J. C., Chen, C. C., & Chen, C. Y. (2018). Fabrication of Fe_3O_4/ZnO magnetite core shell and its application in photocatalysis using sunlight. *Materials Chemistry and Physics*, *216*, 380−386.

Ayodhya, D., & Veerabhadram, G. (2018). A review on recent advances in photodegradation of dyes using doped and heterojunction based semiconductor metal sulfide nanostructures for environmental protection. *Materials Today Energy*, *9*, 83−113.

Augugliaro, V., Bellardita, M., Loddo, V., Palmisano, G., Palmisano, L., & Yurdakal, S. (2012). Overview on oxidation mechanisms of organic compounds by TiO_2 in heterogeneous photocatalysis. *Journal of Photochemistry and Photobiology C: Photochemistry Reviews*, *13*(3), 224−245.

Barsoum, M. W. (2000). The Mn + 1AXn phases: A new class of solids: Thermodynamically stable nanolaminates. *Progress in Solid State Chemistry*, 28(1), 201−281.

Barsoum, M. W. (2014). *The Mn + 1 AXn phases and their properties, ceramics science and technology.* Weinheim: Wiley-VCH Verlag GmbH & Co.

Basha, S., Keane, D., Morrissey, A., Nolan, K., Oelgemöller, M., & Tobin, J. (2010). Studies on the adsorption and kinetics of photodegradation of pharmaceutical compound, indomethacin using novel photocatalytic adsorbents (IPCAs). *Industrial & Engineering Chemistry Research*, 49(22), 11302−11309.

Behara, D. K., Upadhyay, A. P., Sharma, G. P., Krishna Kiran, B. V., Sivakumar, S., & Pala, R. G. (2015). Heterostructures based on TiO2 and silicon for solar hydrogen generation. In A. Tiwari, & L. Uzun (Eds.), *Advanced functional materials* (pp. 219−281). Scrivener Publishing 100 Cummings Center, Suite 541JBeverly, MA 01915-610: John Wiley & Sons, Inc.

Bethi, B., Sonawane, S. H., Bhanvase, B. A., & Gumfekar, S. P. (2016). Nanomaterials-based advanced oxidation processes for wastewater treatment: A review. *Chemical Engineering and Processing: Process Intensification*, 109, 178−189.

Cai, T., Wang, L., Liu, Y., Zhang, S., Dong, W., Chen, H., . . . Luo, S. (2018). Ag_3PO_4/Ti_3C_2 MXene interface materials as a Schottky catalyst with enhanced photocatalytic activities and anti-photocorrosion performance. *Applied Catalysis B: Environmental*, 239, 545−554.

Cao, S., Shen, B., Tong, T., Fu, J., & Yu, J. (2018). 2D/2D heterojunction of ultrathin $MXene/Bi_2WO_6$ nanosheets for improved photocatalytic CO_2 reduction. *Advanced Functional Materials*, 28(21)1800136.

Chaudhari, N. K., Jin, H., Kim, B., San Baek, D., Joo, S. H., & Lee, K. (2017). MXene: An emerging two-dimensional material for future energy conversion and storage applications. *Journal of Materials Chemistry A*, 5(47), 24564−24579.

Cheng, L., Li, X., Zhang, H., & Xiang, Q. (2019). Two-dimensional transition metal MXene-based photocatalysts for solar fuel generation. *The Journal of Physical Chemistry Letters*, 10(12), 3488−3494.

Chen, W., Schmidt, D., Schneider, W. F., & Wolverton, C. (2011). First-principles cluster expansion study of missing-row reconstructions of fcc (110) surfaces. *Physical Review B*, 83(7)075415.

Chen, X., Shen, S., Guo, L., & Mao, S. S. (2010). Semiconductor-based photocatalytic hydrogen generation. *Chemical Reviews*, 110(11), 6503−6570.

Chen, L., Tang, J., Song, L. N., Chen, P., He, J., Au, C. T., & Yin, S. F. (2019). Heterogeneous photocatalysis for selective oxidation of alcohols and hydrocarbons. *Applied Catalysis B: Environmental*, 242, 379−388.

Chen, R., Wang, P., Chen, J., Wang, C., & Ao, Y. (2019). Synergetic effect of MoS_2 and MXene on the enhanced H_2 evolution performance of CdS under visible light irradiation. *Applied Surface Science*, 473, 11−19.

Chowdhury, P., Gomaa, H., & Ray, A. K. (2015). Sacrificial hydrogen generation from aqueous triethanolamine with Eosin Y-sensitized Pt/TiO_2 photocatalyst in UV, visible and solar light irradiation. *Chemosphere*, 121, 54−61.

Collivignarelli, M. C., Abbà, A., Carnevale Miino, M., & Damiani, S. (2019). Treatments for color removal from wastewater: State of the art. *Journal of Environmental Management*, 236, 727−745.

Dahlqvist, M., Lu, J., Meshkian, R., Tao, Q., Hultman, L., & Rosen, J. (2017). Prediction and synthesis of a family of atomic laminate phases with Kagomé-like and in-plane chemical ordering. *Science Advances*, 3(7), e1700642.

Dahlqvist, M., Petruhins, A., Lu, J., Hultman, L., & Rosen, J. (2018). Origin of chemically ordered atomic laminates (i-MAX): Expanding the elemental space by a theoretical/experimental approach. *ACS Nano*, 12(8), 7761−7770.

Dall'Agnese, C., Dall'Agnese, Y., Anasori, B., Sugimoto, W., & Mori, S. (2018). Oxidized Ti_3C_2 MXene nanosheets for dye-sensitized solar cells. *New Journal of Chemistry*, 42(20), 16446−16450.

Debnath, D., Gupta, A. K., & Ghosal, P. S. (2019). Recent advances in the development of tailored functional materials for the treatment of pesticides in aqueous media: A review. *Journal of Industrial and Engineering Chemistry*, 70, 51−69.

Deng, D., Novoselov, K. S., Fu, Q., Zheng, N., Tian, Z., & Bao, X. (2016). Catalysis with two-dimensional materials and their heterostructures. *Nature Nanotechnology*, 11(3), 218−230.

Ding, X., Li, C., Wang, L., Feng, L., Han, D., & Wang, W. (2019). Fabrication of hierarchical g-C_3N_4/ MXene-AgNPs nanocomposites with enhanced photocatalytic performances. *Materials Letters, 247,* 174−177.

Djouadi, L., Khalaf, H., Boukhatem, H., Boutoumi, H., Kezzime, A., Santaballa, J. A., & Canle, M. (2018). Degradation of aqueous ketoprofen by heterogeneous photocatalysis using Bi_2S_3/TiO_2−montmorillonite nanocomposites under simulated solar irradiation. *Applied Clay Science, 166,* 27−37.

dos Santos, J. H. Z., & Cichero, M. C. (2019). MXenes for sensors. In R. B. Inamuddin, & A. M. Asiri (Eds.), *MXenes: Fundamentals and applications* (pp. 1−19). Millersville, PA: Materials Research Forum LLC.

Er, D., Li, J., Naguib, M., Gogotsi, Y., & Shenoy, V. B. (2014). Ti_3C_2 MXene as a high capacity electrode material for metal (Li, Na, K, Ca) ion batteries. *ACS Applied Materials & Interfaces, 6*(14), 11173−11179.

Fang, W., & Shangguan, W. (2019). A review on bismuth-based composite oxides for photocatalytic hydrogen generation. *International Journal of Hydrogen Energy, 44*(2), 895−912.

Finčur, N. L., Krstić, J. B., Šibul, F. S., Šojić, D. V., Despotović, V. N., Banić, N. D., . . . Abramović, B. F. (2017). Removal of alprazolam from aqueous solutions by heterogeneous photocatalysis: Influencing factors, intermediates, and products. *Chemical Engineering Journal, 307,* 1105−1115.

Gao, G., O'Mullane, A. P., & Du, A. (2017). 2D MXenes: A new family of promising catalysts for the hydrogen evolution reaction. *ACS Catalysis, 7*(1), 494−500.

Gao, Y., Wang, L., Zhou, A., Li, Z., Chen, J., Bala, H., . . . Cao, X. (2015). Hydrothermal synthesis of TiO_2/ Ti_3C_2 nanocomposites with enhanced photocatalytic activity. *Materials Letters, 150,* 62−64.

Ghidiu, M., Halim, J., Kota, S., Bish, D., Gogotsi, Y., & Barsoum, M. W. (2016). Ion-exchange and cation solvation reactions in Ti_3C_2 MXene. *Chemistry of Materials, 28*(10), 3507−3514.

Ghidiu, M., Lukatskaya, M. R., Zhao, M. Q., Gogotsi, Y., & Barsoum, M. W. (2014). Conductive two-dimensional titanium carbide 'clay' with high volumetric capacitance. *Nature, 516*(7529), 78−81.

Gogotsi, Y. (2015). Chemical vapour deposition: Transition metal carbides go 2D. *Nature. Materials 14*(11), 1079−1080.

Guo, Z., Zhou, J., Zhu, L., & Sun, Z. (2016). MXene: A promising photocatalyst for water splitting. *Journal of Materials Chemistry A, 4*(29), 11446−11452.

Halim, J., Lukatskaya, M. R., Cook, K. M., Lu, J., Smith, C. R., Näslund, L.-Å., . . . Barsoum, M. W. (2014). Transparent conductive two-dimensional titanium carbide epitaxial thin films. *Chemistry of Materials, 26*(7), 2374−2381.

Halim, J., Palisaitis, J., Lu, J., Thörnberg, J., Moon, E. J., Precner, M., . . . Rosen, J. (2018). Synthesis of two-dimensional Nb1.33C (MXene) with randomly distributed vacancies by etching of the quaternary solid solution (Nb2/3Sc1/3)2AlC MAX phase. *ACS Applied Nano Materials, 1*(6), 2455−2460.

Herder, L. M., Bray, J. M., & Schneider, W. F. (2015). Comparison of cluster expansion fitting algorithms for interactions at surfaces. *Surface Science, 640,* 104−111.

Hu, Q., Sun, D., Wu, Q., Wang, H., Wang, L., Liu, B., . . . He, J. (2013). MXene: A new family of promising hydrogen storage medium. *The Journal of Physical Chemistry A, 117*(51), 14253−14260.

Kanakaraju, D., Glass, B. D., & Oelgemöller, M. (2018). Advanced oxidation process-mediated removal of pharmaceuticals from water: A review. *Journal of Environmental Management, 219,* 189−207.

Khazaei, M., Mishra, A., Venkataramanan, N. S., Singh, A. K., & Yunoki, S. (2019). Recent advances in MXenes: From fundamentals to applications. *Current Opinion in Solid State and Materials Science, 23* (3), 164−178.

Kitsiou, V., Zachariadis, G. A., Lambropoulou, D. A., Tsiplakides, D., & Poulios, I. (2018). Mineralization of the antineoplastic drug carboplatin by heterogeneous photocatalysis with simultaneous synthesis of platinum-modified TiO_2 catalysts. *Journal of Environmental Chemical Engineering, 6*(2), 2409−2416.

Krishnappa, M., Souza, V. S., Ganganagappa, N., Scholten, J. D., Teixeira, S. R., Dupont, J., & Thippeswamy, R. (2015). Mesoporous foam TiO_2 nanomaterials for effective hydrogen production. *Chemistry−A European Journal, 21*(49), 17624−17630.

Liao, J., Sa, B., Zhou, J., Ahuja, R., & Sun, Z. (2014). Design of high-efficiency visible-light photocatalysts for water splitting: MoS$_2$/AlN(GaN) heterostructures. *The Journal of Physical Chemistry C, 118*(31), 17594−17599.

Ling, C., Shi, L., Ouyang, Y., & Wang, J. (2016). Searching for highly active catalysts for hydrogen evolution reaction based on O-terminated MXenes through a simple descriptor. *Chemistry of Materials, 28*(24), 9026−9032.

Liu, Y., Luo, R., Li, Y., Qi, J., Wang, C., Li, J., . . . Wang, L. (2018). Sandwich-like Co$_3$O$_4$/MXene composite with enhanced catalytic performance for bisphenol A degradation. *Chemical Engineering Journal, 347*, 731−740.

Liu, X., Tang, B., Long, J., Zhang, W., Liu, X., & Mirza, Z. (2018). The development of MOFs-based nanomaterials in heterogeneous organocatalysis. *Science Bulletin, 63*(8), 502−524.

Liu, Y., Xiao, H., & Goddard, W. A. (2016). Schottky-barrier-free contacts with two-dimensional semiconductors by surface-engineered MXenes. *Journal of the American Chemical Society, 138*(49), 15853−15856.

Liu, Q., Ai, L., & Jiang, J. (2018). MXene-derived TiO$_2$@C/g-C$_3$N$_4$ heterojunctions for highly efficient nitrogen photofixation. *Journal of Materials Chemistry A, 6*(9), 4102−4110.

Li, Z., & Wu, Y. (2019). 2D early transition metal carbides (MXenes) for catalysis. *Small (Weinheim an der Bergstrasse, Germany), 15*(29)1804736.

Loddo, V., Bellardita, M., Camera-Roda, G., Parrino, F., & Palmisano, L. (2018). Heterogeneous photocatalysis: A promising advanced oxidation process. In A. Basile, S. Mozia, & R. Molinari (Eds.), *Current trends and future developments on (bio-) membranes* (pp. 1−43). Amsterdam: Elsevier, Chapter 1.

Luo, Y., Guo, W., Ngo, H. H., Nghiem, L. D., Hai, F. I., Zhang, J., . . . Wang, X. C. (2014). A review on the occurrence of micropollutants in the aquatic environment and their fate and removal during wastewater treatment. *Science of the Total Environment, 473−474*(0), 619−641.

Lu, J., Thore, A., Meshkian, R., Tao, Q., Hultman, L., & Rosen, J. (2017). Theoretical and experimental exploration of a novel in-plane chemically ordered (Cr2/3M1/3)2AlC i-MAX phase with M = Sc and Y. *Crystal Growth & Design, 17*(11), 5704−5711.

Malato, S., Fernández-Ibáñez, P., Maldonado, M. I., Blanco, J., & Gernjak, W. (2009). Decontamination and disinfection of water by solar photocatalysis: Recent overview and trends. *Catalysis Today, 147*(1), 1−59.

Mamaghani, A. H., Haghighat, F., & Lee, C. S. (2019). Hydrothermal/solvothermal synthesis and treatment of TiO$_2$ for photocatalytic degradation of air pollutants: Preparation, characterization, properties, and performance. *Chemosphere, 219*, 804−825.

Marcone, G. P. S., Oliveira, Á. C., Almeida, G., Umbuzeiro, G. A., & Jardim, W. F. (2012). Ecotoxity of TiO$_2$ to Daphnia similis under irradiation. *Journal of Hazardous Materials, 211−212*, 436−442.

Miklos, D. B., Remy, C., Jekel, M., Linden, K. G., Drewes, J. E., & Hübner, U. (2018). Evaluation of advanced oxidation processes for water and wastewater treatment−A critical review. *Water Research, 139*, 118−131.

Muersha, W., & Pozan Soylu, G. S. (2018). Effects of metal oxide semiconductors on the photocatalytic degradation of 4-nitrophenol. *Journal of Molecular Structure, 1174*, 96−102.

Naguib, M., Come, J., Dyatkin, B., Presser, V., Taberna, P.-L., Simon, P., . . . Gogotsi, Y. (2012). MXene: A promising transition metal carbide anode for lithium-ion batteries. *Electrochemistry Communications, 16*(1), 61−64.

Naguib, M., Halim, J., Lu, J., Cook, K. M., Hultman, L., Gogotsi, Y., & Barsoum, M. W. (2013). New two-dimensional niobium and vanadium carbides as promising materials for Li-ion batteries. *Journal of the American Chemical Society, 135*(43), 15966−15969.

Naguib, M., Kurtoglu, M., Presser, V., Lu, J., Niu, J., Heon, M., . . . Barsoum, M. W. (2011). Two-dimensional nanocrystals produced by exfoliation of Ti$_3$AlC$_2$. *Advanced Materials, 23*(37), 4248−4253.

Naguib, M., Mashtalir, O., Carle, J., Presser, V., Lu, J., Hultman, L., . . . Barsoum, M. W. (2012). Two-dimensional transition metal carbides. *ACS Nano, 6*(2), 1322−1331.

Naguib, M., Mashtalir, O., Lukatskaya, M. R., Dyatkin, B., Zhang, C., Presser, V., ... Barsoum, M. W. (2014). One-step synthesis of nanocrystalline transition metal oxides on thin sheets of disordered graphitic carbon by oxidation of MXenes. *Chemical Communications*, *50*(56), 7420−7423.

Naguib, M., Mochalin, V. N., Barsoum, M. W., & Gogotsi, Y. (2014). Two-dimensional materials: 25th anniversary article: MXenes: A new family of two-dimensional materials. *Advanced Materials*, *26*(7), 982.

Navarro Yerga, R. M., Alvarez-Galván, M. C., Vaquero, F., Arenales, J., & Fierro, J. L. G. (2013). Hydrogen production from water splitting using photo-semiconductor catalysts. In L. M. Gandía, G. Arzamendi, & P. M. Diéguez (Eds.), *Renewable hydrogen technologies* (pp. 43−61). Amsterdam: Elsevier, Chapter 3.

Papoulis, D. (2019). Halloysite based nanocomposites and photocatalysis: A review. *Applied Clay Science*, *168*, 164−174.

Park, J. (2017). Visible and near infrared light active photocatalysis based on conjugated polymers. *Journal of Industrial and Engineering Chemistry*, *51*, 27−43.

Perdew, J. P., Burke, K., & Wang, Y. (1996). Generalized gradient approximation for the exchange-correlation hole of a many-electron system. *Physical Review B*, *54*(23), 16533−16539.

Perdew, J. P., & Wang, Y. (1992). Pair-distribution function and its coupling-constant average for the spin-polarized electron gas. *Physical Review B*, *46*(20), 12947−12954.

Prasad, C., Tang, H., & Bahadur, I. (2019). Graphitic carbon nitride based ternary nanocomposites: From synthesis to their applications in photocatalysis: A recent review. *Journal of Molecular Liquids*, *281*, 634−654.

Ran, J., Gao, G., Li, F. T., Ma, T. Y., Du, A., & Qiao, S. Z. (2017). Ti_3C_2 MXene co-catalyst on metal sulfide photo-absorbers for enhanced visible-light photocatalytic hydrogen production. *Nature Communications*, *8*(1), 13907.

Rasaki, S. A., Bingxue, Z., Guarecuco, R., Thomas, T., & Minghui, Y. (2019). Geopolymer for use in heavy metals adsorption, and advanced oxidative processes: A critical review. *Journal of Cleaner Production*, *213*, 42−58.

Satyapal, S. (2019). *Hydrogen and fuel cell program overview* (p. 22). Crystal City, VA; Fuel Cell Technologies Office.

Shahzad, A., Rasool, K., Nawaz, M., Miran, W., Jang, J., Moztahida, M., ... Lee, D. S. (2018). Heterostructural TiO_2/Ti_3C_2Tx (MXene) for photocatalytic degradation of antiepileptic drug carbamazepine. *Chemical Engineering Journal*, *349*, 748−755.

She, Z. W., Fredrickson, K. D., Anasori, B., Kibsgaard, J., Strickler, A. L., Lukatskaya, M. R., ... Vojvodic, A. (2016). Two-dimensional molybdenum carbide (MXene) as an efficient electrocatalyst for hydrogen evolution. *ACS Energy Letters*, *1*(3), 589−594.

Sillanpää, M., Ncibi, M. C., & Matilainen, A. (2018). Advanced oxidation processes for the removal of natural organic matter from drinking water sources: A comprehensive review. *Journal of Environmental Management*, *208*, 56−76.

Singh, A. K., Mathew, K., Zhuang, H. L., & Hennig, R. G. (2015). Computational screening of 2D materials for photocatalysis. *The Journal of Physical Chemistry Letters*, *6*(6), 1087−1098.

Si, C., Zhou, J., & Sun, Z. (2015). Half-metallic ferromagnetism and surface functionalization-induced metal−insulator transition in graphene-like two-dimensional Cr2C crystals. *ACS Applied Materials & Interfaces*, *7*(31), 17510−17515.

Srivastava, P., Mishra, A., Mizuseki, H., Lee, K.-R., & Singh, A. K. (2016). Mechanistic insight into the chemical exfoliation and functionalization of Ti_3C_2 MXene. *ACS Applied Materials & Interfaces*, *8*(36), 24256−24264.

Sun, Y., Sun, Y., Meng, X., Gao, Y., Dall'Agnese, Y., Chen, G., ... Wang, X.-F. (2019). Eosin Y-sensitized partially oxidized Ti_3C_2 MXene for photocatalytic hydrogen evolution. *Catalysis Science & Technology*, *9*(2), 310−315.

Swaminathan, M. (2018). Semiconductor oxide nanomaterials as catalysts for multiple applications. In C. Mustansar Hussain (Ed.), *Handbook of nanomaterials for industrial applications* (pp. 197−207). Elsevier, Chapter.

Tang, Q., Zhou, Z., & Shen, P. (2012). Are MXenes promising anode materials for Li ion batteries? Computational studies on electronic properties and Li storage capability of Ti_3C_2 and $Ti_3C_2X_2$ (X = F, OH) monolayer. *Journal of the American Chemical Society, 134*(40), 16909—16916.

Tan, C., Cao, X., Wu, X. J., He, Q., Yang, J., Zhang, X., . . . Zhang, H. (2017). Recent advances in ultrathin two-dimensional nanomaterials. *Chemical Reviews, 117*(9), 6225—6331.

Tan, T. L., Jin, H. M., Sullivan, M. B., Anasori, B., & Gogotsi, Y. (2017). High-throughput survey of ordering configurations in MXene alloys across compositions and temperatures. *ACS Nano, 11*(5), 4407—4418.

Tao, Q., Dahlqvist, M., Lu, J., Kota, S., Meshkian, R., Halim, J., . . . Rosen, J. (2017). Two-dimensional Mo1.33C MXene with divacancy ordering prepared from parent 3D laminate with in-plane chemical ordering. *Nature Communications, 8*(1), 14949.

Trawiński, J., & Skibiński, R. (2019). Rapid degradation of clozapine by heterogeneous photocatalysis. Comparison with direct photolysis, kinetics, identification of transformation products and scavenger study. *Science of the Total Environment, 665*, 557—567.

Tzenov, N. V., & Barsoum, M. W. (2000). Synthesis and characterization of Ti_3AlC_2. *Journal of the American Ceramic Society, 83*(4), 825—832.

Urbankowski, P., Anasori, B., Makaryan, T., Er, D., Kota, S., Walsh, P. L., . . . Gogotsi, Y. (2016). Synthesis of two-dimensional titanium nitride Ti_4N_3 (MXene). *Nanoscale, 8*(22), 11385—11391.

Verbruggen, S. W. (2015). TiO_2 photocatalysis for the degradation of pollutants in gas phase: From morphological design to plasmonic enhancement. *Journal of Photochemistry and Photobiology C: Photochemistry Reviews, 24*, 64—82.

Wang, B., de Godoi, F. C., Sun, Z., Zeng, Q., Zheng, S., & Frost, R. L. (2015). Synthesis, characterization and activity of an immobilized photocatalyst: Natural porous diatomite supported titania nanoparticles. *Journal of Colloid and Interface Science, 438*, 204—211.

Wang, X., Garnero, C., Rochard, G., Magne, D., Morisset, S., Hurand, S., . . . Célérier, S. (2017). A new etching environment (FeF3/HCl) for the synthesis of two-dimensional titanium carbide MXenes: A route towards selective reactivity versus water. *Journal of Materials Chemistry A, 5*(41), 22012—22023.

Wang, H., Peng, R., Hood, Z. D., Naguib, M., Adhikari, S. P., & Wu, Z. (2016). Titania composites with 2 D transition metal carbides as photocatalysts for hydrogen production under visible-light irradiation. *ChemSusChem, 9*(12), 1490—1497.

Wang, X., Shen, X., Gao, Y., Wang, Z., Yu, R., & Chen, L. (2015). Atomic-scale recognition of surface structure and intercalation mechanism of Ti_3C_2X. *Journal of the American Chemical Society, 137*(7), 2715—2721.

Wang, H., Wu, Y., Xiao, T., Yuan, X., Zeng, G., Tu, W., . . . Chew, J. W. (2018). Formation of quasi-core-shell In_2S_3/anatase TiO_2@metallic Ti_3C_2Tx hybrids with favorable charge transfer channels for excellent visible-light-photocatalytic performance. *Applied Catalysis B: Environmental, 233*, 213—225.

Wang, H., Zhang, X., & Xie, Y. (2018). Recent progress in ultrathin two-dimensional semiconductors for photocatalysis. *Materials Science and Engineering: R: Reports, 130*, 1—39.

Wang, L., Zhao, J., Liu, H., & Huang, J. (2018). Design, modification and application of semiconductor photocatalysts. *Journal of the Taiwan Institute of Chemical Engineers, 93*, 590—602.

Wang, X. H., & Zhou, Y. C. (2010). Layered machinable and electrically conductive Ti_2AlC and Ti_3AlC_2 ceramics: A review. *Journal of Materials Science & Technology, 26*(5), 385—416.

Wang, H.-J., Cao, Y., Wu, L. L., Wu, S. S., Raza, A., Liu, N., . . . Miyazawa, T. (2018). ZnS-based dual nano-semiconductors (ZnS/PbS, ZnS/CdS or ZnS/Ag2S,): A green synthesis route and photocatalytic comparison for removing organic dyes. *Journal of Environmental Chemical Engineering, 6*(6), 6771—6779.

Wang, P., Guan, Z., Li, Q., & Yang, J. (2018). Efficient visible-light-driven photocatalytic hydrogen production from water by using Eosin Y-sensitized novel g-C_3N_4/Pt/GO composites. *Journal of Materials Science, 53* (1), 774—786.

Wen, C., Yin, A., & Dai, W.-L. (2014). Recent advances in silver-based heterogeneous catalysts for green chemistry processes. *Applied Catalysis B: Environmental 160, 161*, 730—741.

Wetchakun, K., Wetchakun, N., & Sakulsermsuk, S. (2019). An overview of solar/visible light-driven heterogeneous photocatalysis for water purification: TiO_2- and ZnO-based photocatalysts used in suspension photoreactors. *Journal of Industrial and Engineering Chemistry, 71*, 19−49.

Wong, Z. M., Tan, T. L., Yang, S. W., & Xu, G. Q. (2018). Enhancing the photocatalytic performance of MXenes via stoichiometry engineering of their electronic and optical properties. *ACS Applied Materials & Interfaces, 10*(46), 39879−39889.

Xiao, R., Zhao, C., Zou, Z., Chen, Z., Tian, L., Xu, H., . . . Yang, X. (2019). In situ fabrication of 1D CdS nanorod/2D Ti_3C_2 MXene nanosheet Schottky heterojunction toward enhanced photocatalytic hydrogen evolution. *Applied Catalysis B: Environmental,* 118382.

Xie, X., Zhang, N., Tang, Z. R., Anpo, M., & Xu, Y. J. (2018). Ti3C2Tx MXene as a Janus cocatalyst for concurrent promoted photoactivity and inhibited photocorrosion. *Applied Catalysis B: Environmental, 237*, 43−49.

Yasmina, M., Mourad, K., Mohammed, S. H., & Khaoula, C. (2014). Treatment heterogeneous photocatalysis; factors influencing the photocatalytic degradation by TiO_2. *Energy Procedia, 50*, 559−566.

Ying, Y., Liu, Y., Wang, X., Mao, Y., Cao, W., Hu, P., & Peng, X. (2015). Two-dimensional titanium carbide for efficiently reductive removal of highly toxic chromium(VI) from water. *ACS Applied Materials & Interfaces, 7*(3), 1795−1803.

Yin, Q., & Hill, C. L. (2017). Better semiconductor-catalyst interfaces using light. *Joule, 1*(4), 645−646.

Yuan, R., Hu, L., Yu, P., Wang, Z., Wang, H., & Fang, J. (2018). Co_3O_4 nanocrystals/3D nitrogen-doped graphene aerogel: A synergistic hybrid for peroxymonosulfate activation toward the degradation of organic pollutants. *Chemosphere, 210*, 877−888.

Zhang, C., Kim, S. J., Ghidiu, M., Zhao, M. Q., Barsoum, M. W., Nicolosi, V., & Gogotsi, Y. (2016). Layered orthorhombic Nb2O5@Nb4C3Tx and TiO2@Ti3C2Tx hierarchical composites for high performance Li-ion batteries. *Advanced Functional Materials, 26*(23), 4143−4151.

Zhang, H., Yang, G., Zuo, X., Tang, H., Yang, Q., & Li, G. (2016). Computational studies on the structural, electronic and optical properties of graphene-like MXenes (M_2CT_2, M = Ti, Zr, Hf; T = O, F, OH) and their potential applications as visible-light driven photocatalysts. *Journal of Materials Chemistry A, 4*(33), 12913−12920.

Zhang, Q., Teng, J., Zou, G., Peng, Q., Du, Q., Jiao, T., & Xiang, J. (2016). Efficient phosphate sequestration for water purification by unique sandwich-like MXene/magnetic iron oxide nanocomposites. *Nanoscale, 8* (13), 7085−7093.

Zhou, J., Zha, X., Chen, F. Y., Ye, Q., Eklund, P., Du, S., & Huang, Q. (2016). A two-dimensional zirconium carbide by selective etching of Al_3C_3 from nanolaminated $Zr_3Al_3C_5$. *Angewandte Chemie International Edition, 55*(16), 5008−5013.

Zou, J.-P., Chen, Y., Zhu, M., Wang, D., Luo, X. B., & Luo, S. L. (2019). Semiconductor-based nanocomposites for photodegradation of organic pollutants. In X. Luo, & F. Deng (Eds.), *Nanomaterials for the removal of pollutants and resource reutilization* (pp. 25−58). Elsevier, Chapter 2.

CHAPTER 13

Round-the-clock photocatalytic memory systems: Phenomenon and applications

Maryam Mokhtarifar[1,2], **MariaPia Pedeferri**[1], **Maria Vittoria Diamanti**[1],
Mohan Sakar[3] **and Trong-On Do**[2]

[1]*Department of Chemistry, Materials and Chemical Engineering "G. Natta," Polytechnic of Milan, Milan, Italy,* [2]*Department of Chemical Engineering, Laval University, Quebec, QC, Canada,* [3]*Centre for Nano and Material Sciences, Jain University, Bangalore, India*

13.1 Introduction

In the photocatalytic (PC) process, it is clearly known that the photo-activated redox reaction occurs under light irradiation and it is generally limited once the light irradiation is stopped (Zhou et al., 2016). In fact, even if the photocatalyst is highly efficient and photo-stable, the onset of the PC reaction will occur only under light irradiation. Therefore it will be fascinating if PC reactions continue to occur even if the light irradiation is not available. Such PC systems are generally termed as "day–night," "catalytic memory," or "round-the-clock" photocatalysts. To enable such day–night (DN) PC reactions, there should be two distinct processes required, which should also be coupled together as they have to happen simultaneously. These two distinct processes are nothing but (1) the typical redox reactions and (2) the storing of photo-induced electrons. Accordingly, the scheme of DN PC system should contain a PC semiconductor (SC) to generate the charge carriers under the light illumination and an electron acceptor to store the electrons, which should also release them back in the absence of light through cathodic reactions. A memory photocatalyst can work highly efficient with the "engineered" conduction band (CB) and valence band (VB) levels in the SC and the kinetics of the charge carriers between the SC and electron storing material (ESM).

Generally, to develop a memory photocatalyst, some of the fundamental concepts should be considered. First, the CB of the SC should be above the CB edge of the ESM as the electrons should be transferred from SC to ESM. Second, a sufficient connection is required at the interface of SC and ESM for the improved electron transfer and subsequent catalytic reactions under both light irradiation and dark conditions. Alternatively, the other ways to establish PC memory are the (1) peroxidase mimic mechanism (Chiu & Hsu, 2017) and (2) fluorescence-assisted mechanism (Li, Yin, Wang, & Sato, 2012a, 2012b, 2012c, 2017; Li, Yin, Wang, & Sato, 2013). Considering the practical applications, the potential of PC process should be improved in many folds as it holds promising versatile applications in energy production and environmental remediation. Although there are many reviews on conventional photocatalysis (Fang et al., 2018; Jin, Ye, Xie, & Chen, 2017; Knör, 2015; Maeda, 2013), there are only limited reports available on the DN PC systems. In this context, this chapter gives a summary of the advanced improvements in PC memory reactions, describes the PC memory reactions in different systems, presents the possible applications and demonstrates how to combine the theoretical and experimental concepts for analyzing the memory PC systems to understand them and thereby to design a new systems. Accordingly, in the first step, a detailed explanation on DN PC mechanisms is provided. Then, to have a better understanding of the catalytic system mechanism, different types of materials are summarized. Finally, it describes the various applications of memory photocatalysis and followed by the challenges and features are discussed toward the development of such advanced memory photocatalysts for the future.

13.2 Mechanism of dark photocatalyst

The mechanism of memory photocatalysts is varied depending upon the different types of material systems. At present, the most commonly used memory photocatalysts generally consist of two types of materials, PC and ESM, which are based on the carrier storage mechanism. In addition, some substances have the enzymatic effect of peroxidase, which is due to the formation of hydrogen, and cause the catalytic reaction in the dark. Similarly, coupling of an SC with a long afterglow phosphor can also show the DN PC, which is due to the long afterglow phosphor that functions as a light source in the dark. As such, the catalytic system in memory photocatalysts is classified into four types, where their mechanisms are dependent upon the types of materials used to construct the DN PC systems.

13.2.1 Electron storage (reductive) mechanism

The mechanism of electron storage occurs through reductive reaction via the formation of intermediate during the photocatalysis, and it is reversible during the dark catalysis, which ultimately enables the DN PC process. There are range of materials, such as MoO_3 (Takahashi, Ngaotrakanwiwat, & Tatsuma, 2004), carbon nanotube (CNT) (Kongkanand & Kamat, 2007; Yang et al., 2013), V_2O_5 (Wang & Huang, 2008), TiO_2 and WO_3 (Feng, Yang, Gao, Sun, & Li, 2018; Kim, Park, Kim, & Park, 2016; Takahashi & Tatsuma, 2008), C_3N_4 (Kasap et al., 2016; Zeng et al., 2018), and Cu_2O (Yasomanee & Bandara, 2008) developed toward storing the electrons as given in Eqs. (13.1)−(13.3) (Ngaotrakanwiwat & Tatsuma, 2004; Ngaotrakanwiwat, Tatsuma, Saitoh, Ohko, & Fujishim, 2003; Park, Bak,

Jeon, Kim, & Choi, 2012; Tatsuma, Saitoh, Ohko, & Fujishima, 2001). Among these, WO_3 functions as an efficient material to store the electrons and TiO_2 to harvest the required light. As given in the following equations, the light irradiation leads to the formation of electron—hole pairs, wherein the holes will be promoted to TiO_2 surface to act as absorbers of H_2O and/or even an electrolyte reacting with environmental medium, and finally produces oxygen and hydrogen or an alkali metal ion (M^+). The formed alkali ions of the metal can be of Fr, Li, K, Na, Cs, or Rb, which is dependent upon the electrolytes used (Dickens & Whittingham, 1968).

$$TiO_2 + h\nu \rightarrow TiO_2^* \left(e^- + h^+ \right) \rightarrow \text{Carrier generation} \tag{13.1}$$

$$2H_2O \text{ (or electrolyte)} + 4h^+ \rightarrow O_2^* + 4H^+ \left(\text{or } M^+ \right) \tag{13.2}$$

$$e^- (TiO_2) \rightarrow e^- (WO_3) \rightarrow \text{Interfacial electron transfer} \tag{13.3}$$

$$WO_3 + xe^- + xH^+ \left(\text{or } xM^+ \right) \rightarrow H_xWO_3(\text{or } M_x WO_3) \rightarrow \text{Charging} \tag{13.4}$$

$$H_xWO_3(\text{or } M_xWO_3) \rightarrow WO_3 + xe^- + xH^+ \left(\text{or } xM^+ \right) \rightarrow \text{Discharging} \tag{13.5}$$

$$e^- + O_2 \rightarrow {}^\bullet O_2 \rightarrow \text{Dark catalysis} \tag{13.6}$$

As given in the equations, an intermediate compound will be formed upon the injection of the excited electrons to the CB of WO_3 and their subsequent trapping as a result of the formation of H^+ and M^+, which will finally intercalate with them. Later, the release of electrons will occur in the absence of light (Eq. (13.5)), and they will be involved in a reaction with surrounding molecules, such as O_2, and produce radicals and continue the catalytic reactions as shown in Eq. (13.6).

In the absence of light, the release of the stored photoelectrons occurs due to the M^+ ions of the electrolyte, where they will be intercalated into the WO_3 structure during the photoelectrochemical process. Hence, the ability of WO_3 will be extended to function under dark due to the presence of an electrolyte, through which the alkaline metals store the electrons that are generated in the presence of light. Reports have shown that the amount of stored electrons will increase proportionally to the ionic radii of the M^+ ions (for instance, in case of changing H^+ to K^+) due to the limited mobility in the structure and their ability to stay there for a longer period of time (Ng, Ng, Iwase, & Amal, 2011a, 2011b). The mechanism of electron storage through reduction reaction under the light irradiation and their subsequent release in the dark is depicted in Fig. 13.1.

13.2.2 Hole storage (oxidative) mechanism

The storage of holes, commonly referred to as oxidative energy storage, is also possible as similar to the storage of electrons, where it can be explained using "p—n junction" and

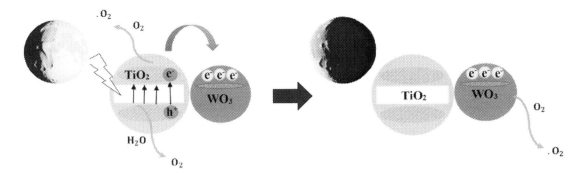

Figure 13.1
The mechanism of reductive electron storage in TiO_2–WO_3 DN PC system. *DN*, day–night; *PC*, photocatalytic.

"mediation" models (Takahashi & Tatsuma, 2005). In the first model, a p–n junction is formed based on the combination of a redox-active p-type SC (such as $Ni(OH)_2$) with an n-type SC (such as TiO_2) (Fig. 13.2A).

In the p–n junction model, upon the irradiation of light, the p–n junction provides a pathway for the transport of the photogenerated holes into the bulk of $Ni(OH)_2$ (Eq. 13.7), through which it will be subsequently oxidized by holes (Eq. 13.8). Subsequently, the surrounding molecules, such as O_2, will consume the photogenerated electrons available in the CB of TiO_2 (Eq. 13.9). Thereby, the formed intermediate ($NiO_x(OH)_{2-x}$) oxidizes the substances in the dark (Eq. 13.10) and reverses to $Ni(OH)_2$. It should be noted that this process leads to the system to be electrically natural due to the intercalation of anions or deintercalation of cations (Eq. 13.8), which finally results in the retention of the oxidative energy.

$$TiO_2{}^*\left(h^+\right) \rightarrow Ni(OH)_2\left(h^+\right) \tag{13.7}$$

$$Ni(OH)_2 + xe^- + xOH^- \rightarrow Ni_x(OH)_{2-x} + xH_2O \tag{13.8}$$

$$O_2 + 2H^+ + 2e^- \rightarrow H_2O_2 \tag{13.9}$$

$$Ni_x(OH)_{2-x} + substance \rightarrow Ni(OH)_2 + products \tag{13.10}$$

In the mediator model, an electron mediator, for example, Fe^{3+}/Fe^{2+} redox couple, is introduced in the system as shown in Fig. 13.2B, which acts as path for the efficient transport of the photogenerated electrons from the $Ni(OH)_2$ to the CB of TiO_2. In this mechanism, the storage of oxidative energy will subsequently happen due to self-oxidation of the remaining photogenerated holes in the VB of $Ni(OH)_2$ (Eq. (13.8)). Separation of the charge carrier is most beneficial of this model, where the high-speed electron transfer is induced through the electron mediator and a built-in electric field can be provided by the p–n junction. Hence, improvement in the efficiency of the hole storage will occur due to

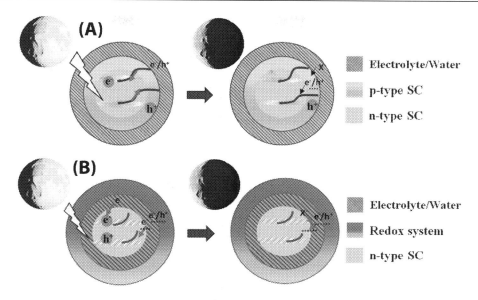

Figure 13.2

Oxidative energy storage through (A) the p—n junction model and (B) mediator model.

the reduction in the loss via carrier recombination. Such mechanism has been usually found in TiO_2—NiO systems (Buama, Junsukhon, Ngaotrakanwiwat, & Rangsunvigit, 2017; Huang et al., 2010), $TiO_2/SiO_2/MnO_x$ systems (Kuroiwa, Park, Sakai, & Tastsuma, 2016), and also TiO_2—$Ni(OH)_2$ systems (Takahashi & Tatsuma, 2005; Yang, Takahashi, Sakai, & Tatsuma, 2010; Yang, Takahashi, Sakai, & Tatsuma, 2011; Zhang et al., 2012).

However, the electron storage efficiency in "electron storage system" (TiO_2—WO_3) is comparably higher than the efficiency of "hole storage systems." This is essentially because of the loss of holes due to the oxidation of the adsorbed water or due to the re-reduction of $Ni(OH)_2$ by the photogenerated electrons. Interestingly, it is realized that the efficiency of hole storage in the p—n junction can be enhanced by tuning the morphological properties of materials, such as introducing the porous TiO_2 (Sakar, Nguyen, Vu, & Do, 2018).

Some studies have reported the interesting combination of phosphorescence materials, such as Ag_3PO_4 (Li, Yin, Wang, Sekino et al., 2013) and $Sr_2MgSi_2O_7$:(Eu, Dy) (SMSO) with photocatalyst systems. After illumination, such systems benefit from the long-lasting phosphorescence in the dark, which excites photocatalysts to have a persistent catalytic reaction, where it acts as a light source once the external light source is stopped available (Fig. 13.3) (Li, Yin, & Sato, 2011; Li et al., 2012c; Li, Yin, Wang, & Sekino et al., 2013; Lu et al., 2018; Yin et al., 2015; Zhou et al., 2016).

In this study, g-C_3N_4 was photo-assisted by SMSO as a long afterglow phosphor. Under light illumination, the onset of energy transfer occurs to the 4f7 ground state of Eu^{2+}.

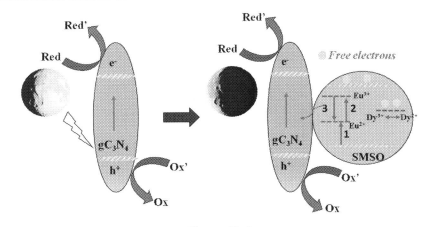

Figure 13.3

DN PC mechanism assisted by long-afterglow phosphorescence in the $Sr_2MgSi_2O_7$:(Eu, Dy) (SMSO)/g-C_3N_4 PC. *DN*, day—night; *PC*, photocatalytic.

Subsequently, Eu^{2+} (arrow 1) is excited to the 4f65d state, which is located beneath the CB (arrow 2). Eu^{3+} is left there due to autoionization, which generates a free electron and the capture of this electron by Dy^{3+} leads to the reduction into Dy^{2+} with a ground state below the bottom of the CB. After cutting the light off, transfer of electron back to the CB happens and finally a 5d-4f emission occurs due to thermally activated release via the recombination with Eu^{3+} (arrow 3). Under such circumstances, the photogenerated carriers can be created by the fluorescence to excite g-C_3N_4. Subsequently, •OH and •O_2^- radicals will be generated due to the transfer of the electrons to the catalyst surface and thereby their reactions with H_2O and/or surrounding molecules. Notably, such systems enable a continuous PC action during round-the-clock. Similar to this system, the silica-based materials have been reported to have the capability of exciting the g-C_3N_4 photocatalyst via blue-light-emitting long-afterglow phosphor (Pan et al., 2008; Sahu, Bisen, & Brahme, 2015). In another study, SMSO/g-C_3N_4 composite PC was employed to degrade rhodamine B and methyl orange (MO) for 6 hours after the removal of the external light source (Lu et al., 2018).

13.3 Applications of the day—night photocatalytic systems

Fujishima et al. developed the TiO_2—WO_3 system in 2001 for the anticorrosion applications, where it was exposed to light and dark conditions (Tatsuma, Saitoh, Ngaotrakanwiwat, Ohko, & Fujishima, 2002). This led to the development of electron storage material (ESM), which could also be considered that it paved a way to materialize the DN/memory photocatalysts. It is noteworthy that memory photocatalysts have developed tremendous interests in many applications, including heavy metal ion removal

(Buama et al., 2017; Huang et al., 2010; Ng et al., 2011a; Takahashi & Tatsuma, 2005), degradation of organic pollutants, disinfectant, and hydrogen production. Memory photocatalysts have high potential in practical applications and therefore a unique PC system needs to be constructed to prepare the continuous day and night catalytic operations. However, it needs a very deeper understanding of their mechanisms and thereby more advanced characterization techniques are required to support the developments in this field.

13.3.1 Corrosion resistance in metals

As mentioned, Fujishima et al. developed the first working electron-storage system using $TiO_2@WO_3$ system for anticorrosion applications in both light and dark conditions (Ohko, Saitoh, Tatsuma, & Fujishima, 2001). When a suitable light source excites the system, the electrons in the VB are excited to the CB and then injected into the metal, and the potential is kept more negative than the potential for corrosion. The ESM partially absorbs the excited electrons and therefore the reductive energy that the excited SC produced can be stored in the ESM. The electrons stored in the electron storage are further injected into the metal after the light is turned off. As a result, it is constantly remained immune against corrosion (Fig. 13.4).

Application of ESMs, that is, WO_3 (Park et al., 2012; Tatsuma et al., 2002) or SnO_2 (Subasri & Shinohara, 2003), along with TiO_2 on the metal surface has given promising results for protecting the material even in the dark. It was shown that when the WO_3 and TiO_2 mixed electrodes were utilized for anticorrosion, there was a flow of photogenerated electrons into the galvanically coupled steel from WO_3. During irradiation time, this flow of photogenerated electrons happened due to the potential cascade of $TiO_2@WO_3@steel$. However, upon turning the light off, the potential cascade of the electron flow was from WO_3 to steel (Park et al., 2012). The further studies also showed that under UV illumination, the corrosion protection level of copper was enhanced by the use of electrode bearing Sn:Ti in the molar ratio 1:1. Notably, even when illumination was stopped for

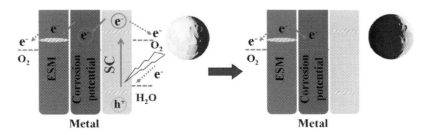

Figure 13.4
Anticorrosion mechanism in a metal coupled with semiconductor-electron storage material system.

several hours, the negative potential of the composite was maintained due to the accumulation of electrons and the intercalation of cations form the electrolyte into SnO_2 (Subasri & Shinohara, 2003).

Generally, it is supposed that the following characteristics of the ESMs attribute toward the anticorrosion properties of the metal (Nawaz et al., 2019): (1) redox activity; (2) more positive redox potential compared with the CB potential of the SC; (3) more negative redox potential in relative to the corrosion potential of the metal to protect it from corrosion; (4) poor oxidizability of the reduced form by oxygen; and (5) long-term stability during different redox cycles. Accordingly, the photocharging and discharging reactions in the $TiO_2@WO_3$ system could be described by Eqs. (13.11)−(13.13).

$$e^-(TiO_2) \rightarrow e^-(WO_3) \quad \text{[Electron transfer]} \tag{13.11}$$

$$xe^- + WO_3 + xH^+ \rightarrow H_xWO_3 \quad \text{[Photocharging]} \tag{13.12}$$

$$H_xWO_3 \rightarrow WO_3 + xe^- + xH^+ \quad \text{[Discharging]} \tag{13.13}$$

13.3.2 Pollutant degradations

Reportedly, the $TiO_2/Ni(OH)_2$ composite was the first system to demonstrate the degradation of multicarbon compounds, such as acetic acid, acetaldehyde, and acetone, into CO_2 even in the absence of light (Kuroiwa et al., 2016). The ability of the bilayer PC system via its oxidatively stored energy was investigated under UV-light irradiation toward the degradation of methanol and formaldehyde. The mechanism of the degradation of methanol and formaldehyde due to the oxidative storage energy of $TiO_2@Ni(OH)_2$ PC is proposed in Eqs. (13.14)−(13.16) (Takahashi & Tatsuma, 2006; Yang et al., 2010):

$$NiO_x(OH)_{2-x} + xH_2O \rightarrow Ni(OH)_2 + xOH^- \tag{13.14}$$

$$CH_3OH + 6OH^- \rightarrow CO_2 + 5H_2O + 6e^- \tag{13.15}$$

$$HCHO + 4OH^- \rightarrow CO_2 + 3H_2O + 4e^- \tag{13.16}$$

Various species, such as formate, aldehydes, phenol, and alcohols, have been oxidized under light irradiation via the oxidative energy stored in the form of holes in the TiO_2-Ni(OH)$_2$ bilayer film. $NiO_x(OH)_{2-x2}$ was realized to be responsible for such oxidation reactions.

In this direction, MnO_x was also demonstrated as an oxidative material for energy storage. However, even though it was not possible to directly integrate MnO_x with TiO_2, use of an interface material, such as fluorine-doped tin oxide (FTO)/MnO_x/nanoporous SiO_2/TiO_2,

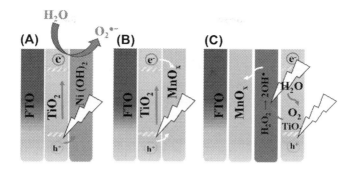

Figure 13.5
Oxidative energy storage in (A) $TiO_2-Ni(OH)_2$, (B) TiO_2-MnO_x, and (C) FTO/MnO_x/nanoporous SiO_2 (or air)/TiO_2 systems.

was used to overcome this limitation. This property was explained as follows: the direct contact of MnO_x with TiO_2 instead of oxidation reaction to store oxidative energy leads to further reduction reaction. As shown in Fig. 13.5, the separation of MnO_x and TiO_2 by SiO_2 with optimum thickness causes oxidation of the MnO_x layer.

As demonstrated in Eqs. (13.17)–(13.21), if TiO_2 is excited, the electrons in CB react with the O_2 and H^+ to produce H_2O_2, where it further transforms into OH radicals under UV light and this OH radicals oxidize the intermediate back to MnO_2. As a result, the developed FTO/MnO_x/nanoporous SiO_2/TiO_2 system was investigated toward the mineralization of methanol, acetaldehyde, acetic acid, and acetone in dark conditions.

$$TiO_2 + h\nu \rightarrow TiO_2(e^- + h^+) \tag{13.17}$$

$$O_2 + 2H^+ + 2e^- \rightarrow H_2O_2 \tag{13.18}$$

$$H_2O_2 + UV \rightarrow 2OH^\bullet \tag{13.19}$$

$$MnOOH + 2OH^\bullet \rightarrow MnO_2 + H_2O \tag{13.20}$$

$$H_2O + h^+ \rightarrow 2OH^\bullet + H^+ \tag{13.21}$$

Interestingly, the levels of formaldehyde degradation over the bare platinum-loaded WO_3 systems (Pt@WO_3) as well as those treated with hydrogen (Pt@H:WO_3) under light irradiation and dark conditions were analyzed (Li et al., 2013c). It was observed that during the 60-minute period of light irradiation, the formaldehyde was degraded into CO_2 by both systems. Subsequently, the degradation process continued up to 360 minutes after the cut-off of the irradiation light. Furthermore, to test the stability of the stored energy, it was observed that even after keeping the samples for 12 hours in the dark, Pt@WO_3 and Pt@H:WO_3 could, respectively, degrade around 20% and 80% of formaldehyde during a period of 360 minutes. Fig. 13.6 depicts the mechanism by which such long, persistent catalytic

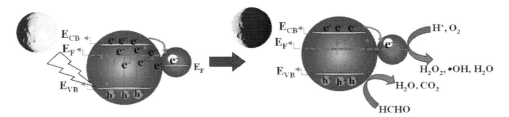

Figure 13.6
Mechanism of electron storage in the Pt@H:WO$_3$ DN PC system. *DN*, day—night; *PC*, photocatalytic.

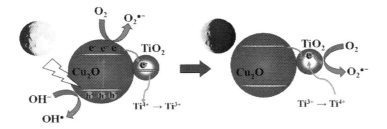

Figure 13.7
Reduction-mediated electron storage and release in the Cu$_2$O/TiO$_2$ DN PC system. *DN*, day—night; *PC*, photocatalytic.

reaction of the hydrogen-treated platinum-loaded WO$_3$ systems occurs. To do so, large amounts of oxygen vacancies are formed due to the hydrogen treatment, which led to the induction of defect band structures in its band gap structures. As a matter of fact, more trapping sites were available for the storage of electrons due to an upshift in the Fermi level caused by the defect bands. In addition to this, a multielectron reduction of O$_2$ on Pt was also facilitated due to Pt nanoparticles (NPs) loaded, which eventually enhanced the formaldehyde degradation.

To investigate the degradation of MO upon irradiation and in dark condition, TiO$_2$ islands were decorated with Cu$_2$O nanospheres to prepare Cu$_2$O/TiO$_2$ composites (Liu, Yang, Li, Gao, & Shang, 2014). The residual percentage of MO under irradiation and in the dark was reported to be about 3% and 80%, respectively. As shown in Fig. 13.7, the transfer of excess electrons from the CB of Cu$_2$O to TiO$_2$ is facilitated due to the difference in band potential and electrostatic field at the interface of TiO$_2$ and Cu$_2$O. Subsequently, through reduction to Ti^{4+} into Ti^{3+} trapped electrons will be stored to be further released in the dark. Reaction of the stored electrons with O$_2$ could then lead to the production of superoxide radicals on one hand, and the reaction of holes in Cu$_2$O produces hydroxyl radicals, and they together involve in the degradation of the dye in the absence of light.

Another interesting material composed of Ag@In@Ni@S was reported for the DN PC degradation of methylene blue (MB) dye molecules (Molla, Sahu, & Hussain, 2015). As it was expected, a rapid dye degradation was observed under tungsten lamp (4 minutes) and sunlight (2 minutes) as compared to the dark conditions (12 minutes). The observed results were attributed to the proper electron-storage properties of silver and hence the developed PC system became more reductive due to the shift of Fermi level of the composite down to a more negative value. As a matter of fact, even in the absence of light, the supply of electrons was continued and maintained the PC activity by producing redox species. Furthermore, it was proposed that the transfer and storage of electrons upon irradiation and their subsequent release in the dark condition were also facilitated by the cascading band structure between the integrated materials.

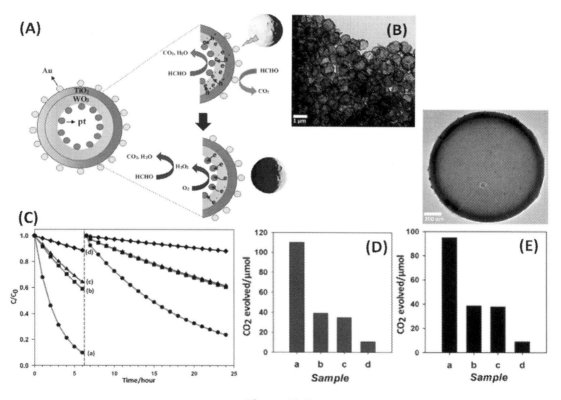

Figure 13.8

(A) Schematic representation of catalytic process of the hollow double-shell $H:Pt-WO_3/TiO_2-Au$ memory PC. (B) TEM and HRTEM images of hollow $H:Pt-WO_3/TiO_2-Au$. (C) Degradation of HCHO. Amount of CO_2 generated over (D) 6 hours in visible light and (E) 18 hours in the dark: (a) hollow $Pt-WO_3/TiO_2-Au$; (b) hollow $H:Pt-WO_3/TiO_2$; (c) hollow $H:Pt-WO_3$; and (d) conventional $H:Pt-WO_3$ prepared from commercial WO_3 (Nguyen et al., 2016). *PC*, photocatalytic.

In another study, the carbon colloidal spheres were exploited as the sacrificial templates to develop a new type of hollow double-shell H:Pt@WO$_3$/TiO$_2$@Au nanospheres and studied its PC potential under both visible-light irradiation and dark conditions to convert formaldehyde (HCHO) into CO$_2$ (Fig. 13.8) (Nguyen, Vu, & Do, 2016).

The obtained results in this study showed the conversion of more than 80% HCHO into CO$_2$ under visible light and dark conditions, respectively, after 6 and 16 hours. Considering approximately the same amount of conversion time, H:Pt@WO$_3$/TiO$_2$ and hollow H: Pt@WO$_3$, respectively, converted 39% and 40% of the HCHO under visible light in 6 hours, while in dark conditions, the percentage of HCHO conversion after 18 hours was 39% and 40%, respectively. It was suggested that the important factors in increasing the PC efficiency of such systems were mainly related to the high surface area, enhanced TiO$_2$/WO$_3$ interfacial contact, contribution of Pt NPs as a cocatalyst, large number of oxygen vacancies in the WO$_{3-x}$ matrix and finally the plasmonic resonance effect induced by Au NPs. The presence of Pt essentially contributed to the possible multielectron reduction reaction on O$_2$, which eventually led to the production hydroxyl radials and superoxide to degrade HCHO molecules in the absence of light. Such reaction is possible due to the trapping of electrons and their subsequent release in dark conditions owing to the presence of large amount of oxygen vacancies in WO$_{3-x}$.

A novel carbon-based metal-free nanocomposite consisting of g-C$_3$N$_4$-carbon nanotubers-graphene (g-C$_3$N$_4$-CNTs-Gr) was developed and explored for their DN PC properties to degrade phenol in the presence and absence of light (Zhang, Wang, Li, Geng, & Leng, 2017) (Fig. 13.9A and B). While g-C$_3$N$_4$ performed the PC function, the storage and discharge of the charges in the presence and absence of light was, respectively, done by CNTs and Gr, which basically functioned as supercapacitors. Their mechanism was proposed to be as follows, which is based on a radical trapping experiment: first, those electrons that are excited to the CB of g-C$_3$N$_4$ will be transferred to the surface and bulk of CNT-Gr. Then, their subsequent reaction with O$_2$ leads to the degradation of phenol under irradiation due to the formation of superoxide radicals (Eqs. 13.22−13.24). On the other hand, those electrons trapped in the bulk will react with O$_2$/H$_2$O and lead to the formation of $^{\bullet}$O$_2^-$ radicals (Eq. 13.25), which perform the degradation of phenol in the absence of light.

$$g\text{-}C_3N_4 + h\upsilon \rightarrow g\text{-}C_3N_4\left(e^- + h^+\right) \tag{13.22}$$

$$g\text{-}C_3N_4 + CNT\text{-}Gr \rightarrow g\text{-}C_3N_4 + CNT\text{-}Gr(e^-) \tag{13.23}$$

$$CNT\text{-}Gr(e^-) + O_2 \rightarrow {}^{\bullet}O_2^- \tag{13.24}$$

$$2{}^{\bullet}O_2^- + 2H^+ + \rightarrow {}^{\bullet}O_2^- + O_2 \tag{13.25}$$

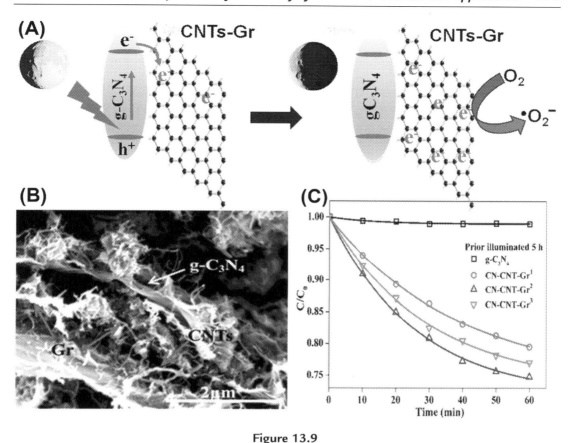

Figure 13.9
(A) Schematic diagram illustrating the removal of phenol in the dark by the CN-CNT-Gr. (B) SEM image. (C) Phenol (C_0 = 5 mg/L) removal in the dark with g-C_3N_4 and different CN-CNT-Gr[1,2,3] photocatalytic samples (Zhang et al., 2017).

Fig. 13.9C depicts the degradation of phenol in 5 hours by g-C_3N_4 and different CN-CNT-Gr concentrations namely CN-CNT-Gr_1, CN-CNT-Gr_2, and CN-CNT-Gr_3 represent the mass ratio of g-C_3N_4:CNT-Gr as 1:1, 2:3, and 1:2, respectively. As seen in Fig. 13.9C, only a negligible postirradiation catalytic effect was observed in the case of bare g-C_3N_4 under dark condition, while the CN-CNT-Gr_2 showed a degradation of 25.3%, which was greater than those of CN-CNT-Gr_3 (22.9%) and CN-CNT-Gr_1 (20.6%).

The ability of Au@Cu_7S_4 yolk@shell nanocrystal-decorated TiO_2 nanowires (TiO_2-Au@Cu_7S_4) to efficiently degrade the MO dye molecules in the round-the-clock time was reported (Chiu & Hsu, 2017). The obtained results of the studies are shown in Fig. 13.10A and B. Among the developed samples, TiO_2-Au@Cu_7S_4 showed the greatest MO degradation activity under irradiation. The Pseudo-first-order approximation was employed to calculate the apparent rate constant of MO degradation (k_{MO}, light) to quantify the degradation efficiency of different

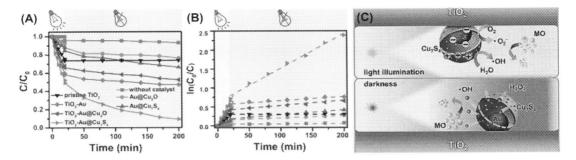

Figure 13.10
(A and B) Photodegradation of MO degradation by TiO_2-Au@Cu_7S_4 system. (C) Mechanism of photo and dark catalytic activity in TiO_2-Au@Cu_7S_4 system (Chiu & Hsu, 2017).

Figure 13.11
Photographic images showing the PC activity of TiO_2-Au@Cu_7S_4 during day–night toward the degradation of MO (for comparison, the specimen without catalyst is also shown) (Chiu & Hsu, 2017). *PC*, photocatalytic.

samples (Shankar et al., 2001). The pronounced charge separation caused by band alignment, abundant active sites equipped with the hollow Cu_7S_4 shell and so increased photodegradation activity due to the decorated $Au@Cu_7S_4$ system attributed to the observed better PC activity of TiO_2-$Au@Cu_7S_4$. As shown in Fig. 13.10C, coupling of PC and peroxidase-mimic provides the photocatalysts to persistently functioning under dark.

Furthermore, it was revealed by the further outdoor environment tests of MO degradation that the developed coupled system of TiO_2-$Au@Cu_7S_4$ offered a new paradigm for designing the more efficient PC systems with DN catalytic functionality (Fig. 13.11). The results showed a 76% MO degradation when TiO_2-$Au@Cu_7S_4$ photocatalyst was used during the day. The remaining residual MO was removed by peroxidase-mimic effect of Au NPs during overnight.

13.3.3 Bactericidal disinfections

$TiO_2@WO_3$ is also intended as a prototypical system to operate in the dark for PC antibacterial applications (Tatsuma, Takeda, Saitoh, Ohko, & Fujishima, 2003). The discovery of $TiO_2@WO_3$ system greatly helped in understanding the possible mechanism of their PC activity in the dark toward the antibacterial applications as well. Accordingly, $TiO_2@WO_3$ was successfully used for antibacterial activity against *Escherichia coli* under dark conditions. The obtained results showed that the observed antibacterial activity of this system was mainly due to the production of H_2O_2 species, which originated via the reaction between O_2 molecules and electrons stored in the dark.

In general, the disinfectant capability of PC SCs originates due to the photogenerated carriers. Their mechanism involves the interaction that are essentially done indirectly by the production of •OH (such as reaction with OH^-) to carry out the chemical transformations on the biomolecules or it can occur directly between the lipids of the membrane and the holes (Sunada, Kikuchi, Hashimoto, & Fujishima, 1998). The electron acceptor (such as H^+) can scavenge the produced electrons in the surrounding media (Wu, Imlay, & Shang, 2010).

The catalytic activity of the TiON/PdO system for the PC disinfectant of *E. coli* in the dark has been investigated (Li, Li, Wu, Xie, & Shang, 2008). It was observed that without the photocatalyst, there would be no antibacterial effect under visible light. The obvious antibacterial effect, which observed on TiON PC fiber under visible light, is consistent with its ability to absorb visible light. TiON/PdO PC fiber (Fig. 13.12) showed much faster antibacterial activity on *E. coli* under visible light as compared to TiON. The addition of PdO NPs increased the PC efficiency induced by visible light. The observed antibacterial effect under dark was attributed to the stored electrons. It is proposed that the electron storage was happened in TiON/PdO system via the transfer of electrons from TiON to PdO

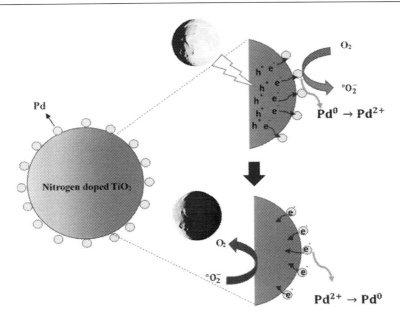

Figure 13.12
Reduction electron storage and release in the TiON/PdO memory photocatalytic toward antibacterial activity.

and the subsequent reduction of PdO was led to the formation of Pd^0 and these Pd^0 NPs effectively captured the electrons or in other words, the electrons fall into the trap. Fig. 13.12 shows the mechanism of electron trapping and their release to the environment to react with O_2 to produce $O_2@/COH$ radicals, which are further involved in killing the bacteria.

The other systems, such as TiO_2/Cu_2O (Liu et al., 2014) and titanium oxide (Wang et al., 2016), have also been investigated for antibacterial activity against *E. coli* and *Enterococcus faecalis* under dark conditions. Liu et al. found that the extremely high antibacterial effect is associated with the ability of the stored electrons, where they found that Ag NPs can store electrons efficiently, while the holes can be accumulated in the TOC (Cao et al., 2013), and this system effectively stimulates oxidation reactions in bacterial cells under dark conditions (Fig. 13.13). They also found that the Ag NPs/TiO$_2$ coatings (TOC) hold the antibacterial effect in the dark. As seen in Fig. 13.13, in the dark, the "bacterial charge" process occurs, which is because of the Schottky barrier effect at the Ag NP/TOC interface that essentially blocks electron—hole recombination (Sze & Ng, 2006) like Helmholtz capacitance effect at the Ag NP/solution interface (Jakob, Levanon, & Kamat, 2003; Subramanian, Wolf, & Kamat, 2004). As a result, holes accumulate at the TOC near the boundaries in Ag NPs/TOC, resulting in significant oxidation reactions and biocide action. Also, these VB holes (h^+) may arouse

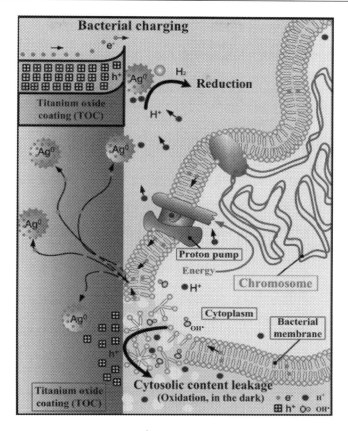

Figure 13.13
Extracellular electron transfer stimulated biocide action of Ag/TOC composites in the dark (Cao et al., 2013).

catalytic oxidation or cause a direct reaction with the membrane lipids due to the electrostatic effects (Maness et al., 1999; Sunada et al., 1998), which involves the formation of pores in the outer membrane and finally cell deterioration.

Similar to the Ag NPs, Lin et al. synthesized Au NPs along with Cu_2O nanocrystals ($Au@Cu_2O$ core@shell nanocrystals) and assessed its ability as a catalyst with a dual function for a continuous activation in light and dark to effectively deactivate *E. coli* (Kuo et al., 2019) (Fig. 13.14A). They attributed the bactericidal mechanism to the Au core peroxidase-mimics and the Fenton reactivity of the Cu_2O shell. As shown in Fig. 13.14B, the catalytic effect of pure Cu_2O continued in the dark for 75 minutes, achieving an additional 2-log reduction of concentration of *E. coli*. In contrast, the remaining cells of *E. coli* were completely inactivated due to the use of $Au@Cu_2O^{-4}$ during the reaction period. The photoexcited Cu_2O electrons are transferred to E_u under light illumination, preferably due to higher E_F of Cu_2O than E_F of Au. Meanwhile, to achieve effective charge separation, the produced holes were displaced at Cu_2O.

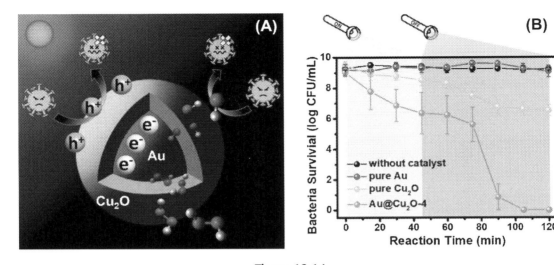

Figure 13.14
(A) Mechanism of antibacterial effect in Au@Cu₂O core@shell nanocrystals. (B) Results of sustainable *Escherichia coli* inactivation under light illumination and darkness conditions by using different catalysts (Kuo et al., 2019).

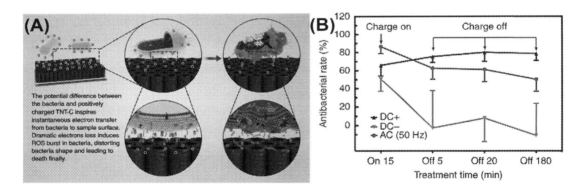

Figure 13.15
(A) The proposed antibacterial mechanism on DC⁺-charged TNT-C. (B) Antibacterial rate of TNT-C-15 on *E. coli* triggered by AC, DC⁺, and DC⁻ (Wang et al., 2018).

A recent study shows that if an external electric current is applied to carbon-doped TiO_2 nanotubes (TNT-C), it will continue to kill bacteria after positive direct current (DC+) is turned off (Wang et al., 2018). The disinfection capacity was considerable because of its intrinsic capacitance and greater discharge. Fig. 13.15A shows that stress in bacteria is produced by the electron transfer at the interface, which increases the generation of intracellular reaction oxygen species and disfigure the morphology of the bacteria, as a result of that the bacteria death occurs. This mechanism is based on the carrier storage and release principle and in fact is similar to the TiON/PdO mechanism that described above. Fig. 13.15B shows the time dependent after charging antibacterial levels of AC, DC+ , and DC groups. The antibacterial level after charging varies slightly from 5 to 180 minutes, which further confirms that the antibacterial effects after charging occur during the initial contact.

13.3.4 Heavy-metal reductions

The WO_3/TiO_2 system was also probed for the reduction of different heavy-metal ions, such as Cr^{6+} into Cr^{3+}, which is relatively nontoxic and stable oxidation state of Cr (Choi et al., 2017). In this process, electrons can be stored in WO_3/TiO_2 under UV light in the presence of oxygen and discharged through a reduction reaction in the absence of light. Finally, electron scavenging released in the dark by the oxygen produces species, such as $O_2 -$ and H_2O_2, which are needed for the reduction process (Zhao, Chen, Yu, Ma, & Zhao, 2009). Accordingly, a faster electron capture rate by WO_3 was observed in experimental investigations. W^{5+} ions will be formed once electrons are excited to the CB of TiO_2 and subsequently transferred to WO_3. In fact, balancing the system's charge neutrality is expected to happen through intercalation of WO_3/TiO_2 by protons due to the stored electrons. In the absence of light, those electrons that were stored in the form Ti^{3+} showed significantly less O_2 reduction as compared to W^{5+}. As a matter of fact, simultaneous storage of electrons and O_2 reduction can happen via the integration of WO_3 to produce more reductive species to reduce the heavy-metal ions.

The system consisting of the flower-like In^{3+}-doped SnS_2 ($InSnS_2$) that synthesized through a microwave-assisted method also showed the ability to reduce the toxic Cr(VI) to less-toxic Cr(II)I (Fig. 13.16A). The spontaneous reduction of Cr(VI) ions was occurred due to the synergistic effect of the released In^{3+} ions from $InSnS_2$ and HCOOH and the subsequent absorbance of reduced Cr(III) ions on the $InSnS_2$ that occurred under dark conditions (Park et al., 2017).

Notably, in the presence of HCOOH, the $InSnS_2$ surface attains a positive charge and PC reduction of Cr(VI) ions occurs under visible light irradiation. However, in the neutral conditions, the spontaneous reduction of $Cr_{(VI)}$ did not occur since no In^{3+} ion was released from $InSnS_2$ and consequently, in the dark condition only 20% of Cr(VI) was adsorbed. Upon the irradiation of visible light, PC reduction of both Cr(VI) ions on the $InSnS_2$ and in the solution was achieved. Hence, a decrease in the reaction time was occurred during the PC reduction in the dark condition, which was due to the spontaneous reduction of Cr(VI), which eventually required further light irradiation in the case when the UV light was used as the source (Fig. 13.16B).

In a similar study, Ag/TiO_2 was used for the photocatalysis-dark degradation of 4-chlorophenol under illumination of UV light, subsequently followed by the reduction of

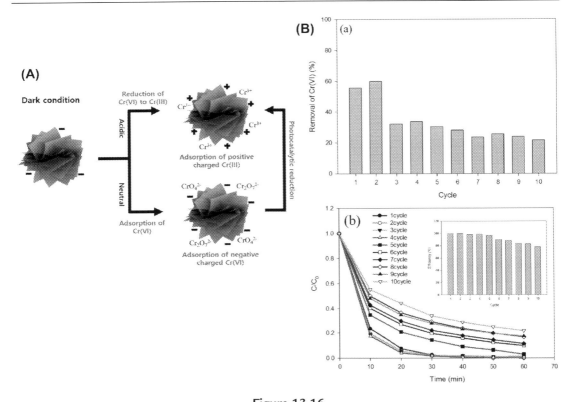

Figure 13.16
(A) Mechanism of the surface charge change of InSnS$_2$ through the reduction of Cr(VI) at different pHs in the dark. (B) PC reduction of Cr(VI) to Cr(III) in the presence of InSnS$_2$ (Park et al., 2017). *PC*, photocatalytic.

Cr$^+$ in the dark. While Ag NPs served as the storage material for the photogenerated electrons during the PC process, where the degradation process of 4-CP was performed by the photogenerated holes (Choi et al., 2017). It this reaction, reduction of Cr^{6+} occurred due to coupling of trapped electrons in Ag NPs with reaction intermediates. In another study, the same Ag/TiO$_2$ was used to degrade the organic pollutants, such as ethanol, humic acid, formic acid, or 4-CP, under irradiation of UV light, followed by the reduction of hexavalent Cr(VI) in the dark. It was also reported that the reduction of Cr(VI) in the absence of light can also occur due to the intermediate compounds produced during the oxidation of 4-CP and also due to the stored electrons in the Ag/TiO$_2$ system. Accordingly, a much higher reduction efficiency for Cr(VI) (about 87%) was achieved through the exploitation of Ag/TiO$_2$ for few hours. The collective effects, such as adsorption, secondary degradation by intermediates, and main reduction by stored electrons, in Ag were expected to play the main roles in achieving such high dark-reduction efficiency in Ag/TiO$_2$ system (Fig. 13.17).

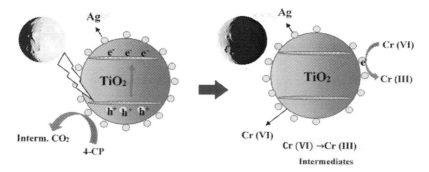

Figure 13.17
Mechanism of electron storage and reduction of metals (Cr^{6+} to Cr^{3+}) in Ag/TiO_2 memory photocatalytic (Choi et al., 2017).

13.3.5 Hydrogen production

The DN PC materials have also been reported to produce hydrogen (H_2) in the dark condition. This system involved the cyanamide-functionalized heptazine polymer-modified g-C_3N_4 ($_{NCN}CN_x$) (Kasap et al., 2016; Lau et al., 2017; Zeng et al., 2018). In this system, the radical species framed in $_{NCN}CN_x$ discharged the trapped electrons in the absence of light toward producing H_2, where these radicals are produced by the preirradiating $_{NCN}CN_x$ suspension in the presence of electron donors (Fig. 13.18A and C). As a result, the hydrogen was generated by keeping the suspension in dark that added with Pt colloids. The obtained results showed that the maximum amount of hydrogen was evolved after 2 hours in dark (where the color of $_{NCN}CN_x$ reached back to yellow) and stopped about 12 hours after illumination was stopped (Fig. 13.18B and E; Kasap et al., 2016). The mechanism of this system was inspirited by the Benson−Calvin cycle in nature to enable hydrogen generation both in light and dark phases.

13.4 Summary and outlook

It is worth remarking that the PC science may decrease concerns in the fields of energy production and environmental remediation. To do this effectively, it is essential to discover more tools to improve the features and properties of the PC materials to rise above their limitations in the PC process. In this direction, the combination of the ESM with a PC material to catalyze the reactions in the presence of light and dark can be considered as one of the promising advancements in this field.

Based on achievements in the development of such DN PCs, it is clear that ESMs should have narrow band gap energy to reduce the hole electron recombination, which is the most

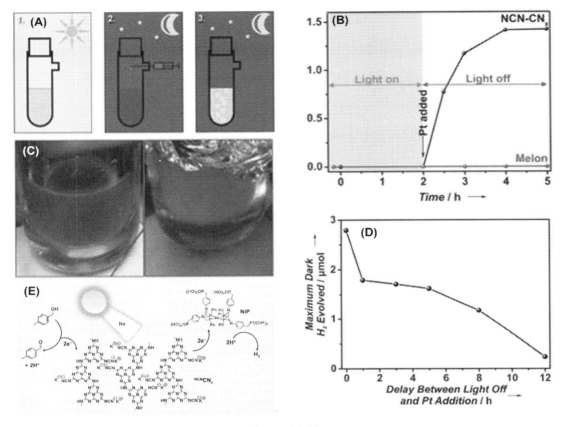

Figure 13.18

(A) Schematic representation of the dark hydrogen evolution process: (1) irradiation of the
$_{NCN}CN_x$ suspension; (2) addition of a solution of hydrogen evolution cocatalyst, and (3) evolution
of hydrogen with the concomitant reversal of suspension color. (B) Rate of hydrogen evolution
with time. (C) Photographs of the "blue radical" (left) and its color reversal in dark (right).
(D) Maximum dark hydrogen evolved as a function of the time between switching off the light
and injection of the Pt colloid (Lau et al., 2017). (E) Schematic representation of irradiation of
surface functionalized carbon nitride, $_{NCN}CN_x$ (Kasap et al., 2016).

important feature needed for increasing the round-the-clock catalytic performance the
system. In addition to this, other properties of the ESMs, such as their electron-storage
ability, rapid charging/gradual discharging, photochemical stability, and durability, should
be improved by means of modification in chemical composition and physical structures to
construct the catalysts with appropriate DN catalytic properties. This can lead to the
materials to possess suitable band edge position with functional intercalation and/or to
reduce properties toward enhancing the overall day—night catalytic properties and thereby
their effective applications in real time.

References

Buama, S., Junsukhon, A., Ngaotrakanwiwat, P., & Rangsunvigit, P. (2017). Validation of energy storage of TiO_2NiO/TiO_2 film by electrochemical process and photocatalytic activity. *Chemical Engineering Journal, 309*, 866−872.

Cao, H., Qiao, Y., Liu, X., Lu, T., Cui, T., Meng, F., & Chu, P. (2013). Electron storage mediated dark antibacterial action of bound silver nanoparticles: Smaller is not always better. *Acta Biomaterialia, 9*(2), 5100−5110.

Chiu, Y.-H., & Hsu, Y.-J. (2017). Au@ Cu_7S_4 yolk@ shell nanocrystal-decorated TiO_2 nanowires as an all-day-active photocatalyst for environmental purification. *Nano Energy, 31*, 286−295.

Choi, Y., Koo, M. S., Bokare, A. D., Kim, D. H., Bahnemann, D. W., & Choi, W. (2017). Sequential process combination of photocatalytic oxidation and dark reduction for the removal of organic pollutants and Cr (VI) using Ag/TiO_2. *Environmental Science & Technology, 51*(7), 3973−3981.

Dickens, P., & Whittingham, M. (1968). The tungsten bronzes and related compounds. *Quarterly Reviews, Chemical Society, 22*(1), 30−44.

Fang, Y., Ma, Y., Zheng, M., Yang, P., Asiri, A. M., & Wang, X. (2018). Metalorganic frameworks for solar energy conversion by photoredox catalysis. *Coordination Chemistry Reviews, 373*, 83−115.

Feng, F., Yang, W., Gao, S., Sun, C., & Li, Q. (2018). Postillumination activity in a single-phase photocatalyst of Mo-doped TiO_2 nanotube array from its photocatalytic "memory". *ACS Sustainable Chemistry & Engineering, 6*(5), 6166−6174.

Huang, H., Jiang, L., Zhang, W. K., Gan, Y. P., Tao, X. Y., & Chen, H. F. (2010). Photoelectrochromic properties and energy storage of TiO_22xNx/NiO bilayer thin films. *Solar Energy Materials and Solar Cells, 94*(2), 355−359.

Jakob, M., Levanon, H., & Kamat, P. V. (2003). Charge distribution between UV-irradiated TiO_2 and gold nanoparticles: Determination of shift in the Fermi level. *Nano Letters, 3*(3), 353−358.

Jin, X., Ye, L., Xie, H., & Chen, G. (2017). Bismuth-rich bismuth oxyhalides for environmental and energy photocatalysis. *Coordination Chemistry Reviews, 349*, 84−101.

Kasap, H., Caputo, C. A., Martindale, B. C. M., Godin, R., Wing-hei Lau, V., Lostsch, B. V., & Reisner, E. (2016). Solar-driven reduction of aqueous protons coupled to selective alcohol oxidation with a carbon nitridemolecular Ni catalyst system. *Journal of the American Chemical Society, 138*(29), 9183−9192.

Kim, S., Park, Y., Kim, W., & Park, H. (2016). Harnessing and storing visible light using a heterojunction of WO_3 and CdS for sunlight-free catalysis. *Photochemical & Photobiological Sciences, 15*(8), 1006−1011.

Knör, G. (2015). Recent progress in homogeneous multielectron transfer photocatalysis and artificial photosynthetic solar energy conversion. *Coordination Chemistry Reviews, 304*, 102−108.

Kongkanand, A., & Kamat, P. V. (2007). Electron storage in single wall carbon nanotubes. Fermi level equilibration in semiconductor−SWCNT suspensions. *ACS Nano, 1*(1), 13−21.

Kuo, M. Y., Hsiao, C. F., Chiu, Y. H., Lai, T. H., Fang, M. J., Wu, J. Y., & Hsu, Y. J. (2019). Au@ Cu_2O core@ shell nanocrystals as dual-functional catalysts for sustainable environmental applications. *Applied Catalysis B: Environmental, 242*, 499−506.

Kuroiwa, Y., Park, S., Sakai, N., & Tastsuma, T. (2016). Oxidation of multicarbon compounds to CO_2 by photocatalysts with energy storage abilities. *Physical Chemistry Chemical Physics, 18*(46), 31441−31445.

Lau, V. W., Klose, D., Kasap, H., Podjaski, F., Pignié, M. C., Reisner, E., Jeschke, G., & Lotsch, B. V. (2017). Dark photocatalysis: Storage of solar energy in carbon nitride for time-delayed hydrogen generation. *Angewandte Chemie International Edition, 56*(2), 510−514.

Li, F., Li, Z., Cai, Y., Zhang, M., Shen, Y., & Wang, W. (2017). Afterglow photocatalysis of Ag_3PO_4 through different afterglow coatings and photocatalysis mechanism. *Materials Letters, 208*, 111−114.

Li, H., Yin, S., & Sato, T. (2011). Novel luminescent photocatalytic deNOx activity of $CaAl_2O_4$:(Eu, Nd)/ $TiO_2 − xNy$ composite. *Applied Catalysis B: Environmental, 106*(3−4), 586−591.

Li, H., Yin, Y., Wang, Y., & Sato, T. (2012a). Blue fluorescence-assisted SrTi 1 − x Cr y O 3 for efficient persistent photocatalysis. *RSC Advances, 2*(8), 3234−3236.

Li, H., Yin, S., Wang, Y., & Sato, T. (2012b). Effect of phase structures of $TiO_2 2xNy$ on the photocatalytic activity of $CaAl_2O_4$: (Eu, Nd)-coupled $TiO_2 2xNy$. *Journal of Catalysis*, *286*, 273−278.

Li, H., Yin, S., Wang, Y., & Sato, T. (2012c). Persistent fluorescence-assisted TiO_2-xNy-based photocatalyst for gaseous acetaldehyde degradation. *Environmental Science & Technology*, *46*(14), 7741−7745.

Li, H., Yin, S., Wang, Y., Sekino, T., Lee, S. W., & Sato, T. (2013). Green phosphorescence-assisted degradation of rhodamine B dyes by Ag_3PO_4. *Journal of Materials Chemistry A*, *1*(4), 1123−1126.

Li, H., Yin, S., Wang, Y., & Sato, T. (2013). Efficient persistent photocatalytic decomposition of nitrogen monoxide over a fluorescence-assisted $CaAl_2O_4$:(Eu, Nd)/(Ta, N)-codoped TiO_2/Fe_2O_3. *Applied Catalysis B: Environmental*, *132*, 487−492.

Li, J., Liu, Y., Zhu, Z., Zhang, G., Zou, T., Zou, Z., & Xie, C. (2013). A full-sunlight-driven photocatalyst with super long-persistent energy storage ability. *Scientific Reports*, *3*, 2409.

Li, Q., Li, Y. W., Wu, P., Xie, R., & Shang, J. K. (2008). Palladium oxide nanoparticles on nitrogen-doped titanium oxide: Accelerated photocatalytic disinfection and post-illumination catalytic "memory". *Advanced Materials*, *20*(19), 3717−3723.

Liu, L., Yang, W., Li, Q., Gao, S., & Shang, J. K. (2014). Synthesis of Cu_2O nanospheres decorated with TiO_2 nanoislands, their enhanced photoactivity and stability under visible light illumination, and their post-illumination catalytic memory. *ACS Applied Materials & Interfaces*, *6*(8), 5629−5639.

Lua, Y., Zhang, X., Chu, Y., Yu, H., Huo, M., Qu, J., Crittenden, J. C., & Yuan, X. (2018). Cu_2O nanocrystals/ TiO_2 microspheres film on a rotating disk containing long-afterglow phosphor for enhanced round-the-clock photocatalysis. *Applied Catalysis B: Environmental*, *224*, 239−248.

Maeda, K. (2013). Z-scheme water splitting using two different semiconductor photocatalysts. *ACS Catalysis*, *3*(7), 1486−1503.

Maness, P. C., Smolinski, S., Blake, D. M., Huang, Z., Wolfrum, E. J., & Jacoby, W. A. (1999). Bactericidal activity of photocatalytic TiO_2 reaction: Toward an understanding of its killing mechanism. *Applied and Environmental Microbiology*, *65*(9), 4094−4098.

Molla, A., Sahu, M., & Hussain, S. (2015). Under dark and visible light: Fast degradation of methylene blue in the presence of Ag−In−Ni−S nanocomposites. *Journal of Materials Chemistry A*, *3*(30), 15616−15625.

Nawaz, A., Kuila, A., Mishra, N. S., Leong, K. H., Sim, L. C., Saravanan, P., & Jang, M. (2019). Challenges and implication of full solar spectrum-driven photocatalyst. *Reviews in Chemical Engineering*, *37*(4), 533−560.

Ng, C., Ng, Y. H., Iwase, A., & Amal, R., (2011a). Charge storage properties of self-photorechargeable flower-shaped WO_3. In Chemeca 2011: Engineering a better world: Sydney Hilton Hotel, NSW, Australia, 18−21 September 2011 (p. 628).

Ng, C., Ng, Y. H., Iwase, A., & Amal, R. (2011b). Visible light-induced charge storage, on-demand release and self-photorechargeability of WO_3 film. *Physical Chemistry Chemical Physics*, *13*(29), 13421−13426.

Ngaotrakanwiwat, P., & Tatsuma, T. (2004). Optimization of energy storage TiO_2−WO_3 photocatalysts and further modification with phosphotungstic acid. *Journal of Electroanalytical Chemistry*, *573*(2), 263−269.

Ngaotrakanwiwat, P., Tatsuma, T., Saitoh, S., Ohko, Y., & Fujishim, A. (2003). Charge discharge behavior of $TiO_2 WO_3$ photocatalysis systems with energy storage ability. *Physical Chemistry Chemical Physics*, *5*(15), 3234−3237.

Nguyen, C.-C., Vu, N.-N., & Do, T.-O. (2016). Efficient hollow double-shell photocatalysts for the degradation of organic pollutants under visible light and in darkness. *Journal of Materials Chemistry A*, *4*(12), 4413−4419.

Ohko, Y., Saitoh, S., Tatsuma, T., & Fujishima, A. (2001). Photoelectrochemical anticorrosion and self-cleaning effects of a TiO_2 coating for type 304 stainless steel. *Journal of the Electrochemical Society*, *148*(1), B24B28.

Pan, W., Ning, G., Zhang, X., Wang, J., Lin, Y., & Ye, J. (2008). Enhanced luminescent properties of long-persistent $Sr_2MgSi_2O_7$: Eu2 + , Dy3 + phosphor prepared by the co-precipitation method. *Journal of Luminescence*, *128*(12), 1975−1979.

Park, H., Bak, A., Jeon, T. H., Kim, S., & Choi, W. (2012). Photo-chargeable and dischargeable TiO_2 and WO_3 heterojunction electrodes. *Applied Catalysis B: Environmental*, *115*, 74−80.

Park, S., Kim, W., Selvaraj, R., & Kim, Y. (2017). Spontaneous reduction of Cr (VI) using $InSnS_2$ under dark condition. *Chemical Engineering Journal, 321*, 97−104.

Sahu, I. P., Bisen, D. P., & Brahme, N. (2015). Luminescence properties of Eu2 + , Dy3 + -doped $Sr_2MgSi_2O_7$, and $Ca_2MgSi_2O_7$ phosphors by solid-state reaction method. *Research on Chemical Intermediates, 41*(9), 6649−6664.

Sakar, M. M., Nguyen, C., Vu, M. N. H., & Do, T. O. (2018). Materials and mechanisms of photo-assisted chemical reactions under light and dark conditions: Can daynight photocatalysis be achieved? *ChemSusChem, 11*(5), 809−820.

Shankar, M. V., Neppolian, B., Sakthivel, S., Banumathi, A., Palanichamy, M., & Murugesan, V. (2001). Kinetics of photocatalytic degradation of textile dye reactive red 2. *Indian Journal of Engineering & Materials Science, 8*, 104−109.

Subasri, R., & Shinohara, T. (2003). Investigations on $SnO_2−TiO_2$ composite photoelectrodes for corrosion protection. *Electrochemistry Communications, 5*(10), 897−902.

Subramanian, V., Wolf, E. E., & Kamat, P. V. (2004). Catalysis with TiO_2/gold nanocomposites. Effect of metal particle size on the Fermi level equilibration. *Journal of the American Chemical Society, 126*(15), 4943−4950.

Sunada, K., Kikuchi, Y., Hashimoto, K., & Fujishima, A. (1998). Bactericidal and detoxification effects of TiO_2 thin film photocatalysts. *Environmental Science & Technology, 32*(5), 726−728.

Sze, S. M., & Ng, K. K. (2006). *Physics of semiconductor devices*. John wiley & Sons.

Takahashi, Y., & Tatsuma, T. (2005). Oxidative energy storage ability of a $TiO_2 − Ni(OH)_2$ bilayer photocatalyst. *Langmuir, 21*(26), 12357−12361.

Takahashi, Y., & Tatsuma, T. (2006). Remote energy storage in $Ni(OH)_2$ with TiO_2 photocatalyst. *Physical Chemistry Chemical Physics, 8*(23), 2716−2719.

Takahashi, Y., & Tatsuma, T. (2008). Visible light-induced photocatalysts with reductive energy storage abilities. *Electrochemistry Communications, 10*(9), 1404−1407.

Takahashi, Y., Ngaotrakanwiwat, P., & Tatsuma, T. (2004). Energy storage $TiO_2−MoO_3$ photocatalysts. *Electrochimica Acta, 49*(12), 2025−2029.

Tatsuma, T., Saitoh, S., Ohko, Y., & Fujishima, A. (2001). $TiO_2−WO_3$ photoelectrochemical anticorrosion system with an energy storage ability. *Chemistry of Materials, 13*(9), 2838−2842.

Tatsuma, T., Saitoh, S., Ngaotrakanwiwat, P., Ohko, Y., & Fujishima, A. (2002). Energy storage of $TiO_2−WO_3$ photocatalysis systems in the gas phase. *Langmuir, 18*(21), 7777−7779.

Tatsuma, T., Takeda, S., Saitoh, S., Ohko, Y., & Fujishima, A. (2003). Bactericidal effect of an energy storage TiO_2WO_3 photocatalyst in dark. *Electrochemistry Communications, 5*(9), 793−796.

Wang, G., Feng, H., Hu, L., Jin, W., Hao, Q., Gao, A., & Chu, P. K. (2018). An antibacterial platform based on capacitive carbon-doped TiO_2 nanotubes after direct or alternating current charging. *Nature Communications, 9*(1), 2055.

Wang, G., Xing, Z., Zeng, X., Feng, C., McCarthy, D. T., Deletic, A., & Zhang, X. (2016). Ultrathin titanium oxide nanosheets film with memory bactericidal activity. *Nanoscale, 8*(42), 18050−18056.

Wu, P., Imlay, J. A., & Shang, J. K. (2010). Mechanism of Escherichia coli inactivation on palladium-modified nitrogen-doped titanium dioxide. *Biomaterials, 31*(29), 7526−7533.

Yang, F., Takahashi, Y., Sakai, N., & Tatsuma, T. (2010). Oxidation of methanol and formaldehyde to CO_2 by a photocatalyst with an energy storage ability. *Physical Chemistry Chemical Physics, 12*(19), 5166−5170.

Yang, F., Takahashi, Y., Sakai, N., & Tatsuma, T. (2011). Visible light driven photocatalysts with oxidative energy storage abilities. *Journal of Materials Chemistry, 21*(7), 2288−2293.

Yang, Z., Li, L., Luo, Y., He, R., Qiu, L., Lin, H., & Peng, H. (2013). An integrated device for both photoelectric conversion and energy storage based on free-standing and aligned carbon nanotube film. *Journal of Materials Chemistry A, 1*(3), 954−958.

Yasomanee, J., & Bandara, J. (2008). Multi-electron storage of photoenergy using $Cu_2O−TiO_2$ thin film photocatalyst. *Solar Energy Materials and Solar Cells, 92*(3), 348−352.

Yin, H., Chen, X., Hou, R., Zhu, H., Li, S., Huo, Y., & Li, H. (2015). Ag/BiOBr film in a rotating-disk reactor containing long-afterglow phosphor for roundthe- clock photocatalysis. *ACS Applied Materials & Interfaces, 7*(36), 20076−20082.

Zeng, Z., Quan, X., Yu, H., Chen, S., Zhang, Y. S., Zhao, H., & Zhang, S. (2018). Carbon nitride with electron storage property: Enhanced exciton dissociation for highefficient photocatalysis. *Applied Catalysis B: Environmental, 236*, 99−106.

Zhang, L., Xu, L., Wang, J., Cai, J., Xu, J., Zhou, H., & Cao, C. N. (2012). Enhanced energy storage of a UV-irradiated three-dimensional nanostructured $TiO_2Ni(OH)_2$ composite film and its electrochemical discharge in the dark. *Journal of Electroanalytical Chemistry, 683*, 55−61.

Zhang, Q., Wang, H., Li, Z., Geng, C., & Leng, J. (2017). Metal-free photocatalyst with visible-light-driven post-illumination catalytic memory. *ACS Applied Materials & Interfaces, 9*(26), 21738−21746.

Zhao, D., Chen, C., Yu, C., Ma, W., & Zhao, J. (2009). Photoinduced electron storage in WO_3/TiO_2 nanohybrid material in the presence of oxygen and postirradiated reduction of heavy metal ions. *The Journal of Physical Chemistry C, 113*(30), 13160−13165.

Zhou, Q., Peng, F., Ni, Y., Kou, J., Lu, C., & Xu, Z. (2016). Long afterglow phosphor driven round-the-clock g-C3N4 photocatalyst. *Journal of Photochemistry and Photobiology A: Chemistry, 328*, 182−188.

CHAPTER 14

Photocatalytic membranes: Synthesis, properties, and applications

Luisa M. Pastrana-Martínez, Sergio Morales-Torres, Álvaro Pérez-Molina and Francisco José Maldonado-Hódar

Carbon Materials Research Group, Department of Inorganic Chemistry, Faculty of Sciences, University of Granada, Granada, Spain

14.1 Introduction

The abatement of the environmental pollution, and especially the availability of clean water, is an important issue in large cities and industrialized areas, as well as, in less-developed regions, making the necessity to develop new efficient water purification technologies at low cost and energy consumption (Ahn et al., 2012; Elimelech & Phillip, 2011; Humplik et al., 2011; Shannon et al., 2008). Up to date, several organic and inorganic contaminants have been detected in water, particularly in the industrial and hospital waters. These pollutants are oil, dyes, pesticides, and drugs, as well as natural organic matter, including microorganisms, bacteria, and algae, among others. Membrane processes can perform efficient and selective separations of colloids, microorganisms, and organic pollutants in water and wastewater treatments (Pendergast & Hoek, 2011; Peters, 2010). Among others, polymers, ceramics, and carbon materials have been successfully proved to produce different types of membranes (Athanasekou et al., 2014; Athanasekou et al., 2015; Baek et al., 2014; Das, Ali, Hamid, Ramakrishna, & Chowdhury, 2014; Pastrana-Martínez et al., 2013; Pastrana-Martínez, Morales-Torres, Figueiredo, Faria, & Silva, 2015; Ulbricht, 2006; Xu, Chang, Hou, & Gao, 2013).

The advantages of membrane technology are high system compacity (small footprint), low energy consumption, easy operational control, easy upscaling, and combination with other processes, good permeate quality, reliable performance, and so on. The most commonly used membrane technologies for water purification are microfiltration (MF, pore size 0.1–10 μm), ultrafiltration (UF, pore size 0.05–0.1 μm), nanofiltration (NF, pore size 0.1–0.001 μm), and reverse osmosis (Kim & Van der Bruggen, 2010; Le-Clech, Lee, & Chen, 2006). However, the major obstacles in the implantation of the membrane technology are (1) the production of a concentrate stream, which contains the chemicals originally present in the feedwater and should be treated separately, and (2) the membrane

Photocatalytic Systems by Design.
DOI: https://doi.org/10.1016/B978-0-12-820532-7.00018-7

385

deactivation by fouling (Riaz & Park, 2020). Both problems can be solved by providing reactivity to the membrane for the degradation of the pollutants, avoiding their accumulation in stream or on the membrane, with a self-cleaning behavior (Zhang, Ding, Luo, Jaffrin, & Tang, 2016).

The removal and/or detoxification of organic pollutants in water can be achieved with the application of advanced oxidation processes (AOPs). They are based on the production of highly reactive oxygen species like hydroxyl radicals (HO$^•$) from different oxidizing reactants (e.g., air, ozone, hydrogen peroxide), which can not only oxidize and mineralize almost each organic compound but also to damage pathogens, including bacteria and viruses (Tsydenova, Batoev, & Batoeva, 2015). Among AOPs, heterogeneous photocatalysis is a very attractive option because it is possible to use a natural light source (sunlight), nontoxic materials, and no additional chemicals with a high efficiency in degrading a wide range of water pollutants (Michael-Kordatou et al., 2015; Sampaio et al., 2015). In photocatalysis, energy from photons is used to boost chemical reactions (Herrmann, 2005). The photogenerated electrons and holes that migrate to the surface of the semiconductor without recombination can, respectively, reduce and oxidize the reactants adsorbed by the semiconductor.

Some of the most active semiconductors traditionally used as photocatalysts are TiO_2, ZnO, CdS, ZnS, Fe_2O_3, WO_3, and Bi_2WO_4 (Gupta & Tripathi, 2012; Xiang, Yu, & Jaroniec, 2012). Among them, TiO_2 is the most recognized semiconductor for photocatalysis due to its low cost, low toxicity, high efficiency, and chemical stability. However, its practical application is severely compromised by two inherent limitations: the low quantum yield, which is primarily impaired by the fast recombination of photogenerated charge carriers, and the poor light-harvesting ability that is restricted by the wide band gap of TiO_2 to the ultraviolet type-A (UVA) spectral range (band gap energy, 3.2 eV) (Fujishima, Rao, & Tryk, 2000; Konstantinou & Albanis, 2004; Nakata & Fujishima, 2012; Pastrana-Martínez et al., 2012).

Photocatalytic membranes (PMs), in which nanoparticles of such photocatalysts are used to enhance the performance of membranes, have received overwhelming research interests in recent years because they provide a complete treatment of wastewater due to the dual action over pollutants: retention/rejection/repulsion and photodegradation (Athanasekou et al., 2014; Gao, Hu, & Mi, 2014; Gao, Liu, Tai, Sun, & Ng, 2013; Mozia, 2010; Zhang, Wang, & Diniz da Costa, 2014). Compared to conventional separation membranes, PMs possess unique properties, such as antibiofouling behavior, less concentrated retentate effluent, higher flux and often more hydrophilic character, among others. Immobilizing the photocatalytic nanoparticles on PMs also avoids the photocatalyst separation/recovery process needed when used in slurry, which is one of the main difficulties of photocatalytic processes, thus advancing their practical application.

In a photocatalytic membrane reactor (PMR), the membrane is used as an active participant in a chemical reaction in the reactor chamber. The membrane not only plays the role of a separator but also works as catalyst support where the desired reactions take place. Therefore it is important to develop PMs that allow effective removal of environmental pollutants through simultaneous degradation and selective separation process based on molecular sizes. The membrane system should be operated under high water flux conditions, which can be maintained or restored after appropriate cleaning. Recently, UV-responsive PMs developed using TiO_2-based catalysts have gained considerable attention (Albu, Ghicov, Macak, Hahn, & Schmuki, 2007; Pan, Zhang, Du, Sun, & Leckie, 2008; Zhang, Quan, Chen, & Zhao, 2006; Zhang, Wang, Lopez, & Diniz da Costa, 2014). However, the suppression of the recombination of photogenerated charge carriers, as well as the effective utilization of visible light, are some of the main challenges before these membranes become economically feasible.

This chapter focuses on the synthesis, properties, and application of the TiO_2-based PMs in water treatment. The strategies to improve the visible light response of TiO_2 and the possible support for the membranes will also be overviewed. In addition, the prospects and further expected developments of TiO_2-based PMs in water treatment will also be overviewed.

14.2 TiO₂-based photocatalytic membranes

The immobilization of the photocatalyst in the membrane can be: (1) coated on the membrane; (2) blended into the polymeric membrane matrix; and (3) free-standing membrane made of a pure photocatalyst. Several types of organic, inorganic, metallic, and polymeric materials are widely used as support materials for the fabrication of TiO_2-based PMs. These can be fabricated using different methods, such as dip coating (Djafer, Ayral, & Ouagued, 2010; Fischer et al., 2015; Zhang et al., 2016), phase inversion (PI) (Kuvarega, Khumalo, Dlamini, & Mamba, 2018; Salim et al., 2019; Singh, Yadav, & Purkait, 2019), hydrothermal synthesis/filtration (Rao, Zhang, Zhao, Chen, & Li, 2016), self-assembly (Starr, Tarabara, Zhou, Roualdès, & Ayral, 2016), electrospinning and electrospraying (An et al., 2014; Daels, Radoicic, Radetic, Van Hulle, & De Clerck, 2014; Dong et al., 2019; Neubert, Pliszka, Thavasi, Wintermantel, & Ramakrishna, 2011; Zhang et al., 2018), and chemical vapor deposition (CVD) (De Filpo et al., 2018; Starr et al., 2016), among others.

The study of membranes can be divided according to two main categories: (1) TiO_2 polymeric membranes and (2) TiO_2 ceramic/inorganic membranes.

14.2.1 TiO₂ polymeric membranes

Polymeric membranes have been used extensively in water and wastewater treatment processes. Tennakone, Tilakaratne, and Kottegoda (1995), in a pioneering work about the use of polymers as TiO$_2$ support, studied polyethylene film to support TiO$_2$ by thermal treatment. Although TiO$_2$ particles can be deposited onto the polymeric surface, when dispersed in the polymeric matrix, the blended TiO$_2$-based photocatalysts present a reduced leaching. Different polymers have been studied as support of photocatalyst nanoparticles, such as polyaniline (PANI), polyvinylidene fluoride (PVDF), polyamine, polysulfone (PSF), cellulose acetate, polyurethane, polytetrafluoroethylene, and polyacrylonitrile, among others.

Typically, one of the most widely known techniques of the preparation of polymer membranes is PI technique. Pereira, Isloor, Ahmed, and Ismail (2015) reported the synthesis of PSF UF membranes with PANI-TiO$_2$ nanocomposites and polyethylene glycol 1000 by the PI method. The modified TiO$_2$ nanotubes were dispersed in the PSF matrix using *N*-methyl-2-pyrrolidone (NMP) as a solvent. PANI-TiO$_2$ nanocomposite membranes showed better hydrophilicity, improved permeability, enhanced porosity, water uptake and good antifouling ability when compared with neat PSF membranes. The performance of prepared membrane was examined for their capability of rejecting Pb^{+2} and Cd^{+2}.

Neubert et al. (2011) reported the preparation of fibrous PANI-TiO$_2$-based polyethylene oxide (PEO) composite membrane by integrating the electrospinning and electrospraying techniques as shown in Fig. 14.1. It was demonstrated that the TiO$_2$ nanoparticle catalyst-embedded PANI-PEO fibrous membrane degraded the pollutant significantly, which is due to uniform dispersion of the catalysts produced by this methodology.

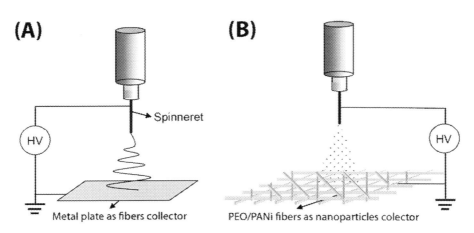

Figure 14.1
Setup for (A) electrospinning and (B) electrospraying. Source: *Reprinted with permission from Neubert, S., Pliszka, D., Thavasi, V., Wintermantel, E., & Ramakrishna, S. (2011). Conductive electrospun PANi-PEO/TiO2 fibrous membrane for photo catalysis.* Materials Science and Engineering: B, 176, 640–646, Copyright 2011, Elsevier.

Other authors reported the combination of electrospinning, cold plasma, and hydrothermal treatments to produce TiO_2-PVDF membranes (Dong et al., 2019). Briefly, PVDF was dissolved in *N,N*-dimethylformamide and acetone and then tetrabutyl titanate (TBT) was added into the prepared gel. The next step was electrospinning of the PVDF-TBT gel through a syringe with a gage needle connected to a high voltage supply. After that, the dry membranes were treated with cold plasma and then placed into stainless steel autoclave. The hydrothermal synthesis was carried out at 90°C for 9 hours.

Some synthesis methods like dip coating normally result in the formation of nanoparticles agglomerates on the membrane having an important effect on membrane activity. Fischer et al. (2015) reported a new and straightforward method for the synthesis of TiO_2 nanoparticles at the surface of two hydrophilic MF membranes (Millipore Express plus membrane, polyethersulfone, PES; Millipore Durapore membrane, PVDF) and one hydrophobic MF membrane (Millipore Durapore membrane and PVDF) via hydrolysis of titanium tetraisopropoxide. To do this, a wet (water) membrane was dipped in titanium tetraisopropoxide/ethanol solution. As the water film on the membrane surface initiates the hydrolysis, TiO_2 nanoparticles were grown on the membrane surface. As a result, a nonaggregated and strongly bonded layer of TiO_2 nanoparticles were built on the membrane surface. By treating the TiO_2 with vapor water under mild conditions, the photoactive phase anatase was generated on the polymer support.

Ong, Lau, Goh, Ng, and Ismail (2015) prepared a series of PVDF hollow fiber UF membranes made up of different TiO_2 concentrations with the presence of polyvinylpyrrolidone (PVP) as additive. The membrane was fabricated by combining the dry-jet-wet spinning and PI techniques. The membrane performances were evaluated for the oily wastewater treatment in a submerged membrane filtration system. The results demonstrated that the composite PVDF membrane showed relatively higher permeate flux and better performance in treating oily solution in comparison with the membrane prepared without TiO_2.

Other authors (Vatanpour et al., 2012) reported the synthesis of TiO_2-embedded mixed-matrix PES NF membrane. The membrane was fabricated by the immersion precipitation and PI. The hydrophilicity of the membrane showed significant enhancement due to the improvement of water permeability of the membrane. The membrane showed better antifouling performance upon the addition of nano-sized TiO_2 nanoparticles and the water flux recovery increased from 56% to 91% by the addition of 4 wt.% TiO_2 nanoparticles.

Pourjafar, Rahimpour, and Jahanshahi (2012) reported the synthesis of polyvinyl alcohol (PVA)-PES composite TiO_2 NF membrane by dip coating of PES membrane in PVA and TiO_2 aqueous suspension. Phase-induced immersion precipitation technique was utilized to fabricate the membrane while glutaraldehyde was used as a cross-linker for the composite polymer membrane to enhance the chemical, thermal, and mechanical stabilities. It was found that the increase in TiO_2 solution concentration can highly affect the surface morphology and filtration performance of coated membranes.

Emadzadeh, Lau, Matsuura, Rahbari-Sisakht, and Ismail (2014) reported the fabrication of the TiO_2-PSF membrane for the desalination of water in forward osmosis. The membrane was fabricated with a prepared solution of TiO_2 in NMP by ultrasounds, followed by the addition of PVP under vigorous stirring. It was found that both hydrophilicity and porosity of the substrate were increased upon addition of TiO_2. The authors claimed that the thin film composite membrane containing 0.5 wt.% TiO_2 nanoparticles showed promising results, such as better water permeability and low reverse solute flux.

Padaki, Emadzadeh, Masturra, and Ismail (2015) reported the antifouling properties of PSF and titanium nanotube mixed matrix UF membrane. PI method was adopted to fabricate the membrane. The authors observed the remarkable improvement of the water reflux recovery, which was attributed to the enhanced hydrophilicity of the membrane. The fouling resistance of the membrane was assessed against bovine serum albumin (NSA); the experimental results demonstrated the improved antifouling performance of the membrane, which results in 100% water reflux recovery.

The fabrication of 3D nanofiber membranes decorated with photocatalytic TiO_2 particles using nylon-6 nanofiber membranes was also reported (An et al., 2014). They use both batch and continuous deposition processes for the fabrication of the membrane following electrospraying methodology (Fig. 14.2). In the batch process, nanoparticles were initially electrosprayed onto a substrate and then nylon-6 nanofibers were electrospun on top of TiO_2. In the continuous process, nylon-6 nanofibers and TiO_2 nanoparticles were deposited simultaneously by electrospinning and electrospraying, respectively, onto a substrate attached to a rotating cylinder, making a truly 3D-structured water purification nanofiber membrane. The membrane fabricated by the continuous means yielded 100% degradation of methylene blue within 90 minutes under UV irradiation.

Trying to increase the efficiency of pure TiO_2, one of the main strategy is the preparation of TiO_2 composites, where different carbon materials can offer many advantages. Thus, different authors have prepared TiO_2-carbon composites that can be thereafter deposited on the polymeric membrane. The combination of TiO_2 and graphene oxide (GO) has sparked off enormous attention in the photocatalysis domain. Typically, GO-TiO_2 composites are efficient photocatalysts under both near-UV/Vis and visible light irradiation, overcoming one of the main limitations of bare TiO_2 (Fan, Lai, Zhang, & Wang, 2011; Long et al., 2013; Pastrana-Martínez et al., 2013; Pastrana-Martínez et al., 2014). However, these materials are usually employed as suspended particles (slurries) in batch reactors and thus a second step is required for catalyst separation from the treated water, limiting its recovery and reuse.

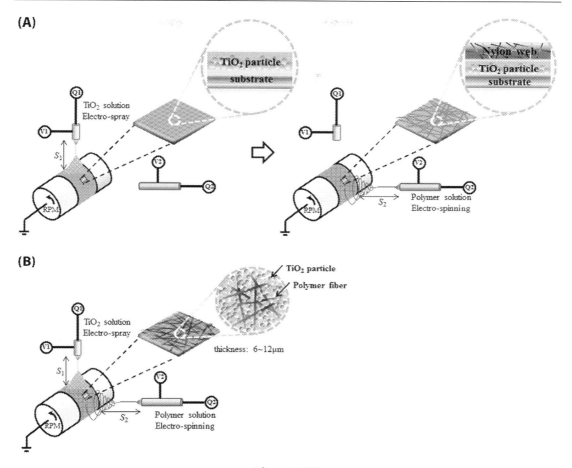

(A)

(B)

Figure 14.2

Schematic representation of (A) batch process (2D structure) and (B) continuous process (3D structure). Source: *Reprinted with permission from An, S., Lee, M. W., Joshi, B. N., Jo, A., Jung, J., & Yoon, S. S. (2014). Water purification and toxicity control of chlorophenols by 3D nanofiber membranes decorated with photocatalytic titania nanoparticles.* Ceramics International, 40, *3305—3313, Copyright 2014, Elsevier.*

Xu et al. (2016) reported the synthesis of an UF PVDF PM based on GO and TiO_2 (GO/TiO_2-PVDF) for water treatment via the PI technique. The GO/TiO_2 nanocomposite was prepared by hydrothermal method. Compared with PVDF membranes supplemented with TiO_2 and GO, respectively, the GO/TiO_2-PVDF membrane displayed significantly improved photodegradation efficiency (improved about 50%—70%) and superior photodegradation kinetics (1.0—1.5 times faster) toward BSA.

Pastrana-Martínez et al. (2015) immobilized several photocatalysts (i.e., TiO_2 and GO-TiO_2, GOT) on a flat sheet UF membranes prepared with laboratory-made TiO_2 and GO-TiO_2

(GOT). The deposition of photocatalysts onto flat sheet membranes was carried out by filtration method and cellulose (MCE) membranes as supports. Cross-sectional images of membranes showed that the corresponding photocatalytic materials were homogeneously deposited on the MCE membrane without appreciable presence of cracks, holes, or another defects, even if considered that these membranes were fractured. The membrane presenting the highest photocatalytic activity (M-GOT) was also modified by intercalating a freestanding GO membrane between the MCE membrane and the GOT photocatalyst layer. For that, a GO dispersion was filtered through an M-GOT membrane, and a homogeneous GO layer was obtained above the MCE membrane and the layer of GOT (labeled as M-GO/GOT, Fig. 14.3A and B). The M-GO/GOT membrane showed higher pollutant removal under dark conditions and good performance under visible and near-UV/Vis irradiation. However, the M-GOT membrane performed better, probably due to the higher compactness in the case of M-GO/GOT as a consequence of the synthesis conditions required for its preparation.

This composite based on GO-TiO$_2$ was also immobilized into the matrix of alginate porous hollow fibers (GO-TiO$_2$−4/APHF) (Pastrana-Martínez et al., 2013). During the synthesis, the composite was added to an aqueous solution of sodium alginate under dynamic stirring. When the homogenization was complete, alginate fibers (containing the composite) were prepared by a dry and wet spinning process. The resulting hydrogel hollow fibers were converted to alcogel via successive immersion in a series of ethanol−water solutions (Papageorgiou et al., 2012). Their internal diameter reaches 500 μm, and their wall thickness is around 41 μm. The external surface exhibits high roughness with GO-TiO$_2$ materials in random orientation (Fig. 14.4). In addition to the high activity in consecutive

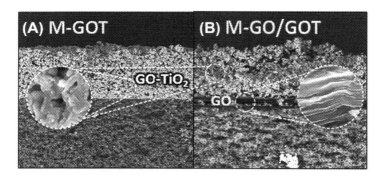

Figure 14.3

Cross-sectional scanning electron microscopy (SEM) micrographs of (A) M-GOT and (B) M-GO/GOT (insets correspond to the GOT composite or the freestanding GO membrane: M-GO).
Source: *Reprinted with permission from Pastrana-Martínez, L. M., Morales-Torres, S., Figueiredo, J. L., Faria, J. L., & Silva, A. M. T. (2015). Graphene oxide based ultrafiltration membranes for photocatalytic degradation of organic pollutants in salty water.* Water Research, *77, 179−190, Copyright 2015, Elsevier.*

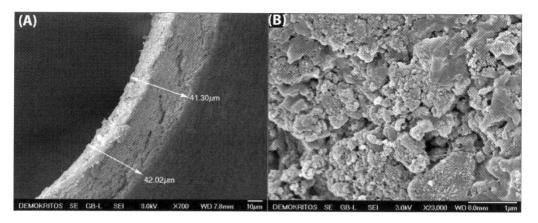

Figure 14.4

SEM micrographs at different magnifications of the GOT composite immobilized into alginate porous hollow fibers (GO-TiO$_2$−4/APHF): (A) wall thickness and (B) external surface. Source: *Reprinted with permission from Pastrana-Martínez, L. M., Morales-Torres, S., Papageorgiou, S. K., Katsaros, F. K., Romanos, G. E., Figueiredo, J. L., . . . Silva, A. M. T. (2013). Photocatalytic behavior of nanocarbon-TiO$_2$ composites and immobilization into hollow fibers.* Applied Catalysis B: Environmental, 142−143, *101−111, Copyright 2013, Elsevier.*

light—dark cycles, the fibers also exhibit high stability. Even so, the authors claimed that possible degradation of the polymer containing the photocatalyst occurs during the experiment.

Similarly, other authors (Dzinun et al., 2017) also reported the effects of UV irradiation on the structure and long-term stability of polymer membranes. They observed that after 30 days of UV irradiation, a tensile strength of hollow fiber TiO$_2$/PVDF membranes significantly decreased, which shows a limited applicability of polymeric PMs.

Thus the disadvantages of the polymeric membrane can be associated to its loss of chemical, thermal, and photo-stability, decreasing durability when used as supports in reactive membranes under UV irradiation. This limits their use in PMRs. The need to fabricate a new type of membrane with chemical and UV resistances has led to interest in the development of inorganic membranes.

14.2.2 TiO$_2$ ceramic/inorganic membranes

In recent years, ceramic membranes have been used increasingly in the water industry including in municipal water and wastewater treatment plants. This is because ceramic membranes have different features, such as high permeability (due to good surface wettability and relatively uniform pore size), good physical stability under high temperature and high-pressure system, good chemical and biological stability, excellent

UV resistance, mechanical strength, high resistance, long working life, easy cleaning, reusability, and so on.

The commonly used materials for making ceramic membrane are Al_2O_3, TiO_2, ZrO_2, and SiO_2, and mixtures of these materials. Among the different types of ceramic porous support materials, α-Al_2O_3 is the most commonly chosen due to its high chemical stability under high pH conditions (up to pH value of 11.0). The pioneering works of ceramic membranes correspond to Choi's and Zhang's groups with the study of permeability and photocatalytic activity of TiO_2/Al_2O_3 (Ulbricht, 2006) and TiO_2/SiO_2 (Xu et al., 2013) membranes, respectively.

Ceramic membranes are an excellent substrate for PM fabricating by coating with TiO_2 photocatalytic layers. Different coating techniques can be used, such as dip coating, spin coating, aerosol spraying, CVD, spray pyrolysis, reactive sputtering, and plasma spraying, resulting in TiO_2 thin films of various thicknesses and morphologies. The goal is to produce a thin TiO_2 layer with strong adherence, good uniformity in thickness and pore size, defect-free surface, and minimal pore blockage of the bare ceramic membrane. The microstructure and the performance of the TiO_2 coating as well as the selective layer strongly depend on the preparation conditions.

Normally, when the photocatalysts are coated on the inorganic membrane, the permeability can decrease. Horovitz et al. (2016) reported that Al_2O_3 coated with N-doped TiO_2 decreased its permeability between 12% and 50% depending on membrane pore size, although Darling (2018) reported that membrane coating can increase the permeability by rendering the interface more hydrophilic.

Alias, Harun, and Latif (2018) reported the synthesis of porous ceramic membranes coated with TiO_2 by dip coating using different concentrations of TiO_2 nanoparticle suspensions before sintering or after the sintering process. The ceramic membranes before sintering showed excellent morphology and strong adhesion of the penetrated TiO_2-coated layer on the surface and within the pores.

Zhang et al. (2016) also used dip-coating technique for coating TiO_2 nanofibers onto the ceramic hollow fiber membrane supports. The ceramic hollow fiber membrane was dipped into a solution containing TiO_2 and water, with one end of the membrane sealed and the other end connected to a pump (50 rpm). Finally, the obtained TiO_2 nanofiber membranes were calcined at 500°C for 2 hours. The obtained results showed a high removal of humic acid (HA) (90%).

Agana, Reeve, and Orbell (2013) evaluated the performance of a commercial TiO_2-ceramic UF membrane to the production of potable water from wastewater. A Membralox T1−70 single channel ceramic membrane with a TiO_2 active layer was used in all the experiments. The researchers investigated the effect of different operating parameters, such as crossflow velocity (CFV), transmembrane pressure (TMP), and the particle size on the performance of the UF membrane. The experimental results showed that the permeate flux and the rejection of contaminants were significantly improved by the higher combinations of CFV and TMP owing to the turbulent flow of suspended particles during the UF process.

Athanasekou et al. (2014) reported the synthesis of a hybrid photocatalytic/UF membrane using UF monochannel monoliths with GO-TiO$_2$ composites stabilized into the pores (Fig. 14.5). This membrane was finally incorporated into an innovative water purification device that combines membrane filtration with semiconductor photocatalysis, being active under visible illumination. An aqueous dispersion (50 g/L) containing the corresponding photocatalyst (GO-TiO$_2$ or bare TiO$_2$) was used to deposit the material on the different ceramic membranes by dip coating. The coated membranes were treated at 200°C, with a heating and cooling rate of 1°C/minute and under a N$_2$ atmosphere. Finally, the ceramic membranes were softly flushed with compressed air to remove the particles not well-adhered before to be used in reaction. Different membranes were developed on the monoliths with the pore size of 1, 5, and 10 nm using the GOT aqueous suspension.

Commercially ceramic MF membranes of Al$_2$O$_3$ were modified by nano-TiO$_2$ coating by an in situ precipitation method (Chang et al., 2014). The membrane was saturated with a TiO$_2$ nanoparticles in suspension placed in the oven at 85°C for 3 hours and washed with distilled water (DW) and dried at 110°C for 12 hours; this cycle was repeated twice. At the

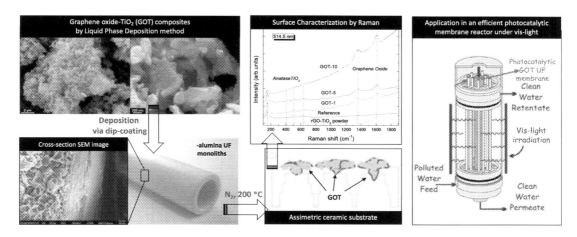

Figure 14.5

Schematic representation of hybrid photocatalytic/ultrafiltration process for water purification under visible light by using GOT stabilized into the pores of ultrafiltration (*UF*) monochannel monoliths. Source: *Adapted with permission from Athanasekou, C. P., Morales-Torres, S., Likodimos, V., Romanos, G. E., Pastrana-Martinez, L. M., Falaras, P., . . . Silva, A. M. T. (2014). Prototype composite membranes of partially reduced graphene oxide/TiO$_2$ for photocatalytic ultrafiltration water treatment under visible light. Applied Catalysis B: Environmental, 158–159, 361–372, Copyright 2014, Elsevier.*

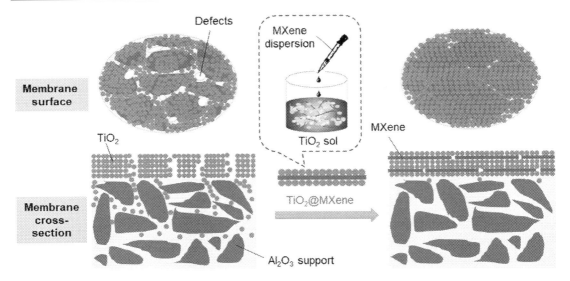

Figure 14.6

Fabrication of TiO₂-MXene membranes to eliminate potential defects. Source: *Reprinted with permission from Xu, Z., Sun, Y., Zhuang, Y., Jing, W., Ye, H., & Cui, Z. (2018). Assembly of 2D MXene nanosheets and TiO2 nanoparticles for fabricating mesoporous TiO₂-MXene membranes.* Journal of Membrane Science, 564, 35−43, Copyright 2018, Elsevier.

final stage, the membrane was calcined at 950°C for 3 hours. The results showed a goof performance of the modified ceramic MF membrane in the treatment of a stable oil−water emulsion.

Xu et al. (2018) proposed a simple and scalable method for the preparation of mesoporous membranes by assembling two-dimensional (2D) MXene nanosheets and TiO₂ nanoparticles on a macroporous support. MXene nanosheets were introduced into TiO₂ hydrosol, and then the sol was transformed into gel for the fabrication of mesoporous TiO₂-MXene membranes (Fig. 14.6). The results revealed that hollow fiber TiO₂-MXene membranes had a narrow pore size distribution, desirable rejection characteristics of dextran, and high water permeate flux.

14.3 Photocatalytic membrane reactors for water treatment

In PMR, the membrane acts both as a selective barrier for species involved in the reaction and the support for the photocatalyst; thus the photocatalytic and the membrane separation processes take place in the same unit (Molinari et al., 2019). The PMR can operate in two main categories: (1) dead-end filtration and (2) cross-flow filtration. In the first mode, the feedwater is forced through the filter surface via an applied pressure. Retained particles stay behind on the filter surface while water flows through. The retained particles accumulate on the filter surface and consequently, the water experiences a greater resistance to passing through the filter. This may result in a decrease in flux. In the cross-flow mode, the feed solution passes along the surface on the membrane. The constant flow along the membrane surface prevents the fouling. A pressure difference across the element drives water though the membrane (permeate) while particles that are retained (concentrate) by the membrane continue to pass along the membrane surface (Molinari, Lavorato, & Argurio, 2017).

PMRs have been widely studied for water and wastewater treatments, removing organic pollutants, oils, or heavy metals as well as for disinfection treatments, among others. In this section, different reactors are briefly overviewed for the degradation of different contaminants.

The most commonly used PMs are based on ceramics membranes with a photoactive layer and the light source placed on the permeate side. Horovitz et al. (2016) successfully prepared a PM based on N-doped TiO_2-coated photocatalytic Al_2O_3 membrane for the degradation of carbamazepine, a widely recalcitrant pharmaceutical. The flow cell was a laboratory-scale system used in recirculating batch mode, operated in two flow configurations, and placed under light from a solar-simulator (Fig. 14.7). The main advantage of this configuration was that the photocatalytic layer of the membrane was not exposed to fouling, and thus the effectiveness of the degradation did not decrease in time. It should also be mentioned that since the treated stream is permeated, the substances exposed to the degradation are smaller than the membrane pores. Recirculating the treated water

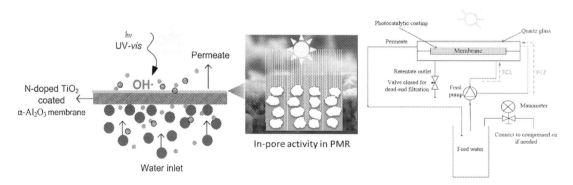

Figure 14.7

(*Left*) Carbamazepine degradation using an N-doped TiO_2-coated photocatalytic membrane reactor. (*Right*) Schematic diagram of a laboratory photocatalytic membrane system and flow configurations used for solar photocatalytic degradation via recirculating batch operation mode. Source: *Reprinted with permission from Horovitz, I., Avisar, D., Baker, M. A., Grilli, R., Lozzi, L., Di Camillo, D., & Mamane, H. (2016). Carbamazepine degradation using a N-doped TiO_2 coated photocatalytic membrane reactor: Influence of physical parameters.* Journal of Hazardous Materials, 310, *98–107, Copyright 2016, Elsevier.*

through the PM (filtration) resulted in a significantly higher carbamazepine reaction rate (~90%). This was attributed to "in-pore" photocatalytic activity due to increased contact of molecules with the active sites caused by the flow through the porous material. The carbamazepine removal rate showed a linear relationship with temperature despite the decrease in dissolved oxygen concentration.

Athanasekou et al. (2014, 2015) demonstrate significant improvement, in particular for visible light efficiency, of a hybrid photocatalysis/UF process for the elimination of methyl orange and methylene blue from water. The purification device allowed the performance of photocatalytic tests in the cross-flow and dead-end filtration mode with irradiation applied on both surfaces of the monochannel monolith. The total flow of polluted water feeding the reactor was performed in the dead-end mode while the membranes were examined in the tangential flow (cross-flow). The total amounts of removed pollutant were obtained from the mass balance between the feed, retentate, and permeate side of membrane. It has been observed that only the membrane developed on the UF monolith with 10-nm pore size exhibited enhanced photocatalytic performance under visible light. The performance of the novel hybrid process was compared with that of standard NF with respect to pollutant removal efficiency and energy consumption, providing firm evidence for its economic feasibility and efficiency.

Wang, Guan, Chen, and Bai (2016) reported the retention and separation of dye 4BS from wastewater by using nitrogen-doped TiO_2 ceramic membranes prepared with dip-coating method as previously reported (Athanasekou et al., 2015). The filtration experiments were carried out in a dead-end mode under UV light obtaining a very good permeation flux (25 L/m^2 per hour, over 96% of pure water flux), very high dye rejection (close to 99%), and good membrane stability after several time reuse was obtained. However, a rapid formation of cake layer due to the dead-end filtration mode was observed.

Agana et al. (2013) evaluated the performance of TiO_2 ceramic UF membrane for the treatment of beverage wastewater. A schematic diagram of the ceramic cross-flow UF rig used in the experiments is shown in Fig. 14.8. Membrane permeate was collected into a container and weighed while reject water was returned into the feed tank to facilitate the increase of feedwater concentration and fast track the rate of membrane fouling. Results show that certain combinations of higher CFV and TMP provide significant improvements in permeate flux and contaminant rejection rates. These improvements are attributed to the turbulent flow regime experienced by suspended particles during UF.

Fischer et al. (2015) evaluated the fouling behavior of the TiO_2 polymeric membranes for the removal of pharmaceuticals ibuprofen and diclofenac as well as methylene blue dye. Due to the on-site synthesis of TiO_2 on the surface of the membrane, no catalyst is lost and the membrane is protected from UV damage. Hydrophilic membranes showed very good photocatalytic activity for ibuprofen and diclofenac. Two or more permeate recycle were

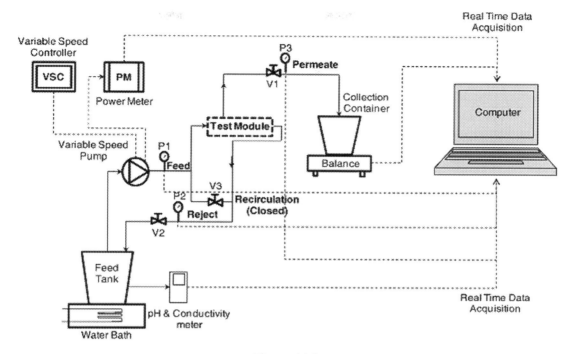

Figure 14.8

Schematic diagram of the ceramic ultrafiltration rig used in the experiments. P1—feed pressure; P2—concentrate pressure; P3—permeate pressure; V1—permeate valve; V2—concentrate valve; V3—recirculation valve. Solid lines represent water flow while broken lines represent real-time data acquisition. Source: *Reprinted with permission from Agana, B. A., Reeve, D., & Orbell, J. D. (2013). Performance optimization of a 5nm TiO₂ ceramic membrane with respect to beverage production wastewater.* Desalination, 311, 162−172, Copyright 2013, Elsevier.

need to obtain acceptable permeate quality (Athanasekou et al., 2015). No fouling studies during photodegradation tests were reported.

Since real effluents usually contain salts and dissolved organic matter together organic pollutants, the effects of these substances on PM performances have also to be considered. Pastrana-Martínez et al. (2015) evaluated the presence of NaCl on the removal of a pharmaceutical compound, diphenhydramine, and an organic dye, methyl orange under both near-UV/Vis and visible light irradiation over flat sheet cellulose membrane modified with TiO_2 and GO. Photocatalytic tests were carried out in a dead-end continuous mode (Fig. 14.9). The system consists of a glass cylindrical reactor with the membrane attached by using a Viton o-ring. The irradiation source was a medium-pressure mercury vapor lamp. DW, simulated brackish water (0.5 g/L NaCl), and seawater (35 g/L NaCl) were considered as aqueous matrixes. The results showed that all the prepared membranes presented high activity and stability in consecutive light−dark cycles under continuous

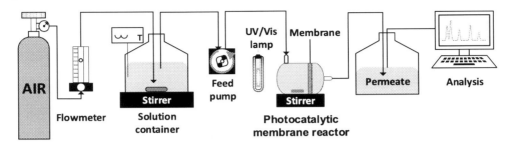

Figure 14.9

Photocatalytic membrane reactor equipped with a polymer membrane with photocatalytically active separation layer. Source: *Reprinted with permission from Pastrana-Martínez, L. M., Morales-Torres, S., Figueiredo, J. L., Faria, J. L., & Silva, A. M. T. (2015). Graphene oxide based ultrafiltration membranes for photocatalytic degradation of organic pollutants in salty water.* Water Research, 77, 179−190, *Copyright 2015, Elsevier.*

mode. During the photocatalytic reaction, the permeate flux increased due to the high hydrophilicity of the membranes and larger contaminant removal by photodegradation. The presence of NaCl (0.5 g/L) leads to a slight decrease in methyl orange degradation, regardless of the membranes employed, since Cl^- ions acted as holes and hydroxyl radical scavengers.

An interesting and promising approach, in view of enhancing the photocatalytic activity of PM, is based on the implementation and use of fiber-based membranes.

Zhang et al. (2016) evaluated the performance of TiO_2 nanofiber membranes for the degradation of HA. The activity of the membranes was evaluated using a photocatalysis system as shown in Fig. 14.10. The membrane module was used as the photocatalytic reactor, which was surrounded by the four mercury lamps. A TiO_2 nanofiber membrane sealed at one end was placed into the quartz reactor. Water was poured into the reactor through a pump and the other end of the unclosed membrane was connected with another pump. Compared with the 29% removal of HA by filtration alone, a HA removal rate of nearly 90% was achieved. The HA removal was enhanced under UV irradiation, likely due to enhanced performance of filtration along with the simultaneous photocatalytic degradation of HA, and the use of TiO_2 nanofibers provided a better reduction of membrane fouling in the presence of UV light irradiation.

14.4 Conclusions and future perspectives

In view of real applications, PMs resulted from the photocatalyst immobilization on/into a ceramic or polymeric membrane present a dual action over pollutants: retention/rejection/repulsion and photodegradation, providing an enhanced treatment of wastewater based on coupling both technologies, photocatalysis and membrane filtration. Compared to

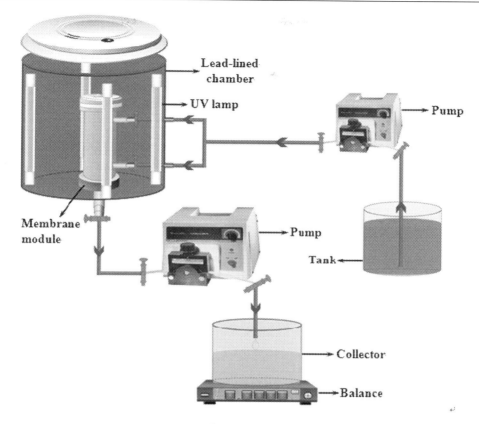

Figure 14.10
Schematic diagram of the TiO$_2$ nanofiber membranes module for water treatment. Source: *Reprinted with permission from Zhang, Q., Wang, H., Fan, X., Lv, F., Chen, S., & Quan, X. (2016). Fabrication of TiO$_2$ nanofiber membranes by a simple dip-coating technique for water treatment.* Surface and Coatings Technology, 298, 45–52, Copyright 2016, Elsevier.

conventional separation membranes, PMs possess unique properties, such as antibiofouling behavior, less concentrated retentate effluent, higher flux, and often more hydrophilic character, among others. Immobilizing the photocatalytic nanoparticles on membranes also avoids the photocatalyst separation/recovery process needed when used in slurry, which is one of the main difficulties of photocatalytic processes, thus advancing their practical applications. The use of TiO$_2$ and TiO$_2$-based catalyst has gained considerable attention to the development of mainly ceramic/inorganic and polymeric PMs using several immobilizing approaches, such as dip coating and PI, among others. Inorganic membranes are preferable for the degradation of different organic pollutants, such as dyes, pharmaceuticals, and other pollutants, owing to their stability during long-term operation under UV−Vis irradiation.

In spite of the considerable fast advancement, there are still several challenges in the synthesis and application of TiO_2-based PMs for highly efficient water treatment. First, different fabrication strategies have diverse effects on the membrane activity. Thus a stable TiO_2 immobilization method without affecting TiO_2 performance is a main goal for the preparation of TiO_2-coated PMs. Second, the studies of the PMs mainly are based on laboratory experiments including their preparation and characterization, as well as the treatment effect of synthetic wastewater in the dead-end or cross-flow filtration systems. Finally, the performance and stability of the PMs should be evaluated during long-term operation conditions, providing the results on the viability of scaling up.

Acknowledgments

This work was supported by the Projects ref. RTI2018-099224-B-I00 from ERDF/Ministry of Science, Innovation and Universities—State Research Agency and ref. PCI2020-112045 from the PRIMA Programme supported by European Union. LMPM (RYC-2016-19347) and SMT (RYC-2019-026634-I/AEI/10.13039/501100011033) acknowledge to the Spanish Ministry of Economy and Competitiveness (MINECO), the State Research Agency and the European Social Found for their Ramon y Cajal research contracts. A.P.-M. is grateful to Ministry of Science, Innovation and Universites for a predoctoral fellowship (ref. PRE2019-087946).

References

Agana, B. A., Reeve, D., & Orbell, J. D. (2013). Performance optimization of a 5nm TiO_2 ceramic membrane with respect to beverage production wastewater. *Desalination, 311*, 162−172.

Ahn, C. H., Baek, Y., Lee, C., Kim, S. O., Kim, S., Lee, S., ... Yoon, J. (2012). Carbon nanotube-based membranes: Fabrication and application to desalination. *Journal of Industrial and Engineering Chemistry, 18*, 1551−1559.

Albu, S. P., Ghicov, A., Macak, J. M., Hahn, R., & Schmuki, P. (2007). Self-organized, free-standing TiO_2 nanotube membrane for flow-through photocatalytic applications. *Nano Letters, 7*, 1286−1289.

Alias, S. S., Harun, Z., & Latif, I. S. A. (2018). Characterization and performance of porous photocatalytic ceramic membranes coated with TiO_2 via different dip-coating routes. *Journal of Materials Science, 53*, 11534−11552.

An, S., Lee, M. W., Joshi, B. N., Jo, A., Jung, J., & Yoon, S. S. (2014). Water purification and toxicity control of chlorophenols by 3D nanofiber membranes decorated with photocatalytic titania nanoparticles. *Ceramics International, 40*, 3305−3313.

Athanasekou, C. P., Morales-Torres, S., Likodimos, V., Romanos, G. E., Pastrana-Martinez, L. M., Falaras, P., ... Silva, A. M. T. (2014). Prototype composite membranes of partially reduced graphene oxide/TiO_2 for photocatalytic ultrafiltration water treatment under visible light. *Applied Catalysis B: Environmental, 158−159*, 361−372.

Athanasekou, C. P., Moustakas, N. G., Morales-Torres, S., Pastrana-Martínez, L. M., Figueiredo, J. L., Faria, J. L., ... Falaras, P. (2015). Ceramic photocatalytic membranes for water filtration under UV and visible light. *Applied Catalysis B: Environmental, 178*, 12−19.

Baek, Y., Kim, C., Seo, D. K., Kim, T., Lee, J. S., Kim, Y. H., ... Yoon, J. (2014). High performance and antifouling vertically aligned carbon nanotube membrane for water purification. *Journal of Membrane Science, 460*, 171−177.

Chang, Q., Zhou, J. E., Wang, Y., Liang, J., Zhang, X., Cerneaux, S., ... Dong, Y. (2014). Application of ceramic microfiltration membrane modified by nano-TiO_2 coating in separation of a stable oil-in-water emulsion. *Journal of Membrane Science, 456*, 128−133.

Daels, N., Radoicic, M., Radetic, M., Van Hulle, S. W. H., & De Clerck, K. (2014). Functionalisation of electrospun polymer nanofibre membranes with TiO_2 nanoparticles in view of dissolved organic matter photodegradation. *Separation and Purification Technology, 133*, 282−290.

Darling, S. B. (2018). Perspective: Interfacial materials at the interface of energy and water. *Journal of Applied Physics, 124*, 030901.

Das, R., Ali, M. E., Hamid, S. B. A., Ramakrishna, S., & Chowdhury, Z. Z. (2014). Carbon nanotube membranes for water purification: A bright future in water desalination. *Desalination, 336*, 97−109.

De Filpo, G., Pantuso, E., Armentano, K., Formoso, P., Di Profio, G., Poerio, T., ... Nicoletta, F. P. (2018). Chemical vapor deposition of photocatalyst nanoparticles on PVDF membranes for advanced oxidation processes. *Membranes, 8*, 35.

Djafer, L., Ayral, A., & Ouagued, A. (2010). Robust synthesis and performance of a titania-based ultrafiltration membrane with photocatalytic properties. *Separation and Purification Technology, 75*, 198−203.

Dong, P., Huang, Z., Nie, X., Cheng, X., Jin, Z., & Zhang, X. (2019). Plasma enhanced decoration of nc-TiO_2 on electrospun PVDF fibers for photocatalytic application. *Materials Research Bulletin, 111*, 102−112.

Dzinun, H., Othman, M. H. D., Ismail, A. F., Puteh, M. H., Rahman, M. A., & Jaafar, J. (2017). Stability study of PVDF/TiO_2 dual layer hollow fibre membranes under long-term UV irradiation exposure. *Journal of Water Process Engineering, 15*, 78−82.

Elimelech, M., & Phillip, W. A. (2011). The future of seawater desalination: Energy, technology, and the environment. *Science (New York, N.Y.), 333*, 712−717.

Emadzadeh, D., Lau, W. J., Matsuura, T., Rahbari-Sisakht, M., & Ismail, A. F. (2014). A novel thin film composite forward osmosis membrane prepared from PSf-TiO_2 nanocomposite substrate for water desalination. *Chemical Engineering Journal, 237*, 70−80.

Fan, W., Lai, Q., Zhang, Q., & Wang, Y. (2011). Nanocomposites of TiO_2 and reduced graphene oxide as efficient photocatalysts for hydrogen evolution. *The Journal of Physical Chemistry C, 115*, 10694−10701.

Fischer, K., Grimm, M., Meyers, J., Dietrich, C., Gläser, R., & Schulze, A. (2015). Photoactive microfiltration membranes via directed synthesis of TiO_2 nanoparticles on the polymer surface for removal of drugs from water. *Journal of Membrane Science, 478*, 49−57.

Fujishima, A., Rao, T. N., & Tryk, D. A. (2000). Titanium dioxide photocatalysis. *Journal of Photochemistry and Photobiology C: Photochemistry Reviews, 1*, 1−21.

Gao, Y., Hu, M., & Mi, B. (2014). Membrane surface modification with TiO_2−graphene oxide for enhanced photocatalytic performance. *Journal of Membrane Science, 455*, 349−356.

Gao, P., Liu, Z., Tai, M., Sun, D. D., & Ng, W. (2013). Multifunctional graphene oxide−TiO_2 microsphere hierarchical membrane for clean water production. *Applied Catalysis B: Environmental, 138−139*, 17−25.

Gupta, S., & Tripathi, M. (2012). An overview of commonly used semiconductor nanoparticles in photocatalysis. *High Energy Chem, 46*, 1−9.

Herrmann, J. M. (2005). Heterogeneous photocatalysis: State of the art and present applications In honor of Pr. R.L. Burwell Jr. (1912−2003), Former Head of Ipatieff Laboratories, Northwestern University, Evanston (Ill). *Topics in Catalysis, 34*, 49−65.

Horovitz, I., Avisar, D., Baker, M. A., Grilli, R., Lozzi, L., Di Camillo, D., & Mamane, H. (2016). Carbamazepine degradation using a N-doped TiO_2 coated photocatalytic membrane reactor: Influence of physical parameters. *Journal of Hazardous Materials, 310*, 98−107.

Humplik, T., Lee, J., O'Hern, S. C., Fellman, B. A., Baig, M. A., Hassan, S. F., ... Wang, E. N. (2011). Nanostructured materials for water desalination. *Nanotechnology, 22*, 292001.

Kim, J., & Van der Bruggen, B. (2010). The use of nanoparticles in polymeric and ceramic membrane structures: Review of manufacturing procedures and performance improvement for water treatment. *Environmental Pollution, 158*, 2335−2349.

Konstantinou, I. K., & Albanis, T. A. (2004). TiO_2-assisted photocatalytic degradation of azo dyes in aqueous solution: Kinetic and mechanistic investigations: A review. *Applied Catalysis B: Environmental, 49*, 1−14.

Kuvarega, A. T., Khumalo, N., Dlamini, D., & Mamba, B. B. (2018). Polysulfone/N,Pd co-doped TiO_2 composite membranes for photocatalytic dye degradation. *Separation and Purification Technology, 191*, 122−133.

Le-Clech, P., Lee, E.-K., & Chen, V. (2006). Hybrid photocatalysis/membrane treatment for surface waters containing low concentrations of natural organic matters. *Water Research, 40*, 323−330.

Long, M., Qin, Y., Chen, C., Guo, X., Tan, B., & Cai, W. (2013). Origin of visible light photoactivity of reduced graphene oxide/TiO_2 by in situ hydrothermal growth of undergrown TiO_2 with graphene oxide. *The Journal of Physical Chemistry C, 117*, 16734−16741.

Michael-Kordatou, I., Iacovou, M., Frontistis, Z., Hapeshi, E., Dionysiou, D. D., & Fatta-Kassinos, D. (2015). Erythromycin oxidation and ERY-resistant Escherichia coli inactivation in urban wastewater by sulfate radical-based oxidation process under UV-C irradiation. *Water Research, 85*, 346−358.

Molinari, R., Lavorato, C., & Argurio, P. (2017). Recent progress of photocatalytic membrane reactors in water treatment and in synthesis of organic compounds. A review. *Catalysis Today, 281*, 144−164.

Molinari, R., Lavorato, C., Argurio, P., Szymański, K., Darowna, D., & Mozia, S. (2019). Overview of photocatalytic membrane reactors in organic synthesis, energy storage and environmental applications. *Catalysts, 9*, 239.

Mozia, S. (2010). Photocatalytic membrane reactors (PMRs) in water and wastewater treatment. A review. *Separation and Purification Technology, 73*, 71−91.

Nakata, K., & Fujishima, A. (2012). TiO_2 photocatalysis: Design and applications. *Journal of Photochemistry and Photobiology C: Photochemistry Reviews, 13*, 169−189.

Neubert, S., Pliszka, D., Thavasi, V., Wintermantel, E., & Ramakrishna, S. (2011). Conductive electrospun PANi-PEO/TiO2 fibrous membrane for photo catalysis. *Materials Science and Engineering: B, 176*, 640−646.

Ong, C. S., Lau, W. J., Goh, P. S., Ng, B. C., & Ismail, A. F. (2015). Preparation and characterization of PVDF−PVP−TiO_2 composite hollow fiber membranes for oily wastewater treatment using submerged membrane system. *Desalination and Water Treatment, 53*, 1213−1223.

Padaki, M., Emadzadeh, D., Masturra, T., & Ismail, A. F. (2015). Antifouling properties of novel PSf and TNT composite membrane and study of effect of the flow direction on membrane washing. *Desalination, 362*, 141−150.

Pan, J. H., Zhang, X., Du, A. J., Sun, D. D., & Leckie, J. O. (2008). Self-etching reconstruction of hierarchically mesoporous F-TiO_2 hollow microspherical photocatalyst for concurrent membrane water purifications. *Journal of the American Chemical Society, 130*, 11256−11257.

Papageorgiou, S. K., Katsaros, F. K., Favvas, E. P., Romanos, G. E., Athanasekou, C. P., Beltsios, K. G., ... Falaras, P. (2012). Alginate fibers as photocatalyst immobilizing agents applied in hybrid photocatalytic/ultrafiltration water treatment processes. *Water Research, 46*, 1858−1872.

Pastrana-Martínez, L. M., Morales-Torres, S., Figueiredo, J. L., Faria, J. L., & Silva, A. M. T. (2015). Graphene oxide based ultrafiltration membranes for photocatalytic degradation of organic pollutants in salty water. *Water Research, 77*, 179−190.

Pastrana-Martínez, L. M., Morales-Torres, S., Kontos, A. G., Moustakas, N. G., Faria, J. L., Doña-Rodríguez, J. M., ... Silva, A. M. T. (2013). TiO_2, surface modified TiO_2 and graphene oxide-TiO_2 photocatalysts for degradation of water pollutants under near-UV/Vis and visible light. *Chemical Engineering Journal, 224*, 17−23.

Pastrana-Martínez, L. M., Morales-Torres, S., Likodimos, V., Falaras, P., Figueiredo, J. L., Faria, J. L., & Silva, A. M. T. (2014). Role of oxygen functionalities on the synthesis of photocatalytically active graphene−TiO_2 composites. *Applied Catalysis B: Environmental, 158−159*, 329−340.

Pastrana-Martínez, L. M., Morales-Torres, S., Likodimos, V., Figueiredo, J. L., Faria, J. L., Falaras, P., & Silva, A. M. T. (2012). Advanced nanostructured photocatalysts based on reduced graphene oxide-TiO_2 composites for degradation of diphenhydramine pharmaceutical and methyl orange dye. *Applied Catalysis B: Environmental, 123−124*, 241−256.

Pastrana-Martínez, L. M., Morales-Torres, S., Papageorgiou, S. K., Katsaros, F. K., Romanos, G. E., Figueiredo, J. L., ... Silva, A. M. T. (2013). Photocatalytic behaviour of nanocarbon-TiO_2 composites and immobilization into hollow fibres. *Applied Catalysis B: Environmental, 142−143*, 101−111.

Pendergast, M. M., & Hoek, E. M. V. (2011). A review of water treatment membrane nanotechnologies. *Energy & Environmental Science, 4*, 1946−1971.

Pereira, V. R., Isloor, A. M., Ahmed, A. A., & Ismail, A. F. (2015). Preparation, characterization and the effect of PANI coated TiO_2 nanocomposites on the performance of polysulfone ultrafiltration membranes. *New Journal of Chemistry, 39*, 703−712.

Peters, T. (2010). Membrane technology for water treatment. *Chemical Engineering & Technology (Elmsford, N.Y.), 33*, 1233−1240.

Pourjafar, S., Rahimpour, A., & Jahanshahi, M. (2012). Synthesis and characterization of PVA/PES thin film composite nanofiltration membrane modified with TiO_2 nanoparticles for better performance and surface properties. *Journal of Industrial and Engineering Chemistry, 18*, 1398−1405.

Rao, G., Zhang, Q., Zhao, H., Chen, J., & Li, Y. (2016). Novel titanium dioxide/iron (III) oxide/graphene oxide photocatalytic membrane for enhanced humic acid removal from water. *Chemical Engineering Journal, 302*, 633−640.

Riaz, S., & Park, S.-J. (2020). An overview of TiO_2-based photocatalytic membrane reactors for water and wastewater treatments. *Journal of Industrial and Engineering Chemistry, 84*, 23−41.

Salim, N. E., Nor, N. A. M., Jaafar, J., Ismail, A. F., Qtaishat, M. R., Matsuura, T., . . . Yusof, N. (2019). Effects of hydrophilic surface macromolecule modifier loading on PES/O-g-C_3N_4 hybrid photocatalytic membrane for phenol removal. *Applied Surface Science, 465*, 180−191.

Sampaio, M. J., Silva, C. G., Silva, A. M. T., Pastrana-Martínez, L. M., Han, C., Morales-Torres, S., . . . Faria, J. L. (2015). Carbon-based TiO_2 materials for the degradation of Microcystin-LA. *Applied Catalysis B: Environmental, 170−171*, 74−82.

Shannon, M. A., Bohn, P. W., Elimelech, M., Georgiadis, J. G., Marinas, B. J., & Mayes, A. M. (2008). Science and technology for water purification in the coming decades. *Nature, 452*, 301−310.

Singh, R., Yadav, V. S. K., & Purkait, M. K. (2019). Cu_2O photocatalyst modified antifouling polysulfone mixed matrix membrane for ultrafiltration of protein and visible light driven photocatalytic pharmaceutical removal. *Separation and Purification Technology, 212*, 191−204.

Starr, B. J., Tarabara, V. V., Zhou, M., Roualdès, S., & Ayral, A. (2016). Coating porous membranes with a photocatalyst: Comparison of LbL self-assembly and plasma-enhanced CVD techniques. *Journal of Membrane Science, 514*, 340−349.

Tennakone, K., Tilakaratne, C. T. K., & Kottegoda, I. R. M. (1995). Photocatalytic degradation of organic contaminants in water with TiO_2 supported on polythene films. *Journal of Photochemistry and Photobiology A: Chemistry, 87*, 177−179.

Tsydenova, O., Batoev, V., & Batoeva, A. (2015). Solar-enhanced advanced oxidation processes for water treatment: Simultaneous removal of pathogens and chemical pollutants. *International Journal of Environmental Research and Public Health, 12*, 9542.

Ulbricht, M. (2006). Advanced functional polymer membranes. *Polymer, 47*, 2217−2262.

Vatanpour, V., Madaeni, S. S., Khataee, A. R., Salehi, E., Zinadini, S., & Monfared, H. A. (2012). TiO_2 embedded mixed matrix PES nanocomposite membranes: Influence of different sizes and types of nanoparticles on antifouling and performance. *Desalination, 292*, 19−29.

Wang, Z.-b, Guan, Y.-j, Chen, B., & Bai, S.-l (2016). Retention and separation of 4BS dye from wastewater by the N-TiO_2 ceramic membrane. *Desalination and Water Treatment, 57*, 16963−16969.

Xiang, Q., Yu, J., & Jaroniec, M. (2012). Graphene-based semiconductor photocatalysts. *Chemical Society Reviews, 41*, 782−796.

Xu, J., Chang, C.-Y., Hou, J., & Gao, C. (2013). Comparison of approaches to minimize fouling of a UF ceramic membrane in filtration of seawater. *Chemical Engineering Journal, 223*, 722−728.

Xu, Z., Sun, Y., Zhuang, Y., Jing, W., Ye, H., & Cui, Z. (2018). Assembly of 2D MXene nanosheets and TiO2 nanoparticles for fabricating mesoporous TiO_2-MXene membranes. *Journal of Membrane Science, 564*, 35−43.

Xu, Z., Wu, T., Shi, J., Teng, K., Wang, W., Ma, M., . . . Fan, J. (2016). Photocatalytic antifouling PVDF ultrafiltration membranes based on synergy of graphene oxide and TiO_2 for water treatment. *Journal of Membrane Science, 520*, 281−293.

Zhang, G., Sheng, H., Chen, D., Li, N., Xu, Q., Li, H., . . . Lu, J. (2018). Hierarchical titanium dioxide nanowire/metal—organic framework/carbon nanofiber membranes for highly efficient photocatalytic degradation of hydrogen sulfide. *Chemistry—A European Journal*, *24*, 15019—15025.

Zhang, H., Quan, X., Chen, S., & Zhao, H. (2006). Fabrication and characterization of silica/titania nanotubes composite membrane with photocatalytic capability. *Environmental Science & Technology*, *40*, 6104—6109.

Zhang, W., Ding, L., Luo, J., Jaffrin, M. Y., & Tang, B. (2016). Membrane fouling in photocatalytic membrane reactors (PMRs) for water and wastewater treatment: A critical review. *Chemical Engineering Journal*, *302*, 446—458.

Zhang, X., Wang, D. K., & Diniz da Costa, J. C. (2014). Recent progresses on fabrication of photocatalytic membranes for water treatment. *Catalysis Today*, *230*, 47—54.

Zhang, X., Wang, D. K., Lopez, D. R. S., & Diniz da Costa, J. C. (2014). Fabrication of nanostructured TiO_2 hollow fiber photocatalytic membrane and application for wastewater treatment. *Chemical Engineering Journal*, *236*, 314—322.

Zhang, Q., Wang, H., Fan, X., Lv, F., Chen, S., & Quan, X. (2016). Fabrication of TiO_2 nanofiber membranes by a simple dip-coating technique for water treatment. *Surface and Coatings Technology*, *298*, 45—52.

CHAPTER 15

Novel photocatalysts for indoor air clean and healthy environments

Vassilios Binas[1,2], Danae Venieri[3], Dimitrios Kotzias[4] and George Kiriakidis[1]

[1]*Institute of Electronic Structure and Laser, Foundation for Research and Technology-Hellas (FORTH-IESL), Irákleion, Greece,* [2]*Department of Physics, University of Crete, Herakleio, Greece,* [3]*School of Environmental Engineering, Technical University of Crete, Chania, Greece,* [4]*European Commission-Joint Research Centre, Institute for Health and Consumer Protection, Ispra, Italy*

15.1 Indoor air environment

Indoor air pollution has been recognized as an emerging environmental health issue. People living in urban areas spend typically 85%–90% of their time indoors and are exposed to a variety of pollutants with known health effects emitted not only from outdoor sources but also from sources in indoor spaces. While air quality guidelines and standards are widely used in outdoor air quality management, systematic science-based approaches for indoor air quality (IAQ) are still in the phase of recommendations. In the last decades, substantial work was done to identify and quantify the main indoor air contaminants, to evaluate human exposure and assess the risk for human health as well as to define strategies to facing the contamination with pollutants in indoor environments. The European Commission with its Environment and Health Action Plan (2004–2010) and the World Health Organization (WHO Guidelines, 2010) significantly contributed to the understanding of the issue and set initiatives to reducing/eliminating the pollution indoors (World Health Organization, Regional Office, 2010).

Chemical and biological compounds are considered to be the relevant factors affecting comfort and well-being in confined spaces. A high number of chemicals and biologically originated compounds have been identified indoors, belonging to different chemical classes (Kotzias, 2012) such as:

- Gaseous inorganic compounds: CO_2, CO, nitrogen oxides, SO_2, ammonia, ozone
- VOCs, including carbonyls.
- Particulate matter (PM10, PM 2,5, PM1, ultra-fine particles/nanoparticles) and compounds bound to PM (SVOCs).
- Asbestos, polycyclic aromatic hydrocarbons (PAHs), pesticides, flame retardants.

Photocatalytic Systems by Design.
DOI: https://doi.org/10.1016/B978-0-12-820532-7.00010-2
407

- Radioactive elements: radon.
- Biological contaminants: substances produced by mold, allergens, and endotoxins.
- Pathogens: viruses, bacteria, and fungi.

In 80%−90% of all investigated confined environments, the indoor air concentrations of chemicals are in the $\mu g/m^3$ range. In many cases, indoor air concentrations exceeding the corresponding air concentrations of some pollutants outdoors were measured.

Toxicological data for individual chemical compounds relevant to the indoor and outdoor environment are well known. However, toxicological data for chemical mixtures are very scarce and mostly for very high doses that do not represent the real indoor environments. Since we are mostly exposed to a mixture of pollutants (cocktail effect), future strategies and technological solutions should follow a holistic approach, that is, not to reduce or eliminate single compounds, but to clean the indoor environment as a whole.

Based on the risk characterization, the European Commission (The INDEX project) and the WHO Guidelines for selected pollutants indoors have established a list of compounds relevant for indoor environments (Koistinen et al., 2008) that includes formaldehyde, nitrogen dioxide, carbon monoxide, benzene, naphthalene, trichloroethylene, tetrachloroethylene, and PAHs (benzo(a)pyrene). Guideline values for these compounds are reported in the WHO Guidelines for selected pollutants.

In the course of the last years, an additional factor—saving energy in buildings—directly affecting IAQ, became important. Saving energy in buildings has partly been achieved with the construction of more air-tight buildings and the appropriate adaptation of elaborate ventilation regimes. In addition, an attempt has been made to promote the development of low emitting products to be used in indoor environments. In this context, relevant Directives/Regulations such as The Energy Performance of Buildings Directive along with the recast Directive on Energy Performance of Buildings, the Construction Products Regulation (CPR), the guidelines for IAQ, and the EU Green Paper on Smoking ban can be applied/integrated at European and national level to facilitate/ensure a healthy indoor environment (Construction Products Regulation CPR, 2011; Energy Performance of Buildings Directive, 2002). Moreover, the recast Directive on Energy Performance of Buildings foresees, that by the end of 2020 (2018 for buildings occupied and owned by public authorities), all new buildings should comply with the Energy Performance of Buildings Directive obligations and thus meet "nearly zero-energy" performance levels using innovative, cost-optimal technologies with integration of renewable energy sources on site or nearby. The performance of innovative technologies and the applied solutions for highly energy-performing buildings should not, however, compromise IAQ, well-being, and comfort.

On the basis of the information available right now, the contribution of poor IAQ to the total burden of disease cannot be quantified with a high degree of certainty. Short-term exposure to single compounds does not reflect all real health risks. Toxicological studies based on chronic low-dose exposure and exposure to mixtures of pollutants would provide more in-depth information on possible long-term effects of air contaminants at concentrations typical for indoor environments. Hence, in future work, the focus should be on combined exposure to chemical mixtures (cocktail effect) and the possibilities, through innovative technological solutions, to effectively clean indoor air as a whole.

This could be achieved by combining epidemiological, chemical, and bio-monitoring studies along with cost-effective technologies and solutions that could be applied in buildings satisfying the energy requirements. Thus, the challenge for the coming years is set to considering, among others, the peculiar economic situation and the herewith emerging societal needs.

Biological contaminants identified in the indoor air environment pose a significant health risk, as they have been associated with the occurrence and transmission of severe human diseases. The most commonly used term for those contaminants is "bioaerosols," which could be liquid, solid, or a mixture, constituted by suspensions of various pathogenic microorganisms, airborne toxins, and other allergens (Zemouri, De Soet, Crielaard, & Laheij, 2017). The diameter of bioaerosols ranges from 0.02 to 100 μm and they can act as infectious agents through inhalation or ingestion, transferring multiple viruses, bacteria, fungi, or even protozoa (Sattar, 2016). Nowadays, there is a general concern regarding diseases caused by bioaerosols due to socio-economic reasons. Up until now, the recorded epidemiological data are referred to even lethal outbreaks, documented worldwide, including the severe acute respiratory syndrome (SARS), influenza A (H1N1), tuberculosis, legionellosis, the recent CORVID19, and many others (Lee et al., 2019; Prussin, Schwake, & Marr, 2017).

The list of airborne pathogenic microorganisms is still growing with multiple consequences towards public health. *Legionella* species are registered in the top, described as a threat to elderly and patients with respiratory complaints. They are Gram-positive bacteria, which may be found in bioaerosols produced by instruments and air-conditioning systems that use tap water (Zemouri et al., 2017). Emerging pathogens like *Clostridium difficile*, *Mycobacterium tuberculosis*, or *Bacillus anthracis* have already been detected in indoor air environments as important vehicles of human diseases (Ijaz, Zargar, Wright, Rubino, & Sattar, 2016). Moreover, antibiotic-resistant bacteria (ARB) are often isolated from such surroundings, being responsible for the spread and prevalence of antibiotic resistance genes (ARGs) (Gupta, Lee, Bisesi, & Lee, 2019). Multi-drug microorganisms, such as methicillin-resistant *Staphylococcus aureus* or *Klebsiella* species, have become one of the major concerns of this century and many efforts are made for their control and successful elimination (Stockwell et al., 2019; Venieri et al., 2017). Other microorganisms contained in bioaerosols are fungi, with *Aspergillus*, *Penicillium*, and *Cladosporium* appearing as the predominant fungal genera in settled dust sampled in buildings and underground spaces (Chatzidiakou, Mumovic, Summerfield, Tàubel, & Hyvärinen, 2015; Radwan & Abdel-Aziz, 2019; Zhang, Wu, et al., 2020). Fungal spores can be spread and travel to long distances, being significantly virulent in immune-compromised patients and vulnerable individuals. Furthermore, different fungal components detected in the air, like mycotoxins, ergosterols, or glucans are of similar importance, as they are able of generating severe symptoms (Ijaz et al., 2016). Viruses, allergens, and endotoxins are included in measurable "bio-components" of aerosols, which are recovered from settled dust and act in a synergistic way with other severe agents, affecting human health (Williams et al., 2016).

All those hazardous airborne contaminants actually verify the initial estimation of WHO, which predicted that in the 21st century, indoor air pollution would be responsible for more diseases outbreaks in comparison with the external air pollution (Radwan & Abdel-Aziz, 2019). Indeed, the emission sources of airborne pathogens indoors (either humans or non-human sources) are much greater than those that can be identified outdoors. Also, ventilation systems harbor microorganisms and contribute in the artificial resuspension of bioaerosols, whose fate is highly influenced by air temperature, relative humidity (RH), and turbulence (Stockwell et al., 2019). Those factors are also responsible for any recorded seasonal variation of microbial concentrations in indoor environments, considering that temperature, heating systems operated in each case, and window-opening behaviors highly affect microbial survival (Chatzidiakou et al., 2015). Furthermore, pathogens should be resistant to the process of aerosolization in order to reach their host and develop symptoms. The infectious dose of each microorganism, as well as the overall health condition of the host, defines the final disease occurrence and its severity (Ijaz et al., 2016).

The increased human exposure to such various pathogens indoors necessitates the exploration and development of new purification technologies, with special emphasis on resistant and virulent microorganisms. The ultimate goal of such methods is to limit down disease-causing microorganisms and to ensure public health protection (Binas, Venieri, Kotzias, & Kiriakidis, 2017). In the last years, the use of modified TiO_2-based photocatalysts under indoor light irradiation have gained enormous attention, as they may serve as valuable novel materials for the degradation of chemical substances and as disinfection agents. This type of catalysts may lead to smart coatings, which act as benchmark materials suitable for indoor applications to degrade/eliminate of priority air pollutants indoors and to inactivate various pathogens with different structural components and resistance levels.

15.2 Indoor air treatment technologies

Indoor air pollution is a rather complex problem due to the variety of pollutants emitted from indoor sources, for example, building and construction materials, carpets, IT-devices, furniture, smoking, cooking, etc. and of pollutants which are emitted from outdoor sources and penetrate indoors.

In Europe, there are about 20,000 materials and products in the market which can be and are used in buildings. However, the market is rather imprecise; there is a distinct lack of investigation for building products and data concerning health and ecological harm. In the last years, in order to bypass existing difficulties in the evaluation of the various building and consumer products, various methodological approaches and technical means have been applied to evaluate the emission behavior of products and their possible impact on health and well-being of building occupants.

Currently, there is no technology which can be applied as a whole to clean confined spaces adequately. There are single treatment devices (air cleaners) which can be applied to eliminate some selected pollutants and particulates. A general overview of various single treatment techniques such as mechanical and electrical filtration, ozone treatment photolysis, photocatalytic oxidation, biological processes, and membrane separation is presented in a relevant report (Luengas, Barona, & Hort, 2015), as well as in other experimental studies (Ciuzas et al., 2016; Martinez, Bertron, Escadeillas, Ringot, & Simon, 2014; Zhong, Lee, & Haghighat, 2012) and in work referred therein (Lee et al., 2019; Prussin et al., 2017; Sattar, 2016; Zemouri et al., 2017). As far as the photocatalytic application for the removal of pollutants (VOCs) is concerned, experimental work until now is done using specifically developed reactors under UV-light conditions at elevated pollutants concentrations, non-typical for indoor spaces.

An efficient way to reduce/eliminate indoor air pollutants is source control. As an alternative, the impact of various sources of pollutants on IAQ can be reduced by increasing the ventilation rate. However, IAQ could also be affected when outdoor and indoor air are mixed through the ventilation system and/or open windows. In such cases, pollutants, for example, ozone produced outside penetrate indoors and may react with chemicals to form compounds with often unknown effects on human health. Since indoor spaces function per se as chemical reactors where a huge number of chemical compounds are present at the same time and may react to each other, indoor air chemistry under these conditions represents an additional factor which might play a substantial role on the deterioration of the air quality indoors.

In analogy to the work done to reducing/eliminating pollutants, for example, NOx and VOC outdoors, heterogeneous photocatalysis using modified titania (TiO_2) with novel photocatalytic properties was applied under indoor light irradiation conditions to eliminate priority air pollutants and inactivate various pathogens in indoor environments.

15.3 Photocatalysis as an effective air treatment technology

Heterogeneous photocatalysis by semiconductors, for example, TiO_2, is a promising technology. Light excitation of TiO_2, with wavelengths between 360 and 380 nm (UV-

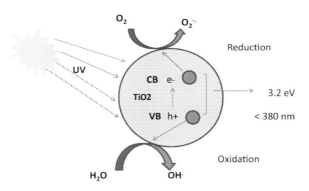

Basic processes onTiO2 surface in the presence of
water molecules and oxygen

Figure 15.1
Basic processes on TiO_2 surface in the presence of water molecule and oxygen.

light), generates electrons and positive holes in the conduction and valence bands (VBs), respectively (Fujishima, Hashimoto, & Watanabe, 1999; Hoffmann, Martin, Choi, & Bahnemannt, 1995). Excited-state conduction band (CB) electrons and VB holes are very reactive and can either recombine or diffuse to semiconductor surface where they are trapped by adsorbed molecules of, for example, water and oxygen. They initiate the formation of hydroxyl radicals and other reactive species that attack pollutant molecules and thus degrade them (Fig. 15.1).

15.4 Novel indoor photocatalysts

There are many photocatalysts reported in the literature. Among these, metal oxides such TiO_2 and ZnO, which are abundant in nature, have also been extensively used as photocatalysts, particularly as heterogeneous catalysts since 1961. In particular, the ability of TiO_2 to generate photocatalytic degradation of organic compounds in the gas phase was announced in 1964 (McLintock & Ritchie, 1965). TiO_2 has been extensively studied as an effective photocatalyst for the degradation of different classes of pollutant gases, such as inorganic (e.g., NO, NO_2, and VOC) and VOCs (e.g., aromatics, alkanes, and odor compounds) (Tsang et al., 2019), along a disinfectant media to degrade a number of pathogens (Etacheri, Michlits, Seery, Hinder, & Pillai, 2013; Fagan, McCormack, Dionysiou, & Pillai, 2016; Zhang, Yuan, et al., 2020), due to its low cost, innoxiousness, chemical inertness, and high photocatalytic performance under UV light. However, the large band gap of TiO_2 (~ 3.2 eV) and the corresponding relatively high recombination rate of the electrons and holes produced under UV illumination inhibited its wide use for indoor applications. Nevertheless, band gap engineering and introduction of suitable dopant materials into the TiO_2 lattice have recently accelerated the research interest resulting in a fair amount of interesting results oriented towards a number of novel photocatalysts suitable both for outdoor and indoor applications (Saqlain et al., 2020; Wang, Ding, & Zhao, 2020). Doping with transition metals with an incomplete d electron shell structure and a capacity to produce cations, mainly with an atomic number between 21 and 30 (i.e., Mn, V, Cr, Fe, Co, Ni, Cu, and Zn), have been proved particularly effective as indoor photocatalysts (Kato & Kudo, 2002; Kiriakidis & Binas, 2014; Osawa, Barrocas, Monteiro, Conceição Oliveira, & Florêncio, 2020; Paul, Chetri, & Choudhury, 2014; Shen, Chuang, Jiang, Liu, & Horng, 2020). In principle, the introduction of transition-metal cations into TiO_2 is changing the coordination environment of titanium in the lattice and is modifying the electronic structure of TiO_2, which leads to a shift in the light absorption edge from UV towards the visible light region (red shift, bathochromic shift). Successful indoor photocatalytic activity has also been reported utilizing various semiconductor composite photocatalysts such as 2D graphene nanosheets-supported TiO_2 (Štengl, Popelková, & Vláčil, 2011) and N-doped anatase TiO_2 (Gao, Zhou, Dai, & Qu, 2009). In the case of graphene nanosheets-supported TiO_2, it enhanced photocatalytic activity in the presence of increasing catalytic active sites on graphene or/and strengthening of interface coupling between the TiO_2 and graphene. On the other hand, in N-doped anatase TiO_2, the enhanced photocatalytic activity attributed to the formation of oxygen vacancies which upon reduction transfer electrons from the Ti^{3+} to the Ti^{4+} ionic state, thus stimulating visible light absorption, promoting electrons from the localized N-impurity states to the CB as demonstrated by EPR analysis.

Although band gap engineering, light energy, and light trapping/absorption along with electron promotion strategies have been widely studied with the development of a large range of material combinations and structures, there is a limited discussion on the striking deviation on the photocatalytic activities reported by different groups even though they may use the same doping or composite combinations. This is largely due to the fact that, in most cases, an additional critical parameter for effective photocatalytic activity, namely, the electron—hole recombination rate, is largely underestimated. Parameters that have been shown to inhibit the electron—hole recombination rate are, among other, the applied physical or chemical doping process, the type of precursor materials, as well as the pH, humidity, and the total amount of dopant concentration. Moreover, different dopant materials have been proved to have different disinfection effectiveness depending on the type of the pathogen interacting with.

Recent examples on the importance of these parameters have been reported using single, and co-doped transition element material combinations synthesizing Mn-, Fe-, Co-, and Mn—Co-doped TiO_2 nanoparticles using the co-precipitation techniques. In the case of single dopant material, it has been demonstrated that both the oxidation state and the total concentration of the dopant material are critical for enhanced photocatalytic activity (Binas, Sambani, Maggos, Katsanaki, & Kiriakidis, 2012).

The photocatalytic degradation of gaseous acetaldehyde on undoped and Mn-doped TiO_2-based materials under visible light irradiation is shown in Fig. 15.2. Clearly the most effective photocatalytic activity is related to the lowest dopant concentration of 0.1% Mn. Studying the role of Mn as a dopant material in the TiO_2 lattice, EPR spectra analysis (Fig. 15.3) has demonstrated that at relatively high concentrations ($>1\%$), Mn is entering into the TiO_2 lattice with a mix of Mn^{4+} and Mn^{2+} oxidation states, which correspond to MnO and Mn_3O_4. However, at lower concentration (i.e., 0.1%), the dominant oxidation state present is that of Mn^{2+} that has a valence electronic configuration of $3d^5$. At this state, trapping an electron, it changes its electronic configuration to d^6 while trapping a hole it changes to d^4, both highly unstable. Consequently, to restore stable electronic configuration, the trapped electron is transferred to an oxygen molecule while the trapped hole transferred to a surface-adsorbed water, thereby it generates superoxide ($O_2^{\bullet-}$) and hydroxyl ($OH^{\bullet}$) radicals, respectively (Gomathi Devi, Girish Kumar, Narasimha Murthy, & Kottam, 2009).

Fig. 15.4 shows the charged carrier distribution in the presence of Mn^{2+} ion-doped TiO_2 and its contribution to the mechanism of radicals' formation that enhances the photodegradation process.

Alternatively, Mn^{3+} can also trap CB electrons, or Mn^+ can trap VB holes to retain the half-filled electronic structure of Mn^{2+}:

$$Mn^+ + h_{VB}^+ \rightarrow Mn^{2+}$$

$$Mn^{3+} + e_{CB}^- \rightarrow Mn^{2+}$$

$$Mn^{2+} + Ti^{3+} \rightarrow Mn^+ + Ti^{4+} (\text{electron trap})$$

$$Mn^{2+} + O^- \rightarrow Mn^{3+} + O^{2-} (\text{hole trap})$$

Because Mn^{3+} and Mn^+ are less stable than Mn^{2+}, there is a tendency for the trapped charges to transfer from Mn^{3+} and Mn^+ to the interface:

$$Mn^+ + O_2 \rightarrow Mn^{2+} + O_2^{\bullet-} (\text{electron release})$$

$$Mn^{3+} + OH^- \rightarrow Mn^{2+} + {}^{\bullet}OH (\text{hole release})$$

In a corresponding study on the heterogeneous photocatalysis using TiO_2 towards the degradation of some of the most common gases such as benzene, toluene, xylene (BTX) in an indoor environment, it was reported that the most effective dopant concentration was found to be 0.1% Mn for the effective gas-degradation under visible light illumination. In addition, in the same study, it was noted at very significant observation that almost zero of the harmful byproduct (CO) formation was taking place under indoor (visible), in direct comparison to analyses on a corresponding material under outdoor/solar (UV) illumination which had shown significant byproduct formation (Binas, Stefanopoulos, Kiriakidis, & Papagiannakopoulos, 2019).

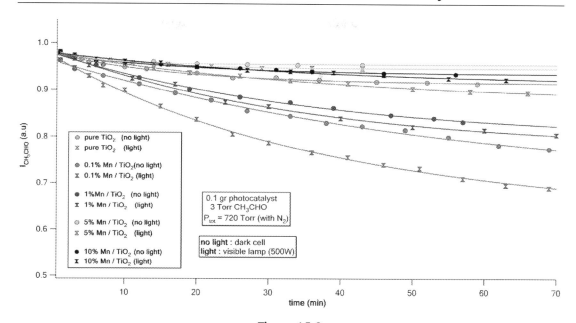

Figure 15.2

CH_3CHO photocatalytic degradation for undoped, 0.1% Mn, 1% Mn, 5% Mn, and 10% Mn TiO_2 materials under dark and visible light conditions.

Further studies on the combined effect of binary transition metal doping (such as Mn with Co) have shown an enhanced effect (under UV) by the bi-metallic presence over that of the corresponding single dopant at the same concentration as far as yield and byproduct formation were concerned (Karafas et al., 2019).

In addition to metal oxides, perovskite materials are also promising visible light-driven photocatalysts due to their electronic properties and crystal structure. The crystal structure and the ABO_3 spatial arrangement of perovskites provide a good framework, which enables the modification of the band gap of the perovskite materials to harvest the maximum portion of visible light. The spatial arrangement and the lattice distortion enhance the quantum efficiency of the resulting photocatalysts by delaying the charge carrier recombination. Titanate perovskites (Alammar, Hamm, Wark, & Mudring, 2015; Murcia-López et al., 2017), ferrite perovskites (Grabowska, 2016; Lam, Sin, & Mohamed, 2017), vanadium- and niobium-based perovskites (Feng, Wang, Luo, Shen, & Lin, 2013), and tantalate perovskites (Thirumalairajan, Girija, Hebalkar, Mangalaraj, & Viswanathan, 2013) are the main visible light-active perovskite photocatalysts that have been reported. Among the various $ATiO_3$ perovskites investigated, the visible light photocatalytic efficiency of Zn, Mg, Co, Ni, Mn, and other metals or binary, ternary, and quaternary systems were found to be very effective on the degradation of pollutants (Bora & Mewada, 2017; Misono, 2009; Sang, Kuai, & Chen, 2014; Shi, Li, & Iwai, 2009; Tanaka & Misono, 2001).

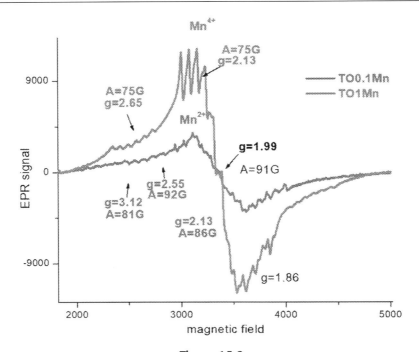

Figure 15.3
EPR spectra of low and high concentration Mn in the titania lattice.

15.5 Heterogeneous photocatalytic degradation of pollutants

15.5.1 Air cleaning

As stated above, indoor and outdoor air pollution is caused by various chemical contaminants including inorganic and organic compounds such as NOx and VOCs. For example, the harmful NOx gases can be oxidized to nitrates on the surface of TiO_2-containing/enriched construction and building materials activated by UV-light. Thus, materials may gain increasing attention, given the wide range of possible applications. They can be used as outer material in the façade of buildings in canyon streets, having the capacity to "clean" the air along the roadways that are polluted by automobile exhaust gases, "using" the solar energy only. Preliminary evidence from experiments carried out in the frame of industry-based research indicated that TiO_2-containing construction materials and paints efficiently destroy the ozone precursors NO and NO_2 up to 80% and 30%, respectively (Maggos, Bartzis, Leva, & Kotzias, 2007). However, TiO_2-containing materials and paints mostly used until now are almost exclusively developed for outdoor use. TiO_2 is the most used photocatalyst for the removal of pollutants due to its highly strong oxidative ability, chemical stability, and non-toxicity. Moreover, TiO_2 is a very common and non-expensive material. Examples of TiO_2 incorporated into buildings' envelope materials are shown in Figs. 15.5 and 15.6.

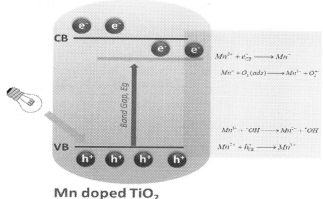

Figure 15.4

The contribution of Mn^{2+} ion-doped TiO_2 on the formation of the superoxide (O^{2-}) and hydroxyl radical (OH) species.

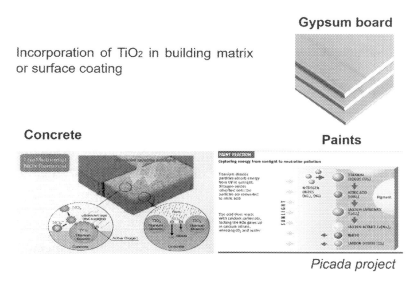

Figure 15.5

Incorporation of TiO_2 in building envelope materials.

In the past years, substantial efforts were made and still are made to further investigate the photocatalytic activity of materials containing (TiO_2) or modified TiO_2 towards priority air pollutants like NO, NO_2, and VOCs frequently accumulated in indoor environments (Binas et al., 2017). A particular asset of the studies was to modify TiO_2 in such way that it can be activated by visible light and therefore to be applied in building materials and paints mostly used indoors (Cacho et al., 2011).

Figure 15.6
Photocatalytic paint with TiO$_2$ on gypsum board. Source: *Photo by Joint Research Centre, Ispra, Italy.*

15.5.1.1 Degradation of NO and toluene in TiO$_2$-containing building materials under UV-light irradiation

The degradation of NO in TiO$_2$-containing photocatalytic materials has been documented in numerous studies (Binas, Papadaki, Maggos, Katsanaki, & Kiriakidis, 2018; Del Cacho, Geiss, Leva, Tirendi, & Barrero-Moreno, 2013; Janus, Zając, Ehm, & Stephan, 2019; Lasek, Yu, & Wu, 2013). Fig. 15.7 shows the UV-induced photocatalytic degradation of NO on the surface of an inorganic (mineral silicate, M1) and an organic (acrylic styrene, M2) paint containing 10% of TiO$_2$, respectively. M1 blk and M2 blk are TiO$_2$-free materials. After about 7 hours of UV irradiation, up to 80% of NO is photochemically degraded on the surface of the inorganic silicate (M1) while degradation of NO using the organic acrylic acrylic paint (M2) reaches up to 90%.

During the aforementioned experiments, we found that high amounts of formaldehyde are produced using the M2 acrylic paint when irradiated with UV. This is because of the degradation of the organic paint itself and represents a serious drawback using this material. Hence, in all subsequent experiments, the mineral silicate paint (M1) was used.

Humidity seems to play a significant role under these experimental conditions in the degradation of NO. Photocatalytic conversion of NO was found to increase with the decrease of RH from 50% to 20% (Fig. 15.8). On contrary, some studies have reported that NO degradation is independent of the RH, while it can significantly affect the degradation

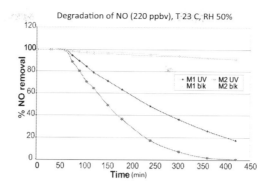

Figure 15.7
Degradation of NO (220 ppbv), *T* 23°C, RH 50. At 60 minutes, UV irradiation was activated.
Source: *Adapted from Maggos, T., Bartzis, J.G., Leva, P., & Kotzias, D. (2007). Application of photocatalytic technology for NOx removal.* Applied Physics A, 89, 81–84.

of VOCs. This probably depends on the concentration of contaminants on the TiO_2 surface and the competition with water molecules (which are a source in the photooxidation process for OH radicals and other reactive species) for adsorption sites.

A selection of basic reactions that occur in the air/semiconductor interface in the presence of H_2O, O_2, NO and NO_2 are:

$$TiO_2 \xrightarrow{h\nu} (TiO_2)^* \longrightarrow h^+ + e^-$$

$$h^+ + H_2O \longrightarrow OH^\bullet + H^+$$

$$H^+ + e^- + O_2 \longrightarrow HO_2^\bullet$$

$$2HO_2^- \longrightarrow H_2O_2 + O_2$$

$$2e^- + O_2 + 2H^+ \longrightarrow H_2O_2$$

$$2h^+ + 2H_2O \longrightarrow H_2O_2 + 2H^+$$

$$NO + HO_2 \longrightarrow NO_2 + OH^-$$

$$NO_2 + NO + H_2O \longrightarrow 2HONO$$

$$NO_2 + OH^- \longrightarrow HNO_3$$

$$NO + OH^- \longrightarrow HNO_2 \longrightarrow H^+ + NO_2^-$$

$$HNO_2 + OH^- \longrightarrow NO_2 + H_2O$$

Degradation of NO

NO 220 ppb
T: 23 C

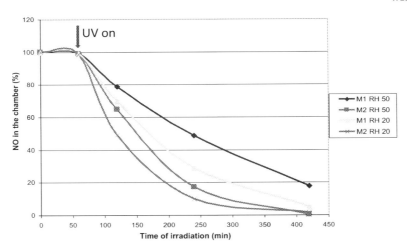

Photocatalytic conversion of NO increases while decreasing relative humidity from 50%
(10.3 g m-3) to 20% (4.1 g m-3) at 23 C.

Figure 15.8
Impact of relative humidity (RH 50%, RH 20%) on the degradation of NO, 220 ppb), T 23°C.
Source: *Adapted from Maggos, T., Bartzis, J.G., Leva, P., & Kotzias, D. (2007). Application of
photocatalytic technology for NOx removal.* Applied Physics A, 89, 81–84.

The produced radicals, for example, OH and HO_2 react with NO forming NO_2 as
intermediate, which further reacts to the final product HNO_3.

15.5.1.2 Degradation of NO under visible light conditions

The activity of TiO_2 depends on the lifetime of charge carriers—positive holes and
electrons—produced on its surface. Recombination of positive holes and electrons, as
mentioned above, occurs in an extremely short time with most charge carriers recombining
at the surface of the photocatalyst semiconductor before undergoing redox reactions. In this
case, no reaction takes place. Therefore, the primary challenge for an efficient
photocatalytic process is to reduce or inhibit the recombination of charge carriers to
maintain the photocatalyst activity at a high level.

One way to reduce or inhibit recombination is to blend/dope TiO_2 with transition metal
cations, which create traps for electrons and/or positive holes and block the charge carriers
by reducing the recombination rate. Doping (change/modification of the crystalline structure
of TiO_2) causes a bathochromic (red) shift, which results in a reduction in the energy gap,
leading to increased absorption in the visible light region. Many transition metals such as

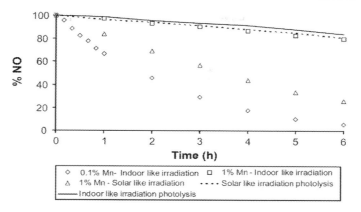

Figure 15.9

Comparison of photochemical degradation of NO (200 ppbv) with manganese (0.1% and 1% Mn) doped TiO_2 catalysts under sun and indoor-like irradiation. Source: *Adapted from Cacho, C., Geiss, O., Barrero-Moreno, J., Binas, V., Kiriakidis, G., Botalico, L., et al. (2011). Studies on photo-induced NO removal by Mn-doped TiO_2 under indoor-like illumination conditions.* Journal of Photochemistry and Photobiology A: Chemistry, 222, 304–306.

V, Cr, Fe, Mn, Ni, Co, Cu, and Zn have been explored to reduce the energy gap and facilitate the transfer of electrons to the CB and thus extend the spectral range of modified TiO_2 to the area of visible light (Fig. 15.4).

In our experiments, 0.1% (w/w) and 1% (w/w) Mn-TiO_2 admixtures were prepared and the ability of the modified photocatalysts to degrade NO by both solar and indoor illumination was evaluated (Figs. 15.9 and 15.10).

Fig. 15.4 illustrates the interaction of manganese (Mn) ions with positive and negative charges and the active radicals produced, a process that ultimately leads to a lowering of the energy gap and greater movement and availability of electric charges.

Experimental results (Fig. 15.9) clearly have shown that the 0.1% Mn-modified/doped photocatalyst is capable of degrading NO up to 95% under indoor illumination conditions, while the 1% Mn-modified/doped photocatalyst remains active only under solar radiation. According to our results, not only the transition metal (Mn) but also its concentration in the crystalline structure plays an important role in the ability of a particular photocatalyst to be active in the visible region of the spectrum. However, further research in this area will be needed to understand how the concentration of the doping compounds affects the energy gap.

Following the confirmed photocatalytic activity of TiO_2 by Mn admixtures under indoor illumination, panels were prepared containing 5% and 10% (w/w) of 0.1% Mn-modified (doped) TiO_2 photocatalyst in calcareous building material. Fig. 15.10 shows that, even

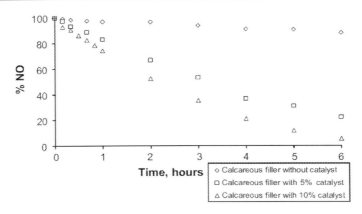

Figure 15.10
Photocatalytic decomposition of NO by calcareous filler panels containing 0.1% Mn—TiO$_2$ photocatalyst. Influence of the amount of the catalyst. Source: *Adapted from Cacho, C., Geiss, O., Barrero-Moreno, J., Binas, V., Kiriakidis, G., Botalico, L., et al. (2011). Studies on photo-induced NO removal by Mn-doped TiO$_2$ under indoor-like illumination conditions.* Journal of Photochemistry and Photobiology A: Chemistry, 222, 304—306.

when added to building material (matrix), the photocatalyst remains active for NO degradation. The activity of these preparations is directly related to the amount of catalyst added to the building material. After exposure to indoor lighting for 6 hours, NO is effectively degraded in the presence of panels containing 5% and 10% photocatalyst up to 80% and 95%, respectively.

15.5.1.3 Degradation of toluene as a model compound for indoor priority pollutants

Photocatalysis of toluene at lower ppbv level (alone and in mixtures) typical for indoor environments was investigated using a commercial gypsum-based plaster containing TiO$_2$. Degradation of toluene after irradiation with UV-light was affected by both the RH and the presence of other substances (NO, benzene). Figs. 15.11 and 15.12 show the conversion of toluene at humidity levels of 20% and 60%, respectively. After 6 hours of UV-irradiation, toluene was almost totally degraded (20%, RH), while in the presence of NO and benzene, conversion of toluene up to 100% after 5 hours was recorded. At higher humidity level (60%, RH), degradation of toluene was almost independent of the presence of NO or benzene. Degradation up to 75% after 6 hours was recorded, indicating the impact of higher humidity levels on toluene conversion (with and without additions).

Fig. 15.13 shows possible reaction pathways for the degradation of toluene. Based on the results obtained from the above experiments, an attempt was made to calculate the OH radical formation at the interface air/TiO$_2$-surface by the UV-irradiation of toluene on TiO$_2$-containing surfaces. For these calculations, the photocatalytic conversion of toluene without any other additions was taken into consideration. As expected, it was found that the

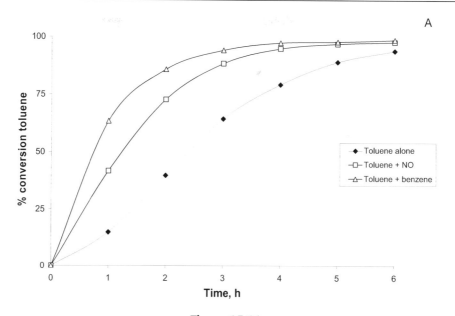

Figure 15.11
Conversion of toluene (alone and in a mixture with NO and benzene), T 23°C, RH 20%.

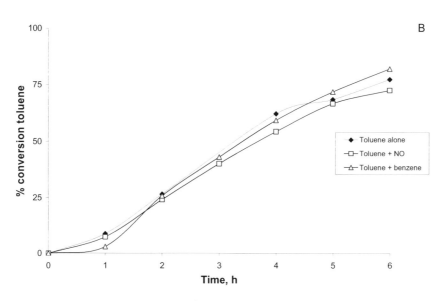

Figure 15.12
Conversion of toluene (alone and in a mixture with NO and benzene), T 23°C, RH 60%.

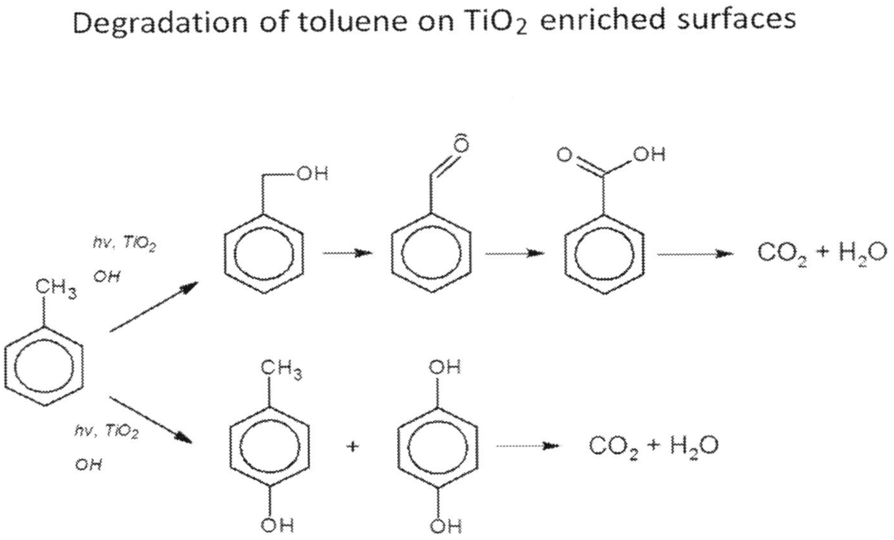

Degradation of toluene on TiO$_2$ enriched surfaces

Figure 15.13
Possible degradation pathways for toluene on TiO$_2$ surface.

OH-radical formation/concentration is higher at low humidity levels (20%) reaching values up to 1.6×10^7 molecules/cm^3, while at higher RH levels (60%), the correspond value was 6.2×10^6 molecules/cm^3. Calculations were done using the following OH-toluene reaction rate constant: k [OH-toluene]: 5.78×10^{-12} cm^3/mol s.

The development of innovative building materials and coatings containing TiO$_2$ and, in particular, the development of catalysts activated by visible light remain the fields of innovative research; further investigations should be conducted in the near future addressing critical parameters involved.

Moreover, targeted studies are needed addressing the efficiency of photocatalytic materials on the degradation of chemical mixtures, including the determination of eventually formed, toxicologically relevant byproducts to evaluate the applicability of these materials and coatings particularly for indoor applications.

15.5.1.4 Byproducts

The necessity of energy saving in buildings has significantly increased the expectations on the potential role of the construction industry to maintain the quality and comfort of confined environments developing innovative materials for various applications indoors. The use of heterogeneous photocatalysis for the degradation of pollutants indoors demands further industrial efforts in research and innovation to developing products sensitive to artificial light, highly efficient in degrading priority indoor air pollutants and free of undesirable side effects (e.g., formation of secondary emissions and byproducts). The efficiency of such products as clean technological solutions has to be proved under real-world settings conditions of temperature, humidity, illumination and pollutant concentrations.

In the frame of our research, byproducts originating from the degradation of the paint matrix during the irradiation with UV-light were detected and the paint ingredients responsible for the formation of these products were identified (Geiss, Cacho, Barrero-Moreno, & Kotzias, 2012). The photocatalytic experiments were carried out in an environmental chamber (the INDOORTRON facility at the JRC, Ispra, Italy) of 30 m^3 under real-world setting conditions of temperature, humidity and illumination. The photocatalytic materials containing semiconductors, for example, TiO$_2$, were irradiated for several days. The formation of byproducts (carbonyl compounds) during the photocatalytic experiments was monitored. It was found that carbonyl compounds, for example, formaldehyde, acetaldehyde and higher carbonyl compounds, are formed from the decomposition of organic constituents present in the supporting material containing the photocatalyst (Fig. 15.14).

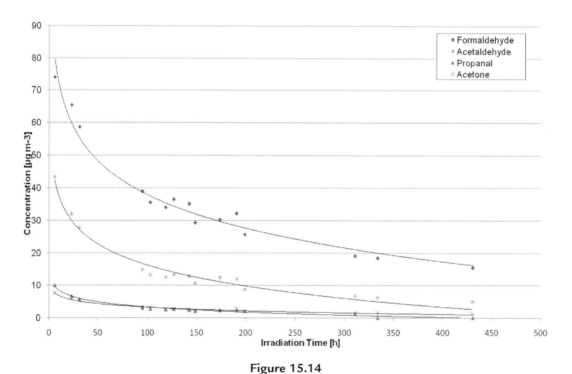

Figure 15.14

Irradiation of TiO$_2$ containing commercially available construction materials in the Indoortron (30 m^3) environmental chamber (Joint Research Centre, Ispra, Italy). Formation of byproducts.
Source: *Adapted from Geiss, O., Cacho, C., Barrero-Moreno, J., & Kotzias, D. (2012). Photocatalytic degradation of organic paint constituents-formation of carbonyls.* Building and Environment, 48, *107–112.*

Longer term irradiation of the matrix has evidenced a continuous decrease of the release of carbonyls formed. Thus, the formation of secondary emissions could be reduced or eliminated by establishing a conditioning period of continuous irradiation. The length of this conditioning period depends on the nature of the material. Hence, the unwanted formation of byproducts depends on several factors such as the history of the sample and the type and intensity of irradiation. Accordingly, photocatalytic formation of carbonyls under indoor-like irradiation is in general lower compared to the creation of byproducts when the solar-like irradiation was used. These findings can be applied by the construction industry to develop new, innovative, and consumer-friendly building materials for indoor and outdoor applications.

In order to evaluate the contribution of the paint additions, that is, cohesion agents and defoamers, to the overall formation of byproducts under UV-irradiation, several experiments were conducted with the individual compounds/components of the paint matrix, with and without TiO_2. These experiments were done using a 0.45 m^3 environmental chamber under defined conditions of humidity and temperature in the static mode. Irradiation time was 5 hours (Fig. 15.15).

The results indicated that all components presented in the paint matrix were contributing to the formation of byproducts when irradiated with UV-light in the presence of TiO_2.

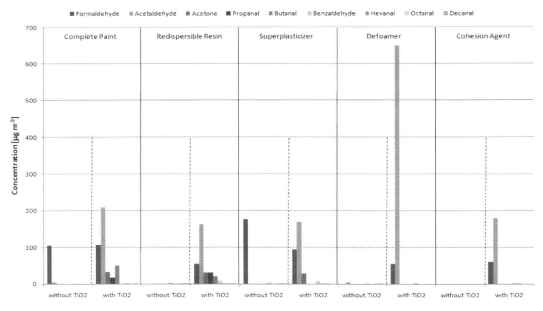

Figure 15.15

Irradiation (5 h) of the paint matrix and the individual components with and without TiO_2 temperature: T 23°C, RH: 50% Source: *Adapted from Clear-up project: https://www.clear-up.eu (Clear-up, project).*

Formaldehyde was also formed by irradiation of the paint matrix without addition of TiO_2, which shows that some of the paint components are not stable enough under UV-light conditions. Interesting are the findings in the case of super plasticizer, which produced formaldehyde without TiO_2 addition, and of de-foamer, which produced high amounts of acetaldehyde after irradiation with TiO_2 addition.

15.5.2 Disinfection

The significant presence of various microorganisms in bioaerosols imposes the application of effective disinfection methods with the view to purify air in indoor environments and to protect human health. Disinfection involves mainly the inactivation of pathogens and their elimination to safe and tolerable limits, according to the current epidemiological data and applicable legislation. Up until now, the basic approaches for such purposes are filtration and adsorption with considerable handling and operational cost and questionable results in terms of the removal of hardy microbes (Ren, Koshy, Chen, Qi, & Sorrell, 2017). In this context, ozonation and UV irradiation have been applied frequently as alternative techniques with major biocidal potential (Hwang, Jung, Jeong, & Lee, 2010; Pham & Lee, 2014a). However, their harmful effects towards human health and some action limitations, such as the ability of some bacteria to restore damages induced by UV irradiation, have led to the exploration of newly developed disinfection technologies.

Advanced oxidation processes (AOPs) have emerged as a promising group of methods, capable of reducing or even destroying completely the hazardous microbial load of bioaerosols. Particular interest has gained photocatalysis as a simple, reliable, and well-established technique that is often applied as a compliment to other air disinfection technologies (e.g., filtration/ventilation, thermal or chemical treatments, etc.) (Sánchez et al., 2012). The overall performance of photocatalysis relies on the in situ generation of reactive oxygen species (ROS), which interact with microorganisms, causing them oxidative stress and ultimate inactivation. The high biocidal potential of photocatalysis is verified by its capacity to eliminate different types of pathogens, including the more tolerant ones, like bacterial spores, viruses, fungi, and their toxins. What is even more important is the achievement of the degradation of any byproducts that may occur during decomposition of microbial organic compounds (Ren et al., 2017). Although the mechanisms of disinfection during photocatalysis have not been fully elucidated, the basic steps during inactivation include (1) the interaction between the catalysts nanoparticles and the microorganism, (2) the cause of destructions in the cellular outer layers, that is, the cell wall, cytoplasmic membrane, or the protein capsid of viruses, (3) the alteration in cell permeability, and (4) the oxidation of Coenzyme A (CoA) inside the cell that suspends metabolic pathways (Binas et al., 2017; Markowska-Szczupak, Ulfig, & Morawski, 2011).

The level of microbial resistance during photocatalysis depends highly on the special characteristics of each pathogen, the structural properties, and the employed operational conditions. The precedence order of bacteria is usually defined by the composition of their cell wall and their cellular from. Although there are no general rules, the resistance order that is most commonly observed is bacterial spores > Gram positive bacteria > Gram negative bacteria. Gram positive bacteria possess a thick cell wall that provides certain protection from the reactive transitory species, which are produced during photocatalysis (e.g., hydroxyl radicals—$HO^{\bullet}$, superoxide radicals—$O_2^{-\bullet}$, hydro-peroxyl radicals—$HO_2^{\bullet}$) (Yoo et al., 2015). On the other hand, Gram-negative bacteria have an extra outer membrane, which serves as an additional "protection shield." Once ROS penetrate cell wall, they reach the cytoplasmic membrane and finally the internal components, whose degradation leads to cell death. Fungi (filamentous forms and yeasts) follow the same trend-line, especially given that they also contain cell wall with chitin. This feature makes them quite persistent during photocatalytic treatment, as they require intense conditions for complete elimination from indoor environments (Markowska-Szczupak et al., 2011; Pigeot-Remy et al., 2014b; Sánchez et al., 2012). Finally, viruses are well known for their significant resistance, despite their simple and noncellular structure. The only targets for ROS are their capsid made by proteins and the inner genetic material (DNA or RNA). Yet, more oxidizing power is required for viral inactivation, which begins either with the adsorption of viruses onto the nanoparticles of the catalysts, or with their interaction with the free radicals in the bulk phase (Gerrity, Ryu, Crittenden, & Abbaszadegan, 2008; Misstear & Gill, 2012). The degradation of the protein capsid decreases the viral infectivity and further lesions occurred in the genetic material lead to complete virus destruction (Ren et al., 2017).

Air purification/disinfection systems based on photocatalysis include a broad variety of catalysts, whose major types are binary metal oxides, complex metal oxides, metal sulfides, and metal-free materials (Ren et al., 2017). The activity and disinfection performance of those catalysts are primarily influenced by the resistant nature and the concentration of the pathogens contained in bioaerosols. Table 15.1 shows the photocatalysts that have been used in recent studies for the inactivation of pathogenic microorganisms in bioaerosols and in some aquatic matrices. As it may be seen, the vast majority of the catalysts are titania-based materials in different forms, like films/coatings, foams and powders. The efficiency of those materials regarding photodisinfection is proven by the high inactivation rates of the tested microorganisms. For example, Josset et al. (2010), who worked with a 3D-structured reticulated polyurethane foam, as a support of a TiO_2 particle coating, succeeded in the elimination of *Legionella pneumophila* with a logarithmic reduction in viability of 1.3 and a survival probability of 6%. Those results are considered satisfactory, given the virulence of *L. pneumophila* and its significant adaptability to extreme conditions, like the ones induced by oxidation processes. Moreover, in another study, researchers developed photocatalytic filters with high filtration efficiency attempting to improve contact between microorganisms and TiO_2. Indeed, this approach led to the inactivation of *Aspergillus*

Table 15.1: Photocatalysts used recently for the inactivation of pathogenic microorganisms.

Pathogenic microorganism		Photocatalyst	Reference
Bacteria	*Escherichia coli* K12	TiO_2 PC500 coated on filter fibers	Pigeot-Remy et al. (2014)
	Escherichia coli	Ag-doped TiO_2/glass fibers (Ag–TiO_2/GF)	Pham and Lee (2014)
	Escherichia coli	Cu-doped TiO_2/glass fibers (Cu-TiO_2/GF)	Pham and Lee (2014)
	Escherichia coli	TiO_2/PU (polyurethane) and Ag–TiO_2/PU	Pham and Lee (2015)
	Escherichia coli	Ag-doped TiO_2 nanofiber	Wu et al. (2019)
	Escherichia coli	Metal-organic frameworks (MOFs)	Li et al. (2019)
	Escherichia coli and Klebsiella pneumoniae	Mn-, Co-, and binary Mn/Co doped TiO_2	Venieri et al. (2014); Venieri et al. (2017)
	Pseudomonas aeruginosa and *Staphylococcus aureus*	Anatase thin films on cellulose acetate monolithic (CAM) structures	Rodrigues-Silva et al. (2018)
	Staphylococcus aureus	Ag-doped TiO_2 nanofiber	Wu et al. (2019)
	Staphylococcus aureus	Fe-, Al-, and Cr-doped TiO_2	Venieri et al. (2017)
	Staphylococcus aureus (MRSA)	Dye-sensitized TiO_2 nanopore thin films	Perillo and Getz (2016)
	Staphylococcus aureus	Cu–TiO_2 deposited on glass fiber	Pham and Lee (2015)
	Legionella pneumophila	TiO_2 deposited on reticulated alveolar polyurethane foam	Josset et al. (2010)
	Legionella pneumophila	Cu^{2+}/TiO_2-coated cordierite foam	Ishiguro et al. (2013)
	Bacillus subtilis	Commercial TiO_2 filter and immobilized TiO_2 slide	Lin and Li (2003)
	Escherichia coli, Pseudomonas aeruginosa, Legionella pneumophila, Klebsiella pneumoniae, and *Staphylococcus aureus* (MRSA)	TiO_2/cordierite foam	Yao et al. (2011)
Fungi	*Aspergillus niger*	TiO_2 coating	Pigeot-Remy et al. (2013)
	Penicillium citrinum	Commercial TiO_2 filter and immobilized TiO_2 slide	Lin and Li (2003)
Viruses	Aerosol-associated influenza virus A/PR/8/1934 (H1N1)	Nanosized TiO_2-coated aluminum plate	Shiraki et al. (2017)
	Airborne Ms2 viruses	Pd-TiO_2 catalysts	Kim and Jang (2018)
	Airborne T2 bacteriophage	TiO_2/β-SiC solid alveolar foams	Doss et al. (2018)

niger spores, causing them irreversible destructions and eliminating the possibility for germination of the residual fungal spores.

Photodisinfection becomes even more attractive, as it offers the prospect of using the solar spectral range, becoming thus environmentally friendly and cost-effective method. In this sense, current research is focused on the development of catalysts that are active in the visible light region, extending the applications of photocatalysis as a purification process. The most common catalysts used in solar photocatalysis are modifications of titania, involving doping with nonmetals or/and noble and transition metals and modification of the substrates of the catalyst. Such materials show remarkable results under solar irradiation, in terms of the inactivation of various pathogens, including viruses and other tolerant microorganisms (Venieri et al., 2014, 2017). Fig. 15.16 shows the effect of Co-doped TiO_2 towards the bacteriophage Ms2, which is a viral indicator, under solar light. Capsid proteins are destructed and degraded in the course of treatment, resulting in the complete inactivation of the virus and the loss of its infectivity. Within the framework of solar photodisinfection, Wu, Lin, Hsu, and Hsu (2019) synthesized metal-doped TiO_2 nanofibers through a hydrothermal procedure, using metal dopants like Ag, Au, Co, Cr, Cu, etc. The acquired visible-light bioactivity of the catalysts led to excellent biocidal effects against *Escherichia coli* and *Staphylococcus aureus* under visible light (Wu et al., 2019). Furthermore, Li et al. (2019) worked with a series of metal-organic frameworks (MOFs), which exhibited bactericidal properties under simulated solar irradiation. In particular, a

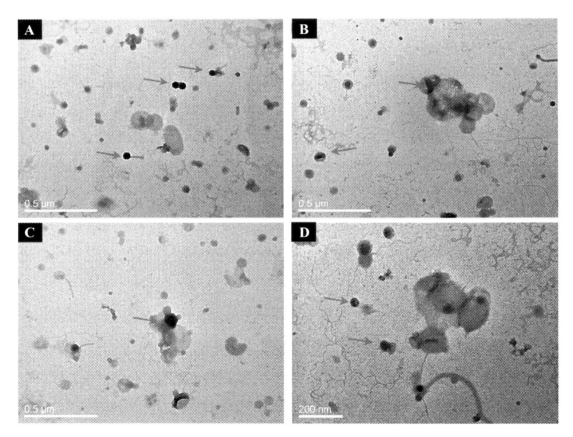

Figure 15.16

TEM images of Ms2 phages without treatment (A—negative control) and after photocatalytic treatment in the presence Co-doped TiO_2 (B—D). Arrows point out Ms2 phages in the samples.

zinc-imidazolate MOF exhibited almost complete decay of *E. coli* ($>99.9999\%$ inactivation efficiency) after 2 hours of exposure under simulated solar irradiation.

Screening the current literature, it is very clear that photodosinfection and the use of newly developed catalysts constitute an attractive technology for air purification and for the inactivation of pathogens in bioaerosols. However, certain microorganisms, such as fungi or even viruses, are merely mentioned in recent research studies, making thus the information regarding their resistance/sensitivity during treatment very limited. On the other hand, bacterial response has been reported frequently (Table 15.1), elucidating several aspects concerning the mechanisms of inactivation caused by photodisinfection. Nevertheless, with respect to the air purification and to the public health protection, it is widely accepted that (solar) photocatalysis is a promising method with high potential of development and evolution. The synthesis of new and "state-of-the-art" catalysts makes this approach appealing and effective in small- and large-scale applications.

15.6 Real-scale photocatalytic applications

Solar light-activated PC coatings have been applied in a number of large-scale projects such as the Church Dives in Misericordia, and "Umberto I" tunnel in Rome (Italy), TiO_2 paving road blocks Osaka Japan, Camden town, London UK, etc. Visible active photocatalytic materials and in particular Mn-doped TiO_2 have also been synthesized in up-scaled lab quantities, and an example of about 100 kg of photocatalytic powder was first used for the internal application in a road tunnel in Greece. In this particular field test, it was demonstrated that the material retained the same structural characteristics and photocatalytic activity as the original lab tests had shown. The full amount of the powder synthesized was used for the production of 1000 lit (1 tn) of photocatalytic paint (Fig. 15.17) developed in

Figure 15.17
Up-scaled production of "PhotoCatalyticNano − PCN Materials" photocatalytic powder for the tunnel application in Greece.

collaboration with a local paints company and FORTH (Foundation for Research and Technology) and has subsequently applied in December 2014 as the first commercial-scale internal application of this photocatalytic material in Greece. The photocatalytic paint produced was used to coat a total surface area of 4000 m^2 (Fig. 15.18). Five years later, the qualitative characteristics of the applied paint have remained unchanged with the passing car drivers noticed a better atmosphere inside the tunnel. As a consequence, the Region of Crete has saved more than 57,000 Euro per year in lighting expenses due to the retention of the lighting levels at the original specified margin values of the construction company.

Figure 15.18
The first real application in a tunnel in Greece. Tunnel in Stalida—Crete (Dec. 2014).

In this field test, it was demonstrated that the material retained the same structural characteristics and photocatalytic activity towards various air pollutants as the original lab tests had shown too.

In a second field test, an amount of about 80 kg of photocatalytic paint was produced early in 2017, in close collaboration with the 691 Army's Industrial Factory (Machaira Camp, Avlonas) of the Hellenic Army and subsequently under the permission of the GFSA (Army General Staff). It was applied at the medical center of the cadets training camp (SEAP) in Crete in an effort to improve the quality of air in their buildings (Fig. 15.19). This area is very closed to the harbor, the industrial zone, and the airport with an accumulation of more than 400 tn/year nitrous oxides. As the result of a month-long tests during the summer of 2017, there is a substantial decrease both in NO and bacteria concentration (Figs. 15.20 and 15.21).

In particular, after 1-year study, the observed results on the effect of a photocatalytic paint on the elimination of air pollutants, and more specifically NO and toluene eliminations, for application in indoor environments are summarized as follows. The physicochemical properties (including the mechanical parameters) of the photo-paint did not seem to be affected from the introduction of the photocatalytic material in the synthesis route. The photocatalytic efficiency of the paint on NO removal was

Figure 15.19
Location of the Army training camp (SEAP) in Heraklion, Crete.

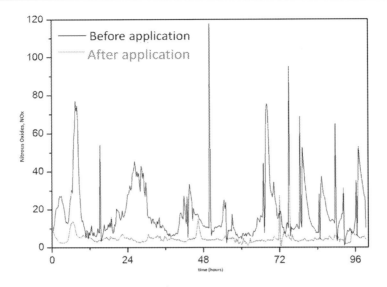

Figure 15.20
Field results of the photocatalytic paint jointly produced with the Hellenic Army.

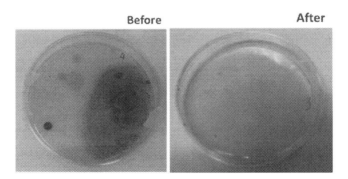

Figure 15.21
Field results of the photocatalytic paint jointly produced with the Hellenic Army.

significantly higher than toluene. Also, the potential of a pollutant removal depended on the intrinsic properties of gas and the chemical nature of the paint in which the TiO_2 particles are embedded (Maggos et al., 2019).

References

Alammar, T., Hamm, I., Wark, M., & Mudring, A.-V. (2015). Low-temperature route to metal titanate perovskite nanoparticles for photocatalytic applications. *Applied Catalysis B: Environmental*, *178*, 20–28. Available from https://doi.org/10.1016/j.apcatb.2014.11.010.

Binas, V., Papadaki, D., Maggos, Th, Katsanaki, A., & Kiriakidis, G. (2018). Study of innovative photocatalytic cement based coatings: The effect of supporting materials. *Construction and Building Materials, 168,* 923−930. Available from https://doi.org/10.1016/j.conbuildmat.2018.02.106.

Binas, V., Venieri, D., Kotzias, D., & Kiriakidis, G. (2017). Modified TiO_2 based photocatalysts for improved air and health quality. *Journal of Materials, 3.* Available from https://doi.org/10.1016/j.jmat.2016.11.002.

Binas, V., Stefanopoulos, V., Kiriakidis, G., & Papagiannakopoulos, P. (2019). Photocatalytic oxidation of gaseous benzene, toluene and xylene under UV and visible irradiation over Mn-doped TiO_2 nanoparticles. *Journal of Materiomics, 5,* 56−65. Available from https://doi.org/10.1016/j.jmat.2018.12.003.

Binas, V. D., Sambani, K., Maggos, T., Katsanaki, A., & Kiriakidis, G. (2012). Synthesis and photocatalytic activity of Mn-doped TiO_2 nanostructured powders under UV and visible light. *Applied Catalysis B: Environmental, 113−114,* 79−86. Available from https://doi.org/10.1016/j.apcatb.2011.11.021.

Bora, L. V., & Mewada, R. K. (2017). Visible/solar light active photocatalysts for organic effluent treatment: Fundamentals, mechanisms and parametric review. *Renewable and Sustainable Energy Reviews, 76,* 1393−1421.

Cacho, C., Geiss, O., Barrero-Moreno, J., Binas, V., Kiriakidis, G., Botalico, L., et al. (2011). Studies on photo-induced NO removal by Mn-doped TiO_2 under indoor-like illumination conditions. *J Photochem Photobiol A Chem, 222,* 304−306.

Chatzidiakou, L., Mumovic, D., Summerfield, A. J., Tàubel, M., & Hyvärinen, A. (2015). Indoor air quality in London schools. Part 2: Long-term integrated assessment. *Intelligent Buildings International, 7,* 130−146. Available from https://doi.org/10.1080/17508975.2014.918871.

Ciuzas, D., Prasauskas, T., Krugly, E., Jurelionis, A., Seduikyte, L., & Martuzevicius, D. (2016). Indoor air quality management by combined ventilation and air cleaning: An experimental study. *Aerosol and Air Quality Research, 16,* 2550−2559.

Clear-up project: https://www.clear-up.eu.

Construction Products Regulation (CPR) [305/2011/CEE].

Del Cacho, C., Geiss, O., Leva, P., Tirendi, S., & Barrero-Moreno, J. (2013). *15 - Nanotechnology in manufacturing paints for eco-efficient buildings. Woodhead Publishing Series in Civil and Structural Engineering, Nanotechnology in Eco-Efficient Construction* (pp. 343−363). Woodhead Publishing. Available from https://doi.org/10.1533/9780857098832.3.343.

Energy Performance of Buildings Directive [91/2002/EPBD], [The recast Directive on Energy Performance of Buildings (EPBD 31/2010].

Etacheri, V., Michlits, G., Seery, M. K., Hinder, S. J., & Pillai, S. C. (2013). A highly efficient $TiO_{2-x}C_x$ nano-heterojunction photocatalyst for visible light induced antibacterial applications. *ACS Applied Materials and Interfaces, 5,* 1663−1672. Available from https://doi.org/10.1021/am302676a.

Fagan, R., McCormack, D. E., Dionysiou, D. D., & Pillai, S. C. (2016). A review of solar and visible light active TiO_2 photocatalysis for treating bacteria, cyanotoxins and contaminants of emerging concern. *Materials Science in Semiconductor Processing, 42,* 2−14. Available from https://doi.org/10.1016/j.mssp.2015.07.052.

Feng, Y. -N., Wang, H. -C., Luo, Y. -D., Shen, Y., & Lin, A. (2013). Ferromagnetic and photocatalytic behaviors observed in Ca-doped $BiFeO_3$ nanofibres. *Journal of Applied Physics, 113,* 146101.

Fujishima, A., Hashimoto, K., & Watanabe, T. (1999). *TiO_2 photocatalysis: Fundamentals and applications.* Japan: BKC Inc.

Gao, H., Zhou, J., Dai, D., & Qu, Y. (2009). Photocatalytic activity and electronic structure analysis of N-doped anatase TiO_2: A combined experimental and theoretical study. *Chemical Engineering & Technology, 32,* 867. Available from https://doi.org/10.1002/ceat.200800624.

Geiss, O., Cacho, C., Barrero-Moreno, J., & Kotzias, D. (2012). Photocatalytic degradation of organic paint constituents-formation of carbonyls. *Building and Environment, 48,* 107−112.

Gerrity, D., Ryu, H., Crittenden, J., & Abbaszadegan, M. (2008). Photocatalytic inactivation of viruses using titanium dioxide nanoparticles and low-pressure UV light. *Journal of Environmental Science and Health.*

Part A, Toxic/Hazardous Substances & Environmental Engineering, 43, 1261−1270. Available from https://doi.org/10.1080/10934520802177813.

Gomathi Devi, L., Girish Kumar, S., Narasimha Murthy, B., & Kottam, Nagaraju (2009). Influence of Mn^{2+} and Mo^{6+} dopants on the phase transformations of TiO_2 lattice and its photo catalytic activity under solar illumination. *Catalysis Communications, 10,* 794−798. Available from https://doi.org/10.1016/j.catcom.2008.11.041.

Grabowska, E. (2016). Selected perovskite oxides: Characterization, preparation and photocatalytic properties— A review. *Applied Catalysis B: Environmental, 186,* 97−126. Available from https://doi.org/10.1016/j.apcatb.2015.12.035.

Gupta, M., Lee, S., Bisesi, M., & Lee, J. (2019). Indoor microbiome and antibiotic resistance on floor surfaces: An exploratory study in three different building types. *International Journal of Environmental Research and Public Health, 16,* 12−15. Available from https://doi.org/10.3390/ijerph16214160.

Hoffmann, M. R., Martin, Scot T., Choi, Wonyong, & Bahnemannt, D. W. (1995). Environmental applications of semiconductor photocatalysis. *Chemical Reviews, 95,* 69−96.

Hwang, G. B., Jung, J. H., Jeong, T. G., & Lee, B. U. (2010). Effect of hybrid UV-thermal energy stimuli on inactivation of *S. epidermidis* and *B. subtilis* bacterial bioaerosols. *The Science of the Total Environment, 408,* 5903−5909. Available from https://doi.org/10.1016/j.scitotenv.2010.08.016.

Ijaz, M. K., Zargar, B., Wright, K. E., Rubino, J. R., & Sattar, S. A. (2016). Generic aspects of the airborne spread of human pathogens indoors and emerging air decontamination technologies. *American Journal of Infection Control, 44,* S109−S120. Available from https://doi.org/10.1016/j.ajic.2016.06.008.

Janus, M., Zając, K., Ehm, C., & Stephan, D. (2019). Fast method for testing the photocatalytic performance of modified gypsum. *Catalysts, 9,* 693. Available from https://doi.org/10.3390/catal9080693.

Josset, S., Hajiesmaili, S., Begin, D., Edouard, D., Pham-Huu, C., Lett, M. C., . . . Keller, V. (2010). UV-A photocatalytic treatment of Legionella pneumophila bacteria contaminated airflows through three-dimensional solid foam structured photocatalytic reactors. *Journal of Hazardous Materials, 175,* 372−381. Available from https://doi.org/10.1016/j.jhazmat.2009.10.013.

Karafas, E. S., Romanias, M. N., Stefanopoulos, V., Binas, V., Zachopoulos, A., Kiriakidis, G., & Papagiannakopoulos, P. (2019). Effect of metal doped and co-doped TiO_2 photocatalysts oriented to degrade indoor/outdoor pollutants for air quality improvement. A kinetic and product study using acetaldehyde as probe molecule. *Journal of Photochemistry and Photobiology A: Chemistry, 371,* 255−263. Available from https://doi.org/10.1016/j.jphotochem.2018.11.023.

Kato, H., & Kudo, A. (2002). Visible-light-response and photocatalytic activities of TiO_2 and $SrTiO_3$ photocatalysts codoped with antimony and chromium. *The Journal of Physical Chemistry B, 106,* 5029−5034. Available from https://doi.org/10.1021/jp0255482.

Kiriakidis, G., & Binas, V. (2014). Metal oxide semiconductors as visible light photocatalysts. *Journal of the Korean Physical Society, 65,* 297−302.

Koistinen, K., Kotzias, D., Kephalopoulos, S., Schlitt, C., Carrer, P., Jantunen, M., . . . Seifert, B. (2008). The INDEX project: Executive summary of a European Union project on indoor air pollutants. *Allergy, 63,* 810−819.

Kotzias, D. (2012). Clearing the pollution indoors. *Fresenius Environmental Bulletin, 21*(11), 3212−3214.

Lam, S.-M., Sin, J.-C., & Mohamed, A. R. (2017). A newly emerging visible light-responsive $BiFeO_3$ perovskite for photocatalytic applications: A mini review. *Materials Research Bulletin, 90,* 15−30. Available from https://doi.org/10.1016/j.materresbull.2016.12.052.

Lasek, J., Yu, Y.-H., & Wu, J. C. S. (2013). Removal of NO_x by photocatalytic processes. *Journal of Photochemistry and Photobiology C: Photochemistry Reviews, 14,* 29−52. Available from https://doi.org/10.1016/j.jphotochemrev.2012.08.002.

Lee, J. H., Kim, J. Y., Cho, B. B., Anusha, J. R., Sim, J. Y., Raj, C. J., & Yu, K. H. (2019). Assessment of air purifier on efficient removal of airborne bacteria, *Staphylococcus epidermidis*, using single-chamber method. *Environmental Monitoring and Assessment, 191,* 1−10. Available from https://doi.org/10.1007/s10661-019-7876-3.

Li, P., Li, J., Feng, X., Li, J., Hao, Y., Zhang, J., . . . Wang, B. (2019). Metal-organic frameworks with photocatalytic bactericidal activity for integrated air cleaning. *Nature Communications*, *10*, 1−10. Available from https://doi.org/10.1038/s41467-019-10218-9.

Luengas, A., Barona, A., Hort, C., et al. (2015). A review of indoor air treatment technologies. *Reviews in Environmental Science and Biotechnology*, *14*, 499. Available from https://doi.org/10.1007/s11157-015-9363-9.

Maggos, T., Bartzis, J. G., Leva, P., & Kotzias, D. (2007). Application of photocatalytic technology for NOx removal. *Applied Physics A*, *89*, 81−84.

Maggos, T., Binas, V., Siaperas, V., Terzopoulos, A., Panagopoulos, P., & Kiriakidis, G. (2019). A promising technological approach to improve indoor air quality. *Applied Science*, *9*, 4837. Available from https://doi.org/10.3390/app9224837.

Markowska-Szczupak, A., Ulfig, K., & Morawski, A. W. (2011). The application of titanium dioxide for deactivation of bioparticulates: An overview. *Catalysis Today*, *169*, 249−257. Available from https://doi.org/10.1016/j.cattod.2010.11.055.

Martinez, T., Bertron, A., Escadeillas, G., Ringot, E., & Simon, V. (2014). BTEX abatement by photocatalytic TiO2-bearing coatings applied to cement mortars. *Building and Environment*, *71*, 186−192.

McLintock, I. S., & Ritchie, M. (1965). Reactions on titanium dioxide; photo-adsorption and oxidation of ethylene and propylene. *Transactions of the Faraday Society.*, *61*, 1007−1016.

Misono, M. (2009). Recent progress in the practical applications of heteropolyacid and perovskite catalysts: Catalytic technology for the sustainable society. *Catalysis Today*, *144*, 285−291.

Misstear, D. B., & Gill, L. W. (2012). The inactivation of phages MS2, ΦX174 and PR772 using UV and solar photocatalysis. *Joutnal of Photochemistry and Photobiology B: Biology*, *107*, 1−8. Available from https://doi.org/10.1016/j.jphotobiol.2011.10.012.

Murcia-López, S., Moschogiannaki, M., Binas, V., Andreu, T., Kiriakidis, G., & Morante, J. R. (2017). Insights into the performance of CoxNi1−xTiO$_3$ solid solutions as photocatalysts for sun-driven water oxidation. *ACS Applied Materials and Interfaces*, *9*, 40290−40297. Available from https://doi.org/10.1021/acsami.7b12994.

Osawa, R. A., Barrocas, B. T., Monteiro, O. C., Conceição Oliveira, M., & Florêncio, M. Helena (2020). Visible light photocatalytic degradation of amitriptyline using cobalt doped titanate nanowires: Kinetics and characterization of transformation products. *Journal of Environmental Chemical Engineering*, *8*, 103585. Available from https://doi.org/10.1016/j.jece.2019.103585.

Paul, S., Chetri, P., & Choudhury, A. (2014). Effect of manganese doping on the optical property and photocatalytic activity of nanocrystalline titania: Experimental and theoretical investigation. *Journal of Alloys and Compounds*, *583*, 578−586. Available from https://doi.org/10.1016/j.jallcom.2013.08.209.

Pham, T. D., & Lee, B. K. (2014a). Effects of Ag doping on the photocatalytic disinfection of *E. coli* in bioaerosol by Ag-TiO$_2$/GF under visible light. *Journal of Colloid and Interface Science*, *428*, 24−31. Available from https://doi.org/10.1016/j.jcis.2014.04.030.

Pigeot-Remy, S., Lazzaroni, J. C., Simonet, F., Petinga, P., Vallet, C., Petit, P., . . . Guillard, C. (2014b). Survival of bioaerosols in HVAC system photocatalytic filters. *Applied Catalysis B: Environmental*, *144*, 654−664. Available from https://doi.org/10.1016/j.apcatb.2013.07.036.

Pigeot-Remy, S., Real, P., Simonet, F., Hernandez, C., Vallet, C., Lazzaroni, J. C., . . . Guillard, C. (2013). Inactivation of *Aspergillus niger* spores from indoor air by photocatalytic filters. *Applied Catalysis B: Environmental*, *134*, 167−173. Available from https://doi.org/10.1016/j.apcatb.2013.01.023.

Prussin, A. J., Schwake, D. O., & Marr, L. C. (2017). Ten questions concerning the aerosolization and transmission of Legionella in the built environment. *Building and Environment*, *123*, 684−695. Available from https://doi.org/10.1016/j.buildenv.2017.06.024.

Radwan, S. M. A., & Abdel-Aziz, R. A. (2019). Evaluation of microbial content of indoor air in hot arid climate. *International Journal of Environmental Science and Technology*, *16*, 5429−5438. Available from https://doi.org/10.1007/s13762-018-2068-1.

Ren, H., Koshy, P., Chen, W. F., Qi, S., & Sorrell, C. C. (2017). Photocatalytic materials and technologies for air purification. *Journal of Hazardous Materials*, *325*, 340−366. Available from https://doi.org/10.1016/j.jhazmat.2016.08.072.

Sánchez, B., Sánchez-Muñoz, M., Muñoz-Vicente, M., Cobas, G., Portela, R., Suárez, S., ... Amils, R. (2012). Photocatalytic elimination of indoor air biological and chemical pollution in realistic conditions. *Chemosphere*, *87*, 625−630. Available from https://doi.org/10.1016/j.chemosphere.2012.01.050.

Sang, Y., Kuai, L., Chen, C., et al. (2014). Fabrication of a visible-light-driven plasmonic photocatalyst of AgVO3@AgBr@Ag nanobelt heterostructures. *ACS Applied Materials and Interfaces*, *6*, 5061−5068.

Saqlain, S., Cha, B. J., Kim, S. Y., Ahn, T. K., Park, C., Oh, J.-M., ... Kim, Y. D. (2020). Visible light-responsive Fe-loaded TiO_2 photocatalysts for total oxidation of acetaldehyde: Fundamental studies towards large-scale production and applications. *Applied Surface Science*, *505*, 144160. Available from https://doi.org/10.1016/j.apsusc.2019.144160.

Sattar, S. A. (2016). Indoor air as a vehicle for human pathogens: Introduction, objectives, and expectation of outcome. *American Journal of Infection Control*, *44*, S95−S101. Available from https://doi.org/10.1016/j.ajic.2016.06.010.

Shen, J.-H., Chuang, H.-Y., Jiang, Z.-W., Liu, X.-Z., & Horng, J.-J. (2020). Novel quantification of formation trend and reaction efficiency of hydroxyl radicals for investigating photocatalytic mechanism of Fe-doped TiO_2 during UV and visible light-induced degradation of acid orange 7. *Chemosphere*, *251*, 126380. Available from https://doi.org/10.1016/j.chemosphere.2020.126380.

Shi, H., Li, X., Iwai, H., et al. (2009). 2-Propanol photodegradation over nitrogen-doped $NaNbO_3$ powders under visible-light irradiation. *The Journal of Physics and Chemistry of Solids*, *70*, 931−935.

Štengl, V., Popelková, D., & Vláčil, P. (2011). TiO_2−graphene nanocomposite as high performace photocatalysts. *Journal of Physical Chemistry C*, *115*, 25209−25218. Available from https://doi.org/10.1021/jp207515z.

Stockwell, R. E., Ballard, E. L., O'Rourke, P., Knibbs, L. D., Morawska, L., & Bell, S. C. (2019). Indoor hospital air and the impact of ventilation on bioaerosols: A systematic review. *The Journal of Hospital Infection*, *103*, 175−184. Available from https://doi.org/10.1016/j.jhin.2019.06.016.

Tanaka, H., & Misono, M. (2001). Advances in designing perovskite catalysts. *Current Opinion in Solid State & Materials Science*, *5*, 381−387.

Thirumalairajan, S., Girija, K., Hebalkar, N. Y., Mangalaraj., & Viswanathan, C. (2013). Shape evolution of perovskite $LaFeO_3$ nanostructures: A systematic investigation of growth mechanism, properties and morphology dependent photocatalytic activities. *RSC Advances*, *3*, 7549−7561.

Tsang, C. H. A., Li, K., Zeng, Y., Zhao, W., Zhang, T., Zhan, Y., ... Huang, H. (2019). Titanium oxide based photocatalytic materials development and their role of in the air pollutants degradation: Overview and forecast. *Environment International*, *125*, 200−228. Available from https://doi.org/10.1016/j.envint.2019.01.015.

Venieri, D., Gounaki, I., Bikouvaraki, M., Binas, V., Zachopoulos, A., Kiriakidis, G., & Mantzavinos, D. (2017). Solar photocatalysis as disinfection technique: Inactivation of *Klebsiella pneumoniae* in sewage and investigation of changes in antibiotic resistance profile. *Journal of Environmental Management*, *195*. Available from https://doi.org/10.1016/j.jenvman.2016.06.009.

Venieri, D., Gounaki, I., Binas, V., Zachopoulos, A., Kiriakidis, G., & Mantzavinos, D. (2014). Inactivation of MS2 coliphage in sewage by solar photocatalysis using metal-doped TiO_2. *Applied Catalysis B: Environmental*, *178*, 54−64. Available from https://doi.org/10.1016/j.apcatb.2014.10.052.

Venieri, D., Tournas, F., Gounaki, I., Binas, V., Zachopoulos, A., Kiriakidis, G., & Mantzavinos, D. (2017). Inactivation of *Staphylococcus aureus* in water by means of solar photocatalysis using metal doped TiO_2 semiconductors. *Journal of Chemical Technology and Biotechnology (Oxford, Oxfordshire: 1986)*, *92*. Available from https://doi.org/10.1002/jctb.5085.

Wang, S., Ding, H., Zhao, Y., et al. (2020). Fabrication of protective textile with N-doped TiO_2 embedded citral microcapsule coating and its air purification properties. *Fibers and Polymers*, *21*, 334−342. Available from https://doi.org/10.1007/s12221-020-9352-7.

Williams, D. L., McCormack, M. C., Matsui, E. C., Diette, G. B., McKenzie, S. E., Geyh, A. S., & Breysse, P. N. (2016). Cow allergen (Bos d2) and endotoxin concentrations are higher in the settled dust of homes proximate to industrial-scale dairy operations. *Journal of Exposure Science & Environmental Epidemiology*, *26*, 42−47. Available from https://doi.org/10.1038/jes.2014.57.

World Health Organization, Regional Office. (2010). WHO guidelines for indoor air quality-selected poll. ISBN: 9789289002134. Retrieved from http://www.euro.who.int.

Wu, M. C., Lin, T. H., Hsu, K. H., & Hsu, J. F. (2019). Photo-induced disinfection property and photocatalytic activity based on the synergistic catalytic technique of Ag doped TiO_2 nanofibers. *Applied Surface Science*, *484*, 326−334. Available from https://doi.org/10.1016/j.apsusc.2019.04.028.

Yoo, S., Ghafoor, K., Kim, S., Sun, Y. W., Kim, J. U., Yang, K., . . . Park, J. (2015). Inactivation of pathogenic bacteria inoculated onto a BactoTM agar model surface using TiO_2-UVC photocatalysis, UVC and chlorine treatments. *Journal of Applied Microbiology*, *119*, 688−696. Available from https://doi.org/10.1111/jam.12877.

Zemouri, C., De Soet, H., Crielaard, W., & Laheij, A. (2017). A scoping review on bio-Aerosols in healthcare & the dental environment. *PLoS One*, *12*, 1−25. Available from https://doi.org/10.1371/journal.pone.0178007.

Zhang, Y., Wu, D., Kong, Q., Li, A., Li, Y., Geng, S., . . . Chen, P. (2020). Exposure level and distribution of airborne bacteria and fungi in an urban utility tunnel: A case study. *Tunnelling and Underground Space Technology*, *96*, 103215. Available from https://doi.org/10.1016/j.tust.2019.103215.

Zhang, J., Yuan, M., Liu, X., Wang, X., Liu, S., Han, B., . . . Shi, H. (2020). Copper modified Ti^{3+} self-doped TiO_2 photocatalyst for highly efficient photodisinfection of five agricultural pathogenic fungus. *Chemical Engineering Journal*, *387*, 124171. Available from https://doi.org/10.1016/j.cej.2020.124171.

Zhong, L., Lee, C.-S., & Haghighat, F. (2012). Adsorption performance of titanium dioxide (TiO_2) coated air filters for volatile organic compounds. *Journal of Hazardous Materials*, 340−349.

CHAPTER 16

Oxyhalides-based photocatalysts: the case of bismuth oxyhalides

Y.N. Teja[1], K. Gayathri[1], C. Ningaraju[1], Adhigan Murali[2] and Mohan Sakar[1]

[1]Centre for Nano and Material Sciences, Jain University, Bangalore, India, [2]School for Advanced Research in Polymers (SARP)-ARSTPS, Central Institute of Plastics Engineering & Technology (CIPET), Chennai, India

16.1 Introduction

Environmental pollution due to rapid industrialization and ever-increasing usage of fossil fuels is a rising global concern and thus the effective strategies to curb or convert these pollutants in an eco-friendly way has become a real-time necessity (Ankush et al., 2019; Ebhota & Jen, 2020). One such greener and cost-effective method is photocatalysis (Lakshmi, Varadharajan, & Kadirvelu, 2020) in which inexhaustible solar light can be used for various environmental remediation, such as the (1) degradation of pollutants, including pharmaceuticals (Akkaria et al., 2018), plastics (de Assis et al., 2018), and organic dyes (Zhu & Zhou, 2019); (2) conversion of CO_2 into nontoxic fuels (Khalil, Gunlazuardi, Ivandini, & Umar, 2019); (3) generation of H_2 and O_2 through water splitting (Wang, Li, & Domen, 2019); (4) reduction of toxic heavy metals to nontoxic (Zhang et al., 2018); and (5) disinfection of dangerous bacterial species (Ganguly, Byrne, Breen, & Pillai, 2018). To conveniently achieve these applications, the efficiency of photocatalysts is the key requirement. The efficiency of photocatalysts depends on various parameters, such as band-gap energy, separation and recombination of charge carriers, charge-transfer characteristics, and size and the shape of the semiconductor photocatalyst. Accordingly, extensive research efforts have been put into practice to investigate different semiconductor photocatalysts, such as TiO_2 (Xing et al., 2018), ZnO (Ong, Ng, & Mohammad, 2018), WO_3 (He, Meng, Cheng, Ho, & Yu, 2020), CdS (Cheng, Xiang, Liao, & Zhang, 2018), among which, TiO_2 has received enormous attention till date. But most of these photocatalysts are active in ultraviolet (UV) region, which does not serve the main purpose of photocatalysis, that is, making use of solar energy, which constitutes 45% of visible light spectrum. In order to make the photocatalysts into visible light active, various strategies, such as tuning the band-gap energy of photocatalysts by doping with metals and nonmetals, compositing with metal oxides and carbonaceous materials, and forming heterojunctions and Z-schemes with different photocatalysts have been developed.

In such a quest for visible light active photocatalysts, the layered nanomaterials, such as double-layered hydroxides (Abazari et al., 2019), graphitic carbon nitride (g-C_3N_4) (Zhang et al., 2019), graphene (Li et al., 2018), transition-metal dichalcogenides (Rosman et al., 2018), perovskites (Pan et al., 2019), and oxyhalides (Ogawa et al., 2019) have been explored and investigated. In this direction, bismuth oxyhalides (BiOX, X = Cl, Br, and I) are one such series of layered nanomaterials belonging to the family of V−VI−VII ternary compound semiconductors. Bismuth oxyhalides are chemically stable, cheap, anticorrosive, and nontoxic nanomaterials with band gap falling in between UV and visible light region making them highly interesting materials for their applications in visible light based photocatalysis. The first study on photocatalytic properties of bismuth oxyhalides was carried out by Lin, Huang, Huang, Wang, and Shi (2006), where they synthesized bismuth oxychloride (BiOCl) and investigated towards photocatalytic degradation of methyl orange under UV irradiation in comparison with TiO_2 and reported that BiOCl showed higher efficiency than that of TiO_2. Followed by the synthesis and investigations on new bismuth oxyhalides, namely, bismuth oxybromide (BiOBr) and bismuth oxyiodide (BiOI) showed appreciable visible-light-driven photocatalytic activity. Since then the bismuth oxyhalide compounds have been widely investigated for various photocatalytic applications particularly for the purpose of pollutant removal.

16.2 Structure and photocatalytic mechanism of BiOX

As mentioned earlier, bismuth oxyhalides are part of the family of V−VI−VII ternary compounds with a crystalline structure of tetragonal matlockite (Bannister, 1934). A typical bismuth oxyhalide structure consists of Bi_2O_2 single layers intertwined with double layers of halide atoms (X) in an arrangement of layer-by-layer stacking as shown in Fig. 16.1, where the layers are strongly bonded via covalent bonds, while a weak nonbonding van der

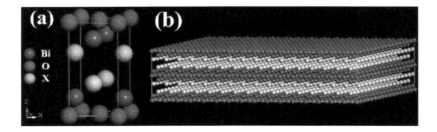

Figure 16.1

(A) Schematic structural representation of BiOX and (B) 2D representation of layered structure of BiOX (Di et al., 2017).

Waals interactions are existing among the layers (Di, Xia, Li, Guo, & Dai 2017). An electric field is induced inside the system of BiOX perpendicular to the layers of Bi_2O_2 and X, which plays an important role in the photocatalytic activity of BiOX. The formation of electric field is fundamentally due to the uneven charge distribution among the layers of Bi_2O_2 and X, which results in the polarization of atoms and orbitals, finally leading to the formation of electric field.

The valance band maximum (VBM) and conduction band minimum (CBM) of BiOX were studied through DFT calculations in which it was concluded that the O 2p and X np states contribute to the VBM, while the CBM is dominated by the Bi 6p states (Bhachu et al., 2016). The X np states contributing to the VBM that alters as $n = 3$, 4, and 5 for Cl, Br, and I, respectively. It is observed that the increasing atomic number of halogen atoms leads to the narrowing of band-gap structure. This fascinating feature helps the BiOX series to have a band gap that works within the range of UV to visible region. Correspondingly, the band gaps of BiOCl, BiOBr, and BiOI are estimated to be 3.2, 2.6, and 1.7 eV, respectively, which are schematically represented in Fig. 16.2A. BiOCl with indirect band gap ranging between 3.2 and 3.4 eV has excellent photocatalytic efficiencies under UV light irradiation. Even though the activity of BiOCl is limited to UV region, it has shown more efficacy than that of TiO_2, which makes it more preferential in industrial applications. BiOBr is found to have limited visible light absorption with a band gap of 2.6−2.8 eV and BiOI with a band gap of 1.7−1.9 eV showed convenient visible light photocatalytic activity.

Photocatalysis typically includes three steps, in which the first step involves the excitation of electrons (e^-) from the valance band (VB) to the conduction band (CB) upon light irradiation and leaving holes (h^+) in the VB, leading to the formation of electron−hole pairs. In the second step, charge carrier separation is taken place, which plays a crucial role for better photocatalytic activity. In the final step, powerful radical species, such as •OH, •O_2, and H_2O_2, are produced and involved in the redox reactions with the surrounding

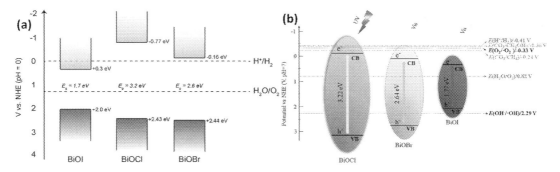

Figure 16.2

(A) Schematic structural representation of bang gap (Bhachu et al., 2016). (B) Schematic representation of photocatalysis activity of BiOCl, BiOBr, and BiOI (Sharma et al., 2019).

species onto the surface of photocatalyst and desired photocatalytic processes are carried out. In the case of BiOX, the induced-electric field in the system helps towards the effective charge carrier separation and suppression of the undesired recombination of electron−hole pairs, leading to excellent photocatalytic efficiency. Among the various photocatalytic applications, BiOX are mainly used for pollutant degradation and taking it into the consideration, the mechanism of photocatalytic degradation involves the reaction steps as shown in Eqs. (16.1−16.9).

$$PC + h\nu \rightarrow e^- + h^+ \tag{16.1}$$

$$h^+ + H_2O \rightarrow H^+ + OH^\bullet \tag{16.2}$$

$$h^+ + OH^- \rightarrow OH^\bullet \tag{16.3}$$

$$e^- + O_2 \rightarrow O_2^- \tag{16.4}$$

$$2e^- + O_2 + 2H^+ \rightarrow H_2O_2 \tag{16.5}$$

$$2e^- + H_2O_2 \rightarrow OH^\bullet + OH^- \tag{16.6}$$

$$R + OH^\bullet \rightarrow \text{Degradation products} \tag{16.7}$$

$$R + h^+ \rightarrow \text{Oxidation degraded products} \tag{16.8}$$

$$R + e^- \rightarrow \text{Reduction degraded products} \tag{16.9}$$

Even though BiOX is considered as one of the high-efficient photocatalysts, still few drawbacks hinder its application to the fullest. One of the main drawbacks is the limitation of the usage of BiOCl due to its absorption wavelength falling in the UV region. The other main drawback involves the fast recombination and the unsuitable band-edge potentials of BiOBr and BiOI towards the desired reduction and oxidation reactions even after having the optical absorptions falling in the visible light region. In order to overcome these drawbacks and to improve the efficacy of BiOX photocatalytic activities, there have been various strategies proposed and put into the practices, which include crystal facet engineering, creating oxygen vacancies (OV), elemental doping and coupling among BiOX with different halogen groups, compositing with other semiconductors through heterojunctions or Z-schemes. Among the various strategies, the crystal facet engineering has gained enormous attention than the other approaches (Liu, Yu, Lu, & Cheng, 2011). In general, the efficiency of a photocatalyst highly depends on the reactions that take place on the surface of semiconductor, such as adsorption of reactant species, transfer of excited electrons to reactant species, and desorption of products. All these activities take place on the surface of the semiconductor, where the surface structure, atomic arrangement and coordination of atoms on the surface affect the photocatalytic process heavily. The surface structure and configuration of surface atoms can be tailored by changing the crystal facet orientation.

Exposing certain crystal facets increases the surface energy of the nanostructures resulting in the enhancement of photocatalytic efficiency. In the case of BiOX nanostructures, the crystal facet engineering is chosen most ardently due to the fact that BiOX normally possesses layered two-dimensional (2D) nanosheet morphology, which is beneficial to tailor the desired facets by controlling the number of layers of nanosheets during the synthesis process. But it is difficult to get desired facet-exposed nanostructures as often during the synthesis process as the facets with high surface energy are unstable and highly reactive resulting in decrease of total crystal surface energy. In order to avoid such obstacles, capping agents are used to control the free surface energy of facets and obtain the desired nanostructures with large number of exposed reactive facets without losing the total surface energy. Various BiOX nanostructures with exposed facets, such as {001}, {110}, and {200}, have been synthesized, which showed the drastically improved photocatalytic performance and are discussed in the following sections.

16.3 Synthesis of BiOX nanostructures

16.3.1 Hydro/solvothermal method

Hydro/solvothermal synthesis process is the most commonly employed synthesis method for the synthesis of bismuth oxyhalides nanostructures. Accordingly, various BiOX nanostructures with different morphologies, such as nanobelts (Wang et al., 2017), nano/microspheres (Mera, Rodriguez, Melendrez, & Valdes, 2017), nanoflowers (Jiang et al., 2018), and nanoplates (Hu, Xu, Zhu, Hua, & Zhu, 2016), have been successfully synthesized using hydrothermal process. Other composites and heterojunctions of BiOX, such as BiOCl/BiOBr (Zhang, Lv, Dai, Liang, & Liu, 2018), SnO_2/BiOBr (Liu et al., 2018), BiOBr/Bi_2WO_6 (Dumrongrojthanath et al., 2018), BiOBr/$FeWO_4$ (Gao et al., 2018), Pd/BiOCl/BiOI (Ren et al., 2017), P-BiOI/$ZnFe_2O_4$ (Yosefi, Haghighi, & Allahyari, 2017), BiOI/g-C_3N_4/CeO_2 (Jiang et al., 2019), have also been synthesized using hydro/solvothermal process.

BiOX nanosheets with particular facets exposed as shown in Fig. 16.3A—D (Jiang, Zhao, Xiao, & Zhang, 2012) were majorly synthesized through hydrothermal process. Generally, for facet tailoring, capping agents are used during the synthesis process, which will be adsorbed onto the specific facets and will suppress the growth of the material along that particular axis and results in the high exposure of the capped facet on the surface of material. Accordingly, versatile facet exposed nanostructures have been synthesized using various capping agents with different ratios. In the case of BiOX nanostructures, the photocatalytic activity of BiOCl was found to be highly dependent on the facet exposure.

It has been reported that the H^+ ions are used as capping agent at pH $= 1$ for the synthesis of BiOCl nanosheets with the {001} exposed facets. In this process, the reaction between the capping agent and the terminated oxygen atoms took place as the H^+ ions possess

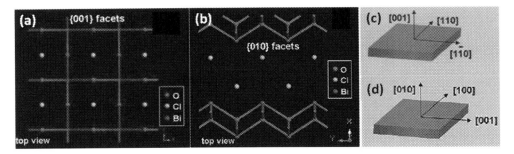

Figure 16.3

Schematic representation of atomic structure of (A) {001} facet and (B) {010} facet, and the crystal orientation of (C) {001} facet and (D) {010} facet of BiOCl nanosheets (Jiang et al., 2012).

strong binding interaction with terminated oxygen atoms that are most dominant on the (001) plane and in turn hinder the growth along that axis making {001} facet highly exposed as shown in Fig. 16.3A and C. Depending on different parameters, such as pH value, type of capping agent, and quantity of capping agent, the BiOX nanostructures with various facets exposed can be obtained. When OH⁻ ions are added as capping agent at pH = 6, the concentration of H^+ ions decreases and leads to the decrement in the interaction between the terminated oxygen atoms and H^+ ions, leading to the growth of {001} facet and resulting in the suppression of growth in (010) plane, which finally results in the formation of BiOCl nanosheets with {010} facet exposure as shown in Fig. 16.3B and D (Jiang et al., 2012).

16.3.2 Precipitation method

Typically, the precipitation synthesis of BiOX nanostructures involves three steps. In the first step, two types of solutions one with bismuth metal salt precursor and another one with a chosen halide source are prepared separately by dissolving in water. In second step, these two prepared solutions are added and stirred vigorously and a basic solution such as NaOH or KOH is added in a drop-wise manner under stirring and a precipitate is formed. In the final step, the precipitate is obtained by centrifuging followed by washing the precipitate with ethanol and waters several times to balance the pH value and finally the obtained product is dried in an oven. Several types of bismuth oxyhalide nanostructures, such as BiOI platelets (Mahmoodi, Ahmadpour, Bastami, & Mousavian, 2018), nanorose-shaped $Bi_4O_5I_2$ (Hou et al., 2019); BiOBr nanocomposites, such as BiOBr-graphene oxide (Zhang, Dong, Xiong, & Zhang, 2014) and CdS/BiOBr composites (Cui, An, Liu, Hu, & Liang, 2014); $BiOCl/Ag_2CO_3$ heterojunction (Fang et al., 2016); and hybrids ultrathin iodine-doped BiOCl nanosheets (Bi et al., 2019) were successfully synthesized using precipitation method and explored for photocatalytic applications.

16.3.3 Microwave-assisted synthesis

Microwave-assisted method is a green synthesis method that provides various advantages over conventional heating methods, which include speed reactions, higher yields, lesser byproducts, and homogenous heating of precursors, leading to the formation of uniform size crystal growth. This method gives the freedom to control the reaction parameters, to carry out reactions at high temperatures and pressures. In a typical process of microwave-assisted synthesis of BiOX nanostructures, a solution of selected precursors will be made and then transferred into a microwave reactor for the subsequent chemical reactions. The resulted product will be washed several times and the final sample will be obtained after drying. Various BiOX nanostructures with different morphologies, such as nanoflowers, nanoplates, and microspherers, have been synthesized using this method along with synthesis of heterojunctions with other semiconductors. In a study, BiOI microspheres were synthesized with different microwave irradiation reaction timings ranging from 5 to 120 minutes (Mera, Cruz, Pérez-Tijerinac, Meléndrezd, & Valdése, 2018). This study showed that at lower reaction time rate, the BiOI nanostructures possesses nanoflake-like morphology with irregular size distribution, while with increasing reaction timing the morphology was changed to microspheres with homogeneous size distribution and less agglomeration. Similarly, the microwave-assisted solvothermal method has been used for the synthesis of novel hierarchical nanostructures BiOX, such as BiOI/rGO composites (Niu, Dai, Zhang, Yao, & Yu, 2018), $Bi_4O_5I_2$ architectures (He, Cao, Yu, & Yang, 2016), and BiOBr microspheres (Zhang, Cao, Chen, & Xue, 2011). Other examples include preparation of BiF_3/BiOBr heterojunctions (Zhang, Chen, & Song, 2019), surfactant-free microwave synthesis of BiOX nanoplates (Patiphatpanya, Ekthammathat, Phuruangrat, Thongteme, & Thongtem, 2018), and BiOBr/BiOCl flowerlike composites (Maisang et al., 2020).

16.3.4 Sonochemical synthesis

Sonochemical method is a facile and cost-effective approach for the synthesis of photocatalytic nanostructures. Ultrasonic waves are used as the source of energy to carry out the chemical reactions during the synthesis process. Accordingly, sonochemical method is employed to synthesize various BiOX nanostructures but mostly for the synthesis of BiOX-based heterojunctions to improve the photocatalytic activity. In this direction, the BiOBr-GO heterojunction was synthesized using sonochemical method (Patil et al., 2016). The morphology of the obtained BiOBr−GO heterojunction nanostructures was found to be flower-like morphology with wavy GO nanosheets well distributed in the BiOBr matrix, which expected to help for the smooth electron transfer to result in highly enhanced photocatalytic activity as compared to the bare BiOBr and GO nanostructures. Similarly, BiOCl nanoflowers (Lei, Wang, Song, Fan, & Zhang, 2009), BiOI−TiO_2 heterojunction (Cai, Yao, Li, Zhang, & Wei, 2019), and ZnO/BiOBr functionalized cotton fabrics

(Yang et al., 2018) of BiOX nanostructures have also been successfully synthesized using sonochemical method.

16.3.5 Template-assisted method

Template synthesis is a very effective method, where preexisting guide is used to synthesize the controlled nanostructures. Nanomaterials synthesized by templating strategies hold a well-defined size, shape, and configuration, which usually benefits from the direct effect of the templates. As a result, template synthesis is capable of producing nanostructures with unique structures, morphologies, and properties. Accordingly, various BiOX nanostructures are synthesized using template-assisted method. One such example is synthesis of nanowire arrays of bismuth oxychloride (BiOCl) using anodic aluminum oxide (AAO) template-assisted sol−gel method (Wu et al., 2010). In this study, BiOCl nanowires of 100 nm diameter were successfully assembled in the porous structure of AAO templates and studied for dye degradation. Other template-based BiOX nanostructures include the synthesis of BiOBr microspheres with ethanol as self-template (Li et al., 2018), belt-like $Bi_4O_5Br_2$ hierarchical nanostructures using belt-like bismuth subsalicylate as the morphological template and bismuth source (Mi, Li, Zhang, & Hou, 2019), and preparation of BiOBr lamellar structures using CTAB as Br source and template (Shang, Wang, & Zhang, 2009).

16.3.6 Mechanochemical synthesis

Mechanochemical synthesis is a simple one-step process of milling the powders during which the chemical reactions occur by the direct absorption of mechanical energy. Mechanochemical synthesis possesses various advantages over the traditional wet-chemical synthesis methods, such as usage of less or no solvents, requires less temperature and pressure, and cost effectiveness. By selecting suitable chemical reaction paths, stoichiometry of starting materials and milling conditions, mechanochemical processing have been successfully used to synthesize a wide range of nanocrystalline particles. In a typical mechanochemical synthesis of BiOX nanostructures, suitable amount of precursor mixture is subjected to grinding in a planetary ball milling machine, which is used as a mechanical energy source and the milling process is carried out for a selective amount of time at desired rpm. If required, postmilling process will also be performed, where the initial reagents will be removed by washing with water and ethanol and finally dried to obtain the desired products. Accordingly, bismuth oxybromide ($Bi_xO_yBr_z$) nanostructures were synthesized using mechanochemical process by grinding a mixture of 2 g of Bi_2O_3 and $BiBr_3$ precursors using a planetary ball mill at 700 rpm for 2 hours and the obtained samples were used without any further treatments (Wu et al., 2019). Interestingly, due to large amount of mechanical energy introduced to the sample and a high-speed operation of the medium balls in the pot resulted in the formation of heavy agglomerated irregular particles with different O/Br ratios, which demonstrated some specific features in the materials. Other examples include the ionic liquid induced mechanochemical synthesis of BiOBr ultrathin nanosheets (Jiang et al., 2020), synthesis of oxygen-rich bismuth oxychlorides $Bi_xO_yCl_z$ (Wu, Li, Zhang, & Liu, 2019), and BiOCl nanoplates (Tadjarodi, Akhavan, Bijanzad, & Khiavi, 2016).

16.3.7 Other methods

Microemulsion is an interesting method for the synthesis of photocatalysts, in which two immiscible liquids, most preferably oil (nonpolar liquid) and water (polar liquid), are dispersed and stabilized with the addition of a surfactant. By adjusting the polar and nonpolar proportions, the reverse microemulsion (water-in-oil) and normal emulsion (oil-in-water) systems are obtained. Mao et al. (2017) used the ionic liquid-in-water microemulsion method to synthesize the novel Bi-rich $Bi_4O_5Br_2$ with tunable morphologies of 2D nanosheets, 3D monodisperse layered microspheres, and hollow spheres. The same group also synthesized the ultrathin BiOCl half-shells using same ionic liquid-in-water microemulsion method, applying the liquid-liquid boundary of the emulsion system as a template (Mao et al., 2016). In both cases, TX-100 was chosen as surfactant, whereas ionic liquids [Omin] Br and [Bmin] Cl were chosen as oil phase liquids. In another report, reverse microemulsion method was used for synthesis of BiOX nanoparticles, where marlophen NP5 was chosen as surfactant (Henle, Simon, Frenzel, Scholz, & Kaskel, 2007). Similarly, the ion-exchange synthesis method is another important technique used for the synthesis of complex and metastable nanostructures, which are typically not possible through conventional synthesis methods. Using ion-exchange method, the heterojunctions, such as Bi_2MoO_6/BiOI (Fan et al., 2016), BiOI/CH_3COO(BiO) (Han et al., 2017), and hybrid architectures such as Bi_2S_3/BiOCl (Cheng, Huang, Qin, Zhang, & Dai, 2012) have been successfully synthesized and demonstrated for photocatalytic applications.

16.4 Photocatalytic applications of BiOX systems

Bismuth oxyhalides as efficient photocatalysts have gained enormous attention due to their tunable band-gap structures. In the recent years, BiOX photocatalysts have been employed for various photocatalytic applications, such as pollutant degradation, oxygen evolution, hydrogen production, nitrogen fixation, CO_2 reduction, and antimicrobial activities. Accordingly, this section discusses these various photocatalytic applications of BiOX and their composite materials.

16.4.1 Dye degradation

Lin et al. (2006) were the first to report the photocatalytic activity of bismuth oxychloride (Bi_3O_4Cl) towards its ability to degrade methyl orange (MO). In this study, the polycrystalline Bi_3O_4Cl nanopowders were synthesized using Bi_2O_3 and $BiOCl$ through high-temperature calcinations at $700°C$ for 24 hours with intermediate grinding. In order to increase the performance of Bi_3O_4Cl, silver metal was loaded on the surface by a wet chemical method. The prepared Bi_3O_4Cl powder showed an optical band gap of 2.79 eV. Accordingly, the photocatalytic activity was carried out under both UV and 300 W Xe lamp using cut-off by a light filter with >420 nm. They reported that the degradation time taken by pure Bi_3O_4Cl under UV irradiation was 55 minutes, which was a less duration as compared to TiO_2, which took 70 minutes. Upon the dispersion of Ag, the performance of Bi_3O_4Cl became even better and the degradation time was found decreasing with increasing concentration of Ag as displayed in Fig. 16.4A. Under visible light irradiation, it took 24 hours for 80% removal of MO, while 3% addition of Ag took 21-hours illumination for the complete degradation. The proposed photocatalytic mechanism is depicted in Fig. 16.4B.

In another work, the heterostructured BiOCl/AgBr composites were synthesized by solvothermal method and investigated for the degradation of rhodamine B (RhB) (Fig. 16.5A; Chang, Zhong, Hu, Luo, & Wang, 2018). BiOCl typically possesses a band gap of 3.2−3.4 eV, which is closer to that of TiO_2 and can be only used under UV irradiation. In order to tune its band gap to absorb visible light and to promote the charge carrier separation, another photocatalyst AgBr was coupled with BiOCl at various molar ratios. Accordingly, the band gap of BiOCl, AgBr, BiOCl/AgBr (1/2), BiOCl/AgBr (1/1), and BiOCl/AgBr (2/1) was found to be tuned as 3.07, 2.41, 2.58, 2.49, and 2.81 eV, respectively, as shown in Fig. 16.5B. This clearly showed that with the heterostructure formation facilitated the band gap reduction towards visible light region, where the best photocatalytic performance was achieved with BiOCl/AgBr (1/1) composites. The schematic representation of photocatalytic mechanism is shown in

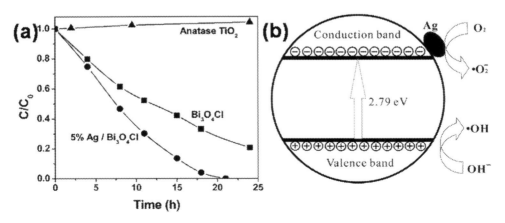

Figure 16.4

(A and B) Visible-light-driven photocatalytic degradation of MO and its mechanism.

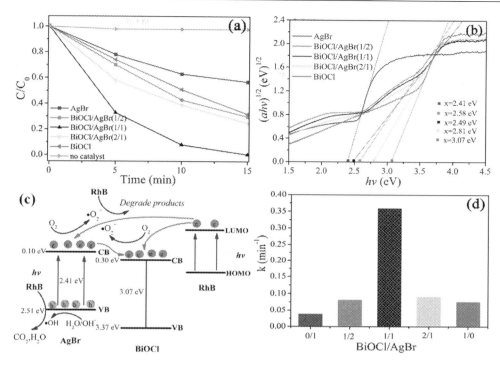

Figure 16.5

(A) Degradation curves of BiOCl/AgBr composites under visible light irradiation, (B) calculated band gap of BiOCl/AgBr composites, (C) schematic representation of photocatalytic mechanism, and (D) comparison of apparent reaction rate constants (Chang et al., 2018).

Fig. 16.5C. Further, the degradation efficiencies of AgBr, BiOCl, BiOCl/AgBr (1/2), BiOCl/AgBr (1/1), and BiOCl/AgBr (2/1) found to be 40%, 60%, 65%, 100%, and 70% in 15 minutes under the visible light irradiation, respectively, as shown in Fig. 16.5D.

Interestingly, Z-scheme based BiOCl−Au−CdS heterostructure was developed and used for the degradation of multiple pollutants, such as anionic dye methyl orange, cationic dye rhodamine B, colorless pollutant phenol, and antibiotics sulfadiazine (Li et al., 2017). Initially, BiOCl nanosheets with {001} and {110} exposed facets were fabricated using solvothermal method, then the step-wise facet-selective photoreduction process was employed for the deposition of Au nanoparticles on the {001} facets of BiOCl nanosheets. Finally, chemical bath deposition method was used for the deposition of CdS nanoparticles onto the BiOCl−Au nanosheets and formed the Z-scheme BiOCl−Au−CdS heterostructure as shown in Fig. 16.6A and B.

During the photocatalytic process, the electrons on the CB of CdS react with the absorbed O_2 on the surface and produce $^\bullet O_2^-$ radical species, while the h^+ on the VB of BiOCl

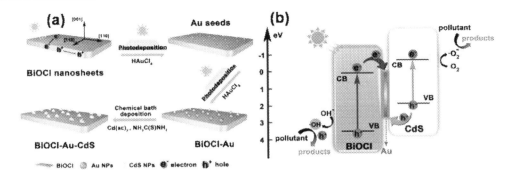

Figure 16.6
Schematic illustration of (A) synthetic route and formation mechanism of Z-scheme BiOCl−Au−CdS and (B) its photocatalytic charge-transfer and degradation mechanism (Li et al., 2017).

produce •OH radical species by oxidizing the hydroxyl groups. As a result, the radical species degraded around 100% of MO within 180 minutes and 100% of RhB within 30 minutes under solar irradiation. This developed Z-scheme BiOCl−Au−CdS showed enhanced photocatalytic activity as compared to the bare BiOCl, BiOCl−Au, and BiOCl−CdS heterostructure. Similarly, Table 16.1 highlights some of the BiOX systems and their photocatalytic dye degradation efficiencies.

16.4.2 Other organic pollutants degradation

Another major application of BiOX photocatalysts is in the removal of organic pollutants released from chemical industries. In this direction, the ZnO/Nd-BiOBr composite was developed and investigated for the visible light photocatalytic degradation of phenol as shown in Fig. 16.7A (Sin, Lim, Lam, Mohamed, & Zeng, 2019). Even though ZnO possesses band-gap energy of 3.3 eV, it has been investigated for various photocatalytic applications, but its large band gap limits its usage only to the UV light region. On the other hand, BiOBr possesses narrow band-gap energy and visible light activity, but its photocatalytic performance is affected by short charge carrier lifetime. Hence, to overcome these drawbacks of these individual materials, these two photocatalysts were coupled together forming a composite, where ZnO nanoflakes were decorated with rare-earth ion neodymium (Nd) doped BiOBr nanosheets.

The mechanism of photocatalytic phenol degradation by ZnO/Nd−BiOBr composite is illustrated in Fig. 16.7B and is explained as follows. Upon light irradiation, photogeneration of electron−hole pairs takes place and the photogenerated h^+ transfer from VB of BiOBr to VB of ZnO as the CB and VB of BiOBr are placed at a more positive position than that of ZnO. The h^+ reacts with the adsorbed water and produce •OH radical species and they

Table 16.1: Photocatalytic degradation of various dyes using BiOX-based photocatalysts.

Photocatalyst	Dye	Degradation time (min)	Efficiency of degradation (%)	Reference
CdS–BiOBr	MO	120	90.1	Liu, Wu, Zhu, Wang, and Wang (2013)
BiOI–GR	MO	240	88	Liu, Cao, Su, Chen, and Wang (2013)
AR–BiOI nanosheets	MO/RhB/MB/CR/AO II	120	99.8/100/99.9/99.6/98.6	Huang et al. (2014)
$Bi_7O_9I_3/Bi_5O_7I/$ g-C_3N_4	CV	48 hours	99	Chou, Chen, Dai, Lin, and Lee (2016)
N–BiOCl nanosheets	MO	120	93	Zhang et al. (2017)
G–BiOBr nanocomposite	RhB	70	100	Alansi, Al-qunaibit, Alade, Qahtan, and Saleh (2018)
t-$PbBiO_2I/$ Bi_5O_7I/g-C_3N_4	CV	24 hours	95	Lee, Wang, and Chen (2019)
Bi_2O_4/BiOBr	MO	60	99.3	Wang, Liu, Guo, Fan, and Tao (2018)
BiOBr/$NiFe_2O_4$	RhB	30	99.8	Li et al. (2018)
BiO_pF_q/BiO_xI_y	CV	24 hours	99	Chen et al. (2018)
BiOBr-x microflowers	RhB	40	100	Song et al. (2019)
$BiVO_4$/BiOBr	RhB	180	90	Liu, Chen, Liu, Shan, and Zhang (2019)
G-$BiOBr_xI_{1-x}$	MO	300	100	Yadav, Garg, Chandra, and Hernadi (2019)

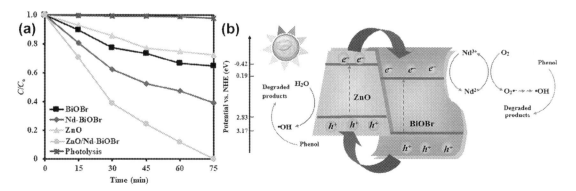

Figure 16.7
(A) Photocatalytic phenol degradation by various photocatalysts and (B) schematic illustration of photocatalytic mechanism of ZnO/Nd–BiOBr composites (Sin et al., 2019).

participate in the phenol degradation. On the other hand, the excited electrons migrate from CB of ZnO to the CB of Nd−BiOBr, where the Nd acts as an electron scavenger and trapped the electrons resulting in the enhancement of charge separation and increasing the lifetime of charge carriers. Then the trapped electrons are further transferred to the O_2 on the surface and producing $\bullet O_2^-$ radical species and consequently transformed to $\bullet OH$ radicals, thereby resulting in an efficient degradation of phenol.

In another study, bismuth oxybromide/oxyiodide ($Bi_4O_5Br_xI_{2-x}$) photocatalysts were investigated for the degradation of resorcinol, o-phenylphenol, and 4-$tert$-butylphenol under visible light irradiation (Sun et al., 2018). The $Bi_4O_5Br_xI_{2-x}$ ($x = 0, 0.2, 0.6, 1.0, 1.4, 1.8,$ and 2) photocatalysts were prepared using simple precipitation method with varying B/I molar ratios. The highest photocatalytic performance was observed on $Bi_4O_5Br_{0.6}I_{1.4}$, which showed around 2.8 and 1.8 times higher degradation rate than that of pure $Bi_4O_5Br_2$ and $Bi_4O_5I_2$ towards degradation of resorcinol as shown in Fig. 16.8A. It is predicted that the CB and VB potentials of $Bi_4O_5Br_{0.6}I_{1.4}$ is positioned at -0.64 and $+1.49$ eV, respectively, and the band-gap energy was estimated to be 2.13 eV as shown in Fig. 16.8B. Upon excitation, the electrons migrate to the CB leaving the holes in the VB. As the CB potential of $Bi_4O_5Br_{0.6}I_{1.4}$ is more negative than that of $O_2/\bullet O_2^-$ (-0.33 V vs NHE), electrons in the CB will continuously react with the O_2 producing $\bullet O_2^-$ radical species which would further react with h^+ to produce $\bullet OH$ that participates in the degradation of phenol. On the other hand, the holes in the VB do not oxidize OH^- to produce $\bullet OH$ as the VB potential is too

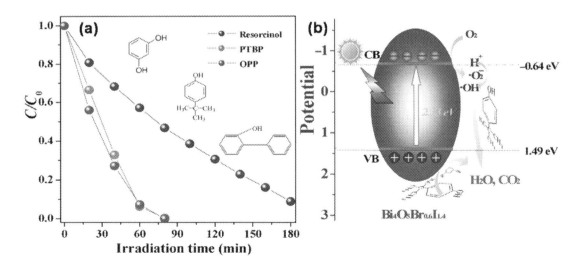

Figure 16.8

(A) Degradation efficiency of $Bi_4O_5Br_xI_{2-x}$ photocatalyst towards resorcinol, o-phenylphenol, and 4-$tert$-butylphenol under visible light irradiation and (B) schematic illustration of the proposed photocatalytic mechanism (Sun et al., 2018).

Table 16.2: Photocatalytic degradation of various pollutants using BiOX photocatalysts.

Photocatalyst	Pollutant	Degradation time (min)	Degradation (%)	Reference
Ag/AgBr/BiOBr	*p*-nitrophenol	210	96	Li, Luo, and Yang (2013)
B−BiOBr	Phenol	120	78.3	Liu, Liu, Wang, Cao, and Niu (2016)
BiOI-110	BPA	350	95	Pan, Zhang, Gao, Liu, and Chen (2015)
BCOI-1−350	Phenol	30	80+	Long et al. (2016)
Co−BiOCl nanosheets	BPA	120	95	Wang et al. (2018)
$Bi_{12}O_{17}Cl_2$ nanobelts	BPA	120	95+	Wang et al. (2017)
Bi−BiOI	BPA	240	86	Luo et al. (2017)
$Bi_{24}O_{31}Br_{10}$ nanobelts	BPA	120	92	Wang et al. (2018)
Pt/BiOBr	BPA	120	90	Zhang et al. (2018)
$Bi_4O_5Br_2$ nanobelts	Resorcinol	8 hours	97	Mi et al. (2019)
$BiFeO_3$/BiOI	BPA	120	71	Malathia, Arunachalam, Kirankumar, Madhavana, and Al-Mayouf (2018)
CQDs/BiOCl/BiOBr	Tetracycline	120	77	Hu et al. (2018)
	Ciprofloxacin		~80	
	BPA		~80	
BiOCl/BiOBr/BiOI/GO	HBA	24 hours	96.9	Siao et al. (2019)

negative as compared with •OH/OH$^-$ potential ($+1.99$ V vs NHE). Furthermore, the performed trapping experiments revealed that even though all three radical species, such as •OH, •O$_2{}^-$, and h$^+$, were generated, the •O$_2{}^-$ and h$^+$ species played major role for the enhanced photocatalytic efficiency of the system. Table 16.2 shows a list of BiOX systems and their photocatalytic pollutant degradation efficiencies.

16.4.3 CO$_2$ reduction

Photocatalytic CO$_2$ conversion could be a greener approach to convert CO$_2$ into carbon containing fuels and BiOX photocatalysts have been studied for this application lately. BiOX photocatalysts have not been employed for CO$_2$ conversion and hydrogen evolution as much as they are used for pollutant degradation due to their less photocatalytic performance towards the respective applications. The main reason for the poor photocatalytic performance towards CO$_2$ conversion is due to their lower position CB minimum. One of the effective ways to improve the photocatalytic performance of BiOX photocatalysts towards CO$_2$ conversion is to synthesize the BiOX photocatalysts with rich bismuth concentration as the CB of BiOX is mainly dependent on the Bi 6p states. For instance, the oxygen vacancy-induced thin BiOCl nanoplates with exposed {001} facets were studied for the CO$_2$ reduction (Zhang, Wang, Jiang, Gao, & Sun, 2015). Under

8 hours of light irradiation, CO yield of 8.1 μmol/g and CH$_4$ yield of 1.2 μmol/g was obtained, which is still not satisfactory. After this, bismuth-rich ultrathin BiOX nanosheets were investigated for CO$_2$ reduction applications, where it was found that the thickness of nanosheets and the bismuth contents were found to be important factors that influence the band edge potentials of BiOX and thereby influence the CO$_2$ reduction performance (Ding et al., 2016; Ye et al., 2016). Even though these studies did not show satisfactory results, they opened doors for new studies based on these reports.

Accordingly, in one of the studies, Bai et al. (2017) synthesized the solid solutions of bismuth-rich Bi$_4$O$_5$Br$_x$I$_{2-x}$ ($x = 0$, 0.5, 1, 1.5, and 2) through molecular precursor method and studied for CO$_2$ reduction to CO, CH$_4$ formation as shown in Fig. 16.9A. From the valance band XPS spectra as shown in Fig. 16.9B, the VB positions of Bi$_4$O$_5$Br$_2$, Bi$_4$O$_5$BrI, and Bi4O$_5$I$_2$ were found to be 1.48, 1.64, and 1.99 eV, respectively, and based on this data, the CB edge potentials were calculated to be -0.55, -0.70, and -0.70 eV, respectively, as shown in Fig. 16.9C. From the band energy diagram as represented in Fig. 16.9c, it can be seen clearly that the CB edges of Bi$_4$O$_5$BrI and Bi$_4$O5I$_2$ are higher than that of Bi$_4$O$_5$Br$_2$,

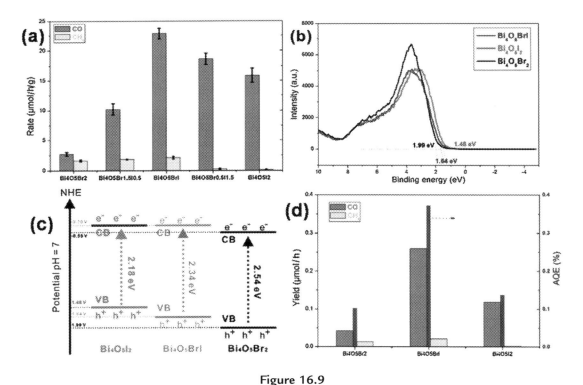

Figure 16.9

(A) CO$_2$ rate of for Bi$_4$O$_5$Br$_x$I$_{2-x}$ ($x = 0$, 0.5, 1, 1.5, and 2) under visible light irradiation for 1 hour, (B) valence band XPS spectrum, (C) schematic representation of band energies, and (D) Quantum efficiency and product yield of Bi$_4$O$_5$Br$_2$, Bi$_4$O$_5$BrI, and Bi4O$_5$I$_2$ (Bai et al., 2017).

which resulted in the enhancement of photocatalytic CO_2 reduction. The highest photocatalytic performance was observed from $Bi_4O_5Br_xI_{2-x}$ ($x = 1$) solid solution with CO as the major final products. The photocatalytic yield of CO and the apparent quantum yield percentage was found to be larger for Bi_4O_5BrI as compared to other samples as shown in Fig. 16.9D.

The same group later investigated the ultrathin $Bi_4O_5Br_2$ nanosheets ($Bi_4O_5Br_2$–UN) for CO_2 reduction and found excellent enhancement in photocatalytic performance (Bai et al., 2019). Precursor method was used to synthesize $Bi_4O_5Br_2$ and $Bi_4O_5Br_2$–UN with Bi:Br with 1:1 and 2.55:1 ratios, respectively. The obtained UV−visible (UV−vis) diffuse reflectance spectra (DRS) of the samples are displayed in Fig. 16.10A, which confirmed the light absorption ability of the photocatalyst and the flat band potentials was estimated to be −0.27 and −0.06 V versus Ag/AgCl electrode, which corresponds to −0.05 and 0.16 V versus the normal hydrogen electrode (NHE).

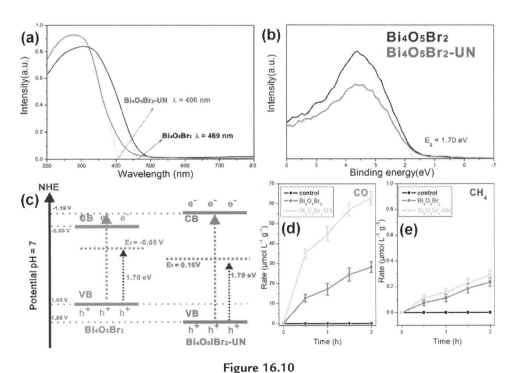

Figure 16.10
(A) UV−vis diffuse reflectance spectra, (B) valence band XPS spectra of $Bi_4O_5Br_2$ and $Bi_4O_5Br_2$–UN, (C) schematic representation of the band structure of $Bi_4O_5Br_2$ and $Bi_4O_5Br_2$–UN and their rate of photocatalytic reduction of CO_2 to (D) CO and (E) CH_4 under UV−vis light irradiation (Bai et al., 2019).

From valance band XPS spectra shown in Fig. 16.10B, the energy gap between VB positions and flat band potential was found to be 1.70 V for $Bi_4O_5Br_2$ and $Bi_4O_5Br_2-UN$. Using the XPS data, the VB and CB potentials were calculated to be 1.65, 1.86 and -0.99, -1.19 V for $Bi_4O_5Br_2$ and $Bi_4O_5Br_2-UN$, respectively, as shown in Fig. 16.10C. It can be clearly seen that the CB potential of $Bi_4O_5Br_2-UN$ is more negative than the CB potential of $Bi_4O_5Br_2$, accordingly, $Bi_4O_5Br_2-UN$ showed high photocatalytic performance as compared to $Bi_4O_5Br_2$ with yield of CO and CH_4 as final products upon CO_2 reduction as shown in Fig. 16.10D and E. $Bi_4O_5Br_2-UN$ was able to show the CO yield of 63.13 μmol/g as major product, which was about 2.3 times higher than that of $Bi_4O_5Br_2$ (27.56 μmol/g) after 2 hours of UV−vis light irradiation. Similarly, the photocatalytic CO_2 reduction by various other BiOX-based photocatalysts is shown in Table 16.3.

16.4.4 Water splitting

BiOX-based photocatalysts have been demonstrated for water splitting in recent years, but their performance has not been satisfactory yet due to their lower position CB minimum. For a photocatalyst to be employed for water splitting applications, the bottom level of CB

Table 16.3: Photocatalytic CO_2 reduction by various BiOX-based photocatalysts.

Photocatalyst	Final products	Yield (in μmol/h/g)	Reference
BiOCl nanoplates	CO	8.1	Zhang et al. (2015)
	CH_4	1.2	
BiOBr	CO	2.67	Ye et al. (2016)
	CH_4	0.16	
$Bi_4O_5I_2$	CO	19.82	Ding et al. (2016)
Bi_5O_7I		1.73	
BiOI	CO	0.51	Ye et al. (2016)
	CH_4	0.18	
$Bi_4O_5Br_xI_{2-x}$	CO	22.85	Bai et al. (2017)
$Bi_4O_5Br_2-UN$	CO	63.13	Bai et al. (2019)
BiOI-001	CO/ CH_4	0.259/0.089	Ye et al. (2016)
BiOI-100	CO/ CH_4	0.076/0.075	
BiOBr	CH_4	4.86	Kong, Lee, Ong, Chai, and Mohamed (2016)
BiOBr−IL	CO	47.1	Wang et al. (2016)
BiOBr−NaBr		21.7	
BOC−OV	CO	16.76	Ma et al. (2017)
$BiOBr_xCl_{1-x}$	CO	15.86	Gao et al. (2019)
$BiOI/g-C_3N_4$	CO	4.86	Wang et al. (2016)
	CH_4	0.18	
$g-C_3N_4/Bi_4O_5I_2$	CO	45.6	Bai et al. (2016)
	CH_4	5.63	
$Au/BiOI/MnO_x$	CO	42.9	Bai et al. (2016)
	CH_4	1.36	

should be more negative than the redox potential of H^+/H_2, while the top level of VB should be more positive than the redox potential of O^+/HO_2. In the early studies conducted by Huizhong et al. (2008), the BiOCl and BiOBr showed O_2 evolution with a rate of 9.36 μmol/h and 11.71, respectively under 10 hours of UV irradiation, while BiOI did not show any O_2 evolution. No H_2 formation was also achieved in the same study even after loading Pt onto the BiOX photocatalysts. From this report, it was found that the CB and VB of BiOCl are formed by Cl 3p and Bi 6p states, respectively. Accordingly, the CB of BiOX was estimated to be 0.09 eV, which is little more positive than the H^+/H_2 potential [0 eV vs standard hydrogen electrode (SHE), pH = 0] because of which no H_2 evolution was observed with the bulk BiOX photocatalysts. Later, various studies have been conducted to improve the water splitting photocatalytic performance of BiOX photocatalysts. One of the studies carried out by Zhang et al. (2013) reported the overall photocatalytic water splitting by BiOCl/CuPc composites from methanol–H_2O–RhB solution under simulated solar irradiation. The same group also reported a study on BiOCl@Au/MnO_x hierarchical structures with selective transport of electron and hole among {001} and {110} facets of BiOCl for pure water splitting, where Au and MnO_x were deposited on the {001} and {110} crystal facets of BiOCl respectively (Zhang, Wang, Sun, Jiang, & Gao, 2015). The highest O_2 evolution rate was observed on BiOCl@Au/MnO_x system as compared to individual performances of BiOCl, BiOCl@Au, and BiOCl@MnO_x with a very small trace of H_2 production. This study also revealed that the main reason for the enhancement was due to the increased spatial separation and transport of charge carriers in the material, which resulted because of the internal electric field along the C-axis in the BiOCl crystal and the strong local electric field induced by the Au nanoparticles along {001} direction.

In another study, Zn-doped BiOBr photocatalyst was studied for visible-light-driven water splitting and dye degradation applications, where Zn was doped into BiOBr through coprecipitation method (Guo et al., 2019). From the obtained UV–vis DRS data, the band gap of the photocatalysts was calculated using Kubelka–Munk equation as shown in Fig. 16.11A and estimated to be 2.42 eV for BiOBr, 2.74 eV for 1/16 Zn–BiOBr, and 2.87 eV for 1/8 Zn–BiOBr from which it can be concluded that the Zn doping showed significant impact on tuning the band-gap structure of BiOBr.

Further in the study, the VB potentials were calculated using XPS data and found to be 2.37, 2.60, and 2.65 eV for BiOBr, 1/16 Zn–BiOBr and 1/8 Zn–BiOBr, respectively, and CB potential were found to be −0.05, −0.16, and −0.22 eV for BiOBr, 1/16 Zn–BiOBr, and 1/8 Zn–BiOBr, respectively, and the band structure was depicted as displayed in Fig. 16.11B. From these CB and VB potential calculations, it can be clearly seen that with increasing Zn doping, the levels of VBM position has become more positive while the CBM position has become more negative, which eventually helped for the enhancement of water splitting performance. DFT calculations were also carried out to understand the band structure, which showed that initially the CB of BiOBr was dominated by Bi 6p states while

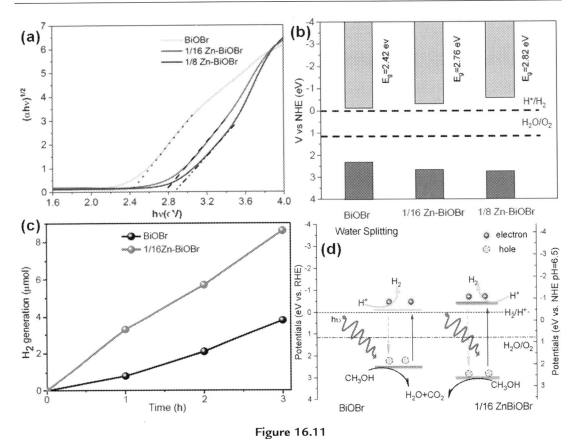

Figure 16.11

(A) Tauc plot of BiOBr and Zn-doped BiOBr photocatalysts, (B) energy-band diagram of BiOBr and Zn-doped BiOBr, (C) H_2 generation over BiOBr and 1/16 Zn−BiOBr, and (D) schematic representation of water splitting mechanism (Guo et al., 2019).

the VB was dominated by Br 4p state along with hybridization of O 2p and Bi 6s electron states. On the other hand the Zn−BiOBr CBM was dominated by stronger density Zn 4s electron states along with hybridization of low-density Bi 6p states, which are slightly above the CBM. The VBM was mainly occupied by slightly lower density Br 4p, O 2p, and Bi 6s states than those of BiOBr. Best photocatalytic water splitting performance was observed by 1/16 Zn−BiOBr, where 29 µmol/g per hour of H_2 evolution rate was obtained, which was much higher as compared with BiOBr that produced the H_2 at the rate of 13 µmol/g per hour. From the performed DFT calculations and other experimental data, the main reason for the enhancement of photocatalytic performance was attributed to the energy difference between CBM and H^+/H_2 for Zn−BiOBr, which is more negative than CBM and H^+/H_2 for BiOBr, which limited the back reaction of H^+ reduction and the recombination of electron−hole as the mechanism depicted in Fig. 16.11D.

One of the examples of BiOX photocatalysts for oxygen evolution through water splitting was reported by Ji et al. (2019) in which they reported the oxygen evolution through Bi-rich bismuth oxyiodide microspheres with rich oxygen vacancy ($Bi_7O_9I_3$−OVR). Even though bismuth oxyiodide showed visible light responsive band gap, it exhibited insufficient redox abilities due to the unmatched band edge potentials, which decreased its possibilities to be directly used for water splitting applications. One of the strategies used for tuning of band gap is by increasing the concentration of bismuth. Accordingly, in this study, in addition to increasing the bismuth content, OV were also introduced as they help to alter the electronic band structure of oxygen framework and also act as adsorption sites leading to the enhancement of overall photocatalytic performance. One-pot ionic liquid assisted solvothermal method was used for synthesis of rich oxygen vacancy $Bi_7O_9I_3$−OVR and partial oxygen vacancy $Bi_7O_9I_3$−OVL microspheres and investigated for their photocatalytic O_2 evolution performance as shown in Fig. 16.12A.

Through XPS analysis and UV−vis diffuse reflection spectra, the band gap of $Bi_7O_9I_3$−OVR and $Bi_7O_9I_3$−OVL was calculated to be 1.79 and 1.84 eV, respectively. The VBM and CBM potentials of $Bi_7O_9I_3$−OVR and $Bi_7O_9I_3$−OVL were found to be 0.95, 1.69 eV and −0.84, −0.15 eV, respectively, as depicted in Fig. 16.12B. The upshifting of VBM position helped for the enhancement of separation of charge carriers and also better photooxidation ability of the holes by increasing the mobility of holes. On the other hand, the migrated electrons with more negative potential on the CB possessed high reduction ability and produced more radical species thus suppressing the undesired recombination of electron−hole pairs. After 6-hours light irradiation, the observed O_2 evolution rate of $Bi_7O_9I_3$−OVR was found to be 2.66 μmol/m²/h (199.26 μmol/g/h), while the rate of $Bi_7O_9I_3$−lVR was measured to be 1.94 μmol/m²/h (129.96 μmol/g/h), from which it can be clearly seen that the increased OV in the system helped in increasing the oxygen activation

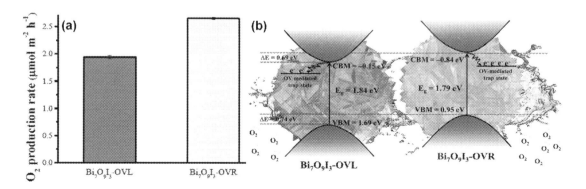

Figure 16.12

Photocatalytic (A) O_2 evolution efficiency and (B) the mechanism of $Bi_7O_9I_3$−OVR and $Bi_7O_9I_3$−OVL microsphere photocatalysts (Ji et al., 2019).

Table 16.4: Photocatalystic H_2 and O_2 evolution by various BiOX photocatalysts.

Photocatalyst	Final product	Yield	Reference
BiOCl/CuPc	H_2	20 μmol/h/g	Zhang et al. (2013)
BU–BiOCl	H_2	2.51 μmol/h	Ye et al. (2015)
BU–BiOCl/Pt		3.96 μmol/h	
BOC–S	H_2	8.4 μmol/h	Zhang et al. (2015)
NiOx@BOC–S		2.6 μmol/h	
BiOCl-{001}–NiO$_x$	H_2	0.24 μmol/h/g	Li, Zhao, Yu, and Zhang (2015)
BiOCl-{010}–NiO$_x$		0.42 μmol/h/g	
$Bi_4O_5Br_2$	H_2	6.81 μmol/h	Bai, Chen, Wang, Wang, and Ye (2016)
BiOBr/α-Fe$_2$O$_3$	H_2	193.0 μmol/g	Si et al. (2017)
CdS QDs/BiOI	H_2	203.0 μmol/h	Kandi, Martha, Thirumurugan, and Parida (2017)
CdS–BiOCl/PAN	H_2	288.65 μmol/h/g	Zhang, Li, Bai, He, and Li (2019)
BiOC–C1	O_2	25.0 μmol/L	Li, Cai, Shang, Yu, and Zhang (2016)
BiOC–C2		55.0 μmol/L	
BiOC–C3		90.0 μmol/L	
Bi_3O_4Cl/BiOCl	O_2	58.6 μmol/h/g	Ning et al. (2019)
BiOI/BiOCl	O_2	7.57 μmol/h	Liu et al. (2017)
Defect engineered BiOCl nanosheets	O_2	56.85 μmol/h/g	Di et al. (2017)

capacity, thereby led to an enhanced oxygen evolution in the system. Table 16.4 presents some of the selective BiOX-based systems and their photocatalytic H_2 and O_2 evolution efficiencies.

16.4.5 Nitrogen fixation

Photocatalytic nitrogen fixation promises both environment-friendly, simple, and cost-effective approach to convert N_2 to NH_3. N_2 fixation is much more challenging as compared to CO_2 reduction and H_2 evolution due to the high energy of N_2 intermediates involved during the fixation process along with the poor binding of N_2 with the photocatalysts. Moreover, high negative reduction potential such as -4.2 V [vs NHE via $N_2 + e^- \rightarrow N_2^-$ (aq)] is required to produce N_2^-, which is not typically possible with the traditional photocatalysts. Even with proton-coupled electron transfer (i.e., $N_2 + H^+ + e^- \rightarrow N_2H$), the reduction potential is still as high as -3.2 V versus NHE, which cannot be achieved via organic scavengers or metal cocatalysts loading. Also, the other biggest challenge in the N_2 fixation process is the activation and breaking of the highly stable N–N triple bonds. Hence, it is generally very challenging to design a visible-light-responsive photocatalyst with enough activation centers to absorb N_2 onto its surface along with the capability of activating and breaking the N–N triple bonds. Notably, the BiOX photocatalysts possess high positive CB edge potential that makes it even more difficult to be employed for N_2 fixation, but their visible light narrow band gap and superior electrical properties make BiOX prospective photocatalysts for N_2 fixation.

For the first time, Li, Shang, Ai, and Zhang (2015) investigated the BiOBr nanosheets with OV exposed on the {001} facets for N_2 fixation without using organic scavengers or any precious metal cocatalysts. Firstly, they investigated the possibility of N_2 fixation to NH_3 through theoretical calculations, which revealed that the Baders charges of two OV-connected Bi atoms (Bi_1, Bi_2) have been increased from 2.87 e to 3.34 e and 3.32 e after the formation of oxygen vacancy on the surface, which indicated that there was a partial reduction of Bi ions by the localized electrons, and through these partially reduced ions as well as with the coordination of OV-connected Bi atoms, the N_2 molecules can be effectively absorbed onto the sites of OV. To study the electron transfer behavior, the charge density difference was employed and it revealed that the major exchange and transfer of electrons took place between the OVs and the adsorbed N_2. From the electron depletion on the OV and the electron accumulation on the adsorbed N_2, it can be estimated that the back transfer of charges from OV to N_2 takes place and results in the bond activation of N—N triple bond. In the next step, it was predicted that the coordinated and activated N_2 will lead to the formation of high-energy intermediates and finally leading to the formation of NH_3 or N_2H_4. Accordingly, a simple solvothermal approach was used to synthesize the {001} facet exposed BiOBr nanosheets with OVs (BOB-001-OV) as well as a reference sample without OVs was also synthesized and labeled as BOB-001-H. After 60 minutes of visible light irradiation, there was no NH_3 traced over BOB-001-H sample, while BOB-001-OV showed a N_2 fixation rate of 104.2 μmol/h/g of catalyst. Under UV—vis light irradiation, the N_2 fixation rate was increased to 223.3 μmol/h in BOB-001-OV. This study essentially paved ways for new investigations and studies on the possibility of employing the BiOX-based photocatalysts for N_2 fixation applications.

In another study, the OV induced Fe-doped BiOCl nanosheets [BiOCl NSs-Fe-x% (x = 1.5, 2.5, 5, 7.5, 10)] with varying Fe ratios were reported for the photocatalytic N_2 fixation (Zhang et al., 2019). Hydrothermal method was adopted to synthesize BiOCl NSs-Fe-x% and the photocatalytic activity was carried out under a full spectrum light irradiation. The photocatalytic NH_3 production with varying Fe ratios as well as under varying environmental conditions is shown in Fig. 16.13A and B. After one hour of light irradiation, BiOCl NSs-Fe-5% showed the highest NH_3 production rate of 1.022 mmol/g/h, which is around 2.53 times higher than that of the bare-BiOCl nanosheets. Further, the CB and VB positions were calculated to be −1.60 and 1.37 eV, respectively, as shown in Fig. 16.13C, where it can be observed that the CB potential is more negative and it was favorable towards NH_3 production. The schematic photocatalytic N_2 fixation is shown in Fig. 16.13D and the photocatalytic mechanism was explained in four steps. The first step involves the escape of few O atoms from the surface of the catalyst with OVs generated on the surface. Next steps involve the chemisorption and activation of N_2 by the OVs followed by the injection of photoinduced electrons into the antibonding π-orbitals of the activated N_2,

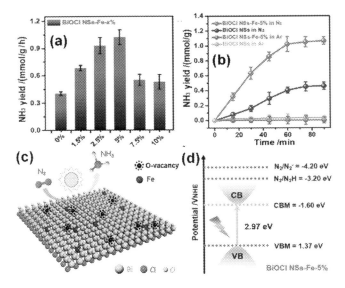

Figure 16.13

(A) Photocatalytic NH_3 production efficiency of BiOCl NSs-Fe-x% in the first 1 hour, (B) quantitative determination of NH_3 yields over the BiOCl NSs and BiOCl NSs-Fe-5% in N_2 and Ar, (C) schematic illustration of the proposed photocatalytic N_2 fixation mechanism, and (D) band energy-level diagram for the BiOCl NSs-Fe-5% (Zhang et al., 2019).

Table 16.5: Photocatalystic N_2 fixation to NH_3 production by various BiOX photocatalysts.

Photocatalyst	Scavenger	NH_3 yield (μmol/h/g)	Reference
BOB-001-OV	—	104.2	Li et al. (2015)
BiOCl NSs—Fe	$AgNO_3$	1.022	Zhang et al. (2019)
Bi_5O_7I—001	CH_3OH	111.5 μmol/L/h	Bai et al. (2016)
Bi_5O_7I—100		47.6 μmol/L/h	
Bi_5O_7Br	—	1.38	Wang et al. (2017)
C—BiOI	Ethanol	311	Zeng et al. (2019)
SUC Bi_3O_4Br	—	380.0	Di et al. (2019)

which leads to the reduction of N_2. In the final step, the OVs are refilled by the vicinal O atoms from H_2O or O_2 and thereby the system showed the enhanced N_2 fixation efficiency. Similar to such results, the N_2 fixations to NH_3 productions by some of the selective BiOX-based photocatalysts are presented in Table 16.5.

16.4.6 Other photocatalytic applications of BiOX materials

BiOX-based materials have also been investigated for the photocatalytic heavy metal reduction and microbial disinfection applications. In this direction, a Z-scheme based BiOX composite consisting of bismuth oxyiodide/reduced graphene oxide/bismuth sulfide (BiOI/RGO/Bi$_2$S$_3$) was developed and investigated for the removal of Cr(VI) and phenol contaminants under visible light irradiation (Chen, Bian, Xu, Xin, & Wang, 2017). Electrostatic self-assembly method was used for the synthesis of this Z-scheme. XPS analysis and UV−vis diffuse reflection spectra were used for calculating VBM and CBM potentials of BiOI and Bi$_2$S$_3$. The VBM positions of BiOI and Bi$_2$S$_3$ were found to be 1.43 and 1.09 eV, respectively, while the CBM position was calculated to be −0.39 and −0.74 eV, respectively, from which it can be observed that BiOI possesses strong oxidative activity due to its higher VBM potential and Bi$_2$S$_3$ possesses strong reduction ability because of its lower CBM value. In both cases of removal of Cr(VI) and degradation of phenol, the BiOI/RGO/Bi$_2$S$_3$ composite showed higher photocatalytic performance than the other individual counter parts as shown in Fig. 16.14A and B. Due to Z-scheme mechanism, the electrons from the CB of BiOI recombine with the holes of VB of Bi$_2$S$_3$ through the surface of rGO that acted as an electron mediator in this case. The electrons on the CB of Bi$_2$S$_3$ reduced Cr(VI) to Cr(III), while the holes on the VB of BiOI oxidized the phenol molecules as depicted in Fig. 16.14C.

In another study, the graphene-like boron nitride-doped bismuth oxychloride (BN/BiOCl) with varying ratios of BN/BiOCl materials with the mass ratios 0.2%, 0.5%, 1%, and 2%

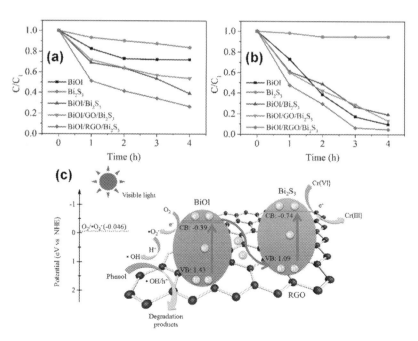

Figure 16.14

Photocatalytic (A) Cr(VI) reduction, (B) phenol oxidation, and (C) schematic illustration of the mechanism of Z-scheme-based BiOI/RGO/Bi$_2$S$_3$ photocatalyst (Chen et al., 2017).

was studied for Cr(VI) reduction. Among the various samples, the BN/BiOCl-1% showed the highest removal rate, which was around 2.39 times higher than that of pure BiOCl (Xu, Wu, Ding, & Gao, 2017). Similarly, another study reported the Cr(VI) removal using bismuth oxybromide ($Bi_{24}O_{31}Br_{10}$), which showed complete removal of Cr(VI) ions within 40 minutes of irradiation (Shang et al., 2014).

Interestingly, the Bi_2WO_6/BiOI-based p−n heterojunction was used for the application of antifouling activity under visible light irradiation, in which 99.99% of *Pseudomonas aeruginosa* (*P. aeruginosa*), *Escherichia coli* (*E. coli*) and *Staphylococcus aureus* (*S. aureus*) bacteria were killed within 60 minutes in the presence of 30% Bi_2WO_6/BiOI (Xiang et al., 2016). Similarly, the CuI−BiOI/Cu films were used for the photocatalytic inactivation of *E. coli* under visible light irradiation, which showed almost complete inactivation within 2 hours of irradiation (Zhang et al., 2018). In another study, a ternary composite composed of g-C_3N_4/BiOI/BiOBr was developed and investigated for the inactivation of *E. coli* under visible light irradiation. The obtained results showed that almost all ($\sim$100%) *E. coli* bacteria were killed over g-C_3N_4/BiOI/BiOBr composite within 3 hours of visible light irradiation, while only 29.4%, 34.2%, and 83.7% of *E. coli* inactivation was happened over the bare-BiOBr, BiOI, and g-C_3N_4 photocatalysts, respectively (Liu et al., 2018). The observed efficiency was attributed to the synergistic properties of the composite towards producing highly reactive oxygen species for the effective killing of the bacteria.

16.5 Summary and outlook

Bismuth oxyhalides due to their unique layered structure and tunable electronic properties along with the other advantages such as high stability and tunable band gaps have gained great attention for photocatalytic applications especially for environmental remediation, such as pollutant degradation, CO_2 reduction, and N_2 fixation. However, BiOX often exhibits few drawbacks, such as limited visible light absorption, high recombination of electron−hole pairs, and the positive conduction band minimum potentials, which hinder its photocatalytic applications. Few strategies, such as making the CBM of BiOX more negative by increasing the bismuth content, inner coupling of BiOX with other materials, formation of heterojunctions and Z-scheme with other semiconductors, varying the synthesis processes, and creating OV, have been largely explored and found to be useful to some extent for photocatalytic applications. In order to further enhance the photocatalytic performance of BiOX, more understanding towards its photocatalytic mechanism is needed. More focus has to be devoted on understanding the role of internal electric field induced inside the BiOX during the photocatalytic process, which is expected to be helpful in effectively tuning their photocatalytic performances. Upon exploring and devising such possibilities, it can be expected that the

BiOX-based photocatalysts can be potentially used for their large-scale applications in the field of photocatalysis.

Acknowledgments

M. Sakar gratefully acknowledges the Department of Science and Technology, Government of India for the funding support through the DST-INSPIRE Faculty Award (DST/INSPIRE/04/ 2016/002227, 14−02−2017).

References

Abazari, R., Mahjoub, A. R., Sanati, S., Rezvani, Z., Hou, Z., & Dai, H. (2019). Ni − Ti layered double hydroxide@graphitic carbon nitride nanosheet: A novel nanocomposite with high and ultrafast sonophotocatalytic performance for degradation of antibiotics. *Inorganic Chemistry, 58*, 1834−1849.

Akkaria, M., Arandaa, P., Belverc, C., Bediac, J., Amarab, A. B. H., & Hitzky, E. R. (2018). ZnO/sepiolite heterostructured materials for solar photocatalytic degradation of pharmaceuticals in wastewater. *Applied Clay Science, 156*, 104−109.

Alansi, A. M., Al-qunaibit, M., Alade, I. O., Qahtan, T. F., & Saleh, T. A. (2018). Visible-light responsive BiOBr nanoparticles loaded on reduced graphene oxide for the photocatalytic degradation of dye. *Journal of Molecular Liquids, 253*, 297−304.

Ankush., Mandal, M. K., Sharma, M., Khushboo., Pandey, S., & Dubey, K. K. (2019). Membrane technologies for the treatment of pharmaceutical industry wastewater. *Water and Wastewater Treatment Technologies*, 103−116.

Bai, Y., Chen, T., Wang, P., Wang, L., & Ye, L. (2016). Bismuth-rich $Bi_4O_5X_2$ (X = Br, and I) nanosheets with dominant {101} facets exposure for photocatalytic H_2 evolution. *Chemical Engineering Journal, 304*, 454−460.

Bai, Y., Yang, P., Wang, L., Yang, B., Xie, H., Zhou, Y., & Ye, L. (2019). Ultrathin $Bi_4O_5Br_2$ nanosheets for selective photocatalytic CO_2 conversion into CO. *Chemical Engineering Journal, 360*, 473−482.

Bai, Y., Ye, L., Chen, T., Wang, L., Shi, X., Zhang, X., & Chen, D. (2016). Facet-dependent photocatalytic N_2 fixation of bismuth-rich Bi_5O_7I nanosheets. *ACS Applied Materials & Interfaces, 8*, 27661−27668.

Bai, Y., Ye, L., Chen, T., Wang, P., Wang, L., Shi, X., & Wong, P. K. (2017). Synthesis of hierarchical bismuth-rich $Bi_4O_5Br_xI_{2-x}$ solid solutions for enhanced photocatalytic activities of CO_2 conversion and Cr(VI)reduction under visible light. *Applied Catalysis B: Environmental, 203*, 633−640.

Bai, Y., Ye, L., Wang, L., Shi, X., Wang, P., & Bai, W. (2016). Dual-cocatalysts loading $Au/BiOI/MnO_x$ system for enhanced photocatalytic greenhouse gases conversion into solar fuels. *Environmental Science: Nano, 3*, 902−909.

Bai, Y., Ye, L., Wang, L., Shi, X., Wang, P., Bai, W., & Wong, P. K. (2016). g-C_3N_4/$Bi_4O_5I_2$ heterojunction with I_3^-/I^- redox mediator for enhanced photocatalytic CO_2 conversion. *Applied Catalysis B: Environmental, 194*, 98−104.

Bannister, F. A. (1934). The crystal-structure and optical properties of matlockite (PbFCl). *Mineralogical Magazine and Journal of the Mineralogical Society, 23*, 587−597.

Bhachu, D. S., Moniz, S. J. A., Sathasivam, S., Scanlon, D. O., Walsh, A., Bawaked, S. M., . . . Carmalt, C. J. (2016). Bismuth oxyhalides: Synthesis, structure and photoelectrochemical activity. *Chemical Science, 7*, 4832−4841.

Bi, Q., Li, Q., Su, Z., Chen, R., Shi, C., & Chen, T. (2019). Room temperature synthesis of ultrathin iodine-doped BiOCl nanosheets. *Colloids and Surfaces, 582*, 123899.

Cai, L., Yao, J., Li, J., Zhang, Y., & Wei, Y. (2019). Sonochemical synthesis of $BiOI$-TiO_2 heterojunction with enhanced visible-light-driven photocatalytic activity. *Journal of Alloys and Compounds, 783*, 300−330.

Chang, J. Q., Zhong, Y., Hu, C. H., Luo, J. L., & Wang, P. G. (2018). Synthesis and significantly enhanced visible light photocatalytic activity of BiOCl/AgBr heterostructured composites. *Inorganic Chemistry Communications*, *96*, 145−152.

Chen, A., Bian, Z., Xu, J., Xin, X., & Wang, H. (2017). Simultaneous removal of Cr(VI) and phenol contaminants using Z-scheme bismuth oxyiodide/reduced graphene oxide/bismuth sulfide system under visible-light irradiation. *Chemoshere*, *188*, 659−666.

Chen, C. C., Fu, J. Y., Chang, J. L., Huang, S. T., Yeh, T. W., Hung, J. T., . . . Chen, L. W. (2018). Bismuth oxyfluoride/bismuth oxyiodide nanocomposites enhance visible-light-driven photocatalytic activity. *Journal of Colloid and Interface Science*, *532*, 375−386.

Cheng, H., Huang, B., Qin, X., Zhang, X., & Dai, Y. (2012). A controlled anion exchange strategy to synthesize Bi_2S_3 nanocrystals/BiOCl hybrid architectures with efficient visible light photoactivity. *ChemComm*, *48*, 97−99.

Cheng, L., Xiang, Q., Liao, Y., & Zhang, H. (2018). CdS-based photocatalysts. *Energy & Environmental Science*, *11*, 1362−1391.

Chou, S. Y., Chen, C. C., Dai, Y. M., Lin, J. H., & Lee, W. W. (2016). Novel synthesis of bismuth oxyiodide/graphitic carbon nitride nanocomposite with enhanced visible-light photocatalytic activity. *RSC Advances*, *6*, 33478−33491.

Cui, W., An, W., Liu, L., Hu, J., & Liang, Y. (2014). Synthesis of CdS/BiOBr composite and its enhanced photocatalytic degradation for rhodamine B. *Applied Surface Science*, *319*, 298−305.

de Assis, G. C., Skovroinski, E., Leite, V. D., Rodrigues, M. O., Galembeck, A., Alves, M. C. F., . . . de Oliveira, R. J. (2018). Conversion of "waste plastic" into photocatalytic nanofoams for environmental remediation. *ACS Applied Materials & Interfaces*, *10*, 8077−8085.

Di, J., Chen, C., Yang, S. Z., Ji, M., Yan, C., Gu, K., . . . Liu, Z. (2017). Defect engineering in atomically-thin bismuth oxychloride towards photocatalytic oxygen evolution. *Journal of Materials Chemistry A*, *5*, 14144−14151.

Di, J., Xia, J., Chisholm, M. F., Zhong, J., Chen, C., Cao, X., . . . Dai, S. (2019). Defect-tailoring mediated electron−hole separation in single-unit-cell Bi_3O_4Br nanosheets for boosting photocatalytic hydrogen evolution and nitrogen fixation. *Advanced Materials*, *31*, 1807576.

Di, J., Xia, J., Li, H., Guo, S., & Dai, S. (2017). Bismuth oxyhalide layered materials for energy and environmental applications. *Nano Energy*, *41*, 172−192.

Ding, C., Ye, L., Zhao, Q., Zhong, Z., Liu, K., Xie, H., . . . Huang, Z. (2016). Synthesis of $Bi_xO_yI_z$ from molecular precursor and selective photoreduction of CO_2 into CO. *Journal of CO_2 Utilization*, *14*, 135−142.

Dumrongrojthanath, P., Phuruangrat, A., Doungarno, K., Thongtem, T., Patiphatpanya, P., & Thongtem, S. (2018). Microwave-hydrothermal synthesis of $BiOBr/Bi_2WO_6$ nanocomposites for enhanced photocatalytic performance. *Ceramics International*, *44*, 148−151.

Ebhota, W. S., & Jen, T. C. (2020). Fossil fuels environmental challenges and the role of solar photovoltaic technology advances in fast tracking hybrid renewable energy system. *International Journal of Precision Engineering and Manufacturing Green Tech*, *7*, 97−117.

Fan, L., Wei, B., Xu, L., Liu, Y., Cao, W., Ma, N., & Gao, H. (2016). Ion exchange synthesis of $Bi_2MoO_6/BiOI$ heterojunctions for photocatalytic degradation and photoelectrochemical water splitting. *Nano*, *11*, 1650095.

Fang, S., Ding, C., Liang, Q., Li, Z., Xu, S., Peng, Y., & Lu, D. (2016). In-situ precipitation synthesis of novel $BiOCl/Ag_2CO_3$ hybrids with highly efficient visible-light-driven photocatalytic activity. *Journal of Alloys and Compounds*, *684*, 230−236.

Ganguly, P., Byrne, C., Breen, A., & Pillai, S. C. (2018). Antimicrobial activity of photocatalysts: Fundamentals, mechanisms, kinetics and recent advances. *Applied Catalysis B: Environmental*, *225*, 51−75.

Gao, J., Gao, Y., Sui, Z., Dong, Z., Wang, S., & Zou, D. (2018). Hydrothermal synthesis of $BiOBr/FeWO_4$ composite photocatalysts and their photocatalytic degradation of doxycycline. *Journal of Alloys and Compounds*, *732*, 43−51.

Gao, M., Yang, J., Sun, T., Zhang, Z., Zhang, D., Huang, H., . . . Wang, X. (2019). Persian buttercup-like $BiOBr_xCl_{1-x}$ solid solution for photocatalytic overall CO_2 reduction to CO and O_2. *Applied Catalysis B: Environmental, 243,* 734−740.

Guo, J., Liao, X., Lee, M. H., Hyett, G., Huang, C. C., Hewak, D. W., . . . Jiang, Z. (2019). Experimental and DFT insights of the Zn-doping effects on the visible-light photocatalytic water splitting and dye decomposition over Zn-doped BiOBr photocatalysts. *Applied Catalysis B: Environmental, 243,* 502−512.

Han, Q., Yang, Z., Wang, L., Shen, Z., Wang, X., Zhu, J., & Jiang, X. (2017). An ion exchange strategy to $BiOI/CH_3COO(BiO)$ heterojunction with enhanced visible-light photocatalytic activity. *Applied Surface Science, 403,* 103−111.

He, F., Meng, A., Cheng, B., Ho, W., & Yu, J. (2020). Enhanced photocatalytic H_2-production activity of WO_3/TiO_2 step-scheme heterojunction by graphene modification. *Chinese Journal of Catalysis, 41,* 9−20.

He, R., Cao, S., Yu, J., & Yang, Y. (2016). Microwave-assisted solvothermal synthesis of $Bi_4O_5I_2$ hierarchical architectures with high photocatalytic performance. *Catalysis Today, 264,* 221−228.

Henle, J., Simon, P., Frenzel, A., Scholz, S., & Kaskel, S. (2007). Nanosized BiOX (X = Cl, Br, I) particles synthesized in reverse microemulsions. *Chemistry of Materials, 19,* 366−373.

Hou, J., Tang, J., Feng, K., Idrees, F., Tahir, M., Sun, X., & Wang, X. (2019). The chemical precipitation synthesis of nanorose-shaped $Bi_4O_5I_2$ with highly visible light photocatalytic performance. *Materials Letters, 252,* 106−109.

Hu, Q., Ji, M., Di, J., Wang, B., Xia, J., Zhao, Y., & Li, H. (2018). Ionic liquid-induced double regulation of carbon quantum dots modified bismuth oxychloride/bismuth oxybromide nanosheets with enhanced visible-light photocatalytic activity. *Journal of Colloid and Interface Science, 519,* 263−272.

Hu, X., Xu, Y., Zhu, H., Hua, F., & Zhu, S. (2016). Controllable hydrothermal synthesis of BiOCl nanoplates with high exposed {001} facets. *Materials Science in Semiconductor Processing, 41,* 12−16.

Huang, Y., Li, H., Balogun, M. S., Liu, W., Tong, Y., Lu, X., & Ji, H. (2014). Oxygen vacancy induced bismuth oxyiodide with remarkably increased visible-light absorption and superior photocatalytic performance. *ACS Applied Materials & Interfaces, 6,* 22920−22927.

Huizhong, A. N., Yi, D. U., Tianmin, W., Cong, W., Weichang, H., & Junying, Z. (2008). Photocatalytic properties of BiOX (X = Cl, Br, and I). *Rare Metals, 27,* 243−250.

Ji, M., Chen, R., Di, J., Liu, Y., Li, K., Chen, Z., . . . Li, H. (2019). Oxygen vacancies modulated Bi-rich bismuth oxyiodide microspheres with tunable valence band position to boost the photocatalytic activity. *Journal of Colloid and Interface Science, 553,* 612−620.

Jiang, J., Zhao, K., Xiao, X., & Zhang, L. (2012). Synthesis and facet-dependent photoreactivity of BiOCl single-crystalline nanosheets. *Journal of the American Chemical Society, 134,* 4473−4476.

Jiang, M., Shi, Y., Huang, J., Wang, L., She, H., Tong, J., . . . Wang, Q. (2018). Synthesis of flower-like g-C_3N_4/BiOBr with enhanced visible light photocatalytic activity for degradation of dyes. *European Journal of Inorganic Chemistry, 17,* 1834−1841.

Jiang, Q., Ji, M., Chen, R., Zhang, Y., Li, K., Meng, C., . . . Xia, J. (2020). Ionic liquid induced mechanochemical synthesis of BiOBr ultrathin nanosheets at ambient temperature with superior visible-light-driven photocatalysis. *Journal of Colloid and Interface Science, 574,* 131−139.

Jiang, X., Lai, S., Xu, W., Fang, J., Chen, X., Beiyuan, J., . . . Guan, G. (2019). Novel ternary BiOI/g-C_3N_4/CeO_2 catalysts for enhanced photocatalytic degradation of tetracycline under visible-light radiation via double charge transfer process. *Journal of Alloys and Compounds, 809,* 151804.

Kandi, D., Martha, S., Thirumurugan, A., & Parida, K. M. (2017). Modification of BiOI microplates with CdS QDs for enhancing stability, optical property, electronic behavior towards rhodamine B de-colorization and photocatalytic hydrogen evolution. *The Journal of Physical Chemistry C, 121,* 4834−4849.

Khalil, M., Gunlazuardi, J., Ivandini, T. A., & Umar, A. (2019). Photocatalytic conversion of CO_2 using earth-abundant catalysts: A review on mechanism and catalytic performance. *Renewable and Sustainable Energy Reviews, 113,* 109246.

Kong, X. Y., Lee, W. P. C., Ong, W. J., Chai, S. P., & Mohamed, A. R. (2016). Oxygen-deficient BiOBr as a highly stable photocatalyst for efficient CO_2 reduction into renewable carbon-neutral fuels. *ChemCatChem, 8,* 1−9.

Lakshmi, K., Varadharajan, V., & Kadirvelu, K. G. (2020). *Photocatalytic decontamination of organic pollutants using advanced materials* M. Oves, M. Ansari, M. Z. Khan, M. Shahadat, M. I. Ismail (Eds.) Modern age waste water problems (pp. 195−212). Cham: Springer.

Lee, A. H., Wang, Y. C., & Chen, C. C. (2019). Composite photocatalyst, tetragonal lead bismuth oxyiodide/bismuth oxyiodide/graphitic carbon nitride: Synthesis, characterization, and photocatalytic activity. *Journal of Colloid and Interface Science, 533,* 319−332.

Lei, Y., Wang, G., Song, S., Fan, W., & Zhang, H. (2009). Synthesis, characterization and assembly of BiOCl nanostructure and their photocatalytic properties. *CrystEngComm, 11,* 1857−1862.

Li, H., Shang, J., Ai, Z., & Zhang, L. (2015). Efficient visible light nitrogen fixation with BiOBr nanosheets of oxygen vacancies on the exposed {001} facets. *Journal of the American Chemical Society, 137,* 6393−6399.

Li, J., Cai, L., Shang, J., Yu, Y., & Zhang, L. (2016). Giant enhancement of internal electric field boosting bulk charge separation for photocatalysis. *Advanced Materials, 28,* 4059−4064.

Li, J., Zhao, K., Yu, Y., & Zhang, L. (2015). Facet-level mechanistic insights into general homogeneous carbon doping for enhanced solar-to-hydrogen conversion. *Advanced Materials, 25,* 2189−2201.

Li, Q., Guan, Z., Wu, D., Zhao, X., Bao, S., Tian, B., & Zhang, J. (2017). Z-scheme BiOCl-Au-CdS heterostructure with enhanced sunlight-driven photocatalytic activity in degrading water dyes and antibiotics. *ACS Sustainable Chemistry & Engineering, 5,* 6958−6968.

Li, R., Ren, H., Ma, W., Hong, S., Wu, L., & Huang, Y. (2018). Synthesis of BiOBr microspheres with ethanol as self-template and solvent with controllable morphology and photocatalytic activity. *Catalysis Communications, 106,* 1−5.

Li, T., Luo, S., & Yang, L. (2013). Microwave-assisted solvothermal synthesis of flower-like Ag/AgBr/BiOBr microspheres and their high efficient photocatalytic degradation for p-nitrophenol. *Journal of Solid State Chemistry, 206,* 308−316.

Li, X., Xu, H., Wang, L., Zhang, L., Cao, X. F., & Guo, Y. C. (2018). Spinel $NiFe_2O_4$ nanoparticles decorated BiOBr nanosheets for improving the photocatalytic degradation of organic dye pollutants. *Journal of the Taiwan Institute of Chemical Engineers, 85,* 257−264.

Li, Y., Wang, X., Gong, J., Xie, Y., Wu, X., & Zhang, G. (2018). Graphene based nanocomposites for efficient photocatalytic hydrogen evolution: Insight into the interface toward separation of photogenerated charges. *ACS Applied Materials & Interfaces, 10,* 43760−43767.

Lin, X., Huang, T., Huang, F., Wang, W., & Shi, J. (2006). Photocatalytic activity of a Bi-based oxychloride Bi_3O_4Cl. *The Journal of Physical Chemistry B, 110,* 24629−24634.

Liu, B., Han, X., Wang, Y., Fan, X., Wang, Z., Zhang, J., & Shi, H. (2018). Synthesis of g-C_3N_4/BiOI/BiOBr heterostructures for efficient visible light-induced photocatalytic and antibacterial activity. *Journal of Materials Science: Materials in Electronics, 29,* 14300−14310.

Liu, G., Yu, J. C., Lu, G. Q. M., & Cheng, H. M. (2011). Crystal facet engineering of semiconductor photocatalysts: Motivations, advances and unique properties. *ChemComm, 47,* 6763−6783.

Liu, H., Cao, W. R., Su, Y., Chen, Z., & Wang, Y. (2013). Bismuth oxyiodide−graphene nanocomposites with high visible light photocatalytic activity. *Journal of Colloid and Interface Science, 398,* 161−167.

Liu, H., Du, C., Li, M., Zhang, S., Bai, H., Yang, L., & Zhang, S. (2018). One-pot hydrothermal synthesis of SnO_2/BiOBr heterojunction photocatalysts for the efficient degradation of organic pollutants under visible light. *ACS Applied Materials & Interfaces, 10,* 28686−28694.

Liu, S., Chen, J., Liu, D., Shan, L., & Zhang, X. (2019). Improved visible light photocatalytic performance through an in situ composition-transforming synthesis of $BiVO_4$/BiOBr photocatalyst. *Journal of Nanoparticle Research, 21,* 191.

Liu, Y., Xu, J., Wang, L., Zhang, H., Xu, P., Duan, X., . . . Wang, S. (2017). Three-dimensional BiOI/BiOX (X = Cl or Br) nanohybrids for enhanced visible-light photocatalytic activity. *Nanomaterials, 7,* 64.

Liu, Z. S., Liu, J. L., Wang, H. Y., Cao, G., & Niu, J. N. (2016). Boron-doped bismuth oxybromide microspheres with enhanced surface hydroxyl groups: Synthesis, characterization and dramatic photocatalytic activity. *Journal of Colloid and Interface Science, 463*, 324−331.

Liu, Z. S., Wu, B. T., Zhu, Y. B., Wang, F., & Wang, L. G. (2013). Cadmium sulphide quantum dots sensitized hierarchical bismuth oxybromide microsphere with highly efficient photocatalytic activity. *Journal of Colloid and Interface Science, 392*, 337−342.

Long, M., Hu, P., Wu, H., Cai, J., Tan, B., & Zhou, B. (2016). Efficient visible light photocatalytic heterostructure of nonstoichiometric bismuth oxyiodide and iodine intercalated $Bi_2O_2CO_3$. *Applied Catalysis B: Environmental, 184*, 20−27.

Luo, S., Xu, J., Li, Z., Liu, C., Chen, J., Min, X., . . . Huang, Z. (2017). Bismuth oxyiodide coupled with bismuth nanodots for enhanced photocatalytic bisphenol A degradation: Synergistic effects and mechanism insight. *Nanoscale, 9*, 15484−15493.

Mahmoodi, V., Ahmadpour, A., Bastami, T. R., & Mousavian, M. T. H. (2018). Facile synthesis of BiOI nanoparticles at room temperature and evaluation of their photoactivity under sunlight irradiation. *Photochemistry and Photobiology, 94*, 4−16.

Maisang, W., Promnopas, S., Kaowphong, S., Narksitipan, S., Thongtem, S., Wannapop, S., . . . Thongtem, T. (2020). Microwave-assisted hydrothermal synthesis of BiOBr/BiOCl fowerlike composites used for photocatalysis. *Research on Chemical Intermediates, 46*, 2117−2135.

Malathia, A., Arunachalam, P., Kirankumar, V. S., Madhavana, J., & Al-Mayouf, A. M. (2018). An efficient visible light driven bismuth ferrite incorporated bismuth oxyiodide (BiFeO$_3$/BiOI) composite photocatalytic material for degradation of pollutants. *Optical Materials, 84*, 227−235.

Ma, Z., Li, P., Ye, L., Zhou, Y., Su, F., Ding, C., . . . Wong, P. K. (2017). Oxygen vacancies induced exciton dissociation of flexible BiOCl nanosheets for effective photocatalytic CO_2 conversion. *Journal of Materials Chemistry A, 5*, 24995−25004.

Mao, D., Ding, S., Meng, L., Dai, Y., Sun, C., Yang, S., & He, H. (2017). One-pot microemulsion-mediated synthesis of Bi-rich $Bi_4O_5Br_2$ with controllable morphologies and excellent visible-light photocatalytic removal of pollutants. *Applied Catalysis B: Environmental, 207*, 153−165.

Mao, D., Yu, A., Ding, S., Wang, F., Yang, S., Sun, C., . . . Yu, K. (2016). One-pot synthesis of BiOCl half-shells using microemulsion droplets as templates with highly photocatalytic performance for the degradation of ciprofloxacin. *Applied Surface Science, 389*, 742−750.

Mera, A. C., Cruz, A. M., Pérez-Tijerinac, E., Meléndrezd, M. F., & Valdése, H. (2018). Nanostructured BiOI for air pollution control: Microwave-assisted synthesis, characterization and photocatalytic activity toward NO transformation under visible light irradiation. *Materials Science in Semiconductor Processing, 88*, 20−27.

Mera, A. C., Rodriguez, C. A., Melendrez, M. F., & Valdes, H. (2017). Synthesis and characterization of BiOI microspheres under standardized conditions. *Journal of Materials Science, 52*, 944−954.

Mi, Y., Li, H., Zhang, Y., & Hou, W. (2019). Synthesis of belt-like bismuth-rich bismuth oxybromide hierarchical nanostructures with high photocatalytic activities. *Journal of Colloid and Interface Science, 534*, 301−311.

Ning, S., Shi, X., Zhang, H., Lin, H., Zhang, Z., Long, J., . . . Wang, X. (2019). Reconstructing dual-induced {0 0 1} facets bismuth oxychloride nanosheets heterostructures: An effective strategy to promote photocatalytic oxygen evolution. *Sol. RRL, 3*, 1900059.

Niu, J., Dai, P., Zhang, Q., Yao, B., & Yu, X. (2018). Microwave-assisted solvothermal synthesis of novel hierarchical BiOI/rGO composites for efficient photocatalytic degardation of organic pollutants. *Applied Surface Science, 430*, 165−175.

Ogawa, K., Nakada, A., Suzuki, H., Tomita, O., Higashi, M., Saeki, A., . . . Abe, R. (2019). Flux synthesis of layered oxyhalide Bi_4NbO_8Cl photocatalyst for efficient Z-scheme water splitting under visible light. *ACS Applied Materials & Interfaces, 11*, 5642−5650.

Ong, C. B., Ng, L. Y., & Mohammad, A. W. (2018). A review of ZnO nanoparticles as solar photocatalysts: Synthesis, mechanisms and applications. *Renewable and Sustainable Energy Reviews, 81*, 536−551.

Pan, A., Ma, X., Huang, S., Wu, Y., Jia, M., Shi, Y., . . . Liu, Y. (2019). $CsPbBr_3$ perovskite nanocrystal grown on MXene nanosheets for enhanced photoelectric detection and photocatalytic CO_2 reduction. *The Journal of Physical Chemistry Letters*, *10*, 6590−6597.

Pan, M., Zhang, H., Gao, G., Liu, L., & Chen, W. (2015). Facet-dependent catalytic activity of nanosheets assembled BiOI microspheres in degradation of bisphenol A. *Environmental Science & Technology*, *49*(15), 6240−6248.

Patil, S. P., Patil, R. P., Mahajan, V. K., Sonawane, G. H., Shrivastava, V. S., & Sonawane, S. (2016). Facile sonochemical synthesis of BiOBr-graphene oxide nanocomposite with enhanced photocatalytic activity for the degradation of Direct green. *Materials Science in Semiconductor Processing*, *52*, 55−61.

Patiphatpanya, P., Ekthammathat, N., Phuruangrat, A., Thongteme, S., & Thongtem, T. (2018). BiOX (X = Cl, Br, and I) nanoplates prepared by surfactant-free microwave synthesis and their photocatalytic performance. *Russian Journal of Physical Chemistry A*, *92*, 2289−2295.

Ren, L. Z., Zhang, D. E., Hao, X. Y., Xiao, X., Jiang, Y. X., Gong, J. Y., . . . Tong, Z. W. (2017). Facile synthesis of flower-like Pd/BiOCl/BiOI composites and photocatalytic properties. *Materials Research Bulletin*, *94*, 183−189.

Rosman, N. N., Yunus, R. M., Minggu, L. J., Arifin, K., Salehmin, M. N. I., Mohamed, M. A., & Kassim, M. B. (2018). Photocatalytic properties of two-dimensional graphene and layered transition-metal dichalcogenides based photocatalyst for photoelectrochemical hydrogen generation: An overview. *International Journal of Hydrogen Energy*, *43*, 18925−18945.

Shang, J., Hao, W., Lv, X., Wang, T., Wang, X., Du, Y., . . . Wang, J. (2014). Bismuth oxybromide with reasonable photocatalytic reduction activity under visible light. *ACS Catalysis*, *4*, 954−961.

Shang, M., Wang, W., & Zhang, L. (2009). Preparation of BiOBr lamellar structure with high photocatalytic activity by CTAB as Br source and template. *Journal of Hazardous Materials*, *167*, 803−809.

Sharma, K., Dutta, V., Sharma, S., Raizadaa, P., Hosseini-Bandegharaei, A., Thakur, P., & Singh, P. (2019). Recent advances in enhanced photocatalytic activity of bismuth oxyhalides for efficient photocatalysis of organic pollutants in water: A review. *Journal of Industrial and Engineering Chemistry*, *78*, 1−20.

Si, H. Y., Mao, C. J., Xie, Y. M., Sun, X. G., Zhao, J. J., Zhou, N., . . . Li, Y. T. (2017). P−N depleted bulk BiOBr/ α-Fe_2O_3 heterojunctions applied for unbiased solar water splitting. *Dalton Transactions*, *46*, 200−206.

Siao, C. W., Lee, W. L. W., Dai, Y. M., Chung, W. H., Hung, J. T., Huang, P. H., . . . Chen, C. C. (2019). $BiO_xCl_y/BiO_mBr_n/BiO_pI_q$/GO quaternary composites: Syntheses and application of visible-light-driven photocatalytic activities. *Journal of Colloid and Interface Science*, *544*, 25−36.

Sin, J. C., Lim, C. A., Lam, S. M., Mohamed, A. R., & Zeng, H. (2019). Facile synthesis of novel ZnO/Nd-doped BiOBr composites with boosted visible light photocatalytic degradation of phenol. *Materials Letters*, *248*, 20−23.

Song, M. X., Du, M., Liu, Q., Xing, F., Huang, C., & Qiu, X. (2019). Enhancement of photocatalytic activities in hierarchical BiOBr microflowers induced by oxygen vacancies. *Catalysis Today*, *335*, 193−199.

Sun, M., Bi, Y., Yan, T., Zhang, Y., Wu, T., Shao, Y., . . . Du, B. (2018). Room-temperature fabrication of bismuth oxybromide/oxyiodide photocatalyst and efficient degradation of phenolic pollutants under visible light. *Journal of Hazardous Materials*, *358*, 20−32.

Tadjarodi, A., Akhavan, O., Bijanzad, K., & Khiavi, M. M. (2016). Mechanochemically prepared BiOCl nanoplates for removal of rhodamine B and pentachlorophenol. *Monatshefte für Chemie*, *147*, 685−696.

Wang, C. Y., Zhang, X., Qiu, H. B., Wang, W. K., Huang, G. X., Jiang, J., & Yu, H. Q. (2017). Photocatalytic degradation of bisphenol A by oxygen-rich and highly visible-light responsive $Bi_{12}O_{17}Cl_2$ nanobelts. *Applied Catalysis B: Environmental*, *200*, 659−665.

Wang, C. Y., Zhang, X., Zhang, Y. J., Chen, J. J., Huang, G. X., Jiang, J., . . . Yu, H. Q. (2018). Direct generation of hydroxyl radicals over bismuth oxybromide nanobelts with tuned band structure for photocatalytic pollutant degradation under visible light irradiation. *Applied Catalysis B: Environmental*, *237*, 464−472.

Wang, C. Y., Zhang, Y. J., Wang, W. K., Pei, D. N., Huang, G. X., Chen, J. J., . . . Yu, H. Q. (2018). Enhanced photocatalytic degradation of bisphenol A by Co-doped BiOCl nanosheets under visible light irradiation. *Applied Catalysis B: Environmental*, *221*, 320−328.

Wang, H. Y., Liu, Z. S., Guo, L. T., Fan, H. L., & Tao, X. Y. (2018). Novel $Bi_2O_4/BiOBr$ heterojunction photocatalysts: In-situ preparation, photocatalytic activity and mechanism. *Materials Science in Semiconductor Processing*, *77*, 8−15.

Wang, J., Yao, H. C., Fan, Z. Y., Zhang, L., Wang, J., Zang, S. Q., & Li, Z. (2016). Indirect Z-scheme BiOI/g-C_3N_4 photocatalysts with enhanced photoreduction CO_2 activity under visible light irradiation. *ACS Applied Materials & Interfaces*, *8*, 3765−3775.

Wang, P., Yang, P., Bai, Y., Chen, T., Shi, X., Ye, L., & Zhang, X. (2016). Synthesis of 3D BiOBr microspheres for enhanced photocatalytic CO_2 reduction. *Journal of the Taiwan Institute of Chemical Engineers*, *68*, 295−300.

Wang, S., Hai, X., Ding, X., Chang, K., Xiang, Y., Meng, X., . . . Ye, J. (2017). Light-switchable oxygen vacancies in ultrafine Bi_5O_7Br nanotubes for boosting solar-driven nitrogen fixation in pure water. *Advanced Materials*, *29*, 1701774.

Wang, Zheng, Li, Can, & Domen, Kazunari (2019). Recent developments in heterogeneous photocatalysts for solar-driven overall water splitting. *Chemical Society Reviews*, *48*, 2109−2125.

Wu, L., Li, Z., Li, Y., Hu, H., Liu, Y., & Zhang, Q. (2019). Mechanochemical syntheses of bismuth oxybromides $Bi_xO_yBr_z$ as visiblelight responsive photocatalyts for the degradation of bisphenol A. *Journal of Solid State Chemistry*, *270*, 458−462.

Wu, L., Li, Z., Zhang, Q., & Liu, X. (2019). Mechanochemical syntheses of oxygen-rich bismuth oxychlorides $Bi_xO_yCl_z$ to enhance ciprofoxacin degradation under visible light irradiation. *Catalysis Letters*, *149*, 2247−2255.

Wu, S., Wang, C., Cui, Y., Wang, T., Huang, B., Zhang, X., . . . Brault, P. (2010). Synthesis and photocatalytic properties of BiOCl nanowire arrays. *Materials Letters*, *64*, 115−118.

Xiang, Y., Ju, P., Wang, Y., Sun, Y., Zhang, D., & Yu, J. (2016). Chemical etching preparation of the Bi_2WO_6/BiOI p−n heterojunctions with enhanced photocatalytic antifouling activity under visible light irradiation. *Chemical Engineering Journal*, *288*, 264−275.

Xing, Z., Zhang, J., Cui, J., Yin, J., Zhao, T., Kuang, J., . . . Zhou, W. (2018). Recent advances in floating TiO_2-based photocatalysts for environmental application. *Applied Catalysis B: Environmental*, *225*, 452−467.

Xu, H., Wu, Z., Ding, M., & Gao, X. (2017). Microwave-assisted synthesis of flower-like BN/BiOCl composites for photocatalytic Cr(VI) reduction upon visible-light irradiation. *Materials and Design*, *114*, 129−138.

Yadav, M., Garg, S., Chandra, A., & Hernadi, K. (2019). Fabrication of leaf extract mediated bismuth oxybromide/oxyiodide ($BiOBr_xI_{1−x}$) photocatalysts with tunable band gap and enhanced optical absorption for degradation of organic pollutants. *Journal of Colloid and Interface Science*, *555*, 304−314.

Yang, H., Zhang, Q., Chen, Y., Huang, Y., Yang, F., & Lu, Z. (2018). Ultrasonic-microwave synthesis of ZnO/BiOBr functionalized cotton fabrics with antibacterial and photocatalytic properties. *Carbohydrate Polymers*, *201*, 162−171.

Ye, L., Jin, X., Ji, X., Liu, C., Su, Y., Xie, H., & Liu, C. (2016). Facet-dependent photocatalytic reduction of CO_2 on BiOI nanosheets. *Chemical Engineering Journal*, *291*, 39−46.

Ye, L., Jin, X., Leng, Y., Su, Y., Xie, H., & Liu, C. (2015). Synthesis of black ultrathin BiOCl nanosheets for efficient photocatalytic H_2 production under visible light irradiation. *Journal of Power Sources*, *293*, 409−415.

Ye, L., Jin, X., Liu, C., Ding, C., Xie, H., Chu, K. H., & Wong, P. K. (2016). Thickness-ultrathin and bismuth-rich strategies for BiOBr to enhance photoreduction of CO_2 into solar fuels. *Applied Catalysis B: Environmental*, *187*, 281−290.

Ye, L., Wang, H., Jin, X., Su, Y., Wang, D., Xie, H., . . . Liu, X. (2016). Synthesis of olive-green few-layered BiOI for efficient photoreduction of CO_2 into solar fuels under visible/near-infrared light. *Solar Energy Materials and Solar Cells*, *144*, 732−739.

Yosefi, L., Haghighi, M., & Allahyari, S. (2017). Solvothermal synthesis of flowerlike p-BiOI/n-ZnFe$_2$O$_4$ with enhanced visible light driven nanophotocatalyst used in removal of acid orange 7 from wastewater. *Separation and Purification Technology, 178*, 18−28.

Zeng, L., Zhe, F., Wang, Y., Zhang, Q., Zhao, X., Hu, X., . . . He, Y. (2019). Preparation of interstitial carbon doped BiOI for enhanced performance in photocatalytic nitrogen fixation and methyl orange degradation. *Journal of Colloid and Interface Science, 539*, 563−574.

Zhang, J., Lv, J., Dai, K., Liang, C., & Liu, Q. (2018). One-step growth of nanosheet-assembled BiOCl/BiOBr microspheres for highly efficient visible photocatalytic performance. *Applied Surface Sciences, 430*, 639−646.

Zhang, G., Chen, D., Li, N., Xu, Q., Li, H., He, J., & Lu, J. (2018). SnS$_2$/SnO$_2$ heterostructured nanosheet arrays grown on carbon cloth for efficient photocatalytic reduction of Cr(VI). *Journal of Colloid and Interface Science, 514*, 306−315.

Zhang, L., Cao, X. F., Chen, X. T., & Xue, Z. L. (2011). BiOBr hierarchical microspheres: Microwave-assisted solvothermal synthesis, strong adsorption and excellent photocatalytic properties. *Journal of Colloid and Interface Science, 354*(2), 630−636.

Zhang, L., Han, Z., Wang, W., Li, X., Su, Y., Jiang, D., . . . Sun, S. (2015). Solar-light-driven pure water splitting with ultrathin BiOCl nanosheets. *Chemistry—A European Journal, 21*, 18089−18094.

Zhang, L., Wang, W., Jiang, D., Gao, E., & Sun, S. (2015). Photoreduction of CO$_2$ on BiOCl nanoplates with the assistance of photoinduced oxygen vacancies. *Nano Research, 8*, 821−831.

Zhang, L., Wang, W., Sun, S., Jiang, D., & Gao, E. (2015). Selective transport of electron and hole among {0 0 1} and {1 1 0} facets of BiOCl for pure water splitting. *Applied Catalysis B: Environmental, 162*, 470−474.

Zhang, L., Wang, W., Sun, S., Sun, Y., Gao, E., & Xu, J. (2013). Water splitting from dye w..stewater: A case study of BiOCl/copper(II) phthalocyanine composite photocatalyst. *Applied Catalysis B: Environmental, 132−133*, 315−320.

Zhang, N., Li, L., Shao, Q., Zhu, T., Huang, X., & Xiao, X. (2019). Fe-doped BiOCl nanosheets with light-switchable oxygen vacancies for photocatalytic nitrogen fixation. *ACS Applied Energy Materials, 2*, 8394−8398.

Zhang, P., Qiu, Y., Yang, S., Jiao, Y., Ji, C., Li, Y., . . . Fan, H. (2017). Oxygen-deficient bismuth oxychloride nanosheets: Superior photocatalytic performance. *Materials Research Bulletin*, 478−484.

Zhang, Q., Li, G., Bai, J., He, R., & Li, C. (2019). Enhanced photocatalytic activities of CdS-BiOCl/PAN composites towards photocatalytic hydrogen evolution. *Materials Research Bulletin, 1117*, 9−17.

Zhang, S., Chen, X., & Song, L. (2019). Preparation of BiF$_3$/BiOBr heterojunctions from microwave-assisted method and photocatalytic performances. *Journal of Hazardous Materials, 367*, 304−315.

Zhang, S., Gu, P., Ma, R., Luo, C., Wen, T., Zhao, G., . . . Wang, X. (2019). Recent developments in fabrication and structure regulation of visible-lightdriven g-C$_3$N$_4$-based photocatalysts towards water purification: A critical review. *Catalysis Today, 335*, 65−77.

Zhang, W., Dong, F., Xiong, T., & Zhang, Q. (2014). Synthesis of BiOBr−graphene and BiOBr−graphene oxide nanocomposites with enhanced visible light photocatalytic performance. *Ceramic International, 40*, 9003−9008.

Zhang, Y., Lin, C., Lin, Q., Jin, Y., Wang, Y., Zhang, Z., . . . Wang, X. (2018). CuI-BiOI/Cu film for enhanced photo-induced charge separation and visible light antibacterial activity. *Applied Catalysis B: Environmental, 235*, 238−245.

Zhang, Z., Wang, Y., Zhang, X., Zhang, C., Wang, Y., Zhang, H., & Fan, C. (2018). Optimized design of novel Pt decorated 3D BiOBr flower-microsphere synthesis for highly efficient photocatalytic properties. *Chemical Papers, 72*, 2413−2423.

Zhu, D., & Zhou, Q. (2019). Action and mechanism of semiconductor photocatalysis on degradation of organic pollutants in water treatment: A review. *Environmental Nanotechnology, Monitoring & Management, 12*, 100255.

CHAPTER 17

Design of photocatalysts for the decontamination of emerging pharmaceutical pollutants in water

K. Rokesh[1], Mohan Sakar[1,2] and Trong-On Do[1]

[1]Department of Chemical Engineering, Laval University, Quebec, QC, Canada, [2]Centre for Nano and Material Sciences, Jain University, Bangalore, India

17.1 Introduction

Water is the major resources for human beings as well as for the survival of the ecological systems on earth. Nowadays, the demand for water has been increased tremendously with the rapid growth of population and vigorous industrial developments. Similarly, the growing world's population also leads to the extensive consumption of chemicals on various levels; hence there is a rapid increment in the number of chemical industries as well as increasing release of chemicals into the environment. Therefore, maintaining the quality water sources is highly necessary towards keeping healthy ecosystems and assurance of safe drinking water. The common water pollutants include dyes, pesticides, polyaromatic compounds, polychlorinated compounds, and persistent organic pollutants, and they have been identified and removed by appropriate techniques (Pal, Gina, Lin, & Reinhard, 2010). Recently, scientific communities have identified the presence of pharmaceutical residues in water/ wastewater and accordingly, the harmful effects on human life and the living ecosystems have received huge attention worldwide. Up to date, the pharmaceutical compounds are not covered under any pollution regulations, and their effects on the biological and human system are also poorly studied (Küster & Adler, 2014). Therefore, these pharmaceutical products are considered new "emerging micropollutants," where their rising occurrence in global water bodies is shown in Fig. 17.1 (aus der Beek et al., 2016).

In addition, the pharmaceutical pollutants are generating serious toxicity and producing long-term chronic effects to humans as well as the ecosystems. According to the World Health Organization (WHO), the antimicrobial resistance is an emerging problem that essentially generates multidrug-resistant infections to humans and animals. The antibiotic residual in the environment could result into various adverse effects and generate stable organic byproducts,

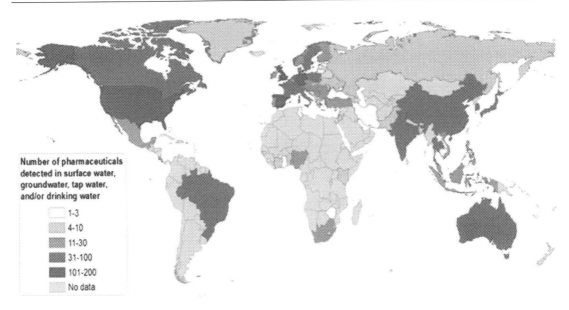

Figure 17.1
The number of pharmaceutical compounds detected in surface waters, groundwater, and tap/drinking water worldwide (aus der Beek et al., 2016).

which are difficult to degrade wastewater treatments and could generate secondary pollution, as well as lead to antibiotic resistance in bacterial populations (Khetan & Collins, 2007; Martínez, 2008). Therefore, there is an urgent need to address this issue and to develop an efficient technique to remove these pollutants from water/wastewater. There are various techniques available to remove pharmaceutical residues from water, such as adsorption, microbial degradation, photocatalysis, ozonolytic, electrocatalysis, and membrane filtration process (Gadipelly et al., 2014). However, the semiconductor-based photocatalytic degradation process has been indentified as an ideal technique thanks to its potential redox properties, which offer an effective mineralization of pollutants, nonharmful byproduct/end product, low cost, recyclability, and an eco-friendly process (Bagheri, TermehYousefi, & Do, 2017). Therefore, this chapter outlines the photocatalytic technology with designing of nanostructured and nanocomposite photocatalysts for the effective degradation of emerging pharmaceutical pollutants (antibiotics) in aqueous environment.

17.2 Photocatalysis and pharmaceutical degradation

Photocatalysis utilizes energy from light and carries out the redox catalytic reactions. A numerous catalytic reaction has been explored, which are potentially useful for energy generation and environmental remediation. As a part clean energy generation, photocatalysis has received great interest thanks to the applications, such as hydrogen production from water,

CO_2 conversion into hydrocarbon fuels, besides reducing the CO_2 emission in the environment (Wang, Li, & Xu, 2017). Similarly, photocatalysis has been found to be promising in environmental applications, such as degradation of organic compounds, pesticides, dyes from water treatment, antimicrobial inactivation, and removal of NO_x/SO_x from air (Hoffmann, Martin, Choi, & Bahnemann, 1995). The understanding of the wide application of photocatalysis reactions is potentially important to apply this technique into emerging environmental and energy issues.

Photocatalysis defined as photoinduced catalytic reaction by acceleration of catalytic process in the presence of light (photon) and the catalyst is called as photocatalyst and the process can be represented as photocatalysis. Generally, as it is known that a photocatalyst is a semiconductor material with the band gap energy (E_g) separated by valance band (VB) and conduction band (CB). The photocatalytic process mainly involves photoabsorption by semiconductor material, then the photoexcited electrons move from VB to CB of the semiconductor, followed by the generation of holes in VB and accumulation of electrons in CB. This electron–hole pair generation and separation process is essential for photocatalytic reaction. The generated free electrons migrated to surface of the semiconductor can undergo the reduction reaction and the holes undergo the oxidation reactions (Banerjee et al., 2014). The oxidation and reduction reactions are the basic mechanisms of the photocatalytic process, which could also lead to secondary reactions under appropriate conditions.

The photocatalytic oxidation occurs in direct and indirect pathways, where the primary reaction involves direct hole oxidization of the adsorbed organic pollutants and the secondary reaction involves oxidization water molecules by holes to produce hydroxyl radicals, which subsequently initiate a chain reaction leading to the oxidation of organic pollutant. On the other hand, the electron can be entrapped by the oxygen molecule, leading to the formation of superoxide radical and oxidizing the pollutants. However, hydroxyl radical (OH$^\bullet$) has attracted more attention due to its strong oxidation potential and can react with complex organic molecules and break into small and nonhazardous end products, such as water and carbon dioxide. In addition, the hydroxyl radicals also offer nonselective destruction as compared to other oxidants, which will be able to react with a wide range of contaminants without any additives (Gaya & Abdullah, 2008). Moreover, the heterogeneous photocatalytic reaction with free radicals generation is able to oxidize and completely mineralize the pharmaceutical pollutants (antibiotics) as depicted in Fig. 17.2.

The degradation rate of pharmaceutical compounds is mainly influenced by many operational parameters, such as loading of catalyst, concentration of the pollutant, pH of the solution, addition of hydrogen peroxide, dissolved oxygen, presence of inorganic ions, light source, and designing of photoreactor. These different optimum parameter conditions are usually studied towards achieving maximum removal efficiency. There are also some limitations associated with the photocatalysis technique. The foremost limitation of this method is its inability to

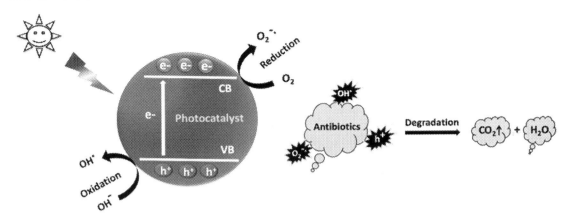

Figure 17.2
Schematic illustration of photocatalytic degradation of antibiotics.

better perform in turbid water and/or high-concentration pollutants. Because of higher concentrations, the pollutant reflects the incident light and diminishes light penetration to reach catalyst surface in the liquid medium. Likewise, the incident light absorbed by pollutants probably allows lower intensity of light to excite the photocatalyst. Therefore, heterogeneous photocatalysis treatment is unable to integrate with primary and secondary treatment process, but it is suggested to be used as a tertiary treatment process, particularly for the elimination of pharmaceutical compounds. The major challenges of photocatalytic process are the limitations of semiconductor photocatalyst. Most of the semiconductor photocatalysts are only UV active due to their wide band gap (> 3.0 eV); however, solar radiation holds only 4% of UV spectrum. Hence, they are unable to utilize the maximum amount of solar energy during the photocatalytic process. Then, the poor photogenerated charge separation leads to higher rate of charge recombination, leading to reduced active species generation as well as photocatalytic activity. Then, the poor surface charge transfer also leads to more charge recombination. In addition, photostability, particle aggregation and recoverability also lead to an inadequate photocatalytic performance. Finally, it is important to achieve the complete mineralization of pollutants, which is very much important, because the intermediates that are generated during the treatment process can even be more toxic than the parent compounds (Gaya & Abdullah, 2008; Hoffmann, Martin, Choi, & Bahnemann, 1995).

17.3 Photocatalysts by design for pharmaceutical pollutant degradation

17.3.1 Semiconductor photocatalysts

Semiconductor photocatalyst material is a key factor in photocatalytic process. The three steps, photoabsorption, charge separation, and charge transfer are the most important steps in semiconductor photocatalysis. The primary limitations of semiconductor photocatalyst is that most of the semiconductor photocatalysts are only UV responsive due to its wide band gap but solar light contains only few percentages of the UV spectrum. Hence, it is important to do maximum utilization of solar radiation during the photocatalytic process. In other words, the photogenerated electron–hole charge recombination is the major limitation of photocatalysts. The electrons–hole pair charge separation occurs after photon absorption to initiate the reduction and oxidation process. Unfortunately, the short lifetime of photogenerated electrons significantly promoted the electron–hole charge recombination processes and limited photocatalytic activity. Therefore, the development of photocatalysis that suppresses charge recombination is the foremost challenge in photocatalyst development. Then, the photogenerated electrons undergo the surface or bulk recombination process, which leads to poor surface charge transfer and reduce the photocatalytic performance. In addition, the photocorrosion of the semiconductor could limit the stability of the photocatalyst as well (Colmenares & Luque, 2014; Gaya & Abdullah, 2008). Therefore, the development of nanostructured and nanocomposites-based semiconductor is a promising strategy to overcome the limitation of simple-semiconductor photocatalysts.

17.3.2 Nanophotocatalysts

Nanotechnology has made ground-breaking impact on photocatalytic material design and development. Nanomaterials offer large surface areas, diverse morphologies, enriched surface, catalytic and physiochemical properties, and easy fabrication (Tong et al., 2012). All these properties are beneficial for the significant development of new nano/nanostructured photocatalyst. Exclusively, nanostructured photocatalyst is offering diverse and potential route towards enhancing the photocatalytic efficiencies in many folds. Nanoscale designing generally favors the surface-dependent photocatalytic activity as it provides more specific surface area and reactive sites. Also, the nanostructured photocatalysts facilitate the separation of a large number of photogenerated electron–hole pairs, which essentially reduces the charge recombination in the system and improves the overall photocatalytic activity. The nanoparticles facilitate the fast charge carrier migration as the distance from the core to the surface of the catalyst is shorter than that in bulk material (Tong et al., 2012). The migration of charge carriers requires a potential gradient for fast and effective charge transfer, which is closely associated with morphology, structure, and surface properties of semiconductor photocatalyst (Fig. 17.3). The notable development has been made in the morphology and size-controlled synthesis, where considerable investigations have been carried out on the relationship between the morphological or structural characteristics and the photocatalytic properties.

For example, the nanoaggregates have accomplished significant advances in tailoring the photocatalytic performances. The synthesis of aggregates is mainly dependent on the

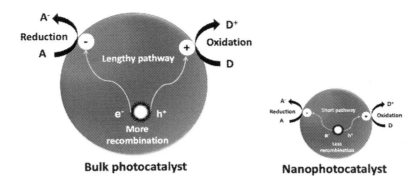

Figure 17.3
Charge carrier migration and recombination process of bulk photocatalyst and nanophotocatalyst.

templates, surfactants, and additives. For instance, ZnO nanoaggregates were synthesized using polyol-aqueous medium by optimizing the polyol and water composition. The photocatalytic performance of ZnO nanoaggregates was examined on ciprofloxacin photodegradation under simulated solar light irradiation and was also compared with different nanostructures (Li et al., 2018). Similarly, the Zn_2GeO_4 hollow spheres were fabricated via in situ template-engaged approach, and with increasing reaction time Zn_2GeO_4 nanorods (400–600 nm size) were gradually transformed into Zn_2GeO_4 nanobundles. As compared to bulk Zn_2GeO_4 (solid state synthesis), the template-assisted Zn_2GeO_4 hollow spheres were found to show efficient photocatalytic activity and excellent photostability on metronidazole degradation due to their hollow structure and high specific surface area (Liu, Zhang, Yu, & Guo, 2013). Then, the feathery-shaped BiOCl nanostructure photocatalyst with large specific surface area and higher oxygen vacancies initiated an efficient visible photocatalytic activity and improved tetracycline hydrochloride degradation. The BiOCl nanostructure photocatalytic efficiency was found to be approximately three times higher than that using bulk BiOCl (Sun et al., 2018).

17.3.3 Nanocomposite photocatalysts

In general, a single-component semiconductor photocatalyst shows the limited solar light absorption, poor charge separation, higher rate of charge recombination, low charge transfer efficiency, and photocorrosion. To develop the nanocomposite photocatalysts, a nanostructured semiconductor can be combined with other components, such as metal/metallic, semiconductor, and carbon nanomaterials, which can potentially overcome these drawbacks of single-component semiconductor photocatalyst and also offer more advantages. Therefore, the construction of nanocomposite semiconductor photocatalyst seems to be particularly important in photocatalysis. Nanocomposite photocatalyst offers improved photogenerated electron–hole charge separation and reduced charge recombination, higher charge transfer, extended light absorption, reduced photocorrosion, and increased photostability. It is realized that the surface and structural properties of semiconductor−metal, semiconductor-semiconductor, semiconductor-carbon systems are the key features to construct potential heterojunction composites. The large surface area and highly efficient interface with effective contacts are important tools for the improved charge transfer and reduced recombination in the system. Likewise, the optimal band gap and band edge alignment also determines the separation and transport efficiency of photogenerated charge carriers, which should be considered during the fabrication of heterojunction nanocomposites. Also, suitable materials should be chosen to couple them in order to achieve highly efficient and stable photocatalysts (Petronella et al., 2017). Therefore, nanocomposite materials play a vital role on improving photocatalytic performance and so longer many nanocomposite photocatalysts have been developed and applied for emerging pharmaceutical pollutant degradation.

17.3.3.1 Semiconductor−metal/metallic composites

Most of the current nanocomposite photocatalysts are generally obtained by loading the metal/metallic nanoclusters onto the surface of semiconductor nanostructure because the incorporation of metal/metallic nanoparticles with semiconductors is one of the most effective routes to achieve higher photocatalytic performance. The semiconductor−metal/ metallic nanocomposite photocatalysts offer two benefits: (1) improved charge separation at the semiconductor−metal interface by the formation of Schottky barrier and (2) increased visible light absorption due to the surface plasmon resonance (SPR) effect (Fig. 17.4; Khan, Chuan, Yosuf, Chowdhury, & Cheng, 2015).

1. Schottky photocatalysts

The development of semiconductor−metal composite established a Schottky barrier via Fermi-level equilibration between metals and semiconductor interface. Due to the existence of

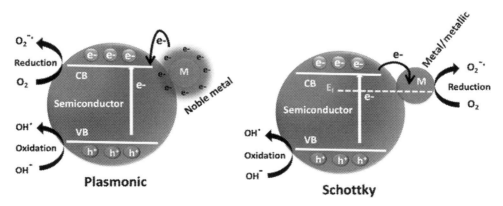

Figure 17.4
Schematic illustrations of working mechanism of plasmonic and Schottky photocatalyst.

the Schottky barrier, the Fermi-level equilibrium alignment of metal and semiconductor leads to the creation of a 'built-in electric field' near the interface, which favors the rapid charge carrier separation in the system (Khan, Chuan, Yosuf, Chowdhury, & Cheng, 2015). Notably, the efficiency of Schottky barrier is also governed by the difference in the work function of metal and semiconductor involved in the system. For instance, the silicate-supported CdS/Pt heterostructure composite was fabricated and Pt nanoparticles were photodeposited on silicate-integrated CdS nanosphere. The Schottky barrier formation between Pt and CdS enhanced the electron—hole pairs separation and suppressed reverse migration of electrons from photoexcitation, thereby improved photocatalytic activity and degradation of tetracycline was achieved (Fang et al., 2017). Similarly, the bimetallic Ag/Pt@TiO_2 composite showed much higher activity than the Ag@TiO_2 and Pt@TiO_2 due to the establishment of Schottky barrier and synergistic effects between the metals and TiO_2. The incorporation of bimetallic nanoparticles on the surface of TiO_2 have become more prominent, since it has drastically improved the photocatalytic activity than in the single metal incorporation. The as-prepared Ag/Pt@TiO_2 composite enhanced the photocatalytic degradation of ciprofloxacin due to the Schottky barrier that increased their efficient charge separation and diffusion (Jiang et al., 2014). On the other hand, the metal carbides are demonstrated to be the promising candidates for photocatalytic applications because of their metal-like properties and good electrical—thermal stability. In this direction, the fabrication of Ag_3PO_4/Ti_3C_2 Schottky photocatalyst has greatly improved the photocatalytic activity and photostability of Ag_3PO_4 towards the degradation of tetracycline. In consequence, the surface hydrophilic functional groups of Ti_3C_2 tend to construct a strong interfacial contact with Ag_3PO_4, and it facilitates to have a strong Schottky junction between Ag_3PO_4 and Ti_3C_2 towards enhancing the charge carrier separation and stability of the photocatalyst (Fig. 17.5). In addition, the strong redox

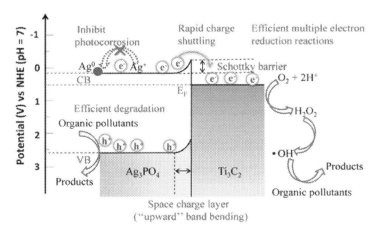

Figure 17.5

The Schottky barrier formation and photocatalytic degradation mechanism of Ag_3PO_4/Ti_3C_2 (Cai et al., 2018).

ability of surface titanium sites has encouraged multiple electron reduction and induced more hydroxyl radical production (Cai et al., 2018).

2. Plasmonic photocatalysts

On the other hand, the metallic components (i.e., noble metals) in the metal/semiconductor heterostructure could also enhance the light absorption of semiconductors through a SPR effect. The SPR effect of noble metals could be described as the collective oscillation of surface electrons; when the frequency of photons matches with the frequency of surface electrons as the oscillation is established against the restoring force of positive nuclei, then it would be transferred to the semiconductor. Thereby, enhanced photoabsorption and photocatalytic activity of the nearby semiconductor photocatalyst can be observed. In such plasmonic photocatalysts, the resonance photon wavelength varies with different metals. Also, the surface plasmon absorption is strongly dependent on the particle size and morphology of metals. In this direction, gold, silver, and platinum have been widely investigated plasmonic metal for improving the photocatalytic performance with various semiconductors. Among them, gold and silver are most commonly used plasmonic metals as they show tunable plasmonic resonance in the visible region (Jiang et al., 2018; Wang et al., 2017). For example, the WO_3/Ag plasmonic composites were developed by photodeposition of Ag NPs onto the WO_3 nanoplate surface. The silver metal particle assigned to SPR effect essentially broadened the visible absorption range of WO_3/Ag composites and improved the photocatalytic degradation of antibiotic sulfanilamide. The photocatalytic results showed that the WO_3/Ag composites accomplished much better activity than pure WO_3, which reached the highest removal efficiency of 96.2% in 5 hours. In addition, the plasmonic composite showed a 100% photocatalytic inhibition efficiency against *Escherichia coli* and *Bacillus subtilis* bacteria under visible light irradiation in 2 hours (Zhu, Liu, Yu, Zhou, & Yan, 2016). More interestingly, the metallic Ag incorporation offered ideal SPR effect and efficient ternary charge transfer mechanism to Ag@PCNS/$BiVO_4$ photocatalyst system. The nanocomposite showed 92.6% of ciprofloxacin removal in 120 minutes under visible light irradiation ($\lambda > 420$ nm), and also presented higher photocatalytic activity than single component or binary composites under near-infrared (NIR) light irradiation ($\lambda > 760$ nm). The excellent photocatalysis was attributed to the strong SPR effect of Ag nanoparticles, establishing local electric field that offered enhanced photoabsorbance. Then, the synergistic effect established between the PCNS and $BiVO_4$ interface via the metallic silver offered an efficient ternary charge transfer mechanism in the system. Therefore, the metallic silver delivered efficient photoinduced charge generation, as well as promoted photogenerated electron–hole charge separation and charge transfer in the Ag@PCNS/$BiVO_4$ photocatalytic system (Deng et al., 2018). Also, this mechanism was found to be suitable for the plasmonic metal and semiconductor composite of Au/Pt/g-C_3N_4 photocatalyst. Under visible light, Au nanoparticles established charge separation and transfer the electrons into g-C_3N_4

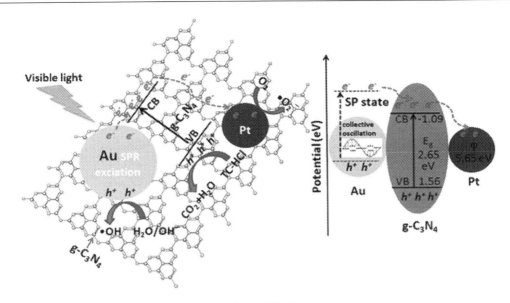

Figure 17.6
The enhancement of photocatalytic activity of Au/Pt/g-C$_3$N$_4$ nanocomposites was attributed by SPR effect of gold under visible light irradiation (Xue, Ma, Zhou, Zhang, & He, 2015).

semiconductor. The SPR effect of Au and electron-sink function of Pt nanoparticles were found to improve the optical absorption property and photogenerated charge carrier separation in g-C$_3$N$_4$ (Fig. 17.6), which synergistically supported photocatalytic degradation of tetracycline antibiotic and observed 93.0% degradation efficiency in 180 minutes (Xue, Ma, Zhou, Zhang, & He, 2015).

Similarly, the SPR effect of Au nanoparticles enhanced photocatalytic activity of Au-TiO$_2$/pDA/PVDF composite membranes. The photocatalytic activity of nanocomposite membranes was investigated by degrading tetracycline under visible-light irradiation and 92% degradation was achieved within 120 minutes, where the degradation efficiency of nanocomposite membranes was found to be increased around 26% and 51% as compared to Au-TiO$_2$ powder and TiO$_2$/pDA/PVDF nanocomposite membranes, respectively. Since the SPR effect of Au played a positive role in improving the photoinduced charge generation and fabrication of Au-TiO$_2$/pDA/PVDF heterostructure, the heterostructure construction improved the overall photoinduced charge transfer for the photocatalytic reaction. Also, the pDA (polydopamine) acted as photosensitizer was broadened by photoabsorbance of TiO$_2$ (Wang et al., 2017).

17.3.3.2 Semiconductor—semiconductor composites

The semiconductor—semiconductor architecture greatly improves the efficiencies of the photocatalytic system because it offers highly distinctive charge collection and separation.

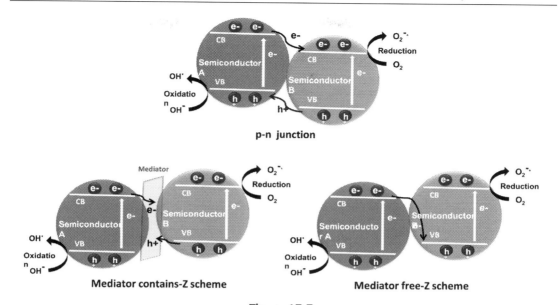

Figure 17.7
Schematic illustration of p-n junction-photocatalyst and Z-scheme-photocatalyst (mediator and mediator-free) mechanism.

The semiconductor—semiconductor heterojunction composite has mainly established two different types of effective architecture correspondingly p-n junction and Z-scheme heterojunction photocatalyst for improving charge separation and photocatalytic activity (Fig. 17.7). The fabrication of semiconductor—semiconductor heterojunction composite includes two steps: firstly, the main component which acts as host, and then the second component, which will be well distributed on the surfaces of host to fabricate the semiconductor—semiconductor effective heterostructure composite. The designing and construction of semiconductor—semiconductor composite offers several benefits: (1) efficient photoinduced charge separation and charge transfer, (2) longer charge carrier lifetime and less charge recombination, (3) provides separate active sites, and (4) extended light absorbance range (Li, Zhou, Tu, Ye, & Zou, 2015; Wang, 2014).

1. p-n junction photocatalysts

The p-n heterojunction photocatalysts can be established by the coupling of p-type and n-type semiconductors. In p-n junction semiconductor composite, semiconductors with different band levels interface could form integral potential with easy charge separation, which can promote photogenerated electron—hole pairs separation and their transfer to the surroundings. The p-n junction mechanism follows the transfer of photogenerated electrons from a semiconductor with more negative CB into a semiconductor with less negative CB. In the meantime, the photogenerated holes from the semiconductor with a lower valence band will jump into the semiconductor with higher valence band. The generated p-n junction formation can efficiently promote the charge separation and inhibit the charge recombination towards enhancing the photocatalytic performance (Li, Salvador, & Rohrer, 2014). For instance, a novel flower-like three-dimensional $BiOBr/Bi_2SiO_5$ p-n heterojunction nanocomposite was developed and demonstrated to have an improved charge separation efficiency and it offered 3.6 times higher visible light photocatalytic activity than BiOBr towards the degradation of tetracycline. In addition, the p-n junction formation improved active radical generation and photostability of the catalyst (Wang, Zhang, Li, & Wang, 2018). The $CoO/g-C_3N_4$ p-n heterojunction photocatalysts were prepared via CoO nanoparticles were uniformly distributed on the wrinkled $g-C_3N_4$ surface to form an efficient p-n heterojunction photocatalyst. As-fabricated $CoO/g-C_3N_4$ p-n heterojunction photocatalyst performance was explored on tetracycline degradation (90% in 60 minutes) and showed a superior visible-light photocatalytic activity as compared to individual $g-C_3N_4$ and CoO. The superior photocatalytic activity was ascribed to the fast charge separation attributed to the inner electric field created by the formation of a p-n heterojunction between CoO and $g-C_3N_4$ (Fig. 17.8). Moreover, the CoO nanoparticle aggregation was repelled by the introduction of $g-C_3N_4$, which improved the photocatalytic stability of $CoO/g-C_3N_4$ p-n heterojunction (Guo et al., 2017).

Recently, a novel organic—inorganic polyaniline/silver molybdate ($PANI/Ag_2MoO_4$) p-n heterojunction nanocatalyst was successfully fabricated via in situ deposition method over Ag_2MoO_4 on PANI. The p-n heterojunction consequently induced efficient photogenerated charge separation and migration due to the formation of internal electric field at heterojunction. This developed composite completely degraded the neurotoxic

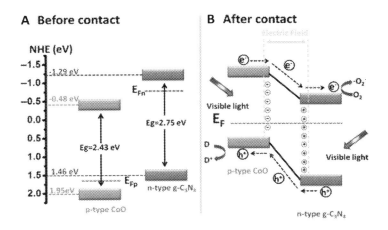

Figure 17.8

The band alignment and charge transfer mechanism of $CoO/g-C_3N_4$ p-n heterojunction (A) before contact and (B) after contact (Guo et al., 2017).

fluoroquinolone antibiotic of ciprofloxacin under UV light within 40 minutes. In addition, PANI was beneficial towards improving both the photocatalytic activity and stability of Ag_2MoO_4 (Mondal et al., 2018).

2. Z-scheme photocatalysts

The construction of Z-scheme photocatalysts exhibits more benefits on the photocatalytic performance, which possess more advantages, such as increased light harvesting, spatial separation of reduction and oxidation sites, and strong redox ability. Z-scheme heterojunctions were constructed to overcome the limitation of the lower redox potential of the p-n heterojunction system. As compared to p-n heterojunction, Z-scheme photocatalysts possess the same band structure configuration but opposite charge transfer mode. During light irradiation, the photogenerated electrons get transferred from semiconductors with lower CB into semiconductors with higher valence band holes, and continuous redox reaction occurs at semiconductor interfaces. Meantime, the photogenerated electrons with strong reduction abilities in the CB of semiconductors and holes with strong oxidation abilities in the valence band of the semiconductors are preserved. As a result, Z-scheme photocatalyst is found to have separate reductive and oxidative sites and lesser charge recombination for driving the photocatalytic reactions (Low, Jiang, Cheng, Wageh, & Al-Ghamdi, 2017; Zhou, Yu, & Jaroniec, 2014).

a. Mediator contains Z-scheme photocatalysts

The mediator contains Z-scheme photocatalyst or all-solid-state Z-scheme photocatalytic system constructed by two different semiconductor photocatalysts with solid electron mediator at the interface of semiconductors. The higher charge carrier separation and transfer both are succeeded by solid-state mediator at the interface of the semiconductors. Mostly, the noble metals (Ag, Au) and reduced graphene oxide (RGO) were used as solid-state electron mediator (Zhou et al., 2014). For example, the construction of Ag_2O-supported RGO-enwrapped TiO_2 nanobelt nanocomposite led to the formation of a solid-state Z-scheme photocatalytic system, which significantly suppressed the photocorrosion and promoted charge separation in the composite. The photocatalytic performance of the developed Z-scheme RGO-Ag_2O/TiO_2 composite was studied on tetracycline degradation under UV light, visible light, NIR light, and simulated solar light irradiation. The RGO incorporation between Ag_2O and TiO_2 potentially improved the photogenerated electron transfer from Ag_2O to TiO_2 in Z-scheme system, which prolonged the lifetime of the photogenerated charge carrier (Fig. 17.9). In addition, the reasonable spatial configuration and hole protective mechanism of Z-scheme photocatalytic system were realized to prevent the Ag_2O nanoparticles from photocorrosion (Hu, Liu, Tian, Li, & Cui, 2017).

Likewise, the nitrogen-doped graphene quantum dots (N-GQDs) modified Ag_2CrO_4@g-C_3N_4 core-shell Z-scheme composite succeeded in complete degradation of antibiotic doxycycline under full-spectrum light from UV to NIR region. The large interface contact area between

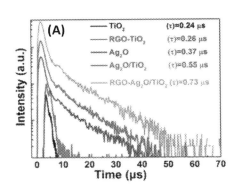
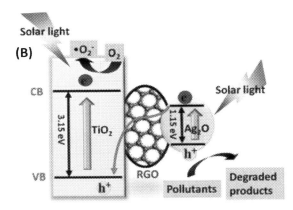

Figure 17.9

(A) Time-resolved transient photoluminescence (PL) spectra and (B) Z-scheme charge transfer mechanism of RGO-Ag$_2$O/TiO$_2$ nanocomposite (Hu, Liu, Tian, Li, & Cui, 2017).

Ag$_2$CrO$_4$ and g-C$_3$N$_4$ and higher conductivity of N-GQDs both effectively promoted Z-scheme charge transfer; moreover, core-shell fabrication predominantly prevented the photocorrosion and improved the photostability of Ag$_2$CrO$_4$ (Feng et al., 2018). Further, the development of mediator containing Z-scheme Ag/FeTiO$_3$/Ag/BiFeO$_3$ photocatalysts with different silver loadings demonstrated the degradation of norfloxacin under visible light and 96.5% removal efficiency was observed in 150 minutes. The Z-scheme system showed higher photocatalytic activity than FeTiO$_3$/BiFeO$_3$, since the loading of Ag mediator improved the charge separation and photocatalytic activity. In addition, Z-scheme catalysts held good photocatalytic stability and reusability during the degradation of norfloxacin (Tang et al., 2018). Similarly, the CdS-Au-BiVO$_4$ heterostructures presented much higher photocatalytic activity on tetracycline removal as compared to BiVO$_4$, Au-BiVO$_4$, and CdS-BiVO$_4$, which is essentilly because of their potential Z-scheme-like structure that improved the separation of photogenerated electron—hole pairs and reduced charge recombination and thereby offered higher redox properties in the system. Correspondingly, the Au NPs were effectively anchored on BiVO$_4$(010) crystals and CdS was further selectively deposited on Au NPs surface via the strong S—Au interaction, and the selective loading of Au NPs resulted into electrons accumulation on BiVO$_4$ (010) facets (Bao, Wu, Chang, Tian, & Zhang, 2017). Interestingly, a novel magnetic Z-scheme photocatalyst WO$_3$/Fe$_3$O$_4$/C$_3$N$_4$ was potentially developed, and upon investigation on tetracycline degradation it was found to degrade the tetracycline by around 90% in 120 minutes. The introduction of Fe$_3$O$_4$ offered the magnetic properties, where it can also act as a conducting mediator between the WO$_3$ and g-C$_3$N$_4$ components, which can also participate on photogenerated charge carrier separation and transfer. Accordingly, the percentage of Fe$_2$O$_3$ and WO$_3$ in the WO$_3$/Fe$_3$O$_4$/C$_3$N$_4$ Z-scheme system and their impact on tetracycline degradation were systematically studied (Pan, Ma, & Wang, 2016).

b. Mediator-free Z-scheme photocatalysts

The Z-scheme photocatalysts gained great interests because of its stronger redox capacity and enriched photocatalytic performance. However, the mediator containing Z-scheme photocatalyst system was usually constructed using two different semiconductors with electron mediators at the interface, but the backward reactions and poor stability are significantly affecting the photocatalytic performance of this system. Therefore, the mediator-free Z-scheme (direct Z-scheme) system designing relatively established more promising practical applications compared with mediator-containing Z-scheme system. The direct Z-scheme photocatalyst is usually constructed with two semiconductor photocatalysts without mediator, one acts as strong oxidation photocatalyst and the other acts as strong reduction photocatalyst. The oxidation photocatalysts possess low VB position and exhibit strong oxidation ability, while the reduction photocatalysts usually have a high CB position and display strong reduction ability. Moreover, the mediator-free Z-scheme photocatalyst possesses several advantages, including separate and potential reductive and oxidative active sites, higher photogenerated charge carrier separation and wide range of spectrum for efficient visible/solar photocatalytic reactions (Low, Jiang, Cheng, Wageh, & Al-Ghamdi, 2017). For example, mediator-free Z-scheme γ-Fe_2O_3/g-C_3N_4 heterojunctions were developed over mesoporous γ-Fe_2O_3 nanospheres anchored on g-C_3N_4 nanosheet, and the mesoporous property and enhanced specific surface area offered large number of active sites to γ-Fe_2O_3/g-C_3N_4 heterojunction system. Furthermore, Z-scheme heterostructure construction between γ-Fe_2O_3 and g-C_3N_4 potentially extends and speeds up the photoinduced charge carrier separation and transfer, which is more beneficial for boosting the photocatalytic activity and degradation of tetracycline (73.8% in 120 minutes) under visible light (Li et al., 2018). Likewise, the redox-mediator-free direct Z-scheme marigold-flower-like $CaIn_2S_4$/TiO_2 composite was prepared via facile wet-impregnation method. The redox-mediator-free direct Z-scheme fabrication significantly suppressed photoinduced charge recombination rate and also improved charge transfer thereby achieved higher photocatalytic efficiency on degradation of pharmaceutical compounds isoniazid and metronidazole. Then, the developed direct Z-scheme $CaIn_2S_4$ marigold-flower-like/TiO_2 showed a higher photocatalytic degradation efficiency of isoniazid (71.9%) and metronidazole (86.5%) as compared to the individual TiO_2 and $CaIn_2S_4$ (Jo & Natarajan, 2015). More interestingly, the nitrogen-doped graphene quantum dots (N-GQDs) are accomplished as effective active sites and charge carrier collectors in modified Z-scheme g-C_3N_4/Bi_2WO_6 heterojunctions photocatalyst. The efficient charge carriers' collection properties and up-conversion luminescence properties of NGQDs potentially enhanced Z-scheme electron—hole charge separation and light-harvesting ability of g-C_3N4/Bi_2WO_6 heterojunction photocatalyst system (Fig. 17.10). Moreover, NGQD-modified Z-scheme g-C_3N_4/Bi_2WO_6 heterojunctions exhibited greater photocatalytic activity on degradation of various antibiotics (e.g., tetracycline, ciprofloxacin, and oxytetracycline) under visible and NIR light irradiation (Che et al., 2018).

Then, the simple oxygen vacancy involved in Z-scheme photocatalytic system potentially enhanced photocatalytic activity. Typically, the development of Z-scheme $BiO_{1-x}Br$/$Bi_2O_2CO_3$

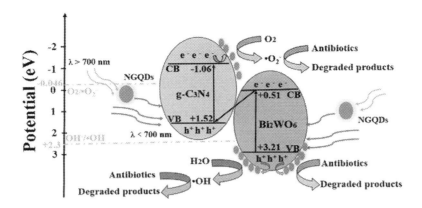

Figure 17.10

The possible photocatalytic mechanism of NGQD-modified g-C₃N₄-Bi₂WO₆ Z-scheme heterojunctions under visible light and NIR light irradiation (Che et al., 2018).

composite photocatalytic system was found to have rich oxygen vacancies, and it enhanced visible photocatalytic activity on degradation ciprofloxacin. The oxygen vacancies offered more redox ability and significantly promoted the photoinduced charge carrier separation, which benefited for the production of superoxide radical on $BiO_{1-x}Br$ and hydroxyl radical on $Bi_2O_2CO_3$ towards the antibiotic degradation (Ding et al., 2017). A new insight into the establishment of a nonmetal plasmonic effect in Z-scheme system was demonstrated to be highly efficient photocatalyst. The novel Z-scheme $W_{18}O_{49}/g$-C_3N_4 nanograss composites were synthesized via hierarchical assembly of nonstoichiometric tungsten oxide ($W_{18}O_{49}$) on g-C_3N_4 nanosheets. This developed Z-scheme composites presented a high photocatalytic activity on removal of 93.5% ciprofloxacin in 120 minutes under full-spectrum light ($\lambda > 365$ nm) irradiation and 44.7% in 240 minutes under NIR light ($\lambda > 700$ nm) irradiation, respectively. The boosted photocatalytic activity of $W_{18}O_{49}/g$-C_3N_4 composites was originated from the establishment of Z-scheme structure with improved charge carrier separation and plasmon resonance effect induced hot electron injection. In addition, the unique nanograss structure was found to show more light reflection and refraction process, which ultimately improved the maximum utilization of light energy, where it was also found that the oxygen vacancies greatly influenced photocatalytic activity of $W_{18}O_{49}/g$-C_3N_4 nanograss composites (Deng et al., 2018).

3. Heterojunction photocatalysts

The construction of semiconductor/semiconductor-based heterojunction photocatalysts is also considered one of the most promising approaches for improving the photocatalytic performance of the system. The heterojunction photocatalysts offer effective spatial electron—hole pairs separation and charge transfer due to the interfacial contacts between two semiconductors with dissimilar band structures. In fact, this heterojunction charge transfer mechanism is different from the conventional p-n junction and Z-scheme heterostructure photocatalytic system

(Low, Yu, Jaroniec, Wageh, & Al-Ghamdi, 2017). For example, the Ag_3VO_4 hybridized WO_3 (Ag_3VO_4/WO_3) heterojunction photocatalyst was designed essentially to address the lower photocatalytic activity of WO_3 nanoparticles. This formation of heterojunction greatly improved the charge separation and prolonged the lifetime of photoexcited charge carriers followed by increased photocatalytic activity. Therefore, Ag_3VO_4/WO_3 heterojunction nanocomposite founded superior photocatalytic performance on tetracycline degradation and achieved best photodegradation rate of 71.2% in 30 minutes (Wu, Zhu, Zhu, Hua, & Shi, 2016). Then, the CdS/SnO_2 heterojunction photocatalyst presented a lattice mismatch between CdS and SnO_2, which enhanced the interfacial contact areas to speed up electron migration. Also, CB offset of CdS and SnO_2 offered excellent conductivity for SnO_2 which promoted photogenerated electron transfer and reduced the electron and hole recombination process of CdS/SnO_2 heterojunction catalyst. In addition, the smaller CdS nanoparticles with strong interface interaction remarkably enhanced the visible photoabsorption and surface charge transfer of CdS/SnO_2 heterostructure nanocomposite. As developed, the heterostructure photocatalyst showed 85.4% tetracycline degradation in 60 minutes and it retained the high-level degradation efficiency after several cycles. Then, the inductively coupled plasma mass spectrometry (ICP-MS) results revealed the detection of very small quantity of Cd^{2+} ions during photocatalytic reaction, which represented that the photocorrosion of CdS was inhibited by SnO_2 (Zhang et al., 2018).

The fabrication of Ta_3N_5 nanoparticles (0D) anchored TiO_2 hollow nanosphere (3D) heterojunction nanocomposites and were systematically studied on antibiotics levofloxacin, ciprofloxacin and tetracycline degradation under solar light. As developed, the 0D/3D Ta_3N_5/TiO_2 heterojunction composites exhibited a higher photocatalytic activity as compared to TiO_2 hollow spheres. The efficient charge transfer and separation between the Ta/Ti (0D/3D) heterojunction interfaces, expanded optical response, and improved specific surface area were mainly accountable for the superior photocatalytic performance of the Ta_3N_5/TiO_2 nanocomposite. Subsequently, the metal-like properties of Ta_3N_5 offered narrow band gaps and good stability to heterojunction composite (Jiang et al., 2018). Likewise, 3D $CdIn_2S_4$ nanooctahedron/2D ZnO nanosheet embedded heterostructure composite offered greater photocatalytic activity for tetracycline hydrochloride degradation ($\sim$94% in 40 minutes) under visible light irradiation, where it was found to be 1.95 and 4.74 times higher than ZnO and $CdIn_2S_4$. The outstanding photocatalytic performance of $CdIn_2S_4/ZnO$ heterostructure photocatalyst was established by the construction of 3D/2D heterojunction, which offered efficient interface contact and superior photogenerated electron—hole separation. In addition, the 3D/2D heterostructure fabrication remarkably improved the photostability of the system (Peng et al., 2018).

17.3.3.3 Semiconductor—carbon composites

The semiconductor—carbon composite photocatalysts have been recently paid more attention due to their higher surface area, good conductivity, chemical stability, and special

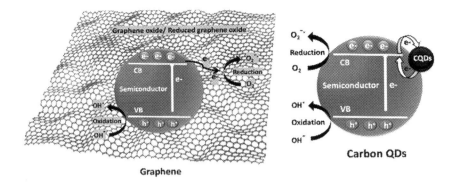

Figure 17.11
Schematic illustration of graphene—photocatalyst and CQD—photocatalyst mechanism.

electronic properties. The carbonaceous materials, such as graphene, carbon quantum dots (CQDs), and carbon sphere, are widely explored in photocatalysis (Fig. 17.11). Also, there are numerous graphene, CQDs, and carbon sphere-based nanocomposite photocatalysts developed via different preparation routes and explored through photocatalytic applications, the semiconductor—carbon was established to have potential photoinduced charge separation and charge transfer, thereby enhancing the overall photocatalytic performances of the system. Moreover, semiconductor—carbon composite offers: (1) efficient charge separation, (2) higher charge mobility, (3) extended photoabsorbance, (4) large specific surface area, and (5) good chemical and photostability (Phang & Tan, 2019; Xiang, Yu, & Jaroniec, 2012; Zhao, Wang, Yang, & Yang, 2012).

Graphene photocatalysts: Graphene has received great interest in photocatalysis since past several years. Graphene possesses many unique properties, such as large specific surface area, high charge mobility, and good mechanical strength due to their special two-dimensional (2D) hybridized carbon structure (Li, Yu, Wageh, & Al-Ghamdi, 2016; Xiang, Cheng, & Yu, 2015). Therefore, it can be a potential candidate for designing semiconductor—carbon composite photocatalyst and pharmaceutical degradation. For example, a 2D graphene and reduced-grapheme oxide (r-GO) sheets have attracted much attention in photocatalysis because of their unique structure properties. A highly efficient and stable $Ag_2MoO_4/Ag/AgBr/GO$ heterostructure photocatalyst demonstrated an excellent charge separation and photocatalytic performance. Consequently, the antibiotic tetracycline hydrochloride was completely degraded within 75 minutes by $Ag_2MoO_4/Ag/AgBr/GO$ heterostructure, which is 1.36 times higher photocatalytic efficiency than that of without GO-based $Ag_2MoO_4/Ag/AgBr$ heterostructure composite because of the graphene oxide that acted as an efficient electron acceptor, where it also prevented the metallic silver to exchange into silver ions, thereby improving the stability of the heterostructure photocatalyst (Bai, Wang, & Liu, 2016).

Similarly, the RGO offered good conductivity with outstanding electron transporting properties, which inhibited the photoexcited electron−hole pairs recombination in Fe_3O_4@g-C_3N_4/RGO ternary photocatalyst. Furthermore, graphene oxide is found to increase the photostability to Fe_3O_4@g-C_3N_4/RGO and showed enhanced photocatalytic performance on tetracycline degradation (Fan et al., 2018). Further, the incorporation of RGO into WO_3 nanoparticles improved the photostability of the semiconductor photocatalyst and RGO with large surface area provided more adsorption sites as well as catalytic sites for photocatalytic reaction. In addition, RGO acted as electron trapper and hence the photogenerated electrons are effectively transferred to RGO, leading to efficient electron−hole separation, thereby improving photocatalytic activity and achieving complete degradation of sulfamethoxazole antibiotic in 3 hours (Zhu, Sun, Goei, & Zhou, 2017). Then, the visible-light-responsive reduced graphene oxide−zinc telluride (RGO-ZnTe) hybrid photocatalyst was synthesized by one-step hydrothermal technique. The distinctive ZnTe nanoparticles were well spread onto a 2D wrinkled graphene sheet surface, where the RGO functioned as a solid support as well as a nucleation center of ZnTe nanocrystals. RGO-ZnTe composite exhibited around 2.6 times higher photocatalytic efficiency than ZnTe towards the degradation of tetracycline antibiotic under visible light. The enhanced catalytic performance was attributed to better interaction and synergy between RGO and ZnTe nanoparticles (Chakraborty, Pal, & Ghosh, 2018).

Carbon quantum dots photocatalysts: In recent years, the semiconductor−carbon quantum dot composites have also received more interest in photocatalysis due to their distinct electronic properties. In this direction, the CQDs possess many unique properties, such as: (1) excellent charge transport properties, which can enhance photogenerated electron−hole separation and reduce recombination process; (2) the extension of the excitation wavelength into visible light region, acting as photosensitizers; and (3) can act as either electron acceptor or electron donor. Therefore, fabrication of semiconductor−CQDs composites would provide enhanced photocatalytic activity (Liu et al., 2019; Sharma, Umar, Mehta, Ibhadon, & Kansal, 2018). For example, CQDs-loaded mesoporous g-C_3N_4 (mpg-C_3N_4/CQDs) was found to show synergistic absorption effect and higher visible photodegradation of fluoroquinolones antibiotic ofloxacin (90.1% degradation). The mpg-C_3N_4/CQDs composite introduced a larger number of adsorption sites, exclusive up-converted photoluminescence properties, and efficient charge separation in order to improve the fluoroquinolones antibiotics ofloxacin degradation. The scavenger experiments showed that the superoxide anion radical ($O_2^{\bullet-}$) and photoinduced hole (h^+) were identified as primarily active species that initiated the effective degradation of ofloxacin (Wang et al., 2018). Similarly, 2−4 nm sized CQDs were well dispersed on dandelion-like ZnS surface and found to have higher photocatalytic activity on ciprofloxacin removal under simulated sunlight. The optical and electrochemical results proved the enhancement of photocatalytic activities assigned to improve electron−hole charge separation efficiency by CQDs.

Because of the smaller size, CQDs can be uniformly distributed and can well establish contact onto the semiconductor surface, which build bulk-to-small surface channels and excellent charge transport ability (Ming, Hong, Xu, & Wang, 2016).

Moreover, the CQDs are considered an ideal supporting material for semiconductor photocatalyst as they possess an excellent photoinduced electron transfer property. In this context, towards improving the photocatalytic activity and antiphotocorrosion property, the Ag_3PO_4 was modified with CQDs and benzoxazine to obtain the 3D core-shell CQDs/Ag_3PO_4@benzoxazine tetrapod's composite system. The composite system showed excellent photocatalytic performance on sulfamethoxazole antibiotic degradation, which was found to have 95% of removal efficiency within 15 minutes. In this case, CQDs induced more photocarrier generation and accelerated the electron transfer from Ag_3PO_4 to CQDs. In addition, the 3D core-shell structure effectively decreased the photocorrosion of Ag_3PO_4 followed by improved stability and reusability of the catalyst (Shao, Hou, Zhu, Wang, & François-Xavier, 2018). More interestingly, Ag^+-CDs-Bi_2WO_6 ternary composites with excellent solar-light-driven photocatalyst activity were studied on tetracycline degradation. The carbon dots (CDs) were uniformly spread on Bi_2WO_6 nanosheets surface, where it also potentially performed as the efficient electron acceptors. The coexistence of CDs and Ag^+ significantly improved the photogenerated electron−hole pair separation and distinctly improved visible light absorbance (Fig. 17.12). It facilitated a superior photocatalytic activity towards the tetracycline degradation with 91.33% removal efficiency in 10 minutes and complete degradation was achieved in 20 minutes (Li, Zhu, Wu, Wang, & Qiang, 2016a).

Carbon sphere photocatalysts: The carbon spheres recently regained great interest in photocatalyst due to their remarkable characteristic properties, such as uniform geometry, controllable porosity, surface functionality, tunable particle size distribution, and excellent

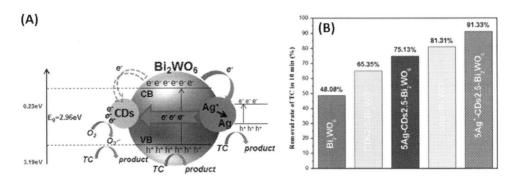

Figure 17.12
(A) Photocatalytic mechanism of Ag^+-CDs-Bi_2WO_6 and (B) photocatalytic removal efficiency of tetracycline with different catalyst in 10 minutes under solar light (Li, Zhu, Wu, Wang, & Qiang, 2016a).

chemical and photostability. Therefore, carbon spheres served as efficient supporting materials for the immobilization or encapsulation of semiconductor photocatalytic particles, and it also acted as a template for the formation of hollow structures to enrich the surface properties and charge transfer of semiconductor—carbon composite. In addition, their low cost and simple synthesis procedure offered feasible ways for material synthesis (Li, Pasc, Fierro, & Celzard, 2016; Wu, 2017). For example, carbon spheres were developed through hydrothermal treatment using different carbon precursors, such as glucose, β-cyclodextrin, and sucrose, with/without addition of surfactant CTAB. The CTAB-modified carbon spheres showed a higher degree of carbonation, and then the modified carbon spheres/g-C_3N_4 composite's photocatalytic performance was examined on the degradation of antibiotic tylosin under simulated solar light. The photocatalytic results displayed that the modified carbon spheres/g-C_3N_4 composite found to degrade 98% of tylosin in 45 minutes, which was higher than that of pure carbon spheres and g-C_3N_4. Similarly, the improved photocatalytic activity was found in the carbon spheres embedded g-C_3N_4 structure, which formed a potential heterojunction with closer interface contact and the boosted visible light photoabsorption, enlarged surface areas, and facilitated to the segregation of electron—hole pairs of g-C_3N_4 in the carbon sphere/g-C_3N_4 composites. In addition, the radical experiments presented that the superoxide radical ($O_2^{\bullet-}$) and hydroxyl radical ($OH^{\bullet}$) were the main active species in the photodegradation (Guo et al., 2018).

Further, the magnetic core-shell Fe_3O_4@carbon sphere tailored Co_3O_4 nanochains (Fe_3O_4@C/Co_3O_4) photocatalyst was explored on the tetracycline degradation. The photocatalytic performance was significantly boosted under visible light irradiation because of the synergetic effect of carbon layer and Co_3O_4 nanochains, which was beneficial for the higher photogenerated charge carrier transfer and separation. In addition, the introduction of magnetic materials effectively improved recyclability and reusability of the catalyst. The semiconductor photocatalyst tailored with carbon layer is proved to possess a potential advantage in dramatically improving the photocatalytic activity (Zhao et al., 2018).

17.3.4 Prospective photocatalytic materials

The development of new potential photocatalysts with proper materials and design is highly required for effective removal of antibiotic from water because these new photocatalysts offer significant advantages over traditional photocatalytic materials, such as interesting visible-/solar-light-driven-tunable narrow bandgap, fast and efficient charge transfer, higher redox potential, and higher surface and catalytic properties.

17.3.4.1 Ferroelectric MBiO₃ materials

$MBiO_3$ (M = Ca, Cd, Mg, and Zn) is theoretically proposed single-phase perovskite-structured ferroelectric oxide, which can be operated under visible light irradiation. The

$MBiO_3$ exhibits small band gaps (~ 2.0 eV), strong visible-light absorption, small carrier effective masses, and large electrical polarizations. $MBiO_3$ materials are taking more advantage towards photocatalytic application due to their effective Bi^{3+}/Bi^{5+} charge disproportionation and small radii of M-site cations in the $MBiO_3$ compounds. Besides, the designing of small band gap ABO_3 ferroelectric oxides by combining small A-site cations, such as Mg^{2+}, Zn^{2+}, Cd^{2+}, and Ca^{2+}, with larger B-site cation of bismuth was favoring higher charge ordering and better light absorption, specifically disproportionation of Bi^{4+} into Bi^{3+}/Bi^{5+}. Also, $MBiO_3$ offers the ferroelectric compounds with smaller tolerance factor (τ) and the band edges formed by the 6s-like states derived from the Bi^{3+} and Bi^{5+} cations. The spatially extended 6s orbitals of Bi^{3+} and Bi^{5+} cations should guarantee a high photogenerated carrier mobility as observed in similar material. The band gap reduction in these systems is mainly due to the stereochemically active lone pair of Bi^{3+} 6s shifting the top of the valence band to higher energy, which does not affect the redox processes during the photocatalytic process. Therefore, the development of new bivalent $MBiO_3$ ferroelectric perovskite materials are possible potential candidates for photocatalytic oxidation of pharmaceutical pollutants (Tang, Zou, & Ye, 2007). Recently, we have developed calcium bismuthate ($CaBiO_3$) perovskite material with distinct Bi^{3+} and Bi^{5+} multicharge disproportion via glycine-complexation and ion-exchange methods for the first time. The as-developed $CaBiO_3$ is found to have an efficient Bi^{3+}/Bi^{5+} charge disproportion and be well established in BiO_6 octahedral crystal arrangement, where it also offered an efficient visible photoabsorbance and higher photogenerated charge carrier generation and separation. Moreover, the effective Ca^{2+} incorporation into bismuthate (BiO_6) crystal lattices potentially modified the band structure and band potential of $CaBiO_3$, which offered more suitable photocatalytic redox properties. Also, the developed $CaBiO_3$ materials in their nanostructures with higher surface area provided enhanced surface properties for the improved catalytic reactions. Furthermore, the developed $CaBiO_3$ materials have also been potentially explored on the degradation of emerging pharmaceutical contaminants of antibiotics under solar light. In this direction, the other ferroelectric materials, such as $CdBiO_3$, $MgBiO_3$, and $MgBiO_3$, can also be prepared and investigated for their potential application in photocatalytic antibiotics degradation (Rokesh, Sakar, & Do, 2020).

17.3.4.2 Black metal oxides

Recently, black metal oxides, such as black titanium dioxide (TiO_{2-x}), black $BiVO_4$, and tungstic oxide (WO_{3-x}) have been found to show great performances in photoelectrocatalytic applications because these materials offer moderate oxygen vacancies and increase the donor concentration, which ultimately favors efficient charge separation and transport. Furthermore, their lower valence band helps achieving the narrow band gap and broaden photoabsorption range. Therefore, black metal oxides tend to show enhanced solar energy utilization and charge carrier migration.

The black-TiO_2(TiO_{2-x}) materials have also received great attention in photocatalytic and photoelectrochemical (PEC) performance. The black-TiO_2 shows unique properties, such as extended photoabsorption, narrow band gap, structural and surface disorder, oxygen vacancies, generation of Ti^{3+} species, offering Ti-H species or Ti-OH bonds, and modified electron density (Chen, Liu, Yu, & Mao, 2011; Naldoni et al., 2018; Zhu, Xu, Zhao, & Huang, 2016). For example, the controllable synthesis of black titania via metal-reduction without hydrogen treatment showed five times higher hydrogen production rate than TiO_2 P25 in PEC water treatment (Tian, Yin, Lin, & Huang, 2017). The black titania nanotubes showed higher photocurrent of 3.65 mA cm^{-2} and a high applied bias photon-to-current efficiency of 1.20%, in PEC water splitting, which was about five times higher than that of titanium nanotubes. The improved PEC performance was attributed to the optimum Ti^{3+} and oxygen vacancies, extended visible absorbance, and higher charge separation and transfer (Cui et al., 2014). Similarly, the low-temperature water plasma synthesis based black titania (H-TiO_{2-x}) showed excellent visible-light photocatalytic activity towards the dye degradation; the higher photocatalytic activity was attributed to the visible-light photoabsorption and the existence of oxygen vacancies and Ti^{3+} sites (Panomsuwan, Watthanaphanit, Ishizaki, & Saito, 2015).

Further, the development of black $BiVO_4$@amorphous TiO_{2-x} material provided an effective strategy to prompt solar water splitting, where it showed a high photocurrent density of 6.12 mA and remarkable applied bias photon-to-current efficiency up to 2.5%. Because of the black-$BiVO_4$ with moderate oxygen vacancies, it offered band gap reduction and significant enhancement in solar utilization and provided better charge separation and transfer (Tian et al., 2019). Recently, the pulsed laser irradiation technique derived black-$BiVO_4$ colloidal nanoparticles showed rich oxygen vacancies and outstanding stability for sodium-ion batteries application (Xu et al., 2020). The hydrogen-treated WO_3 (black-WO_3) nanoflakes showed magnitude enhancement in photocurrent density and exhibited extraordinary stability for water oxidation than compared to pristine WO_3. The improved photoactivity of black-WO_3 was attributed by controlled introduction of oxygen vacancies into WO_3 during hydrogen treatment. The enhanced photostability was established to the formation of substoichiometric WO_{3-x}, which highly prevented the re-oxidation (Wang et al., 2012). Similarly, the black WO_{3-x}@TiO_{2-x} core-shell nanosheet photoanode showed around eight times higher photocurrent density than the pure WO_3 photoanode. The improved photoelectrocatalytic water oxidation performance was originated from the enhanced light absorption by black WO_{3-x} and the efficient charge separation and transfer occurred by the well-established band alignment in of the WO_{3-x}@TiO_{2-x} nanosheets (Cao et al., 2017).

17.4 Summary and outlook

The pharmaceutical compounds especially antibiotics are considered an emerging organic contaminant due to their serious public health issues of generation of antimicrobial resistivity. The antibiotic molecules are very difficult to remove completely by the conventional water purification techniques. On the other hand, the photocatalytic technique can potentially degrade the pharmaceutical/antibiotic molecules in water using natural solar energy. The pharmaceutical removal efficiency mainly depends upon effectiveness of the photocatalytic system, which essentially includes the photoabsorbance potential, photoinduced charge separation, charge transfer, morphology, and surface properties of the materials. Recently, several photocatalytic systems have already been potentially developed and well studied for the photocatalytic degradation of antibiotics pollutant. However, the traditional photocatalysts still exist with several limitations and, therefore, the development of new potential photocatalysts with proper materials design is highly required. In this direction, the identified hybrid-based, nanocomposite-based photocatalysts, new ferroelectric perovskites, black-TiO_2, black-$BiVO_4$, and black-WO_3 can be explored as the new visible-light-driven photocatalysts towards the effective degradation of antibiotic pollutants in water environments.

References

aus der Beek, T., Weber, F.-A., Bergmann, A., Hickmann, S., Ebert, I., Hein, A., et al. (2016). Pharmaceuticals in the environment—Global occurrences and perspectives. *Environmental Toxicology and Chemistry*, 35(4), 823–835.

Bagheri, S., TermehYousefi, A., & Do, T.-O. (2017). Photocatalytic pathway toward degradation of environmental pharmaceutical pollutants: Structure, kinetics and mechanism approach. *Catalysis Science & Technology*, 7(20), 4548–4569.

Bai, Y.-Y., Wang, F.-R., & Liu, J.-K. (2016). A new complementary catalyst and catalytic mechanism: Ag_2MoO_4/Ag/AgBr/GO heterostructure. *Industrial & Engineering Chemistry Research*, 55(37), 9873–9879.

Banerjee, S., Pillai, S. C., Falaras, P., O'Shea, K. E, Byrne, J. A, & Dionysiou, D. D (2014). New insights into the mechanism of visible light photocatalysis. *The Journal of Physical Chemistry Letters*, 5(15), 2543–2554.

Bao, S., Wu, Q., Chang, S., Tian, B., & Zhang, J. (2017). Z-scheme CdS−Au−$BiVO_4$ with enhanced photocatalytic activity for organic contaminant decomposition. *Catalysis Science & Technology*, 7(1), 124–132.

Cai, T., Wang, L., Liu, Y., Zhang, S., Dong, W., Chen, H., et al. (2018). Ag_3PO_4/Ti_3C_2 MXene interface materials as a Schottky catalyst with enhanced photocatalytic activities and anti-photocorrosion performance. *Applied Catalysis B: Environmental*, 239, 545–554.

Chakraborty, K., Pal, T., & Ghosh, S. (2018). RGO-ZnTe: A graphene based composite for tetracycline degradation and their synergistic effect. *ACS Applied Nano Materials*, 1(7), 3137–3144.

Che, H., Liu, C., Hu, W., Hu, H., Li, J., Dou, J., et al. (2018). NGQD active sites as effective collectors of charge carriers for improving the photocatalytic performance of Z-scheme gC_3N_4/Bi_2WO_6 heterojunctions. *Catalysis Science & Technology*, 8(2), 622–631.

Chen, X., Liu, L., Yu, P. Y, & Mao, S. S (2011). Increasing solar absorption for photocatalysis with black hydrogenated titanium dioxide nanocrystals. *Science (New York, N.Y.), 331*(6018), 746−750.

Colmenares, J. C., & Luque, R. (2014). Heterogeneous photocatalytic nanomaterials: Prospects and challenges in selective transformations of biomass-derived compounds. *Chemical Society Reviews, 43*(3), 765−778.

Cui, H., Zhao, W., Yang, C., Yin, H., Lin, T., Shan, Y., et al. (2014). Black TiO_2 nanotube arrays for high-efficiency photoelectrochemical water-splitting. *Journal of Materials Chemistry A, 2*(23), 8612−8616.

Deng, Y., Tang, L., Feng, C., Zeng, G., Wang, J., Zhou, Y., et al. (2018). Construction of plasmonic Ag modified phosphorous-doped ultrathin g-C_3N_4 nanosheets/$BiVO_4$ photocatalyst with enhanced visible-near-infrared response ability for ciprofloxacin degradation. *Journal of Hazardous Materials, 344*, 758−769.

Deng, Y., Tang, L., Feng, C., Zeng, G., Chen, Z., Wang, J., et al. (2018). Insight into the dual-channel charge-charrier transfer path for nonmetal plasmonic tungsten oxide based composites with boosted photocatalytic activity under full-spectrum light. *Applied Catalysis B: Environmental, 235*, 225−237.

Ding, J., Dai, Z., Qin, F., Zhao, H., Zhao, S., & Chen, R. (2017). Z-scheme $BiO_{1-x}Br/Bi_2O_2CO_3$ photocatalyst with rich oxygen vacancy as electron mediator for highly efficient degradation of antibiotics. *Applied Catalysis B: Environmental, 205*, 281−291.

Fan, W., Zhu, Z., Yu, Y., Liu, Z., Li, C., Huo, P., et al. (2018). Fabrication of magnetic gC_3N_4 for effectively enhanced tetracycline degradation with RGO as mediator. *New Journal of Chemistry, 42*(19), 15974−15984.

Fang, J., Wang, W., Zhu, C., Fang, L., Jin, J., Ni, Y., et al. (2017). CdS/Pt photocatalytic activity boosted by high-energetic photons based on efficient triplet−triplet annihilation upconversion. *Applied Catalysis B: Environmental, 217*, 100−107.

Feng, C., Deng, Y., Tang, L., Zeng, G., Wang, J., Yu, J., et al. (2018). Core-shell Ag_2CrO_4/N-GQDs@ g-C_3N_4 composites with anti-photocorrosion performance for enhanced full-spectrum-light photocatalytic activities. *Applied Catalysis B: Environmental, 239*, 525−536.

Gadipelly, C., Pérez-González, A., Yadav, G. D., Ortiz, I., Ibáñez, R., Rathod, V. K., et al. (2014). Pharmaceutical industry wastewater: Review of the technologies for water treatment and reuse. *Industrial & Engineering Chemistry Research, 53*(29), 11571−11592.

Gaya, U. I., & Abdullah, A. H. (2008). Heterogeneous photocatalytic degradation of organic contaminants over titanium dioxide: A review of fundamentals, progress and problems. *Journal of Photochemistry and Photobiology C: Photochemistry Reviews, 9*(1), 1−12.

Guo, F., Shi, W., Wang, H., Han, M., Li, H., Huang, H., et al. (2017). Facile fabrication of a CoO/gC_3N_4 p−n heterojunction with enhanced photocatalytic activity and stability for tetracycline degradation under visible light. *Catalysis Science & Technology, 7*(15), 3325−3331.

Guo, X., Dong, H., Xia, T., Wang, T., Jia, H., & Zhu, L. (2018). Highly efficient degradation toward tylosin in the aqueous solution by carbon spheres/g-C_3N_4 composites under simulated sunlight irradiation. *ACS Sustainable Chemistry & Engineering, 6*(10), 12776−12786.

Hoffmann, M. R., Martin, S. T., Choi, W., & Bahnemann, D. W (1995). Environmental applications of semiconductor photocatalysis. *Chemical Reviews, 95*(1), 69−96.

Hu, X., Liu, X., Tian, J., Li, Y., & Cui, H. (2017). Towards full-spectrum (UV, visible, and near-infrared) photocatalysis: Achieving an all-solid-state Z-scheme between Ag_2O and TiO_2 using reduced graphene oxide as the electron mediator. *Catalysis Science & Technology, 7*(18), 4193−4205.

Jiang, E., Liu, X., Che, H., Liu, C., Dong, H., & Che, G. (2018). Visible-light-driven Ag/Bi_3O_4Cl nanocomposite photocatalyst with enhanced photocatalytic activity for degradation of tetracycline. *RSC Advances, 8*(65), 37200−37207.

Peng, Z. Y., Wu, F., Xiao, Y., Jing, X., Wang, L., Liu, Z., et al. (2018). A novel 3D/2D $CdIn_2S_4$ nano-octahedron/ZnO nanosheet heterostructure: Facile synthesis, synergistic effect and enhanced tetracycline hydrochloride photodegradation mechanism. *Dalton Transactions, 47*(26), 8724−8737.

Jiang, Y., Jing, x., Zhu, K., Peng, Z., Zhang, J., Liu, Y., et al. (2018). Ta_3N_5 nanoparticles/TiO_2 hollow sphere (0D/3D) heterojunction: Facile synthesis and enhanced photocatalytic activities of levofloxacin degradation and H_2 evolution. *Dalton Transactions, 47*(37), 13113−13125.

Jiang, Z., Zhu, J., Liu, D., Wei, W., Xie, J., & Chen, M. (2014). In situ synthesis of bimetallic Ag/Pt loaded single-crystalline anatase TiO_2 hollow nano-hemispheres and their improved photocatalytic properties. *CrystEngComm*, *16*(12), 2384–2394.

Jo, W.-K., & Natarajan, T. S. (2015). Facile synthesis of novel redox-mediator-free direct Z-scheme $CaIn_2S_4$ marigold-flower-like/TiO_2 photocatalysts with superior photocatalytic efficiency. *ACS Applied Materials & Interfaces*, *7*(31), 17138–17154.

Khan, M. R., Chuan, T. W., Yosuf, A., Chowdhury, M. N. K, & Cheng, C. K. (2015). Schottky barrier and surface plasmonic resonance phenomena towards the photocatalytic reaction: Study of their mechanisms to enhance photocatalytic activity. *Catalysis Science & Technology*, *5*(5), 2522–2531.

Khetan, S. K., & Collins, T. J. (2007). Human pharmaceuticals in the aquatic environment: A challenge to green chemistry. *Chemical Reviews*, *107*(6), 2319–2364.

Küster, A., & Adler, N. (2014). Pharmaceuticals in the environment: Scientific evidence of risks and its regulation. *Philosophical Transactions of the Royal Society B: Biological Sciences*, *369*(1656), 20130587.

Li, C., Yu, S., Che, H., Zhang, X., Han, J., Mao, Y., et al. (2018). Fabrication of Z-scheme heterojunction by anchoring mesoporous γ-Fe_2O_3 nanospheres on g-C_3N_4 for degrading tetracycline hydrochloride in water. *ACS Sustainable Chemistry & Engineering*, *6*(12), 16437–16447.

Li, H., Zhou, Y., Tu, W., Ye, J., & Zou, Z. (2015). State-of-the-art progress in diverse heterostructured photocatalysts toward promoting photocatalytic performance. *Advanced Functional Materials*, *25*(7), 998–1013.

Li, L., Salvador, P. A., & Rohrer, G. S. (2014). Photocatalysts with internal electric fields. *Nanoscale*, *6*(1), 24–42.

Li, S., Pasc, A., Fierro, V., & Celzard, A. (2016). Hollow carbon spheres, synthesis and applications – A review. *Journal of Materials Chemistry A*, *4*(33), 12686–12713.

Li, X., Yu, J., Wageh, S., & Al-Ghamdi, A. A. (2016). Graphene in photocatalysis: A review. *Small (Weinheim an der Bergstrasse, Germany)*, *12*(48), 6640–6696.

Li, Z., Zhu, L., Wu, W., Wang, S., & Qiang, L. (2016a). Highly efficient photocatalysis toward tetracycline under simulated solar-light by Ag + -CDs-Bi2WO6: Synergistic effects of silver ions and carbon dots. *Applied Catalysis B: Environmental*, *192*, 277–285.

Liu, J., Zhang, G., Yu, J. C, & Guo, Y. (2013). In situ synthesis of Zn_2GeO 4 hollow spheres and their enhanced photocatalytic activity for the degradation of antibiotic metronidazole. *Dalton Transactions*, *42* (14), 5092–5099.

Liu, W., Li, Y., Liu, F., Jiang, W., Zhang, D., & Liang, J. (2019). Visible-light-driven photocatalytic degradation of diclofenac by carbon quantum dots modified porous g-C_3N_4: Mechanisms, degradation pathway and DFT calculation. *Water Research*, *150*, 431–441.

Low, J., Jiang, C., Cheng, B., Wageh, S., & Al-Ghamdi, A. A. (2017). A review of direct Z-scheme photocatalysts. *Small Methods*, *1*(5), 1700080.

Low, J., Yu, J., Jaroniec, M., Wageh, S., & Al-Ghamdi, A. A. (2017). Heterojunction photocatalysts. *Advanced Materials*, *29*(20), 1601694.

Martínez, J. L. (2008). Antibiotics and antibiotic resistance genes in natural environments. *Science (New York, N.Y.)*, *321*(5887), 365–367.

Ming, F., Hong, J., Xu, X., & Wang, Z. (2016). Dandelion-like ZnS/carbon quantum dots hybrid materials with enhanced photocatalytic activity toward organic pollutants. *RSC Advances*, *6*(37), 31551–31558.

Mondal, P., Satra, J., Ghorui, U. K., Saha, N., Srivastava, D. N., & Adhikary, B. (2018). Facile fabrication of novel hetero-structured organic–inorganic high-performance nanocatalyst: A smart smart system for enhanced catalytic activity toward ciprofloxacin degradation and oxygen reduction. *ACS Applied Nano Materials*, *1*(11), 6015–6026.

Naldoni, A., Altomare, M., Zoppellaro, G., Liu, N., Kment, S., Zbořil, R., et al. (2018). Photocatalysis with reduced TiO_2: From black TiO_2 to cocatalyst-free hydrogen production. *ACS Catalysis*, *9*(1), 345–364.

Pal, A., Gina, K. Y.-H., Lin, A. Y.-C., & Reinhard, M. (2010). Impacts of emerging organic contaminants on freshwater resources: Review of recent occurrences, sources, fate and effects. *Science of the Total Environment*, *408*(24), 6062–6069.

Pan, Z., Ma, W., & Wang, L. (2016). Construction of a magnetic Z-scheme photocatalyst with enhanced oxidation/reduction abilities and recyclability for the degradation of tetracycline. *RSC Advances*, 6(115), 114374−114382.

Panomsuwan, G., Watthanaphanit, A., Ishizaki, T., & Saito, N. (2015). Water-plasma-assisted synthesis of black titania spheres with efficient visible-light photocatalytic activity. *Physical Chemistry Chemical Physics*, 17 (21), 13794−13799.

Petronella, F., Truppi, A., Ingrosso, C., Placido, T., Striccoli, M., Curri, M. L., et al. (2017). Nanocomposite materials for photocatalytic degradation of pollutants. *Catalysis Today*, 281, 85−100.

Phang, S. J., & Tan, L.-L. (2019). Recent advances in carbon quantum dot (CQD)-based two dimensional materials for photocatalytic applications. *Catalysis Science & Technology*, 9(21), 5882−5905.

Rokesh, K., Sakar, M., & Do, T.-O. (2020). Calcium bismuthate (CaBiO$_3$): A potential sunlight-driven perovskite photocatalyst for the degradation of emerging pharmaceutical contaminants. *ChemPhotoChem*, 4(5), 373−380.

Shao, N., Hou, Z., Zhu, H., Wang, J., & François-Xavier, C. P. (2018). Novel 3D core-shell structured CQDs/ Ag$_3$PO$_4$@Benzoxazine tetrapods for enhancement of visible-light photocatalytic activity and anti-photocorrosion. *Applied Catalysis B: Environmental*, 232, 574−586.

Sharma, S., Umar, A., Mehta, S. K., Ibhadon, A. O., & Kansal, S. K. (2018). Solar light driven photocatalytic degradation of levofloxacin using TiO$_2$/carbon-dot nanocomposites. *New Journal of Chemistry*, 42(9), 7445−7456.

Sun, J., Xu, H., Li, D., Zou, Z., Wu, Q., Liu, G., et al. (2018). Ultrasound-assisted synthesis of a feathery-shaped BiOCl with abundant oxygen vacancies and efficient visible-light photoactivity. *New Journal of Chemistry*, 42(24), 19571−19577.

Tang, J., Wang, R., Liu, M., Zhang, Z., Song, Y., Xue, S., et al. (2018). Construction of novel Z-scheme Ag/ FeTiO$_3$/Ag/BiFeO$_3$ photocatalyst with enhanced visible-light-driven photocatalytic performance for degradation of norfloxacin. *Chemical Engineering Journal*, 351, 1056−1066.

Tang, J., Zou, Z., & Ye, J. (2007). Efficient photocatalysis on BaBiO$_3$ driven by visible light. *The Journal of Physical Chemistry C*, 111(34), 12779−12785.

Tian, Z., Zhang, P., Qin, P., Sun, D., Zhang, S., Guo, X., et al. (2019). Novel black BiVO$_4$/TiO$_2$ − x photoanode with enhanced photon absorption and charge separation for efficient and stable solar water splitting. *Advanced Energy Materials*, 9(27), 1901287.

Tong, H., Ouyang, S., Bi, Y., Umezawa, N., Oshikiri, M., & Ye, J. (2012). Nano-photocatalytic materials: Possibilities and challenges. *Advanced Materials*, 24(2), 229−251.

Wang, C., Wu, Y., Lu, J., Zhao, J., Cui, j., Wu, X., et al. (2017). Bioinspired synthesis of photocatalytic nanocomposite membranes based on synergy of Au-TiO$_2$ and polydopamine for degradation of tetracycline under visible light. *ACS Applied Materials & Interfaces*, 9(28), 23687−23697.

Wang, F., Li, Q., & Xu, D. (2017). Recent progress in semiconductor-based nanocomposite photocatalysts for solar-to-chemical energy conversion. *Advanced Energy Materials*, 7(23), 1700529.

Wang, G., Ling, Y., Wang, H., Yang, X., Wang, C., Zhang, J. Z., et al. (2012). Hydrogen-treated WO$_3$ nanoflakes show enhanced photostability. *Energy & Environmental Science*, 5(3), 6180−6187.

Wang, H., Zhang, L., Chen, Z., Hu, J., Li, S., Wang, Z., et al. (2014). Semiconductor heterojunction photocatalysts: Design, construction, and photocatalytic performances. *Chemical Society Reviews*, 43(15), 5234−5244.

Wang, J., Zhang, G., Li, J., & Wang, K. (2018). Novel three-dimensional flowerlike BiOBr/Bi$_2$SiO$_5$ p−n heterostructured nanocomposite for degradation of tetracycline: Enhanced visible light photocatalytic activity and mechanism. *ACS Sustainable Chemistry & Engineering*, 6(11), 14221−14229.

Wang, Y., Wang, F., Feng, Y., Xie, Z., Zhang, Q., Jin, X., et al. (2018). Facile synthesis of carbon quantum dots loaded with mesoporous gC$_3$N$_4$ for synergistic absorption and visible light photodegradation of fluoroquinolone antibiotics. *Dalton Transactions*, 47(4), 1284−1293.

Wu, H., Wu, X.-L., Wang, Z.-M, Aoki, H., Kutsuna, S., Jimurac, K., et al. (2017). Anchoring titanium dioxide on carbon spheres for high-performance visible light photocatalysis. *Applied Catalysis B: Environmental*, 207, 255−266.

Xiang, Q., Cheng, B., & Yu, J. (2015). Graphene-based photocatalysts for solar-fuel generation. *Angewandte Chemie International Edition, 54*(39), 11350−11366.

Xiang, Q., Yu, J., & Jaroniec, M. (2012). Graphene-based semiconductor photocatalysts. *Chemical Society Reviews, 41*(2), 782−796.

Li, W., Gan, Y., Wei, Y., Cheng, G., Dou, S., Li, Z., et al. (2018). Extremely rapid engineering of zinc oxide nanoaggregates with structure-dependent catalytic capability towards removal of ciprofloxacin antibiotic. *Inorganic Chemistry Frontiers, 5*(10), 2432−2444.

Tian, Z., Yin, G., Lin, T., & Huang, F. (2017). Controllable reduced black titania with enhanced photoelectrochemical water splitting performance. *Dalton Transactions, 46*(4), 1047−1051.

Xu, X., Xu, Y., Xu, F., Jiang, G., Jian, J., Yu, H., et al. (2020). Black $BiVO_4$: Size tailored synthesis, rich oxygen vacancies, and sodium storage performance. *Journal of Materials Chemistry A, 8*(4), 1636−1645.

Xue, J., Ma, S., Zhou, Y., Zhang, Z., & He, M. (2015). Facile photochemical synthesis of $Au/Pt/g$-C_3N_4 with plasmon-enhanced photocatalytic activity for antibiotic degradation. *ACS Applied Materials & Interfaces, 7* (18), 9630−9637.

Wu, Y., Zhu, F., Zhu, F., Hua, Y., & Shi, W. (2016). The fabrication of a novel Ag_3VO_4/WO_3 heterojunction with enhanced visible light efficiency in the photocatalytic degradation of TC. *Physical Chemistry Chemical Physics, 18*(4), 3308−3315.

Cao, Q., Lu, H.-L, Zhong, M., Zheng, X., Chen, H.-Y., Wang, T., et al. (2017). Oxygen-deficient $WO_3 − x@TiO_2 − x$ core−shell nanosheets for efficient photoelectrochemical oxidation of neutral water solutions. *Journal of Materials Chemistry A, 5*(28), 14697−14706.

Zhang, L., Niu, C.-G., Liang, C., Wen, X.-J, Huang, D.-W., Guo, H., et al. (2018). One-step in situ synthesis of CdS/SnO_2 heterostructure with excellent photocatalytic performance for Cr (VI) reduction and tetracycline degradation. *Chemical Engineering Journal, 352*, 863−875.

Zhao, W., Wang, Y., Yang, Y., & Yang, Y. (2012). Carbon spheres supported visible-light-driven CuO-$BiVO_4$ heterojunction: Preparation, characterization, and photocatalytic properties. *Applied Catalysis B: Environmental, 115*, 90−99.

Zhao, X., Lu, Z., Wei, M., Zhang, M., Dong, H., Yi, C., et al. (2018). Synergetic effect of carbon sphere derived from yeast with magnetism and cobalt oxide nanochains towards improving photodegradation activity for various pollutants. *Applied Catalysis B: Environmental, 220*, 137−147.

Zhou, P., Yu, J., & Jaroniec, M. (2014). All-solid-state Z-scheme photocatalytic systems. *Advanced Materials, 26*(29), 4920−4935.

Zhu, G., Xu, J., Zhao, W., & Huang, F. (2016). Constructing black titania with unique nanocage structure for solar desalination. *ACS Applied Materials & Interfaces, 8*(46), 31716−31721.

Zhu, W., Liu, J., Yu, S., Zhou, Y., & Yan, X. (2016). Ag loaded WO_3 nanoplates for efficient photocatalytic degradation of sulfanilamide and their bactericidal effect under visible light irradiation. *Journal of Hazardous Materials, 318*, 407−416.

Zhu, W., Sun, F., Goei, R., & Zhou, Y. (2017). Facile fabrication of RGO-WO_3 composites for effective visible light photocatalytic degradation of sulfamethoxazole. *Applied Catalysis B: Environmental, 207*, 93−102.

CHAPTER 18

Magnetic photocatalytic systems

Jagadeesh Babu Sriramoju, Chitrabanu C. Paramesh, Guddappa Halligudra, Dinesh Rangappa* and Prasanna D. Shivaramu*

Department of Applied Sciences, Visvesvaraya Technological University, Center for Postgraduate Studies, Bengaluru Region, Muddenahalli, Chikkaballapur, India
*Corresponding author. e-mail addresses: dineshrangappa@gmail.com, prasuds@gmail.com

18.1 Introduction

The environment and all living species on the earth are threatened by uncontrolled release of harmful compounds in water, air, and earth. In the past few decades, due to rapid industrialization and population growth, many toxic chemicals have been introduced in the environment. Most of these are organic, inorganic, and biopollutants on surfaces, soil, sewage, and drinking water (Pal, He, Jekel, Reinhard, & Gin, 2014). This situation has increased demand for new technologies for sustainable clean water production, avoiding large amounts of energy consumption. Therefore, development of eco-friendly, low-cost, and high-efficiency water purification are a great global challenge today (Shannon et al., 2008). Waste water generated at various steps of many industrial processes contains large amount mixture of organic and inorganic pollutants. These pollutants are highly toxic to living organisms, including human beings, and difficult to degrade due to their stability, toxicity, and nonbiodegradable nature. This causes devastating effects on the ecosystem. The greatest environmental concerns with dyes, organic, and inorganic compounds are that, they absorb and reflect sunlight entering the water and this interferes with the photosynthetic activity of aquatic plants (Lellis, Fávaro-Polonio, Pamphile, & Polonio, 2019). Therefore, the concentration of the toxic pollutants needs to be reduced, degraded, or transformed to nontoxic forms before releasing them into the aquatic system. So this is a most serious global environmental issue today to be addressed on priority.

Over the decades, many efficient and effective technique shave been invented and developed to reduce the concentration of pollutants such as adsorption and coagulation, membrane technologies, chemical, biological, and physico-chemical methods, electrochemical methods, and advanced oxidation processes (AOPs) (Hamzezadeh-Nakhjavani et al., 2015; Jadhav et al., 2015; Saiz, Bringas, & Ortiz, 2014; Saiz, Bringas, & Ortiz, 2014; San Román, Bringas, Ibañez, & Ortiz, 2010; Uribe, Mosquera-Corral, Rodicio, & Esplugas, 2015). Even though the presently available techniques showed impressive level of success, they often suffer a number of drawbacks, like high operating costs, inefficiency, or both. For example, adsorption and coagulation are associated with incomplete elimination or destruction of the pollutant into a nontoxic or biodegradable form. The chemical process and membrane technologies involve high operating cost and sometimes form other toxic secondary pollutants (Mamba & Mishra, 2016). In addition, in most of these processes, to transmit contaminants between different stages of fluid, use of excess chemicals or large amounts of energy in the process, and the subsequent steps produce waste products that need to be processed further. In addition, some biological methods are not suitable for the removal of many organic contaminants and pollutants, in emerging concerns such as personal protective products, drugs, pesticides, endocrine pests, and materials. Hence, these compounds are accumulated in the atmosphere due to extreme stability (Anglada, Urtiaga, & Ortiz, 2009; Fernández-Castro, Vallejo, San Román, & Ortiz, 2015; Ribeiro, Nunes, Pereira, & Silva, 2015). Therefore, innovative water treatment technologies must be developed to ensure water quality. The most attractive and popular method for sewage, groundwater, or surface water treatment is the AOP (Pérez, Ibáñez, Urtiaga, & Ortiz, 2012; Rivero et al., 2014). AOP is a group of processes, which offers the degradation of organic, inorganic compounds, and even microorganisms into less harmful substances (Buthiyappan, Abdul Aziz, & Wan Daud, 2016). During the process, strong oxidative species produced on surfaces such as hydroxy radicals (OH) leads to many reactions to break down organic contaminants into one or more double bonds with low, less harmful substances such as CO_2, H_2O, and organic chain acids (Blanco et al., 2009; Kumar & Bansal, 2013). The advantages of AOPs include their versatility, ambient operating temperature and pressure, and complete mineralization of the parent compound and their intermediates without leaving secondary pollutants, thereby eliminating the problem of formation of sludge and offering low operating costs (Wang, Li, Lv, Zhang, & Guo, 2014). In recent years, semiconductor photocatalysis is the most attractive and widely used AOPs, which include ZnO, TiO_2, SnO_2, and WO_3 and two-dimensional (2D) transition metal chalcogenides like ZnS, CdS, WS_2, SnS2, etc. (Satheesh, Vignesh, Suganthi, & Rajarajan, 2014; Wang et al., 2014).

Despite the numerous advantages, the typical drawback of photocatalysis using semiconductors is that most of the photocatalysts normally used are not very photo-stable under the operating conditions and the illumination of these catalysts in aqueous media usually leads to their corrosion, which leads to the migration of metal ions into water and further complete dissolution of the solid catalysts (Wang et al., 2014). Since photocatalysts are usually applied in powdered form, the separation and recovery of the photocatalyst after completion of the treatment becomes an expensive and time-consuming process. This leads to the incomplete recovery of the photocatalyst, which hinders the recyclability and increases the cost of treatment (Li et al., 2009). Also, the commonly used photocatalyst, that is, titanium dioxide (TiO_2), has a band gap energy, 3.2 eV for anatase TiO_2 and 3.02 eV for rutile TiO_2. This material can utilize only 4%–6% of solar light, which limits their practical applications (He, Li, Jiang, & Chen, 2012). The recent scientific community has attracted attention because of its unique physical and chemical properties that overtake nanoparticles as photocatalytic materials (Xu et al., 2012; Zhou, Fang, Li, & Wang, 2015). These materials represent different behaviors from bulk materials due to large surface and quantum effects that improve mechanical, optical, electrical, and magnetic properties, as well as chemical reactivity (Buzea, Pacheco, & Robbie, 2007; Udom, Ram, Stefanakos, Hepp, & Goswami, 2013). The advantages of using nano-photocatalysts are associated with better characteristics such as high surface area, chemical stability, high decomposition capacity, etc. This means lower dose and faster kinetics than conventional catalytic materials, lower process costs, and increased photocatalytic decomposition efficiency. The use of nanoscale catalysts has been shown to improve the oxidative potential compared with micro-catalysts by effectively producing oxidative species on the surface of the substance (Chong, Jin, Chow, & Saint, 2010; Gupta & Tripathi, 2011; Lee & Park, 2013).

On the other hand, the recovery of nanocatalyst from the solution is still difficult, limiting a wide range of photocatalysts (Lazar, Varghese, & Nair, 2012; McCullagh, Skillen, Adams, & Robertson, 2011; Udom et al., 2013). To overcome these shortcomings, it is proposed to fix the material on the inert substrate, while avoiding the step of material restoration. However, this configuration of the system reduces the photocatalytic efficiency of such nanocatalyst, as it reduces the amount of catalytic active area (Chong et al., 2010; Gaya & Abdullah, 2008). To solve the problem of photocatalyst recovery problems is incorporating magnetic nano-photocatalysts in water purification processes is proposed. With the adoption of magnetic nanoparticles, the separation can be easily carried out through the application of external magnetic fields and also this enhances the feasibility of multiple recycling of nanophotocatalysts for more efficient, economical, and environmentally related water purification process (Linley, Leshuk, & Gu, 2013). Therefore, the ultimate goal of the research in this field is to engineer nanomaterials with well-designed morphology and multiple functions that can be easily separated and recovered. Since the magnetic nanoparticles can be easily recovered with an external magnetic field, they attracted researchers in developing photocatalysts using magnetic nanoparticles as integral ingredients (Cao et al., 2015; Li et al., 2015).

The magnetic properties were explored in a few elements, which include iron (Fe), nickel (Ni), cobalt (Co), etc. Among all, Fe-based materials exhibit a higher magnetic nature compared with other elements. Existing iron-based magnetic materials are categorized into four major types: hematite (Fe_2O_3), maghemite (Fe_2O_3), magnetite (Fe_3O_4), and ferrites (MFe_2O_3, where M could be Mg, Ni, Cu, Co, Cd, etc.). These MNPs are directly used as catalysts or as supports for further modification or functionalization with the catalytic species (Mamba & Mishra, 2016).

Among the MNPs, magnetite (Fe_3O_4) has been widely used as an ideal support in catalysis because of its easy preparation, low cost, biocompatibility, and good magnetic and adsorption properties (Satheesh et al., 2014). Hematite (Fe_2O_3) is an abundantly available magnetic form, which possesses ferromagnetic behavior. It has a narrow band gap, high chemical resistance, low conductivity, good visible light response, and high resistance to corrosion. Hematite exhibits poor photocatalytic activity due to rapid electron hole recombination and poor conductivity. The narrow band gap of hematite makes it an ideal sensitizer for improving the visible light response of various other semiconductors (Mamba & Mishra, 2016; Venugopal, Thangavel, & Raghavan, 2019). Maghemite is another important form of iron oxide that possesses good magnetic properties with approximately 76 emu/g of saturation magnetization. This behavior grabbed attention in designing magnetic semiconductor nanocomposite for photocatalytic applications (Venugopal et al., 2019). Ferrites have also received considerable attention due to their ferromagnetic nature, good chemical and thermal stability, and a narrow band gap (Mamba & Mishra, 2016; Satheesh et al., 2014; Wang & Astruc, 2014).

18.2 Mechanism of magnetic materials in photocatalysis

The process of photocatalysis occurs when photocatalyst absorbs light energy, which is equal to or greater than its band gap energy and excites electron from the valence band into the conduction band. The process of excitation leaves holes (h^+) in the valence band and creates electrons (e^-) in conduction band. They then react with water and dissolved oxygen to form hydroxyl radical (•OH) and superoxide anion (O^{2-}) anion. These generated radical species having higher oxidation potential contribute to the degradation of organic pollutant into readily biodegradable compounds (Ajmal, Majeed, Malik, Idriss, & Nadeem, 2014).

The chemical representation of the reactions that occur during photocatalysis can be written as follows [Eqs. (18.1)−(18.4)]:

$$\text{Photocatalysts} \xrightarrow{h\nu} h^+ + e^- \tag{18.1}$$

$$\text{Photocatalyste}^-(\text{CB}) + O_2 \rightarrow O_2^- \tag{18.2}$$

$$\text{Photocatalysth}^+(\text{VB}) + H_2O \rightarrow H^+ + \bullet\text{OH} \tag{18.3}$$

$$O_2^- + \bullet\text{OH} + \text{Pollutant} \rightarrow \text{Degraded Pollutant} + CO_2 + H_2O \tag{18.4}$$

A schematic representation of the photocatalytic process is shown in Fig. 18.1.

In case of photocatalysis using iron oxide, generations of radicals by the ferrous and ferric ion surfaces are also to be considered apart from the light excitation. During the process, traces of iron ions may enter the coating of passive layer. The iron ions, which are present in the aqueous solution, can promote the degradation of organic pollutant by Fenton (Fe^{2+}/H_2O_2) and Fenton-like (Fe^{3+}/H_2O_2) reaction mechanism (Pang, Lim, Ong, & Chong, 2016). In Fenton

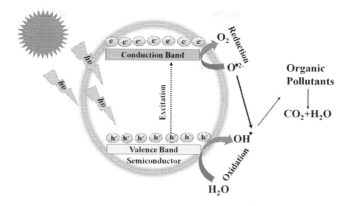

Figure 18.1
Schematic mechanism of photocatalytic degradation.

process, Fe^{2+} is oxidized to Fe^{3+} to generate •OH (Eq. 18.5). Then, Fe^{2+} is regenerated through the reduction of Fe^{3+} by H_2O_2 [Eq. (18.6)].

$$Fe^{2+} + H_2O_2 \rightarrow Fe^{3+} + OH^- + \bullet HO \qquad (18.5)$$

$$Fe^{3+} + H_2O_2 \rightarrow Fe^{2+} + \bullet O_2H + H^+ \qquad (18.6)$$

The Fe^{2+} on iron oxide surface can further react with the oxygen dissolved in the reaction mixture to generate Fe^{3+} ions and reactive oxygen species (ROS) including •O_2, H_2O_2 and •OH as shown in reactions (18.7)−(18.9). These generated radicals are highly reactive and powerful oxidizing reagents that can nearly break down all the organic pollutants and thus serve significant role in the catalytic degradation of organic pollutants (Pang et al., 2016). The degradation mechanism is as shown below.

$$Fe^{2+} + O_2 \rightarrow \bullet O_2^- + Fe^{3+} \qquad (18.7)$$

$$Fe^{2+} + \bullet O_2^- + 2H^+ \rightarrow Fe^{3+} + H_2O_2 \qquad (18.8)$$

$$\bullet O_2^- + H^+ \rightarrow \bullet O_2H \qquad (18.9)$$

$$2\bullet O_2^- + 2H^+ \rightarrow H_2O_2 + O_2 \qquad (18.10)$$

$$\bullet O_2^- + H_2O_2 + H^+ \rightarrow O_2 + \bullet OH + H_2O \qquad (18.11)$$

18.3 Synthesis of magnetic photocatalytic systems

The synthesis of magnetic nanoparticles can be achieved by physical or chemical processes. The bottom-up method is a chemical process that includes coprecipitation, thermal decomposition, solvothermal, hydrothermal, sol—gel, sonochemical, microwave, and emulsion techniques. The top-down approach is a physical method where the bulk material/macromolecule is transformed into nanoparticles. Here, the ions or atoms are responsible for the growth of the nanoparticle. Mechanical milling, vapor deposition, and patterning are the common methods of top-down approach. However, in this section, we are discussing only few chemical processes for magnetic nanoparticle synthesis.

Coprecipitation is the commonly used method to synthesize either Fe_3O_4 or "—Fe_2O_3." It is an oldest, simple, and widely used method to obtain bigger quantity of magnetic materials. The method includes the precipitation of iron precursors (Fe^{2+} and Fe^{3+}) at a molar proportion of 1:2 in presence of base such as sodium hydroxide (NaOH) or ammonium hydroxide (NH_4OH) under inert atmosphere, at room or high temperature (Pang et al., 2016).

The reaction is expressed as

$$Fe^{2+} + 2Fe^{3+} + 8OH^- \rightarrow Fe_3O_4 + 4H_2O \tag{18.12}$$

The size, shape, and composition of the obtained iron oxide nanoparticles rely on the concentration of Fe^{2+}/Fe^{3+}, reaction temperature, pH, type of salts (e.g., sulfates, chlorides, nitrates), stabilizing agent used, ionic strength of the reaction media, and other parameters (Lu, Salabas, & Schüth, 2007). Also, it should be noted that, the various synthesis processes involved will contribute to different morphologies of iron oxide nanoparticles as shown in Fig. 18.2. The process of coprecipitation does not require any special stabilizing agents and the products obtained are water soluble (Gogate & Pandit, 2004). It is difficult to avoid nucleation and subsequent growth of the particles during coprecipitation reaction, which leads to the formation of particles with broad size distribution and irregular morphology. Hence, a size selection process after coprecipitation synthesis has to be incorporated to stabilize colloidal solution or precipitate bigger particles (Pang et al., 2016). Several types of magnetic nanoparticles are prepared using this method that includes Fe_3O_4, $NiFe_2O_4$, Ni-doped $CoFe_2O_4$ and their composite $Fe_3O_4@Fe_2O_3/Al_2O_3$, Fe_3O_4/hydroxyapatite, and $BiOBr/BiOI/Fe_3O_4$ (Venugopal et al., 2019).

Another most commonly used solution method is hydrothermal/solvothermal process. This is a method of synthesizing nanoparticles, where the chemical reaction is performed in a closed vessel in presence of solvents at a temperature above than the critical points. The process is referred as hydrothermal when water is used as a solvent and solvothermal when organic solvents such as methanol, ethanol, or polyol are used as solvent (Pang et al., 2016). A schematic representation of the method is shown in Fig. 18.3. The hydrothermal technique provides unique morphology, controlled dimensions (0D, 1D, and 2D), and good interactions between the counterparts, which strongly influence the photocatalyst efficiency. For example, 1D pearl chains Fe_3O_4/C core-shell, $BiOBr/NiFe_2O_4$ nanosheets/nanorods, $CoFe_2O_4/CeO_2$ spherical nanocomposite, thin-layer MnO_2 nanosheet-coated Fe_2O_3 nanocomposites, and

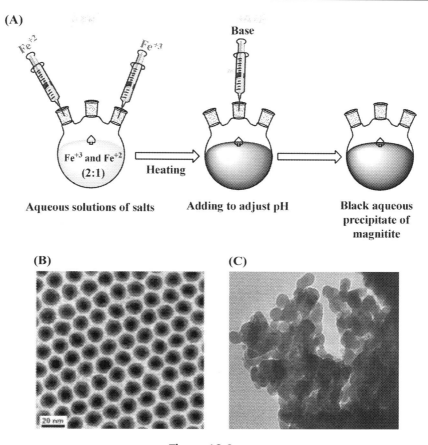

Figure 18.2

(A) Schematic representation of coprecipitation method for synthesis of iron oxide nanoparticles. (B) Transmission electron microscope (TEM) images of hollow Fe_3O_4 nanoparticles synthesis by coprecipitation method and (C) spherical $ZnO/\gamma\text{-}Fe_2O_3$. Source: *(B) Reprinted with permission from Peng and Sun (2007), Synthesis and characterization of monodisperse hollow Fe3O4 nanoparticles. Angewandte Chemie, 119, 4233−4236. https://doi.org/10.1002/ange.200700677. John Wiley and Sons. (C) Reprinted with permission from Lee and Abdullah (2015), Synthesis and characterization of zinc oxide/ maghemite nanocomposites: Influence of heat treatment on photocatalytic degradation of 2,4-dichlorophenoxyacetic acid. Materials Science and Semiconductor Processing, 30, 298−306. https://doi.org/ 10.1016/j.mssp.2014.100.017. Elsevier.*

C-Fe_2O_3@SiO_2@N,La-TiO_2 composite photocatalysts have been successfully prepared byhydrothermal methods (Venugopal et al., 2019). It has been reported that, at higher reaction temperature, solvothermal synthesis favors the formation of nanoparticles with high crystallinity, monodispersity, and nonaggregated nanoparticles (Pang et al., 2016).

Sol−gel method is the other widely preferred technique for the preparation of nanocomposite specifically used to design the core-shell structure. It is a wet chemical approach to synthesize

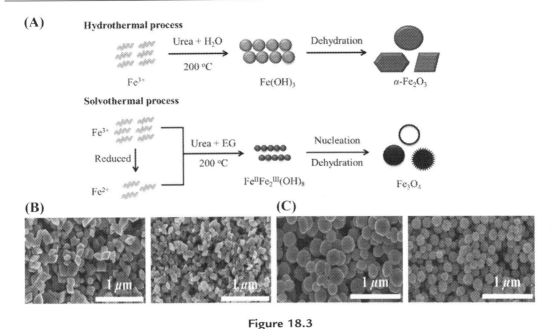

Figure 18.3

(A) Schematic representation of synthesis of iron oxide nanoparticles by hydrothermal/ solvothermal methods. (B) Scanning Electron Microscope (SEM) images of rhomb-like nano particles obtained from urea—water hydrothermal (UWH) system. (C) SEM images of sphere-like nano particles obtained from urea—ethylene glycol solvothermal (UEGS) system. Source: *Reprinted with permission from Su, He, & Shih (2016), Facile synthesis of morphology and size-controlled α-Fe2O3 and Fe3O4 nano- and microstructures by hydrothermal/solvothermal process: The roles of reaction medium and urea dose. Ceramics International, 42, 14793—14804. https://doi.org/10.1016/j.ceramint.2016.060.111. Elsevier.*

metal oxide nanoparticles, which is a low-cost, low-temperature process, and allows homogeneous blending at a molecular level (Venugopal et al., 2019). A schematic representation of the sol—gel method is shown in Fig. 18.4. By adopting this method, a core-shell structured nanoparticles can be obtained. For instance, Fe_3O_4-$NaYF_4$:Yb,Tm@TiO_2 (Lv et al., 2018), Fe_3O_4@TiO_2@Ag (Zhao et al., 2016), TiO_2-SiO_2/$CoFe_2O_4$ (Harraz, Mohamed, Rashad, Wang, & Sigmund, 2014), etc., where iron oxide acts as core with other components in the shell. $BiFeO_3$ nanofiber was synthesized using the combination of electrospinning and a sol—gel method (Wang et al., 2013). Electrospinning is a simple route to fabricate continuous 1D nanocomposites at a large scale, and the prepared material is typically called a nanofiber.

Thermal decomposition is another technique for preparation of magnetic nanoparticles. The technique is based on decomposition of metal precursors in high boiling point organic solvents such as benzyl ether oroctadecene. The iron precursors can be either acetylacetonates, acetates,

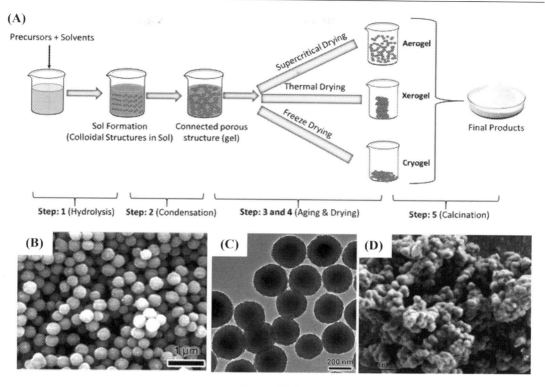

(A)

Precursors + Solvents

Sol Formation
(Colloidal Structures in Sol)

Connected porous
structure (gel)

Supercritical Drying

Thermal Drying

Freeze Drying

Aerogel

Xerogel

Cryogel

Final Products

Step: 1 (Hydrolysis) Step: 2 (Condensation) Step: 3 and 4 (Aging & Drying) Step: 5 (Calcination)

Figure 18.4

(A). Schematic representation of synthesis of iron oxide nanoparticles by sol−gel method. (B, C) Scanning Electron Microscope (SEM) and TEM images of core-shell $Fe_3O_4/SiO_2/TiO_2$. (D) Grape-like Fe_3O_4/SiO_2 nanospheres. Source*: (A) Reprinted with permission from Parashar, Shukla, & Singh (2020). Metal oxides nanoparticles via sol−gel method: A review on synthesis, characterization and applications. Journal of Materials Science: Materials in Electronics, 31, 3729−3749. https://doi.org/ 10.1007/s10854-020-02994-8. Springer Nature. (B, C) Reprinted with permission from M. Ye, Q. Zhang, Y. Hu, J. Ge, Z. Lu, L. He, Z. Chen, Y. Yin, Magnetically Recoverable Core-Shell Nanocomposites with Enhanced Photocatalytic Activity,* Chemistry - A European Journal, *16 (2010) 6243−6250. https://doi. org/10.1002/chem.200903516. John Wiley and Sons. (D) Reprinted with permission from R. Wang, X. Wang, X. Xi, R. Hu, G. Jiang, Preparation and Photocatalytic Activity of Magnetic $Fe_3O_4/SiO_2/TiO_2$ Composites,* Advances in Materials Science and Engineering, *2012 (2012) 1−8. https://doi.org/ 10.1155/2012/409379. Hindawi Publishing Corporation.*

oleates, carbonyl, oxalates, or ferrocene (Pang et al., 2016). In this method, using different suitable stabilizing surfactants, a uniform and monodisperse iron oxide nanocrystals can be obtained. A schematic representation of the method is displayed in Fig. 18.5.

Microemulsion or reverse micelles are another attractive method of nanoparticle synthesis among other reported methods due to their potential benefits. Microemulsions are the

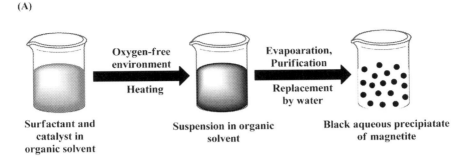

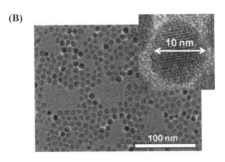

Figure 18.5

(A) Schematic representation of thermal decomposition method for synthesis of iron oxide nanoparticles. (B) Transmission Electron Microscope (TEM) images of iron oxide NPs synthesized by thermal decomposition. Source: *Reprinted with permission from Basly, Felder-Flesch, Perriat, Pourroy, & Bégin-Colin (2011). Properties and suspension stability of dendronized iron oxide nanoparticles for MRI applications,* Contrast Media & Molecular Imaging, *6, 132–138. https://doi.org/10.1002/cmmi0.416. John Wiley and Sons.*

thermodynamically stable dispersions of two immiscible phases (water and oil) in the presence of surfactant and cosurfactant molecules at the interface (Majidi, Sehrig, Farkhani, Goloujeh, & Akbarzadeh, 2016). In case of water in oil emulsion, water is dispersed as small nanodroplets (typically 1–100 nm in diameter) in oil phase; these droplets contain metal precursors inside, which are stabilized by suitable surfactant molecules, forming the micelles. Thus, formed micelles function as nanoreactors, where the synthesis of the MNPs takes place (Pang et al., 2016). Some of the morphologies obtained by the microemulsion process include hollow Fe_3O_4 (Zhang, Tong, Li, Zhang, & Ying, 2008), well-defined grass-like Fe_3O_4 (Li et al., 2014), and flower-like Fe_3O_4 (Li, Jiang, Liu, Lv, & Zhang, 2014). Schematic representation of the method is presented in Fig. 18.6.

Among the top-down processes, ball milling/mechanical milling is considered as one of the best top-down approaches for the synthesis of nanocomposites. It is a high-energy mechanical milling process, where nanoparticles of different natures in enormous quantities

(A)

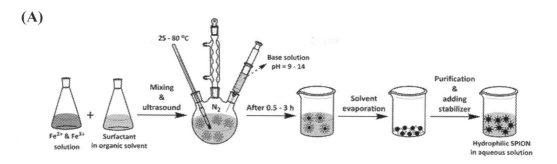

(B)

(C)

Figure 18.6

(A) Schematic representation of microemulsion method for synthesis of iron oxide nanoparticles. (B) Field Emission Scanning Electron Microscope (FESEM) images of the hollow Fe$_3$O$_4$ nanospheres. (C) Transmission Electron Microscope (TEM) and Field Emission Scanning Electron Microscope (FESEM) images of flower-like iron oxide. *Source: (A) Reprinted with permission from S.M. Dadfar, K. Roemhild, N.I. Drude, S. von Stillfried, R. Knüchel, F. Kiessling, T. Lammers, (2019). Iron oxide nanoparticles: Diagnostic, therapeutic and theranostic applications, Advanced Drug Delivery Reviews 138, 302−325. https://doi.org/10.1016/j.addr.2019.010.005. Elsevier. Reprinted with permission from D.E. Zhang, Z.W. Tong, S.Z. Li, X.B. Zhang, A. Ying, (2008). Fabrication and characterization of hollow Fe3O4 nanospheres in a microemulsion, Materials Letter. 62 (2008) 4053−4055. https://doi.org/10.1016/j.matlet.2008.050.023. Elsevier. Reprinted with permission from Y. Li, R. Jiang, T. Liu, H. Lv, X. Zhang, Single-microemulsion-based solvothermal synthesis of magnetite microflowers, Ceramics International, 40 (2014) 4791−4795. https://doi.org/10.1016/j.ceramint.2013.09.025. Elsevier.*

can be obtained. For example, $NiFe_2O_3$-rGO and $CoFe_2O_3$-rGO nanocomposites were synthesized through an one-pot ball milling technique. The process is simple, cost-effective along with high efficiency and good yield. Since no harsh chemicals are used during mechanical milling, it is also environmentally friendly (Venugopal et al., 2019).

Coating and functionalization of magnetic nanoparticles is the current ongoing research trend to get photocatalyst with desired functionalities. By doing so, some of the most commonly recognized limitations of MNPs can be overcome. At the same time, they can also provide added capabilities for multifunctional applications. The commonly prepared semiconductor magnetic nanocomposite includes magnetic core coated with semiconductor cells, magnetic core-coated semiconductor cells with an interlayer, and magnetic core-coated doped semiconductor cells with an insulator interlayer (Venugopal et al., 2019).

A core-shell structure consists of two different materials, where one acts as a core and other one as the shell, which exhibits the properties of both the materials. Inorganic materials such as silica (SiO_2), gold, or gadolinium are being employed as coating materials. In addition, nonpolymer organic stabilizers such as oleic acid, stearic acid, and phosphates, and polymer stabilizers such as dextran, polyethylene glycol, polyvinyl alcohol, and other are also used as coating materials (Laurent et al., 2008). The coating provides a nanoparticle with modified surface that can be further functionalized to alter the magnetic properties. In addition, coating makes nanocomposite chemically and colloidally stable. Some of the TiO_2-coated magnetic materials are $BaFe_2O_4$-TiO_2, $SrFe_{12}O_{19}$-TiO_2, Fe_3O_4@C@TiO_2, and $ZnFe_2O_4$-TiO_2 (Chen et al., 2014; Venugopal et al., 2019). On the other method, to prevent the direct physical contact between the semiconductor and magnetic materials, an insulating layer was induced between them. The incorporated insulator inhibits the charge carrier transformation from shell to core, thereby increasing the efficiency of the photocatalyst. The commonly used intermediate layer is SiO_2, which provides a good insulation and strong inhibition to the charge carrier transformation inhibitor and also associated with a wide band gap compared with TiO_2 (Pang et al., 2016).

The third class of structures is the magnetic core-coated semiconductor cell with a doped semiconductor, where there is an attempt to overcome the rapid electron—hole recombination, which limits the efficiency of photocatalyst. Introduction of various types of dopant species (metals, noble-metals, and nonmetals) into iron oxide structure have been tried over the years to improve its photocatalytic activity (Pang et al., 2016). Some of the dopants explored are platinum (Pt), palladium (Pd), molybdenum (Mo), chromium (Cr), silver (Ag), gold (Ag), gallium (Ga), Ba—Cd—Sr—Ti, etc. (Pang et al., 2016; Venugopal et al., 2019). A few repoted metal-doped magnetically separable photocatalysts are TiO_2-coated Ag@Fe_3O_4@C-Au microspheres, Fe_3O_4@C—Au@void@TiO_2—Pd hierarchical microspheres, iodine/nitrogen-doped TiO_2, and activated carbon-based magnetic photocatalysts (Venugopal et al., 2019).

18.4 Applications

18.4.1 Dye degradation

In the textile industry dyeing processes produce large amounts of colored and contaminated wastewater, which is discharged into the environment. The dyeing process releases large amount of sodium, sulfate, chloride, heavy metals, and carcinogenic dyes, which leads to high biochemical oxygen requirements and complete dissolution of solids. The most complex contaminants that need to be treated are toxic micronutrients, acids, alkaline, azo, and diazo. The complex aromatic molecular structures and nonbiodegradable properties of dyes with a high presence of carcinogenic aromatic amines can be considered as most complex pollutants (Tünay, Kabdasli, Eremektar, & Orhon, 1996). In both dumping and wastewater generation, the textile industry is considered to be the most polluted among all industries. Therefore, there is a great need for dye/color removal techniques that work properly in the above situations and are cost-effective. Different techniques such as membrane separation, absorption (Mohamed et al., 2017; You, Semblante, Lu, Damodar, & Wei, 2012), fluid extraction, adsorption, and biological treatment (Saroj, Dubey, Agarwal, Prasad, & Singh, 2015) have been proposed for dye removal from time to time. However, all these methods are struggling with one or another restriction and have not been able to completely remove the color from the wastewater. AOP using the UV system has become an alternative to the oxidation process for degradation of more dye waste. Under ultraviolet (UV) radiation, excited electron in photocatalysts produces hydroxyl radicals, which is a powerful oxidizer that tends to break down the color pigments and give complete discoloration.

In general, the developed photocatalyst is suspended in a dye-contaminated water sewage and irradiated under the visible/UV light. The properties of the photocatalyst with maximized surface area and reactive site were studied. The high concentration of dye molecules in solution leads to accumulation of dye molecules on the surface of active site. This in turn results in the decrease in the efficiency of the catalyst, where it leads an additional process to restore/remove the catalyst.

To solve this problem, a self-removable photocatalytic system was introduced in the photocatalysts of particulate matter. Many research groups have begun to focus on self-removing photocatalysts for wastewater treatment, demonstrating the value of the special properties of magnetic materials. These magnetic properties are studied in several elements, including iron (Fe), nickel (Ni), and cobalt (Co). Among all these, the Fe-based material has higher magnetic properties than other factors. Magnetic nanoparticles and their nanocomposites have proven to be good in photocatalytic performance under light irradiation for dye degradation (Sun, Han, & Chen, 2019). In detail, $Fe_3O_4@SiO_2$ nanocomposites delivered a Fenton-like catalytic activity along with H_2O_2 and achieved a high degree of Methylene Blue degradation at neutral pH. This confirmed the enhanced photocatalytic activity of nanocomposite material when compared with pristine Fe_3O_4 (Yang et al., 2015). In the reported study, a very stable $Fe_3O_4@TiO_2$ sample with almost 100% degradation of Methylene Blue was reported with addition of H_2O_2 in 5 minutes under UV cutoff visible light (Abbas, Rao, Reddy, & Kim, 2014). Similarly, Fe_3O_4/chitosan/TiO_2 nanocomposites reported 93% of Methylene Blue dye within 40 minutes (Xiang, Wang, He, & Song, 2015). In another study, enhanced photocatalytic dye degradation activity of Fe_3O_4/ZnO/$CuWO_4$ nanocomposites was reported for degradation of Rhodamine-B (Rh-B) compared with binary composite Fe_3O_4/ZnO or Fe_3O_4/$CuWO_4$ under visible light irradiation (Shekofteh-Gohari & Habibi-Yangjeh, 2016). A recent study showed an effective elimination of Congo Red dye with $Fe_3O_4@SiO_2$ nanocomponents (Wang et al., 2016).

Improved photocatalytic dye degradation activity is observed in degradation of Rh-B with prepared Fe_3O_4/SiO_2/TiO_2 nanocomposite that was prepared by sol−gel method when compared to Degussa P25 and SiO_2 with high chemical stability, and rapid magnetic separation lasting 18 cycles (Ye et al., 2010).

Similarly, Fe_3O_4/SiO_2/TiO_2 nanocomposites synthesized by sol−gel method shows photocatalytic activity on Methylene Blue dye in an aqueous solution maintained at pH of 10 at ambient temperature. Under UV irradiation, the nanocomposites displayed enhanced photocatalytic degradation of Methylene Blue by 78% in a short time of 5 minutes (Wang et al., 2012).

A core-shell-type nanocomposite $SrFe_{12}O_{19}$/$Zn_{0.65}Ni_{0.25}Cu_{0.1}Fe_2O_4$ (1:1 ratio) showed degradation rate 100% of Methyl Violet within 3 minutes in H_2O. The hybrid complex of magnetic nanoparticles and graphite carbon nitride (g-C_3N_4) has been most reported nanocomposites for photocatalytic dye degradation. For example, photocatalytic effect and Fe_3O_4 stability are significantly improved when the 8 nm size Fe_3O_4 nanoparticles were deposited on the g-C_3N_4 sheet. Visible light photocatalytic degradation activity of Rh-B dye was significantly improved with nanocomposites compared with pristine materials. In addition, it was reported that a nanocomposite demonstrated good recycling and highly stable activity after six cycles (Kumar, Surendar, Kumar, Baruah, & Shanker, 2013). This synergy is usually observed when other nanomaterials dispersed into magnetic nanoparticles. Similarly, compared to the usual bare g-C_3N_4, the photocatalytic activity of $Fe_2O_3@$ g-C_3N_4 over rh-B dye was observed 1.8 times higher for Rh-B dye degradation when visible light was exposed (Ye et al., 2013).

Core-shell nanostructures $CuFe_2O_4@$g-C_3N_4 of dose 0.1 g/L significantly enhanced the photocatalytic activity on Orange-II (0.28 mM) dye along with 0.01 M of H_2O_2 as oxidant, where Orange-II dye was degraded within 90 minutes at 65°C under visible light (Yao et al., 2015). The composite showed excellent activity than pristine $CuFe_2O_4$ and bare g-C_3N_4. In addition, the study showed the significance of increasing temperatures to minimize the activation energy barrier of H_2O_2, which led to an increase in the production of hydroxy radicals. In general, these conclusions evidently demonstrated the synergetic effect of each material in prepared nanocomposites, which makes the application of the $CuFe_2O_4$@g-C_3N_4 core-shell photocatalyst a viable and effective material for industrial wastewater treatment. In all kind of dyes, the hardly recyclable organic complexes are azo dyes, due to their high stability under light irradiation and microbial attack resistance. However, recently reported nanocomposites such as $MnFe_2O_4$@PANI@Ag (Amir, Kurtan, Baykal, & Sözeri, 2016) and $Co_{0.53}Mn_{0.31}Fe_{2.16}O_4$@$TiO_2$ (Neris et al., 2018) have demonstrated considerable degradation capability with improved proficiency towards azo dyes. Studies comparing the decomposition ability of Ni^{2+}-incorporated $MnFe_2O_4$ ($M_{1-x}Ni_xFe_2O_4$, where $x = 0.1, 0.2...$to 0.5 of Ni), among all stoichiometric compositions, $Mn_{0.5}Ni_{0.5}Fe_2O_4$ nanoparticles were found to be an excellent photocatalyst by decomposing indigo carmine synthetic dye (Jesudoss et al., 2016). In another study, sulfur-modified $CoFe_2O_4$ (S-$CoFe_2O_4$) nanomaterial dosage of 0.1 g/L showed improved degradation activity within 15 minutes of 50 mg/L Acid Orange-II (AO-II) with 0.1 g/L sodium bicarbonate ($NaHCO_3$)-activated 0.3 mM of hydrogen peroxide (Guo, Li, & Zhao, 2015). In general, these reports suggest the importance of magnetic nanomaterials for enhanced photocatalytic degradation of organic pollutants in waste water. The $CoFe_2O_4$/g-C_3N_4 (2:1 ratio) nanocomposites demonstrated degradation of several dyes with good stability, high rate of recovery, and good recyclability.

18.4.2 Nondye-based pollutants degradation (phenolic compounds etc)

Phenol and phenolic by-products are one of the major and most serious pollutants in environment. This is mainly due to the industrial-scale production of various antibacterial and antifungal materials which results in the release of lot of phenolic and related compounds (Krastanov, Alexieva, & Yemendzhiev, 2013). These phenolic compounds were treated as severe pollutants by the United States Environment protection agency, in part because of their toxic, carcinogenic, and mutant properties. Low concentrations of these pollutants can cause serious damage to human liver and animals, lungs, and blood cells. Furthermore, the removal of these pollutants is difficult because of its resistance. Thus, they remain longer time in environment (Wang, Hsieh, Chou, & Chang, 1999). Subsequent sections summarize the photocatalytic degradation of the most commonly used forms of phenols such as chlorophenols and nitrophenols, with a brief review.

18.4.2.1 Phenols

Phenols are the contaminants of highly harmful and hardly biodegradable chemicals which are found in wastewater. This can create an enormous damage to anthropoid health even at less exposure (Stoia, Muntean, & Militaru, 2017). In recent years, the use of magnetic nanoparticles in photocatalytic applications for the complete decomposition and removal of phenols from wastewater has become an essential practice. For example, the magnetic Fe_3O_4 nanomaterial, added with H_2O_2 in wastewater containing of phenolics, was removed 85% of phenols within 3 hours at $16°C$ without any secondary contamination (Zhang et al., 2008). The nanocomposites with semiconducting materials and magnetic nanoparticles also increase the rate of phenolic decomposition. The decomposition of phenols through the photocatalytic process using Fe_3O_4-ZnO nanocomposites was recently explored (Feng, Guo, Patel, Zhou, & Lou, 2014). The stability and restoration of hybrid nanocomposites was also assessed during the study period. As a result, about 89% of used photocatalyst of 89% was restored even after three repeated cycles. The percentage of phenol reduction for 1, 2, and 3 cycles were 82.8%, 72.4%, and 65.1%, respectively. These values are much higher than the values observed for pristine ZnO with 52% of phenolic decomposition, indicating an improvement in photocatalytic activity, stability, and the ability to restore synthesized hybrid photocatalytic. Improved photocatalytic performance is synergistic between magnetic nanoparticles and semiconductors, which quickly increase the charge separation efficiency, and reduce the e^-/h^+ recombination, using maximum electron-holes for oxidation (Ahmed, El-Katori, & Gharni, 2013; Byrne, Subramanian, & Pillai, 2018; Casbeer, Sharma, & Li, 2012; Feng et al., 2014).

18.4.2.2 Chlorophenols

Depending on the location and the size of chlorine, there are many types of chlorophenol compounds such as 2,4-dichlorophenol, para-chlorophenol, penta-chlorophenol, and 2-chlorophenol. They are mostly used in many industries in the manufacture of dyes, pesticides, fungicides, pharmaceuticals, explosives, pesticides, and preservatives. Widespread use of these chlorophenols has led to pollution from industrial emissions. They are suspected of being one of the mutagenic and carcinogenic chemicals (Taleb, 2014). Magnetic nanocomposites are one of the very efficient catalytic materials for chlorophenol breakdown. Recently, the enhanced degradation of 4-chlorophenol complex with $Cu_{0.5}Mn_{0.5}Fe_2O_4-TiO_2$ photocatalyst was reported than the pure $CuFe_2O_4$ (Manikandan et al., 2015).

Improved photocatalytic performance of the nanocomposite is attributed to TiO_2 semiconductors in the interconnected $Cu_{0.5}Mn_{0.5}Fe_2O_4$. In an another study, 88.6% of degradation of the 4-chlorophenol using ZrO_2/Fe_3O_4/chitosan nanocomposite was reported in 3 hours under natural sun light (Kumar et al., 2016). The semiconductor hetero-compound metal oxide reduces the recombination of the electron-hole, and improves charge separation and electronic transmission. Chitosan increases the absorption and stability, and prevents particle aggregation. Similarly, there is a significant improvement in photocatalytic performance of the chitosan/$CoFe_2O_4$ composites for the complete decomposition of 2-chlorophenol in sewage water when exposed to sunlight (Taleb, 2014). This exceptional effectiveness of magnetic nanoparticles and magnetic nanocomposites in reliable degradation of hazardous pollutants in environment might be an indicator of importance in addressing wastewater treatment issues.

18.4.2.3 Nitrophenol

Organic nitro compounds are the most dangerous contaminants and nonbiodegradable organic pollutants due to their carcinogenic properties. Usually these are found in discharged wastewater from industrial and agricultural waste. Thus, the elimination of nitrophenol from wastewater is very necessary by decomposing or converting beneficial amino-aromatic components. In fact, conversion to amino-aromatics has more benefits because it is one of the important organic compounds for the manufacture of some biologically active compounds, pharmaceuticals, photo developers, corrosion inhibitors, and several industrial dyes (Ibrahim, Ali, Salama, Bahgat, & Mohamed, 2016; Min et al., 2008). For the conversion of beneficial amino-aromatic acids from nitro-phenol compounds, there are several methods reported such as heterogeneous catalytic hydrogenation, photo-reduction (Gazi & Ananthakrishnan, 2011; Rashad, Mohamed, Ibrahim, Ismail, & Abdel-Aal, 2012), and so on (Chen, Wang, Du, Xing, & Xu, 2009).

However, the abovementioned methods have few limitations, like recovery of catalyst, more energy consumption, and laborious difficulties. In fact, all of these restrictions can be reduced by using pristine or substituted magnetic nanoparticles to reduce nitro-phenol (Goyal, Bansal, & Singhal, 2014; Ibrahim et al., 2016). For example, among all the synthesized spinel ferrites using hydrothermal sol−gel technology, 10.6 nm $MnFe_2O_4$ nanoparticles showed effective performance with 100% conversion of 4-nitrophenol, 4-nitroaniline, and 2,4,6-trinitrophenol into corresponding amino derivates within a short reaction time of 4.5 minutes. The significance of $MnFe_2O_4$ was demonstrated by means of increasing the pore size and volume (Ibrahim et al., 2016). In other study, Fe_3O_4−3-aminopropyl-triethoxysilane (APTES)-Pd(0) (Karaoglu, Summak, Baykal, Sözeri, & Toprak, 2013), $NiFe_2O_4$-APTESPd(0), and $NiFe_2O_4$-Pd performed an improved catalytic activity for 4-nitroaniline and 1.3-dinitrobenzene hydrogenation in liquid phase (Karaoğlu et al., 2012). Also, the most effective photocatalyst for 4-nitroaniline and 1.3-dinnitro-cyclohexanone hydrogenation was synthesized, $CoFe_2O_4$-APTES-Pd(0) (Demirelli, Karaoğlu, Baykal, Sözeri, & Uysal, 2014).

Moreover, all these findings demonstrated the significance of attaining high stability, and ease to restore photocatalyst with the external magnetic field. The importance of achieving reusability without significantly reducing the catalytic activity of synthesized magnetic nanocomposites was also demonstrated. The catalyst Fe_3O_4 is also capable of the possible conversion of nitrophenol to advantageous amino-aromatic acids. Recently, the Au-Fe_3O_4 organic layer-coated nanocomposites [Au-Fe_3O_4 @ metal-organic framework (MOF)] catalyst showed the excellent performance. This surpasses the 4-nitrophenol reduction with support of

NaBH$_4$. Self-processing and recycling are reinforced with Fe$_3$O$_4$ and Au nanoparticles, which serve as effective sites for the catalytic reduction of 4-nitrophenol. Similarly, Fe$_3$O$_4$@SiO$_2$-Ag nanocomposite has demonstrated a better catalytic activity on p-nitrophenol to p-aminophenol conversion with 95% in 15 minutes in the presence of NaBH$_4$. Now this has become the potential photocatalytic material for various applications in the future. The Pd-loaded titanium nanotubes have shown an effective conversion of p-nitrophenol to p-amino phenols with the support of NaBH$_4$. However, it showed poor performance in the second and third cycles of recovery and recyclability (Kalarivalappil et al., 2016). The dispersion of magnetic nanoparticles in catalytic materials is expected to improve recovery and reuse of this material and this requires further research in this direction.

18.4.3 Pharmaceutical pollutants degradation

Pharmaceutical compounds found in wastewater and environments are antibiotics used for human and veterinary. Environmental impacts involve the development of antibiotic-resistant microbes for water treatment and underwater environments (Kümmerer, 2001; Wise et al., 1998); retardation of nitrate oxidation (Dokianakis, Kornaros, & Lyberatos, 2004), and methane production (Dokianakis et al., 2004; Fountoulakis, Drillia, Stamatelatou, & Lyberatos, 2004); and potential increase in toxicity of chemical combinations and metabolites (Eljarrat, 2003). However, this potential effects on humans and aquatic ecosystems, however, are not clearly understood. The release of antibiotics into the environment can lead to the development of bacterial resistance, which can greatly affect the effectiveness of antibiotics. In addition, these antibiotics are almost noneliminated through conventional water purification methods. Various advanced processing options are available, including membrane filtration technology (Wintgens, Gallenkemper, & Melin, 2004; Yoon, Westerhoff, Snyder, & Wert, 2006), AOPs (Petrovic, 2003), UV irradiation, and hybrid systems (Yoon et al., 2006). Heterogeneous photocatalysts with semiconductor metal oxides among advanced oxidation options are rapidly growing in the area of fundamental and applied research and are a promising tool for water treatment. Effective decomposition of essential medicines has been observed in some cases without the formation of harmful intermediates (Augugliaro et al., 2005; Doll & Frimmel, 2005). Other benefits include low chemical input, energy efficiency, and renewable and nonpollution of solar energy (Kabra, Chaudhary, & Sawhney, 2004).

The solar-light active CNT−TiO$_2$ magnetic complex is used for photocatalytic decomposition of carbamazepine and sulfamethoxazole. The maximum efficiency of photocatalytic degradation of pharmaceutical pollutants was achieved with the optimal mass ratio of MCNT and TiO$_2$. Compared with the pristine TiO$_2$, photocatalytic activity of MCNT-TiO$_2$ was much higher (Awfa, Ateia, Fujii, & Yoshimura, 2019).

Magnetic nanoparticles MnFe$_2$O$_4$-based graphite carbon sand complex (GSC) and bentonite (BT) nanocomposites photocatalyst have been used in degradation of ampicillin (AMP) and antibiotics like oxytetracycline (OTC) under sunlight. The output results of nanocomposites demonstrated about 99% and 90% of AMP and 96% and 83% of OTC. over-the-counter reduction within 60 and 120 minutes using $MnFe_2O_4$/GSC and $MnFe_2O_4$/BT under the exposure of natural sunlight, respectively (Gautam et al., 2017). Furthermore, under solar radiation, the two used photocatalysts were separated magnetically from the reaction mixture in 10 seconds and also showed significant recycling efficiency even after 10 consecutive cycles without effecting their activity. Likewise, the $Fe_3O_4-GO-Ce-TiO_2$ hybrid complex nanocomposite has showed an enhanced degradation and good adsorption capacity of tetracycline (Cao et al., 2016). It is considered as one of the promising photocatalysts for the decomposition of antibiotics. Similarly, the pyrrole-imprinted $CoFe_2O_4$/MWCNTs nanocomposites demonstrated an efficient photocatalytic performance for decomposition of 2-mercaptobenzothiazol (Lu et al., 2015).

A magnetic Fe_2O_3@ZnO core-shell structural photocatalyst synthesized through hydrothermal process in combination with ALD has provided the benefits of magnetic separation. More attractively, the precisely controlled deposition of the 3D ZnO layer can grow evenly on the surface of α-Fe_2O_3, improving the level of photocatalytic activity. Photocatalytic degradation efficiency of the ciprofloxacin was increased from 11.7% of α-Fe_2O_3 to 92.5% of Fe_2O_3@ZnO composite under simulated sunlight irradiation. In addition, the composite photocatalyst showed stability and reprocessing after six cycles. It has been proven that the improvement of the photocatalyst comes from the type II heterogeneous structure of α-Fe_2O_3 and ZnO, which can effectively inhibit the recombination of photogenerated electronic holes. Holes and hydroxy radicals have been identified as the main contribution to the decomposition of ciprofloxacin (Li et al., 2017). The ZnO/α-Fe_2O_3 nanocomposite photocatalyst using the precipitation–calcination method showed efficient photocatalytic degradation over 2,4-dichlorophenoxyacetic acid. and reached to 40% efficiency after four cycles in reusable experiments (Lee & Abdullah, 2015).

The ternary nanocomposite Ag_2WO_4@ZnO@Fe_3O_4 is one of these photocatalysts used to break down tetracycline hydrochloride. Compared to Ag_2WO_4 and pristine ZnO nanoparticles, ternary nanocomposites improve decomposition rates of approximately 152% and 143%, respectively. In addition, the ternary nanocomposites completely decomposed tetracycline hydrochloride for 1.75 hours under sun exposure (Shen, Lu, Liu, & Yang, 2016). The photocatalytic activity of the nanocomplex α-Fe_2O_3-ZnO can be used to decompose cefixime trihydrate drugs in UV–visible range intensities by varying different parameters, such as the initial concentration of photocatalyst, the concentration of the drug, the UV–vis intensity and irradiation time, and the appropriate conditions. Results obtained suggest that the intensity of light is an important factor. 99.1% of cefiximetrihydrate is decomposed in optimized conditions. Shooshtari and Ghazi (2017). Diphenhydramine photodegradation under visible light was studied by Fe_2O_3-ZnO supported on clinoptilolite natural zeolite. In optimized conditions, approximately 95% of the drug has been degraded for Fe_2O_3—ZnO—Zeolite composite, which was more than the Fe_2O_3—TiO_2—Zeolite (80%) (Davari, Farhadian, Nazar, & Homayoonfal, 2017).

The Fe_3O_4/Bi_2WO_6 nanohybrid composite is an efficient photocatalyst for photocatalytic decomposition of ibuprofen (IBP). IBP, 2-[4-(2-methylprofile) phenyl] propanoic acid belongs to the category of nonsteroidal antiinflammatory drugs widely used in the treatment of muscle pain, inflammatory diseases, fever, migraines, and tooth pain. The test results of IBP decomposition showed the high photocatalytic activity of Fe_3O_4/Bi_2WO_6 nanohybrid under sunlight conditions with the addition of hydrogen peroxide to improve the efficiency of IBP decomposition. According to the results, the light generation hole (h^+) was found to be the main active species for the decomposition of IBP by Fe_3O_4/Bi_2WO_6 hybrid nanocomposite. The solar-Fenton degradation of IBP deterioration took place on the surface of the photocatalyst in the presence of H_2O_2 (Rohani Bastami, Ahmadpour, & Ahmadi Hekmatikar, 2017). In addition, the photocatalytic activity and magnetization properties of magnetic photocatalyticnanohybrids provide a promising solution for the decomposition of water contaminants and photocatalytic recovery.

18.4.4 Heavy metal reduction

Among the several types of waste water treatment practices discovered by researchers, photocatalyst remains one of the best practices, which is capable of destroying or reducing contamination rather than just reducing, trapping, or isolating. The use of various semiconductor materials, such as titanium dioxide (TiO_2) and zinc oxide (ZnO), as a photocatalyst for waste water treatment, has attracted the attention of many researchers. These semiconductors can produce powerful free radicals that can destroy various organic pollutants and reduce heavy metal ions with the exposure of light sources. These semiconductors absorb photons and thus the electrons move to high energy levels leaving the holes in valence band. Then, these electrons and holes are further promoted to the surface of the semiconductor, which eventually produce the ROS such as OH and O_2^-.

Since semiconductors can effectively decompose natural organic matter, much research focuses on fine-tuning and modification of these semiconductors to effectively dissociate organic contaminants. Since it is impossible to decompose metal ions, photocatalysts may not be used to remove heavy metals. However, photocatalysts have shown that they can be used as a tool to reduce the harmful effects of heavy metal ions in wastewater by reducing them to less harmful byproducts. Reduction of heavy metal ions is a restraint tool for the treatment of heavy metal contamination in wastewater. Particularly, CR(VI) is significantly harmful to water living organisms as compared to Cr(II). Hence, the major practice for calibration of Cr(VI) is to reduce it to Cr(II). Photocatalysts play an important role in the treatment of heavy metal ions in wastewater. Based on current research trends, the employment of photocatalysts is used to reduce Cr(VI) into Cr(II), which is usually found in contaminated water released from industries (Yu, Lv, Wang, & Tan, 2018). The photo reduction of Cr(IV) is explained in Eqs. (18.13) and (18.14) (Du et al., 2019).

$$Cr_2O_7^{2-} + 14H^+ + 6e^- \rightarrow 2Cr^{3+}7H_2O \tag{18.13}$$

$$2H_2O + 2h^+ \rightarrow H_2O_2 + 2H^+ \tag{18.14}$$

Cr(VI) has been widely studied because hexagon equivalents are far more toxic than its insoluble variants, and correspond to CR(III). When photocatalysts are exposed to photons, they absorb photons and stimulate electrons to the photocatalyst surface. CR(VI) consumes these electrons, allowing for a single reduction in light. This is a reactive catalyst because of the subsequent presence of e/h + pairs. To use this mechanism, research on photoreduction of Cr(VI) ions combined with organic pollutants and organic pollutants has caused a chain reaction to the continuous reduction of Cr(IV) and organic pollutants under light source. This mechanism improves versatility by providing additional evidence that photocatalysts have on the capability of wastewater treatment, including heavy metal and organic pollutants degradation. Recently, iron oxide (II) bismuth carbonate photocatalyst showed an excellent photocatalytic activity towards the reduction of carcinogenic and mutagenic Cr(VI) to nontoxic Cr(II) (Kar, Jain, Kumar, & Gupta, 2019).

The effects of titanium dioxide with silicon layer ($NiFe_2O_4$-SiO_2-TiO_2) nanocompositephotocatalyst to reduce Cr(VI) in a water solution were evaluated. Further, the photocatalytic activity of the composites has been compared to determine whether the introduction of silica has improved Cr(VI) reduction in wastewater. TiO_2 was recognized as the best photocatalyst for reducing Cr(VI) in simulated wastewater samples with a 96.7% reduction rate within 240 minutes. $NiFe_2O_4-SiO_2-TiO_2$ has a 96.5% reduction efficiency within 300 minutes, and $NiFe_2O_4-TiO_2$ has a 60% reduction in UV radiation within 300 minutes. The integration of the silica layer between titanium dioxide and magnetic core has shown significant improvement and efficient photocatalytic properties without affecting the magnetic recovery of magnetic titanium nanocomposites. This creates a magnetic photocatalyst suitable for wastewater decontamination (Ojemaye, Okoh, & Okoh, 2017).

Nanotitania-cobalt ferrite modified by silver nanoparticles reduces Cr(VI) via photocatalytic activity. Ag/TCF composites provide higher photocatalytic efficiency as titania and ferrite improved the reduction process only by UV and solar light or in combination. The high efficiency is due to improved light absorption (plasma effect) and more efficient charge separation, as well as a significant static attraction between new photocatalysts and chromium cations. The rate constant was not affected throughout five consecutive photocatalytic cycles, confirming the reliability of photocatalyst Ag/TCF under UV radiation. The new composite also represents interesting magnetic properties and can be easily separated from the reaction solution using magnets (Ibrahim et al., 2020).

18.4.5 Microbial disinfection

Microbial pollution has always been harmful to human health since the existence of human society. Many types of bacteria can lead to serious diseases and even death in humans. Therefore, it is necessary to develop an important and effective, low-cost, and environmentally friendly methods. Platinum doping TiO_2-mediated photon-based disinfection method is widely researched since 1985, and the photocatalytic disinfection from UV light to visible light is widely researched in particular, TiO_2-mediated bacterial growth control seems to be preferable compared with UV and chlorination methods. Chlorination is environmental unfriendly and not resistant to the presence of some free chlorine-resistant bacteria such as *Mycobacterium avium* and carcinogenic disinfection byproducts (Shannon et al., 2008). In addition, UV rays are not effective against some UV-resistant bacteria and the risk of direct and intense use of UV rays limits the application (Wang et al., 2013).

In recent years, several magnetic nanoparticles functionalized with biological proteins (Liu, Tsai, Lee, & Chen, 2008), amino acids (Jin, Liu, Shan, Tong, & Hou, 2014), antibodies (Ryan et al., 2009), and carbohydrates (Liu, Dietsch, Schurtenberger, & Yan, 2009) were utilized to capture and detect targeted bacteria. Therefore, a promising solution to separate used photocatalyst after the application is to combine with magnetic materials such as Fe_3O_4, TiO_2/Fe_3O_4, Au-decorating $Fe_3O_4@mTiO_2$ core-shell (Li et al., 2014), $Ag_3PO_4/TiO_2/Fe_3O_4$ heterogeneous structure (Xu, Gao, Han, Liu, & Song, 2014), and Fe_3O_4/TiO_2 core-shell magnetic nanoparticles etc. Some similar strategies have been reported to deactivate pathogenic bacteria via photocatalytic root.

The ultra-thin Fe_3O_4—itanium Oxide nano sheets (TNS) nanocomposite synthesized using solvothermal method was demonstrated best photocatalytic disinfection activity. In the presence or absence of light exposure, magnetic separation of photocatalysts was systematically investigated. Due to a larger surface area, higher absorption rate in visible light and a lower recombination ratio of photogenerated pairs of electrons and holes, Fe_3O_4-TNS nanocomposite exhibited improved photocatalytic sterilization and bacterial removal efficiency compared with individual TNS and Fe_3O_4 (Ai, Zhang, & Chen, 2011).

Restoration and reuse of nanostructured photocatalyst in the treatment of water was significant. Quick and convenient separation of Fe_3O_4-TNS from water after use is important not only for reuse but also to prevent secondary contamination. Material was collected by magnets after the photocatalytic disinfection study for Fe_3O_4—TNS recyclability study. Wasting the photocatalytic during separation and purification was caused by contamination of the photocatalytic surface with bacterial residues, which can reduce the effectiveness of bacterial removal by increasing processing time. Although the removal of bacteria indicates a slight decrease in cycle efficiency, more than 90% of the *E. coli* strain can be removed from Fe_3O_4—TNS within five cycles (Ai et al., 2011). A review of the iron oxide based photocatalytic systems for the

Table 18.1: Review of iron oxide photocatalytic systems for degradation of dyes, nondye-based and pharmaceutical pollutants, and reduction of heavy metals and microbial disinfection.

Material	Method of synthesis	Morphology of the material	Application	Reference
$CoFe_2O_4$	Wet chemistry	Near spheroidal	Dye degradation (CR, MO, MB, RhB, 4-N, 4-C)	Sun et al. (2019)
Fe_3O_4/TiO_2	Sono-chemical	Core/shell nanocubes	Photocatalytic MB degradation	Abbas et al. (2014)
$Fe_3O_4/chitosan/TiO_2$	Solvents thermal reduction	Spherical	Photocatalytic MB degradation	Xiang et al. (2015)
$Fe_3O_4/ZnO/CuWO_4$	Chemical coprecipitation	Nanoparticles	Photocatalytic Rh-B degradation	Shekofteh-Gohari and Habibi-Yangjeh (2016)
$Fe_3O_4@SiO_2$	Hydrothermal method	Core-shell microspheres	Photocatalytic degradation of congo Red dye	Wang et al. (2016)
$Fe_3O_4/SiO_2/TiO_2$	Sol–gel method	Core-shell	Photocatalytic Rh-B degradation	Ye et al. (2010)
$Fe_3O_4/SiO_2/TiO_2$	Sol–gel method	Core-shell	Photocatalytic MB degradation	Wang et al. (2012)
$g-C_3N_4—Fe_3O_4$	Precipitation method	Nanoparticles on sheets	Photocatalytic Rh-B degradation	Kumar et al. (2013)
$g-C_3N_4—Fe_3O_4$	Heat treatment	Nanoparticles on porous g-C3N4	Photocatalytic Rh-B degradation	Ye et al. (2013)
$CuFe_2O_4@g-C_3N_4$	Self-assembly process	Core-shell	Photocatalytic Orange-II degradation	Yao et al. (2015)
$MnFe_2O_4@PANI@Ag$	Reflux/precipitation	Spherical particles on PANI layer	Catalytic reduction of MB, MO, eosin Y, RhB	Amir et al. (2016)
$Mn_{0.5}Ni_{0.5}Fe_2O_4$	Microwave combustion method	Particle	Photocatalytic degradation of Indigo caramel	Jesudoss et al. (2016)
$Fe_3O_4—ZnO$	Chemical coprecipitation	Irregular	Phenol degradation	Feng et al. (2014)
$Chitosan/CoFe_2O_4$	Gamma-radiation	Agglomerated	2-Chlorophenol degradation	Taleb (2014)
$Cu_{0.5}Mn_{0.5}Fe_2O_4-TiO_2$	One-pot microwave combustion method	Spherical	Degradation of 4-chlorophenol	Manikandan et al. (2015)
$ZrO_2/Fe_3O_4/chitosan$	Coprecipitation	Spherical	Cr(vi), 4-chlorophenol	Kumar et al. (2016)
$CoFe_2O_4$	Sol–gel-hydrothermal technique	Spherical	Catalytic reduction of 4-nitrophenols, 2,4,6-trinitrophenol and 4-nitroaniline	Ibrahim et al. (2016)

(Continued)

Table 18.1: (Continued)

Material	Method of synthesis	Morphology of the material	Application	Reference
$MnFe_2O_4$	Sol–gel-hydrothermal technique	Spherical	Catalytic reduction of 4-nitrophenols, 2,4,6-trinitrophenol and 4-nitroaniline	Ibrahim et al. (2016)
$ZnFe_2O_4$	Sol–gel-hydrothermal technique	Spherical	Catalytic reduction of 4-nitrophenols, 2,4,6-trinitrophenol and 4-nitroaniline	Ibrahim et al. (2016)
Fe_3O_4–APTES	Chemical coprecipitation	Polygonic and spherical	Catalytic reduction of 4-nitroaniline and 1, 3-dinitrobenzene	Karaoglu et al. (2013)
$NiFe_2O_4$–Pd	Sonochemical method	Polygonic	Catalytic activity of $NiFe_2O_4$–Pd	Karaoğlu et al. (2012)
$CoFe_2O_4$-APTES-Pd(0)	Chemical coprecipitation	Polygonic and spherical	Catalytic activity of CoFe2O4-APTES-Pd(0)	Demirelli et al. (2014)
Pd-TiO_2	Hydrothermal method	Nanotubes	Catalytic reduction of p-Nitrophenol	Kalarivalappil et al. (2016)
MCNT–TiO_2	Simple mixing method	Particles-nanotubes	Photocatalytic degradation of carbamazepine and sulfamethoxazole	Awfa et al. (2019)
$MnFe_2O_4$, $MnFe_2O_4$/GSC	Hydrolysis method	Agglomeration over sheets	Photocatalytic activity for the mineralization of ampicillin (AMP) and oxytetracycline (OTC) antibiotics	Gautam et al. (2017)
Ce-doped TiO_2-MGO	Simple mixing method	Spherical particles/sheets	Photocatalyst in the degradation of tetracyclines	Cao et al. (2016)
$CoFe_2O_4$/MWCNTs	Hydrothermal method with suspension polymerization	Aggregated particles on CNTs	Photodegradation of 2-mercaptobenzothiazole	Lu et al. (2015)
γ-Fe_2O_3@ZnO	Hydrothermal process, ALD	Core-shell	Ciprofloxacin photodegradation	Li et al. (2017)
ZnO/γ-Fe_2O_3	Coprecipitation	Spherical	Photodecomposition of 2,4-dichlorophenoxyacetic acid (2,4-D)	Lee and Abdullah (2015)
Ag_2WO_4@ZnO@Fe_3O_4	Hydrothermal method	Spherical	Photodegradation of tetracycline hydrochloride	Shen et al. (2016)
α-Fe_2O_3/ZnO	Hydrothermal and total reflux method	Nano-flower	Photodegradation of Cefixime trihydrate	Shooshtari and Ghazi (2017)
ZnO/Fe_2O_3 and TiO_2/Fe_2O_3	Mechanical milling	Agglomerated	Photodegradation of diphenhydramine	Davari et al. (2017)
Fe_3O_4/Bi_2WO_6	Modified solvothermal method	Nanospheres	Photocatalytic degradation of pharmaceutical ibuprofen	Rohani Bastami et al. (2017)
$BiVO_4$ with {010}-orientation growth	Hydrothermal process	Fishbone-like morphology	Photocatalytic treatment of Cr (VI) and phenol	Yu et al. (2018)

(Continued)

Table 18.1: (Continued)

Material	Method of synthesis	Morphology of the material	Application	Reference
Fe_2O_3@BOC	Wet-chemical	Nanosheet	Degradation of toxic dye (MB) and reduction of toxic metal (Cr6 +)	Kar et al. (2019)
$NiFe_2O_4-SiO_2-TiO_2$	Sol–gel	Spherical	Photocatalytic reduction of Cr (VI)	Ojemaye et al. (2017)
$TiO_2/CoFe_2O_4/Ag$	Hydrothermal process	Clusters	Photocatalytic reduction of hexavalent chromium	Ibrahim et al. (2020)
$CdIn_2S_4$	A modified USP method	Microsphere	Photocatalyst in inactivation of *Escherichia coli* (*E. coli*) K-12	Wang et al. (2013)
Pigeon ovalbumin-bound Fe_3O_4@Al_2O_3	Coprecipitation	Nanoparticles	Capture of uropathogenic *E. coli*	Liu et al. (2008)
Au-decorated Fe_3O_4@$mTiO_2$	Hydrothermal process	Core-shell	Degradation of toxic dye (MB *E. coli*)	Li et al. (2014)
$Ag_3PO_4/TiO_2/Fe_3O_4$	Hydrolysis, precipitation	Nanospheres	Bactericidal activity of *E. coli*	Xu et al. (2014)

degradation of dyes, nondye-based and pharmaceutical pollutants, and reduction of heavy metals and microbial disinfection is provided in (Table 18.1).

18.5 Conclusion

In this chapter, we have summarized the advance magnetic photocatalytic systems for degradation of dyes, nondye-based pollutants such as phenolics, pharmaceutical pollutants degradation, reduction of heavy metal ions in the aqueous system, and microbial disinfections. Especially we have discussed about various iron oxides-based magnetic systems, such as hematite, magnetite, meghamite, and ferrites. Synthesis methods such as hydrothermal, solvo thermal, coprecipitation, thermal decomposition, sol–gel, microemulsion, and other coating methods used for the preparation of these magnetic photocatalytic systems are discussed. Synthesis methods are very important, which decide the shape, size, and property of the systems to get the desired activity. Composites of iron oxides with TiO_2 and doped iron oxides have been widely employed for the degradation of both dye-based and pharmaceutical waste-based pollutants. In addition, these have been effectively employed in reduction of metal ions, and silver-based composite systems are used for microbial disinfection purposes.

References

Abbas, M., Rao, B. P., Reddy, V., & Kim, C. (2014). Fe_3O_4/TiO_2 core/shell nanocubes: Single-batch surfactantless synthesis, characterization and efficient catalysts for methylene blue degradation. *Ceramics International, 40,* 11177−11186. Available from https://doi.org/10.1016/j.ceramint.2014.030.148.

Ahmed, M. A., El-Katori, E. E., & Gharni, Z. H. (2013). Photocatalytic degradation of methylene blue dye using Fe_2O_3/TiO_2 nanoparticles prepared by sol−gel method. *Journal of Alloys and Compounds, 553,* 19−29. Available from https://doi.org/10.1016/j.jallcom.2012.100.038.

Ai, L., Zhang, C., & Chen, Z. (2011). Removal of methylene blue from aqueous solution by a solvothermal-synthesized graphene/magnetite composite. *Journal of Hazardous Materials, 192,* 1515−1524. Available from https://doi.org/10.1016/j.jhazmat.2011.060.068.

Ajmal, A., Majeed, I., Malik, R. N., Idriss, H., & Nadeem, M. A. (2014). Principles and mechanisms of photocatalytic dye degradation on TiO_2 based photocatalysts: A comparative overview. *RSC Advances, 4,* 37003−37026. Available from https://doi.org/10.1039/c4ra06658h.

Amir, M., Kurtan, U., Baykal, A., & Sözeri, H. (2016). MnFe2O4@PANI@Ag heterogeneous nanocatalyst for degradation of industrial aqueous organic pollutants. *Journal of Material Science and Technology, 32,* 134−141. Available from https://doi.org/10.1016/j.jmst.2015.120.011.

Anglada, Ángela, Urtiaga, A., & Ortiz, I. (2009). Contributions of electrochemical oxidation to waste-water treatment: Fundamentals and review of applications. *Journal of Chemical Technology and Biotechnology (Oxford, Oxfordshire: 1986), 84,* 1747−1755. Available from https://doi.org/10.1002/jctb.2214.

Augugliaro, V., García-López, E., Loddo, V., Malato-Rodríguez, S., Maldonado, I., Marcì, G., ... Palmisano, L. (2005). Degradation of lincomycin in aqueous medium: Coupling of solar photocatalysis and membrane separation. *Solar Energy, 79,* 402−408. Available from https://doi.org/10.1016/j.solener.2005.020.020.

Awfa, D., Ateia, M., Fujii, M., & Yoshimura, C. (2019). Novel magnetic carbon nanotube-TiO_2 composites for solar light photocatalytic degradation of pharmaceuticals in the presence of natural organic matter. *Journal of Water Process Engineering, 31,* 100836. Available from https://doi.org/10.1016/j.jwpe.2019.100836.

Basly, B., Felder-Flesch, D., Perriat, P., Pourroy, G., & Bégin-Colin, S. (2011). Properties and suspension stability of dendronized iron oxide nanoparticles for MRI applications. *Contrast Media & Molecular Imaging, 6,* 132−138. Available from https://doi.org/10.1002/cmmi0.416.

Blanco, J., Malato, S., Fernández-Ibañez, P., Alarcón, D., Gernjak, W., & Maldonado, M. I. (2009). Review of feasible solar energy applications to water processes. *Renewable and Sustainable Energy Reviews, 13,* 1437−1445. Available from https://doi.org/10.1016/j.rser.2008.080.016.

Buthiyappan, A., Abdul Aziz, A. R., & Wan Daud, W. M. A. (2016). Recent advances and prospects of catalytic advanced oxidation process in treating textile effluents. *Reviews in Chemical Engineering, 32,* 1−47. Available from https://doi.org/10.1515/revce-2015-0034.

Buzea, C., Pacheco, I. I., & Robbie, K. (2007). Nanomaterials and nanoparticles: Sources and toxicity. *Biointerphases, 2,* MR17−MR71. Available from https://doi.org/10.1116/1.2815690.

Byrne, C., Subramanian, G., & Pillai, S. C. (2018). Recent advances in photocatalysis for environmental applications. *Journal of Environmental Chemical Engineering, 6,* 3531−3555. Available from https://doi.org/10.1016/j.jece.2017.070.080.

Cao, M., Wang, P., Ao, Y., Wang, C., Hou, J., & Qian, J. (2015). Photocatalytic degradation of tetrabromobisphenol A by a magnetically separable graphene-TiO_2 composite photocatalyst: Mechanism and intermediates analysis. *Chemical Engineering Journal, 264,* 113−124. Available from https://doi.org/10.1016/j.cej.2014.100.011.

Cao, M., Wang, P., Ao, Y., Wang, C., Hou, J., & Qian, J. (2016). Visible light activated photocatalytic degradation of tetracycline by a magnetically separable composite photocatalyst: Graphene oxide/magnetite/cerium-doped titania. *Journal of Colloid and Interface Science, 467,* 129−139. Available from https://doi.org/10.1016/j.jcis.2016.010.005.

Casbeer, E., Sharma, V. K., & Li, X.-Z. (2012). Synthesis and photocatalytic activity of ferrites under visible light: A review. *Seperation and Purification Technology, 87,* 1−14. Available from https://doi.org/10.1016/j.seppur.2011.110.034.

Chen, L., Li, L., Wang, T., Zhang, L., Xing, S., Wang, C., & Su, Z. (2014). A novel strategy to fabricate multifunctional Fe_3O 4@C@TiO_2 yolk−shell structures as magnetically recyclable photocatalysts. *Nanoscale, 6,* 6603−6608. Available from https://doi.org/10.1039/c4nr00175c.

Chen, R., Wang, Q., Du, Y., Xing, W., & Xu, N. (2009). Effect of initial solution apparent pH on nano-sized nickel catalysts in p-nitrophenol hydrogenation. *Chemical Engineering Journal, 145,* 371−376. Available from https://doi.org/10.1016/j.cej.2008.070.042.

Chong, M. N., Jin, B., Chow, C. W. K., & Saint, C. (2010). Recent developments in photocatalytic water treatment technology: A review. *Water Research, 44,* 2997−3027. Available from https://doi.org/10.1016/j.watres.2010.020.039.

Dadfar, S. M., Roemhild, K., Drude, N. I., von Stillfried, S., Knüchel, R., Kiessling, F., & Lammers, T. (2019). Iron oxide nanoparticles: Diagnostic, therapeutic and theranostic applications. *Advanced Drug Delivery Reviews, 138,* 302−325. Available from https://doi.org/10.1016/j.addr.2019.010.005.

Davari, N., Farhadian, M., Nazar, A. R. S., & Homayoonfal, M. (2017). Degradation of diphenhydramine by the photocatalysts of ZnO/Fe_2O_3 and TiO_2/Fe_2O_3 based on clinoptilolite: Structural and operational comparison. *Journal of Environmental Chemical Engineering, 5,* 5707−5720. Available from https://doi.org/10.1016/j.jece.2017.100.052.

Demirelli, M., Karaoğlu, E., Baykal, A., Sözeri, H., & Uysal, E. (2014). Synthesis, characterization and catalytic activity of $CoFe_2O_4$-APTES-Pd magnetic recyclable catalyst. *Journal of Alloys and Compounds, 582,* 201−207. Available from https://doi.org/10.1016/j.jallcom.2013.070.174.

Dokianakis, S. N., Kornaros, M. E., & Lyberatos, G. (2004). On the effect of pharmaceuticals on bacterial nitrite oxidation. *Water Science and Technology: A Journal of the International Association on Water Pollution Research, 50,* 341−346. Available from https://doi.org/10.2166/wst.2004.0347.

Doll, T. E., & Frimmel, F. H. (2005). Removal of selected persistent organic pollutants by heterogeneous photocatalysis in water. *Catalysis Today, 101,* 195−202. Available from https://doi.org/10.1016/j.cattod.2005.030.005.

Du, X.-D., Yi, X.-H., Wang, P., Zheng, W., Deng, J., & Wang, C.-C. (2019). Robust photocatalytic reduction of Cr(VI) on UiO-66-NH_2(Zr/Hf) metal-organic framework membrane under sunlight irradiation. *Chemical Engineering Journal, 356,* 393−399. Available from https://doi.org/10.1016/j.cej.2018.090.084.

Eljarrat, E. (2003). Priority lists for persistent organic pollutants and emerging contaminants based on their relative toxic potency in environmental samples. *TrAC Trends in Analytical Chemistry, 22,* 655−665. Available from https://doi.org/10.1016/S0165-9936(03)01001-X.

Feng, X., Guo, H., Patel, K., Zhou, H., & Lou, X. (2014). High performance, recoverable Fe_3O_4ZnO nanoparticles for enhanced photocatalytic degradation of phenol. *Chemical Engineering Journal, 244,* 327−334. Available from https://doi.org/10.1016/j.cej.2014.010.075.

Fernández-Castro, P., Vallejo, M., San Román, M. F., & Ortiz, I. (2015). Insight on the fundamentals of advanced oxidation processes. Role and review of the determination methods of reactive oxygen species. *Journal of Chemical Technology and Biotechnology (Oxford, Oxfordshire: 1986), 90,* 796−820. Available from https://doi.org/10.1002/jctb.4634.

Fountoulakis, M., Drillia, P., Stamatelatou, K., & Lyberatos, G. (2004). Toxic effect of pharmaceuticals on methanogenesis. *Water Science and Technology: A Journal of the International Association on Water Pollution Research, 50,* 335−340. Available from https://doi.org/10.2166/wst.2004.0346.

Gautam, S., Shandilya, P., Priya, B., Singh, V. P., Raizada, P., Rai, R., ... Singh, P. (2017). Superparamagnetic $MnFe_2O_4$ dispersed over graphitic carbon sand composite and bentonite as magnetically recoverable photocatalyst for antibiotic mineralization. *Seperation and Purification Technology, 172,* 498−511. Available from https://doi.org/10.1016/j.seppur.2016.090.006.

Gaya, U. I., & Abdullah, A. H. (2008). Heterogeneous photocatalytic degradation of organic contaminants over titanium dioxide: A review of fundamentals, progress and problems. *Journal of Photochemistry and Photobiology C: Photochemistry Reviews, 9,* 1−12. Available from https://doi.org/10.1016/j.jphotochemrev.2007.120.003.

Gazi, S., & Ananthakrishnan, R. (2011). Metal-free-photocatalytic reduction of 4-nitrophenol by resin-supported dye under the visible irradiation. *Applied Catalysis B: Environmental, 105,* 317−325. Available from https://doi.org/10.1016/j.apcatb.2011.040.025.

Gogate, P. R., & Pandit, A. B. (2004). A review of imperative technologies for wastewater treatment I: Oxidation technologies at ambient conditions. *Advances in Environmental Research*, 8, 501–551. Available from https://doi.org/10.1016/S1093-0191(03)00032-7.

Goyal, A., Bansal, S., & Singhal, S. (2014). Facile reduction of nitrophenols: Comparative catalytic efficiency of MFe_2O_4 (M = Ni, Cu, Zn) nano ferrites. *International Journal of Hydrogen Energy*, 39, 4895–4908. Available from https://doi.org/10.1016/j.ijhydene.2014.010.050.

Guo, X., Li, H., & Zhao, S. (2015). Fast degradation of Acid Orange II by bicarbonate-activated hydrogen peroxide with a magnetic S-modified $CoFe_2O_4$ catalyst. *Journal of the Taiwan Institute of Chemical Engineering*, 55, 90–100. Available from https://doi.org/10.1016/j.jtice.2015.030.039.

Gupta, S. M., & Tripathi, M. (2011). A review of TiO_2 nanoparticles. *Chinese Science Bulletin*, 56, 1639–1657. Available from https://doi.org/10.1007/s11434-011-4476-1.

Hamzezadeh-Nakhjavani, S., Tavakoli, O., Akhlaghi, S. P., Salehi, Z., Esmailnejad-Ahranjani, P., & Arpanaei, A. (2015). Efficient photocatalytic degradation of organic pollutants by magnetically recoverable nitrogen-doped TiO_2 nanocomposite photocatalysts under visible light irradiation. *Environmental Science and Pollution Research*, 22, 18859–18873. Available from https://doi.org/10.1007/s11356-015-5032-3.

Harraz, F. A., Mohamed, R. M., Rashad, M. M., Wang, Y. C., & Sigmund, W. (2014). Magnetic nanocomposite based on titania-silica/cobalt ferrite for photocatalytic degradation of methylene blue dye. *Ceramics International*, 40, 375–384. Available from https://doi.org/10.1016/j.ceramint.2013.060.012.

He, M., Li, D., Jiang, D., & Chen, M. (2012). Magnetically separable γ-Fe 2O 3@SiO 2@Ce-doped TiO 2 core-shell nanocomposites: Fabrication and visible-light-driven photocatalytic activity. *Journal of Solid State Chemistry*, 192, 139–143. Available from https://doi.org/10.1016/j.jssc.2012.040.004.

Ibrahim, I., Ali, I. O., Salama, T. M., Bahgat, A. A., & Mohamed, M. M. (2016). Synthesis of magnetically recyclable spinel ferrite (MFe_2O_4, M = Zn, Co, Mn) nanocrystals engineered by sol gel-hydrothermal technology: High catalytic performances for nitroarenes reduction. *Applied Catalysis B: Environmental*, 181, 389–402. Available from https://doi.org/10.1016/j.apcatb.2015.080.005.

Ibrahim, I., Kaltzoglou, A., Athanasekou, C., Katsaros, F., Devlin, E., Kontos, A. G., ... Falaras, P. (2020). Magnetically separable $TiO_2/CoFe_2O_4$/Ag nanocomposites for the photocatalytic reduction of hexavalent chromium pollutant under UV and artificial solar light. *Chemical Engineering Journal*, 381, 122730. Available from https://doi.org/10.1016/j.cej.2019.122730.

Jadhav, S. V., Bringas, E., Yadav, G. D., Rathod, V. K., Ortiz, I., & Marathe, K. V. (2015). Arsenic and fluoride contaminated groundwaters: A review of current technologies for contaminants removal. *Journal of Environmental Management*, 162, 306–325. Available from https://doi.org/10.1016/j.jenvman.2015.070.020.

Jesudoss, S. K., Vijaya, J. J., Kennedy, L. J., Rajan, P. I., Al-Lohedan, H. A., Ramalingam, R. J., ... Bououdina, M. (2016). Studies on the efficient dual performance of $Mn_{1-x}Ni_xFe_2O_4$ spinel nanoparticles in photodegradation and antibacterial activity. *Journal of Photochemistry and Photobiology B: Biology*, 165, 121–132. Available from https://doi.org/10.1016/j.jphotobiol.2016.100.004.

Jin, Y., Liu, F., Shan, C., Tong, M., & Hou, Y. (2014). Efficient bacterial capture with amino acid modified magnetic nanoparticles. *Water Research*, 50, 124–134. Available from https://doi.org/10.1016/j.watres.2013.110.045.

Kabra, K., Chaudhary, R., & Sawhney, R. L. (2004). Treatment of hazardous organic and inorganic compounds through aqueous-phase photocatalysis: A review. *Industrial & Engineering Chemistry Research*, 43, 7683–7696. Available from https://doi.org/10.1021/ie0498551.

Kalarivalappil, V., Divya, C. M., Wunderlich, W., Pillai, S. C., Hinder, S. J., Nageri, M., ... Vijayan, B. K. (2016). Pd loaded TiO_2 nanotubes for the effective catalytic reduction of p-nitrophenol. *Catalysis Letters*, 146, 474–482. Available from https://doi.org/10.1007/s10562-015-1663-8.

Kar, P., Jain, P., Kumar, V., & Gupta, R. K. (2019). Interfacial engineering of Fe_2O_3@BOC heterojunction for efficient detoxification of toxic metal and dye under visible light illumination. *Journal of Environmental Chemical Engineering*, 7, 102843. Available from https://doi.org/10.1016/j.jece.2018.102843.

Karaoğlu, E., Özel, U., Caner, C., Baykal, A., Summak, M. M., & Sözeri, H. (2012). Synthesis and characterization of $NiFe_2O_4$−Pd magnetically recyclable catalyst for hydrogenation reaction. *Materials Research Bulletin, 47*, 4316−4321. Available from https://doi.org/10.1016/j.materresbull.2012.090.011.

Karaoglu, E., Summak, M. M., Baykal, A., Sözeri, H., & Toprak, M. S. (2013). Synthesis and characterization of catalytically activity Fe_3O_4−3-aminopropyl-triethoxysilane/Pd nanocomposite. *Journal of Inorganic and Organometallic Polymers and Materials, 23*, 409−417. Available from https://doi.org/10.1007/s10904-012-9796-3.

Krastanov, A., Alexieva, Z., & Yemendzhiev, H. (2013). Microbial degradation of phenol and phenolic derivatives. *Engineering in Life Science, 13*, 76−87. Available from https://doi.org/10.1002/elsc.201100227.

Kumar, A., Guo, C., Sharma, G., Pathania, D., Naushad, M., Kalia, S., & Dhiman, P. (2016). Magnetically recoverable ZrO_2/Fe_3O_4/chitosan nanomaterials for enhanced sunlight driven photoreduction of carcinogenic Cr(VI) and dechlorination & mineralization of 4-chlorophenol from simulated waste water. *RSC Advances, 6*, 13251−13263. Available from https://doi.org/10.1039/C5RA23372K.

Kumar, J., & Bansal, A. (2013). Photocatalysis by nanoparticles of titanium dioxide for drinking water purification: A conceptual and state-of-art review. *Materials Science Forum, 764*, 130−150. Available from https://doi.org/10.4028/www.scientific.net/MSF.764.130.

Kumar, S., Surendar, T., Kumar, B., Baruah, A., & Shanker, V. (2013). Synthesis of magnetically separable and recyclable g-C_3N_4 −Fe_3O_4 hybrid nanocomposites with enhanced photocatalytic performance under visible-light irradiation. *Journal of Physical Chemistry C, 117*, 26135−26143. Available from https://doi.org/10.1021/jp409651g.

Kümmerer, K. (2001). Drugs in the environment: Emission of drugs, diagnostic aids and disinfectants into wastewater by hospitals in relation to other sources − A review. *Chemosphere, 45*, 957−969. Available from https://doi.org/10.1016/S0045-6535(01)00144-8.

Laurent, S., Forge, D., Port, M., Roch, A., Robic, C., Vander Elst, L., & Muller, R. N. (2008). Magnetic iron oxide nanoparticles: Synthesis, stabilization, vectorization, physicochemical characterizations and biological applications. *Chemical Reviews, 108*, 2064−2110. Available from https://doi.org/10.1021/cr068445e.

Lazar, M., Varghese, S., & Nair, S. (2012). Photocatalytic water treatment by titanium dioxide: Recent updates. *Catalysts, 2*, 572−601. Available from https://doi.org/10.3390/catal2040572.

Lee, K. M., & Abdullah, A. H. (2015). Synthesis and characterization of zinc oxide/maghemite nanocomposites: Influence of heat treatment on photocatalytic degradation of 2,4-dichlorophenoxyacetic acid. *Materials Science and Semiconductor Processing, 30*, 298−306. Available from https://doi.org/10.1016/j.mssp.2014.100.017.

Lee, S.-Y., & Park, S.-J. (2013). TiO_2 photocatalyst for water treatment applications. *Journal of Industrial and Engineering Chemistry, 19*, 1761−1769. Available from https://doi.org/10.1016/j.jiec.2013.070.012.

Lellis, B., Fávaro-Polonio, C. Z., Pamphile, J. A., & Polonio, J. C. (2019). Effects of textile dyes on health and the environment and bioremediation potential of living organisms. *Biotechnology Research and Innovation, 3*, 275−290. Available from https://doi.org/10.1016/j.biori.2019.090.001.

Li, C., Younesi, R., Cai, Y., Zhu, Y., Ma, M., & Zhu, J. (2014). Photocatalytic and antibacterial properties of Au-decorated Fe_3O_4@mTiO$_2$ core-shell microspheres. *Applied Catalysis B: Environmental, 156−157*, 314−322. Available from https://doi.org/10.1016/j.apcatb.2014.030.031.

Li, G., Wong, K. H., Zhang, X., Hu, C., Yu, J. C., Chan, R. C. Y., & Wong, P. K. (2009). Degradation of Acid Orange 7 using magnetic AgBr under visible light: The roles of oxidizing species. *Chemosphere, 76*, 1185−1191. Available from https://doi.org/10.1016/j.chemosphere.2009.060.027.

Li, N., Zhang, J., Tian, Y., Zhao, J., Zhang, J., & Zuo, W. (2017). Precisely controlled fabrication of magnetic 3D γ-Fe_2O_3 @ZnO core-shell photocatalyst with enhanced activity: Ciprofloxacin degradation and mechanism insight. *Chemical Engineering Journal, 308*, 377−385. Available from https://doi.org/10.1016/j.cej.2016.090.093.

Li, W., Tian, Y., Li, P., Zhang, B., Zhang, H., Geng, W., & Zhang, Q. (2015). Synthesis of rattle-type magnetic mesoporous Fe_3O_4@mSiO_2@BiOBr hierarchical photocatalyst and investigation of its photoactivity in the degradation of methylene blue. *RSC Advances, 5*, 48050−48059. Available from https://doi.org/10.1039/c5ra06894k.

Li, Y., Jiang, R., Liu, T., Lv, H., Zhou, L., & Zhang, X. (2014). One-pot synthesis of grass-like Fe_3O_4 nanostructures by a novel microemulsion-assisted solvothermal method. *Ceramics International, 40*, 1059−1063. Available from https://doi.org/10.1016/j.ceramint.2013.060.104.

Li, Y., Jiang, R., Liu, T., Lv, H., & Zhang, X. (2014). Single-microemulsion-based solvothermal synthesis of magnetite microflowers. *Ceramics International, 40*, 4791−4795. Available from https://doi.org/10.1016/j.ceramint.2013.090.025.

Linley, S., Leshuk, T., & Gu, F. X. (2013). Magnetically separable water treatment technologies and their role in future advanced water treatment: A patent review. *CLEAN - Soil, Air, and Water, 41*, 1152−1156. Available from https://doi.org/10.1002/clen.201100261.

Liu, J.-C., Tsai, P.-J., Lee, Y. C., & Chen, Y.-C. (2008). Affinity capture of uropathogenic *Escherichia coli* using pigeon ovalbumin-bound Fe_3O_4 @Al_2O_3 magnetic nanoparticles. *Analytical Chemistry, 80*, 5425−5432. Available from https://doi.org/10.1021/ac800487v.

Liu, L.-H., Dietsch, H., Schurtenberger, P., & Yan, M. (2009). Photoinitiated coupling of unmodified monosaccharides to iron oxide nanoparticles for sensing proteins and bacteria. *Bioconjugate Chemistry, 20*, 1349−1355. Available from https://doi.org/10.1021/bc900110x.

Lu, A. H., Salabas, E. L., & Schüth, F. (2007). Magnetic nanoparticles: Synthesis, protection, functionalization, and application. *Angewandte Chemie International Edition, 46*, 1222−1244. Available from https://doi.org/10.1002/anie.200602866.

Lu, Z., He, M., Yang, L., Ma, Z., Yang, L., Wang, D., . . . Hua, Z. (2015). Selective photodegradation of 2-mercaptobenzothiazole by a novel imprinted $CoFe_2O_4$/MWCNTs photocatalyst. *RSC Advances, 5*, 47820−47829. Available from https://doi.org/10.1039/C5RA08795C.

Lv, Y., Yue, L., Li, Q., Shao, B., Zhao, S., Wang, H., . . . Wang, Z. (2018). Recyclable (Fe_3O_4-NaYF_4:Yb,Tm) @TiO_2 nanocomposites with near-infrared enhanced photocatalytic activity. *Dalton Transactions, 47*, 1666−1673. Available from https://doi.org/10.1039/c7dt04279e.

Majidi, S., Sehrig, F. Z., Farkhani, S. M., Goloujeh, M. S., & Akbarzadeh, A. (2016). Current methods for synthesis of magnetic nanoparticles. *Artificial Cells, Nanomedicine, and Biotechnology, 44*, 722−734. Available from https://doi.org/10.3109/21691401.2014.982802.

Mamba, G., & Mishra, A. (2016). Advances in magnetically separable photocatalysts: Smart, recyclable materials for water pollution mitigation. *Catalysts, 6*, 1−34. Available from https://doi.org/10.3390/catal6060079.

Manikandan, A., Hema, E., Durka, M., Seevakan, K., Alagesan, T., & Arul Antony, S. (2015). Room temperature ferromagnetism of magnetically recyclable photocatalyst of $Cu_{1−x}Mn_xFe_2O_4$-TiO_2 ($0.0 \leq x \leq 0.5$) nanocomposites. *Journal of Superconductivity and Novel Magnetism, 28*, 1783−1795. Available from https://doi.org/10.1007/s10948-014-2945-x.

McCullagh, C., Skillen, N., Adams, M., & Robertson, P. K. J. (2011). Photocatalytic reactors for environmental remediation: A review. *Journal of Chemical Technology and Biotechnology (Oxford, Oxfordshire: 1986), 86*, 1002−1017. Available from https://doi.org/10.1002/jctb.2650.

Min, K.-I., Choi, J.-S., Chung, Y.-M., Ahn, W.-S., Ryoo, R., & Lim, P. K. (2008). p-Aminophenol synthesis in an organic/aqueous system using Pt supported on mesoporous carbons. *Applied Catalysis A: General, 337*, 97−104. Available from https://doi.org/10.1016/j.apcata.2007.120.004.

Mohamed, M. A., Salleh, W. N. W., Jaafar, J., Ismail, A. F., Abd Mutalib, M., Mohamad, A. B., . . . Hir, Z. A. Mohd (2017). Physicochemical characterization of cellulose nanocrystal and nanoporous self-assembled CNC membrane derived from Ceiba pentandra. *Carbohydrate Polymers, 157*, 1892−1902. Available from https://doi.org/10.1016/j.carbpol.2016.110.078.

Neris, A. M., Schreiner, W. H., Salvador, C., Silva, U. C., Chesman, C., Longo, E., & Santos, I. M. G. (2018). Photocatalytic evaluation of the magnetic core@shell system (Co,Mn)Fe_2O_4 @TiO_2 obtained by the modified Pechini method. *Materials Science and Engineering B, 229*, 218−226. Available from https://doi.org/10.1016/j.mseb.2017.120.029.

Ojemaye, M. O., Okoh, O. O., & Okoh, A. I. (2017). Performance of $NiFe_2O_4-SiO_2-TiO_2$ magnetic photocatalyst for the effective photocatalytic reduction of Cr(VI) in aqueous solutions. *Journal of Nanomaterials, 2017*, 1−11. Available from https://doi.org/10.1155/2017/5264910.

Pal, A., He, Y., Jekel, M., Reinhard, M., & Gin, K. Y.-H. (2014). Emerging contaminants of public health significance as water quality indicator compounds in the urban water cycle. *Environment International, 71*, 46−62. Available from https://doi.org/10.1016/j.envint.2014.050.025.

Pang, Y. L., Lim, S., Ong, H. C., & Chong, W. T. (2016). Research progress on iron oxide-based magnetic materials: Synthesis techniques and photocatalytic applications. *Ceramics International, 42*, 9−34. Available from https://doi.org/10.1016/j.ceramint.2015.080.144.

Parashar, M., Shukla, V. K., & Singh, R. (2020). Metal oxides nanoparticles via sol−gel method: A review on synthesis, characterization and applications. *Journal of Materials Science: Materials in Electronics, 31*, 3729−3749. Available from https://doi.org/10.1007/s10854-020-02994-8.

Peng, S., & Sun, S. (2007). Synthesis and characterization of monodisperse hollow Fe_3O_4 nanoparticles. *Angewandte Chemie, 119*, 4233−4236. Available from https://doi.org/10.1002/ange.200700677.

Pérez, G., Ibáñez, R., Urtiaga, A. M., & Ortiz, I. (2012). Kinetic study of the simultaneous electrochemical removal of aqueous nitrogen compounds using BDD electrodes. *Chemical Engineering Journal, 197*, 475−482. Available from https://doi.org/10.1016/j.cej.2012.050.062.

Petrovic, M. (2003). Analysis and removal of emerging contaminants in wastewater and drinking water. *TrAC Trends in Analytical Chemistry, 22*, 685−696. Available from https://doi.org/10.1016/S0165-9936(03) 01105-1.

Rashad, M. M., Mohamed, R. M., Ibrahim, M. A., Ismail, L. F. M., & Abdel-Aal, E. A. (2012). Magnetic and catalytic properties of cubic copper ferrite nanopowders synthesized from secondary resources. *Advances in Powder Technology, 23*, 315−323. Available from https://doi.org/10.1016/j.apt.2011.040.005.

Ribeiro, A. R., Nunes, O. C., Pereira, M. F. R., & Silva, A. M. T. (2015). An overview on the advanced oxidation processes applied for the treatment of water pollutants defined in the recently launched Directive 2013/39/EU. *Environment International, 75*, 33−51. Available from https://doi.org/10.1016/j.envint.2014.100.027.

Rivero, M. J., Alonso, E., Dominguez, S., Ribao, P., Ibañez, R., Ortiz, I., & Irabien, A. (2014). Kinetic analysis and biodegradability of the Fenton mineralization of bisphenol A. *Journal of Chemical Technology and Biotechnology (Oxford, Oxfordshire: 1986), 89*, 1228−1234. Available from https://doi.org/10.1002/jctb.4376.

Rohani Bastami, T., Ahmadpour, A., & Ahmadi Hekmatikar, F. (2017). Synthesis of Fe_3O_4/Bi_2WO_6 nanohybrid for the photocatalytic degradation of pharmaceutical ibuprofen under solar light. *Journal of Industrial and Engineering Chemistry, 51*, 244−254. Available from https://doi.org/10.1016/j.jiec.2017.030.008.

Ryan, S., Kell, A. J., van Faassen, H., Tay, L.-L., Simard, B., MacKenzie, R., ... Tanha, J. (2009). Single-domain antibody-nanoparticles: Promising architectures for increased *Staphylococcus aureus* detection specificity and sensitivity. *Bioconjugate Chemistry, 20*, 1966−1974. Available from https://doi.org/10.1021/bc900332r.

Saiz, J., Bringas, E., & Ortiz, I. (2014). New functionalized magnetic materials for As 5 + removal: Adsorbent regeneration and reuse. *Industrial & Engineering Chemistry Research, 53*, 18928−18934. Available from https://doi.org/10.1021/ie500912k.

Saiz, J., Bringas, E., & Ortiz, I. (2014). Functionalized magnetic nanoparticles as new adsorption materials for arsenic removal from polluted waters. *Journal of Chemical Technology and Biotechnology (Oxford, Oxfordshire: 1986), 89*, 909−918. Available from https://doi.org/10.1002/jctb.4331.

San Román, M. F., Bringas, E., Ibañez, R., & Ortiz, I. (2010). Liquid membrane technology: Fundamentals and review of its applications. *Journal of Chemical Technology and Biotechnology (Oxford, Oxfordshire: 1986), 85*, 2−10. Available from https://doi.org/10.1002/jctb.2252.

Saroj, S., Dubey, S., Agarwal, P., Prasad, R., & Singh, R. P. (2015). Evaluation of the efficacy of a fungal consortium for degradation of azo dyes and simulated textile dye effluents. *Sustainable Water Resources Management, 1*, 233−243. Available from https://doi.org/10.1007/s40899-015-0027-2.

Satheesh, R., Vignesh, K., Suganthi, A., & Rajarajan, M. (2014). Visible light responsive photocatalytic applications of transition metal (M = Cu, Ni and Co) doped α-Fe_2O_3 nanoparticles. *Journal of Environmental Chemical Engineering, 2*, 1956−1968. Available from https://doi.org/10.1016/j.jece.2014.080.016.

Shannon, M. A., Bohn, P. W., Elimelech, M., Georgiadis, J. G., Mariñas, B. J., & Mayes, A. M. (2008). Science and technology for water purification in the coming decades. *Nature, 452*, 301−310. Available from https://doi.org/10.1038/nature06599.

Shekofteh-Gohari, M., & Habibi-Yangjeh, A. (2016). Fabrication of novel magnetically separable visible-light-driven photocatalysts through photosensitization of Fe_3O_4/ZnO with $CuWO_4$. *Journal of Industrial and Engineering Chemistry, 44*, 174−184. Available from https://doi.org/10.1016/j.jiec.2016.080.028.

Shen, J., Lu, Y., Liu, J.-K., & Yang, X.-H. (2016). Design and preparation of easily recycled Ag_2WO_4@ZnO@Fe_3O_4 ternary nanocomposites and their highly efficient degradation of antibiotics. *Journal of Materials Science, 51*, 7793−7802. Available from https://doi.org/10.1007/s10853-016-0063-9.

Shooshtari, N. M., & Ghazi, M. M. (2017). An investigation of the photocatalytic activity of nano α-Fe_2O_3/ZnO on the photodegradation of cefixime trihydrate. *Chemical Engineering Journal, 315*, 527−536. Available from https://doi.org/10.1016/j.cej.2017.010.058.

Stoia, M., Muntean, C., & Militaru, B. (2017). $MnFe_2O_4$ nanoparticles as new catalyst for oxidative degradation of phenol by peroxydisulfate. *Journal of Environmental Science, 53*, 269−277. Available from https://doi.org/10.1016/j.jes.2015.100.035.

Su, M., He, C., & Shih, K. (2016). Facile synthesis of morphology and size-controlled α-Fe_2O_3 and Fe_3O_4 nano- and microstructures by hydrothermal/solvothermal process: The roles of reaction medium and urea dose. *Ceramics International, 42*, 14793−14804. Available from https://doi.org/10.1016/j.ceramint.2016.060.111.

Sun, M., Han, X., & Chen, S. (2019). Synthesis and photocatalytic activity of nano-cobalt ferrite catalyst for the photo-degradation various dyes under simulated sunlight irradiation. *Materials Science and Semiconductor Processing, 91*, 367−376. Available from https://doi.org/10.1016/j.mssp.2018.120.005.

Taleb, M. F. A. (2014). Adsorption and photocatalytic degradation of 2-CP in wastewater onto CS/$CoFe_2O_4$ nanocomposite synthesized using gamma radiation. *Carbohydrate Polymers, 114*, 65−72. Available from https://doi.org/10.1016/j.carbpol.2014.070.061.

Tünay, O., Kabdasli, I., Eremektar, G., & Orhon, D. (1996). Color removal from textile wastewaters. *Water Science and Technology: A Journal of the International Association on Water Pollution Research, 34*. Available from https://doi.org/10.1016/S0273-1223(96)00815-3.

Udom, I., Ram, M. K., Stefanakos, E. K., Hepp, A. F., & Goswami, D. Y. (2013). One dimensional-ZnO nanostructures: Synthesis, properties and environmental applications. *Materials Science and Semiconductor Processing, 16*, 2070−2083. Available from https://doi.org/10.1016/j.mssp.2013.060.017.

Uribe, I. O., Mosquera-Corral, A., Rodicio, J. L., & Esplugas, S. (2015). Advanced technologies for water treatment and reuse. *AIChE Journal. American Institute of Chemical Engineers, 61*, 3146−3158. Available from https://doi.org/10.1002/aic.15013.

Venugopal, G., Thangavel, Sakthivel, & Raghavan, Nivea (2019). Magnetically separable iron oxide-based nanocomposite photocatalytic materials for environmental remediation. *Photocatalytic Functional Materials for Environmental Remediation*, 243−265.

Wang, C. C., Li, J. R., Lv, X. L., Zhang, Y. Q., & Guo, G. (2014). Photocatalytic organic pollutants degradation in metal-organic frameworks. *Energy and Environmental Science, 7*, 2831−2867. Available from https://doi.org/10.1039/c4ee01299b.

Wang, D., & Astruc, D. (2014). Fast-growing field of magnetically recyclable nanocatalysts. *Chemical Reviews, 114*, 6949−6985. Available from https://doi.org/10.1021/cr500134h.

Wang, K.-H., Hsieh, Y.-H., Chou, M.-Y., & Chang, C.-Y. (1999). Photocatalytic degradation of 2-chloro and 2-nitrophenol by titanium dioxide suspensions in aqueous solution. *Applied Catalysis B: Environmental, 21*, 1−8. Available from https://doi.org/10.1016/S0926-3373(98)00116-7.

Wang, P., Wang, X., Yu, S., Zou, Y., Wang, J., Chen, Z., . . . Wang, X. (2016). Silica coated Fe_3O_4 magnetic nanospheres for high removal of organic pollutants from wastewater. *Chemical Engineering Journal, 306,* 280−288. Available from https://doi.org/10.1016/j.cej.2016.070.068.

Wang, R., Wang, X., Xi, X., Hu, R., & Jiang, G. (2012). Preparation and photocatalytic activity of magnetic $Fe_3O_4/SiO_2/TiO_2$ composites. *Advances in Materials Science and Engineering, 2012,* 1−8. Available from https://doi.org/10.1155/2012/409379.

Wang, W., Li, N., Chi, Y., Li, Y., Yan, W., Li, X., & Shao, C. (2013). Electrospinning of magnetical bismuth ferrite nanofibers with photocatalytic activity. *Ceramics International, 39,* 3511−3518. Available from https://doi.org/10.1016/j.ceramint.2012.100.175.

Wang, W., Ng, T. W., Ho, W. K., Huang, J., Liang, S., An, T., . . . Wong, P. K. (2013). $CdIn_2S_4$ microsphere as an efficient visible-light-driven photocatalyst for bacterial inactivation: Synthesis, characterizations and photocatalytic inactivation mechanisms. *Applied Catalysis B: Environmental, 129,* 482−490. Available from https://doi.org/10.1016/j.apcatb.2012.090.054.

Wintgens, T., Gallenkemper, M., & Melin, T. (2004). Removal of endocrine disrupting compounds with membrane processes in wastewater treatment and reuse. *Water Science and Technology: A Journal of the International Association on Water Pollution Research, 50,* 1−8. Available from https://doi.org/10.2166/wst.2004.0301.

Wise, R., Hart, T., Cars, O., Streulens, M., Helmuth, R., Huovinen, P., & Sprenger, M. (1998). Antimicrobial resistance. *British Medical Journal (Clinical Research Edition), 317,* 609−610. Available from https://doi.org/10.1136/bmj.317.71590.609.

Xiang, Y., Wang, H., He, Y., & Song, G. (2015). Efficient degradation of methylene blue by magnetically separable $Fe_3O_4/chitosan/TiO_2$ nanocomposites. *Desalination and Water Treatment, 55,* 1018−1025. Available from https://doi.org/10.1080/19443994.2014.922441.

Xu, J.-W., Gao, Z.-D., Han, K., Liu, Y., & Song, Y.-Y. (2014). Synthesis of magnetically separable $Ag_3PO_4/TiO_2/Fe_3O_4$ heterostructure with enhanced photocatalytic performance under visible light for photoinactivation of bacteria. *ACS Applied Materials and Interfaces, 6,* 15122−15131. Available from https://doi.org/10.1021/am5032727.

Xu, P., Zeng, G. M., Huang, D. L., Feng, C. L., Hu, S., Zhao, M. H., . . . Liu, Z. F. (2012). Use of iron oxide nanomaterials in wastewater treatment: A review. *The Science of the Total Environment, 424,* 1−10. Available from https://doi.org/10.1016/j.scitotenv.2012.020.023.

Yang, S.-T., Zhang, W., Xie, J., Liao, R., Zhang, X., Yu, B., . . . Guo, Z. (2015). $Fe_3O_4@SiO_2$ nanoparticles as a high-performance Fenton-like catalyst in a neutral environment. *RSC Advances, 5,* 5458−5463. Available from https://doi.org/10.1039/C4RA10207J.

Yao, Y., Lu, F., Zhu, Y., Wei, F., Liu, X., Lian, C., & Wang, S. (2015). Magnetic core-shell $CuFe_2O_4@C_3N_4$ hybrids for visible light photocatalysis of Orange II. *Journal of Hazardous Materials, 297,* 224−233. Available from https://doi.org/10.1016/j.jhazmat.2015.040.046.

Ye, M., Zhang, Q., Hu, Y., Ge, J., Lu, Z., He, L., . . . Yin, Y. (2010). Magnetically recoverable core-shell nanocomposites with enhanced photocatalytic activity. *Chemistry, 16,* 6243−6250. Available from https://doi.org/10.1002/chem.200903516.

Ye, S., Qiu, L. G., Yuan, Y. P., Zhu, Y. J., Xia, J., & Zhu, J. F. (2013). Facile fabrication of magnetically separable graphitic carbon nitride photocatalysts with enhanced photocatalytic activity under visible light. *Journal of Materials Chemistry A, 1,* 3008−3015. Available from https://doi.org/10.1039/c2ta01069k.

Yoon, Y., Westerhoff, P., Snyder, S. A., & Wert, E. C. (2006). Nanofiltration and ultrafiltration of endocrine disrupting compounds, pharmaceuticals and personal care products. *Journal of Membrane Science, 270,* 88−100. Available from https://doi.org/10.1016/j.memsci.2005.060.045.

You, S.-J., Semblante, G. U., Lu, S.-C., Damodar, R. A., & Wei, T.-C. (2012). Evaluation of the antifouling and photocatalytic properties of poly(vinylidene fluoride) plasma-grafted poly(acrylic acid) membrane with self-assembled TiO_2. *Journal of Hazardous Materials, 237−238,* 10−19. Available from https://doi.org/10.1016/j.jhazmat.2012.070.071.

Yu, T., Lv, L., Wang, H., & Tan, X. (2018). Enhanced photocatalytic treatment of Cr(VI) and phenol by monoclinic BiVO$_4$ with {010}-orientation growth. *Materials Research Bulletin, 107*, 248−254. Available from https://doi.org/10.1016/j.materresbull.2018.070.033.

Zhang, D. E., Tong, Z. W., Li, S. Z., Zhang, X. B., & Ying, A. (2008). Fabrication and characterization of hollow Fe$_3$O$_4$ nanospheres in a microemulsion. *Materials Letter, 62*, 4053−4055. Available from https://doi.org/10.1016/j.matlet.2008.050.023.

Zhang, J., Zhuang, J., Gao, L., Zhang, Y., Gu, N., Feng, J., . . . Yan, X. (2008). Decomposing phenol by the hidden talent of ferromagnetic nanoparticles. *Chemosphere, 73*, 1524−1528. Available from https://doi.org/10.1016/j.chemosphere.2008.050.050.

Zhao, Y., Tao, C., Xiao, G., Wei, G., Li, L., Liu, C., & Su, H. (2016). Controlled synthesis and photocatalysis of sea urchin-like Fe$_3$O$_4$@TiO$_2$@Ag nanocomposites. *Nanoscale, 8*, 5313−5326. Available from https://doi.org/10.1039/c5nr08624h.

Zhou, Q., Fang, Z., Li, J., & Wang, M. (2015). Applications of TiO$_2$ nanotube arrays in environmental and energy fields: A review. *Microporous and Mesoporous Materials, 202*, 22−35. Available from https://doi.org/10.1016/j.micromeso.2014.090.040.

Index

Note: Page numbers followed by "*f*" and "*t*" refer to figures and tables, respectively.

Printed in the United States
by Baker & Taylor Publisher Services